Advanced Structured Materials

Volume 16

Series Editors
Andreas Öchsner
Lucas F. M. da Silva
Holm Altenbach

For further volumes:
http://www.springer.com/series/8611

Andreas Öchsner · Lucas F. M. da Silva
Holm Altenbach
Editors

Materials with Complex Behaviour II

Properties, Non-Classical Materials and New Technologies

Springer

Prof. Dr.-Ing. Andreas Öchsner
Department of Solid Mechanics and Design
Faculty of Mechanical Engineering
University of Technology Malaysia (UTM)
Skudai, Johor
Malaysia
e-mail: andreas.oechsner@gmail.com

Prof. Dr. Holm Altenbach
Otto-von-Guericke-Universität Magdeburg
Institut für Mechanik
Universitätsplatz 2
30106 Magdeburg, Sachsen-Anhalt
Germany
e-mail: Holm.Altenbach@ovgu.de

Prof. Dr. Lucas F. M. da Silva
Departamento de Engenharia Mecânica
Faculdade de Engenharia da
Universidade do Porto
Rua Dr. Roberto Frias s/n
4200-465 Porto
Portugal
e-mail: lucas@fe.up.pt

ISSN 1869-8433
e-ISSN 1869-8441
ISBN 978-3-642-44800-3
ISBN 978-3-642-22700-4 (eBook)
DOI 10.1007/978-3-642-22700-4
Springer Heidelberg Dordrecht London New York

Printed on acid-free paper

Springer is part of Springer Science+Business Media (www.springer.com)

Preface

Common engineering materials reach in many demanding applications such as automotive or aerospace their limits and new developments are required to fulfill increasing demands on performance and characteristics. The properties of materials can be increased for example by combining different materials to achieve better properties than a single constituent or by shaping the material or constituents in a specific structure. Many of these new materials reveal a much more complex behavior than traditional engineering materials due to their advanced structure or composition. Furthermore, the classical applications of many engineering materials are extended to new ranges of applications and to more demanding environmental conditions such as elevated temperatures. All these tendencies require in addition to the synthesis of new materials, proper methods for their manufacturing and extensive programs for their characterization. In many fields of application, the development of new methods and processes must be accomplished by accurate and reliable modeling and simulation techniques. Only the interaction between these new developments with regard to manufacturing, modeling, characterization, further processing and monitoring of materials will allow to meet all demands and to introduce these developments in safety-relevant applications.

The 4th International Conference on Advanced Computational Engineering and Experimenting, ACE-X 2010, was held in Paris, France, from 05 to 07 July 2010 with a strong focus on the above-mentioned developments. This conference served as an excellent platform for the engineering community to meet with each other and to exchange the latest ideas. This volume contains 45 revised and extended research articles written by experienced researchers participating in the conference. The book will offer the state-of-the-art of tremendous advances in engineering technologies of materials with complex behavior and also serve as an excellent reference volume for researchers and graduate students working with advanced materials. The covered topics are related to *Materials and Properties, Non-classical Materials and Structures and New Technologies*.

The organizers and editors wish to thank all the authors for their participation and cooperation which made this volume possible. Finally, we would like to thank the team of Springer-Verlag, especially Dr. Christoph Baumann, for the excellent cooperation during the preparation of this volume.

November 2011

Andreas Öchsner
Lucas F. M. da Silva
Holm Altenbach

Contents

Part I
Materials and Properties

A Finite Element Simulation of Longitudinal Impact Waves in Elastic Rods

Hesham A. Elkaranshawy and Nasser S. Bajaba

Abstract In this chapter, wave propagation in a thin rod struck by a rigid mass is considered and a finite element simulation of the system is developed. Both cases of free–free and fixed-free rods are considered. Though impact generates a propagating stress wave in both cases, the free–free rod is going to have a rigid-body motion. The analytical equations of motion are presented and the corresponding finite element equations are derived. A numerical scheme is constructed and solutions are obtained using Newmark implicit integration method and Newton–Raphson iterative technique. Solutions include time histories of displacement, velocity, stress, and contact force. The contact force is calculated, according to St. Venant's impact model. Numerical results of the simulation are compared to traditional analytical results. A simulated visualization of the propagation of the stress wave in the rod is presented, which enhances the understanding of this complicated physical phenomenon. The achieved results are accurate enough to have confidence in using this model for practical applications in wave propagation simulation and analysis.

Keywords Longitudinal impact · Stress and strain analysis · Wave propagation · Finite element simulation

H. A. Elkaranshawy (✉)
Department of Engineering Mathematics and Physics,
Faculty of Engineering, Alexandria University,
P.O. Box 21544, Alexandria, Egypt
e-mail: hesham_elk@yahoo.ca

N. S. Bajaba
Department of Mechanical Engineering Technology,
Yanbu Industrial College, Yanbu, Saudi Arabia
e-mail: nbajabaa@gmail.com

A. Öchsner et al. (eds.), *Materials with Complex Behaviour II*,
Advanced Structured Materials 16, DOI: 10.1007/978-3-642-22700-4_1,

Abbreviations

A	Cross-sectional area (m^2)
c	Wave propagation velocity (m/s)
E	Young's modulus (N/m^2)
F	Contact force (N)
$\{f\}$	Global force vector (N)
$\{f\}_{el}$	Force vector for element in contact with the rigid mass (N)
$[K]$	Global stiffness matrix (N/m)
l	Length (m)
L	Lagrangean (J)
m	Mass of the rod (kg)
m_0	Mass of the rigid mass (kg)
$[M]$	Global mass matrix (kg)
$[N]$	Finite element shape functions (m/m)
q	Displacement of the rigid mass (m)
$\dot{q}$	Velocity of the rigid mass (m/s)
t	Time (s)
t_c	Contact period (s)
T	Kinetic energy (J)
u	Displacement of the rod at position x(m)
$\{U\}$	Nodal displacement vector (m)
$\{\dot{U}\}$	Velocity vector (m/s)
$\{\ddot{U}\}$	Acceleration vector (m/s^2)
$\{U\}_{el}$	Nodal displacement vector element in contact with the rigid mass (m)
U_s	Strain energy (J)
υ_0	Initial velocity of the rigid mass (m/s)
W	Work done (J)
x	Position in the rod (m)
ε	Strain in the rod (m/m)
Π	Potential energy (J)
ρ	Density (kg/m^3)
σ	Stress in the rod (N/m^2)
τ	Time for the wave to travel across the rod from one end to the other end (s)
$()_N$	Value at the previous time step ($N = 0,1,2,\ldots$)
$()_{N+1}$	Value at the current time step ($N = 0,1,2,\ldots$)

1 Introduction

Investigation of wave propagation in a rod due to impact has a long history. Bernolli, Navier, Poisson, and St. Venant are among the great researchers who investigated this problem. Good reviews of the treatments of longitudinal waves in rods produced by impact are offered in [1, 2]. Recently, due to the presence of powerful computers, new computational methods are applied to solve this classical problem. An analytical simulation, symbolic solution, and a solution using time delay method have been developed [3–5]. Both theoretical and experimental researches were conducted [6–8] and a review of the experimental studies is offered in [9].

One practical device which utilizes longitudinal wave propagation in rods is the Hopkinson (or Davies) bar. The device is used to calibrate shock accelerometers under high acceleration levels and a wide frequency bandwidth. The bar is a long, thin, and elastic rod, in which a stress wave is generated at one end by a projectile impact. The projectile is a rigid mass or a striker bar. At the other end of the bar the generated wave can be used in many applications. The propagation of the shock wave in a Hopkinson bar is modeled [10, 11]. Insertion of a deformable disk between the projectile and the bar can decrease the wave dispersion, hence, a commercial finite element code was utilized to investigate dispersion in the bar and to find the optimum characteristics of the inserted deformable disk [12].

Wave propagation can be used in the determination of mechanical properties of materials. Some dynamic strength material constants were obtained using the split Hopkinson pressure bar [13]. The split tensile Hopkinson bar tests are interrupted to evaluate the damage in the materials at high strain rate [14]. The evaluation of the coefficient of restitution, through numerical simulation of impact of a rigid mass and a slender elastic rod, was investigated [15, 16]. Furthermore, there is an increasing interest in using wave propagation in crack detection, for example, wave propagations in cracked beams and plates were examined [17, 18].

Some machine elements are rod-like bodies that are subjected to impact loading during their functional operations. Examples are encountered in piling, percussive drilling and hydraulic hammering. Due to the elasticity of these axial elements, waves propagate through them while they are in translational motion. At the same time, it is obvious that wave propagation is gaining more potential in non-destructive testing methods. Therefore, reliable finite element models are needed to be used in the simulation of the propagation of waves. In this chapter, a finite element model is constructed to represent impact of a rigid mass on a flexible rod. The model overcomes the limitations in the previously reviewed works where the impact forces were assumed, see [17, 18], or calculated using methods that are highly time consuming, see [15, 16]. Hence, the contact force is calculated using an efficient approach utilizing the St.Venant's classical impact model. The two cases of free–free and free-fixed elastic rods are investigated. A numerical scheme is formulated depending upon Newmark implicit time stepping method and Newton–Raphson iterative method. The contact force is calculated and the wave propagation in the rod is simulated. To enhance the understanding of the complicated physical phenomenon, a simulated visualization of the propagation of the impact wave through the bar is monitored.

2 Mathematical Modeling

It is assumed that the rod has mass m, Young's modulus E, density ρ, cross-sectional area A and length l. The rod is initially at rest and is struck on the right end $x = l$ at the initial time $t = 0$ by a moving rigid mass m_0 with initial velocity υ_0. The displacement of the rigid mass at time t is donated by $q(t)$ and the displacement of the rod at position x and time t is given by $u(x, t)$, as in Fig. 1.

The governing equation for the longitudinal wave in the rod is

$$\frac{\partial^2 u(x,t)}{\partial t^2} = c^2 \frac{\partial^2 u(x,t)}{\partial x^2} \tag{1}$$

where c is the wave propagation velocity

$$c = \sqrt{\frac{E}{\rho}} \tag{2}$$

The strain $\varepsilon(x, t)$ in the rod is given by

$$\varepsilon(x,t) = \frac{\partial u(x,t)}{\partial x} \tag{3}$$

For an elastic rod, the stress is proportional to strain, or

$$\sigma(x,t) = E \frac{\partial u(x,t)}{\partial x} \tag{4}$$

As contact is established between the mass and the rod, both the mass and the contact end of the rod ($x = l$) are assumed to have the same velocity υ_0. Therefore, a compression wave is created in the rod. The wave travels along the rod and is reflected at the other end ($x = 0$). During the contact period, displacement $q(t)$ and velocity $\dot{q}(t)$ of the mass are the same as those of the contact end of the rod ($x = l$).

$$q(t) = u(l,t) \quad \text{and} \quad \dot{q}(t) = \frac{\partial u}{\partial t}(l,t), \quad 0 < t < t_c \tag{5}$$

where t_c is the contact period.

The contact persists as long as the contact force between the mass and contact end of the rod does not vanish. The contact force equals the stress at the contact end times the rod's cross-sectional area, i.e.

$$F(t) = EA \frac{\partial u(l,t)}{\partial x} \tag{6}$$

The motion of the rigid mass is governed by

$$m \frac{d\dot{q}}{dt} = F(t) \tag{7}$$

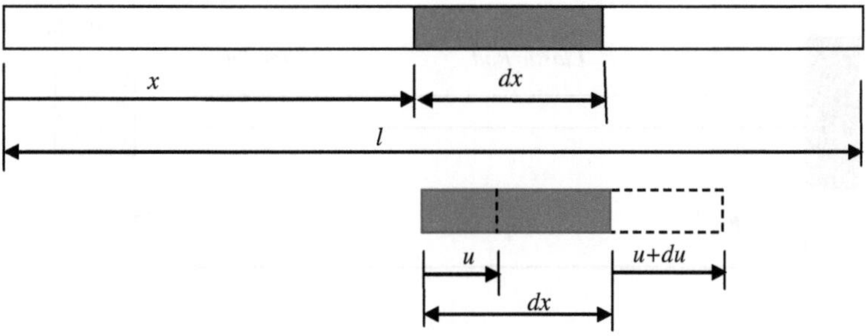

Fig. 1 Displacement of the rod at a general position x

Equations (1), (6), and (7) are the equations of motion of the rod and the rigid mass during the impact period. After the cease of impact the motion of the rod is controlled by Eq. (1). In the same time, since $F(t)$vanishes, Eq. (7) declares that the rigid mass moves with a constant velocity.

3 Finite Element Solutions

The pre-mentioned differential formulation of the equations of motion is equivalent to integral formulation, which requires the application of Lagrange's equation of motion. First, one defines Lagrangean 'L' by

$$L = T - \pi \tag{8}$$

where 'T' is the kinetic energy and 'π' is the potential energy defined by

$$\pi = U_s - W \tag{9}$$

U_s and W are the strain energy and the work done, respectively, that are given by

$$U_s = \sum\nolimits_e \frac{1}{2} \int\limits_e EA \left(\frac{\partial u}{\partial x} \right)^2 dx \tag{10}$$

$$T = \sum\nolimits_e \frac{1}{2} \int\limits_e \rho A \left(\frac{\partial u}{\partial t} \right)^2 dx + \frac{1}{2} m_0 \left[\frac{\partial u}{\partial t} (x = l) \right]^2 \tag{11}$$

$$W = F(t) u(l, t) \tag{12}$$

The finite element shape functions $[N(x)]$ link the displacement 'u' to the nodal displacement vector $\{U\}$ through

$$u(x, t) = [N]\{U\} \tag{13}$$

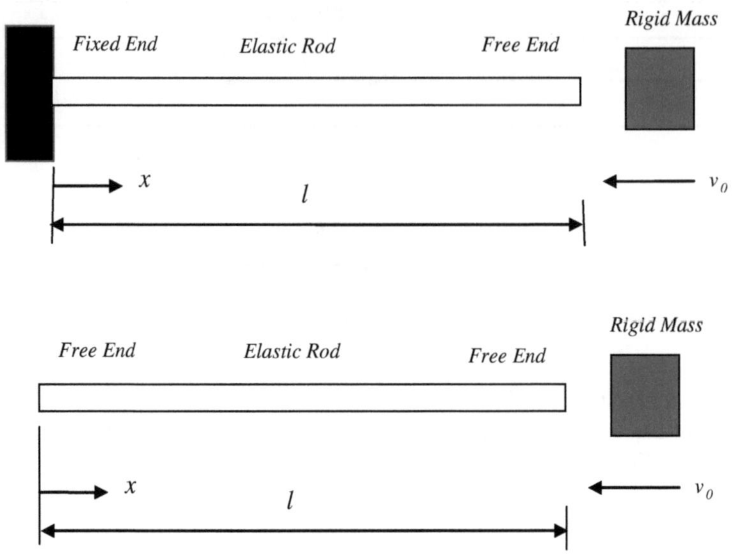

Fig. 2 Longitudinal impact of a mass on a rod (free-fixed and free–free)

Consequently;

$$L = \frac{1}{2}\{\dot{U}\}^T[M]\{\dot{U}\} - \frac{1}{2}\{U\}^T[K]\{U\} + \{f(t)\}^T\{U\} \tag{14}$$

$[M]$ and $[K]$ are the global mass and stiffness matrices and $\{f(t)\}$ is the global force vector. $\{f(t)\}$ contains only the nodal forces of the last element, $\{f(t)\}_{el}$, due to the contact force. The rest of the global force vector is full of zeros. $\{f(t)\}_{el}$ is given by

$$\{f(t)\}_{el} = F(t)[N(x = l)]^T \tag{15}$$

The Lagrange's equation of motion is given by

$$\frac{d}{dt}\left(\frac{\partial L}{\partial\{\dot{U}\}}\right) - \frac{\partial L}{\partial\{U\}} = 0 \tag{16}$$

this leads to

$$[M]\{\ddot{U}\} + [K]\{U\} = \{f(t)\} \tag{17}$$

Equations (6) and (13) give

$$F(t) = EA\left[\frac{\partial N}{\partial x}(x = l)\right]\{U\}_{el} \tag{18}$$

where $\{U\}_{el}$is the nodal displacement vector for the last element, which is the element in contact with the rigid mass.

Equations (17) and (18) are the finite element equations of motion during impact. These equations are applied for both cases of free- free rod and free-fixed rod, see Fig. 2. In the case of free-fixed bar both $[M]$ and $[K]$ are positive definite matrices. For the free–free bar, though $[M]$is positive definite matrix, $[K]$ is positive semi-definite matrix due to the existence of rigid body modes.

Newmark implicit time stepping method (Bathe [19]) is used to express the current velocity $\{\dot{U}\}_{N+1}$ and acceleration$\{\ddot{U}\}_{N+1}$ in terms of the current displacement $\{U\}_{N+1}$ and previously determined values of displacement $\{U\}_N$, velocity $\{\dot{U}\}_N$, and acceleration$\{\ddot{U}\}_N$. Combining these equations with the equations of motion (17) and (18) yields a system of algebraic equations in terms of $\{U\}_{N+1}$ and $F(t)_{N+1}$. The Newton–Raphson iterative method (Bathe [19]) is used to solve the resulting equations to find the current displacement and contact force. The displacement and other variables' distributions in the rod at the end of impact serve as the initial conditions for the subsequent free vibrations of the bar, which are governed by the solution of Eq. (17) while Eq. (18) is no longer relevant.

4 Numerical Simulation

Numerical simulations, for a rigid mass collides with a free–free elastic rod and with a free-fixed elastic rod, are presented in this section, see Fig. 2. The rod in both cases is an aluminum rod with a 3 mm × 25 mm cross section, 200 mm length, 70 GN/m^2 Young's modulus, and 2,710 kg/m^3 mass density. The rigid mass has the same mass as the rod. The mass is moving towards the rod with a velocity of 1 m/s. Fifty elements are used to model the rod in the finite element model. The elements are two-nodes and one-dimensional linear elastic elements. The velocity of the created wave is $c = 5082.35$ m/s and the time for the wave to travel across the rod from one end to the other end is $\tau = \frac{l}{c} = 3.935 \times 10^{-5}$ s.

According to St. Venant's principle, as contact starts the velocity of the contact end becomes immediately equals to the rigid mass velocity and right away a compression wave is created at the contact end and travels across the rod with velocity 'c'. The initial compression stress at the contact end is $\sigma_0 = v_0\sqrt{E\rho}$ and the stress at that end starts to decrease with time until the reflected wave reaches that end.

For the bar with the other end free, the stress at the free end is always zero, therefore, the traveling compression stress wave is reflected at the free end as a tension wave and whenever that tension wave reaches the contact end at time τ, it cancels out the stress at that end and contact is terminated. Following the cease of contact, the wave is reflected from the contact end as compression wave and periodic cycles start with a period equals to τ.

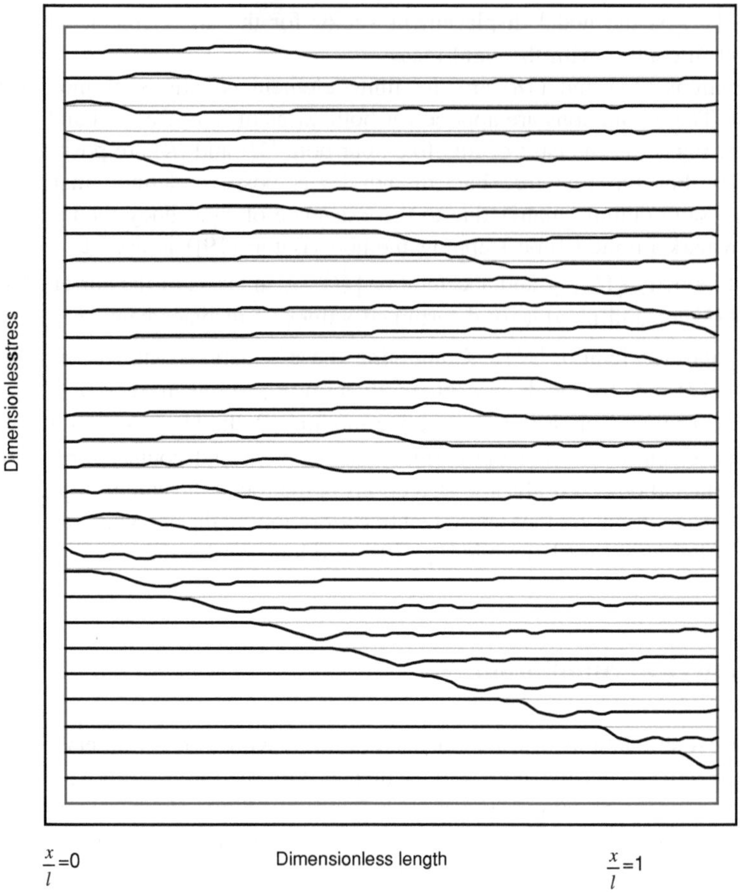

Fig. 3 Stress wave propagation in the free–free bar; waves of dimensionless stress $\frac{-c}{v_0 E}\sigma$ verses dimensionless length $\frac{x}{l}$ are shown at times 0.125 $\frac{c}{l}t$ apart

The finite element solutions successfully predict this phenomenon as can be seen in Figs. 3, 4, 5, 6, 7.

Figures 4, 5, 6, 7 show the dimensionless displacement $\frac{c}{v_0 l}u$, velocity $\frac{v}{v_0}$, stress $\frac{-c}{v_0 E}\sigma$, and contact force $\frac{1}{v_0 A\sqrt{E\rho}}F$, respectively, with respect to dimensionless time $\frac{c}{l}t$. Slight numerical damping is introduced to reduce the oscillations in solutions. Dimensionless analytical solutions are given in [4]. Very good agreement is found between the solutions of the proposed finite element model and the analytical solutions [1, 2]. Figure 3 shows the distribution of the dimensionless stress over the dimensionless length $\frac{x}{l}$, at equal dimensionless time steps of 0.125. Therefore, the wave propagation can be visualized in that figure. Figures 3 and 7 show that the arrival of the reflected tension wave into the contact end nullifies the contact

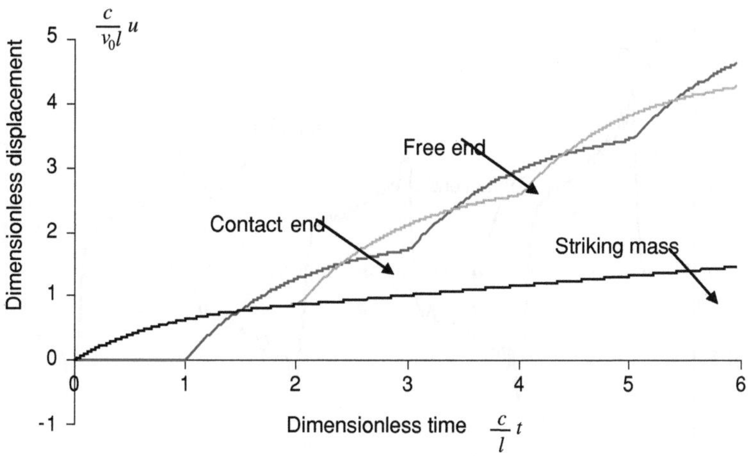

Fig. 4 Dimensionless displacements of free end, contact end and striking mass (free–free bar)

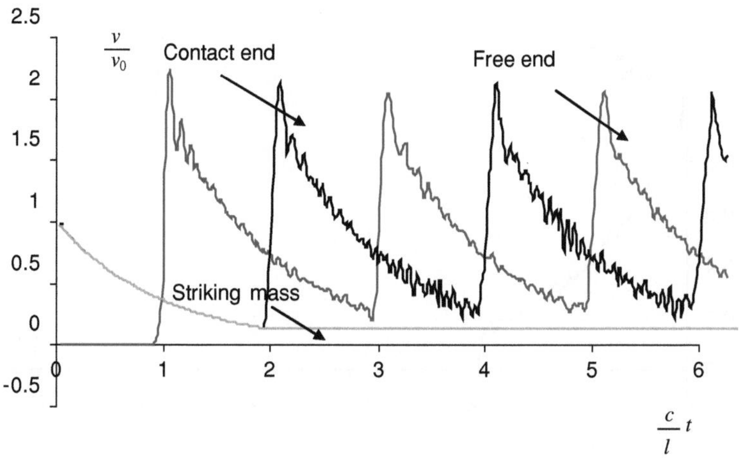

Fig. 5 Dimensionless velocities of free end, contact end and striking mass (free–free bar); $(\frac{v}{v_0})$ is the dimensionless velocity and $(\frac{c}{l}t)$ is the dimensionless time

force. Therefore, it marks the end of impact. Most of the time, a portion of the rod is in tension while the other portion is in compression, as can be seen in Fig. 3. Therefore, the mid-point stress alternate between compression and tension marked by the arrival of the wave at that point, see Figs. 3 and 6. Though the displacements of the bar ends are continuous, see Fig. 4, the slope of each displacement history suffers discontinuity corresponding to the arrival of the wave at that end,

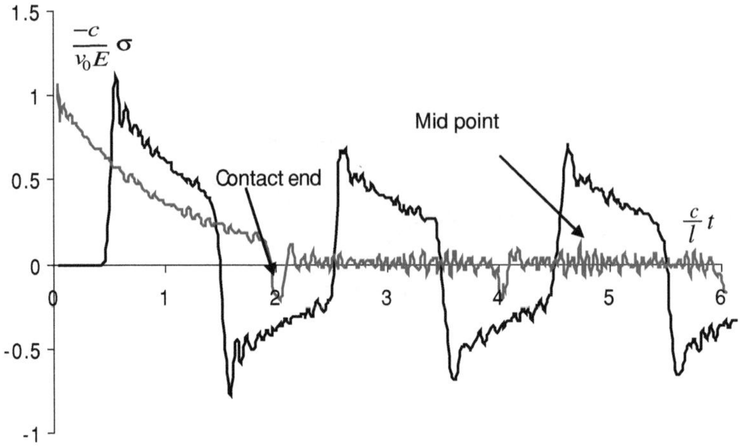

Fig. 6 Dimensionless stresses at mid point and contact end (free–free bar); $\frac{-c}{v_0 E}\sigma$ is the dimensionless stress and $\frac{c}{l}t$ is the dimensionless time

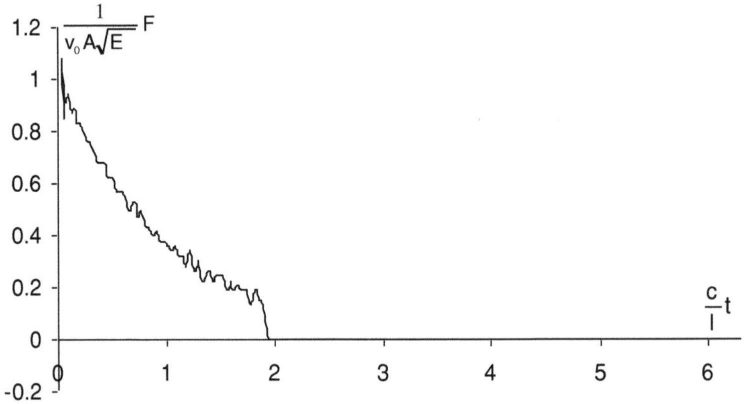

Fig. 7 Dimensionless contact force (free–free bar); $\frac{1}{v_0 A\sqrt{E\rho}}F$ is the dimensionless force and $\frac{c}{l}t$ is the dimensionless time

which is reflected in the discontinuity of the velocities, Fig. 5. The time history of velocity in Fig. 5 indicates that after the end of impact, the striking mass does not change its original moving direction and the bar starts a continued free vibration. The bar has an average rigid body motion velocity and for each end, the velocity is varying between two limits. The arrival of the wave at each bar end increases the velocity of that end impulsively to its maximum value. The analytical solutions given by Goldsmith [1] predict that the final dimensionless velocity of the rigid

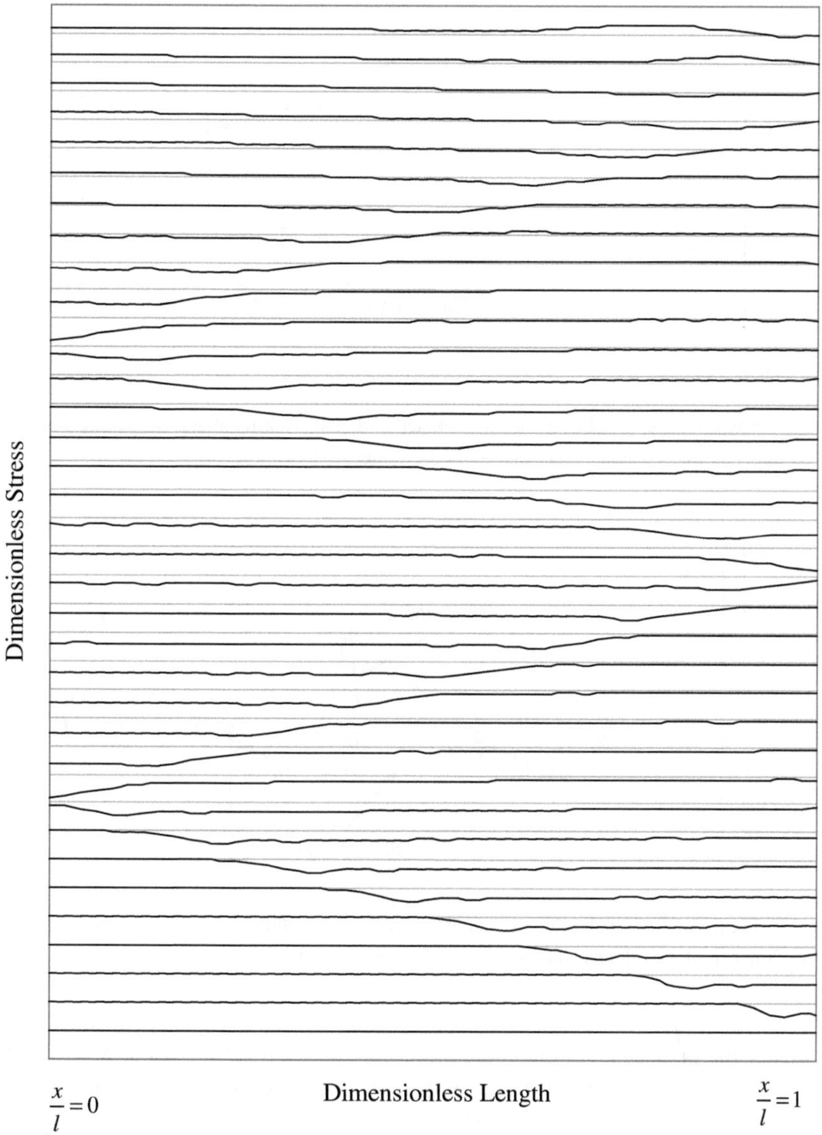

Fig. 8 Stress wave propagation in the fixed-free bar; waves of dimensionless stress $\frac{-c}{v_0 E}\sigma$ verses dimensionless length $\frac{x}{l}$ are shown at times $0.125\,\frac{c}{l}t$ apart

mass to be 0.1353 and the present simulation predicts 0.1469, see Fig. 5. At the arrival of the reflected wave to the contact end, the analytical dimensionless stress is 0.1353 and in Fig. 6 the finite element calculates 0.1325.

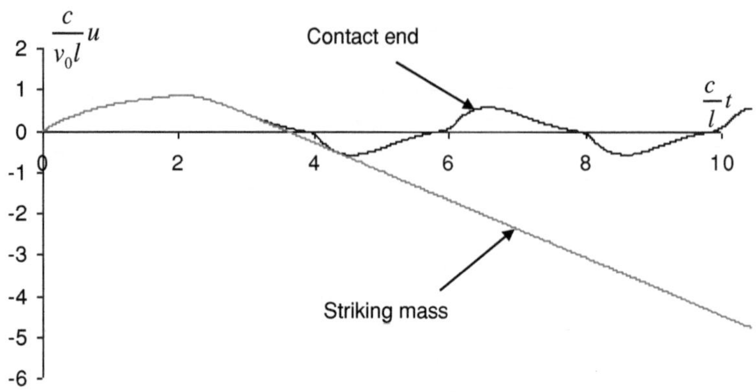

Fig. 9 Dimensionless displacements of contact end and striking mass (fixed-free bar); $\frac{c}{v_0 l}u$ is the dimensionless displacement and $\frac{c}{l}t$ is the dimensionless time

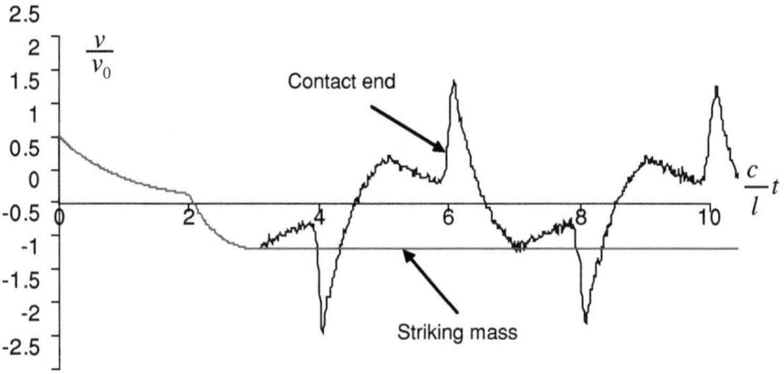

Fig. 10 Dimensionless velocities of contact end and striking mass (fixed-free bar); $(\frac{v}{v_0})$ is the dimensionless velocity and $(\frac{c}{l}t)$ is the dimensionless time

Using the present finite element simulation, a visualization of the wave motion is illustrated in Fig. 8 for the bar with one fixed end. The figure illustrates the distribution of the dimensionless stress over the dimensionless length $\frac{x}{l}$, at equal dimensionless time steps of 0.125. It shows that the traveling compression stress wave is reflected at the fixed end as a compression wave, as expected. Since the contact is not terminated yet, the contact end operates as a fixed end and the compression wave is reflected from that end as a compression wave again. Once more, the wave is reflected as a compression wave at the fixed end, but shortly after that the contact is terminated. For that reason, during the contact period, the whole rod is under compression all the time, see Fig. 8. Without presenting a similar figure, analytical solutions given in [1, 2] predict the same phenomena. After the end of impact, the subsequent free vibration of the bar has periodic cycles

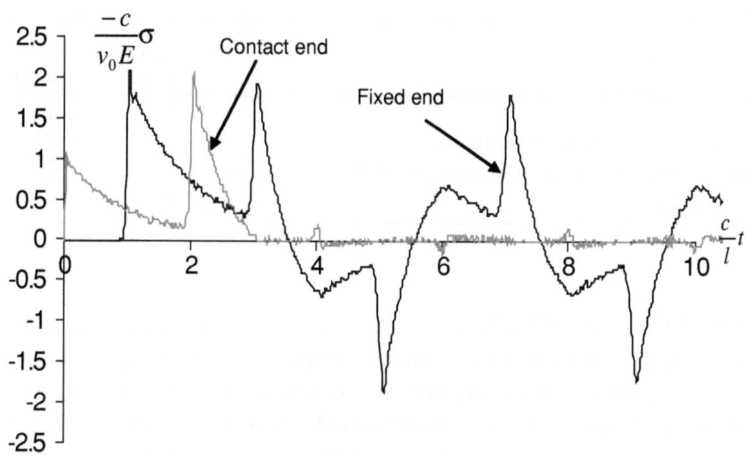

Fig. 11 Dimensionless stresses of contact end and fixed end (fixed-free bar); $\frac{-c}{v_0 E}\sigma$ is the dimensionless stress and $\frac{c}{l}t$ is the dimensionless time

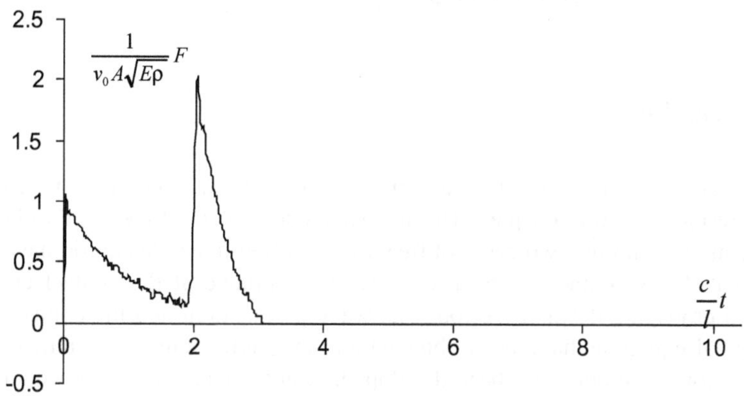

Fig. 12 Dimensionless contact force (fixed-free bar), $\frac{1}{v_0 A\sqrt{E\rho}}$ $F$is the dimensionless force and $\frac{c}{l}t$ is the dimensionless time

with a period equals to 2τ, as can be seen in Figs. 8, 9, 10, and 11. As anticipated, the free end reflects the wave with opposite polarity. Therefore, during these periods, most of the time a part of the rod is in compression while the other part is in tension, see Fig. 8. The time histories of the bar displacement, velocity, stress, and contact force in dimensionless forms are shown in Figs. 9, 10, 11, and 12. The figures illustrate that the displacement is continuous while velocity, stress, and contact force suffer discontinuities. At any location in the bar, the discontinuities occur at intervals corresponding to the arrival of the waves to that location; see

Table 1 Comparison between the analytical and the finite element results (free-fixed rod)

Dimensionless values	Analytical results [1, 2]	Proposed finite element results
Duration time	3.068	3.075
Displacement of contact end at separation	0.375	0.371
Maximum value of displacement of contact end after separations	0.576	0.576
Maximum contact force	2.135	2.045

Figs. 8, 9, 10, 11, and 12. Figures 8, 11 and 12 confirm that the arrival of the reflected compression wave to the contact end raises the stress at the contact end, and accordingly the contact force, to its maximum value. Next, contact force starts to decrease and impact is terminated when the stress at the contact end vanishes. The analytical solutions given in [1, 2] predict the dimensionless duration time, displacement of contact end at separation and its maximum value after separation, and maximum contact force. Both the analytical results and the corresponding results of the current finite element simulation are given in Table 1.

It has to be noticed that slight numerical damping is introduced to reduce the oscillations in the numerical solutions.

5 Conclusion

A finite element simulation for the impact of a rigid mass on an elastic rod has been presented in this chapter. The impact model utilizes St.Venant's classical impact model, and the two cases of free–free and free-fixed elastic rod have been investigated. As contact established, a wave is initiated at the contact end and starts to propagate through the rod. The wave propagation and the contact force differential equations have been obtained and the finite element discretization of the equations of motion has been developed. A numerical solution procedure has been proposed along the lines of Newmark implicit integration method and Newton–Raphson iterative technique. The current simulation calculates the contact force accurately and efficiently which is a significant advantage over other simulations, which just assume the contact force or calculate it inefficiently.

Results show the variation of contact force, displacements, velocities, and stresses with respect to time for both cases of free and fixed far end of the bar. Very good agreement has been found between numerical results and the well-known analytical results. A simulated visualization of the propagation of the stress wave through the bar has been developed. This visualization enhances the understanding of the physical phenomena of impact and wave propagation including the reflection of the wave at free and fixed ends as well as at the contact end. The results demonstrate that the proposed finite element simulation is accurate enough for further investigation in wave analysis and simulation.

References

1. Goldsmith, W.: Impact: The Theory and Physical Behaviour of Colliding Solids. Edward Arnold, Ltd, London (1960)
2. Timoshenko, S.P., Goodier, J.N.: Theory of Elasticity. McGraw Hill, New York (1970)
3. Shi, P.: Simulation of impact involving an elastic rod. Comput. Met. Appl. Mech. Eng. **151**, 497–499 (1997)
4. Hu, B., Eberhard, P.: Symbolic computation of longitudinal impact waves. Comput. Met. Appl. Mech. Eng. **190**, 4805–4815 (2001)
5. Hu, B., Eberhard, P.: Simulation of longitudinal impact waves using time delayed systems. J. Dyn. Syst., Meas. Control (Special Issue) **126**(3), 644–649 (2004)
6. Keskinen, E., Kuokkala, V.-T., Vuoristo, T., Martikainen, M.: "Multi-body wave analysis of axially elastic rod systems," Proc. Instn. Mech. Engrs, Part K. J. Multi-body Dyn. **221**, 417–428 (2007)
7. Maekawa, I., Tanabe, Y., Suzuki, M.: Impact stress in a finite rod. JSME Int. J. Ser. I **31**, 554–560 (1988)
8. Hu, B., Schiehlen, W., Eberhard, P.: Comparison of analytical and experimental results for longitudinal impacts elastic rods. J. Vib. Control **9**, 157–174 (2003)
9. Al-Mousawi, M.M.: On experimental studies of longitudinal and flexural wave propagations: an annotated bibliography. Appl. Mech. Rev **39**, 853–864 (1986)
10. Ueda, K., Umeda, A.: Characterization of shock accelerometers using Davies bar and strain-gages. Exp. Mech. **33**(3), 228–233 (1993)
11. Rusovici, R.: "Modeling of shock wave propagation and attenuation in viscoelastic structure," Ph.D. Dissertation, Virginia polytechnic institute and state university (1999)
12. Ramirez, H., Rubio-Gonzalez, C.: Finite-element simulation of wave propagation and dispersion in Hopkinson bar test. Mater. Des. **27**, 36–44 (2006)
13. Allen, D.J., Rule, W.K., Jones, S.E.: Optimizing material strength constants numerically extracted from Taylor impact data. Exp. Mech. **37**(3), 333–338 (1997)
14. El-Saeid Essa, Y., Lopez-Puente, J., Perez-Castellanos, J.L.: Numerical simulation and experimental study of a mechanism for Hopkinson bar test interruption. J. Strain Anal. Eng. Des. **42**(3), 163–172 (2007)
15. Seifried, R., Schiehlen, W., Eberhard, P.: Numerical and experimental evaluation of the coefficient of restitution for repeated impacts. Int. J. Impact Eng. **32**(1–4), 508–534 (2005)
16. Seifried, R., Eberhard, P.: Comparison of numerical and experimental results for impacts. ENOC-2005, pp. 399–408. Eindhoven, Netherlands, (2005) 7–12 August
17. Krawczuk, M.: Application of spectral beam finite element with crack and iterative search technique for damage detection. Finite Elements Anal. Des. **38**(6), 537–548 (2002)
18. Krawczuk, M., Palacz, M., Ostachowicz, W.: Wave propagation in plate structures for crack detection. Finite Elements Anal. Des. **40**(9–10), 991–1004 (2004)
19. Bathe, K.-J.: Finite Element Procedures. Prentice Hall, Upper Saddle River (1996)

Hamiltonian Formalisms Applied to Continuum Mechanics: Potential Use for Fracture Mechanics

N. Recho

Abstract The first part of this chapter deals with several Hamiltonian formalisms in elasticity. The formalisms of Zhong ((1995) Dalian Science & Technology University Press, Liaoning, China) and Bui ((1993) Introduction aux problèmes inverses en mécaniques des matériaux, Editions Eyrolles, Paris), which resolve respectively the two-end problem and the Cauchy problem in elasticity, are presented briefly. Then we propose a new Hamiltonian formalism, which resolves simultaneously the two problems mentioned above and shows the link between the two formalisms. The potential use for fracture mechanics purposes is then mentioned. In fact, when traditional theories in fracture mechanics are used, asymptotic analyses are often carried out by using high-order differential equations governing the stress field near the crack tip. The solution of the high-order differential equations becomes difficult when one deals with anisotropic or multilayer media etc. The key of our idea was to introduce the Hamiltonian system, usually studied in rational mechanics, into continuum mechanics. By this way, one can obtain a system of first-order differential equations, instead of the high-order differential equation. This method is very efficient and quite simple to obtain a solution of the governing equations of this class of problems. It allows dealing with a large range of problems, which may be difficult to resolve by using traditional methods. Also, recently we developed another new way to resolve fracture mechanics problems with the use of ordinary differential equations (ODEs) with respect to the circumferential coordinate θ around the crack (or notch) tip. This method presents the opportunity to be coupled with finite element analysis and then allows resolving more complicated geometries.

Keywords Hamiltonian analysis · Stress-singular fields · V-notch · Boundary element method

N. Recho (✉)
ERMESS/EPF-Ecoles d'Ingénieurs, Sceaux and LaMI, University Blaise Pascal, Clermont II, France
e-mail: naman.recho@epf.fr; recho@moniut.univ-bpclermont.fr

A. Öchsner et al. (eds.), *Materials with Complex Behaviour II*, Advanced Structured Materials 16, DOI: 10.1007/978-3-642-22700-4_2,

1 Introduction

Recently, an important effort has been made in the reform of the classical theory of continuum mechanics in the frame of the Hamiltonian system. In these new approaches, the principle of Hamilton is applied in a special manner, i.e., by considering a dimensional parameter as "time". In this topic, we can distinguish two formalisms: the formalism of Bui [1] and the formalism of Zhong [2]. By seeking the variations of the couple (displacements, traction forces) on an arbitrary front in a solid when this front virtually moves from an initial position to a neighbor one, a first-order differential equation system governing the mechanical fields was explicitly established. That is the Cauchy problem in elasticity resolved by Bui. On the other hand, the formalism of Zhong looks more classical. In simple words, he established an analogy between quantities in rational mechanics and those in continuum mechanics. For example, a dimensional coordinate in continuum mechanics is considered as time in rational mechanics; the displacement vector as the generalized coordinates; the strain energy density as the Lagrange function and so on. This analogy leads to the canonical equations of Hamilton governing the mechanical fields in elastic bodies. The main advantage of these approaches is that the fundamental equations can directly be resolved. The traditional semi-inverse method is then replaced by a direct, systematic and more structural resolution method.

2 Zhong's Formalism: The Two-End Problem

Let us consider a solid V described by a coordinate system Z in which z is one chosen coordinate. Let us consider now q the displacements in the Z system associated to neighbor displacements, $q + \delta q$. One notes $\dot{\boldsymbol{q}} = \frac{\partial \boldsymbol{q}}{\partial z}$. If we suppose that the displacements are imposed at $z = z_0$ and $z = z_1$, named the two end points, then we have:

$$\delta \boldsymbol{q}(z = z_0) = \delta \boldsymbol{q}(z = z_1) = \boldsymbol{0} \tag{1}$$

Let us write the total potential energy Π of the solid:

$$\Pi = \int_{z_0}^{z_1} \int_{S} (U_0 - W) dS dz = \int_{z_0}^{z_1} L dz \quad \text{avec} \quad L = \int_{S} (U_0 - W) dS \tag{2}$$

where U_0 is the strain energy density and W is the work density of the external forces. We define the Lagrange function as the integral over S. If S is constant along z and we neglect the body forces and we just consider a volume element inside the solid, we can write $L = U_0 - W$. In general, L is a function of $\boldsymbol{q}$ and $\dot{q}$. Following the principle of the minimum of total potential energy, $\delta \Pi = 0$ with respect to δq and using the conditions (1), one obtains the Euler equation in L:

$$\frac{\partial L}{\partial \boldsymbol{q}} - \frac{\partial}{\partial z}\frac{\partial L}{\partial \dot{q}} = 0 \tag{3}$$

In rational mechanics, L is named Lagrange's function, and (3) Lagrange's equation. Then we construct the Hamilton function $H(\boldsymbol{p}, \boldsymbol{q})$ through the Legendre's transformation:

$$\begin{aligned} \boldsymbol{p} &= \frac{\partial L(\boldsymbol{q}, \dot{\boldsymbol{q}})}{\partial \dot{q}} \\ H(p, \boldsymbol{q}) &= \boldsymbol{p}^T \dot{q} - L(\boldsymbol{q}, \dot{\boldsymbol{q}}) \end{aligned} \tag{4}$$

From (3) and (4), one deduces immediately the canonical equations of Hamilton:

$$\frac{\partial H}{\partial \boldsymbol{q}} = -\frac{\partial L}{\partial \boldsymbol{q}} = -\dot{\boldsymbol{p}} \quad \frac{\partial H}{\partial p} = \dot{q} \tag{5}$$

$\boldsymbol{q}$ and $\boldsymbol{p}$ are dual conjugate variables. Differently from rational mechanics, these two variables represent respectively the displacement vector and the normalized stress vector.

3 Bui's Formalism: Cauchy's Problem in Elasticity

Bui [1] has solved the Cauchy problem in elasticity by seeking the variations of the mechanical quantities ($\boldsymbol{q}$ as a displacement vector, $\boldsymbol{p}$ as a traction vector) at an arbitrary front in the solid when it moves from an initial position Γ_t to a neighbour position Γ_{t+dt}, where t defines the movement of the front in the solid. This approach leads to an explicit system of first-order differential equations.

Let us consider a domain divided into two parts Ω and Ω_t by a contour Γ_t. Suppose that mechanical fields are known at the interior of the contour; consequently $\boldsymbol{q}$ and $\boldsymbol{p}$ are known at the contour Γ_t. Suppose $\boldsymbol{q}'$ a virtual compatible displacement. The virtual work principle leads to:

$$\int_{\Omega_t} \nabla \boldsymbol{q}.\boldsymbol{\Lambda}.\nabla q' d\Omega = \int_{\Gamma_t} \boldsymbol{q} \cdot q' d\Gamma \tag{6}$$

Λ is the elastic tensor. Let us consider now an evolution of Γ_t to Ω_t, i.e. at $t + dt$, the contour Γ_t reaches Γ_{t+dt}. It's suitable to consider that Γ_{t+dt} is deduced from Γ_t following the normal to Γ_t with a quantity $\psi \boldsymbol{n} dt$ where $\boldsymbol{n}$ is a unit vector normal to the contour and ψ is a positive scalar field describing the velocity of the contour evolution. The derivation of (6) with respect to dt gives:

$$\frac{d}{dt}\int_{\Omega_t} \nabla \boldsymbol{q}.\boldsymbol{\Lambda}.\nabla q' d\Omega = \frac{d}{dt}\int_{\Gamma_t} \boldsymbol{p} \cdot q' d\Gamma \tag{7}$$

If introducing the following notations of tangential operators:

$$\mathrm{grad}_\Gamma(\cdot) := \nabla(\cdot) - \mathbf{n}\tfrac{\partial}{\partial n}(\cdot) \quad \mathrm{div}_\Gamma(\cdot) := \mathrm{div}(\cdot) - \mathbf{n}\cdot\tfrac{\partial}{\partial n}(\cdot) \tag{8}$$

equation (7) leads to:

$$\int_{\Gamma_t} \nabla\mathbf{q}.\mathbf{\Lambda}.\nabla\mathbf{q}'\psi d\Gamma - \int_{\Gamma_t} \left\{\frac{d\mathbf{p}}{dt} + \mathrm{div}_\Gamma(\psi\mathbf{n})\mathbf{p}\right\}\cdot\mathbf{q}' d\Gamma - \int_{\Gamma_t} \psi\mathbf{p}\cdot\frac{\partial\mathbf{q}'}{\partial n} d\Gamma = 0 \tag{9}$$

After rearrangement and integration by parts, one can deduce the following differential equations:

$$\begin{aligned} \frac{d\boldsymbol{q}}{dt} &= \boldsymbol{B}_q(\boldsymbol{q},\boldsymbol{p},\psi) \\ \frac{d\boldsymbol{p}}{dt} &= \boldsymbol{B}_p(\boldsymbol{q},\boldsymbol{p},\psi) \end{aligned} \tag{10}$$

$\boldsymbol{B}_q$ and $\boldsymbol{B}_p$ are expressed as function of quantities defined on the contour Γ_t. Their explicit expressions are given in the [1].

4 Unified Description of the Two Formalisms

Here we describe a formalism unifying the two precedents within the frame of minimization of the total potential energy of the structure [3].

4.1 Hamilton Principle Written as Variation of Total Potential Energy

Following (2) and (4), the total potential energy is written as:

$$\Pi = \int_{z_1}^{z_2} L dz = \int_{z_1}^{z_2} (\boldsymbol{p}\cdot\dot{\boldsymbol{q}} - H)\, dz \tag{11}$$

u is a parameter describing the solid's evolution. The description of a solid between an event a and an event b could be done under parametrical form of six functions in 2D media: two displacements $\boldsymbol{q}(u)$, three normalised stresses $\boldsymbol{p}(u)$ and one coordinate $z(u)$. Consider u_1 and u_2 as values of u corresponding to events a and b. For $z_1 = u_1$ and $z_2 = u_2$, the total potential energy is re-written as:

$$\Pi(u) = \int_{u_1}^{u_2} \left(\boldsymbol{p}\cdot\frac{\partial\boldsymbol{q}}{\partial u} - H\frac{\partial z}{\partial u}\right) du \tag{12}$$

And its variation becomes:

$$\delta\Pi = \frac{\partial \Pi}{\partial u}\delta u = \left\{ \begin{array}{l} \int\limits_{u_1}^{u_2} \left[\frac{\partial \mathbf{p}}{\partial u}\cdot\frac{\partial \mathbf{q}}{\partial u} - \frac{\partial \mathbf{p}}{\partial u}\cdot\frac{\partial \mathbf{q}}{\partial u} - \frac{\partial H}{\partial u}\cdot\frac{\partial z}{\partial u} + \frac{\partial H}{\partial u}\frac{\partial z}{\partial u} \right] du \\ \qquad\qquad\qquad + \left[\mathbf{p}\cdot\frac{\partial \mathbf{q}}{\partial u} - H\frac{\partial z}{\partial u} \right]_{u_1}^{u_2} \end{array} \right\} \delta u \tag{13}$$

One notes:

$$\frac{\partial \boldsymbol{q}}{\partial u}\delta u = \delta \boldsymbol{q};\quad \frac{\partial \boldsymbol{p}}{\partial u}\delta u = \delta \boldsymbol{p};\quad \frac{\partial H}{\partial u}\delta u = \delta H;\quad \frac{\partial z}{\partial u}\delta u = \delta z \tag{14}$$

When u represents the coordinate z, (13) is written as follow:

$$\delta\Pi = \int\limits_{z_1}^{z_2} \left[\dot{\boldsymbol{q}}.\delta\boldsymbol{p} - \dot{\boldsymbol{p}}.\delta\boldsymbol{q} - \delta H + \dot{H}\delta z \right] dz + \left[\boldsymbol{p}\cdot\delta\boldsymbol{q} - H\delta z \right]_{z_1}^{z_2} \tag{15}$$

So we have $\delta\Pi$ divided into two parts, the first one is an integral; the second one is in the square bracket.

4.2 Application to the Two-End Problem

Consider now the variation of $\boldsymbol{q}$ and z are zero at z_1 and z_2, $\delta\,\mathbf{q} = \mathbf{0}$ and $\delta\, z = 0$. This means we have fixed boundaries and fixed displacement boundary conditions at the two-ends, so we have got the so-called two end point problem. In this case, the quantities in the square bracket of equation (15) vanish. According to the principle of minimum total potential energy, we directly obtain the canonical equations of Hamilton. This is the problem resolved by the formalism of Zhong.

$$\delta\Pi = \int\limits_{z_1}^{z_2} \left[\dot{\mathbf{q}}.\delta\mathbf{p} - \dot{\mathbf{p}}.\delta\mathbf{q} - \delta H + \dot{H}\delta z \right] dz = 0 \tag{16}$$

δH being:$\delta H = \frac{\partial H}{\partial \boldsymbol{q}}\delta\mathbf{q} + \frac{\partial H}{\partial p}\delta\boldsymbol{p} + \frac{\partial H}{\partial z}\delta z$, one obtains:

$$\delta\Pi = \int\limits_{z_1}^{z_2} \left[\dot{\mathbf{q}}.\delta\mathbf{p} - \dot{\mathbf{p}}.\delta\mathbf{q} - \frac{\partial H}{\partial \mathbf{q}}\delta\mathbf{q} - \frac{\partial H}{\partial \mathbf{p}}\delta\mathbf{p} - \frac{\partial H}{\partial z}\delta z + \dot{H}\delta z \right] dz = 0 \tag{17}$$

This equation is available for arbitrary $\delta\boldsymbol{q}$, $\delta\boldsymbol{p}$ and δz. Consequently, we deduce the Hamilton canonical equations:

$$\frac{\partial H}{\partial z} = \dot{H};\quad \frac{\partial'}{\partial \boldsymbol{p}} = \dot{\boldsymbol{q}};\quad \frac{\partial H}{\partial \boldsymbol{q}} = -\dot{\boldsymbol{p}} \tag{18}$$

4.3 Application to Cauchy's Problem

Now consider a natural evolution of the structure, this means that the Hamilton canonical equations are satisfied everywhere in the structure, but with possible variations of $(\boldsymbol{q}, z)$ at $z = z_1$ and $z = z_2$. In this case, we have no fixed boundaries neither fixed boundary conditions at the two ends but we have *natural* evolution everywhere, this is the so called Cauchy problem. In this case, the integral in equation (15) vanishes i.e.:

$$\delta\Pi = \boldsymbol{p}_2 \cdot \delta\boldsymbol{q}_2 - H_2 \cdot \delta z_2 - \boldsymbol{p}_1 \cdot \delta\boldsymbol{q}_1 + H_1 \cdot \delta z_1 \tag{19}$$

For a small displacement of events a and b, the variation of the total potential energy is:

$$\delta\Pi = \delta\boldsymbol{q}_1 \cdot \frac{\partial\Pi}{\partial\boldsymbol{q}_1} + \delta z_1 \frac{\partial\Pi}{\partial z_2} + \delta\boldsymbol{q}_2 \cdot \frac{\partial\Pi}{\partial\boldsymbol{q}_2} + \delta z_2 \frac{\partial\Pi}{\partial z_2} \tag{20}$$

The variables $\boldsymbol{q}_1, z_1, \boldsymbol{q}_2, z_2$ are independent. By identification between (19) and (20), we have got the Hamilton–Jacobi equations:

$$\frac{\partial\Pi}{\partial\boldsymbol{q}_2} = p_2 \quad \frac{\partial\Pi}{\partial z_2} = -H_2 \quad \frac{\partial\Pi}{\partial\boldsymbol{q}_1} = -\boldsymbol{p}_1 \quad \frac{\partial\Pi}{\partial z_1} = H_1 \tag{21}$$

This is the problem resolved by Bui. We know that the Hamilton canonical equations and the Hamilton–Jacobi equations are equivalent. So we can say the formalism of Zhong and that of Bui are equivalent in the differential point of view, even they look quite different. Now, dealing with Bui's formalism, it's obvious that the virtual work principle (6) could be written as a total potential energy by replacing $\boldsymbol{q}'$ by virtual displacements $\delta\boldsymbol{q}$.: (Note that $d\Omega = d\Gamma dt$)

$$\delta\left[\int_t \frac{1}{2}\int_{\Gamma_t} \nabla\boldsymbol{q}.\boldsymbol{\Lambda}.\nabla\boldsymbol{q} d\Gamma dt - \int_{\Gamma_t} \boldsymbol{p}\cdot\boldsymbol{q} d\Gamma\right] = \delta\Pi = 0 \tag{22}$$

If we define:

$$L = \frac{1}{2}\int_{\Gamma_t} \nabla\boldsymbol{q}.\boldsymbol{\Lambda}.\nabla\boldsymbol{q} d\Gamma - \frac{d}{dt}\int_{\Gamma_t} \boldsymbol{p}\cdot\boldsymbol{q} d\Gamma$$

equation (22) becomes:

$$\delta\int_t L dt = 0 \tag{23}$$

The partial derivation of (6) with respect to *t,* which represents the variation of virtual works due to virtual displacements during the evolution of the contour is equivalent to equation (23) if we consider a natural evolution.

5 Hamiltonian Formalism Applied to Fracture Mechanics

We can actually write the equations governing the crack tip fields under the form of (5). The main idea [4, 5], is to consider one coordinate in the polar system as "time" and take the total potential energy as the Lagrange function. For example, we can consider the radial coordinate r or the angular coordinate θ as time and take the variational principles established in continuum mechanics as the Hamilton variational principle. Then all the procedures currently used in rational mechanics can be translated into continuum mechanics. In the following, the angular coordinate θ will be substituted to time.

5.1 Governing Equations of the Problem

Consider a notch formed from several elastic materials. We establish a cylindrical coordinate system with their origins at the notch tip and the z-axis representing the notch front. Material 1 occupies domain $[\theta_0,\theta_1]$, named zone 1; Material 2 occupies zone 2, bounded by $[\theta_1,\theta_2]$, and so on. Under remote loading, the stress concentration at the notch tip will take a mixed mode nature due to the anisotropy of the materials.

First, we write the stress components in the polar coordinate system as:$\sigma = \{\sigma_r \quad \sigma_\theta \quad \tau_{r\theta}\}^T$. The corresponding strain components are $\varepsilon = \{\varepsilon_r \quad \varepsilon_\theta \quad \gamma_{r\theta}\}^T$. The linear elastic stress–strain relationship is:

$$\boldsymbol{\sigma} = \mathbf{C}\,\boldsymbol{\varepsilon}. \tag{24}$$

$\boldsymbol{C}$ is the stiffness matrix of the material. All its components are constant.

We write now the fundamental equations of linear elasticity in the polar system:

(a) Equilibrium equations:

$$\frac{\partial \sigma_r}{\partial r} + \frac{1}{r}\frac{\partial \tau_{r\theta}}{\partial \theta} + \frac{\sigma_r - \sigma_\theta}{r} = 0 \quad \frac{\partial \tau_{r\theta}}{\partial r} + \frac{1}{r}\frac{\partial \sigma_\theta}{\partial \theta} + \frac{2\tau_{r\theta}}{r} = 0 \tag{25}$$

We perform the following variable changes:

$$\xi = \ln r \quad r = \exp(\xi); \tag{26}$$

and

$$\begin{aligned} S_r = r\sigma_r \quad \sigma_r = S_r/r; \quad S_\theta = r\sigma_\theta \quad \sigma_\theta = S_\theta/r; \\ S_{r\theta} = r\tau_{r\theta} \quad \tau_{r\theta} = S_{r\theta}/r; \ldots \text{etc} \end{aligned} \tag{27}$$

Then, by using the notation $(\dot{\cdot}) = \frac{\partial}{\partial\theta}$, the equilibrium equations (25) can be rewritten as:

$$\dot{S}_{r\theta} = S_\theta - \frac{\partial S_r}{\partial \xi} \quad \dot{S}_\theta = -\frac{\partial S_{r\theta}}{\partial \xi} - S_{r\theta} \tag{28}$$

We define the following variable vectors:

$$\mathbf{p} = \{S_\theta \quad S_{r\theta}\}^T \tag{29}$$

Hence, the equilibrium equations (28) can be rewritten as:

$$\dot{\mathbf{p}} = \mathbf{E}_1\mathbf{p} + \mathbf{E}_2\frac{\partial \mathbf{p}}{\partial \xi} \tag{30}$$

where

$$\mathbf{E}_1 = \begin{bmatrix} 0 & -1 \\ 1 & 0 \\ & \end{bmatrix} \mathbf{E}_2 = \begin{bmatrix} 0 & -1 \\ 0 & 0 \\ & \end{bmatrix}$$

(b) Displacement-stress relationship:

$$\begin{gathered} \varepsilon_r = \frac{\partial u_r}{\partial r} \quad \varepsilon_\theta = \frac{1}{r}\left(u_r + \frac{\partial u_\theta}{\partial \theta}\right) \\ \gamma_{r\theta} = \frac{1}{r}\frac{\partial u_r}{\partial \theta} + \frac{\partial u_\theta}{\partial r} - \frac{u_\theta}{r} \end{gathered} \tag{31}$$

By substituting (31) into (24) and by using the variable changes (26) and (27), one obtains:

$$\begin{Bmatrix} S_r \\ S_\theta \\ S_{r\theta} \\ \end{Bmatrix} = \begin{bmatrix} c_{12} & c_{14} \\ c_{22} & c_{24} \\ c_{42} & c_{44} \\ & \end{bmatrix} \begin{Bmatrix} \frac{\partial u_\theta}{\partial \theta} \\ \frac{\partial u_r}{\partial \theta} \end{Bmatrix} + \begin{bmatrix} -c_{14} & c_{12} \\ -c_{24} & c_{22} \\ -c_{44} & c_{42} \\ & \end{bmatrix} \begin{Bmatrix} u_\theta \\ u_r \end{Bmatrix} + \begin{bmatrix} c_{14} & c_{11} \\ c_{24} & c_{21} \\ c_{44} & c_{41} \\ & \end{bmatrix} \begin{Bmatrix} \frac{\partial u_\theta}{\partial \xi} \\ \frac{\partial u_r}{\partial \xi} \end{Bmatrix} \tag{32}$$

Similarly, we define a displacement vector

$$\{\mathbf{q}\} = \{u_\theta \quad u_r\}^T \tag{33}$$

By using the definitions (29) and (33), the relationship (32) can be rewritten as:

$$\mathbf{p} = \mathbf{C}_d\dot{\mathbf{q}} + \mathbf{C}_e\mathbf{q} + \mathbf{C}_f\frac{\partial\mathbf{q}}{\partial\xi} \tag{34}$$

or

$$\dot{\mathbf{q}} = \mathbf{C}_d^{-1}\left(\mathbf{p} - \mathbf{C}_e\mathbf{q} - \mathbf{C}_f\frac{\partial\mathbf{q}}{\partial\xi}\right) \tag{35}$$

with:

$$\mathbf{C}_d = \begin{bmatrix} c_{22} & c_{24} \\ c_{42} & c_{44} \end{bmatrix} \quad \mathbf{C}_e = \begin{bmatrix} -c_{24} & c_{22} \\ -c_{44} & c_{42} \end{bmatrix} \quad \mathbf{C}_f = \begin{bmatrix} c_{24} & c_{21} \\ c_{44} & c_{41} \end{bmatrix} \tag{36}$$

The strain energy in solids is always positive, consequently, $\boldsymbol{C}_d$ is a positively definite matrix. Therefore, the inversion of the matrix $\boldsymbol{C}_d$ is permitted

(c) *Governing equations:* By substituting Eq. (34) into the equilibrium equation (30), the variable vector $\boldsymbol{p}_t$ is eliminated. Then, we obtain, from (30) and (35), the following dual equations that govern the problem:

$$\dot{\boldsymbol{q}} = H_{11}\boldsymbol{q} + H_{12}\boldsymbol{p} \quad \dot{\boldsymbol{p}} = \boldsymbol{H}_{21}\boldsymbol{q} + \boldsymbol{H}_{22}\boldsymbol{p} \tag{37}$$

with:

$$\begin{array}{ll} \mathbf{H}_{11} = \mathbf{E}_1 - \mathbf{C}_d^{-1}\mathbf{C}_f\frac{\partial}{\partial\xi} & \mathbf{H}_{12} = \mathbf{C}_\mathbf{d}^{-1} \\ \mathbf{H}_{21} = \mathbf{E}_3\left(\mathbf{C}_d^{-1}\mathbf{C}_f\right)\frac{\partial^2}{\partial\xi^2} & \mathbf{H}_{22} = \mathbf{E}_1 + \left(\mathbf{E}_2 + \mathbf{E}_3\mathbf{C}_d^{-1}\right)\frac{\partial}{\partial\xi} \end{array} \tag{38}$$

with:

$$\mathbf{E}_1 = \begin{bmatrix} 0 & -1 \\ 1 & 0 \end{bmatrix} \mathbf{E}_2 = \begin{bmatrix} 0 & -1 \\ 0 & 0 \end{bmatrix} \mathbf{E}_3 = \begin{bmatrix} 0 & 0 \\ -1 & 0 \end{bmatrix}$$

In fact, it is more convenient to define a total vector $\boldsymbol{v}$ as variables in the state space:

$$v = \left\{\boldsymbol{q}^T\boldsymbol{p}^T\right\}^T \tag{39}$$

such that the governing equations (37) become:

$$\dot{\boldsymbol{v}} = \boldsymbol{H}\boldsymbol{v} \tag{40}$$

with:

$$\boldsymbol{H} = \begin{vmatrix} \boldsymbol{H}_{11} & \boldsymbol{H}_{12} \\ \boldsymbol{H}_{21} & \boldsymbol{H}_{22} \end{vmatrix} \tag{41}$$

(d) *Boundary conditions and continuity conditions:* Referring to Fig. 1, we adopt the superscript $^{(i)}$ to indicate the quantities in zone i. For example, $\boldsymbol{v}^{(i)}$, $\boldsymbol{H}^{(i)}$, etc.

The boundary conditions at the two free surfaces of the notch are:

$$\boldsymbol{p}^{(1)}(\theta = \theta_0) = \boldsymbol{0} \quad \boldsymbol{p}^{(n)}(\theta = \theta_n) = \boldsymbol{0} \tag{42}$$

The continuity conditions across the interfaces are:

$$\boldsymbol{v}^{(1)}(\theta = \theta_1) = \boldsymbol{v}^{(2)}(\theta = \theta_1) \quad \cdots\cdots \quad \boldsymbol{v}^{(n-1)}(\theta = \theta_{n-1}) = \boldsymbol{v}^{(n)}(\theta = \theta_{n-1}) \tag{43}$$

These relations show the advantage of the choice of the dual variables in the present study: the multi-material problem can be dealt with as a single material problem since the variable vector $\boldsymbol{v}$ is continuous across all the interfaces. This makes much easier the resolution of governing equation (40).

By adapting this new stiffness matrix, all formulations deduced for generalized plane strain can directly be used for plane stress problems.

5.2 Resolution Method

By examining governing equation (40), it is self-evident to try to solve it by using the variable separation method. We suppose that the variable vector $\boldsymbol{v}(\xi, \theta)$ can be written under separable form:

$$\mathbf{v}(\xi, \theta) = \exp(\lambda\xi)\boldsymbol{\psi}(\theta) \tag{44}$$

where λ is an undetermined eigenvalue and $\boldsymbol{\psi}(\theta)$ is a variable vector depending exclusively on θ. Then, equation (40) becomes:

$$\dot{\boldsymbol{\psi}}(\theta) = \boldsymbol{H}(\theta)\boldsymbol{\psi}(\theta) \tag{45}$$

In (45), $\boldsymbol{H}$ is function of θ only,

$$\mathbf{H}(\theta) = \begin{vmatrix} \mathbf{E}_1 - \mathbf{C}_d^{-1}\mathbf{C}_f\lambda & \mathbf{C}_\mathbf{d}^{-1} \\ \mathbf{E}_3\left(\mathbf{C}_d^{-1}\mathbf{C}_f\right)\lambda^2 & \mathbf{E}_1 + \left(\mathbf{E}_2 + \mathbf{E}_3\mathbf{C}_d^{-1}\right)\lambda \end{vmatrix} \tag{46}$$

The continuity conditions across the interfaces become:

$$\boldsymbol{\psi}^{(1)}(\theta = \theta_1) = \boldsymbol{\psi}^{(2)}(\theta = \theta_1) \quad \cdots \quad \boldsymbol{\psi}^{(n-1)}(\theta = \theta_{n-1}) = \boldsymbol{\psi}^{(n)}(\theta = \theta_{n-1}) \tag{47}$$

Any numerical method providing a good accuracy can be used for solving this problem and the eigenvectors $\boldsymbol{\psi}$ can straightforwardly be given with all stress and displacement components.

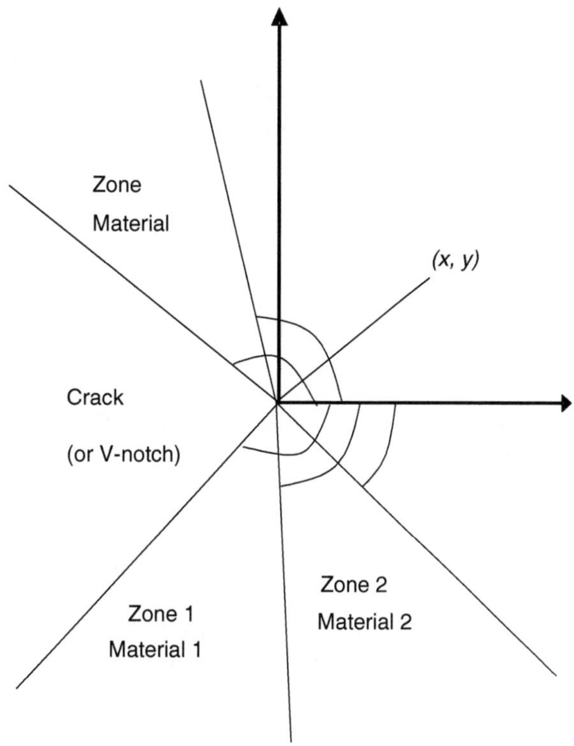

Fig. 1 V-Notch with various materials

6 Future Extensions

In this paragraph, a new way [6] is proposed in order to determine the orders of singularity for two dimensional V-notch problems. Firstly, on the basis of an asymptotic stress field in terms of radial coordinates at the V-notch tip, the governing equations of the elastic theory are transformed into an eigenvalue problem of ordinary differential equations (ODEs) with respect to the circumferential coordinate θ around the notch tip. Then, the singularity orders of the V-notch problem are determined through solving the corresponding ODEs by means of the interpolating matrix method. Meanwhile, the associated eigenvectors of the displacement and stress fields near the V-notches are also obtained. This method is also available to deal with the plane V-notch problems in bonded orthotropic multi-material.

Firstly, let us consider a V-notch of isotropic material with opening angle $2\pi - \theta_1 - \theta_2$ as shown in Fig. 2.

A polar coordinate system (ρ, θ) is defined taking the notch tip as origin. In the linear elastic analysis, it has been verified that the displacement field in the notch tip region can be expressed as a series expansion with respect to the radial

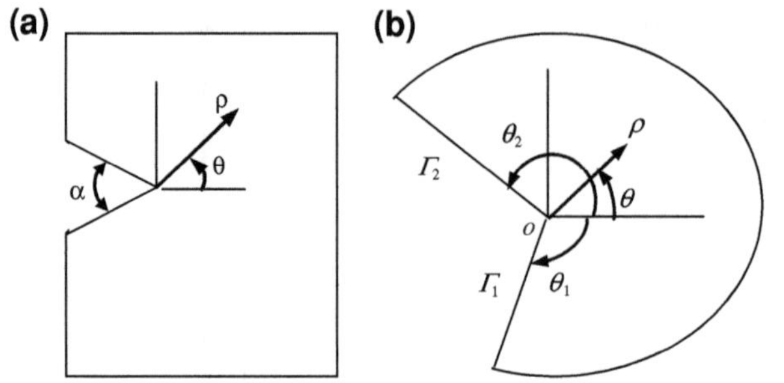

Fig. 2 : Geometry and local field **a** A V-notch with opening angle α **b** Geometry near a V-notch

coordinate ρ originating from the notch tip [7]. One typical term of the series can be written in the following form:

$$u_\rho(\rho,\theta) = \rho^{\lambda+1}\tilde{u}_\rho(\theta) \tag{48a}$$

$$u_\theta(\rho,\theta) = \rho^{\lambda+1}\tilde{u}_\theta(\theta) \tag{48b}$$

where λ, $\tilde{u}_\rho(\theta)$and $\tilde{u}_\theta(\theta)$are eigenpairs. Introducing Eqs. (3) into the strain–displacement relations of linear elastic theory yields the strain components as:

$$\varepsilon_{\rho\rho} = (1+\lambda)\rho^\lambda\tilde{u}_\rho(\theta) \tag{49a}$$

$$\varepsilon_{\theta\theta} = \rho^\lambda\tilde{u}_\rho(\theta) + \rho^\lambda\tilde{u}'_\theta(\theta) \tag{49b}$$

$$\gamma_{\rho\theta} = \rho^\lambda\tilde{u}'_\rho(\theta) + \lambda\rho^\lambda\tilde{u}_\theta(\theta) \tag{49c}$$

where $(\cdots)' = d(\cdots)/d\theta$. From linear elastic behavior law (Hooke's law) of plane stress problems, the plane stresses are expressed as:

$$\sigma_{\rho\rho} = \frac{E}{1-\nu^2}\rho^\lambda[(1+\lambda)\tilde{u}_\rho + \nu\,\tilde{u}_\rho + \nu\,\tilde{u}'_\theta] \tag{50a}$$

$$\sigma_{\theta\theta} = \frac{E}{1-\nu^2}\rho^\lambda[(1+\lambda)\nu\,\tilde{u}_\rho + \tilde{u}_\rho + \tilde{u}'_\theta] \tag{50b}$$

$$\sigma_{\rho\theta} = \frac{E}{2(1+\nu)}\rho^\lambda\left(\lambda\,\tilde{u}_\theta + \tilde{u}'_\rho\right) \tag{50c}$$

where E is the Young's modulus and ν the Poisson's ratio. Neglecting the body forces, the equilibrium equations are:

$$\frac{\partial\sigma_{\rho\rho}}{\partial\rho} + \frac{1}{\rho}\frac{\partial\sigma_{\rho\theta}}{\partial\theta} + \frac{\sigma_{\rho\rho}-\sigma_{\theta\theta}}{\rho} = 0 \tag{51a}$$

$$\frac{1}{\rho}\frac{\partial\sigma_{\theta\theta}}{\partial\theta}+\frac{\partial\sigma_{\rho\theta}}{\partial\rho}+\frac{2\sigma_{\rho\theta}}{\rho}=0 \tag{51b}$$

Substituting Eqs. (50a, 50b, 50c) into Eqs. (51a, 51b) gives:

$$\tilde{u}_\rho''+(\frac{1+v}{1-v}\lambda-2)\tilde{u}_\theta'+\frac{2}{1-v}\lambda(\lambda+2)\tilde{u}_\rho=0\,,\quad \theta\in(\theta_1,\ \theta_2) \tag{52a}$$

$$\tilde{u}_\theta''+\left[2+\frac{1}{2}(1+v)\lambda\right]\tilde{u}_\rho'+\frac{1}{2}(1-v)\lambda(\lambda+2)\tilde{u}_\theta=0,\quad \theta\in(\theta_1,\ \theta_2) \tag{52b}$$

Assume that all the tractions on the two edges, $\Gamma_1$and Γ_2, near the notch tip are zero. That is:

$$\left\{\begin{matrix}\sigma_{\theta\theta}\\ \sigma_{\rho\theta}\end{matrix}\right\}_{\theta=\theta_1}=\left\{\begin{matrix}\sigma_{\theta\theta}\\ \sigma_{\rho\theta}\end{matrix}\right\}_{\theta=\theta_2}=\left\{\begin{matrix}0\\ 0\end{matrix}\right\} \tag{53}$$

Hence, substitution of Eqs. (50a, 50b, 50c) into Eq. (53) yields:

$$\tilde{u}_\theta'+(1+v+v\lambda)\tilde{u}_\rho=0\,,\ \theta=\theta_1 \text{ and } \theta_2 \tag{54a}$$

$$\tilde{u}_\rho'+\lambda\,\tilde{u}_\theta=0\,,\qquad \theta=\theta_1 \text{ and } \theta_2 \tag{54b}$$

Considering that the appearance of λ^2 in Eqs. (52a, 52b) leads to nonlinear eigenanalysis if Eqs. (52a, 52b) are directly solved, an alternative approach is adopted in this paper to transfer the equation into a linear eigenvalue problem. To this end, two new field variables are introduced as follows:

$$g_\rho(\theta)=\lambda\,\tilde{u}_\rho(\theta)\,,\qquad \theta\in(\theta_1,\ \theta_2) \tag{55a}$$

$$g_\theta(\theta)=\lambda\,\tilde{u}_\theta(\theta)\,,\qquad \theta\in(\theta_1,\ \theta_2) \tag{55b}$$

Thus, Eqs. (55a, 55b), Eqs. (52a, 52b) can been rewritten as:

$$\tilde{u}_\rho''+\left(\frac{1+v}{1-v}\lambda-2\right)\tilde{u}_\theta'+\frac{2}{1-v}(\lambda+2)g_\rho=0\,,\quad \theta\in(\theta_1,\ \theta_2) \tag{56a}$$

$$\tilde{u}_\theta''+\left[2+\frac{1}{2}(1+v)\lambda\right]\tilde{u}_\rho'+\frac{1}{2}(1-v)(\lambda+2)g_\theta=0\,,\quad \theta\in(\theta_1,\ \theta_2) \tag{56b}$$

By following the above procedure, the evaluation of the singularity orders near a V-notch tip is transformed to solving a linear eigenvalue problem of the ODEs governed by Eqs. (55a, 55b), (56a, 56b) subjected to the boundary condition of

Fig. 3 A V-notch of bonded bi-material

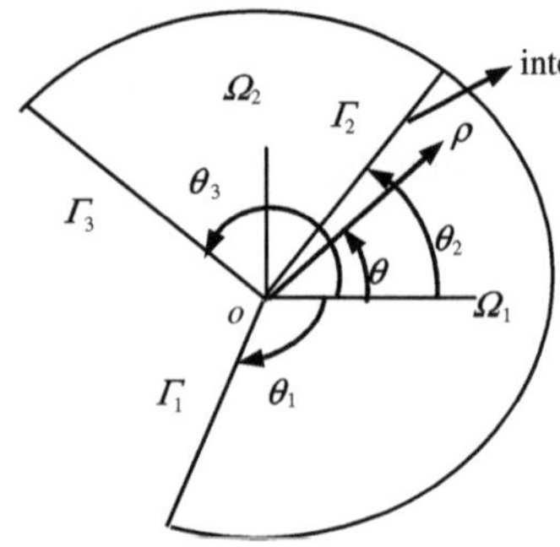

Eqs. (54a, 54b). In the solutions, the associated eigenfunctions $\tilde{u}_\rho$ and $\tilde{u}_\theta$ can also be obtained and can be used to determine the stresses in the vicinity of the notch tip.

In Fig. 3 we show an example of solution using this method applied in the case of bounded dissimilar linear elastic materials containing a V-notch tip.

Table 1 shows the comparison between the singularity degrees obtained by this method for various mesh levels of the used interpolating matrix method (IMMEI) and those of the literature. Reference [8] gives only one singularity degree λ_1 (one term in Eq. 48). Reference [9] gives two singularity degrees λ_1 and λ_2 as the present method noted (IMMEI) in the table. The value of n in the table indicates the discritization level considered in the IMMEI.

7 Concluding Remarks

In this chapter, we give a new Hamiltonian formalism resolving simultaneously the two-end problem and the problem of Cauchy and as a consequence, showing the relationship between the formalisms of Bui and Zhong which look so different. The key idea is to write the total potential energy of a solid as an integral along a special axis z, then over a section S normal to it. Using integration by part, the variation of the total potential energy can be written as two parts [see Eq. (15)]. The first part is an integral along z, and the second one is an integrated quantity depending on the two ends z_1 and z_2. For the two end problem, the displacements are imposed at the two ends; so their variations vanish. According to the minimum principle of the total potential energy, the canonical equations of Hamilton are immediately obtained, [see Eqs. (16)–(18)]. On the other hand, for a natural evolution of the structure (i.e., the canonical equations of Hamilton are satisfied everywhere in the solid), but with possible variations of the two ends, the first part in the variation of the total potential energy vanishes [see Eq. (19)]. This corresponds to the Cauchy problem in elasticity. In this case, the equations of Hamilton–Jacobi can be deduced [see Eq. (21)]. Since the canonical equations of

Table 1 The eigenvalues of the V-notch with $\theta_1 = 0^\circ$, $\theta_2 = 90^\circ$, $\theta_3 = 270^\circ$ Ref [8]

Ref [9]		Ref [10]		IMMEI					
				n = 20		*n* = 40		*n* = 80	
E2/E1	λ_1	λ_1	λ_2	λ_1	λ_2	λ_1	λ_2	λ_1	λ_2
0.33	−0.498	−0.49805	−0.22194	−0.4980360	−0.2218669	−0.4980532	−0.2219326	−0.4980545	−0.2219375
0.5	−0.488	−0.48756	−0.16731	−0.4875405	−0.1672198	−0.4875569	−0.1673009	−0.4575581	−3.1673069
1.0	−0.450	−0.45074	−0.09252	−0.4507273	−0.0923977	−0.4507428	−0.0925077	−4.4507440	−0.0925159
2.14	−0.390	−0.39034	−0.04170	−0.3903270	−0.0415509	−0.3903419	−0.0416593	−0.3903430	−0.0416992
4.0	−0.337	−0.33611	−0.01944	−0.3360963	−0 0192742	−0.3361104	−0.0194300	−0 3361115	−0.0194415
10.0	−0.270	−0.26966	−0.00529	−0.2696441	−0 0051051	−0.2696558	−0.0052500	−0.2696567	−0.0052930

Hamilton and the equations of Hamilton–Jacobi are fundamentally equivalent, we can see that the formalisms of Bui and Zhong are equivalent too.

Zhong's formalism has been successfully applied to Fracture Mechanics in order to determine the asymptotic mechanical fields near the crack tip [4]. This work has shown that the Hamiltonian approach provides a systematic method in asymptotic analysis near the crack tip. It leads to a first order differential equation system, which is easy to deal with. We insist on the fact that this approach is not only a new formalism other than the traditional methods, but it can be used as a powerful tool in asymptotic analysis of fracture mechanics.

By using this approach, we have resolved various problems. Some of them have been solved previously and some not yet. For example, we can calculate the stress singularities for an interfacial crack between two elastic and isotropic materials. The results are completely identical as those obtained by using the well-known theoretical formula. Similar example is a crack tip normally touching an interface has been resolved see Ref. [4]. For a crack in a generally anisotropic material, we obtained a near tip field identical to theoretical results [5].The comparison shows no difference between these two stress distributions. Another example consists in finding stress singularities near a notch tip formed from two generally anisotropic materials and stress singularities near an inclined crack tip touching an interface between two generally anisotropic materials [5]. From this work, we see that the present method is particularly efficient for resolving multi-material problems. This is because the selected dual variables are continuous across all the interfaces. So the multi-material problem can be resolved as a single material problem through the construction of the transfer matrix.

We believe that a large domain can be found in applying this new approach into fracture mechanics.

Nevertheless, the connection between the local obtained solution of the stress field and the far field is still a tremendous problem. That is why we investigate a new way transforming the fracture mechanics problem into an eigenvalue problem. That allows us to compute more terms in the stress expansion and then to connect the local field easily to the far field. The far field could be the finite element solution. This way will allow more efficiency to deal with various structural geometries and boundary conditions.

References

1. Bui, H. D.: Introduction aux problèmes inverses en mécaniques des matériaux, Editions Eyrolles (1993)
2. Zhong, W. X.: A New Systematically Methodology in Elasticity Theory, Dalian Science & Technology University Press, Liaoning, China (1995)
3. Li, J., Recho, N.: Hamiltonian formalisms in elasticity—potential use for fracture mechanics. In: 11th International Conference on Fracture (ICF 11), mars , Turin (Italie)(2005)
4. Li J., Recho N.: Méthodes asymptotiques en mécanique de la rupture, Editions Hermes Science, ISBN: 2-7462-0366-9 (2002)

5. Li, J., Zhang, X.B., Recho, N.: Stress singularities near the tip of a two-dimensional notch forded from several anisotropic materials. Int. J. Fract. **107**, 379–395 (2001)
6. Niu, Z.R., Cheng, C., Ye, J., Recho, N.: A new boundary element approach of modeling singular stress fields of plane V-notch porblems. Int. J. Solids Struct. **46**(16), 2999–3008 (2009)
7. Yosibash, Z., Szabó, B.A.: A note on numerically computed eigenfunctions and generalized stress intensity factors associated with singular points. Eng. Frac. Mech. **54**, 593–595 (1996)
8. Niu, Z.R., Ge, D., Cheng, C.Z., Ye, J., Recho, N.: Evaluation of the stress singularity degrees of plane V-notches in bonded dissimilar materials. J. Appl. Math. Model. **33**(3), 1776–1792 (2009)
9. Bogy, D.B.: Two edge-bonded elastic wedges of different materials and angles under surface traction. J. Appl. Mech. **38**(2), 377–386 (1971)
10. Jian, Z., Huang, S.H., Hu, L.M.: Computation of stress factors for biomaterial V-notch and heel of concrete dam on rock foundation. Chin. J. Hydraul. Eng. **6**, 77–81 (1998) (in Chinese)

A Finite Element Solution for Transient Wave Propagation in an Infinite Circular Plate With Nonlinear Viscoelastic Model

Saeid Arabnejad, Manouchehr Salehi and Hamed Farahmand

Abstract In this study, stress wave propagation in an infinite nonlinear viscoelastic plate is investigated by the Finite Element method. In addition, explicit Galerkin Finite Element approach is employed with Lagrangian strain and Gaussian stresses. For viscoelasticity modeling, Kelvin's model with a nonlinear dashpot is considered. The resulted relations for deflections and deflection angles are derived during a specified period of time. Result show that second order shear deformation theory predicts higher speed for bending wave than the first order shear deformation theory. But both theories predict the same speed for shear waves.

Keywords Stress wave propagation · Plate's shear deformation theories · Nonlinear viscoelastic

S. Arabnejad (✉) · M. Salehi
Mechanical Engineering Department,
Amirkabir University of Technology,
Tehran, Iran
e-mail: sarabnezhad@yahoo.com

M. Salehi
e-mail: msalehi@aut.ac.ir

H. Farahmand
Mechanical Engineering Department, Kerman Branch,
Islamic Azad University, Kerman, Iran
e-mail: hamed_1256@yahoo.com

A. Öchsner et al. (eds.), *Materials with Complex Behaviour II*,
Advanced Structured Materials 16, DOI: 10.1007/978-3-642-22700-4_3,

1 Introduction

The problem of wave propagation in a solid medium that exhibits the attenuation of a propagating wave is a wide and significant branch of continuum mechanics. In the field of structural engineering, wave propagation phenomena has found increasing applications especially in the area of structural health monitoring and active control of vibrations and noise. Dynamic analysis in structural engineering falls into two different classes: the first one involves low frequency loading and the other one involves high frequency loading. Low frequency problems are categorized as structural dynamics problems whereas those involving high frequency loading fall into the category of wave propagation problems. In structural dynamics problems, the frequency content of the dynamic load is in the order of a few hundred Hertz (Hz) and the designer will be mostly interested in the long-term (or steady-state) effects of the dynamic load on the structures. Most of the dynamic problems in structures will fall into this category. On the other hand, in wave propagation problems, the frequency content of the input loading is very high (in the order of kilo-Hertz (kHz) or higher) and hence, short-term effects (transient response) will become very critical. Furthermore, many higher order modes will participate in amplifying the dynamic response. Impact and blast-type of loading is in this category.

The state of stress in viscoelastic materials is influenced by many factors, yet only strain and strain rate are usually taken into account. The behavior of viscoelastic materials is more often described by relaxation functions. For the beginning, strain and stress tensors are decomposed into the deviatoric and volumetric terms. The first one is connected with changes in shape and the other one represents volumetric relaxation. However, most of viscoelastic materials display negligible bulk modulus changes with respect to the changes of shear modulus so that, a constant bulk modulus and a relaxation function for shear modulus are often assumed.

The study of wave propagation has been initiated with Pochhammer [1] with wave propagation in an elastic rod. Afterward it was followed by Chree [2] and Bancraft [3] in case of elastic rods. Kolsky [4] started wave propagation analysis in viscoelastic rods and proceeded his own work in 1968 [5]. Lundberg and Blanc [6] and Blanc [7] continued estate of viscoelastic rods. Rizzi and Doyle [8] used spectral analysis and Fourier transform to investigate wave propagation in plates with boundaries. Nkemzi and Green [9] studied wave propagation in linear viscoelastic sandwich plates. The plane wave propagation in inhomogeneous nonlinear viscoelastic plates with the finite element method (FEM) was done by Jiang and Haddow [10]. Yang and Yuan [11] published a comprehensive analytical solution for wave propagation in elastic plates using Second order Shear Deformation Theory (SSDT) and 3D Elasticity. In the same year, Adamek et al. [12] utilized the same procedure as that in Yang et al.'s work, however the material was linear viscoelastic and First order Shear Deformation Theory (FSDT) was applied.

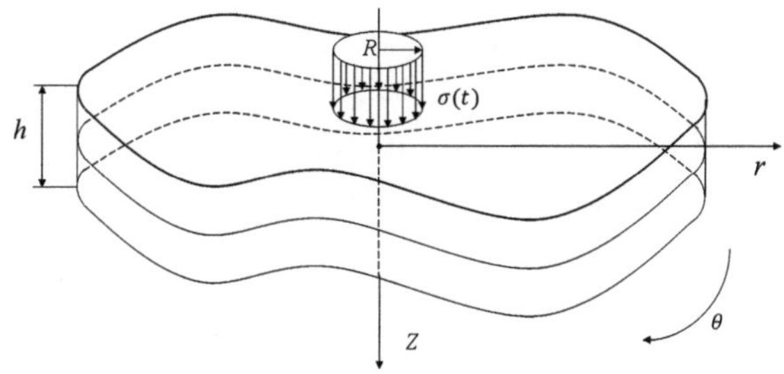

Fig. 1 Geometry definition

Stress wave propagation in structures is usually studied in three ways: by 3D elasticity, structural analysis and numerical methods. Because of the complexity in 3D analysis, structural and numerical methods are utilized to solve this problem.

2 Solution of the Problem

After defining the problem's geometry and external load, finite element equations are obtained from constitutive equations using virtual power method.

2.1 Problem Definition and Initial Assumptions

The problem's geometry consists of an infinite plate with constant thickness, h, loaded by transverse pressure. The loading has been applied on the upper face of the plate (see Fig. 1). As depicted in Fig. 2, the uniformly distributed load has a harmonic amplitude in time only in the circular area with finite radius, R. The geometry of the plate is assumed infinite and with respect to the axial symmetry of the loading applied, the problem is solved as an axisymmetric problem in cylindrical coordinates. The positive orientation of axis is depicted in Fig. 1.

2.2 Constitutive Equations

Using constitutive equation, stress tensor is decomposed into volumetric and deviatoric tensors:

$$\sigma_{ij} = \begin{bmatrix} \sigma_{11} & \sigma_{12} & \sigma_{13} \\ \sigma_{21} & \sigma_{22} & \sigma_{23} \\ \sigma_{31} & \sigma_{32} & \sigma_{33} \end{bmatrix} = \begin{bmatrix} \sigma_{\mathrm{v}} & 0 & 0 \\ 0 & \sigma_{\mathrm{v}} & 0 \\ 0 & 0 & \sigma_{\mathrm{v}} \end{bmatrix} + \begin{bmatrix} s_{11} & s_{12} & s_{13} \\ s_{21} & s_{22} & s_{23} \\ s_{31} & s_{32} & s_{33} \end{bmatrix} \tag{1}$$

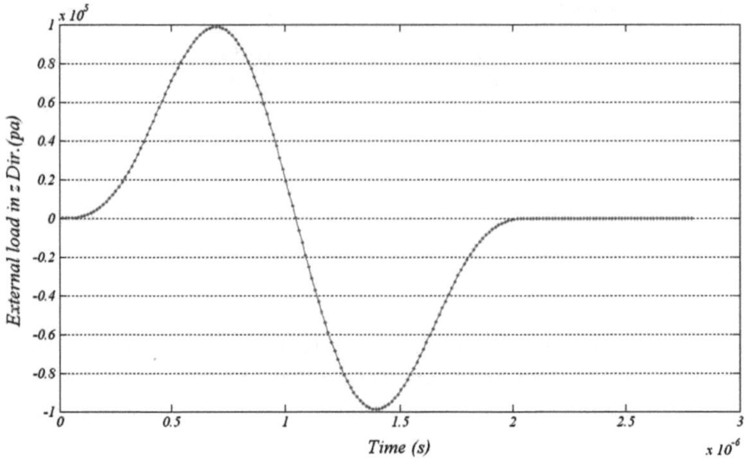

Fig. 2 Harmonic external load applied at plate center

where σ_v is the volumetric stress and s_{ij} is the deviatoric stress tensor. Strain, ε_{ij}, and strain rate, d_{ij}, pursue the same way as stress:

$$\varepsilon_{ij} = \begin{bmatrix} \varepsilon_{11} & \varepsilon_{12} & \varepsilon_{13} \\ \varepsilon_{21} & \varepsilon_{22} & \varepsilon_{23} \\ \varepsilon_{31} & \varepsilon_{32} & \varepsilon_{33} \end{bmatrix} = \begin{bmatrix} \varepsilon_v & 0 & 0 \\ 0 & \varepsilon_v & 0 \\ 0 & 0 & \varepsilon_v \end{bmatrix} + \begin{bmatrix} \varepsilon_{11}^{\mathrm{div}} & \varepsilon_{12}^{\mathrm{div}} & \varepsilon_{13}^{\mathrm{div}} \\ \varepsilon_{21}^{\mathrm{div}} & \varepsilon_{22}^{\mathrm{div}} & \varepsilon_{23}^{\mathrm{div}} \\ \varepsilon_{31}^{\mathrm{div}} & \varepsilon_{32}^{\mathrm{div}} & \varepsilon_{33}^{\mathrm{div}} \end{bmatrix} \tag{2}$$

$$d_{ij} = \begin{bmatrix} d_{11} & d_{12} & d_{13} \\ d_{21} & d_{22} & d_{23} \\ d_{31} & d_{32} & d_{33} \end{bmatrix} = \begin{bmatrix} d_v & 0 & 0 \\ 0 & d_v & 0 \\ 0 & 0 & d_v \end{bmatrix} + \begin{bmatrix} d_{11}^{\mathrm{div}} & d_{12}^{\mathrm{div}} & d_{13}^{\mathrm{div}} \\ d_{21}^{\mathrm{div}} & d_{22}^{\mathrm{div}} & d_{23}^{\mathrm{div}} \\ d_{31}^{\mathrm{div}} & d_{32}^{\mathrm{div}} & d_{33}^{\mathrm{div}} \end{bmatrix} \tag{3}$$

As it was mentioned before, volumetric strain rates regarding the mentioned viscoelastic model are used in order to obtain the volumetric stress.

$$\sigma_v = K\varepsilon_v \tag{4}$$

$$\sigma_v^{t+1} = \sigma_v^t + K\, dt\, d_v^t \tag{5}$$

Stress within each time step is calculated by a forward difference method:

$$S_{ij}^{t+1} = S_{ij}^t + G\, dt\, d_{ij}^{div^t} + F\left(d_{ij}^{div^t}\right) \tag{6}$$

This is the general form for a viscoelastic material. Where, F is represented by Eyring model.

$$F\left(d_{ij}^{div^t}\right) = L\, sinh^{-1}\left(C\, d_{ij}^{div^t}\right) \tag{7}$$

C and L are parameters defined with respect to the material properties.

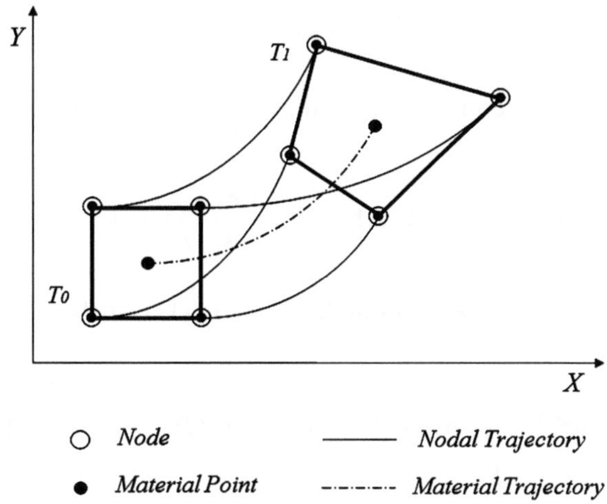

Fig. 3 Lagrangian description

2.3 Finite Element Form of Governing Equations

In stress wave propagation problems, the strains are usually small but strain rates are large. This is the main reason of utilizing Lagrangian a mesh (see Fig. 3). In this case, the material does not pass the mesh borders and also external loads and boundary conditions are on the mesh borders and nodes.

To initiate the finite element formulation, linear momentum conservation equation is first used.

$$\frac{\partial \sigma_{ij}}{\partial x_j} + \rho b_i = \rho \dot{v}_i \quad in \quad \boldsymbol{\Omega} \tag{8}$$

$$n_j \, \sigma_{ij} = \bar{t}_i \quad on \quad \boldsymbol{\Gamma}_{t_i} \tag{9}$$

where $\boldsymbol{\Omega}$ is the volume and $\boldsymbol{\Gamma}_{t_i}$ is its border. The momentum equation is multiplied by a test function and is integrated on its domain, in order to obtain the weak form of the governing equation:

$$\int\limits_{\Omega} \delta v_i \left(\frac{\partial \sigma_{ij}}{\partial x_j} + \rho b_i - \rho \dot{v}_i \right) d\boldsymbol{\Omega} = 0 \tag{10}$$

After simplifying the integral in Eq. 10, the weak form can be expressed as:

$$\int\limits_{\Omega} \left(\frac{\partial(\delta v_i)}{\partial x_j} \right) \sigma_{ji} d\boldsymbol{\Omega} - \int\limits_{\Omega} \delta v_i \rho b_i d\boldsymbol{\Omega} - \sum_{i=1}^{n_{SD}} \int\limits_{\Gamma_{t_i}} \delta v_i \bar{t}_i d\boldsymbol{\Gamma} + \int\limits_{\Omega} \delta v_i \rho \dot{v}_i d\boldsymbol{\Omega} = 0 \tag{11}$$

This is known as the virtual power equation. The definition for each of these terms is specified by the first integral as shown below:

$$\left(\frac{\partial(\delta v_i)}{\partial x_j}\right)\sigma_{ij} = \delta L_{ij}\sigma_{ij} = \left(\delta D_{ij} + \delta W_{ij}\right)\sigma_{ij} = \delta D_{ij}\sigma_{ij} = \delta \boldsymbol{D} : \boldsymbol{\sigma} \tag{12}$$

This term is nominated as the internal virtual power per unit volume. So we have

$$\delta p^{int} = \int_{\Omega} \delta D_{ij}\sigma_{ij} d\boldsymbol{\Omega} = \int_{\Omega} \left(\frac{\delta(\delta v_i)}{\delta x_j}\right)\sigma_{ij} d\boldsymbol{\Omega} = \int_{\Omega} \delta L_{ij}\sigma_{ij} d\boldsymbol{\Omega} = \int_{\Omega} \delta D : s \, d\boldsymbol{\Omega} \tag{13}$$

And the second and third term in (11) can be designated as the external virtual power.

$$\delta p^{ext} = \int_{\Omega} \delta v_i \rho b_i d\boldsymbol{\Omega} + \sum_{i=1}^{n_{SD}} \int_{\Gamma_{t_i}} \delta v_i \bar{t} d\boldsymbol{\Gamma} = \int_{\Omega} \delta v.\rho b \, d\boldsymbol{\Omega} + \sum_{j=1}^{n_{SD}} \int_{\Gamma_{t_i}} \delta \delta v_j e_j \cdot \bar{t} d\boldsymbol{\Gamma} \tag{14}$$

And the last term in (11) is rewritten as the virtual kinetic power:

$$\delta p^{kin} = \int_{\Omega} \delta v_i \rho \dot{v}_i d\boldsymbol{\Omega} \tag{15}$$

Substituting (13),(14),(15) in (12), the virtual power equation is defined as

$$\delta p = \delta p^{int} - \delta p^{ext} + \delta p^{kin} = 0 \quad \forall \delta v_i \in \mathfrak{U}0 \tag{16}$$

In order to transfer the character from the element domain to element nodes, linear shape functions are used.

$$x_i(\boldsymbol{x}, t) = N_I(\boldsymbol{X})x_{iI}(t) \quad or \quad \boldsymbol{x}(\boldsymbol{X}, t) = N_I(\boldsymbol{X})x_I(t) \tag{17}$$

Velocities and accelerations are defined similar to displacements. To differentiate dissimilar terms, differentiating shape functions are employed whilst nodal displacements are considered as constant terms.

$$L_{ij} = v_{i,j} = v_{iI}\left(\frac{\partial N_I}{\partial x_j}\right) = v_{iI}N_{I,j} \quad or \quad \boldsymbol{L} = v_I \nabla N_I = v_I N_{I,x} \tag{18}$$

Therefore, the strain is described as a function of shape functions:

$$D_{ij} = \left(\frac{1}{2}\right)\left(L_{ij} + L_{ji}\right) = \frac{1}{2}\left(v_{iI}N_{I,j} + v_{jI}N_{I,i}\right) \tag{19}$$

Rewriting Eq. 16, the virtual power comes to:

$$\int_{\Omega} \left(\frac{\partial N_I}{\delta x_j}\right)\sigma_{ij} d\boldsymbol{\Omega} - \int_{\Omega} N_I \rho b_i d\boldsymbol{\Omega} - \sum_{j=1}^{n_{SD}} \int_{\Gamma_{t_i}} N_I \bar{t}_i d\boldsymbol{\Gamma} + \int_{\Omega} N_I \rho \dot{v}_i d\boldsymbol{\Omega} = 0 \quad \forall (I, i) \notin \boldsymbol{\Gamma}_{v_i} \tag{20}$$

Fig. 4 Problem algorithm

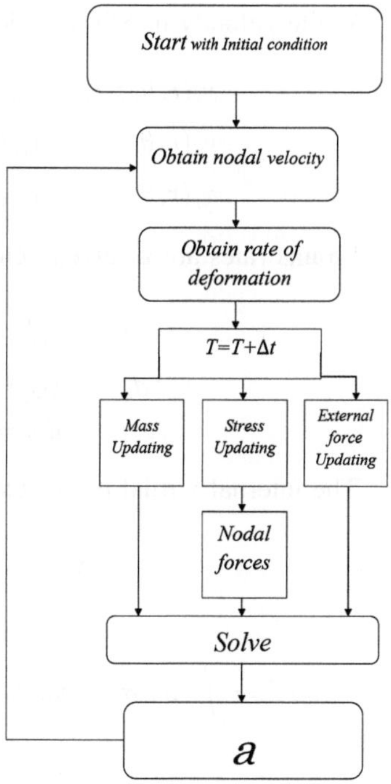

The velocity increments are factored out from integrals as constants.

$$\delta v_{iI}(f_{iI}^{int} - f_{iI}^{ext} + M_{ijIJ}\dot{v}_{jJ}) = 0 \quad \forall \delta v_{iI} \notin \Gamma_{v_i} \tag{21}$$

The algorithm for solving this equation is illustrated in Fig. 4.

The nodal forces are obtained from the strain definition. The axisymmetric linear strains in cylindrical coordinate are expressed as:

$$\begin{aligned} &\varepsilon_{rr} = \tfrac{\partial v_r}{\partial r}, \quad \varepsilon_{\theta\theta} = \tfrac{v_r}{r}, \quad \varepsilon_{zz} = \tfrac{\partial v_z}{\partial z} \\ &\varepsilon_{r\theta} = 0, \quad \varepsilon_{z\theta} = 0, \quad \varepsilon_{rz} = \tfrac{1}{2}\left(\tfrac{\partial u_r}{\partial z} + \tfrac{\partial u_z}{\partial r}\right) \end{aligned} \tag{22}$$

And strain rate is described as:

$$\begin{aligned} &d_{rr} = \tfrac{\partial v_r}{\partial r}, \quad d_{\theta\theta} = \tfrac{v_r}{r}, \quad d_{zz} = \tfrac{\partial v_z}{\partial z} \\ &d_{r\theta} = 0, \quad d_{z\theta} = 0, \quad d_{rz} = \tfrac{1}{2}\left(\tfrac{\partial u_r}{\partial z} + \tfrac{\partial u_z}{\partial r}\right) \end{aligned} \tag{23}$$

Velocity in FSDT is defined as:

$$\begin{aligned} v_r(r,\theta,z,t) &= v_{r_0}(r,\theta,t) + z\dot{\varphi}r(r,\theta,t) \\ v_\theta(r,\theta,z,t) &= v_{\theta_0}(r,\theta,t) + z\dot{\varphi}_\theta(r,\theta,t) \\ v_z(r,\theta,z,t) &= v_{z_0}(r,\theta,t) \end{aligned} \tag{24}$$

While velocity in SSDT is defined as:

$$\begin{aligned} v_r(r,\theta,z,t) &= v_{r_0}(r,\theta,t) + z\dot{\psi}_1(r,\theta,t) + z^2\dot{\varphi}_1(r,\theta,t) \\ v_\theta(r,\theta,z,t) &= v_{\theta_0}(r,\theta,t) + z\dot{\psi}_2(r,\theta,t) + z^2\dot{\varphi}_2(r,\theta,t) \\ v_z(r,\theta,z,t) &= v_{z_0}(r,\theta,t) + z\dot{\psi}_3(r,\theta,t) \end{aligned} \tag{25}$$

Strains rates are as given below for FSDT;

$$\begin{aligned} d_{rr} &= \tfrac{\partial v_{r0}}{\partial r} + z\left(\tfrac{\partial \dot{\varphi}_r}{\partial r}\right) \quad d_{\theta\theta} = \tfrac{v_{r0}}{r} + z\left(\tfrac{\dot{\varphi}_r}{r}\right) \\ d_{rz} &= \tfrac{1}{2}(\dot{\varphi}_r) + \left(\tfrac{(\partial v_{z0})}{\partial r}\right) \quad d_{\theta z} = 0 \\ d_{r\theta} &= 0 \end{aligned} \tag{26}$$

The internal virtual power using strain rate is rewritten as:

$$\delta p^{int} = \int_{v^e} (\delta d'^t \sigma' + k.\delta d''^t \sigma'')dv \tag{27}$$

$$\delta p^{int} = \int_{v^e} (\delta d_r \sigma_r + \delta d_\theta \sigma_\theta + 2\delta d_{r\theta}\sigma_{r\theta} + 2k\delta d_{rz}\sigma_{rz} + 2k\delta d_{\theta z}\sigma_{\theta z})dv \tag{28}$$

where the shear correction factor is $k = 0.88629$, corresponding to the parabolic distribution of τ_{rz} over the plate's cross-section.

Before substituting strain rates, parameters of deformation theories are defined with shape functions;

$$\begin{aligned} v_{r_0} &= N_I(r,\theta)v_{r_{0_I}}, \quad v_{\theta_0} = N_I(r,\theta)v_{\theta_{0_I}}, \quad v_{z_0} = N_I(r,\theta)v_{z_{0_I}} \\ \dot{\varphi}_r &= N_I(r,\theta)\dot{\varphi}_{r_I}, \quad \dot{\varphi}_\theta = N_I(r,\theta)\dot{\varphi}_{\theta_I} \end{aligned} \tag{29}$$

The differentiation of shape functions are;

$$B_{1_I} = \tfrac{(\partial N_I)}{\partial r}, \quad B_{2_I} = \tfrac{(\partial N_I)}{\partial \theta} 30 \tag{30}$$

By substituting in internal virtual power for FSDT, internal virtual power is obtained as;

$$\delta p^{int} = \int_{v^e} \left[B_{1_I}\left(\delta v_{r_{0_I}} + z\delta\dot{\varphi}_{r_I}\right)\sigma_r + k\sigma_{rz}\left(N_I\delta\dot{\varphi}_{r_I} + B_{1_I}\delta v_{z_I}\right)\right]dv \tag{31}$$

$$\begin{aligned} \delta p^{int} &= \delta v_{r_{0_I}} \int_{R_1}^{R_2}\int_0^{2\pi}\int_{-h/2}^{h/2} (B_{1_I}\sigma_r) r\,dr\,d\theta\,dz + \delta v_{z_I} \int_{R_1}^{R_2}\int_0^{2\pi}\int_{-h/2}^{h/2} (kB_{1_I}\sigma_{rz}) r\,dr\,d\theta\,dz \\ &+ \delta\dot{\varphi}_{r_I} \int_{R_1}^{R_2}\int_0^{2\pi}\int_{-h/2}^{h/2} (B_{1_I} z\sigma_r + kN_I\sigma_{rz}) r\,dr\,d\theta\,dz \end{aligned} \tag{32}$$

$$\delta p^{int} = \delta v_{r_{0_I}} \int_{R_1}^{R_2} \int_{0}^{2\pi} (B_{1_I} f_r) r\, dr\, d\theta + \delta v_{z_I} \int_{R_1}^{R_2} \int_{0}^{2\pi} (k B_{1_I} f_{rz}) r\, dr\, d\theta + \delta \dot{\varphi}_{rI} \int_{R_1}^{R_2} \int_{0}^{2\pi} (B_{1_I} m_r + k N_I f_{rz}) r\, dr\, d\theta \tag{33}$$

In which

$$f_r = \int_{-h/2}^{h/2} \sigma_r dz, \quad f_{rz} = \int_{-h/2}^{h/2} \sigma_{rz} dz, \quad m_r = \int_{-h/2}^{h/2} z \sigma_r dz \tag{34}$$

After simplifying internal virtual power, we have

$$\delta p^{int} = \delta v_{r_{0_I}} F_{r_I} + \delta v_{z_I} F_{z_I} + \delta \dot{\varphi}_{r_I} M_{r_I} \tag{35}$$

where

$$F_{r_I} = \int_{-1}^{1} \left(\frac{2}{(R_2 - R_1) B_{1_I} f_r} \right) \pi(\zeta(R_2 - R_1) + (R_1 + R_2)) d\zeta \tag{36}$$

$$F_{z_I} = \int_{-1}^{1} \left(\frac{k2}{(R_2 - R_1) B_{1_I} f_{rz}} \right) \pi(\zeta(R_2 - R_1) + (R_1 + R_2)) d\zeta \tag{37}$$

$$M_{r_I} = \int_{-1}^{1} \left(\frac{2}{(R_2 - R_1) B_{1_I} m_r + k N_I f_{rz}} \right) \pi(\zeta(R_2 - R_1) + (R_1 + R_2)) d\zeta \tag{38}$$

The trail is the same for SSDT so it is exorbitant to be brought here. We start from kinetic virtual power for mass matrix.

$$\delta p^{kin} = \int_{\Omega} \delta v_i \rho \dot{v}_i d\Omega \tag{39a}$$

$$\delta p^{kin} = \delta v_{i_I} \int_{V^e} N_I \rho N_I \dot{v}_{i_I} dV \tag{39b}$$

$$\delta p^{kin} = \delta v_{r_I} \int_{R_1}^{R_2} \int_{0}^{2\pi} \int_{-h/2}^{h/2} N_I \rho N_I \dot{v}_{r_I} r\, dr\, d\theta\, dz + \delta v_{z_I} \int_{R_1}^{R_2} \int_{0}^{2\pi} \int_{-h/2}^{h/2} N_I \rho N_I \dot{v}_{z_I} r\, dr\, d\theta\, dz \tag{39c}$$

For a linear order shape function like;

$$N_I = \frac{1}{2} \begin{bmatrix} 1 - \zeta \\ 1 + \zeta \end{bmatrix} \tag{40}$$

$$\zeta = \frac{(2r - (R_1 + R_2))}{(R_2 - R_1)} \tag{41}$$

where r is the radius, and R_2, R_1 are the inner and outer radius of the element, respectively. Here r is a variable and R_2, R_1 are constants. After integration, internal power is in the following form:

$$\delta p^{kin} = \left\{ [\rho h(\delta v_{r_0I}\dot{v}_{r_0I} + \delta v_{zI}\dot{v}_{zI})] + \left[((\rho h^3)/12)(\delta\dot{\varphi}_{rI}\ddot{\varphi}_{rI})\right] \right\} 2\pi$$

$$\begin{bmatrix} (16/105)(R_2-R_1)(R_1+R_2)-(1/15)(R_2-R_1)^2 & (41/420)(R_2-R_1)(R_1+R_2) \\ (41/420)(R_2-R_1)(R_1+R_2) & (16/105)(R_2-R_1)(R_1+R_2)-(1/15)(R_2-R_1)^2 \end{bmatrix} \tag{42}$$

So, in this case the mass matrix of an element is defined as:

$$M_e = 2\pi \begin{bmatrix} \rho hi & 0 & 0 & \rho hj & 0 & 0 \\ 0 & \rho hi & 0 & 0 & \rho hj & 0 \\ 0 & 0 & (\rho h^3/12)i & 0 & 0 & (\rho h^3/12)j \\ \rho hj & 0 & 0 & \rho hi & 0 & 0 \\ 0 & \rho hj & 0 & 0 & \rho hi & 0 \\ 0 & 0 & (\rho h^3/12)j & 0 & 0 & (\rho h^3/12)i \end{bmatrix} \tag{43}$$

where

$$i = \left(\frac{1}{6}\right)(R_2 - R_1)(R_1 + R_2) - \left(\frac{1}{12}\right)(R_2 - R_1)^2$$

$$j = \left(\frac{1}{12}\right)(R_2 - R_1)(R_1 + R_2)$$

Mass matrix is obtained for SSDT by the same procedure. To verify the assumed method, the finite element method is changed to be consistent with the approach in Ref. [12]. In this case, a step load is used in a circular form at the middle of plate. The deflection results are plotted 2 mm from the plate's center during the first 4.5 μm. For this problem: $h = 0.4$ mm, $\sigma_0 = 50$ Mpa, $R = 0.3$ mm

The constitutive equations for the linear viscoelastic material are in a nonlinear integral form as stated below:

$$\sigma_r = \left(\frac{E_1}{(1-\mu_1^2)}\right)(\varepsilon_r + \mu_1\varepsilon_\theta) + \left(\frac{E_2}{(1-\mu_2^2)}\right)(\varepsilon_r + \mu_2\varepsilon_\theta) - \left(\frac{E_2^2}{\lambda(1-\mu_2^2)}\right)\int_0^t(\varepsilon_r + \mu_2\varepsilon_\theta)exp\left(\frac{-E_2}{\lambda(t-\tau)}\right)d\tau \tag{44}$$

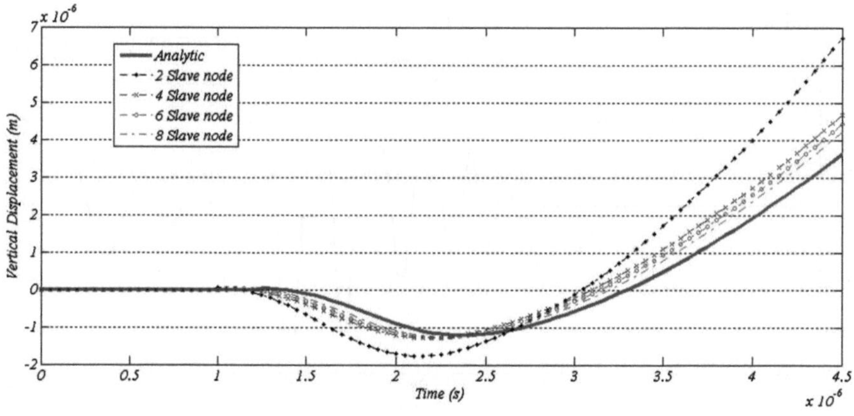

Fig. 5 Plate deflection at a specified point, 2 mm far from center

$$\sigma_\theta = \left(\frac{E_1}{(1-\mu_1^2)}\right)(\varepsilon_\theta + \mu_1\varepsilon_r) + \left(\frac{E_2}{(1-\mu_2^2)}\right)(\varepsilon_\theta + \mu_2\varepsilon_r) - \left(\frac{E_2^2}{\lambda(1-\mu_2^2)}\right)\int_0^t(\varepsilon_\theta + \mu_2\varepsilon_r)exp\left(\frac{-E_2}{\lambda(t-\tau)}\right)d\tau \tag{45}$$

$$\tau_{rz} = (G_1 + G_2)\gamma_{rz} - \left(\frac{G_2^2}{\eta}\right)\int_0^t \gamma_{rz}exp\frac{-G_2}{\eta(t-\tau)}d\tau \tag{46}$$

Finally, the analytical solution is compared with the present method, considering 2, 4, 6 and 8 slave node elements in Fig. 5.

Figure 5 shows that with more slave nodes for integration, response compare better with the analytical solution which uses an exact integration through thickness.

3 Case Study and Results Analysis

After verifying the finite element method, response of the plates with two structural theories is investigated. The problem is programmed and solved by MATLAB 64 bit (2009). For this reason, the nonlinear viscoelastic properties of an Acrylonitrile Butadiene Styrene (ABS) copolymer are considered.

$$\rho = 700\,\frac{\text{kg}}{\text{m}^3} \quad h = 4\,\text{mm} \quad E = 100\,\text{GPa} \quad \nu = 0.33$$

$$L = 100\,\text{kPa} \quad C = 0.01\,\text{s}$$

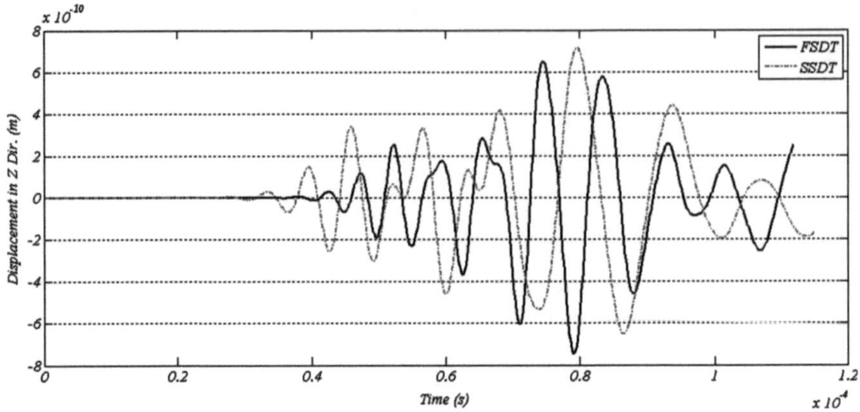

Fig. 6 Plate deflection comparison of FSDT and SSDT

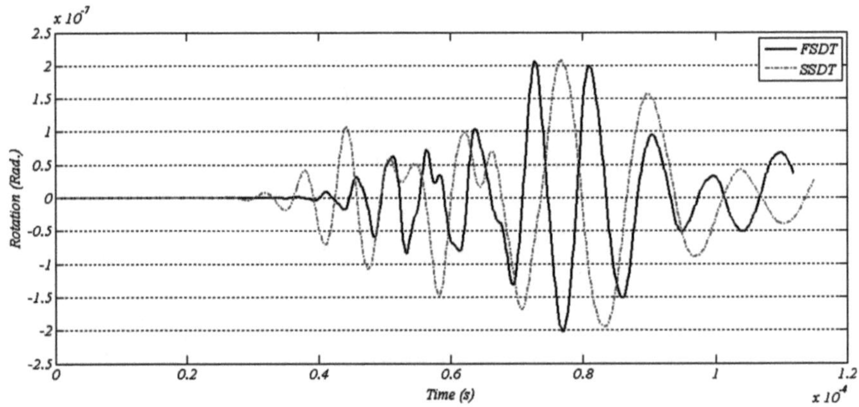

Fig. 7 Plate rotation comparison of FSDT and SSDT

A semi sinusoidal load, as depicted in Fig. 2, with a maximum of 100 kPa is exerted in two microseconds. The loaded area is a 2 cm diameter circle.

Vertical displacements of the plate using both FSDT and SSDT are shown in Fig. 6 at a point 75 cm far from the plate's center. This point is selected since the shear and moment waves are separated and the effects can be analyzed discretely.

In Fig. 6, the wave is divided in tow waves which are nominated shear and bending wave. In case of pure elastic material, the bending wave's speed is $\sqrt{2}$ times larger than the shear wave's. SSDT predicts a higher speed than FSDT, but the same deflection amplitude. In both theories, shear and bending waves have an overlap in the middle of considered time period. In other words, the considered

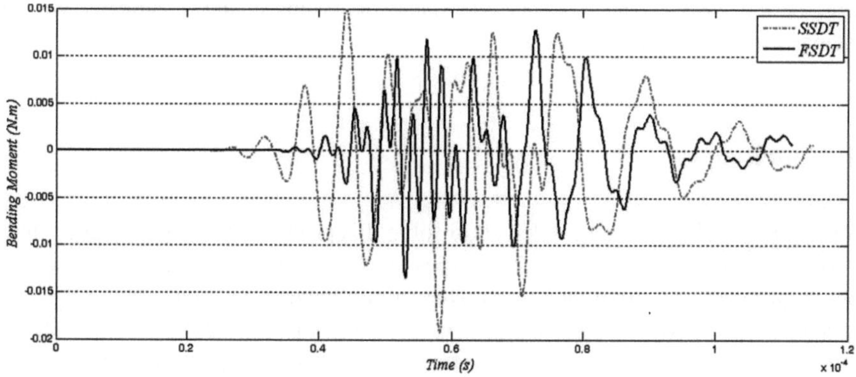

Fig. 8 Bending moment comparison of FSDT and SSDT

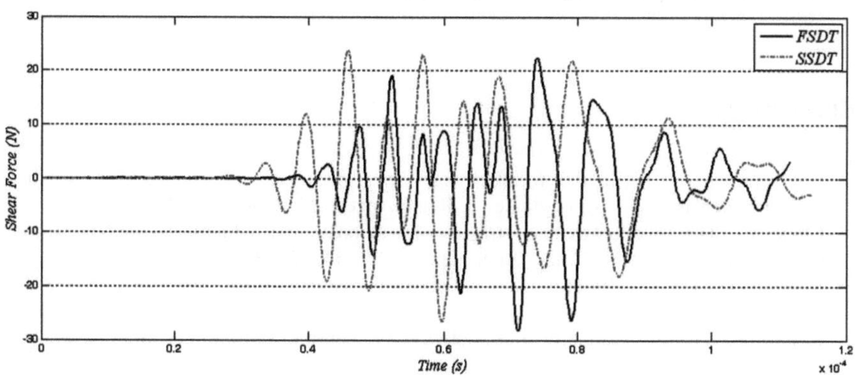

Fig. 9 Shear Force comparison of FSDT and SSDT

point is not far enough from the center so that the waves become completely separated.

Figure 7 depicts plate's response and wave's behavior showing midplane's rotation in both theories. The location is again a point 75 cm far from the plate's center and the investigation time period is selected in order to cover both bending and shear waves.

Again, SSDT predicts higher speed and larger rotations in the case of the bending wave, while for the shear wave, the maximum amplitude, however the time when it takes place is the same for both theories.

Figure 8 illustrates the comparison of the bending moment obtained by the two theories. Both theories show bending wave effects on the bending moment, like high amplitude and fluctuation in bending moment.

Finally, the transverse shear force waves are illustrated in Fig. 9. Figure 9 shows that the bending wave does not have a significant effect on the shear forces.

But for the case of shear waves, the shape of graphs changes radically due to changes in shear stresses.

4 Conclusion

Since structural theories are selected to solve this problem, the waves will deplete into two main types of waves. These types are known as bending and shear waves which propagate in different speeds. Due to nonlinearity in such problems, wave speeds cannot be defined exactly. However, in any case, bending wave speeds are more than shear's. The comparison between these two structural theories (SFSDT and SSDT) shows that SSDT predicts higher speed and larger amplitude than FSDT for bending wave, while they are equal in the case of shear wave. In all cases of both displacements and stress resultants, FSDT is more sensitive and shows more fluctuations encountering waves.

References

1. Pochhammer, L.: Uber die fortplanzungs geschwindigkeiten schwingungen in einem unbegrentzten isotropen kreiscylinder. J. Reine Angew. Math. **81**, 324–336 (1876)
2. Chree, C.: The equation of an isotropic solid in polar and cylindrical coordinates, their solutions and application. Trans. Camb. Philos. Soc. **14**, 250–369 (1889)
3. Bancraft, D.: The velocity of longitudinal waves in cylindrical bars. Phys. Rev. **59**, 588–593 (1941)
4. Kolsky, H.: The propagation of stress pulses in viscoelastic solids. Philos. Mag. **1**, 693–710 (1956)
5. Tsai, Y.M., Kolsky, H.: Surface wave propagation for linear viscoelastic solids. J. Mech. Phys. Solids **16**(2), 99–100 (1968)
6. Lundberg, B., Blanc, R.H.: Determination of mechanical material properties from the two point response of an impact linearly viscoelastic rod specimen. J. Sound Vib. **126**, 97–108 (1988)
7. Blanc, R.H.: Transient wave propagation method for determining the viscoelastic properties of solids. J. Appl. Mech. **60**, 763–768 (1993)
8. Rizzi, S.A., Doyle, J.F.: Spectral analysis of wave motion in plane solids with boundaries. Trans. ASME J. Vib. Acoust. **40**, 114–133 (1992)
9. Nkemzi, D., Green, W.A.: Transient wave propagation in viscoelastic sandwich plate. J. Acta Mech. **82**, 102–167 (1994)
10. Jiang, L., Haddow, J.B.: A finite element solution of plane wave propagation in inhomogeneous linear viscoelastic solids. J. Sound Vib. **184**(3), 429–438 (1995)
11. Yang, S., Yuan, F.G.: Transient wave propagation of isotropic plates using a higher order plate theory. J. Solid Struc. **42**, 4115–4153 (2005)
12. Adamek, V., Vales, F., Las, V.: Analytical and numerical solution of no stationary state of stress in a thin viscoelastic plate. J. Nonlinear Anal. **63**, 955–962 (2005)

Analytical, Numerical and Experimental Study of the Effects of Braking on Single Disc Motorcycle Forks

Dario Croccolo, M. De Agostinis and N. Vincenzi

Abstract This work deals with the development of an analytical model which allows to describe the tensile state arising in *single-disc* motorcycle *forks*, during the brake. Stress and strain trends are computed as functions of some key parameters of the motorcycle (mass and centre of gravity location) and of the fork (lengths and diameters). The fork geometry is represented by a portal frame loaded out of its plane, whose axisymmetric elements represent the legs (pillars) and the wheel pin (transverse beam). Each of the three elements has material and inertia parameters variable along their axis, allowing for the actual mechanical properties of the component. Finally, the stress state of several fork models has been investigated either via *Finite Element Analysis* and with field tests, in order to support the validity of the proposed model.

Keywords Motorcycle · Fork · Stress · FEA · Analytical · Experimental

1 Introduction

Although it may look as a simple component, the front fork has a critical role in the overall dynamic behaviour of motorcycles. It must provide appropriate stiffness characteristics, damping capabilities and the lowest sliding friction values in

---

D. Croccolo (✉) · M. De Agostinis · N. Vincenzi
DIEM, University of Bologna, V.le Risorgimento 2, 40136 Bologna, Italy
e-mail: dario.croccolo@unibo.it

M. De Agostinis
e-mail: m.deagostinis@unibo.it

N. Vincenzi
e-mail: nicolo.vincenzi@unibo.it

A. Öchsner et al. (eds.), *Materials with Complex Behaviour II*,
Advanced Structured Materials 16, DOI: 10.1007/978-3-642-22700-4_4,

Fig. 1 Standard (**a**-left) and Up-Side Down (**b**-right) forks

order to guarantee as much performance, safety and comfort as possible. From a structural mechanics standpoint, the main frame of front suspensions consists of two legs, two steering plates and a steering pin, put together by means of shaft-hub couplings [1–5]. Each leg realises a cylindrical joint, providing a way for its inner and outer tube to translate and rotate freely with respect to each other. The whole front suspension is assembled to the motorcycle frame by means of the steering pin. Whether the inner or outer tubes are coupled with the steering plates, the fork architecture is referred to as 'standard', see Fig. 1a or as 'up-side down', see Fig. 1b.

Inside the tubes are a number of hydraulic and elastic elements whose combined action defines the damping characteristic of the fork. Figure 2 reports a section view of a 'standard' fork, showing the arrangement of hydraulic and elastic components. Motorbike forks can be subdivided further into single-disc and twin-discs architectures. Despite single disc forks were the preferred choice for most of motorbikes in the past, nowadays such solution may be easily found on low cost and 'Offroad' motorbikes only, because the increase in performances and weight of either 'Supersports' and 'Tourism' motorbikes determined the spread of twin discs architectures.

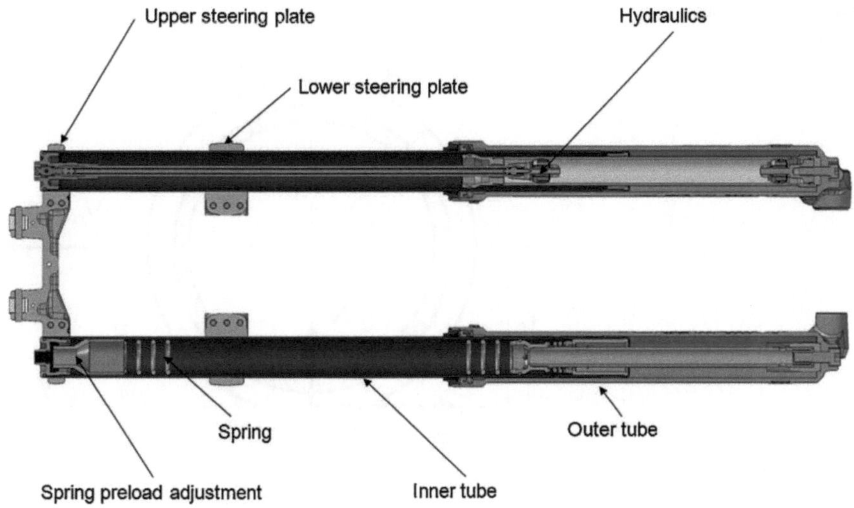

Fig. 2 Inner components of a fork

2 Structural Design of Forks

The design of a motorbike fork is a challenging task which shall take into account both the dynamic response and the structural strength of the suspension. While a good dynamic response can be improved afterwards by a fine-tuning of several manually adjustable parameters (whose choice is quite often based on the rider's feeling on the bike), only a correct design can ensure an adequate structural strength. Moreover, the design phase of a new product shall provide a quick fulfilment of the structural requirements because of the shortening of vehicles lifecycle (about 2:3 years). A deep knowledge of the product lifecycle is the key for a correct design: for that reason, suspensions producers must carry out a number of road/bench testing activities in order to collect all the relevant data. Several road tests carried out by our research group in cooperation with Paioli Meccanica, led to defining the most severe load conditions in terms of mechanical stress on the structural elements of the fork. A hard braking manoeuvre in which the rear tyre looses contact with the ground, and the whole motorcycle weight is transferred to the front wheel, determines the highest flexural stress values on the fork legs. In order to build an analytical model useful for the structural design of forks, such loading condition must be, therefore, deeply understood and described.

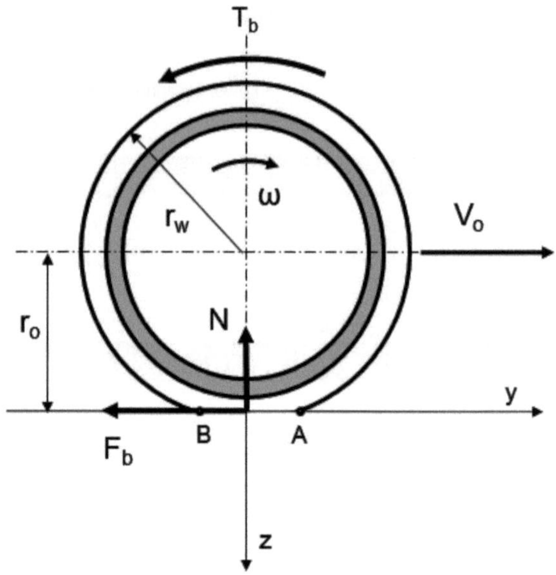

Fig. 3 Tire subject to a brake

2.1 Basics of Tire Dynamics

Tires are flexible elements which provide shock absorption while keeping the wheel in close contact with the ground. Tires grip characteristics play a critical role in the overall traction, brake and cornering performances a motorbike can deliver. Traction and brake forces arise during the ride involving shear forces along the contact area between the tire and the ground. Such forces make the rubber fibres belonging to the tire circumference compress along the tangential direction during the traction phase and extend during the brake. Figure 3 shows a tire travelling with a velocity $V_0 = \omega r_0$, subject to a braking torque T_b which determines a braking force F_b at the interface between the tire and the ground. A vertical load N acts on the tire.

Due to the longitudinal braking force F_b, the rubber fibres on the running circumference elongate when passing through the tire-ground contact segment AB (Fig. 4).

Hence, the circumferential velocity of a point fixed to the aforementioned fibres decreases as it travels from point A to point B: the circumferential velocity V of such point is therefore lower than the travelling velocity V_0.

Now, define the longitudinal slip, k:

$$k = \frac{V - V_0}{V_0} \tag{1}$$

The longitudinal slip takes positive values for traction and negative values for braking. Over the years, tire manufacturers defined a variety of semi-empirical relationships expressing the longitudinal force F (and then the longitudinal friction

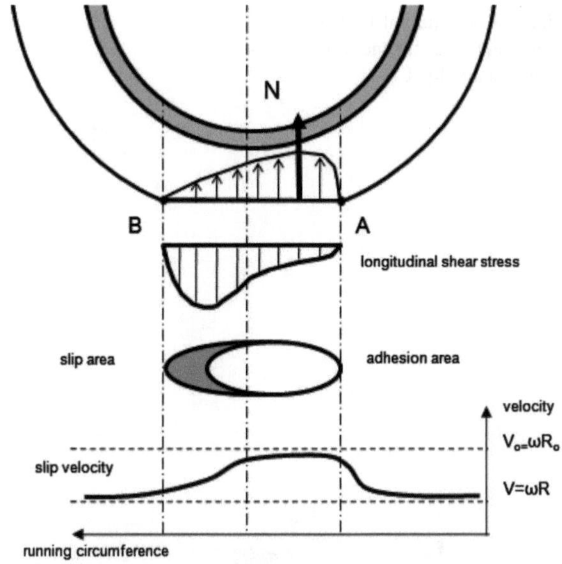

Fig. 4 Slip velocity during a brake

coefficient μ') as a function of vertical load N and longitudinal slip values k. The best-known relationship of such kind is the 'Magic Formula' by Prof. H. B. Pacejka [6], a transcendental function generally expressed in the form:

$$F(k) = D \cdot sin\{C \cdot arctan[B \cdot k - E(B \cdot k - arctan(B \cdot k))]\} \tag{2}$$

$$\mu' = \frac{F}{N} \tag{3}$$

With F being the longitudinal force and k the longitudinal slip, B, C, D, and E being input coefficients which depend on several 'static' parameters (tied to the geometrical and chemical characteristics of the tire) and on two 'dynamic' parameters, namely the longitudinal slip k and the vertical load N. The 'Magic Formula' owes its name to the fact that there is no particular physical basis behind the structure of the equation chosen, but it fits a wide variety of tire constructions and operating conditions. The equation can be plotted as shown in Fig. 5, where μ' is reported as a function of k.

Looking at Fig. 5 it could also be noticed that the curve $\mu'(k)$ is characterized by a peak, which is typically around $k = \pm\ 0.15$.

2.2 Equilibrium During a Brake

When a braking manoeuvre is performed, a load transfer from the rear wheel to the front wheel of the motorbike takes place. The amount of the load transfer depends on the center of mass position, on the motorbike wheelbase and on the deceleration value.

Fig. 5 Longitudinal friction coefficient μ' as a function of longitudinal slip k

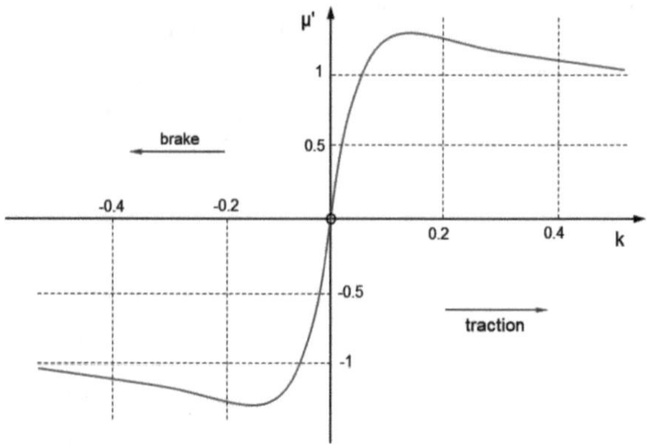

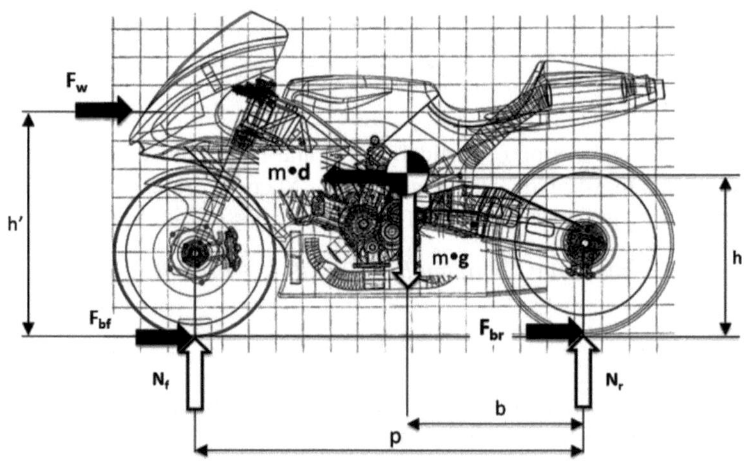

Fig. 6 Rigid body equilibrium during the brake

Referring to Fig. 6, define:

- m: bike + rider mass [kg];
- p: bike wheelbase [mm];
- b: bike + rider centre of mass to rear wheel axis distance [mm];
- h: bike + rider centre of mass to ground distance [mm];
- h': centre of the aerodynamic pressure [mm];
- F_w: aerodynamic force [N];
- F_{bf}, F_{br}: front wheel and rear wheel braking forces [N];
- N_f, N_r: weight on the front and on the rear axle [N].

When the motorbike has its wheels on a flat surface, the vertical load acting on the rear wheel is:

$$N_r = m \cdot g \cdot \frac{(p-b)}{p} \tag{4}$$

Now define the load transfer value due to the brake:

$$\Delta N = m \cdot d \cdot \frac{h}{p} \tag{5}$$

During a brake, vertical loads on the wheels take the values:

$$N'_r = N_r - m \cdot d \cdot \frac{h}{p} \tag{6}$$

$$N'_f = m \cdot g - N'_r \tag{7}$$

The in-plane equilibrium governing relationships during the brake are:

$$\begin{cases} F_w + F_{bf} + F_{br} = m \cdot d \\ m \cdot d \cdot h = m \cdot g \cdot (p-b) + F_w \cdot h' \end{cases} \tag{8}$$

Where:

$$\begin{cases} F_{bf} = \mu' \cdot N'_f \\ F_{br} = \mu' \cdot N'_r \end{cases} \tag{9}$$

Here μ' is the longitudinal friction coefficient, as defined into (3). A braking manoeuvre in which the rear tyre looses contact with the ground, and the whole motorcycle weight is transferred to the front wheel is, therefore, characterized by $N'_r = 0$ [7, 8]. Provided that $N'_r = 0$, (8) transforms as follows:

$$\begin{cases} F_{bf} = m \cdot d - F_w \\ m \cdot d \cdot h = m \cdot g \cdot (p-b) + F_w \cdot h' \end{cases} \tag{10}$$

Then, recalling (7) and (9), (10) gives:

$$\begin{cases} F_{bf} = \mu' \cdot m \cdot g \\ F_{bf} = m \cdot g \cdot \dfrac{(p-b)}{h} + F_w \cdot \dfrac{(h'-h)}{h} \end{cases} \tag{11}$$

Which expresses that the maximum applicable braking force during a hard braking manoeuvre is:

$$F_{bf\_MAX} = min\left[\mu' \cdot m \cdot g; m \cdot g \cdot \frac{(p-b)}{h} + F_w \cdot \frac{(h'-h)}{h}\right] \tag{12}$$

On a dry, clean asphalt surface, it is commonly assumed that motorcycle standard sport touring tires can deliver a maximum longitudinal friction coefficient $\mu' \approx 1.5$ [9]. During a hard brake, motorcycles with a long wheelbase and a low

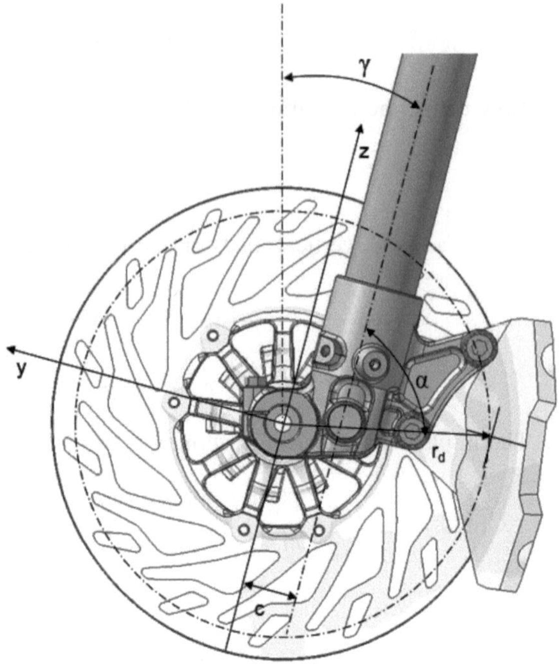

Fig. 7 Coordinate system

center of mass (like Cruisers) would probably reach the skidding condition before capsizing, while Hypersport motorcycles would probably behave the opposite way. The maximum applicable braking force is a key parameter for fork manufacturers, since it is strictly tied to the maximum bending load that the legs must withstand during the product lifecycle.

2.3 Loads Analysis

The tensile state of single disc forks subject to braking loads is worthy of particular interest, because such an asymmetric architecture leads to an unequal distribution of bending stresses between the two legs. Define a Cartesian coordinate system with its origin into the wheel pin centre, x-axis coincident with the wheel pin axis and z-axis oriented upwards along the leg axis. Referring to Fig. 7, r_d is the mean disc radius and α the angle between z-axis and the brake pads centre (brake caliper angle). γ is the caster angle and c is the offset between the wheel pin and the leg axis.

When a hard brake is performed and the external forces shown in Fig. 6 act on the rolling circumference of the front tire, the internal forces shown in Fig. 8 arise. The total vertical load $N = N_f\,(N_r = 0)$ is transmitted to the wheel pin, along with the braking force F_b. Define the braking torque on the disc:

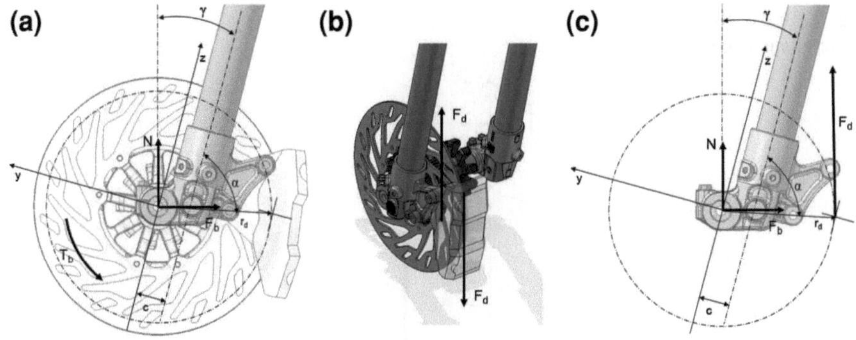

Fig. 8 **a, b, c** Internal forces during the brake

$$T_b = F_b \cdot r_w \tag{13}$$

In order to deliver a braking torque T_b the braking pads must provide a braking force on the disc:

$$F_d = \frac{T_b}{r_d} = F_b \cdot \frac{r_w}{r_d} \tag{14}$$

In order to satisfy the internal equilibrium of the caliper–disc system, the same F_d force shall act either on the disc and on the caliper (Fig. 8 (b)):

Such force is, therefore, transmitted by the brake caliper to the relevant leg (Fig. 8 (c)) and by the disc to the wheel pin (Fig. 9). From now on, F_d will be referred to as 'disc force'.

It must be noticed that any force applied either to the disc or to the wheel is transmitted to the wheel pin through the front wheel hub bearings. Forces belonging to the wheel-ground contact (i.e. N and F_b) act on the fork vertical mid-plane, hence they are equally divided between the two bearings, therefore between the two legs. Since the brake disc cannot be aligned with the leg axis, an offset between z-axis and the disc mid-plane (Ω) always exists, such that F_d actually acts on Ω, as shown in Fig. 9. For that reason, the disc force transmitted by the disc to the wheel pin is unevenly distributed between the two bearings.

2.4 Structural Scheme

In order to retrieve an analytical model able to describe the stress state of the fork under the aforementioned loading conditions, a line body scheme will now be defined. A portal frame made up of three axisymmetric elements, the two pillars representing the legs and the transverse beam representing the wheel pin (Fig. 10), was chosen as a line body scheme [10]. The relevant Cartesian coordinate system is taken as shown in Fig. 10, conforming to what was stated in Sect. 2.2.

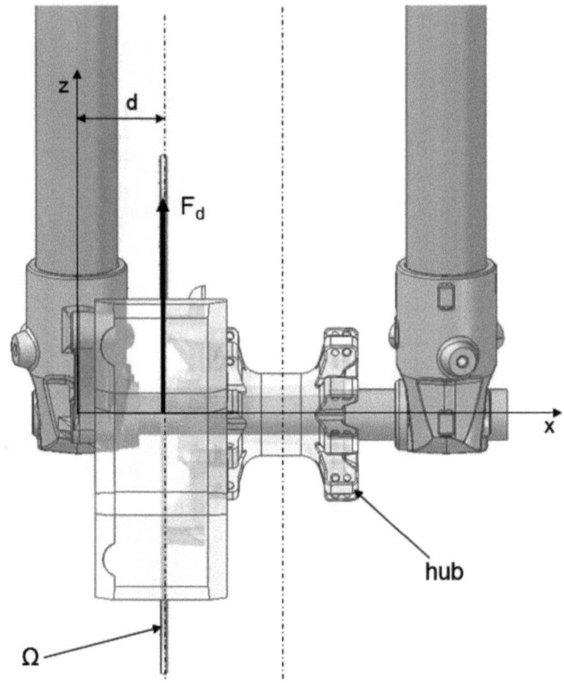

Fig. 9 Brake disc offset along x-axis (d)

The offset along y-axis c between the wheel axle and the leg axis is reported in Fig. 10 as well. The pillars are constrained in C and D by means of two hinges, which allow the sole rotation around z-axis.

The beam and the pillars have material and inertia parameters variable along their axes, in order to achieve a reliable approximation of the mechanical properties of the actual components. Each leg has been subdivided into three segments, E_i and I_i being the Young's moduli and the x-axis moments of inertia of the leg segments respectively. In order to clarify what was above stated, variable thickness lines represent the different properties of the portal frame elements in Fig. 2.8.

This complication is due to the fact that the leg is made up of two elements, the inner and the outer tube, having variable materials and sections along their axes (Fig. 11)

Looking at Fig. 12, it can be appreciated how the leg intermediate section (L_2) is actually composed of two elements (the inner and the outer tube). These elements work together like mechanical springs in parallel, therefore, the overall Young's modulus and the x-axis moment of inertia of the leg intermediate section shall be computed consequently.

Wheel pins usually have their diameter variable along the symmetry axis. In order to provide for that, the portal beam has to be subdivided into four sections with different moments of inertia (I_{pj}, J_{pj}) but with the same material (E_p, G_p, ν_p). As Figs. 10 and 11 report, the overall free length of the legs ($L = L_1 + L_2 + L_3$)

Fig. 10 Line body scheme

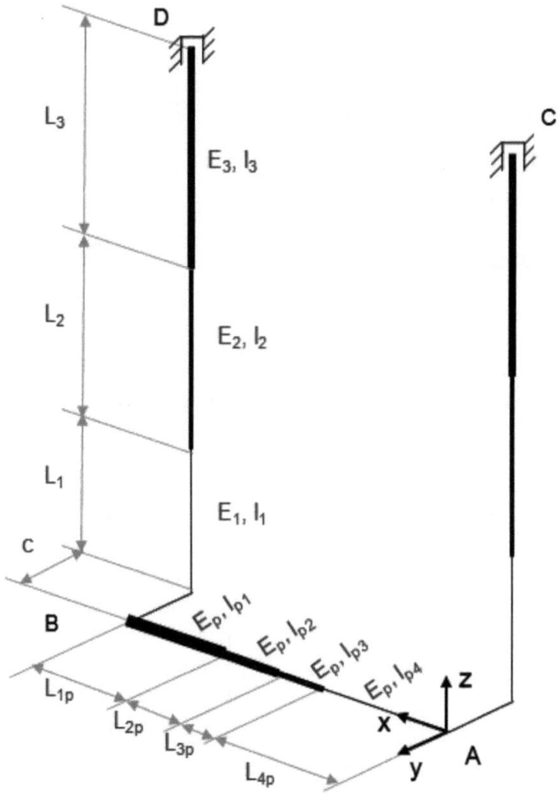

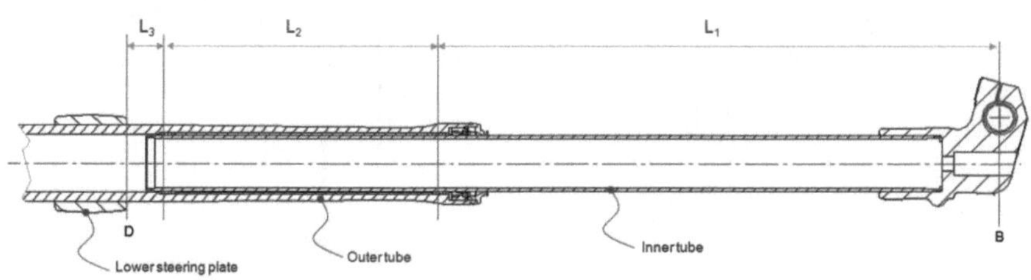

Fig. 11 Leg section

is the distance, along z-axis, from the wheel axle to the lower edge of the lower steering plate. Once the line body scheme has been defined, loads can be applied. Figure 13 (a) represents the portal frame loaded out of its plane with the forces N, F_b and F_d, shown in their actual positions.

Figure 13 (b) represents an equivalent system in which all the forces have been separated into the relevant components along the coordinate system axes. F_d (leg) components have been transferred to the braking leg axis: three concentrated

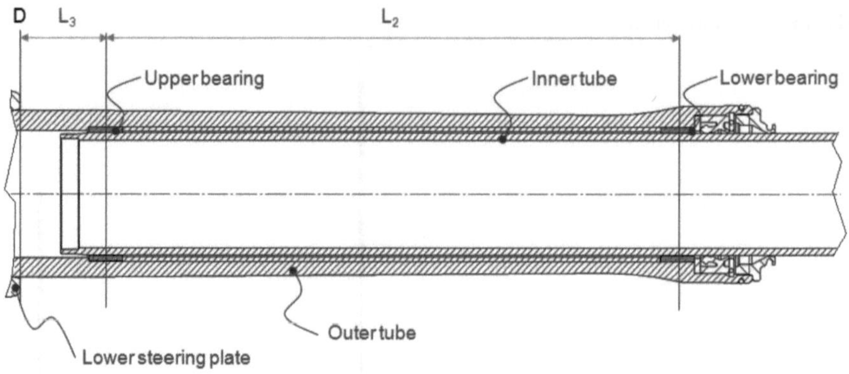

Fig. 12 Leg section enlargement–bearings

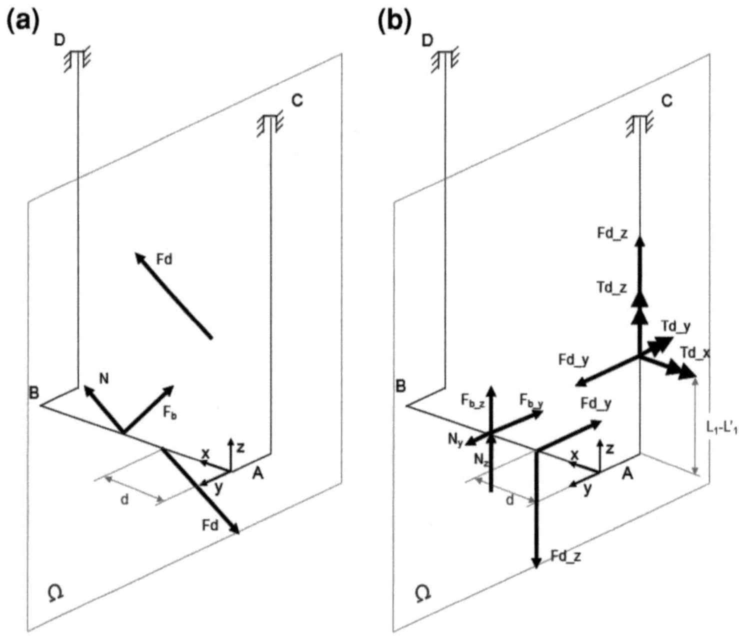

Fig. 13 **a** Forces on the portal frame, **b** Force components on the portal frame

moments T_{d_x}, T_{d_y} and T_{d_z} have been introduced to enforce equilibrium conditions. Referring to Figs. 13 and 14:

$$\begin{cases} F_{d\_y} = F_d \cdot cos\alpha \\ F_{d\_z} = F_d \cdot sin\alpha \end{cases} \tag{15}$$

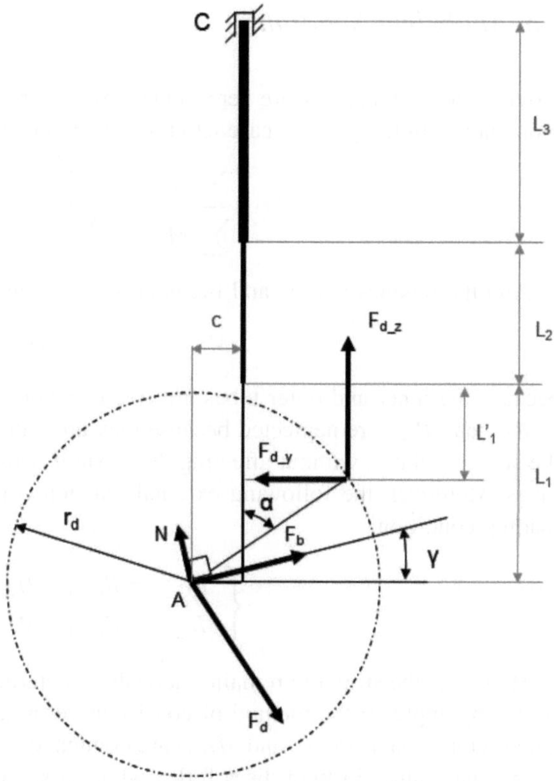

Fig. 14 Force components on the portal frame–side view

T_{d_x} is due to the y-axis distance between the leg axis and the brake pads centre, as shown into the side view of the braking leg reported in Fig. 14.

$$T_{d\_x} = F_{d\_z} \cdot (r_d \cdot sin\ \alpha - c) \tag{16}$$

T_{d_y} can be written as:

$$T_{d\_y} = F_{d\_z} \cdot d \tag{17}$$

And will be neglected because it yields any contribution neither to the legs deflection on yz-plane, nor to the maximum stress of the legs. Finally T_{d_z} is given by:

$$T_{d\_z} = F_{d\_y} \cdot d \tag{18}$$

Referring to Fig. 14, the distance, measured along z-axis, between the F_d projection on y-axis, F_{d_y}, and the constraint C, is $L' = L'_1 + L_2 + L_3$, where:

$$L'_1 = (L_1 - r_d \cos(\alpha)) \tag{19}$$

2.5 Analytical Solution

External force reactions are represented by upper case letter R, while external moment reactions by upper case letter M. The number of equations of statics is six:

$$\begin{cases} \sum R_{x,y,z} = \sum F_{x,y,z} \\ \sum M_{x,y,z} = \sum T_{x,y,z} \end{cases} \tag{20}$$

And the unknown force and moment reactions in C and D are ten, being:

$$M_{C\_z} = M_{D\_z} = 0 \tag{21}$$

because the inner and outer tubes are free to rotate with respect to each other.

R_{C_z} and R_{D_z} are neglected because they are transmitted by the fork springs to the steering plates, without affecting the flexural stress state of the inner and outer tubes. Moreover, the following external reaction components are null due to the loading condition:

$$\begin{cases} R_{C\_x} = R_{D\_x} = 0 \\ M_{C\_y} = M_{D\_y} = 0 \end{cases} \tag{22}$$

However, the structure remains statically indeterminate and shall be, therefore, solved by applying the method of consistent deformations: the reaction moments values along x-axis (M_{C_x} and M_{D_x}) are computed. Such reaction moments would have equal values in twin discs forks, while in single disc forks, their difference expresses the uneven stress distribution between the two legs introduced by the structure asymmetry. The rotation angles around x-axis of points A and B, are defined by φ_A and φ_B respectively. The structure is uncoupled at points A and B and the consistent deformations [11] equation is introduced:

$$\phi_A = \phi_B + \phi' \tag{23}$$

Where φ' is the wheel pin torsion angle between A and B ends. As formerly specified, a number of effects combine to bring about the overall M_{C_x} and M_{D_x} reaction moments: for instance F_b and N determine equal external reaction moments around x-axis while F_d does not. Notwithstanding such effects are subdivided and discussed separately on the following paragraphs, they are eventually superimposed (relying on the fact that the analysis is carried out in the elastic field) in order to compute the overall M_{C_x} and M_{D_x} values.

2.5.1 First Effect: $\mathbf{F_{d\_y}}$ and $\mathbf{T_{d\_x}}$ Contribution

The first effect is due to the combined action of the disc force component F_{d_y} and of the concentrated moment T_{d_x}, both applied to the braking leg axis at $z = L_1 - L'_1$.

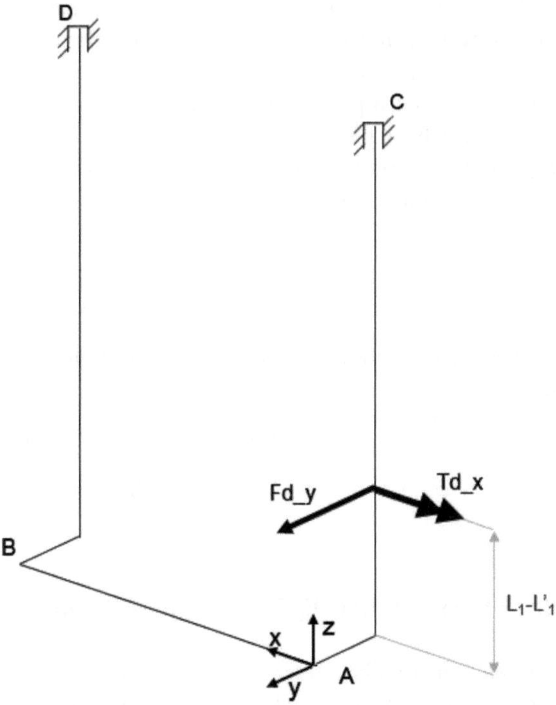

Fig. 15 Force components on the portal frame–first effect

Such effect could be split further into two contributions: the primary one is tied to the bending of the legs under the direct action of F_{d_y}, T_{d_x}.

Referring to Figs. 14 and 15, and recalling (19), it is convenient to define T_1 and T_2 in order to enforce the method of consistent deformations:

$$\begin{gathered} T_1 = F_{d\_y} \cdot \left(L_1' + L_2\right) \\ T_2 = F_{d\_y} \cdot L_1' \end{gathered} \tag{24}$$

Define also:

$$E_2 I_2 = E_1 I_1 + E_3 I_3 \tag{25}$$

Then, the reaction moments due to the primary contribution of the first effect in D and C are:

$$M_{D\_x11} = \frac{\left[\frac{F_{d\_y} \cdot L_3^2}{2 \cdot E_3 I_3} + \frac{T_1 \cdot L_3}{E_3 I_3} + \frac{F_{d\_y} \cdot L_2^2}{2 \cdot E_2 I_2} + \frac{T_2 \cdot L_2}{E_2 I_2} + \frac{T_2 \cdot L_1'}{2 E_1 I_1} - \frac{T_{d\_x} \cdot L_3}{E_3 I_3} - \frac{M \cdot L_2}{E_2 I_2} - \frac{M \cdot L_1'}{E_1 I_1}\right]}{\left(\frac{2 \cdot L_3}{E_3 I_3} + \frac{2 \cdot L_2}{E_2 I_2} + \frac{2 \cdot L_1}{E_1 I_1} + \frac{L_p}{G_p J_p}\right)} \tag{26}$$

$$M_{C\_x11} = F_d cos\alpha \cdot \left(L_1' + L_2 + L_3\right) - F_d \cdot r_d \cdot sin^2\alpha + F_d \cdot c \cdot sin\alpha - M_{D\_x11} \tag{27}$$

Looking at (26) and (27), it can be appreciated how the ends of the two legs (A and B points) undergo different rotations along x-axis due to bending. Provided that the offset c is not null, such different rotations determine different displacements of A and B points, making the wheel pin bending into xz-plane. Ends A and B of the wheel pin react, therefore, with two forces along z-axis: R_{A_z12} and R_{B_z12}. These forces produce a secondary contribution to x-axis moment reactions in C and D equal to:

$$\begin{cases} M_{C\_x12} = \bar{R}_{A\_z\_12} \cdot c \\ M_{D\_x12} = \bar{R}_{B\_z\_12} \cdot c \end{cases} \tag{28}$$

The superposition principle yields the moment reactions due to the first effect as:

$$\begin{cases} M_{C\_x1} = M_{C\_x11} + M_{C\_x12} \\ M_{D\_x1} = M_{D\_x11} + M_{D\_x12} \end{cases} \tag{29}$$

2.5.2 Second Effect: $\mathbf{F_d}$ Loads Applied to the Wheel Pin

For internal equilibrium reasons, F_d acts on the disc as well. Then it is transmitted by the hub to the wheel pin and, therefore, to the legs in A and B points. The legs react on the wheel pin with two forces whose intensity depends on the distance between Ω and the leg axis d and on the length of the wheel pin L_p (Figs. 16 and 17).

Just like the first, the second effect brings about two contributions to the supports moment reactions. The first contribution yields:

$$\begin{cases} M_{D\_x21} = R_{B\_y21} \cdot (L_1 + L_2 + L_3) - \left(R_{B\_y21} - R_{A\_y21}\right) \cdot \dfrac{S_1}{2 \cdot S_2 + S_3} \\ M_{C\_x21} = R_{A\_y21} \cdot (L_1 + L_2 + L_3) + \left(R_{B\_y21} - R_{A\_y21}\right) \cdot \dfrac{S_1}{2 \cdot S_2 + S_3} \end{cases} \tag{30}$$

Where:

$$\begin{aligned} S_1 &= \frac{L_3^2}{2 \cdot E_3 I_3} + \frac{(L_1 + L_2) \cdot L_3}{E_3 I_3} + \frac{L_2^2}{2 \cdot E_2 I_2} + \frac{L_1 \cdot L_2}{E_2 I_2} + \frac{(L_1)^2}{2 \cdot E_1 I_1} \\ S_2 &= \frac{L_3}{E_3 I_3} + \frac{L_2}{E_2 I_2} + \frac{L_1}{E_1 I_1} \\ S_3 &= \frac{L_p}{G_p \cdot J_p} \end{aligned} \tag{31}$$

As above reported for the first effect, since the offset c is not null, the different rotations and displacements determined by R_{A_y21} and R_{B_y21} at A and B points, force the wheel pin to bend into xz-plane. Ends A and B of the wheel pin react,

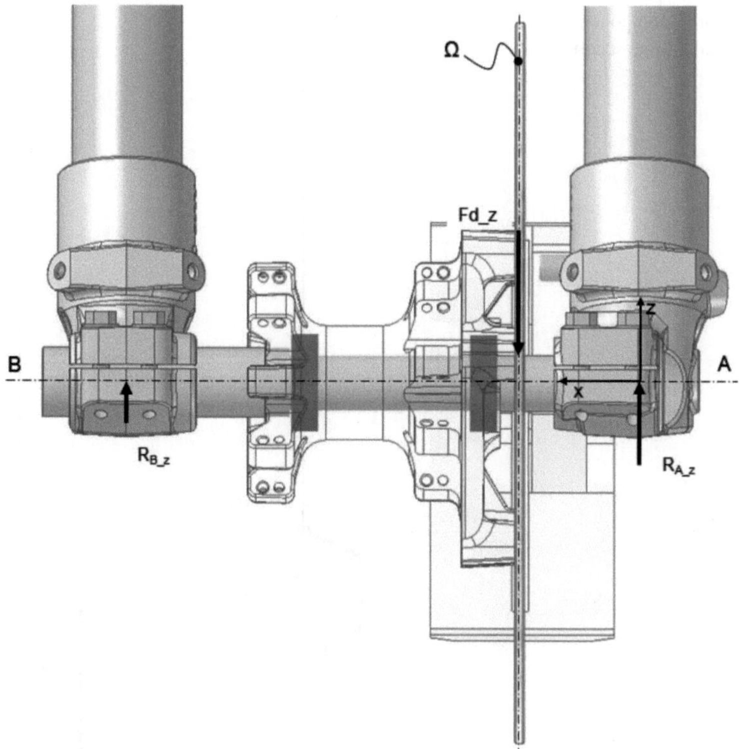

Fig. 16 Wheel pin equilibrium along z-axis–second effect

therefore, with two forces along z-axis: R_{A_z22} and R_{B_z22}. These forces, summed up with the components along z-axis due to the first effect, produce a secondary contribution to x-axis moment reactions in C and D equal to:

$$\begin{cases} M_{D\_x22} = (R_{B\_z21} + R_{B\_z22}) \cdot c \\ M_{C\_x22} = (R_{A\_z21} + R_{A\_z22}) \cdot c \end{cases} \tag{32}$$

The moment reactions due to the second effect can be, therefore, written as:

$$\begin{cases} M_{C\_x2} = M_{C\_x21} + M_{C\_x22} \\ M_{D\_x2} = M_{D\_x21} + M_{D\_x22} \end{cases} \tag{33}$$

2.5.3 Third Effect: T_{d_z} Contribution

Now, the third effect is computed as a superposition of two contributions. Recalling Fig. 13 and (15), T_{d_z} concentrated torque brings about the last

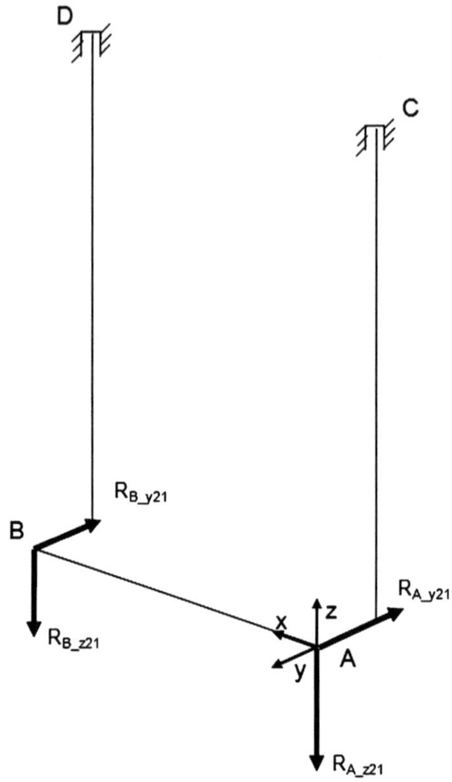

Fig. 17 Force components on the portal frame–second effect

contribution to the overall x-axis moment reactions in C and D due to the braking force. Since the legs are allowed to rotate along z-axis (Figs. 11, 12), they behave as simple supports for the wheel pin on xy-plane. The concentrated torque (Fig. 18) can be transported to the wheel pin centre: the wheel pin can, therefore, be treated as a simply supported beam, loaded by T_{d_z} (Fig. 19).

The two supports react with two forces having the same intensity and directed along y-axis, R_{A_y3} and R_{B_y3}:

$$R_{A\_y3} = R_{B\_y3} = \frac{T_{d\_z}}{L_p} \tag{34}$$

Such forces determine the third effect, first contribution, moment reactions in C and D:

$$\begin{cases} M_{C\_x31} = R_{A\_y3} \cdot (L_1 + L_2 + L_3) + \left(R_{B\_y3} - R_{A\_y3}\right) \cdot \dfrac{S_1}{2 \cdot S_2 + S_3} \\ M_{D\_x31} = R_{B\_y3} \cdot (L_1 + L_2 + L_3) - \left(R_{B\_y3} - R_{A\_y3}\right) \cdot \dfrac{S_1}{2 \cdot S_2 + S_3} \end{cases} \tag{35}$$

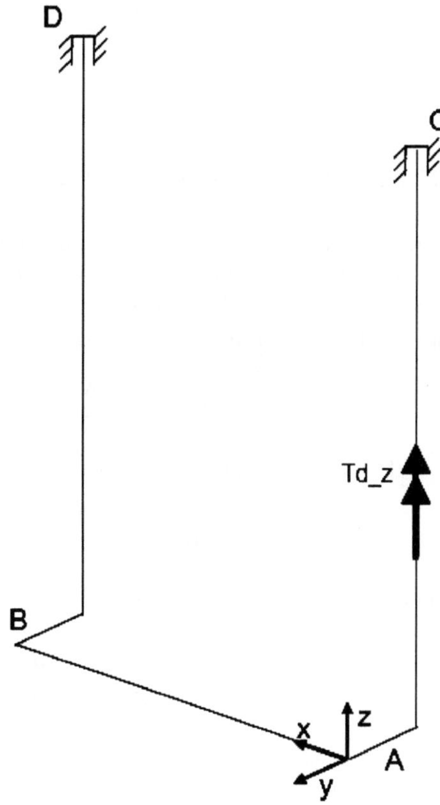

Fig. 18 Force components on the portal frame–third effect

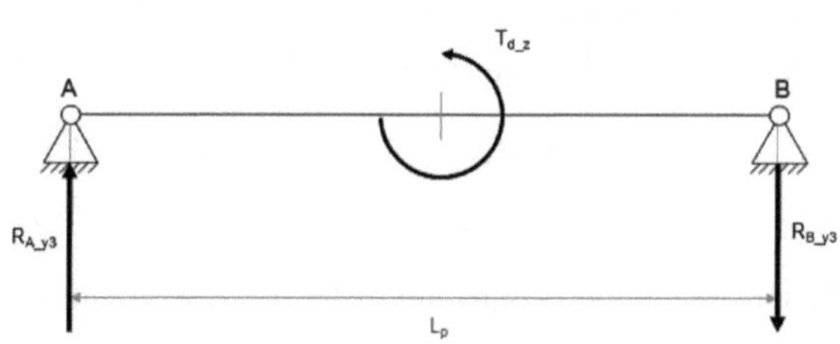

Fig. 19 Wheel pin loaded by $\mathbf{T_{d\_z}}$ on xy-plane

Where S_1, S_2, S_3 are defined into (31).

The second contribution of the third effect can be derived as an analog of the second contributions of first and second effect.

$$\begin{cases} M_{C\_x32} = R_{A\_z32} \cdot c \\ M_{D\_x32} = R_{B\_z32} \cdot c \end{cases} \tag{36}$$

The moment reactions belonging to the third effect are:

$$\begin{cases} M_{C\_x3} = M_{C\_x31} + M_{C\_x32} \\ M_{D\_x3} = M_{D\_x31} + M_{D\_x32} \end{cases} \tag{37}$$

2.5.4 Fourth Effect, Symmetric Contributions $\mathbf{F_b}$ and N

Since the effects, which concur in determining the load unevenness between the legs have been examined, focus now on F_b and N forces. Their effects are equally subdivided between the legs. Referring to Fig. 14 notation, the moment reactions in C and D due to the vertical load N is defined as:

$$M_{C\_x41} = M_{D\_x41} = \frac{N}{2} \cdot [(L_1 + L_2 + L_3) \cdot sin\gamma + c \cdot cos\gamma] \tag{38}$$

While the moment reactions belonging to F_b can be written as:

$$M_{C\_x42} = M_{D\_x42} = \frac{F_b}{2} \cdot [(L_1 + L_2 + L_3) \cdot cos\gamma - c \cdot sin\gamma] \tag{39}$$

Then:

$$\begin{cases} M_{C\_x4} = M_{C\_x41} + M_{C\_x42} \\ M_{D\_x4} = M_{D\_x41} + M_{D\_x42} \end{cases} \tag{40}$$

2.5.5 Overall Moment Reactions Estimation

Recalling (29), (33), (37) and (40), the overall moment reactions in C and D can be obtained by adding up the four aforementioned effects:

$$\begin{cases} M_{C\_x} = M_{C\_x1} + M_{C\_x2} + M_{C\_x3} + M_{C\_x4} \\ M_{D\_x} = M_{D\_x1} + M_{D\_x2} + M_{D\_x3} + M_{D\_x4} \end{cases} \tag{41}$$

Where the major contribution to the overall moment reactions is given by the fourth effect (M_{C_x4}, M_{D_x4}).

2.6 Comparison between Analytical and Finite Element Analysis Results

Once relationships (41) are known, the stress values along the fork legs, or the moment reactions of the constraints C and D can be back calculated. In order to understand the validity of the presented model, a comparison with a Finite

Table 1 Bike parameters

Bike parameter	Value	Unit
m	195	kg
p	1.475	mm
b	650	mm
h	650	mm
r_w	300	mm
r_d	117,5	mm
F_b	2.428	N

Table 2 Fork parameters

Fork parameter	Value	Unit
γ	26	deg
α	85	deg
c	34	mm
d	37	mm
L_1	397	mm
L_2	211	mm
L_3	24	mm
L_p	190	mm
I_1	56.261	mm^4
I_2	378.861	mm^4
I_3	322.600	mm^4
J_p	19.769	mm^4
E_1, E_p	206.000	MPa
E_3	71.000	MPa

Elements Analysis (*FEA*) of a production fork is reported. The motorbike chosen for the comparison is a 125 cc. Enduro model, equipped with a 41 mm upside-down fork. In conforming with Sect. 2.2 notation, the bike parameters are reported below.

Table 2 shows instead the fork parameters, used as inputs for the analytical model.

Given Table 1 and 2 values, Eq. (41) provide the overall moment reaction values in C and D:

$$\begin{cases} M_{C\_x} = M_{C\_x1} + M_{C\_x2} + M_{C\_x3} + M_{C\_x4} \cong -955{,}600\,\text{Nmm} \\ M_{D\_x} = M_{D\_x1} + M_{D\_x2} + M_{D\_x3} + M_{D\_x4} \cong -527{,}400\,\text{Nmm} \end{cases} \tag{42}$$

In the vicinity of C and D constraints, also the normal stress along z-axis on the external tube can be calculated as follows [12]:

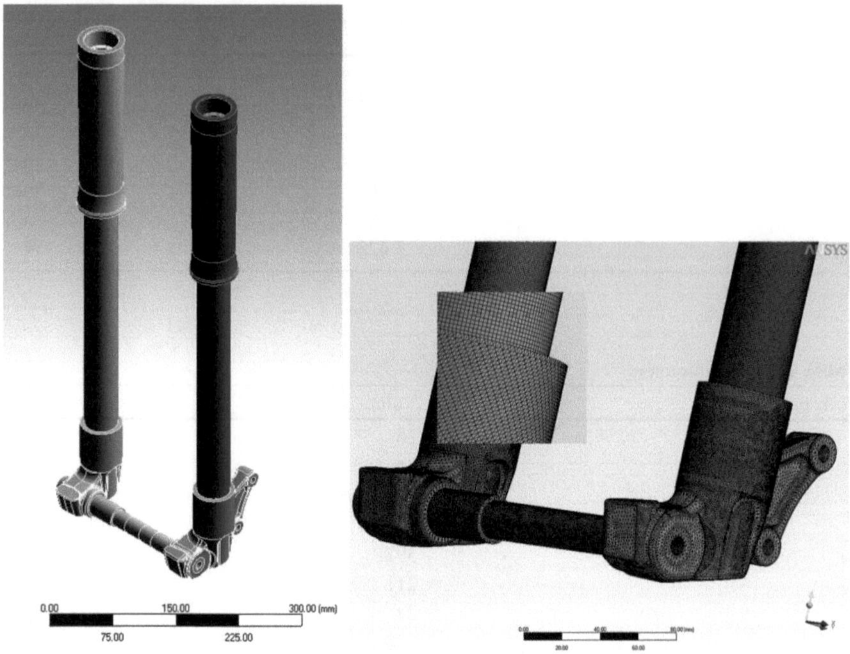

Fig. 20 FEA model and Mesh

$$\begin{cases} \sigma_{C\_z} = \dfrac{M_{C\_x}}{I_3} \cdot \dfrac{D_{e3}}{2} \cong 83\text{MPa} \\ \sigma_{D\_z} = \dfrac{M_{D\_x}}{I_3} \cdot \dfrac{D_{e3}}{2} \cong 46\text{MPa} \end{cases} \tag{43}$$

The *FEA* has been carried out with the ANSYS Workbench R.12 code on the same geometry. The model has been meshed with tetrahedrons and hexahedrons imposing an element size of 1 mm, which resulted into approximately 4,400,000 nodes and allowed to have more than one element across the tube thickness (Fig. 20).

F_d, N and F_b forces have been applied to the structure conveniently, as shown in Fig. 21.

Two cylindrical supports, allowing the tubes to rotate around z-axis, have been applied in C and D. F_d, the force acting on the wheel pin, has been subdivided into two components F_e and F_f. Then, such components have been applied each to a restricted surface of the wheel pin, which corresponds to the contact area between the wheel pin and the wheel bearing internal ring. Results have been obtained for moment reactions in C and D and for normal stress values along z-axis. Table 3 reports a comparison between Analytical and FEA results.

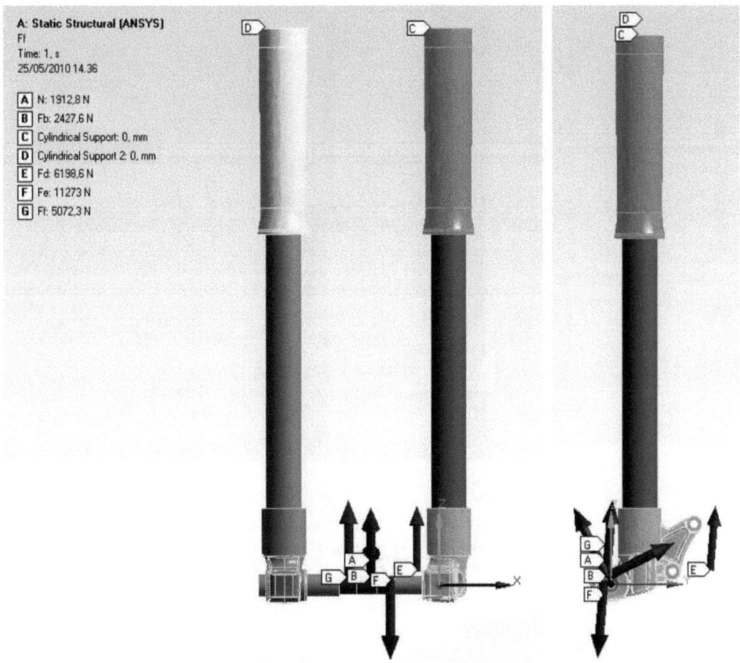

Fig. 21 Loads and constraints

Table 3 Analytical vs. Fea: moment reactions, normal stress (z-axis) and percentage error

	Analytical	FEA	Error (%)
M_{C_x} [N mm]	−955.600	−926.000	3
M_{D_x} [N mm]	−527.400	−538.000	2
σ_{C_z} [MPa]	83	80	4
σ_{D_z} [MPa]	46	47	2

It can be observed that a good convergence between the two methods exists, since errors are always lower than 5%.

Figure 22 shows the normal stress distribution along z-axis: two flags indicate the stress values near the lower steering plate.

2.7 Experimental Results

An experimental campaign has been carried out as a means for assessing the validity of the analytical and numerical models. Results obtained for the same motorbike-fork mentioned into Sect. 2.5 are presented. The fork has been

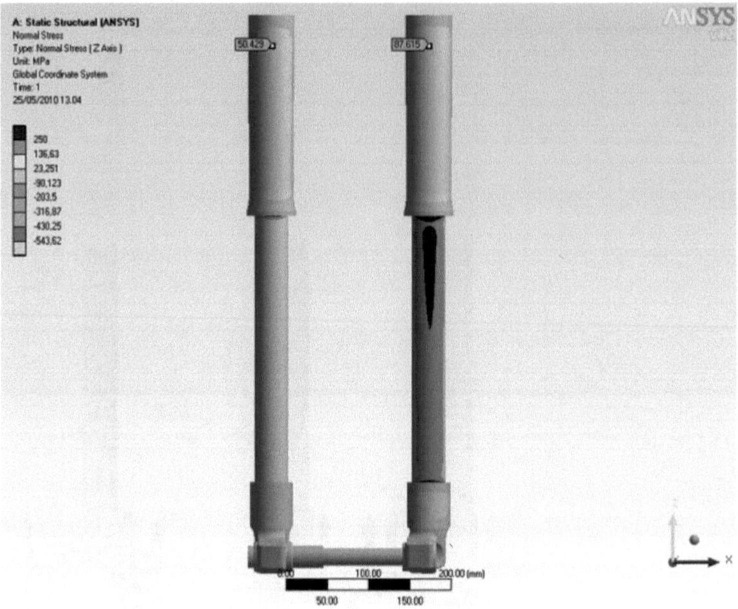

Fig. 22 Z-axis normal stress

Fig. 23 Strain gauges installation

instrumented with four strain gauges placed close to the lower steering plate joint. The strain gauges have been glued to the external surface of the outer tubes, with the main grid axis aligned with the tube axis (Fig. 23).

Each couple of strain gauges, half bridge configured [13], has been connected to a National Instruments 9237 module, plugged into a NI C-Rio portable controller.

Fig. 24 Brake at the capsizing limit

A Labview software has been used to manage all the data acquisition parameters: the data sampling frequency has been set at 100 Hz. Then, a series of hard brakes has been performed by a professional rider on an even asphalt mat (Fig. 24), and the relevant stress values have been recorded for both the legs. It is well known how the asphalt mat conditions deeply influence the longitudinal friction coefficient value μ'. Tire characteristics, as mentioned into Sect. 2.1, play a critical role towards longitudinal friction as well. Another element which impacts on maximum bending stress values is the fork stroke during the brake: as N and F_b forces increase during the brake, taking peak values at the end of the brake, conversely the fork L_l length decreases to become minimum at the same moment. Such behavior determines a significant decrease in flexural stresses on the outer tubes that cannot be taken into account by a static (analytical or *FEA*) model. At the same time, front suspensions outer tubes should be designed at the fully extended position since the sliding between the inner and outer tubes can be locked due to springs or hydraulic failures.

Therefore, the stress values sampled during field-testing can be significantly lower than those calculated either by *FEA* or Analytical models. Experimental peak stress values in C and D are reported in Table 4.

Figure 25 reports the stress tracks, for both the legs, recorded during a series of ten brakes at the capsizing limit.

3 Discussion

Forks legs have identical sections, and all the methods described above allow defining the ratio between the maximum stress values read on the two legs. This parameter conveys the amount of load unbalance between the legs. Therefore,

Table 4 Experimental values: normal stress (z-axis)

	Experimental
σ_{C_z} [MPa]	53
σ_{D_z} [MPa]	26

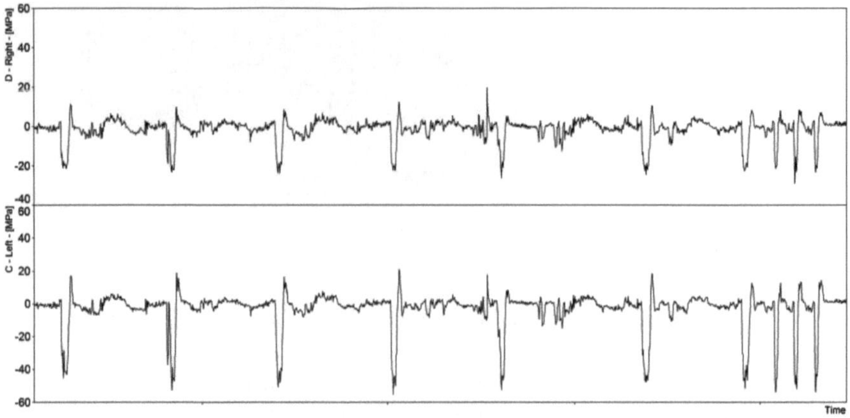

Fig. 25 Series of ten brakes at the capsizing limit

Table 5 Analytical, numerical and experimental values for normal stress (z-axis) and total load percentages on each leg

	Analytical		FEA		Experimental	
σ_{C_z} [MPa]	83	**64%**	80	**63%**	53	**67%**
σ_{D_z} [MPa]	46	**36%**	47	**37%**	26	**33%**

unbalance ratios given by each method can be compared, in order to prove the effectiveness of the analytical model in forecasting the stress unbalance in a new fork, given a limited set of design parameters.

Unless the peak stress values returned by experimental analyses are lower than those given by *FEA* and Analytical models, Table 5 proves the analytical model reliability in predicting the load unevenness between the two legs.

It is important to underline that the global equilibrium is always satisfied so that the sum of M_{c_x} and M_{d_x} is equal to the total bending moment M_{tot_x} produced by external forces applied to the tire (F_b and N). M_{tot_x} can be easily calculated by means of the following equation:

$$M_{tot\_x} = F_b \cdot [r_w + (L_1 + L_2 + L_3) \cdot cos\gamma - c \cdot sin\gamma] - N \cdot [(L_1 + L_2 + L_3) \cdot sin\gamma + c \cdot cos\gamma] \quad (44)$$

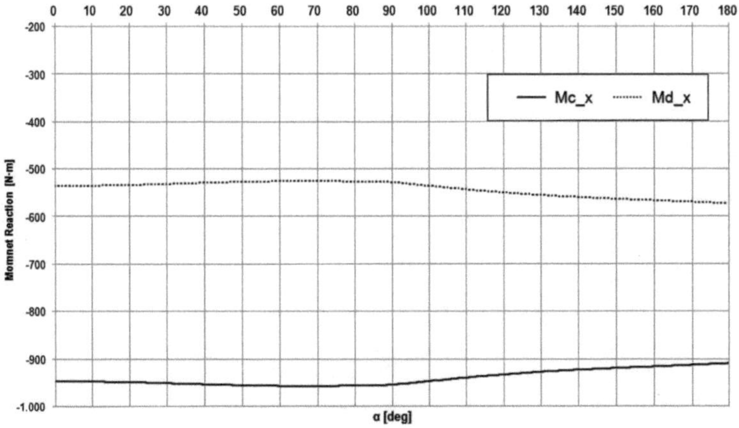

Fig. 26 Overall moment reactions vs. brake caliper angle (flexible wheel pin)

As a matter of fact, for the studied fork, M_{tot_x} results equal to 1.483.000 [Nmm]: in short, the braking leg bears about 2/3 of the total bending moment ($M_{C\_x} = -955.600$ [Nmm]), whereas the non-braking leg about 1/3 ($M_{D\_x} = -527.400$ [Nmm]).

Tests made on different geometries returned similar results, with differences between Analytical, *FEA* and Experimental stress distribution ratios within a few percentage points. Moreover, the newly developed analytical method allows fork designers to estimate the impact on the fork stress state of any design change upfront. For example, the overall moment reactions in C and D (2.41) can be plotted as functions of the brake caliper angle α, in order to evaluate the influence of a different placement of the brake caliper along the disc circumference. Figures 26 and 27 report such trends either for a fork equipped with a flexible wheel pin ($J_p = 19.769$ mm^4) and for the same fork equipped with a stiffer one ($J_p = 101.806$ mm^4). It can be appreciated how for the flexible wheel pin an even load distribution between the legs could never be reached, whatever the angular position of the brake caliper is.

Conversely, if the wheel pin is rigid enough, values of α allowing an equal load distribution between the legs exist, for example around $\alpha = 45°$ in Fig. 27. Similar graphs, displaying for example the overall moment reactions as functions of the wheel pin offset c can be plotted. Moreover, the analytical model allows calculating the bending moment at each section of the legs: it is noticeable that, into the intermediate section L_2 of each leg, two elements (inner and outer tube) work in parallel.

Each element will bear an amount of the total load applied to the leg proportional to its stiffness, according to a relationship of the kind (here written for moments):

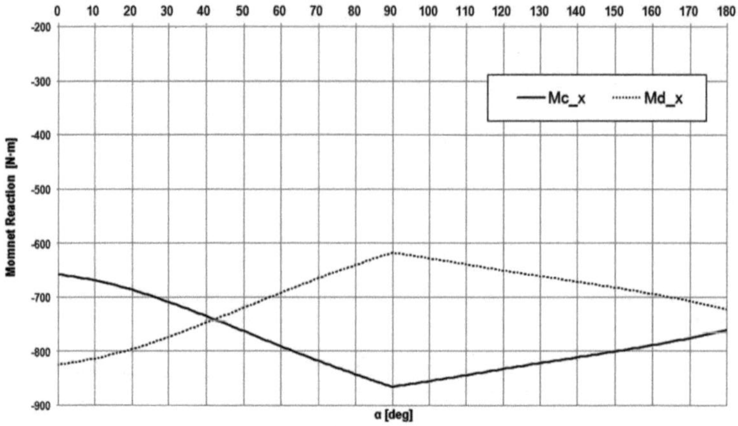

Fig. 27 Overall moment reactions vs. brake caliper angle (rigid wheel pin)

$$\begin{cases} M_{OT\_x} = \dfrac{M_{LEG\_x} \cdot I_3 E_3}{I_1 E_1 + I_3 E_3} \\ M_{IT\_x} = M_{LEG\_x} - M_{OT\_x} \end{cases} \tag{45}$$

Where M_{OT_x} and M_{IT_x} stand for bending moment at a certain section of the leg on outer tube and inner tube namely. M_{LEG_x} is the overall leg bending moment at the same section.

4 Conclusions

The stress field in single disc motorcycle forks, under the effects of a severe brake, has been studied throughout this work. An analytical approach, useful to determine the stress distribution unbalance between the legs, has been proposed and validated either by *FEA* and Experimental Analyses (Errors always within 5%). The analytical model will support fork designers during the concept design phase, as it allows understanding the effects of a structural change on the stress state of a new fork without performing *FEA*.

References

1. Croccolo, D., De Agostinis, M., Vincenzi, N.: Recent improvements and design formulae applied to front motorbike suspensions. Eng. Fail. Anal. (2010). doi:10.1016/j.engfailanal.2010.02.002
2. Croccolo, D., Cuppini, R., Vincenzi, N.: Design improvement of clamped joints in front motorbike suspension based on FEM analysis. Finite Elem. Anal. Des. (2009). doi:10.1016/j.finel.2008.11.007

3. Croccolo, D., Vincenzi, N.: A generalized theory for shaft–hub couplings. Proc. Inst. Mech. Eng. C.: J. Mech. Eng Sci. (2009). doi:10.1243/09544062jmes1437
4. Croccolo, D., Cuppini, R., Vincenzi, N.: The design and optimization of fork–pin compression coupling in front motorbike suspension. Finite Elem. Anal. Des. (2007). doi:10.1016/j.finel.2007.06.016
5. Croccolo, D., Cuppini, R., Vincenzi, N.: Friction coefficient definition in compression-fit couplings applying the DOE method. Strain (2007). doi: 10.1111/j.1475-1305.2007.00363.x
6. Pacejka, H.B.: Tire and Vehicle Dynamics. Butterworth-Heinemann, Oxford (2002)
7. Cossalter, V.: Motorcycle Dynamics, 2nd edn. Lulu, Padova (2006)
8. Cossalter, V., Lot, R., Maggio, F.: On the stability of motorcycle during braking. Small engine technology conference & exhibition SAE paper number: 2004-32-0018/20044305 (2004)
9. Corno M, Savaresi, S.M., Tanelli, M., Fabbri, L.: On optimal motorcycle braking. Control Eng. Pract. (2008). doi:10.1016/j.conengprac.2007.08.001
10. Belluzzi, O.: Scienza delle Costruzioni, vol. 2. Zanichelli, Bologna (1994)
11. Timoshenko, S., Goodier, J.N.: Theory of Elasticity. McGraw-Hill, New York (1951)
12. Niemann, G., Winter, H., Hohn, B.R.: Maschinenelemente. Springer, Berlin (2005)
13. Doyle, J.F.: Modern Experimental Stress Analysis. Wiley, Chichester (2004)

Computational Simulation in Centrifugal Casted Aluminum–Silicon Engine Cylinder Liner

Valter Barragan Neto, Antonio Augusto Couto and Jan Vatavuk

Abstract Automotive cylinder liners are mechanical components with the function of internal coating of the cylinder automotive engines. The replacement of parts made of steel/cast iron by aluminum alloys has been made with advantages not only in reducing weight as well as fuel consumption and emission of pollutants. This study was aided by the finite element software Hyperworks, where the mesh was generated and the simulation was performed in Abaqus. The mesh for the engine block was defined with elements of four nodes of tetrahedrons. The liners were designed with hexahedron elements of six nodes. Due to the manufacturing process of the cylinder liners (centrifugal casting), the finite elements model was created in layers to meet the variation in the amount of silicon along the wall thickness. The variation in the amount of silicon affects the physical properties of the liners along the wall thickness. With this model, it was possible to show the viability of application of aluminum liners in engine blocks made of cast iron. The modal analysis showed that the model does not contain its first natural frequency within the range of work of the engine, approving its application with this concept.

V. Barragan Neto (✉)
General Motors do Brasil and Mackenzie Presbyterian University,
Av. Goias, 2769-CT20, Sao Caetano do Sul, SP, CEP: 09501-970, Brazil
e-mail: valtbn@gmail.com

A. A. Couto · J. Vatavuk
IPEN-CNEN/SP and Mackenzie Presbyterian University,
Rua da Consolação, 930, Sao Paulo, SP, CEP: 01302-907, Brazil
e-mail: acouto@mackenzie.br

J. Vatavuk
e-mail: janvatavuk@uol.com.br

A. Öchsner et al. (eds.), *Materials with Complex Behaviour II*,
Advanced Structured Materials 16, DOI: 10.1007/978-3-642-22700-4_5,

1 Introduction

Automotive cylinder liners are mechanical components assembled to the engine block holes in internal combustion engines. Piston and piston rings run in the internal diameter of the cylinder liners to produce mechanical power due to the combustion action, producing high pressures, heat release, combustion products, friction, promoting wear in the internal engine parts like piston rings, cylinder liners, pistons, with a direct influence on the engine durability. Nowadays, the internal combustion engines cylinder liners are based in gray cast iron, mainly with a perlite matrix. However, the technological evolution in order to reduce vehicle mass enhances lower density materials application. The substitution of steel and cast iron parts by lower density materials promotes lower emissions due to the lower fuel consumption. Besides the lower density materials with good thermal and mechanical properties, the automotive industry has been improving the substitution of heavier materials by aluminum and aluminum based alloys. Steel and gray cast iron density lies between 7.4 and 7.8 g/cm^3 almost three times the aluminum density 2.7 g/cm^3. Like steel and gray cast iron, aluminum is totally recyclable without property losses.

Among the aluminum alloys, the Al–Si system can be classified as the most important cast alloys mainly due to its high fluidity, low thermal contraction, high corrosion resistance, good welding and brazing properties, as well as low thermal expansion coefficient. Aluminum alloys with 5–20% Si (weight percent) are the most used in the industry. The microstructure is based on α aluminum or β silicon primary phase and an aluminum silicon eutectic. Normally, the eutectic β phase has an acicular shape, and the primary silicon based phase consists of big faceted plates.

The main material requirement in the internal combustion engines cylinder liners is the wear resistance, due to the heavy tribological working conditions between the working surfaces piston/piston rings over the inner diameter of the cylinder liners during engine running. Among the aluminum alloys, the Al–Si hypereutectoid system (above 12.6% Si) has been preferred due to silicon effect enhancing wear resistance. The addition of silicon in the aluminum promotes a dispersion of hard silicon particles (proxy. 1,000 HV), that can impair machinability and mechanical properties when added in high proportions.

Between the different foundry casting methods, the centrifugal casting technique is related to a tubular die with a high rotational speed during pouring (1,000 to 1,500 rpm). The dynamic effect of the high rotational speed, promotes a high centrifugal acceleration in the order of 100 g, reducing porosity and promoting some density segregation effect that cannot be noticed in the static foundry processes. The centrifugal casting process applied to the Al–Si alloys for cylinder liners can induce some reduction in the silicon content, working with low hypereutectic compositions (above 14%) as related to the materials used in static processes, around 17%. The high centrifugal acceleration prompts silicon to run mainly in the inner diameter direction, due its lower density as related to aluminum alloy, even in the liquid state. In this way, it becomes easier to guarantee the wear resistance in the cylinder/piston and rings working surface, as well as the mechanical properties requirements, due to the low silicon content in the outer side diameter.

Nowadays, the substitution of heavier materials by lighter ones, obey a validation step, based on wear and mechanical tests, temperature resistance, as well as vibration and noise characteristics. The cost associated with component material change can represent a high validation tests investment. Computational techniques have been developed in order to avoid as much as possible prototypes fabrication associated with a reduction of tests. The industrial strategy departments promote project expertise improving reliability in structural and systems analysis.

The reliability of a project is directly related to the model development according to the product real working conditions. Very accurate hypotheses and consideration must be taken in account in order to develop consistent models. In terms of physical phenomena, as well as constructive aspects, like materials and geometry parts [1]. A high power computational system as well as more than one processing nucleus is required to run representative and reliable modeling systems.

The finite elements method (FEM) was introduced in 1960 by Raymond William Clough, and is still an important science and engineering instrument. All technical procedures can be run in a computational virtual environment. The component must be divided in small and simple geometric elements (line, triangle, square, tetrahedron, pentahedron, and hexahedron). These parts must be interconnected by their nodes. A small part division is necessary to fit the linear equations only applicable to finite elements, based exclusively on linear trajectories.

The main objective of this work is to test the applicability of Al–Si internal combustion engines cylinder liners produced through the centrifugal casting technology. The first step of this project were the manufacturing of a slightly hypereutectoid Al–Si cylinder liner by centrifugal casting. The microstructure and hardness were analyzed through the cylinder liner thickness. A structural finite element numerical system was run by computational Computer Aided Design (CAD) and Computer Aided Engineering (CAE) tools. In order to fit closely the microstructure gradient of the centrifugal casted Al–Si alloy, the procedure was based on the cylinder thickness division in seven slices each one with its own characteristics, to take in account the microstructure gradient produced by the centrifugal casting method in the Al–Si alloys.

2 Manufacturing of Al–Si Cylinder Liners by Centrifugal Casting

This work deals with horizontal centrifugal casting foundry process, represented in Fig. 1. The as cast products were tubes based on Al–Si alloy with 14.7% silicon weight percent. The melting process was carried out in an induction furnace, and

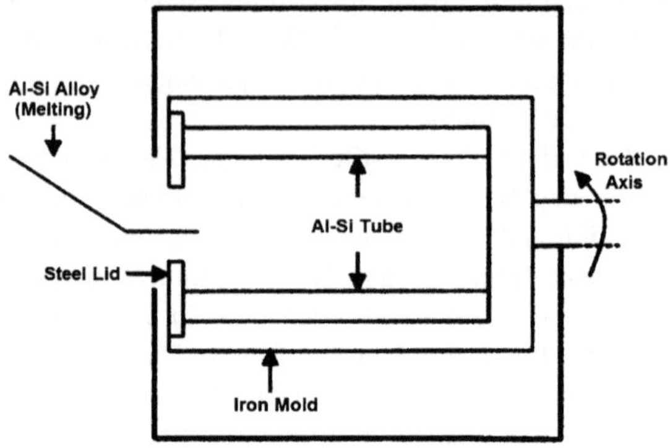

Fig. 1 Scheme of the horizontal centrifugal casting process with cooling die system

Fig. 2 Centrifugal casted Al–Si alloy tube

the pouring temperature between 720 and 770°C. The cylindrical iron mold was pre heated and the rotational speed maintained at 1,500 rpm until solidification processes took place. The hardness tests were carried out by the Vickers indenter technique through a 5 kg load. The sample preparation for optical metallographic observations was carried out by conventional techniques, and the microstructure observed in an Olympus system. The silicon (β phase) area fraction and the medium β primary phase particle size through the wall thickness were determined by an Image Pro-Plus program. The silicon fraction area and the hardness determinations were carried out from the external to the internal wall in 1 mm intervals.

Figure 2 shows tee centrifuged tubes. Figure 3 is related to a cross section for metallographic observations of the tube, 100 mm away from one of its sides. The mean silicon particles size (β phase) were around 25 μm ranging from 12 to 52 μm. The same alloy produced by sand casting promotes silicon particles at list twice larger than the centrifugal casting method. The size of the hard particles must be controlled to optimize cylinder liner tribological working conditions. Large silicon particles typical in sand casted Al–Si hypereutectic alloys can reduce engine performance due to particles detachment while running engine in severe conditions. In this situation, depth scratches on the cylinder liner piston rings working surface can promote an increase in oil consumption as well as combustion gases blow by. The silicon particle size requirement to satisfy the engine working conditions are fulfilled by the centrifugal casting, mainly due to its high efficiency in terms of heat extraction method associated with the high heat extraction.

Fig. 3 Cross section area 100 mm away from the Al–Si tube rear

The micro hardness values for the α (Al) and β (Si) phases were 85 and 1,018 HV. The higher silicon particles hardness can be compared to the Fe_3C iron carbide hardness. Aluminum corresponds to the softer continuous phase responsible for toughness and ductility.

Figure 4 is related to the tube microstructure range through the wall thickness, starting the observations from the external to the internal diameter. A thin layer rich in β phase is observed close to the external tube diameter (Fig. 4a). The high cooling rate due to the molten aluminum contact with the iron mold promotes a solidification front speed higher than the primary β silicon migration in the inner diameter direction. Going forward in the internal diameter direction, the tube presents an eutectic structure. The observations realized close to the internal diameter shows again β silicon primary phase, with an increasing volume (Fig. 4c and d) fraction, reaching values much higher than that typical for a 14.7% silicon alloy.

Figure 5 presents the total silicon volume fraction evolution (eutectic + primary) through the wall thickness from the external to the internal cylinder liner surface in 1 mm intervals. The analyzed tube in Fig. 5 was manufactured with 14.7% Si aluminum alloy. The chemical profile shown in Fig. 5 is in accordance with Fig. 4 observations, a thin layer with a higher silicon content coming from the outside diameter followed by an eutectic composition (lower silicon contend) going forward (inward) the tube wall, in the internal wall direction, and an increase in the silicon content in the internal diameter related to the piston and piston rings working surface. The hardness increases in the higher β silicon content regions.

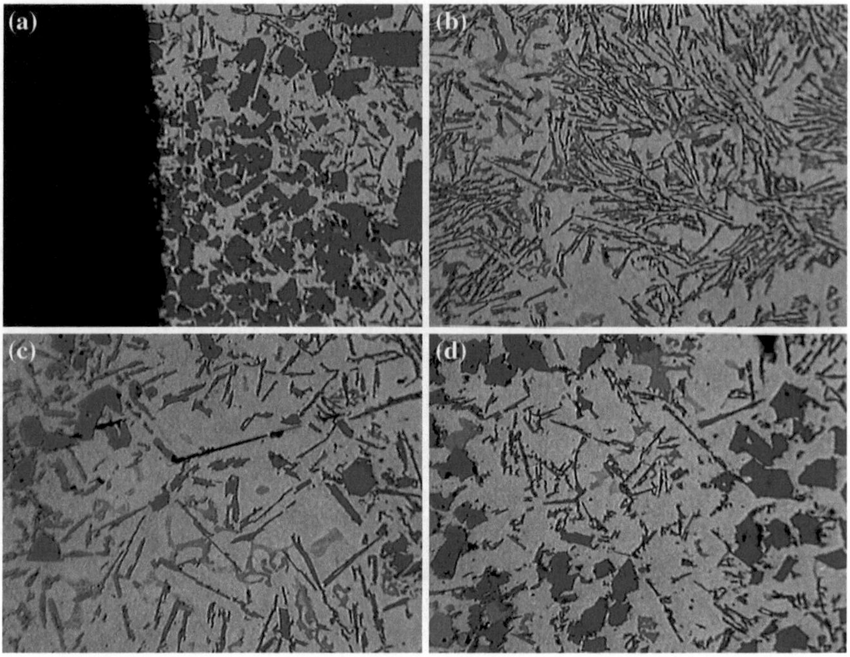

Fig. 4 Al–14.7%Si alloy microstructures from **a** external wall, **b** 3 mm from the external wall, **c** 4 mm from the external wall, and **d** close to the internal wall

Fig. 5 Total silicon contend evolution (primary plus eutectic β phase) from the external wall inward the internal tube diameter in 1 mm steps, in an Al–14.7% Si alloy

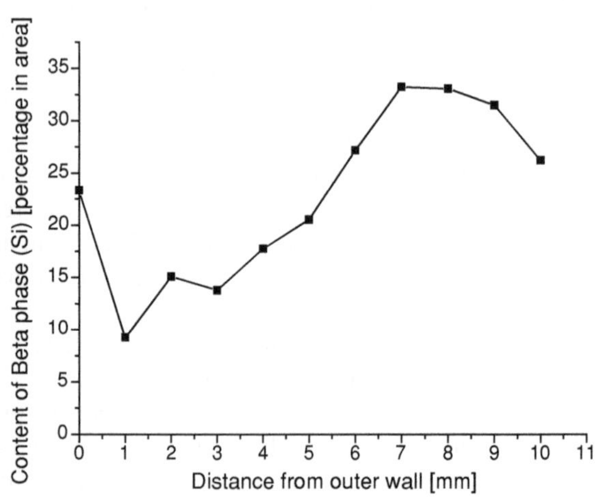

Finite element modal analyses related to the vibration pattern and loading tension related to the engine working conditions discussed above, were carried out taking in account an Al–Si slightly hyper eutectic alloy. A cylinder liner with a

Fig. 6 Degrees of freedom of a solid element [3]

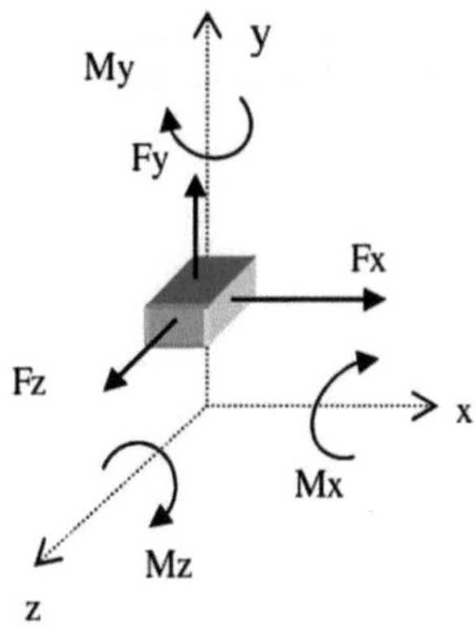

6.5 mm wall, were divided in six slices of 1 mm and one of 0.5 mm. To guarantee the better silicon distribution as well as to reduce the wall thickness of the as cast tube (10 mm) to 6.5, 1.5 and 2 mm material was removed from the internal and external wall [2].

3 Modal Analysis

The best boundary conditions enable the physical model to fit closely the real working conditions. The modal analysis is defined by frequency functions or movement equations from the frequency response functions (FRF). The vibration pattern takes in account a free of constraint components joints. In accordance to Beer and Johnston [3], degree of freedom (DOF) represents the number of independent rigid body possible movements.

Figure 6 is related to the six DOF possibilities, three translations represented by F_x, F_y and F_z in the three X, Y and Z axis, associated to three rotational axis M_x, M_y and M_z. The block geometry was developed using CAD technique associated to ProEngineer, in accordance to a gray cast iron 1,000 cm^3 commercial block.

Figure 7 shows a conventional engine block. This geometry is closely related to reality, with a 33.340 kg approximately mass. The substitution of the internal holes of the block by Ai–Si tubes enables a total mass reduction of 2 kg. The model analysis was carried out by a 6 mm radial mass remotion from the block holes. This procedure was taken in account to allow the (FEM) modeling represent the materials properties change due to the silicon variation in the cylinder wall.

The virtual procedure of cutting of the gray cast iron cylinder liner from the block is represented by the red lines Fig. 8. At left the new block holes taking in account the machined material to introduce the Al–Si cylinder liners. This methodology promotes the direct contact between the new cylinder liners with the liquid cooling system (wet cylinder liner). The direct contact between Al–Si

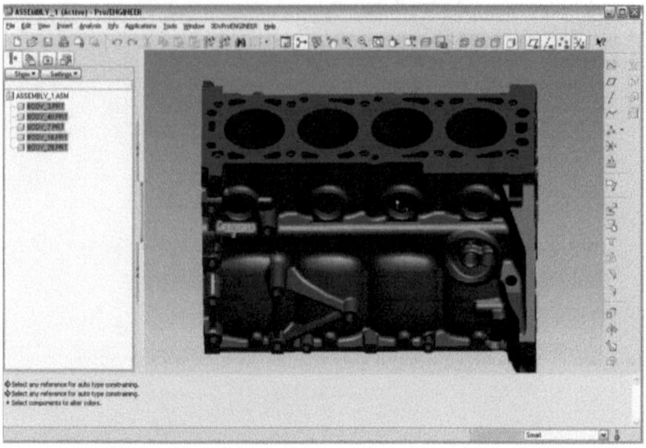

Fig. 7 Computer Aided Design (CAD) block model

Fig. 8 Machined cylinder liner (*red lines*) and the resultant block holes

cylinder liner and the cooling fluid improves efficiency in terms of engine heat dissipation, due to the higher thermal conductivity coefficient of aluminum as related to iron (Al = 205 W/mK; Fe = 79.5 W/mK) [4]. After defining the new block geometry taking in account the new sleeves inserts, the mathematical model construction with the geometric information took place with the HyperMesh (HM) software.

The boundary conditions are responsible for the restrictions imposed in the studied system. Some care must be taken during the choice of the contour conditions to establish a good fitting for the free–free condition representative of some natural vibration characteristics; the rigid body frequency pattern must be verified. In theory, these frequencies must be null or sufficiently low as related to flexural modes, i.e., around 10–20% [5]. A practical solution to deal with this question is a low rigidity suspension. On the other hand, for the set in loading condition, the practical conditions are based on small displacements and inclinations values, that are not taken in account. Most experiments are carried out in the free–free condition, due to the easier contour condition as related to the fixed structure.

The condition free–free was adopted on the modal analysis model, which means the definition of the block vibration frequencies defining the directions and displacement amplitude, the auto vectors and auto values, respectively.

The modal analysis objective lies on the exclusion of the block vibration mode into the working rotational engine speed. The rotational speed adopted in the present work goes from 800 rpm (idle speed) until 7,200 rpm (ignition limited rotation), to avoid collision between the piston and the engine valves. Taking in account that in the four strokes four cylinder engines there are two combustions per rotation cycle, the components excitations per cycle can be determined by the next expression:

$$F_{\text{excitation}} = \frac{RPM \cdot n_{\text{cyl}}}{C_{\text{rev}} \cdot 60} \tag{1}$$

where:

$F_{\text{excitation}}$ = Excitation frequency related to the engine rotation
n_{cyl} = Engine cylinder number
C_{rev} = Engine combustion frequency per cycle

According to Kelly [6], the resonance is responsible for the higher vibration amplitudes, due to the excitation in the natural body frequency. This concept means that the first frequency of the engine parts must be out of the rotational engine range. Some engine parts do not fit this criterion, due to low working load condition, or physical conditions. Considering the engine speed of 7,200 rpm, a speed limit of 7,500 rpm that prevents the collision between the valves and the piston was used in the calculations, giving 300 rpm as safety margin. The next mathematical expression was used for the determination of the minimum engine frequency to avoid its natural vibration frequency:

$$F_{\text{excitation}} = \frac{7,500 \cdot 4}{2 \cdot 60} = 250\,\text{Hz} \tag{2}$$

The frequency values calculated from Eq. 2 shows that the first vibrational mode (natural frequency) must occur above 250 Hz, the maximum rotation engine speed. On the other hand, Eq. 1 must be used as the lower engine rotation to establish the frequency range to be avoided by the engine components.

$$F_{\text{excitation}} = \frac{800 \cdot 4}{2 \cdot 60} = 26.67\,\text{Hz} \tag{3}$$

Adopting this criterion, the range frequency during engine running must be between 26.67 Hz and 250 Hz. Taking this concept in account, the modal analysis is the first to be discussed adopting boundary conditions based in the free–free model, without any DOF restrictions, which means that the parts vibration occurs without restrictions. Physically, the parts vibration reduces as the DOF restriction values increase.

Fig. 9 First vibration mode at 692 Hz related to a normal block (gray cast iron block and cylinder liner)

4 Vibration Modes: Auto Vector and Auto Value

To establish a modeling comparison in the modal analysis, two simulations were run, one with a fully gray cast iron block, and another assembled with the Al–Si alloy cylinder liners. This criterion prompts the system rigidity reduction, the first vibration mode occurred at 692 Hz, and the torsion between block front and rear are represented by the auto vectors, see vectors on orange color in Fig. 9. This figure shows in colors scale from blue to red the auto vectors, being red for the highest displacement. The black lines represent the unloaded block to visualize the block displacement. The natural vibration deformation displacement was amplified ten times to be sensible for human eye. The highest auto value for this vibration mode (fully red), reach an amplitude of 1.26 mm. Only the first frequency is shown regarding the maximum engine frequency, 250 Hz. The second frequency value is 890 Hz and the third 1,087 Hz. Figure 10 shows the result for the block assembled with the Al–Si cylinder liners. As previously expected, the auto vectors and auto values are still unchanged, but the block rigidity diminishes, and first natural frequency reduced to 544 Hz, almost 21% lower than the normal block (only gray cast iron), and the displacement amplitude 1.178 mm.

Figure 11 presents a block top view showing the left first cylinder in the first modeled mode vibration. The vibration promotes a loss in circularity in the 544 Hz excitation frequency. Although out of the engine natural frequency, in these circumstances a 0.7 mm displacement out from the circularity would be produced, with an excessive cylinder liner piston and rings wear, as can be seen in Fig. 11. The second frequency was 766 Hz and its vibration mode affect more directly the block/transmission interface, with little effect in the cylinder liners. An auto vector displacement of 1.025 mm is represented in Fig. 12. The third frequency was 944 Hz.

A block with Al–Si liners promotes similar auto vectors and auto values as a system based on a block with gray cast iron cylinder liners. The block with gray cast iron cylinder liner rigidity (stiffness) is a little bit higher than that assembled

Fig. 10 First vibration mode at 544 Hz related to a normal block (gray cast iron block with Al–Si cylinder liner)

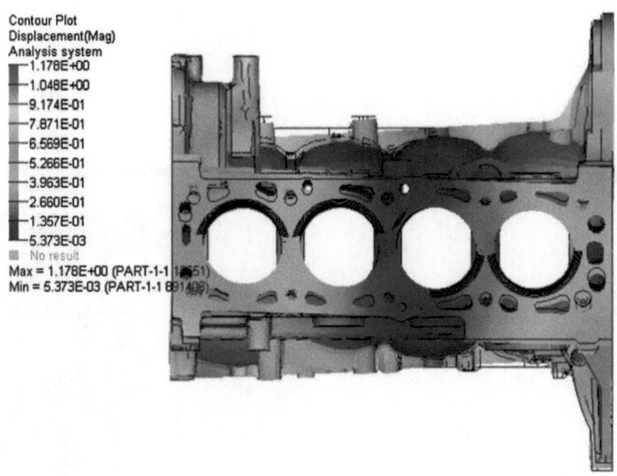

Fig. 11 First vibration mode at 544 Hz. *Top* view

with Al–Si cylinder liner. This result can be attributed to the lower elasticity modulus of the Al–Si alloys as related to gray cast iron. This behavior was expected, due to the block weakening effect of changing from block casted cylinder to assembled Al–Si cylinder liner in machined block holes.

5 Finite Elements Modeling Method

The Al–Si cylinder liners gray cast iron block interaction was assumed as a slight sliding with a 0.2 friction coefficient. The boundary conditions were maintained for the cylinder liners block combustion pressure interaction. According to the sliding, at least one degree of freedom must be fixed. In this study, was adopted the interface between carter block as being clamped to avoid rotation and translation

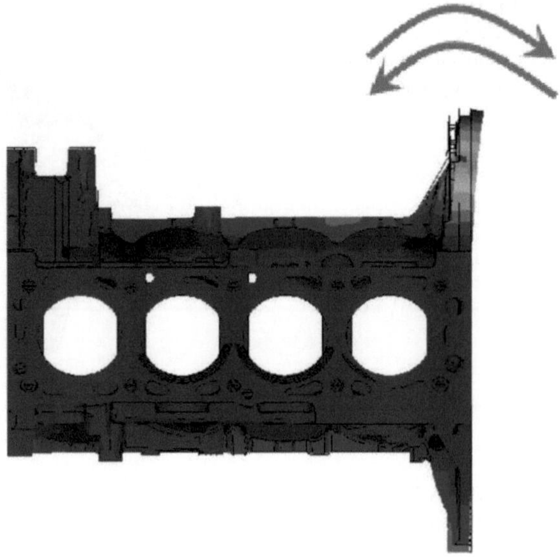

Fig. 12 Second vibration mode at 766 Hz. *Top* view

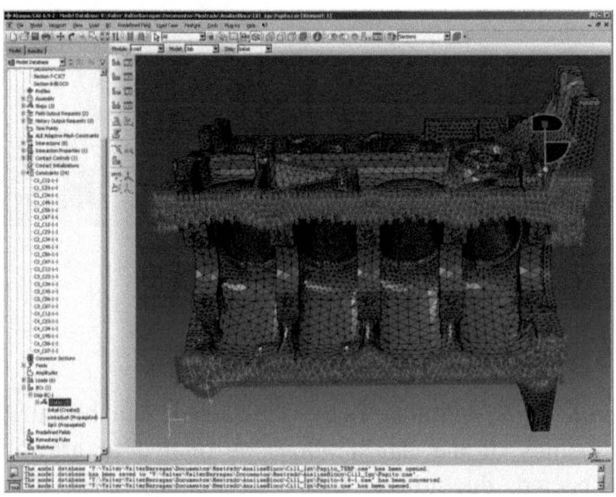

Fig. 13 Engine block constraint

in the three Cartesian axes, as can be seen in Fig. 13. Figure 13 shows the Abaqus graphical user interface (GUI) and the bottom of the engine block with the constrained nodes. The constraints are represented by blue and orange arrows, translational and rotational respectively. This figure also shows the mesh density as it grows from the cylinder liner.

The Abaqus-CAE 6.9.1. software was used to prepare the modeling, taking in account the loading and restriction conditions. Refined hexahedral elements, proxy. 1 mm^3, were applied for the liners taking in account slices, to represent the through thickness properties gradient due to the centrifugal casting method effect

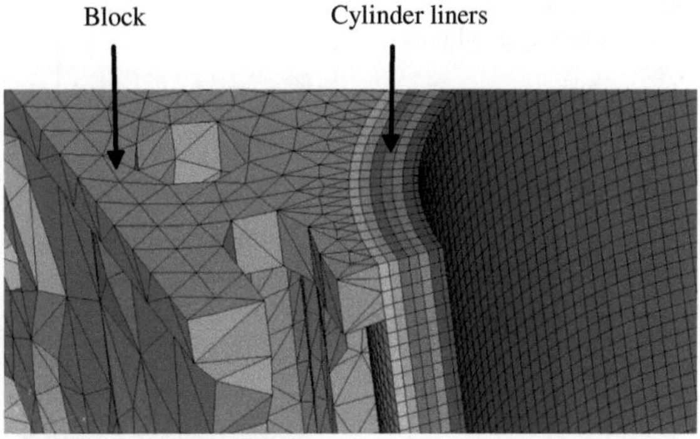

Fig. 14 Cylinder liners represented by slices and its block interation

in the Al–Si alloy. Figure 14 shows a finite element cylinder liner block section, with a seven fold division for the cylinder liner (six with 1 mm and one with 0.5 mm), every one taking in account the silicon stiffness effect. The tetrahedral block elements are larger than the cylinder liner elements, but they are in perfect fit at the interface. This methodology was adopted to adjust the interface results, without the necessity to node fitting in the Abaqus solution, due to the components contact surfaces development. This methodology was adopted in order to improve results at the interface, although not necessary the knots matching to solve the Abaqus, since they created the contact surface between the components under study.

As mentioned before, due to the centrifugal casting process the silicon particles volume fraction varies along the cylinder liner thickness. This was the reason to define a different material in every cylinder liner considered layer. Between these layers, there is an interaction surface, assembled without sliding, producing a heavier and complex model. Figure 15 shows the squares embedded on one hexahedral face, a graphic representation of cylinder liner contact elements. These elements are related to the contact surfaces. They were generated for both surfaces that are in contact, once the mesh was generated in a refined matched manner and nodes of the cylinder liner mesh match perfectly with the engine block.

In terms of the cylinder liner block interaction, even adopting refined elements, a perfect fit cannot be realized, due to the different elements geometry, hexahedral for cylinder liners and tetrahedral for the block, reducing the matching conditions, as shown in Fig. 16. Figure 16 shows the cylinder line surface elements, blue squared, and the engine block surface, green triangles. The surface elements are used to define the interaction between the bodies. This figure shows that the blue squares pertain to the cylinder liner, and the green triangular surface elements are related to the block along the face elements.

The better fitting associated to a refined mesh improves the model resolution reducing simulation interpretation. Nowadays, the contact modeling interpretation is the major problem, and in many circumstances the simulation stops by a fatal error, without a converging solution.

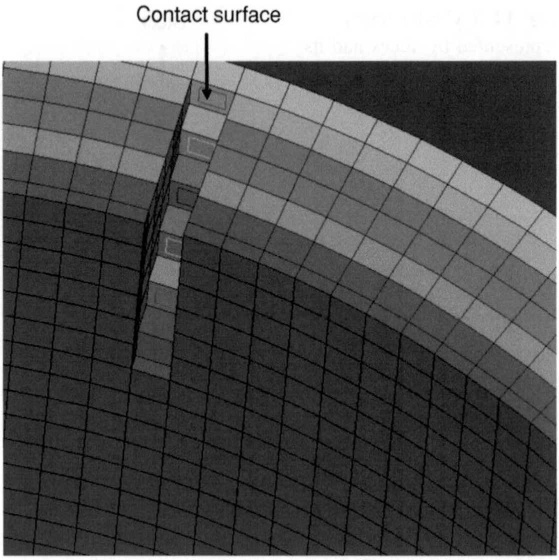

Fig. 15 Contact surface elements between cylinder layers

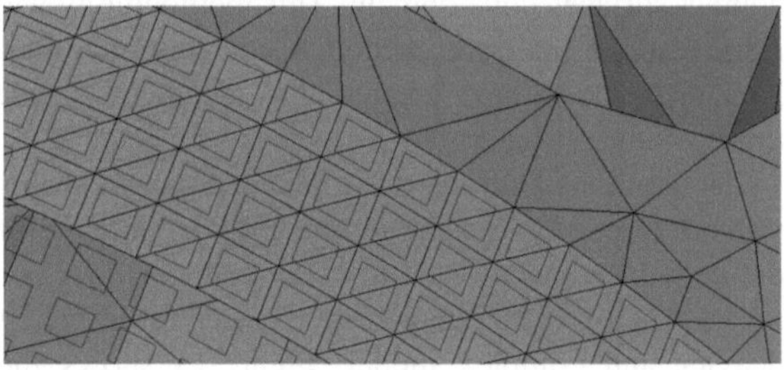

Fig. 16 Surface contact between the cylinder liner block elements, *blue squared*, and the engine block surface, *green triangles*

Initially, many simulation errors occurred during the mathematical modeling, due to the contact tolerances used to run Abaqus. A contact tolerance review was necessary to improve its interaction. To solve this model, a nodes overlapping technique for the closer nodes was adopted, excluding any interpenetration between them. With this model procedure and the modal absence of loading or set in, the model analysis took place. In the pressure working conditions study, some modification in the block took place in order to realize the analysis criteria.

The cylinders pressures were obtained as can be seen in the Fig. 17. The pressure peak occurs some degrees after the top death center (TDC), pointed graphically in Fig. 17 through 90° fazed points. Cylinder 3 pressure was the highest as related to the other cylinders (about 7,500 kPa), and was used in the simulation program as the highest combustion pressure for all cylinders. The pressures at every step of the engine cycle are presented in Table 1.

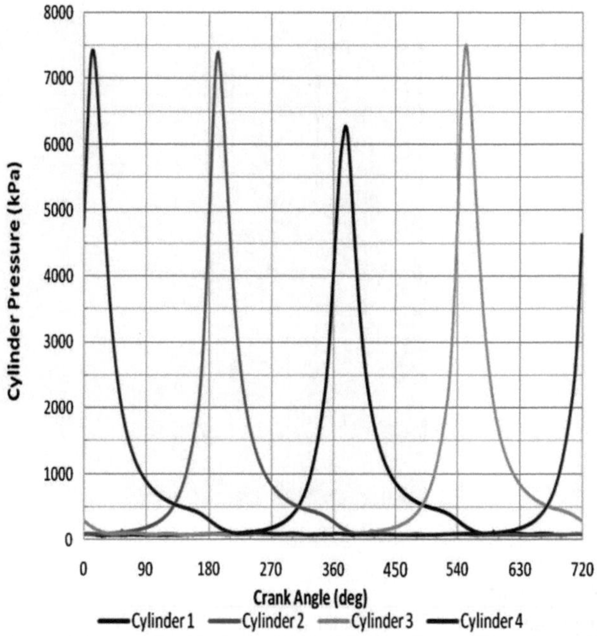

Fig. 17 Cylinder pressures. Complete cycle

Table 1 Cylinder pressures for each engine cycle

Pressure (kPa)	Pressure (kPa)	Pressure (kPa)	Pressure (kPa)
204	83	7,500	92
Exhaust	Intake	Ignition	Compression

The engine cooling fluid medium pressure is about 100 kPa [7]. Due to the flex furl technology in the Brazilian automotive industry, a 160 kPa pressure was adopted for safety, due to the higher running temperature when using ethyl alcohol fuel. Figure 18 shows the elements, in red, where the cooling fluid pressure is applied. The pressure is represented by the orange vectors and is applied in all elements that would be in contact with the cooling fluid. The triangle elements represent the engine block and the squared elements the cylinder liner.

A pressure steps map was developed from Table 1, and can be seen in Table 2 where the engine running step has its pressures mapped. In the piston assembly positioning in order to balance the engine, the pressures were adopted according to Table 2 step one, been applied in the areas pointed in Fig. 19, following Table 1 pressures. The left side of the illustration is related to cylinder 1.

Fig. 18 Cooling system pressure vectors and loaded elements

Table 2 Engine cycle map

	Cylinder 1	Cylinder 2	Cylinder 3	Cylinder 4
Step 1	Ignition	Exhaust	Compression	Intake
Step 2	Exhaust	Intake	Ignition	Compression
Step 3	Intake	Compression	Exhaust	Ignition
Step 4	Compression	Ignition	Intake	Exhaust

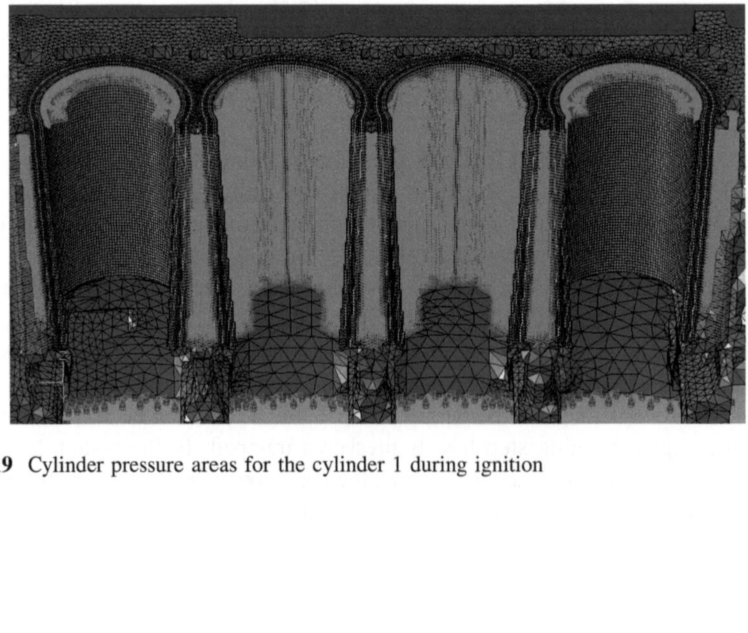

Fig. 19 Cylinder pressure areas for the cylinder 1 during ignition

Fig. 20 Maximum principal tensions—Cylinders 1, 3, 4 and 2 during ignition

After modeling, the loaded and modal conditions were calculated in High Performance Computer (HPC) with 98 processing nucleus, running with eight processors. The modal analysis model runs perfectly, converging to the expected node values.

The loading analysis model showed some contact problems, and the analysis was incomplete even after 3 h simulation. A model adjustment was necessary to run a new calculation. The data convergence was reached after seven modeling procedures. The running time for this analysis was about 13,755 s (229 min or approximately 4 h). Some problems occurred due to the nodes interpenetration during loading. Despite being careful during the contact elements generation, some nodes penetrate elements, impairing the model solution.

The analysis sequences are related to the previously described conditions and assume an information character, with values that do not fit perfectly the real working conditions, due to the virtual modeling. The engine cycle working pressures affect the stresses, as can be seen graphically in Fig. 20. The maximum cylinder liner stress was approximately 20 MPa. This value is lower than the Al–Si alloy yield strength, producing only elastic deformation, in accordance to a perfect engine running condition.

6 Conclusions

The finite elements numerical simulation method applied in Otto Cycle internal combustion engine Al–Si cylinder liners, showed that the modal analysis was fit perfectly, having as consequence the expected converging values for the nodes.

The assembled parts modal analysis showed that the natural first frequency was out of the engine working range. The auto vectors and auto values were unchanged, but the system stiffness is reduced after assembling the Al–Si cylinder liner. The computational simulation showed that the centrifugally casted Al–Si cylinder liners are applicable taking in account the engine loading and vibration working conditions. A prototype is in construction to study the wear behavior of the Al–Si cylinder liner in the piston and piston rings working surface, as well as to validate the simulation technique.

Acknowledgments This work was partially financial supported by the Presbyterian Mackenzie Institute through Mackenzie Research Fund (MackPesquisa) and Conselho Nacional de Desenvolvimento Científico e Tecnológico (CNPq). The authors also thank Magneti Marelli Cofap and GM of Brazil for supporting research.

References

1. Kloutsey, H., Diniz, A., et al.: Influência das Condições de Contorno na Análise Modal Experimental de Vigas. 16° POSMEC Universidade Federal de Uberlândia, Brazil (2006)
2. Couto, A.A., Cabral Neto, A., et al.: Estudo de Ligas Al–Si Fundidas Por Centrifugação Para Aplicação Como Camisa de Cilindro Automotive. 11° Seminário de Metais Não Ferrosos da Associação Brasileira de Metalurgia e Materiais—ABM, São Paulo (2009)
3. Beer, F.P., Johnston, E.R., et al.: Vector Mechanics for Engineers: Statics and Dynamics, 8th edn. McGraw-Hill, New York (2007)
4. Young, H.D., Freedman, R.A.: Solution Manual for University Physics with Modern Physics, 12th edn. Pearson Higher Education, USA (2008)
5. Ewins, D.J.: Modal Testing: Theory, Practice and Application, 2nd edn. Research Studies Press, London (2000)
6. Kelly, S.G.: Fundamentals of Mechanical Vibrations, 2nd edn. McGraw-Hill, Ohio (2000)
7. CelicaTech Toyota. http://www.celicatech.com/imagearchive/bgbonline/93celica/co/description.pdf(2004). Accessed 26 July 2009

Numerical Study of Heavy Oil Flow on Horizontal Pipe Lubricated by Water

Tony Herbert Freire de Andrade, Kelen Cristina Oliveira Crivelaro, Severino Rodrigues de Farias Neto and Antonio Gilson Barbosa de Lima

Abstract This chapter reports information related to multiphase flow with emphasis to core-annular flow. Industrial application has been given to transient water-heavy ultraviscous oil two-phase flow in horizontal pipe. The high viscosity heavy oil transportation is one of the main technological challenges for the oil industry. This fact is related with the high pressure drop due to the viscous effects during the flow. Different techniques for the heavy oil transportation have been cited in the literature, core-flow is one. In this technique, water is injected in the pipe and flows as an annular film near the wall while oil moves in the core region. This way, a smallest amount of energy is required for heavy oil pumping. Mathematical formulation to describe transient and isothermal two-phase flow (water-heavy oil) is presented. Results of the velocity, pressure and volume fraction distributions of the phases were obtained and analyzed. A large reduction of pressure drop by comparison with single phase heavy oil flow (around 59 times) was verified and shows the efficiency of the technique applied to production and transportation of heavy oils.

T. H. F. de Andrade (✉) · S. R. de Farias Neto
Department of Chemical Engineering, Center of Science and Technology,
Federal University of Campina Grande (UFCG), Campina Grande-PB,
58429-900, Brazil
e-mail: tonyherbert2000@yahoo.com.br

S. R. de Farias Neto
e-mail: fariasn@deq.ufcg.edu.br.

K. C. O. Crivelaro · A. G. B. de Lima
Department of Mechanical Engineering, Center of Science and Technology,
Federal University of Campina Grande (UFCG), P.O. Box 10069,
Campina Grande-PB, 58429-900, Brazil
e-mail: kelencrivelaro@yahoo.com.br

A. G. B. de Lima
e-mail: gilson@dem.ufcg.edu.br.

A. Öchsner et al. (eds.), *Materials with Complex Behaviour II*,
Advanced Structured Materials 16, DOI: 10.1007/978-3-642-22700-4_6,

Keywords Heavy oil · Numerical simulation · CFX® · Two-phase flow oil–water · Core-flow

Notation

$A_{\alpha\beta}$	density of interfacial area per unit volume (m^{-1})
C_D	drag coefficient (–)
C_1C_2	empirical constant (–)
c_μ	empirical costant (–)
c_α	empirical constant (–)
$D_{\alpha\beta}$	drag force per unit volume ($N.m^{-3}$)
$d_{\alpha\beta}$	scale length coefficient of the mixture (m)
f	volume fraction (–)
G_α	generation of turbulent kinetic energy ($kg.m^{-1}.s^{-3}$)
k_α	turbulent kinetic energy ($m^2.s^{-2}$)
L_α	spatial length scale (–)
M_α	interfacial forces ($N.m^{-3}$)
N_p	number of phases involved (–)
P	pressure (Pa)
q_α	scale of velocity ($m^4.s^{-4}$)
$S_{MS\alpha}$	mass sources ($kg.s^{-1}.m^{-3}$)
$S_{M\alpha}$	momentum sources ($N.m^{-3}$)
t	time (s)
U	velocity vector ($m.s^{-1}$)

Greek letters

α, β	phases (–)
ε	turbulent dissipation rate ($m^2.s^{-3}$)
$\Gamma_{\alpha\beta}$	mass flow rate per unit volume (–)
μ	viscosity (Pa.s)
$\mu_{t\alpha}$	turbulent viscosity (Pa.s)
ρ	density ($kg.m^{-3}$)

1 Fundamentals of Multiphase Flow

Multiphase flow consists of one or more fluids (continuous phase) and one or more particulate phase (disperse phase) of coexisting matter in motion. The continuous phase can be liquid or gas and the disperse phase can be solid particles, gas bubbles or liquid drops. A more usual definition considers a multiphase system as that in which fluids components are immiscible and separated by interfaces. The use of multiphase flow in pipelines is common in practice in today's petroleum industry.

The occurrence of multiphase flow in the petroleum industry is very common in the transport, production and processing facilities of hydrocarbon of an oil field. Multiphase flow occurs in the transport facilities when the produced fluids are transferred for other areas through pipelines. In the production systems, the multiphase flow happens when the fluids inside the reservoirs move until the surfaces through wells, pipelines and risers (platforms offshore), for example.

Gas–liquid two-phase flow in a pipe can exhibit a variety of spatial distribution, depending on the flow rates and physical properties of the fluids, and also on the geometry and inclination of the pipe [8]. These spatial distributions of the two phases are termed flow patterns. The upward gas–liquid flow in a vertical pipe can exhibit five basic flow patterns: bubbly, slug or plug, churn, annular and dispersed bubble flows. In horizontal pipes, the following regimes can be cited: smooth stratified, wavy stratified, intermittent (slug and bubbles), annular with dispersed liquid and dispersed bubble. For liquid–liquid two-phase flow (oil–water), we can cite the following flow patterns: disperse flow, separated flow (core-annular flow and stratified flow) and intermittent flow [21, 24].

Two-phase flows always involve some relative motion of one phase with respect to the other; therefore, a two-phase flow problem is formulated in terms of two velocity fields. In despite of this comment, the flow pipe is influenced by the velocity, density and volume fraction of each phase, surface tension and shear rate between the phases.

2 Heavy Oil

2.1 General Aspects

In the world, some areas met exceptional features of nature that allowed the emergence of oil. The oil that is extracted from reservoirs on land or at sea is transported for example by pipelines to onshore or offshore platforms. From this place, oil is transported to refineries, where it will be processed to produce gasoline, diesel, gas, fuel oil, lubricants, asphalt etc. Among the different types of produced oil, there is heavy ultraviscous oil. The heavy oil has low degree API (American Petroleum Institute) (between 10 and 20°), high viscosity (between 100 and 10,000 cP) and density close to water. Besides these features, heavy oil have a high ratio carbon/hydrogen, large amounts of carbon residue, asphaltenes, sulfur, nitrogen, heavy metals, aromatic and/or paraffins [15]. However, interests in the production of heavy ultraviscous oils have increased in recent years because of the large amount of reserves available. The estimate of world reserves is difficult, but the order of magnitude of the total volume of heavy oil is the same as conventional oil [7]. However, Rodriguez and Bannwart [24] report heavy oil reserves estimation of 4.6 trillion barrels throughout the world.

2.2 Heavy Viscous Oil Transportation

The research and development to increase heavy oil production is very important to world economy. However, the production of heavy oil imposes a series of technological challenges, especially in transportation due to high viscosity. The difficulty is still more dramatic in deep water production where low sea water temperature is found. So, due to difference between flowing fluids (oil and water) and sea water temperatures, heat transfer across pipe wall occurs. This fact increases oil viscosity and provokes paraffin deposition into the pipe, and a higher pressure drop is verified. According to Bensakhria et al. [7], one solution to ensure the transport of heavy oil is to reduce effects of viscosity, through the addition of heat, dilution of the heavy oil with lighter oil and formation of emulsions. The heating process is very expensive while addition of light oil requires storage in the same production area. An alternative to transport the heavy ultraviscous oils is based in the behavior of annular flow, so called core-flow technique. In this flow pattern, one fluid is placed at the core and another fluid in the annulus. The higher viscosity fluid tends to become encapsulated by the lower viscosity fluid [12].

In the heavy oil–water flow, water is at the pipe surface and lubricates the oil core. This method was created by "Isaacs and Speed" in 1904, described in Patent No 759374 in the United States, citing the ability to transport viscous products by water lubrication. However, only in 1970 a large industrial pipeline was built to transport heavy oil by the company "Shell" around Bakersfield in California with 38 km long and a diameter of 15 cm. For over ten years, a viscous crude oil was transported at a flow rate of 24,000 barrels per day (bbl/d) in a system lubricated with water [7].

This technique has brought attractive results with respect to energy consumption. This fact is due to pressure drop during the oil–water flow to be comparable to pumping water alone at the same throughput, independent of the heavy oil viscosity. So, to keep the core annular flow pattern, it is necessary the lowest pumping power [13].

3 Core-Annular Oil–Water Flow

3.1 Background

For establishing core-annular flow, we have setting flow conditions related mainly for water flow rate to create a continuous water film around the oil, reducing the shear stress on the pipe wall. The water annular film can be very thin and thus requires a small mass flow rate. Besides, the contact surface area and mass transfer between the phases is minimal [17]. Besides, the transport of very viscous oil is possible with pressure close to that obtained when pure water flows alone. Due this fact, petroleum industry has large interest in the technique of water-lubricated transport of heavy oil. Therefore, many theoretical and experimental studies of this topic have been published in recent years such as Arney et al. [1], Bai and Joseph [4], Bannwart [5, 6], Bensakhria et al. [7], Huang et al. [9], Joseph et al. [10], Ko et al. [11], Li and Renardy [12], Oliemans et al. [14], Ooms et al. [17], Prada and Bannwart [19], Renardy and Li [21] Rodriguez and Bannwart [22, 23, 24]. These works include information about model developments, pressure drop, friction factor and stability.

From the literature it is well known which core-annular flow pattern is very stable under established flow conditions. In this sense, two theories have been given: the hydrodynamic stability theory and the levitation theory (based on the classical lubrication theory). Discussions about this topic can be found in the cited works.

Bannwart [6] proposed a theory for the stabilization of the annular pattern when two liquids of different densities and viscosities flow into a horizontal pipe. The theory is based on the analysis of the linear momentum equation in a cross section of the pipeline taking into account the effect of interfacial tension.

Ooms and Poesio [16] analyzed the annular flow in stationary regime in a horizontal pipe and proposed a theoretical model based on the hydrodynamic lubrication theory. According to this model, there was a harmonic motion in annular flow, i.e., the more viscous fluid (oil) moved of the wavy shape in the center of the horizontal pipe; such behavior is well known as wavy core-annular flow (WCAF).

Bannwart [5] investigated the behavior of the annular flow in the horizontal pipe modifying the inner surface of the pipe. This author proposed two mathematical correlations to calculate the pressure gradient by considering a horizontal pipe: (a) with an inner surface oleophobic and (b) with an inner surface oleophilic. The difference between oleophobic and oleophilic surfaces is related to the contact angle between the oil and the inner surface of the pipe. Therefore, it is said that a surface is oleophobic when the contact angle decreases with decreasing inner surface roughness of the pipe, thus facilitating the slip of the oil along the pipe and reduces adhesion problems [25].

Bai et al. [2] presented experimental results of oil flow in vertical pipes lubricated by water flows upward and downward. According to these authors, in the upflow, oil tends to stay afloat in a concentric axis of the tube due to the center of gravity. In upward flow, the pressure gradient and buoyancy have the same direction, waves develop and the force of lubrication with the buoyant force tends to extend the waves. In a horizontal annular flow with density difference between fluids, the core of oil tends to occupy a position eccentric to the axis of the pipe and the presence of waves at the interface between oil and water induces a secondary movement perpendicular to axis of the tube. Ooms and Poesio [16] reports that this motion is not considered secondary in a concentric annular flow.

Ooms et al. [18] investigated theoretically the hydrodynamic counterbalances a buoyant force of the nucleus of oil seeping into the pipe, taking into account the difference in density between the two fluids. During the study, it was assumed that the fluid that forms the core consists of a solid surrounded by a layer of low viscosity fluid. The hydrodynamic lubrication theory used takes into account the flow of an annular layer of a low viscosity liquid and in the center a liquid layer of high viscosity. With this, the development of interfacial waves between the fluids was calculated.

In the literature, it is possible find theoretical and experimental studies that show the behavior of sub-patterns of core-annular flow, known as perfect core-annular flow (PCAF) and wave core-annular flow (WCAF).

The ideal flow or PCAF corresponds to an exact solution of the governing equation applied to concentric fluid flow with different density and viscosity in a pipe of circular cross section. PCAF is a rectilinear flow with a velocity component that varies only with the radial coordinate. The two fluids are organized asymmetrically, with a fluid in the center or core and an adjacent the wall of the tube forming a ring. The core of oil has a perfect cylindrical interface of uniform radius. However, the effect of gravity tends to disorder the flow in horizontal tubes and in this case the PCAF cannot happen unless the gravity is canceled. In vertical tubes this is not true, because gravity is concentric to the two fluids [2]. The PCAF of two fluids with approximated densities flowing in horizontal and vertical pipes is possible, but are rarely stable [2, 20]. According to Bensakhria et al. [7], the ideal or perfect annular flow seems to be very rare and can only exist for the flow of two fluids of equal densities. Bai et al. [2] made several experimental observations showing that waves are formed at the interface between water and oil, leading to a wavy behavior. Bensakhria et al. [7] reported that, for a fixed volume ratio between water and oil, the annular flow is not stable at low velocity. The capillary instability due to interfacial tension becomes visible, causing a disturbance in the core of the oil. However, with increasing velocity, stability is achieved and the flow pattern can then be observed. These authors also mention that for even higher speeds, the PCAF again becomes unstable due to interfacial tension, and then ripples appear in the flow, leading to a wavy annular flow.

According to Bai et al. [2], two types of the oil/water interface are considered. The Rayleigh–Taylor and Kelvin–Helmholtz type instabilities. Rayleigh–Taylor instability is due to the combined effect of interfacial tension and/or unfavorable density gradients in fluids and Kelvin–Helmholtz type instability is related to the imbalance of interfacial velocities of the fluid that manifest by a ripple at the interface of fluids. One factor that has great influence on the stability of annular flow is the interfacial tension. To model the curvature of the oil–water interface helps understand the configuration of the Core-Annular Flow pattern. Bannwart [5] reports that in the field hydrostatic with fully developed flow, an equilibrium condition can be established between surface tension and buoyancy. The lubrication model proposes that the wavy motion of the core of oil with respect to the pipe wall generates a pressure gradient in the annular region exerting forces in the vertical direction, counteracting the forces of buoyancy on the core due to density difference. Thus, the presence of waves is of fundamental importance in the lubrication of the core. If the amplitude of these waves is null, the core reaches the top of the pipe, i.e. the wall of the pipe [17]. The lubrication theory is valid when inertia is neglected (Reynolds Lubrication Theory) and, when the wave amplitude is small and the radial velocity is neglected [3].

3.2 Mathematical Modeling

The modeling procedure consists in the mathematical description of the physical problem to be analyzed. When it is possible to put the variables of the problem in the form of one or more precise quantitative relations, we obtain a well known mathematical models. In the case of fluid flow, the mathematical model is composed by conservation equations (mass, energy and momentum), initial and boundary conditions and mechanical constitutive equations that establish relationship between stress field and velocity field in the flow. However, this model corresponds to a set of data and abstract ideas that allow engineers and researchers to propose an explanation for the phenomenon.

The conservations equations can be written as follows.

(a) Continuity equation

$$\frac{\partial}{\partial t}(f_\alpha \rho_\alpha) + \nabla \bullet (f_\alpha \rho_\alpha U_\alpha) = S_{MS\alpha} + \sum_{\beta=1}^{N_P} \Gamma_{\alpha\beta} \tag{1}$$

(b) Momentum equation

$$\frac{\partial}{\partial t}(f_\alpha \rho_\alpha \mathbf{U}_\alpha) + \nabla \bullet [f_\alpha(\rho_\alpha \mathbf{U}_\alpha \otimes \mathbf{U}_\alpha)] = -f_\alpha \nabla P_\alpha + \nabla \bullet \left\{ f_\alpha \mu_\alpha \left[\nabla U_\alpha + (\nabla U_\alpha)^T \right] \right\} + \\ + \sum_{\beta=1}^{N_P} \left(\Gamma^+_{\alpha\beta} \mathbf{U}_\beta - \Gamma^+_{\beta\alpha} \mathbf{U}_\alpha \right) + S_{M\alpha} + M_\alpha \tag{2}$$

where α and β represent the phases involved (water or oil), f is the volume fraction, ρ is density, **U** is the velocity vector, N_p is the number of phases involved, P is the pressure, μ is viscosity. The term $\mathbf{S}_{MS\alpha}$ describes user specified mass sources, $\Gamma_{\alpha\beta}$ is the mass flow rate per unit volume from phase β to phase α, $\mathbf{S}_{M\alpha}$ describes momentum sources due to external body forces (buoyancy force and rotational force), $\mathbf{M}_\alpha$ describes the interfacial forces acting on phase α due to the presence of other phases (drag force, lift force, wall lubrication force, virtual mass force and interphase turbulent dispersion force) and, the term $\Gamma^+_{\alpha\beta}\mathbf{U}_\beta - \Gamma^+_{\beta\alpha}\mathbf{U}_\alpha$ represents momentum transfer induced by interphase mass transfer.

When we use mixture model, only the total drag force exerted by the phase β in the phase α per unit volume is considered, $D_{\alpha\beta}$ given by:

$$\mathrm{D}_{\alpha\beta} = \mathrm{C_D}\rho_{\alpha\beta}\mathrm{A}_{\alpha\beta}\left|\mathrm{U}_\beta - \mathrm{U}_\alpha\right|\left(\mathrm{U}_\beta - \mathrm{U}_\alpha\right) \tag{3}$$

where C_D is the drag coefficient and $\rho_{\alpha\beta}$ corresponds to the mixture density given by:

$$\rho_{\alpha\beta} = f_\alpha\rho_\alpha + f_\beta\rho_\beta \tag{4}$$

The density of interfacial area per unit volume, $A_{\alpha\beta}$, is given below:

$$A_{\alpha\beta} = \frac{f_\alpha f_\beta}{d_{\alpha\beta}} \tag{5}$$

where $d_{\alpha\beta} = 1\ mm$ is the scale length coefficient of the mixture. The effects of gravity were not taken into account because the closeness of the involved phases densities.

(c) k-ε Turbulence Model

The k-ε turbulence model is a model of turbulent viscosity which assumes that Reynolds stress tensors are proportional the mean velocity gradient, with the constant of proportionality characterized by turbulent viscosity (well known like hypothesis of Boussinesq).

The characteristic of these types of models is that two transport equations modeled separately are solved for the turbulent length and time scale or for any two independent linear combinations of them. The transport equations for turbulent kinetic energy, k and turbulent dissipation rate, ε, are given by:

$$\frac{\partial(\rho_\alpha f_\alpha k_\alpha)}{\partial t} + \nabla\bullet\left\{f_\alpha\left[\rho_\alpha U_\alpha k_\alpha - \left(\mu + \frac{\mu_{t\alpha}}{\sigma_k}\right)\nabla k_\alpha\right]\right\} = f_\alpha(G_\alpha - \rho_\alpha\varepsilon_\alpha) \tag{6}$$

$$\frac{\partial(\rho_\alpha f_\alpha \varepsilon_\alpha)}{\partial t} + \nabla\bullet\left\{f_\alpha\rho_\alpha U_\alpha\varepsilon_\alpha - \left(\mu + \frac{\mu_{t\alpha}}{\sigma_\varepsilon}\right)\nabla\varepsilon_\alpha\right\} = f_\alpha\frac{\varepsilon_\alpha}{k_\alpha}(C_1G_\alpha - C_2\rho_\alpha\varepsilon_\alpha) \tag{7}$$

where G_α is the generation of turbulent kinetic energy inside of the phase α, C_1 and C_2 are empirical constant. In this equation, ε_α is the rate of dissipation of the turbulent kinetic energy of the phase α, defined by:

$$\varepsilon_\alpha = \frac{c_\mu q_\alpha^3}{L_\alpha} \tag{8}$$

and k_α is the turbulent kinetic energy to phase α given by:

$$k_\alpha = \frac{q_\alpha^2}{2} \tag{9}$$

where L_α represents spatial length scale, q_α is the scale of velocity, c_μ is a empirical constant calculated by:

$$c_\mu = 4c_\alpha^2 \tag{10}$$

where c_α is a empirical constant.

In Eqs. 6 and 7, $\mu_{t\alpha}$ corresponds to turbulent viscosity, defined by:

$$\mu_{t\alpha} = c_\mu \rho_\alpha \frac{k_\alpha^2}{\varepsilon_\alpha} \tag{11}$$

In the previous equations: C_1= 1.44; C_2 = 1.92; c_μ = 0.09; σ_k= 1.0 and σ_ε = 1.3.

3.3 *Applications to Heavy Ultraviscous Oil Transport*

3.3.1 Mesh Generation

The geometrical representation of the pipe used to study two-phase flow of water/heavy ultraviscous oil is illustrated in Fig. 1.

Due to the angular symmetry observed in flow in tubes of circular cross section, a study domain in a two-dimensional space on the plane rz was considered, as illustrated in Figs. 2 and 3.

The unstructured mesh was made in the three-dimensional domain in cylindrical coordinates (Fig. 4) and, after different mesh refinement and time (seeking non-dependence of the numerical results with the time and space), resulted in a mesh consisting of 127,210 controls volumes (84,178 tetrahedral, 42,974 prismatic and 58 pyramidal). The grid was obtained by using the CFX® 5.6 Commercial code.

3.3.2 Hydrodynamics Results

The heavy ultraviscous oil and water flow in pipe, using water as a lubricating fluid, is governed by the general laws of conservation. To model core-annular flow in horizontal pipe, the following assumptions were adopted: (a) no mass source and interfacial mass transfer between the phases occurs, (b) gravity effects were

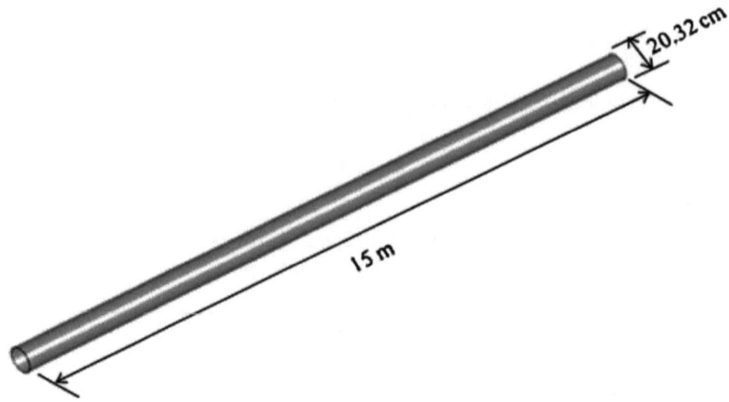

Fig. 1 Geometrical representation of the pipe

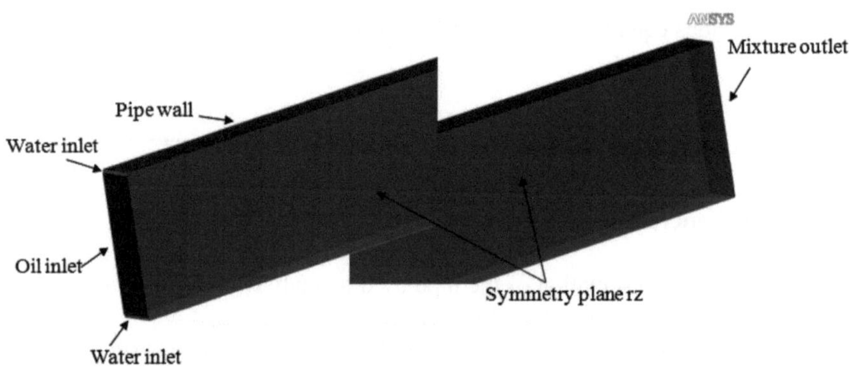

Fig. 2 Detailed illustration of two-dimensional domain rz

Fig. 3 Zoom of the water inlet of the pipe

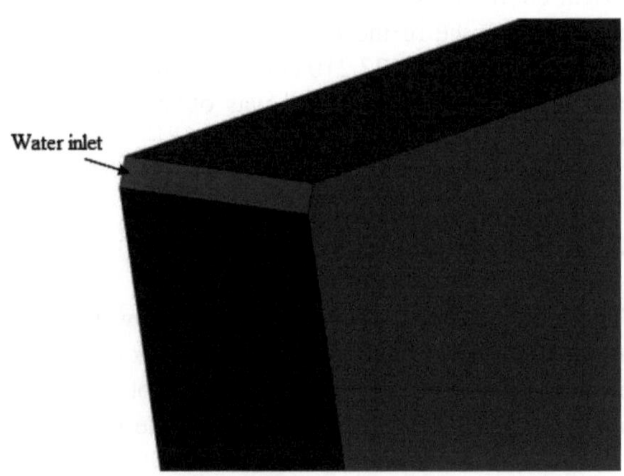

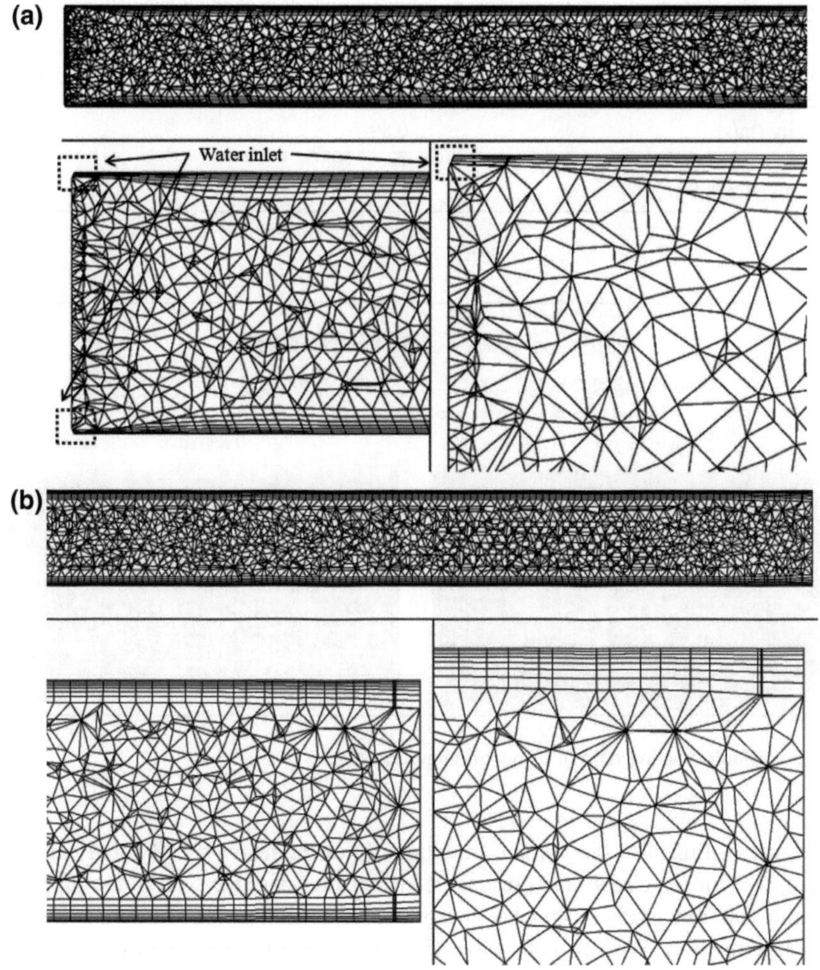

Fig. 4 Grid used in the simulations. **a** zoom pipe inlet and **b** zoom pipe outlet

neglected, (c) chemical reaction effect were neglected, and (d) incompressible and isothermal flow. The effects of gravity were not taken into account because the closeness of the involved phases densities.

The software CFX® 10 was used to obtain numerical solutions of the physical problem. Table 1 summarizes the thermo-physical properties of fluid phases and inlet velocity used on the simulations. For initial and boundary conditions, it was considered that the pipe was initially filled with water and null velocity vector in the pipe wall (no slip condition). In the outlet, a prescribed static pressure (P = 98.1 kPa) was used. The transient numerical results were obtained using a $\Delta t = 0.5$ s for an elapsed time t = 150 s and drag coefficient $C_D = 0.44$.

Table 1 The thermal–physical properties of the water and oil used in all simulations

	Water	Oil
Density (kg/m^3)	997.00	905.00
Viscosity (Pa.s)	88.99×10^{-5}	12.00
Velocity (m/s)	0.80	0.40
Surface tension (N/m)	72×10^{-3}	
Wall roughness (m)	5.00×10^{-5}	

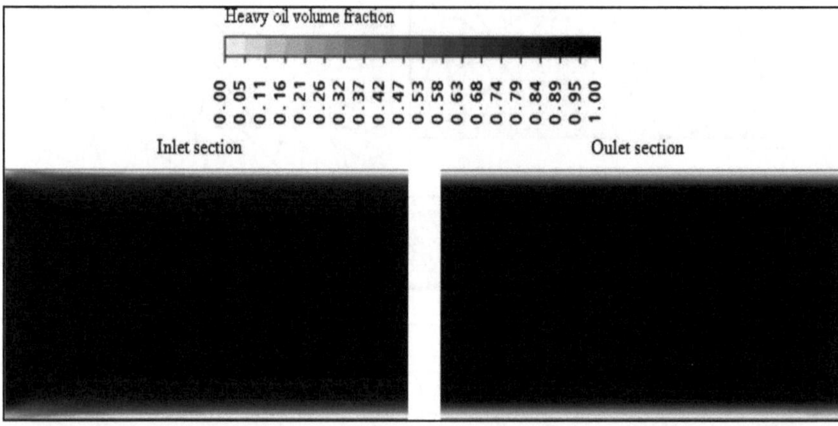

Fig. 5 Volume fraction field of heavy oil along the pipe

Figure 5 illustrates the volume fraction field of the heavy oil. By analyzing this figure, we can see clearly the presence of water flux near the wall of the pipe. The volume fraction gradient can be interpreted like mixture of water/heavy oil or formed emulsion. We can note wave motion of the oil core, mainly at the input region. This phenomenon might be related to the form of water injection, however, does not resemble the "bamboo waves" like reported by Bensakhria et al. [7], Ooms et al. [17, 18], Oliemans et al. [14], Bai [3], Joseph et al. [10], Bai and Joseph [4], Ko et al. [11], Ooms and Poesio [16], Rodriguez and Bannwart [22, 23], among others. According to these authors the presence of waves is very nature for the real cases of heavy oil transportation, and contributes positively in the heavy oil flow by using core-annular flow technique.

A search through the ratio between the velocity of input water/oil adequate, so that the efficiency of lubrication of the wall be maximum (minimum pressure drop), i.e., transport of oil by core-flow technique be optimized, leads that to change water velocity and keeping fixed the oil velocity. Figure 6 depicts the results of pressure drop in the pipe as a function of the ratio between water and oil volumetric flow rates, Q_W/Q_O. This result shows a decrease in pressure drop along the pipe for increased water velocity. This fact can be explained by the reduction

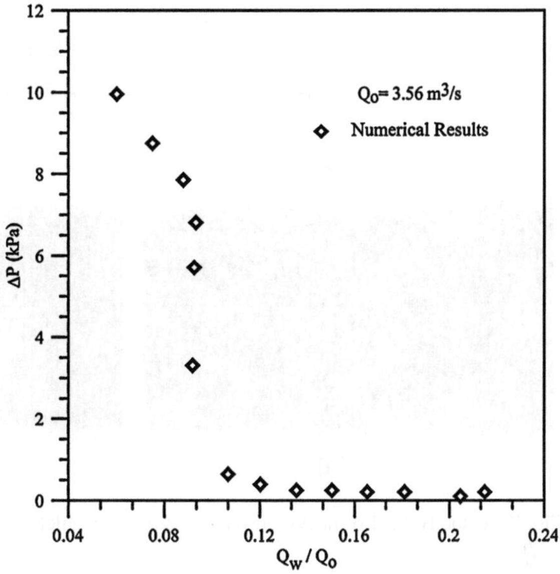

Fig. 6 Pressure drop as a function of water/oil volumetric flow rate relationships for elapsed time t = 150 s

of friction in consequence of the presence of water film, thereby forming a oil core that practically does not touch with the wall of the pipe. It was found that from the ratio Q_w/Q_o= 0.12 there is no practically pressure drop change (ΔP = 388 Pa). This observation suggests that the water flow rate has a maximum value (optimum value) that becomes the process of heavy ultraviscous oil transportation to be optimized in an economic and mechanical perspective.

When water velocity is progressively increased, we have an increase of the pressure drop due to a significant effect of the water turbulent flow regime. Huang et al. [9], Ooms and Poesio [16], Bensakhria et al. [7], Ooms et al. [18] and Vanaparthy and Meiburg [26] by studying the core-flow technique, they reports that due to density difference between phases, the oil tends to rise to the top of the pipe (if it is in horizontal).

However, the lubrication forces tend to repel the oil core to the center of the pipe. A competition between these two effects gives a variation in the position of the oil core inside the pipe, which depends strongly on the ratio between the velocities of water and heavy oil.

Figure 7 illustrates superficial velocity field of heavy oil in the axial direction and details about the inlet and outlet of the pipe. This figure clearly shows the presence of a water stream near the pipe wall (lubricating film of water).

Figure 8 illustrates the evolution of the radial distribution of axial velocity components of heavy oil for different process time placed to 1 m from the pipe input. It can be observed that for an elapsed time of more than four seconds, the velocity profiles are coincident and present a behavior of plug flow in the section of oil core flow. Moreover, it can be said that due to the formation of water ring around the core

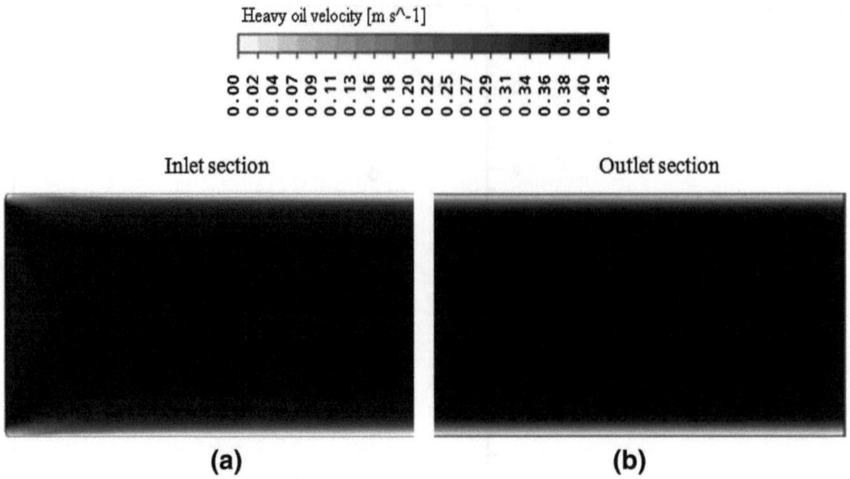

Fig. 7 Velocity field of heavy oil on the rz plane. **a** inlet and **b** outlet

Fig. 8 Radial velocity profiles of oil as a function of radial position at z = 1 m from the inlet of pipe

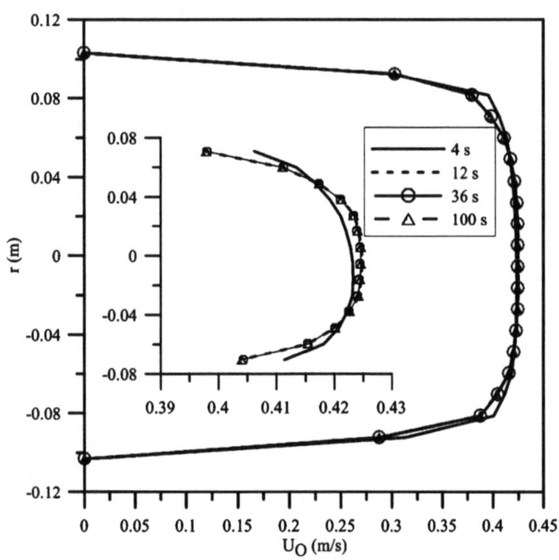

oil, there is a significant reduction in the pressure drop. In this case, the oil presents almost the same velocity of water in heavy oil–water interface.

It is desirable to transport as much heavy ultraviscous crude oil as possible at fixed pressure drop. Then, to evaluate the efficiency of the core-flow technique, we compare the pressure drop in the water–oil two-phase flow with water and oil single-phase flows. In core-annular flow, water always touch the pipe wall, so,

Fig. 9 Comparison of the pressure drop between the single-phase water and oil flows and core annular flow as a function of time

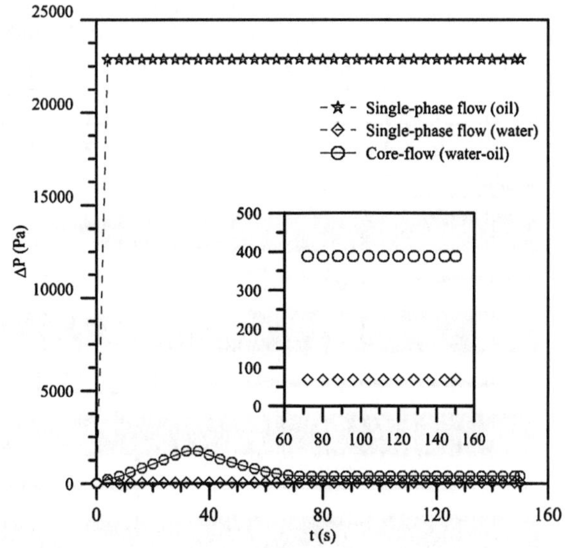

pressure drop should be close to the pressure drop in the water flowing alone at the mass flow rate of the mixture, in accordance with the cited literature.

Figure 9 shows the behavior of pressure drop versus time along the pipe. We can see an increase in pressure drop until 40 s, due to the increased heavy oil–water mixture viscosity. From this moment, reduction in pressure occurs until a stable condition is reached in the elapsed time 80 s. This fact is associated with the lubrication of the pipe wall by the formed water film. By analyzing the Fig. 9 we can see a large reduction of the pressure drop (approximately 59 times). If water flow is stopped, the pressure drop will be increased continuously until it reaches stable conditions of the single-phase oil flow.

These observations can be better understood by observing Fig. 10 where the evolution of the oil as advanced inside the pipe initially full with water is illustrated. The results correspond to the volume fraction of oil on the rz plane for different times. It is very clear that when oil moves through the pipe there is a reduction in heavy oil–water mixture that favors the formation of a oil core in the center of the pipe until a proper lubrication of the tube wall is reached minimizing the pressure drop along the pipe. The stability in the lubrication can be observed in detail in Fig. 10, after an elapsed time of 150 s.

According to Fig. 10, around 28 s, a considerable concentration of oil was observed (average volume fraction of heavy oil equal to 0.662) in this same position (z = 10 m). For an elapsed time more than 28 s, it is observed that oil velocity profile reaches the plug flow behavior.

Figures 11 and 12 illustrate velocity and volume fraction of the oil as a function of the radial position, respectively, in one section located 10 m from inlet of the pipe. We can see that, in the elapsed times of 4 and 12 s, the velocity and volume fraction of the oil are null, i.e., complete absence of oil in the pipe section.

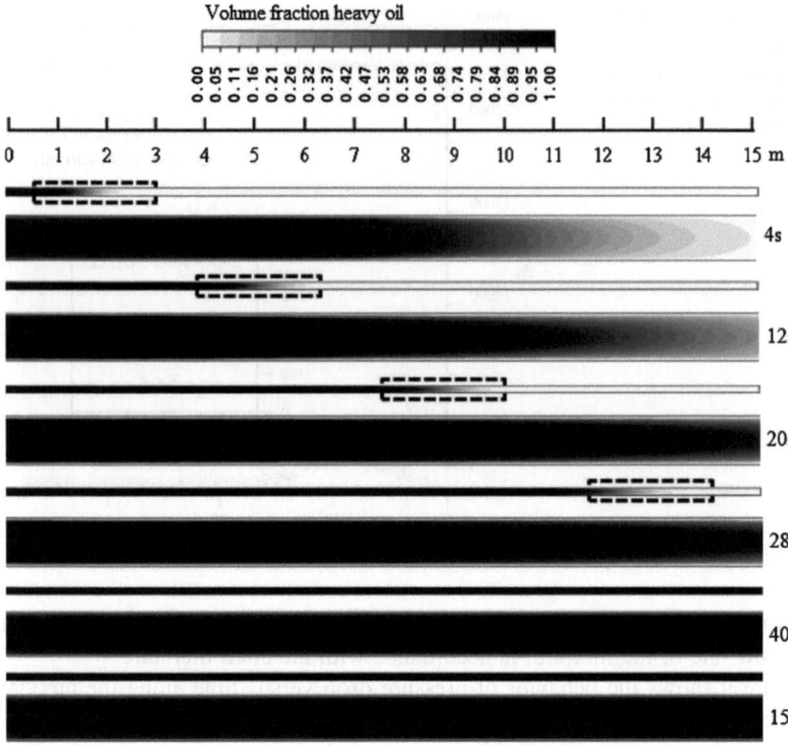

Fig. 10 Oil volume fraction distribution on the rz plane for different elapsed time

Fig. 11 Radial velocity profile of oil as a function of radial position at z = 10 m from the inlet of the pipe

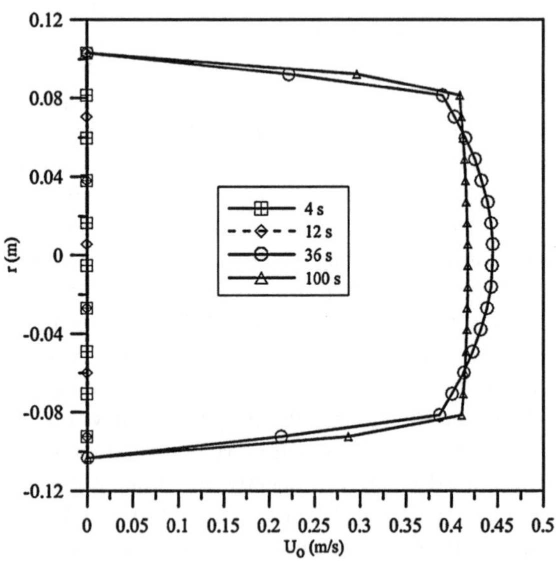

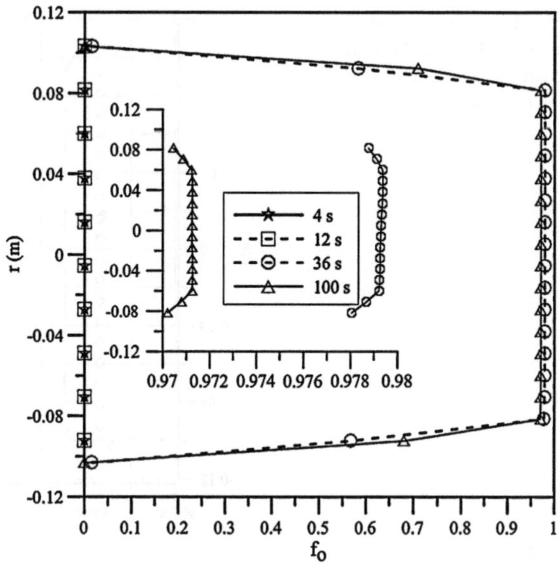

Fig. 12 Oil volume fraction profile as a function of the radial position at z = 10 m from the inlet of the pipe

Joseph et al. [10] recommend lubricating oil core with lower water flow rate as much as possible to minimize the problem of dewatering. On the other side, the probability to oil fouls the wall is large when small water flow rate is used. Here, we can use a 3 mm water thickness layer. Moreover, when oil fouls the wall continuously, pressure drop increases very much, so, it is necessary to stop the flow. The cleaning of the pipe and restart of the flow will be easier if the oil does not strongly stick the pipe wall. Additional discussions about this topic will be found in Arney et al. [1] and, due to importance of this theme, the authors recommend new researches.

Figures 13 and 14 illustrate pressure profiles along the radius at the positions taken z = 2 and 10 m, respectively, from pipe inlet. The results presented here highlight four different situations (it should be noted that the pipe was initially filled with water):

(a) At time t = 4 s at z = 2 m, the heavy oil volume fraction is still very small inside the tube (Fig. 10), keeping a pressure of 98,400 Pa approximately (Fig. 13).
(b) In t = 12 s the oil volume fraction has already achieved the position z = 2 m (Fig. 10), so pressure increases to 98,800 Pa approximately in this position (Fig. 13).
(c) At time t = 36 s, the oil volume fraction reached position z = 10 m (Fig. 10). In this section, the pressure increases immediately to 99,400 Pa (Fig. 14).
(d) For t = 100 s at z = 10 m the annular flow pattern is reached, and the pressure decreases drastically to 98,200 Pa approximately (Fig. 14).

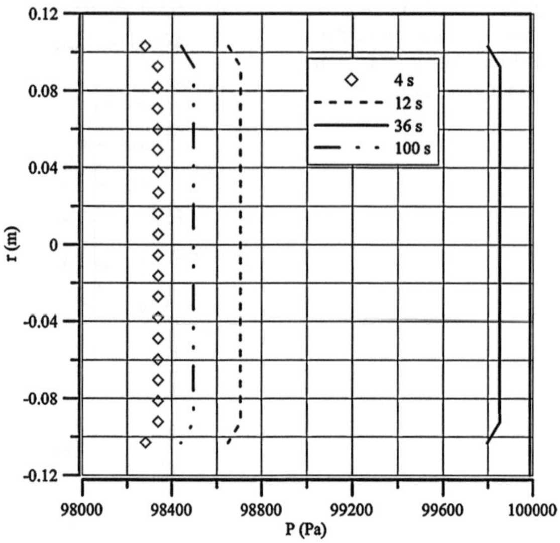

Fig. 13 Pressure distribution as a function of the radial position at z = 2 m from the inlet of the pipe

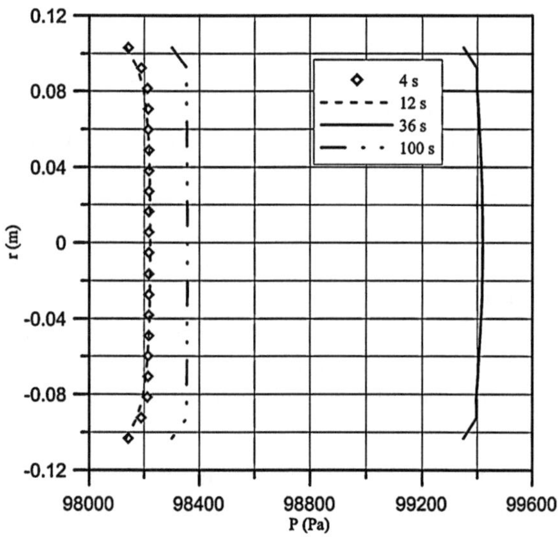

Fig. 14 Pressure profile as a function of the radial position at z = 10 m from the inlet of the pipe

When annular flow is reached, the pressure measured in a section 10 meters from the input (Fig. 14) has a value very close to the situation where water flows alone through this section (single-phase flow water). This fact confirms the results presented and discussed in Figs. 9 and 10. In this case, we conclude that core annular flow technique has a big importance in the transport of heavy oils in horizontal and vertical pipes.

4 Conclusions and Remarks

This chapter provides a discussion about the core-annular flow and its importance to petroleum industry. Information about pressure drop and saving energy pumping are given too. Application is directed to heavy ultraviscous oil flowing in horizontal pipes lubricated by water (20.32 cm internal diameter). From the numerical results of the heavy ultraviscous oil transportation we can conclude that:

(a) There is presence of water stream near the pipe wall forming a film of water that surrounds the oil core flowing into the central region of the pipe, characterizing the annular flow or "core-flow".
(b) A reduction in pressure drop was observed approximately 59 times when compared with single heavy oil flow by using inlet oil and water velocities of 0.4 and 0.8 m/s, respectively.
(c) Core-annular flow pattern is obtained 40 s after water is injected into the tube, and remains stable during the time rest period keeping a relatively low pressure drop by friction, 388 Pa.
(d) For a water/oil volumetric flow rate relationships of 0.12, the core-annular flow was established and, oil does not touch the wall (fouling). Sometimes, when the fouling builds up, progressively, we have a big increase of the pressure drop which leads to block the flow.

Despite the stability of the core-annular flow we notice that pump power, roughness and geometry of the pipe and phase flow conditions must be compatible. For example, at low water velocity, this flow pattern is not stable. Besides, when using pipes with large diameters, the effects of roughness of the pipe wall can be diminished depending on the flow conditions.

Acknowledgments The authors are grateful to CNPq, FINEP, ANP/UFCG-PRH-25, CAPES, PETROBRAS S/A and JBR Engenharia Ltda. for the granted financial support. Sincere thanks to Editors by the opportunity given to present our research in this book. Our gratitude is also directed to the authors cited in the text that helps in the improvement of this chapter.

References

1. Arney, M.S., Bai, R., Guevara, E., Joseph, D.D., Liu, K.: Friction factor and holdup studies for lubricated pipelining. Int. J. Multiph. Flow **19**(6), 1061–1076 (1993)
2. Bai, R., Chen, K., Joseph, D.D.: Lubricated pipelining: stability of core annular-flow. Part 5. Experiments comparison of theory. J. Fluid Mech. **240**, 97–142 (1991)
3. Bai R.: Traveling waves in a high viscosity ratio and axisymmetric core annular flow. Ph.D Thesis, University of Minnesota, USA (1995)
4. Bai, R., Joseph, D.D.: Steady flow and interfacial shapes of a highly viscous dispersed phase. Int. J. Multiph. Flow **26**(9), 1469–1491 (2000)
5. Bannwart, A.C.: Pressure drop in horizontal core-annular flow. Third International Conference on Multiph. Flow (ICMF), Lyon (1998)

6. Bannwart, A.C.: Modeling aspects of oil–water core–annular flows. J. Pet. Sci. Eng. **32**(2–4), 127–143 (2001)
7. Bensakhria, A., Peysson, Y., Antonini, G.: Experimental study of the pipeline lubrication for heavy oil transport. Oil Gas Sci. Technol. **59**(5), 523–533 (2004)
8. Chen, X.T., Brill, J.P.: Slug to churn transition in upward vertical two-phase flow. Chem. Eng. Sci. **52**(23), 4269–4272 (1997)
9. Huang, A., Christodoulou, C., Joseph, D.D.: Friction factor and holdup studies for lubricated pipelining-II, laminar and k–ε models of eccentric core-flow Int. J. Multiph. Flow **20**(3), 481–491 (1994)
10. Joseph, D.D., Bai, R., Chen, K.P., Renardy, Y.Y.: Core-annular flows. Annu. Rev. Fluid Mech. **29**, 65–90 (1997)
11. Ko, T., Choi, H.G., Bai, R., Joseph, D.D.: Finite element method simulation of turbulent wavy core–annular flows using a k-ω turbulence model method. Int. J. Multiph. Flow **28**(7), 1205–1222 (2002)
12. Li, B.J., Renardy, Y.: Direct simulation of unsteady axisymmetric core-annular flow with high viscosity ratio. J. Fluid Mech. **391**, 123–149 (1999)
13. Oliemans, R.V.A., Ooms, G.: Core-annular flow of oil and water through a pipeline. Multiph Sci. Technol. **2**, 427–479 (1986)
14. Oliemans, R.V.A., Ooms, G., Wu, H.L., Duijvestijn, A.: The core-annular oil/water flow turbulent-lubricating-film model and measurements in a 5 cm pipe loop. Int. J. Multiph. Flow **13**(1), 23–31 (1987)
15. Olsen, D.K., Ramzel, E.B.: Heavy oil refining and transportation: effect on The feasibility of increasing domestic heavy oil production. Fuel **71**(12), 1391–1401 (1992)
16. Ooms, G., Poesio, P.: Stationary core-annular flow through a horizontal pipe. Phys. Rev. **68** (2003)
17. Ooms, G., Segal, A., Van Der Wees, A.J., Meerhoff, R., Oliemans, R.V.A.: A theoretical model for core-annular flow of a very viscous oil core and a water annulus through a horizontal Pipe. Int. J. Multiph. Flow. **10**(1), 41–60 (1984)
18. Ooms, G., Vuik, C., Poesio, P.: Core-annular flow through a horizontal pipe: hydrodynamic counterbalancing of buoyancy force on core. Phys. Fluids **19**, 092103 (2007)
19. Prada, J.W.V., Bannwart, A.C.: Modeling of vertical core annular flows and application to heavy oil production. energy for the new millenium. In: Proceedings of ETCE/OMAE, New Orleans, 14–17 February 2000
20. Preziosi, L., Chen, K., Joseph, D.: Lubricated pipelining: stability of core-annular flow. J. Fluid Mech. **201**, 323–356 (1989)
21. Renardy, Y.Y., Li, J.: Merging of props to form bamboo waves. Int. J. Multiph. Flow **27**, 753–763 (2001)
22. Rodriguez, O.M.H., Bannwart, A.C.: Analytical model for interfacial waves in vertical core-flow. J. Pet. Sci. Eng. **54**(3–4), 173–182 (2006a)
23. Rodriguez, O.M.H., Bannwart, A.C.: Experimental study on interfacial waves in vertical core-flow. J. Pet. Sci. Eng **54**(3–4), 140–148 (2006b)
24. Rodriguez, O.M.H., Bannwart, A.C.: Stability analysis of core-annular flow and neutral stability wave number. AIChE J. **54**(1), 20–31 (2008)
25. Silva, R C R.: Wettability alteration of pipeline internal surfaces for use in the transportation of heavy oil via core-flow. Master Thesis, State University of Campinas, Campinas, Brazil (In Portuguese) (2003)
26. Vanaparthy, S.H., Meiburg, E.: Variable density and viscosity, miscible displacements in capillary tubes. Eur. J. Mech. B/Fluids **27**, 268–289 (2008)

Bending Experiments on Thin Cylindrical Shells

Tohid Ghanbari Ghazijahani and Hossein Showkati

Abstract Cylindrical shell structures are highly susceptible to buckling phenomena when they experience compressive stress. In fact, there are few experimental researches that give the real behavior of a cylindrical shell submitted to pure bending, especially thin shells. This is due to the difficulty of pure bending applying to such thin shells and that such structures behavior under bending is frequently considered rather similar to pure compression. This chapter describes an experimental investigation of a procedure including a system for applying pure bending to cylindrical shells with radius to thickness ratio equals 155. The instrumentation consists of a new loading system in which the pure bending is applied using concentrated loads at the ends of the test model. Ultimately, the critical values for moments as well as buckling modes were compared with finite element (FE) results.

Keywords Cylindrical shells · Experimental technique · Pure bending · Buckling load · Failure mode

1 Introduction

Thin-walled cylindrical shells form numerous parts of many deep-seated industrial structures. Therefore, accurate assessments of the maximum load carrying capacity of such structures are of paramount concern to the engineers. Above all, stability

T. Ghanbari Ghazijahani (✉)
Islamic Azad University, Mahabad Branch, Mahabad, Iran
e-mail: tohidghanbari@gmail.com

H. Showkati
Urmia University, Urmia, Iran
e-mail: h.showkati@urmia.ac.ir

A. Öchsner et al. (eds.), *Materials with Complex Behaviour II*,
Advanced Structured Materials 16, DOI: 10.1007/978-3-642-22700-4_7,

behavior of such shell structures exposed to pure bending has been reported mostly related to finite element analyses carried out on the subject and there are few experimental studies on this field.

Generally, for thin-walled metal cylinders with clamped edges, failure mostly occurs similar to those obtained under pure axial compression. The mode of such instability is in diamond-shaped buckling waves at the compression half of the experimental models. In such tubes, which have moderate length, this instability is due to the fact that the boundary conditions prevent the sectional ovalization which is in general not visible prior to failure. A bifurcation type of instability was analyzed in [1] for the first time. Bifurcation stress equal to 1.3 σ_{CL} was also obtained, where σ_{CL} is the buckling stress of a cylindrical shell under axial compression which independently was identified by Timoshenko [2], Lorenz [3] and Southwell [4]. According to a statistical analysis, the experimental buckling stress is from 20 to 60% higher in bending, depending on the radius to thickness ratio [5, 6].

In 1927, for long cylindrical shells, another collapse mechanism was reported by Brazier [7] and in 1933 by Chwalla [8] in which the flattening of cross-section occurs due to the produced curvature developed by bending moment. It should be noted that, this phenomenon can diminish the bending resistance of pipes progressively because for producing a certain curvature, the required bending moment for an oval section is smaller than for a round one. This nonlinear effect leads to limit load-type instability, whereas the snap buckling into diamond-shaped waves is bifurcation instability [9]. Brazier also theoretically calculated the flexural limit moment of cylindrical shells considering section flattening (ovalization). It is worth bearing in mind that as obtained in [5, 10], for long tubes without considering pre-buckling cross-sectional state, a high value of critical bending buckling moment is reached.

Seide and Weingarten [11] studied bifurcation of cylindrical shells subjected to bending, assuming a linear pre-buckling state, and a Ritz-type bifurcation solution. They numerically obtained that for finite length simply supported cylindrical shells, the buckling stress of the shells under bending can be considered similar to the shells under pure compression. Reddy [12] studied steel and aluminum specimens and observed the presence of wave-like ripples on the compression half of the tubes under bending, before collapse occurred in the tests. Instability of an aluminum cylindrical shell subjected to pure bending with a diameter to thickness ratio of 19.5–60.5 and a length to diameter ratio of 18.1–30.1 was studied by Kyriakides and Ju [13] and Ju and Kyriakides [14]; they also observed the presence of wave-like ripples on the compression side of the bent tubes before collapse. The interaction between Brazier's flattening effect and bifurcation instability has been investigated by Libai and Bert [15], Tatting et al. [16], Stephens and Starnes [17] and Fabian [18]. Stephens and Starnes [17] observations indicate that for short cylinders ($L/R < 3$, where L = length and R = radius of the shells) Brazier effect can be excluded.

Generally, the flexural capacity of steel cylindrical shells is a function of the following parameters, which are listed in arbitrary sequences:

- Stress–strain behavior of used material,
- the ratio of D/t (where D is diameter and t is thickness of the shells),
- initial out-of-circularity of the cross sections and other cases of imperfections,
- stress concentrations near the critical points, e.g. the zone of stiffness variation,
- welding (longitudinal as well as circumferential).

Some of the above parameters were investigated to assess the behavior of thin shells subjected to bending. Nonetheless, there are quite few experiments studying the real behavior of plain thin shells (without reinforcement) and also strengthened thin shells under pure bending, so that it has remained almost untouched. The present study investigates the buckling behavior and the load carrying capacity of plain/reinforced cylindrical shells under pure pending. To this end, three cylindrical steel shells were tested to evaluate the buckling capacity and the failure mode of such thin structures. Note that, the efficiency of local end thickening of the cylindrical shells on the strengthening of such shells was detected. Ultimately, acceptable concord was obtained, comparing the experimental results and the numerical outcomes.

2 Experimental Means

2.1 Test Setup

An overall view of the empirical set-up is shown in Fig. 1. The same general procedure was followed for all three tests. A vertical hydraulic jack was used to apply flexural stress to the specimens labeled as CSP1, CSP2 and CSP3 which the two later mentioned specimens were thickened at both ends. The hydraulic jack was mounted on a bearing frame anchored to rigid floor. The bending moment was applied through an IPE 160 section beam which was mounted under the hydraulic jack. The beam was laterally braced to prevent potential torsional instabilities. Two loading arms with sectorial shell segments were connected to the bottom flange of the beam to a proper load application. A digital load cell was placed and fixed between the top flange of the beam and the hydraulic jack, to record the loads applied on the specimens. Linear variable differential transformer (LVDTs) were fixed on the sections of the specimens to determine the relative motion. All data were recorded and processed by means of a digital scanner and data logger.

The test model includes a simply supported system. Two UCP bearing and a connected shaft, as can be seen in Fig. 1b, used as a model of a hinged support and two ball bearings with a proper shaft was used to model the roller support in this set-up. The shafts were fitted and fixed into the internal rings surfaces of the ball bearings. These supports allow the specimen to have free axial displacement and corresponding rotation to which the pure bending is being closely applied.

The pipes (cylindrical models) in these tests consist of two parts: (1) rigid part at both ends, (2) main specimen with thickness of 0.5 mm and corresponding end

Fig. 1 **a** Overall view of test setup; **b** hinged support

thickening, as shown in Fig. 2. The rigid part of the models was designed to apply the net bending moment to the cylindrical specimens. These pipes were braced and stiffened with some cross-shaped reinforcements increasing the stiffness of such pipes and were connected to the specimens by means of eight bolts on each side and friction which connect the two aforementioned parts (the rigid pipe and the main specimen).

The three models, with diameter of 155 mm and length of 900 mm, were carefully assembled by cord-oriented spot welding over the rolled sheet fragment edges. It is worth to mention that CSP1 (plain specimen) was directly connected to the rigid pipe without any end thickening, whereas CSP2 and CSP3 were thickened with stepped wall thickness at both ends, as the length of the thickened zone and

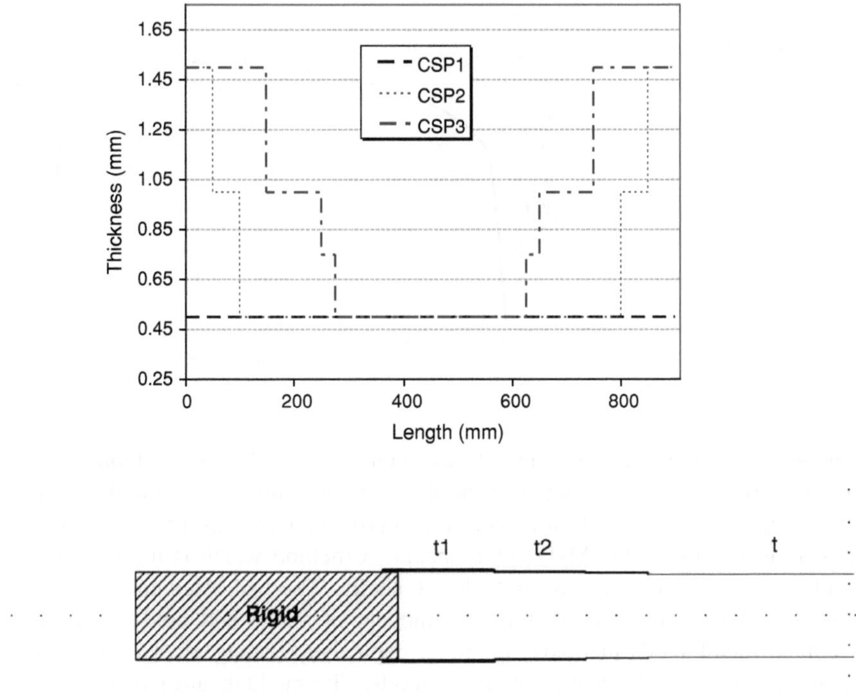

Fig. 2 Thickness variation along the length of the specimens

the thickness variation of the specimens was different from one to another (see Fig. 2).

Tensile coupon tests machined from the experimental models waste material revealed the stress–strain curve of Fig. 3.

2.2 Fabrication Techniques

The fabrication of high quality specimens is a major important issue in shell buckling experiments because their buckling behavior is more sensitive to the details, in particular initial geometrical imperfections, the choice of material and fabrication method. Babcok [19] pointed out that the most considerable point in shell fabrication is that the shell must be made with a method in which the difference between buckling loads and behavior of one nominally identical specimen to another must be less than a parameter in the experiment under investigation.

Fabrication methods have been developed extensively in [19, 20], among which is electroforming which was introduced by Thompson from University College,

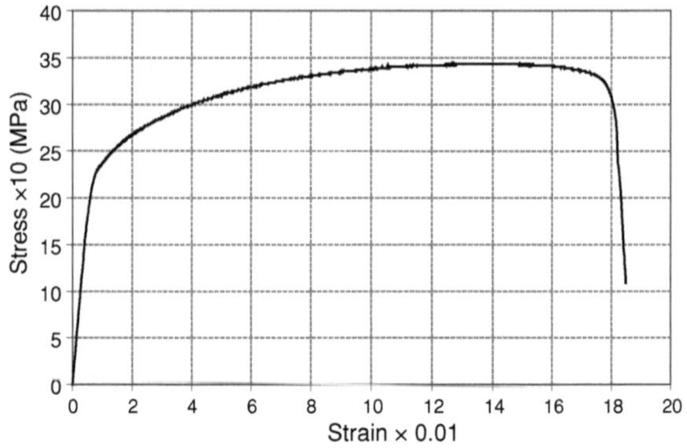

Fig. 3 Stress–strain curve of tensile coupon test

London, in 1960 [21]. This method has been developed making duplicates by electroplating metal onto a mold of an object, then removing the mold in which the intricate surface details are precisely reproduced by this process. Another fabrication method is the Mylar shell specimen method which is most proper to develop the surfaces such as cylindrical shells. Thermal vacuum forming is another widely-used method of plastic models of shells and cold-worked and machined metal shells methods (spinning, explosive forming, or hydroforming), are another realistic fabrication shells methods. The methods are mostly specialized laboratory techniques for making geometrically perfect models. Where the present experiments were aimed at duplicating full-scale steel shell construction as closely as possible, the method of rolling thin steel sheets followed by spot welding has been commonly used [22].

2.2.1 Specimen Fabrication

Sheet cutting was performed using manually controlled shears for a precise shell specimen fabrication. It should be noted that parts were obtained with an accuracy within ±0.1 mm. After cutting the sheets, the plates were cold-rolled into the desired shape. Then, meridional seam connection was performed by means of a proper electrical spot welding machine. Such a connecting process is widely used in industrial shell structures. It should be emphasized that before the main tests, we tested such a connecting method and obtained a good strength at the zone of welding with minor imperfections in the different cases of shell deformations. Therefore, such promising connection was used in pure bending stability tests. Other advantages of such welding in small scale stability tests are as follows:

1. Deformability of the zone of seam welding and proportionality of the stiffness of the welding zone to its adjacent shell body.
2. Minimum residual stresses in comparison to other longitudinal welding techniques.

3. Having a symmetric model in comparison with other welding methods.
4. Minimum geometrical imperfections along the aforementioned connected zone.
5. Prevention of stress concentration development in comparison to the other welding methods near the zone of stiffness differing.

It is of interest to note that when the seam welding process were completed, all the specimens were placed on the rolling machine and rolled again to minimize the geometrical fabrication imperfections. It should be noted that using the described method, nearly geometrically perfect shells could be obtained.

2.3 *Measurement of Geometric Imperfections*

Generally speaking, instability of thin-walled shells is known to be highly sensitive to the different modes of imperfections. In fact, if one wants to get a good agreement between experimental results and theoretical or numerical upshots, then one must take the effects of unavoidable imperfections into account. In addition, all the deviations from the shells perfect shape, so-called geometric imperfections, load eccentricities, supporting system imperfections, used material non-similarities and residual stresses in welded assemblies are all examples of initial imperfections [9]. Thanks to the importance of the measurement of such geometric irregularities, many measuring techniques have been developed [23]. Contacting and non-contacting probes are used for measuring the geometric imperfections. Indeed, non-contacting probes are preferably used for very thin shells (particularly isotropic ones) which have relatively weaker transverse strength, preventing the distortion of the measurements and consequently providing accurate imperfection measurement. In this research, contacting probes were used to a full survey of the imperfections. The details of such measuring technique are available in two previous experimental studies [22, 24]. Topographical layout of initial geometric imperfections of outer surface of CSP2 and CSP3 is shown in Figs. 4 and 5.

2.4 *Shell Thickness Variation*

As we already mentioned, the nominal wall thickness of the steel sheets used for these experiments is 0.5 mm. However, to obtain exact results and conduct reliable analysis, wall thickness variation was also detected and investigated in the specimens. This was done through measuring 18 points in the circumferential direction at both ends of the specimens. A proper micrometer was employed to this purpose. It can be seen from Fig. 6 that thickness variation differs from −8.8 to 8.4% which can be most likely thanks to the manufacturing process, particularly rolling. However, the tolerances are small and we can consider the nominal wall thickness in numerical simulations.

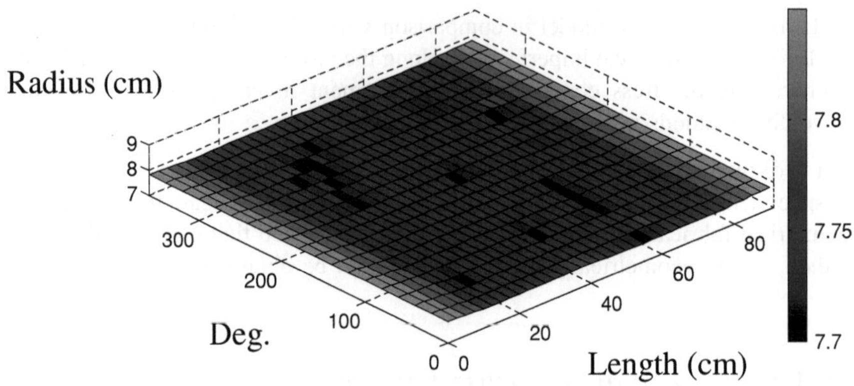

Fig. 4 Topographical layout of initial geometric imperfections of outer surface of CSP2

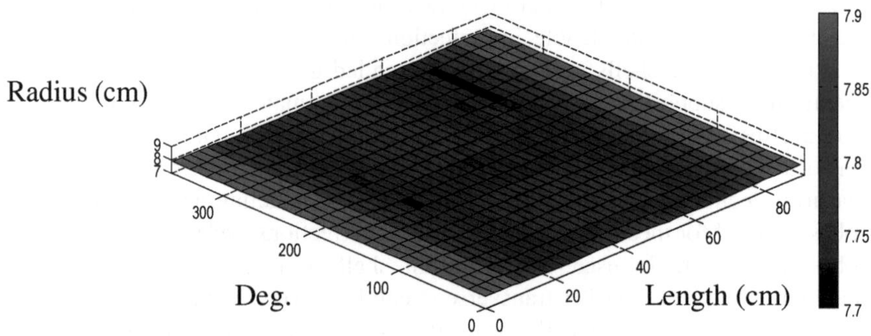

Fig. 5 Topographical layout of initial geometric imperfections of outer surface of CSP3

3 Exhaustive Implementation of the Experiments

Firstly, the instrumental apparatus was installed and calibrated. After full installation of the whole system, all the imperfections of the geometry were gauged through the technique as before mentioned. Then, the bending moment was increasingly applied up to 15% of the numerically anticipated critical load to ensure that the empirical system and the gauging probes are behaving satisfactory. After making certain the system behaviors well, the bending moment was gradually increased up to the first indication of initial buckling. Subsequently load applying was continued to reach the full buckling mode and failure. Finally, a full survey of shell buckling deformations was conducted to evaluate the structural behavior of such thin shells.

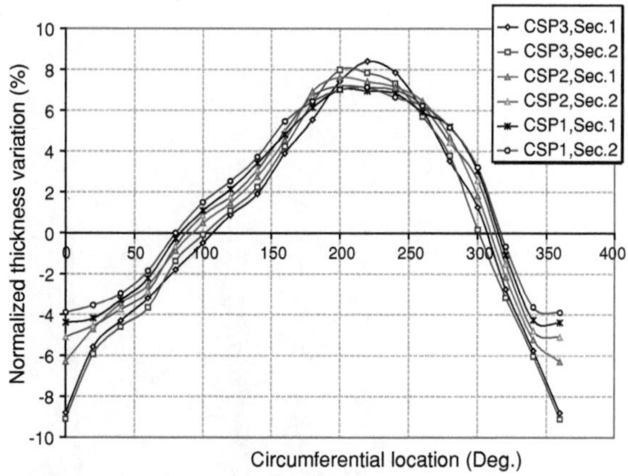

Fig. 6 Thickness variation measurements in the hoop direction of the specimens at two end sections

4 Experimental Observations

In all specimens, as the load reached the critical value, the first evidence of initial buckling appeared at the compression half of the cross section, closely adjacent to the longitudinally thickness variation of the shells. By increasing the bending moment, the buckling propagates circumferentially up to development on the whole of the compression side. Then, one can observe one-tier, sharp-edged diamond pattern buckling mode. Figure 7 shows plainly the global buckling mode of the specimens.

In this pattern of instability, the onset of buckling forms a local unstable pattern. This local deformation, after propagating circumferentially, transforms into a rounded pattern of diamonds of about twice the size of the initial rounded buckles. Initial local buckling is of infinitesimal amplitude and hence cannot be observed by the naked eye. As the buckles grow and deepen, they can be detected by naked eye.

It is of interest to note that along with such rounded diamond shapes to be grown and deepen, shriveling of the outward buckles to narrow ridges simultaneously happens. It should be underlined that such buckling mode is accompanied by a circumferentially bulging state near the intersection of the thickened zone and the thinnest part of the specimens. In contrast, in such mode, inward deformations are larger than the outward ones as obvious in Fig. 8. Moreover, the ratio of the outward deformations to the inward deformations is mostly less than 1:3 in different sections.

It should be pointed out that deformational diamonds are arranged regularly and symmetrically. One may note that the regular and symmetric form of the deformations can indicate the fact that the loading was applied quite vertically and the loading beam properly braced laterally. The remarkable point to be emphasized is that in all the tests, the buckling occurred near the location of the roller support. Therefore, one can consider that the present structures buckle near the side in which axial freedom is applied (near the roller support).

Fig. 7 General buckling mode of the specimens

Fig. 8 Buckling mode features

Fig. 9 Finite element model of CSP3 specimen

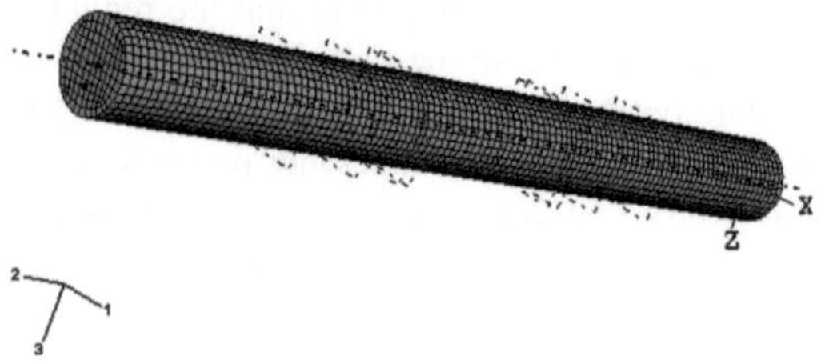

5 Finite Element Modeling

The FEA presented in this chapter was obtained using the general-purpose finite element program ABAQUS designed specifically for advanced structural analysis. The program has been shown to give accurate predictions and has been extensively applied to study buckling behavior of shells of revolution. The derived results are referred to as linear and nonlinear buckling loads respectively in this chapter. Finite element model of CSP3 specimen is shown in Fig. 9.

Generally speaking, buckling stability studies of shell structures require two types of analysis. First, eigenvalue analysis is used to obtain estimates of the buckling loads and modes. Such studies also provide guidance in mesh design because mesh convergence studies are required to ensure that the eigenvalue estimates of the buckling load have converged. This requires that the mesh be adequate to model the buckling modes, which are usually more complex than the pre-buckling deformation mode. The second phase of the study is the performance of load-displacement analyses, usually using the Riks method to handle possible instabilities. These analyses would typically study imperfection sensitivity by perturbing the perfect geometry with different magnitudes of imperfection to assess the effect of such geometric perturbation on the response [25].

The numerical models which have been studied herein are intended for the non-linear elasto-plastic analysis of shells of revolution. Comprehensive modeling of structures requires determination of the entire equilibrium path until collapse occurs. For the instability analysis of these structures, material and geometrical nonlinear analysis should be undertaken. The imperfect geometry of each specimen has been used in finite element analyses. In the present study, the 'Arc-Length-Type Method' has been used which is the most efficient method for this purpose and is now predominantly used in structural nonlinear analysis programs. Use of this technique in structural analysis was originally proposed by Riks [26], in which a constraint equation controls the load increment in order to force the iteration path to follow either a plane normal to the tangent at the starting point of the iteration, or a sphere with its center at the starting point.

The element of S4R has been commonly used for modeling such shell elements. Element S4R has four nodes including five independent degrees of freedom. Note that, these are the three orthogonal translations and the dimensions changing of two independent components of a unit vector normal to the surface of the shell, which is considered as the rotations. It is to be said that the normal vector third component is derived from the condition in which the normal vector length is assumed to be equal to unity. The independent degrees of freedom are all interpolated linearly. Externally, three rotational and three translational degrees of freedom per node are available to the user. This element is well suited for modeling shell structures. Large deflection, stress stiffening and nonlinear analysis are from its capabilities.

The geometric non-linear analysis feature has been taken into account, which is reliable for the large displacements, large rotation, and finite membrane strain

Table 1 Comparison of buckling moment and mode derived by experiments and FEA

Specimen label	Initial buckling moment (experimental study) (N.m)	Initial buckling moment (FEA) (N.m)	Initial buckling difference between experiments and FEA (%)	Number of rounded diamonds (experiments) (half-section)	Number of rounded diamonds (FEA) (half-section)	Difference between diamonds number (experiments and FEA)
CSP1	2,788.7	3,187.5	14.3	3	4	1
CSP2	3,049.6	3,413	11.9	4	5	1
CSP3	3,918.4	4,219.8	7.7	4	5	1

analysis. Two concentrated forces have been modeled at both ends of the models, as follower forces. Boundary conditions are simply supported. It is worth recalling that the model has a degree of freedom in the axial direction. The stress–strain input of the model matched the properties in Fig. 3.

Without doubt, in shell structures, by varying the location and the value of geometrical irregularities, we obtain big differences in the critical buckling load. Therefore, to get realistic results, the imperfections were modeled directly and closely as a table of node numbers and perturbations.

6 Results and Discussion

Table 1 gives a comparison between experimental and numerical values of the initial buckling loads of specimens CSP1, CSP2 and CSP3. Good agreement was obtained which shows that the FEA was able to predict the buckling load of the specimens. Bearing in mind that the test values are in lower range in comparison to the numerical ones owing to the presence of initial imperfections of the geometry, shortcomings of the apparatus and other human and instrument-related parameters. Figures 10 and 11 illustrate the deformed mode for specimens, CSP1 and CSP3 and corresponding deformations derived from the nonlinear finite element analysis.

In Figs. 12, 13, 14, 15, 16 and 17 a comparison is carried out between initial and ultimate geometry of specimen CSP1, CSP2 and CSP3 in which the maximum deformation is located roughly at a distance 2 cm from the end of the thickening region. In Figs. 13, 15 and 17 parts "a" are related to the experimental measurements of the specimens, and parts "b" are related to corresponding deformed sections derived from FEA simulations. As is shown in these figures, three buckling waves were formed and a good agreement between careful experiments and numerical predictions was achieved.

The following points can be concluded, considering both the aforementioned experimental and FEA outcomes:

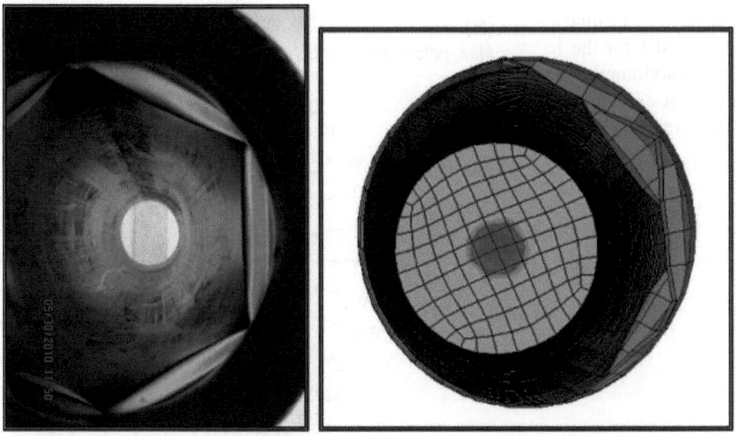

Fig. 10 Final deformed mode of CSP1 specimen and corresponding FE model

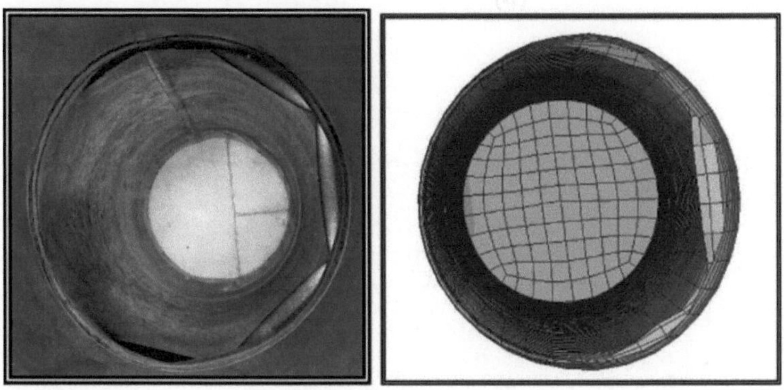

Fig. 11 Final deformed mode of CSP3 specimen and corresponding FE model

Fig. 12 Deformed shape of different sections on the tested specimen CSP1, measured from the end of thickened length toward the center

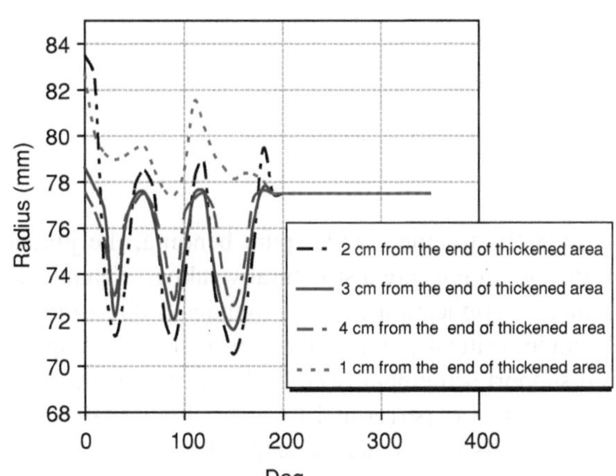

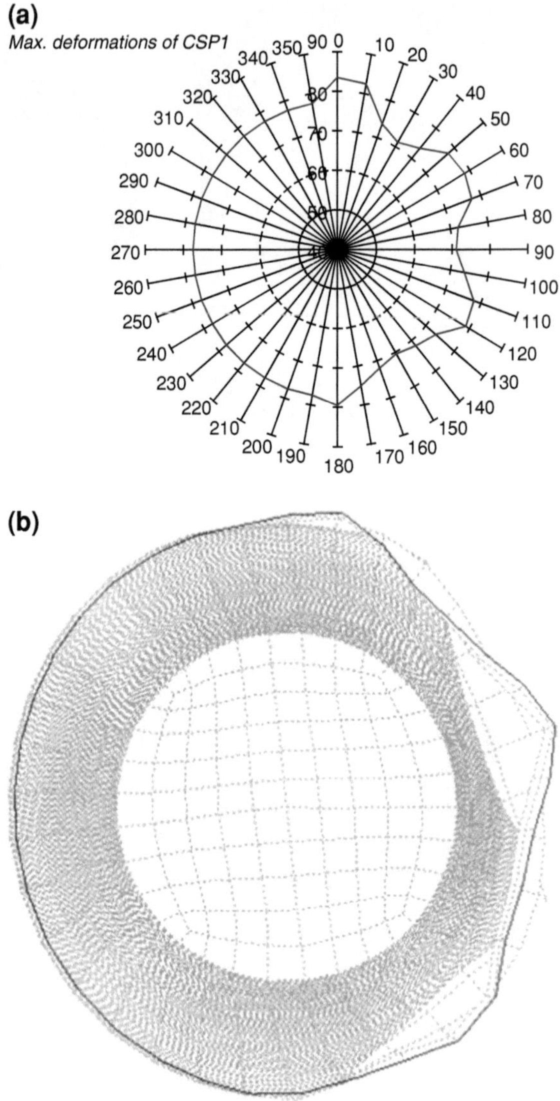

Fig. 13 Polar plot of final geometry of CSP1 for the section with maximum deformations: **a** experimental gauging; **b** corresponding FE model

In shell structures under pure bending, the junctions and generally the points of discontinuities are a structural weakness. Thanks to this fact, in various types of industrial applications, out of roundness due to point loads is to be checked. For example, critical point loads may arise at free-span shoulders, artificial supports and support settlement [27]. In this experimental study, the parameter of the local thickening was performed to overcome negative local effects. This local thickening and its length and the steps of thickness variation, has a dominant influence on the buckling load of the experimental and numerical models. However, we can

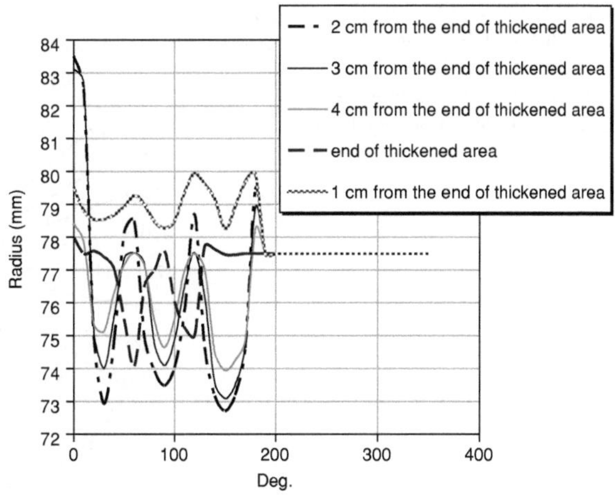

Fig. 14 Deformed shape of different sections on the tested specimen CSP2, measured from the end of thickened length toward the center

define the parameter h as a factor in which the local thickening length (l_i), the thickness of thickened area (t_i) and local thickening step numbers n are effective in this parameter:

$$h = \frac{\sum_{i=1}^{n} l_i.t_i}{\sum_{i=1}^{n} l_i} \tag{1}$$

Note that, l_i and t_i are for each thickening step of half-length of the whole shell, respectively. As is shown in Fig. 18, the considerable point of this issue is that the critical buckling moment varies linearly with the parameter h.

It can be concluded from the experimental and numerical results that, the buckling capacity increases with stepped increasing length of thickening. The reason of this behavior can be due to the negative effect of boundaries, an abrupt changing in the stiffness in the zone of rigid pipe connection to the main specimen (stress concentration). It can also be due to the decrease in the length of the thin specimen which leads to the corresponding decrease in the value of sectional ovalization and, therefore, approaching to the bifurcation type of instability which contain a smaller value of bending buckling load, in comparison to the limit point flattening type of instability. It is noteworthy that the interaction between Brazier's flattening effect and bifurcation buckling has been studied in [17, 18]. The results of [17] show that for short cylinders of ($L/R < 3$), Brazier ovalization effect can be excluded (specimen CSP3).

In the models of this research, as local thickening length and the number of stiffness variation steps from the rigid pipe to the thinnest part increases gradually and step by step, the buckling load increasingly changes. Therefore, for similar cases, with stiffness difference of the parts along the longitudinal direction, gradually thickness variation is recommended as an approach to increase the bending capacity of such structures.

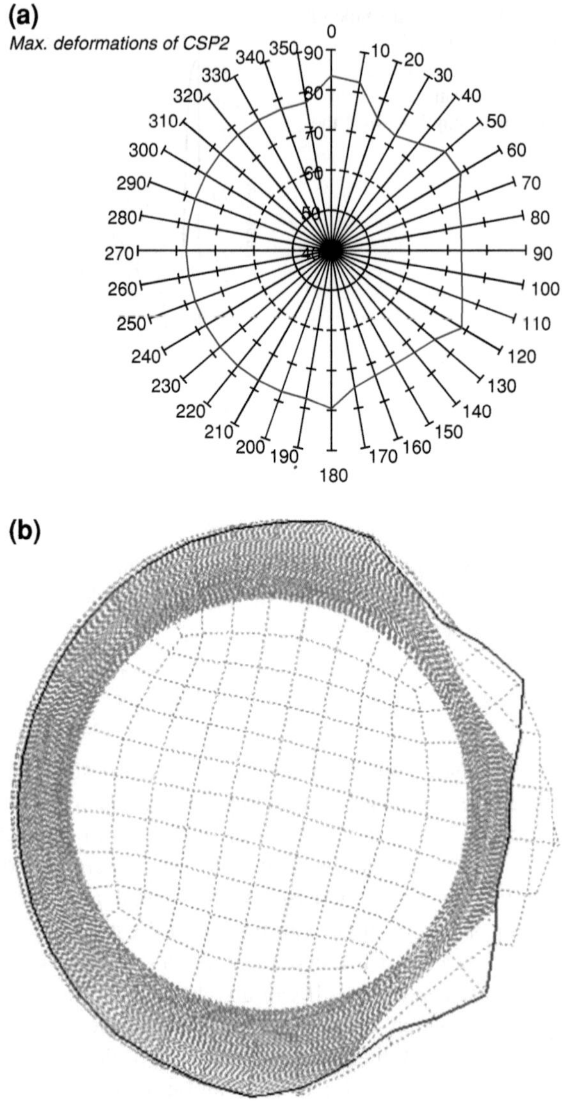

Fig. 15 Polar plot of final geometry of CSP2 for the section with maximum deformations: **a** experimental gauging; **b** corresponding FE model

Generally, a good agreement is observed between the numerical models and experimental results in terms of initial buckling load, buckling waves, and the region of buckling incidence. The numerical and experimental results show that the diamond mode of buckling has been formed in the thinner part of the cylinders and near the intersection of the thinnest part and adjacent local thickened region (Fig. 19). The development of such mode also caused a clear break angle between the two parts of the models. This result correlates with the point which has been

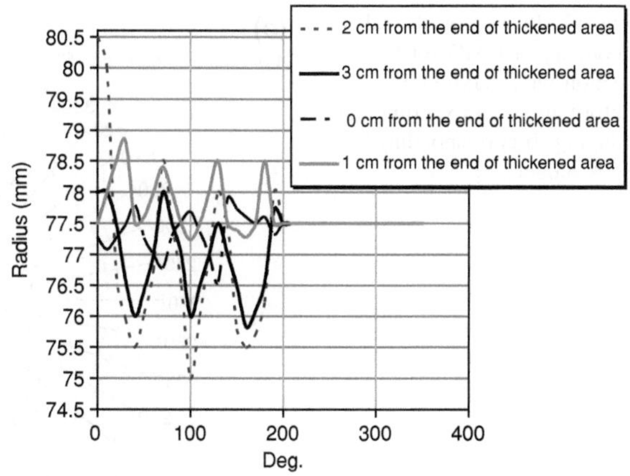

Fig. 16 Deformed shape of different sections on the tested specimen CSP3, measured from the end of thickened length toward the center

pointed out in [28] that thin-walled shells with large radius to thickness ratio possess a low strength for the pure bending condition. However, in some cases, bending conditions cannot be neglected. It turns out that strong severe bending conditions are localized only in a small domain near some discontinuities in loading and geometrical conditions, as well as near supports, etc. As we move away from such a disturbance zone, the bending stresses will decrease rapidly.

All the samples buckled in a diamond pattern in the compression half of the specimens. On the one hand, the local thickening of such models has not a significant effect on the circumferentially wave number of the buckled section. Moreover, it is obvious that the length of buckling development of the waves toward the mid-span of the specimens is rather affected by the steps of thickening. As the local end thickening of the shells become longer, the development of the deformations in the longitudinal direction become shorter, thereby the CSP3 specimen deformation is less developed along the length of the specimen, comparing the two other specimens.

The yield lines were formed in a semicircular pleated shape in the hoop direction of the buckled region. As the shell thickening steps increase, the radius of these semicircles in the longitudinal direction decrease.

In terms of imperfection sensitivity, we can remark that shells under pure bending are less imperfection sensitive than under pure compression, though in both cases, shells must withstand compression stresses. The reason for lower imperfection sensitivity of such structures under pure bending is primarily that in pure bending condition, buckling initiated in a narrow zone of greatest compressive stress, while under pure axial compression any imperfect point on the shell surface can trigger buckling. In this study, the aforementioned narrow region is limited to the joint area of local stiffened section (local thickening in specimens CSP2, CSP3 and rigid pipe in specimen CSP1) and the shell with the lowest

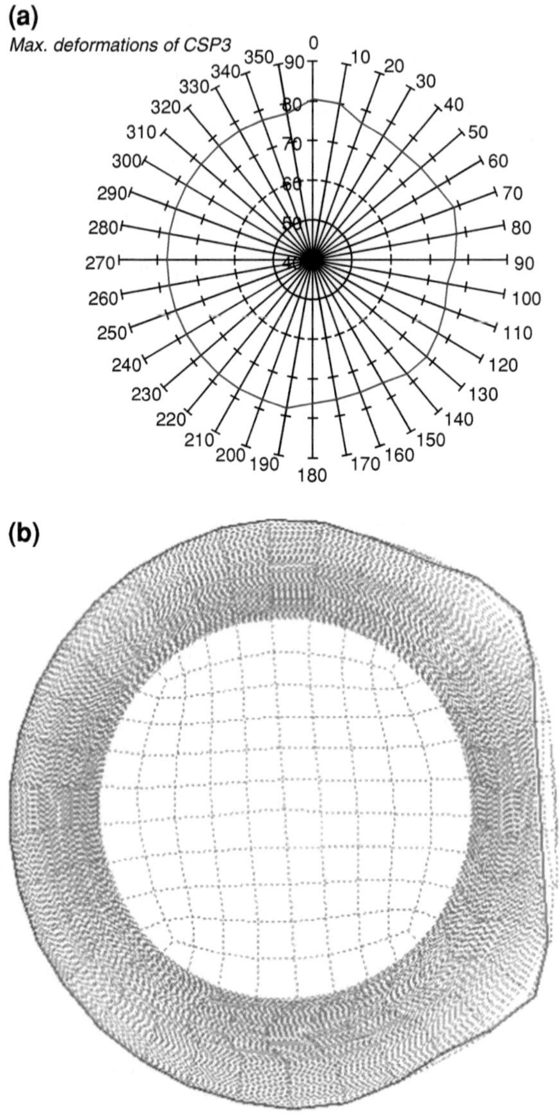

Fig. 17 Polar plot of final geometry of CSP3 for the section with maximum deformations: **a** experimental gauging; **b** corresponding FEA model

thickness. Thus, this region is of major importance which extreme care should be taken to have high quality geometry due to the sensitivity of this area. In our sense, we can emphasize that imperfection sensitivity of such models is less than the other cases of structures under compressive stresses owing to the relatively small target area size of the discontinuity region.

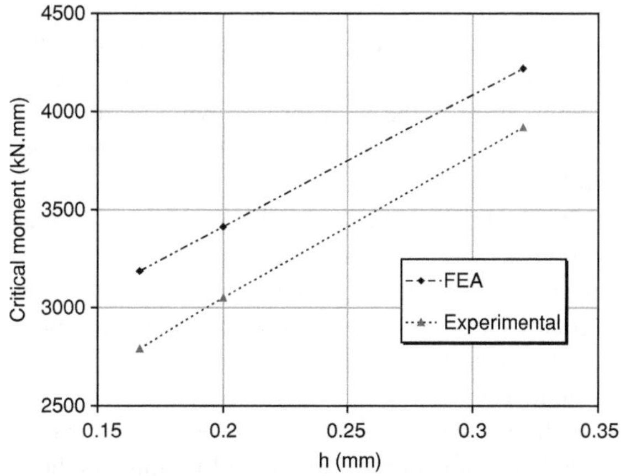

Fig. 18 Buckling moment versus parameter *h*

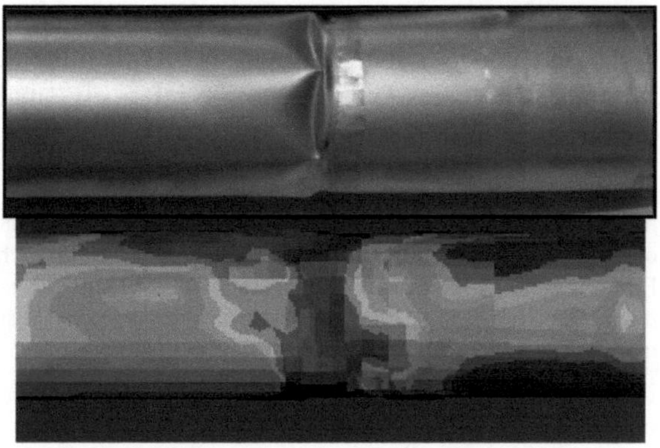

Fig. 19 Final deformed zone of CSP3 specimen and corresponding numerical model

7 Concluding Remarks

The structural buckling behavior of thin-walled cylindrical shells of radius to thickness ratio equal to 155 exposed to pure bending has been investigated with an experimental approach. It has been shown that the general buckling mode of such models can be assimilated to that of pure axial compression. The salient conclusive points to be drawn from the experimental program described herein are as follows:

- In the above specimens, buckling mode formed locally in a diamond pattern circumferentially in the compression half of pipe sections. The diamond mode of buckling formed in the thinner part of the shells and near the intersection of the thinnest part and adjacent local thickened area.
- In all the tests, buckling occurred near the location of the roller support. Consequently, we can consider that the structure buckles near the side in which axial freedom is applied.

- A reasonable agreement is observed between the numerical models and experimental results in terms of initial buckling load, buckling mode, and the region of buckling incidence. However, experimental values are lower in comparison to the FEA results owing to the presence of initial imperfections of the geometry, shortcomings of the apparatus and other human and instrument-related parameters.
- The buckling capacity of the specimens increase as the length and the step number of local thickening increase.
- Under pure bending, some discontinuity zones created by loading and geometrical conditions, supports, etc. are most susceptible to the buckling phenomenon. Thus, these critical regions must be considered in designing such cases.
- In these tests, the yield lines were developed in a semicircular pleated shape in the hoop direction of the bucked region. As the shell thickening steps increase, the radius of these semicircles in the longitudinal direction decrease.
- In all the specimens, imperfection sensitivity is mostly related to the region of joint areas and the points of geometrical discontinuity. Therefore, in these models, imperfection sensitivity is less than the other cases of structures under compressive stresses owing to the relatively small target area of the discontinuity region.

Acknowledgments The authors would like to put across gratitude to the Structures Laboratory of Urmia University specially the technician Mr. Azimzadeh. The authors also want to thank Prof. A. Behravesh for encouraging this work. Lastly but not least the authors would like to thank Dr. Hodjat Shiri from the University of Western Australia to his valuable remarks and many discussions regarding this research.

References

1. Flügge, W.: Die Stabilität der Kreiszylinderschale. Ing. Arch. **3**, 463–506 (1932)
2. Timoshenko, S.: Einige Stabilitatsprobleme der Elastizitatstheorie. Zeitshcrift für Math. Phys. **58**, 337–385 (1910)
3. Lorenz, R.: Achsensymmetrische Verzerrungen in dünnwandigen Hohlzylinder. Z. Vereines Deutscher Ing. **52**, 1706–1713 (1908)
4. Southwell, R.W.: On the general theory of elastic stability. Philos. Trans. R. Soc. Lond. Ser. A **213**, 187–244 (1914)
5. Lundquist, E.E.: Strength tests of thin-walled duralumin cylinders in pure bending. NACA Report 479 (1933)
6. Donnell, L.H.: A new theory for the buckling of thin cylinders under axial compression and bending. Trans. ASME **56**, 795–806 (1934)
7. Brazier, L.G.: On the flexure of thin cylindrical shells and other "thin" sections. Proc. R. Soc. Ser. A **116**, 104–114 (1927)
8. Chwalla, E.: Reine Biegung schlanker, dünnwandiger Rohre mit gerader Achse. Zeitschrift für angewandete Mathematik und Mechanik **13**(1), 48–53 (1933)
9. Singer, J.: Buckling experiments: experimental methode in buckling of thin-walled structures, vol. 2. Wiley, New York (2002)

10. Imperial, F.F.: The criterion of elastic instability of thin duralumin tubes subjected to bending. MS Thesis, Department of Mechanical Engineering, University of California (1932)
11. Seide, P., Weingarten VI.: On the buckling of circular cylindrical shells under pure bending. J. Appl. Mech. ASME **28**, 112–116 (1961)
12. Reddy, B.D.: An experimental study of the plastic buckling of circular cylinders in pure bending. Int. J. Solids Struct. **15**, 669–683 (1979)
13. Kyriakides, S., Ju, G.T.: Bifurcation and localization instabilities in cylindrical shells under bending. I. Experiments. Int. J. Solids Struct. **29**, 1117–1142 (1992)
14. Ju, G.T., Kyriakides, S.: Bifurcation and localization instabilities in cylindrical shells under bending. II. Predictions. Int. J. Solids Struct. **29**, 1143–1171 (1992)
15. Libai, A., Bert, C.W.: A mixed variational principle and its application to the nonlinear bending problem of orthotropic tubes—II: application to nonlinear bending of circular cylindrical tubes. Int. J. Solids Struct. **31**(7), 1019–1033 (1994)
16. Tatting, B.F., Gürdal, Z., Vasiliev, V.V.: The Brazier effect for finite length composite cylinders under bending. Int. J. Solids Struct. **34**, 1419–1440 (1997)
17. Stephens, W.B., Starnes Jr, J.H.: Collapse of long cylindrical shells under combined bending and pressure loads. AIAA J. **13**(1), 20–25 (1975)
18. Fabian, O.: Collapse of cylindrical, elastic tubes under combined bending, pressure and axial loads. Int. J. Solids Struct. **13**, 1257–1270 (1977)
19. Babcock, C.D.: Experiments in shell buckling. In: Fung, Y.C., Sechler, E.E. (eds.) Thin-Shell Structures, Theory; Experiments and Design, pp. 345–369. Prentice-Hall, Englewood Cliffs (1974)
20. Singer, J.: Buckling experiments on shells-a review of recent developments. Solid Mech. Arch. **7**, 213–313 (1982)
21. Thompson, J.M.T.: Making of thin metal shells for model stress analysis. J. Mech. Eng. Sci. **2**(2), 105–108 (1960)
22. Golzan, B.S., Showkati, H.: Buckling of thin-walled conical shells under uniform external pressure. Thin-Walled Struct. **46**, 516–529 (2008)
23. Singer, J., Abramovich, H.: The development of shell imperfection measurement techniques. Thin-Walled Struct. **23**, 379–398 (1995)
24. Showkati, H., Shahandeh, R.: Experiments on the buckling behavior of ring-stiffened pipelines under hydrostatic pressure, ASCE. J. Eng. Mech. **136**, 464–471 (2010)
25. ABAQUS, Documentation and help, Version 6.6
26. Riks, E.: An incremental approach to the solution of snapping and buckling problems. Int. J. Solids Struct. **15**, 529–551 (1979)
27. Bai, Y.: Pipelines and risers, vol. 3. Elsevier Ocean Engineering Book Series, London (2001)
28. Venstel, E., Krauthammer, T.: Thin plates and shells, theory, analysis and application. p. 3. Marcel Dekker, New York (2001)

17. [illegible]
18. [illegible]

Numerical Simulation of Cross Wedge Rolling: Influence of Die Geometry, Process Conditions and Inclusion Content of Two Steels on the Formation of Internal Defects

Mário Luiz Nunes da Silva and Sérgio Tonini Button

Abstract Cross wedge rolling (CWR) is a metal forming process used in the manufacturing of stepped rotational parts. In this process, a cylindrical billet is heated and plastically deformed into an axisymmetric product by the action of wedge-shaped dies moving tangentially relative to one another. Since internal defects in CWR can weaken the integrity of the final product and can ultimately lead to catastrophic failures, it is necessary to investigate the mechanisms of their generation and growth. This defect has its origin in the center of the rolled pieces and its causes are not fully identified yet. Based on the finite element method, numerical simulations of CWR in three dimensions were studied with a commercial software. Numerical simulations can provide useful information helping decision making about die geometry and process conditions and, therefore, is a valuable tool to define ideal process parameters. Aided by this tool, researchers try to understand the role of process variables and die geometric features on the internal defects formation. The purpose of this study was aimed at the variables: rolling speed, relative reduction and forming and stretching angles. Stress, effective plastic strain and damage values at the rolled parts cross section were chosen as analysis criteria in several situations according to a given set of process variables. Earlier practical tests performed by the author showed the significant influence of these variables, and also proved the random behavior of that influence. Due this randomness, further studies were done with two different steels (AISI 1045 and 38MnSiVS5) taking into account their chemical composition and the possibility of the inclusion content has any influence on the crack generation which causes the internal defects.

M. L. N. da Silva (✉) · S. T. Button
School of Mechanical Engineering, University of Campinas, Rua Mendeleiev 200, Barao Geraldo, Campinas, São Paulo 13083-860, Brazil
e-mail: marioluiznunes@gmail.com

S. T. Button
e-mail: sergio1@fem.unicamp.br

A. Öchsner et al. (eds.), *Materials with Complex Behaviour II*,
Advanced Structured Materials 16, DOI: 10.1007/978-3-642-22700-4_8,

Keywords Cross wedge rolling · Hot forming · Finite element method · Internal defects

1 Introduction

In cross wedge rolling (CWR), a forged part or a cylindrical billet is formed by wedge-shaped dies which are fixed in rolling mills or plates. Until this moment, the plastic forming mechanisms of CWR are not totally clear due the complexity of the metal forming and thus experimental research is still dominant. The lack of precise theories lead to experiments with many repetitions in which products defects are hardly controlled, as well as porosity, voids and cracks initiation, and therefore these defects limit the use of the process in large scale [1].

The first objective of this work was to improve the knowledge about the formation of central cavities which are the main defect found in CWR products. Several researches have been dedicated to better know the role of the process variables in the generation of such defects. Equations that correlate geometric variables and try to establish safe process conditions have been defined. Tests performed by the author in a laboratory equipment showed that even in stable conditions, internal defects can occur pointing that other influent mechanisms and variables should be present.

The second objective was to establish the influence of the inclusions present in two commercial steels on the crack generation that origin the internal defects.

2 The CWR Process

CWR is a rotational forming process. As a result of tools movement, a cylindrical workpiece is rolled, obtaining shafts with tapers, steps, shoulders and free elongation in the axial direction. Figure 1 shows the typical design of a CWR tool with its four forming zones: knifing, guiding, stretching and sizing [2].

Due to modification of the billet geometry along the process, plastic forming mechanisms are meaningly different in each of these zones. In the knifing zone, the tool presents a wedge with height starting at zero and that increases to the total reduction of the workpiece diameter. In the guiding zone, tool cross section does not change to obtain a uniform V-shaped groove around the workpiece surface. The stretching zone is the most critical tool section because in it happens the most meaningful plastic deformation of the process. In this zone, the material is stretched and forced to flow to the edge of the workpiece, and therefore the shaft steps can be formed.

In the sizing zone, a small plastic deformation occurs in order to adjust the tolerance and the surface quality of the workpiece [3].

Fig. 1 CWR tool zones

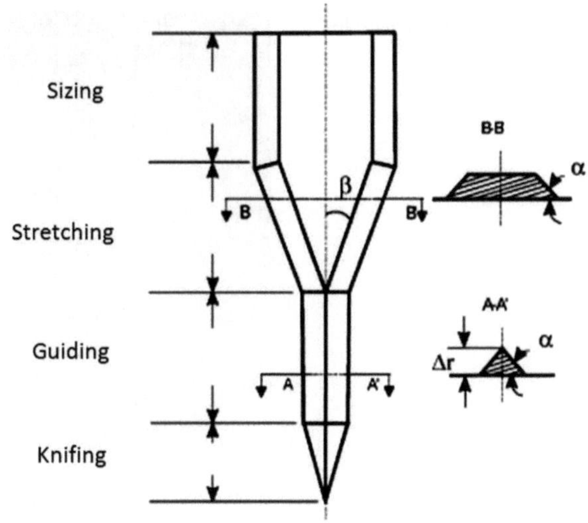

The key parameters of the process related to tool geometry are the forming angle α on the wedge side, the stretching angle β of the wedge, and the relative reduction δ, or the ratio between the initial diameter of the billet and the smaller diameter of the rolled product.

These parameters determine the plastic forming level experimented by the workpiece and they have a relevant role in the probability of internal defects occurrence [4]. The forming angle α, for example, controls the contact area between tools and rolled part. Smaller angles mean less sharp tools and a larger contact area with the part.

The stretching angle β determines the total axial deformation, so bigger stretching angles mean higher plastic deformation and elongation.

The relative reduction δ is a measure of the radial reduction of the rolled part. The bigger the relative reduction, the bigger the radial compression suffered by the part [5].

Besides these geometric variables, the rolling speed (v) was also considered in this research.

3 Numerical Simulation

In the numeric simulations of the CWR process, the commercial software Deform 3D, version 6.1 based on the finite element (FE) method was used. With the purpose of minimize the processing time, the tools were considered as rigid and the friction factor as constant. Figure 2 presents a picture captured from the software showing the tools and its zones and the rolled workpiece. The tools were designed with only three zones, eliminating the guiding zone. The material used in

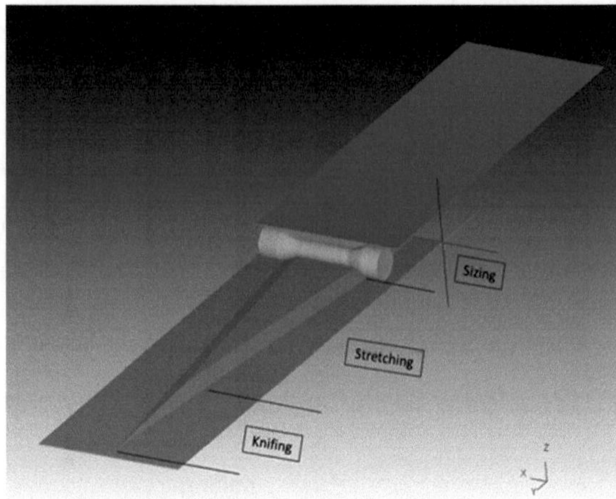

Fig. 2 Workpiece positioned on the lower tool

the numerical models was the steel AISI 1045, available in the FE program library. The pre-heating temperature was 1,100°C for the material and the tools were considered at room temperature, 25°C. The friction factor was assumed as $m = 1.0$.

Workpieces were discretized with 45,000 elements on average. Billet diameters varied between 24 and 34 mm, according to the relative reduction chosen for each simulation. Billet length was assumed as 80 mm.

Considering the process variables, the following conditions were analyzed:

1. forming angle $\alpha = 20°$ with relative reductions $\delta = 1.44$; 1.57 and 1.70;
2. forming angle $\alpha = 25°$ with relative reductions $\delta = 1.41$; 1.51 and 1.61;
3. forming angles $\alpha = 10$ and 30° with relative reduction $\delta = 1.57$;
4. rolling speed (v): 100; 150 and 200 mm/s to the conditions (1) and (2), and 200 mm/s to the condition (3).

Stretching angle β was kept constant in all simulations and equal to 7°. The pre-heating temperature was also kept constant considering preliminary studies which showed that small variations around 1,100°C did not caused meaningful influence on the material plastic behavior.

3.1 Damage

Damage is usually associated with the fracture in a component. Particularly, the damage model developed by Cockcroft–Latham, which is one of the damage models available in the simulation software, has been shown to be a good indicator of ductile fracture under tensile stress in cold forming. Although this research is related to hot rolling, damage results can be used as a reliable option of damage evaluation [6].

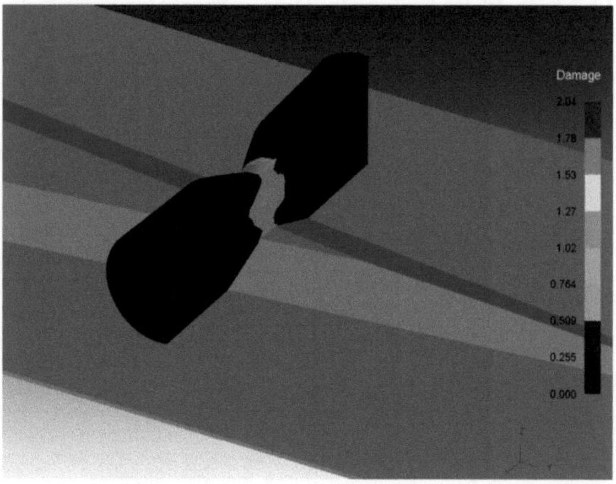

Fig. 3 Damage distribution on the cross section of the rolled part

The numerical analysis was initiated by studying the cross sections of the rolled parts to know the damage distribution along those sections. A typical example of such sections can be seen in Fig. 3, which shows the part right after passing the knifing zone. Preliminary studies done by the author and confirmed by other researchers point the center of the rolled parts as the place where the internal cavities initiate. The damage distribution showed in Fig. 3 confirms that conclusion.

Afterwards, the analysis was chosen by tracking points in some regions of the rolled parts, which means that the values of each variable were studied during the whole rolling process for each chosen point.

The chosen points for the analysis damage variable and the effective stress of the rolled part are shown in Fig. 4. Two points were chosen in the central cross section: $P1$, in the center of the part and $P2$, right under the surface and at the same vertical line of $P1$.

3.1.1 Damage Versus Rolling Speed

Three rolling speeds were chosen: 100; 150 and 200 mm/s. In general, simulation results presented the same trend: the damage levels increase as the rolling speed decreases, as shown in the damage charts with the variables: $\alpha = 20°$ and $\delta = 1.44$ (Fig. 5).

At the points $P1$ and $P2$, damage presented a significant increase from the knifing zone to the initial region of the stretching zone, and then the damage becomes stable. Short baselines in the $P2$ curve show the time when that region loses contact with the tools during the revolutions of the part.

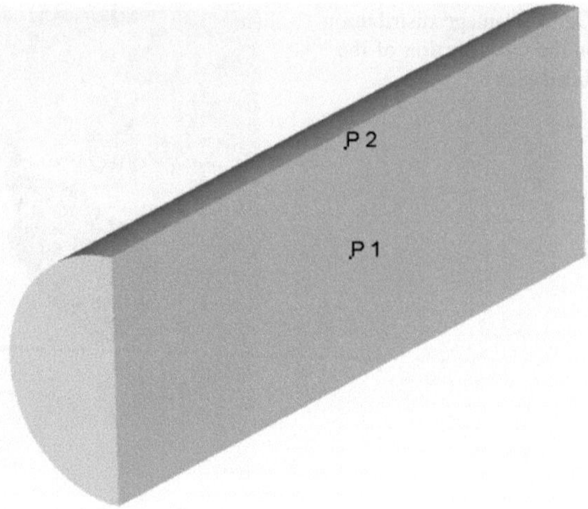

Fig. 4 Position of selected points for tracking

3.1.2 Damage Versus Forming Angle

To analyze the influence of this variable, the process speed and the relative reduction were kept constant. Then, two situations are presented: (a) $\alpha = 10°$; $\delta = 1.53$; v = 200 mm/s and (b) $\alpha = 30°$; $\delta = 1.53$; v = 200 mm/s, and the results are shown in Fig. 6.

These results prove the important influence of the forming angle on the possibility of failure in the rolled part. With smaller angles, damage levels are much higher if compared to the levels of bigger forming angles for the two considered points. With $\alpha = 10°$, damage levels are higher in the center of the part than under the surface during all the process. The damage increases until practically the end of the stretching zone, stabilizing only in the sizing zone. With $\alpha = 30°$, damage levels under the surface are bigger than in the center of the part. For both regions, damage increases in the knifing zone stabilizing along the process.

3.2 Stress

The simulation software adopted the von Mises flow yield criterion and denominates the effective stress as the flow stress σ_0 given by:

$$\sigma_0 = \frac{1}{\sqrt{2}}\sqrt{(\sigma_1 - \sigma_2)^2 + (\sigma_2 - \sigma_3)^2 + (\sigma_3 - \sigma_1)^2} \quad (1)$$

where σ_1, σ_2 and σ_3 are the principal stresses. According to the von Mises criterion, metal flow does not depend on a particular normal stress or shear stress, although depends on a function of the three main shear stresses values. Due to the

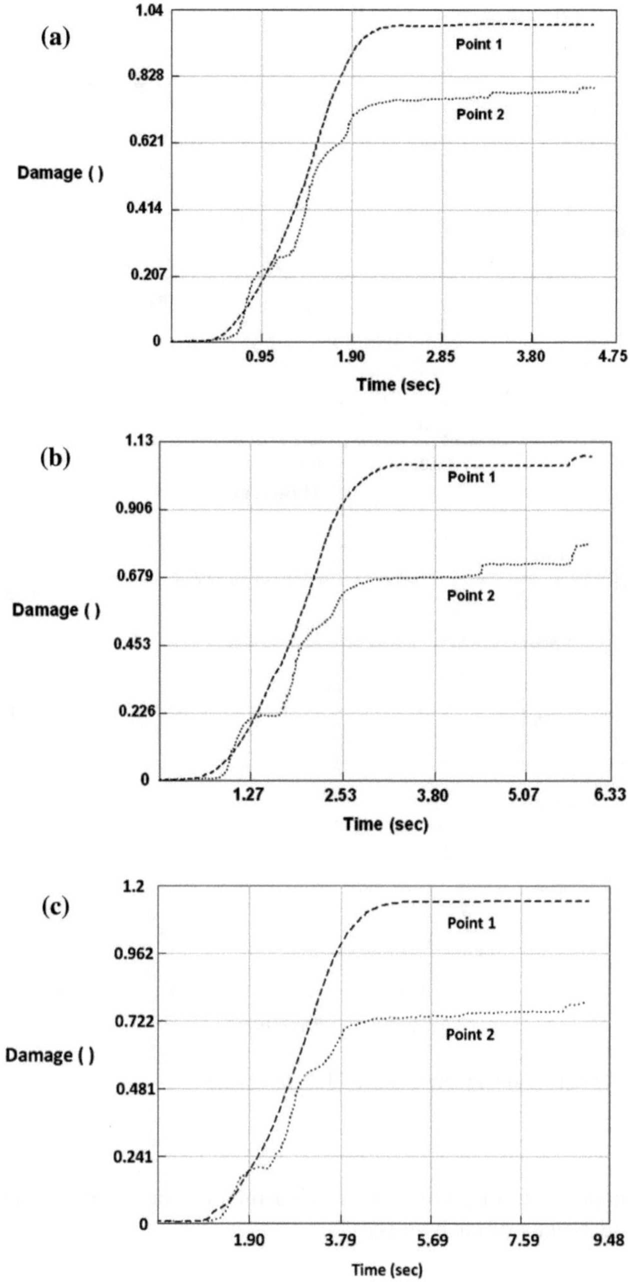

Fig. 5 Damage versus rolling speed ($\alpha = 20°$ and $\delta = 1.44$): **a** 200 mm/s, **b** 150 mm/s, **c** 100 mm/s

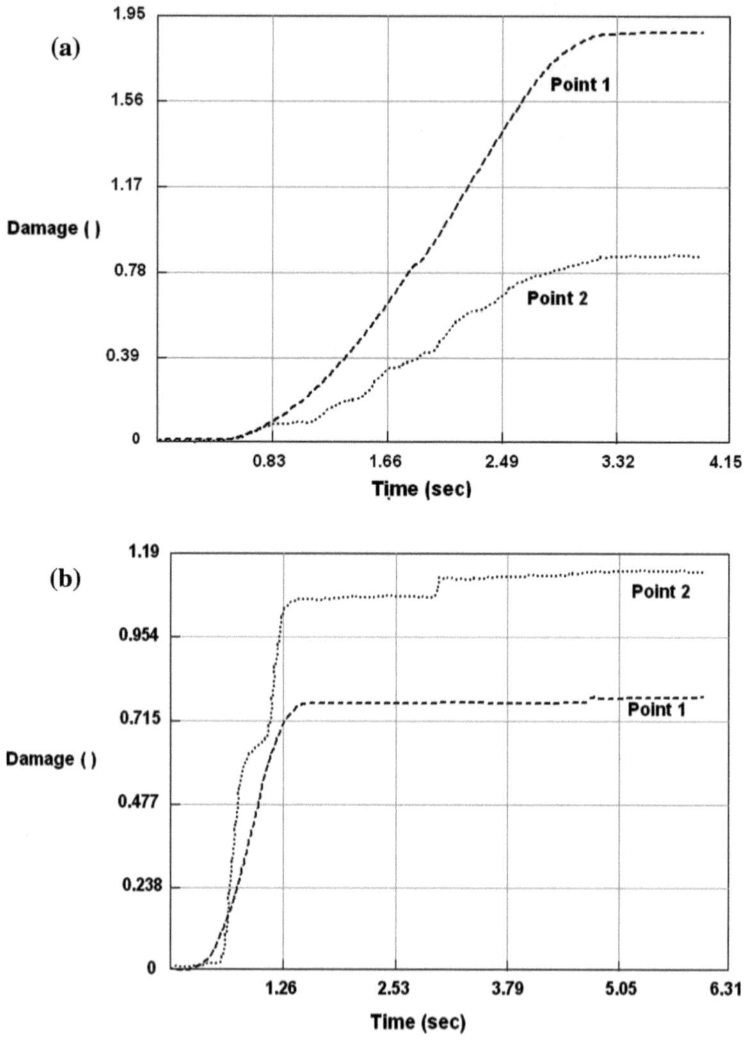

Fig. 6 Damage versus forming angle: **a** $\alpha = 10°$, **b** $\alpha = 30°$

involving quadratic terms, von Mises criterion presents a result that does not depend on the signals of each stress [7].

3.2.1 Effective Stress Versus Relative Reduction

The charts in Fig. 7 show the results of the simulations with $\alpha = 20°$; $v = 200$ mm/s and $\delta = 1.41$; 1.51 and 1.61. As the relative reduction increases, the stress in the center of the part, point $P1$, increases until the beginning of the stretching zone. Till the end of the process, the stress behavior is unstable, however, the trend of higher level stresses with increasing values of reduction is maintained.

The analysis described above can also be done to the region near the surface of the part, point $P2$, taking in account that in the first stages, the stress values present more pronounced peaks and valleys.

Considering the effective stress, it can be observed that the probability for occurrence of internal defects increases with higher relative reductions.

3.2.2 Effective Stress Versus Forming Angle

The influence of the forming angle is shown in the charts of Fig. 8. With $\alpha = 10°$ and in the center of the part, the effective stress was practically constant on the first two zones of the tools, with values between 80 and 74 MPa. In the region under the surface, there was a significant variation with peaks of up to 135 MPa.

With $\alpha = 30°$ the effective stress was stable in the center of the part on the knifing zone and at the beginning of the stretching zone. In the remainder of the stretching zone, there was a marked decrease to stresses around 40 MPa. On the region under the surface, there were peaks of 150 MPa in the knifing zone with a decrease up to 40 MPa in the stretching zone.

Smaller forming angles cause higher effective stress during all the forming process, both in the center and near the surface of the part.

3.3 *Stress Component*

Whereas in the CWR process, material flows mainly in the axial direction of the workpiece, the normal stress component on that direction was also studied. According to the coordinate system presented in Fig. 3, it is the component in "x" direction (from now denominated "x" stress).

3.3.1 Analysis of "x" Stress Versus Rolling Speed

With decreasing rolling speeds, an increase in the "x" stress was noted. The results can be exemplified with $\alpha = 20°$ and $\delta = 1.44$ (Fig. 9). Except for the initial instants of the process when alternating peaks of tensile and compressive stresses can be observed in the center of the part, the stress is tensile, up to about half the stretching zone causing the longitudinal flow of the material. From this point, tensile and compressive stresses alternate again, predominating the compressive stress caused by the tools.

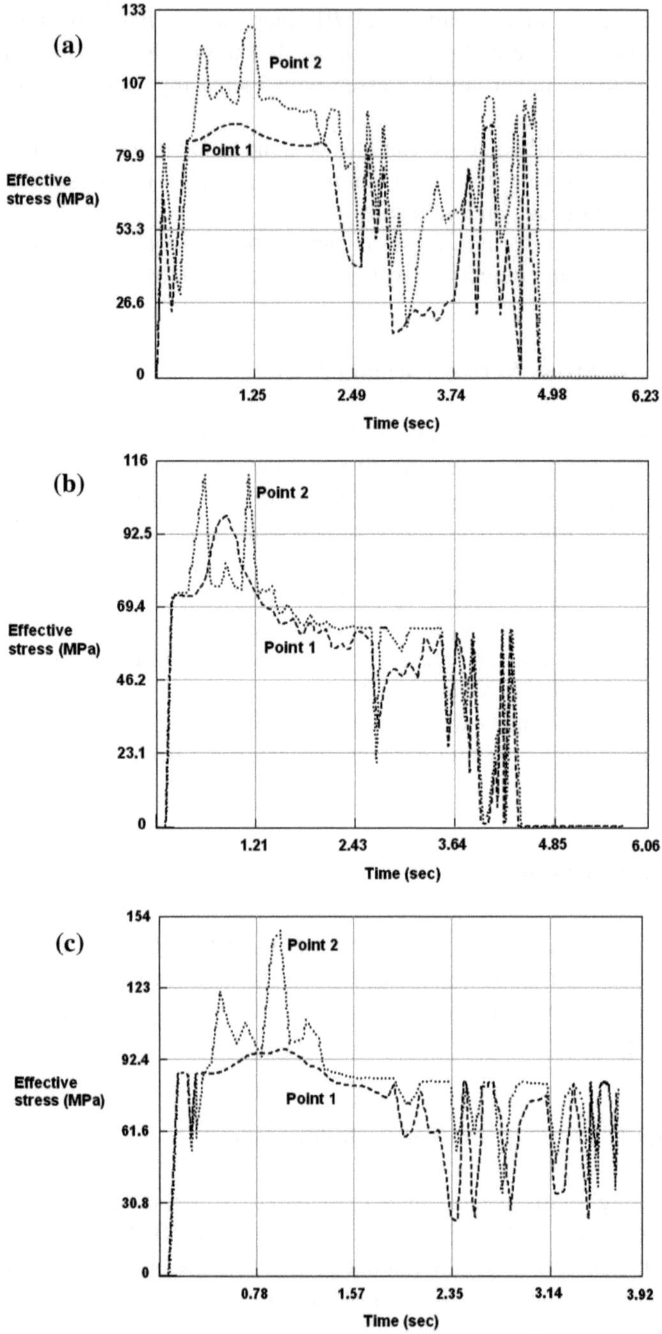

Fig. 7 Effective stress versus relative reduction ($\alpha = 20°$ and $v = 200$ mm/s): **a** $\delta = 1.41$, **b** $\delta = 1.51$, **c** $\delta = 1.61$

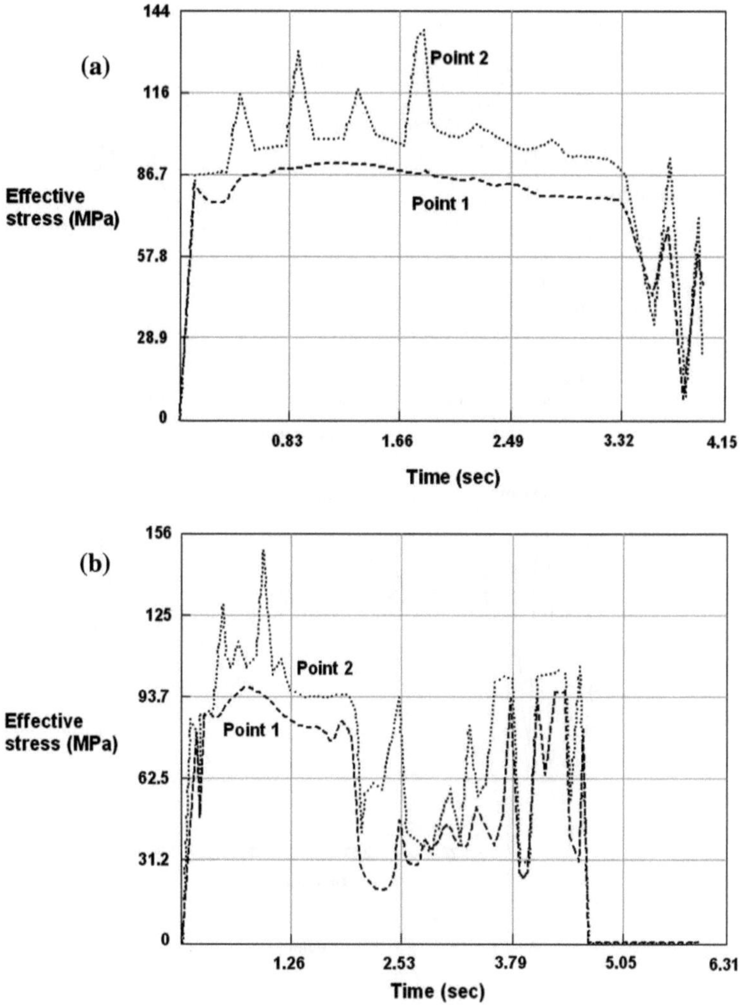

Fig. 8 Effective stress versus forming angle: **a** $\alpha = 10°$; $\delta = 1.53$; $v = 200$ mm/s, **b** $\alpha = 30°$; $\delta = 1.53$; $v = 200$ mm/s

Near the surface of the part, the whole process presents alternating tensile and compressive stresses when that region is in contact or not with the tools. That particular behavior among the regions of the rolled part explains the difficulty of knowing with detail the forming and failure mechanisms of the CWR process.

3.3.2 Analysis of "x" Stress Versus Relative Reduction

The analysis of "x" stress also indicates that increasing relative reductions can be harmful to the CWR process. In the center of the part, stress remains tensile and

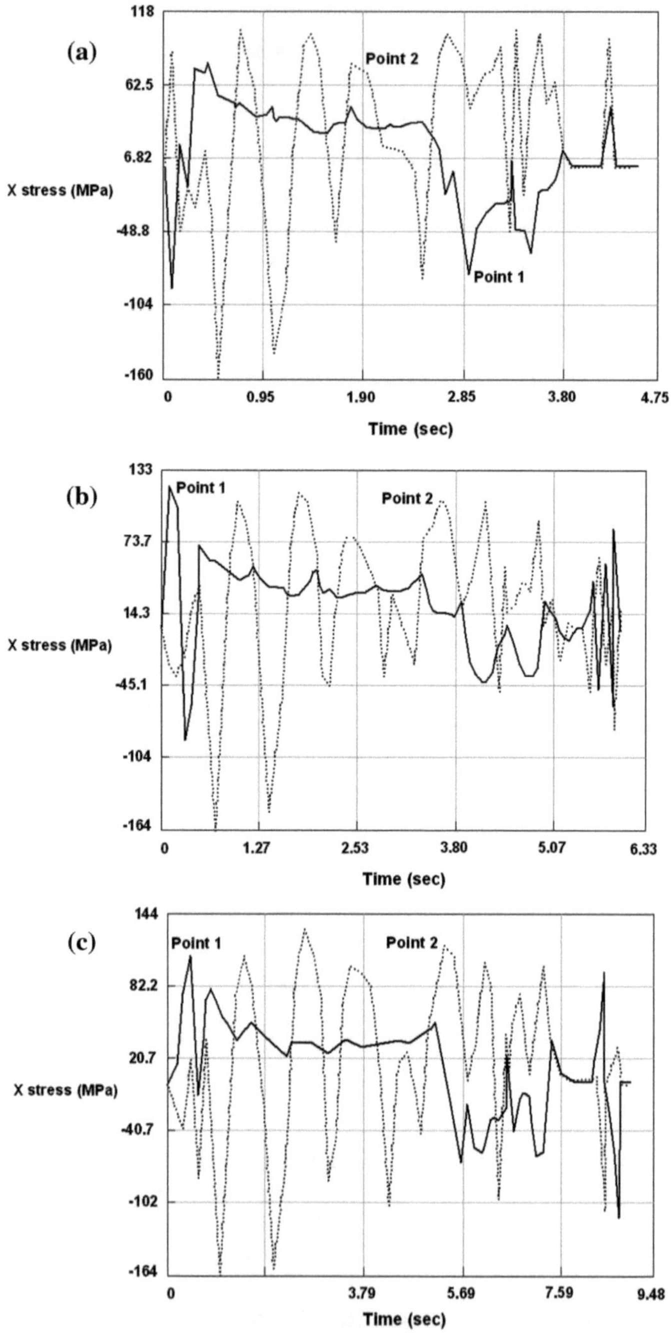

Fig. 9 "*X*" stress versus rolling speed ($\alpha = 20°$ and $\delta = 1.44$)**:** **a** 200 mm/s, **b** 150 mm/s, **c** 100 mm/s

increases as the reduction increases until the beginning of the stretching zone. Under the surface of the part, tensile and compressive stresses alternated practically during the whole process. Those results can be found in the charts of Fig. 10.

3.3.3 Analysis of "x" Stress Versus Forming Angle

Figure 11 shows "x" stress charts for two conditions: $\alpha = 10$ and 30°. In the center of the part, with $\alpha = 10^\circ$, tensile stress is present practically during the whole rolling process, except for the initial instants. With $\alpha = 30^\circ$, stress behavior is more alternated, varying between tensile and compressive stresses, with tensile stresses lower than with $\alpha = 10^\circ$. Therefore, it can be concluded that $\alpha = 10^\circ$ is certainly more favorable for the generation of voids and cracks and subsequent propagation.

However, under the surface, the "x" stress behavior follows the pattern of previous analysis with interchange between tensile and compressive stresses during the entire process.

It can be stated that with conditions favorable to the generation of defects, mainly in those near to the limits of safe conditions, i.e. close to points of overall process failure, an option to consider is to increase the rolling speed. Speed increase is related to lower damage and tensile stress in the axial direction, which contributes to avoid crack generation or minimize the propagation of generated cracks.

An increase of relative reductions is harmful to the process and favors the occurrence of internal defects, since they lead to higher effective stress and axial tensile stress in the center of rolled parts.

Tool designs with small forming angle must be avoided. The smaller the forming angles, the higher are the damage levels, effective stresses and axial tensile stresses in the center of the part.

4 Experimental Tests With AISI 1045 Steel

Several tests were performed with workpieces made with commercial steel AISI 1045 bars. That steel has the typical chemical analysis described in Table 1.

All workpieces were 80 mm in length and with diameters equal to 25.0; 27.5 and 29.5 mm. Other process and geometric variables of the tests are shown in Table 2.

The workpieces were cut from the steel bars withdrawing small slices before and after each workpiece for micrographic analysis of the inclusions (Fig. 12).

This analysis showed the variability of the quantity and size of inclusions along the bars and even though it was difficult to prevent the presence of those inclusions in the center of the workpieces, it was possible to observe the regions of higher and lower concentration of inclusions.

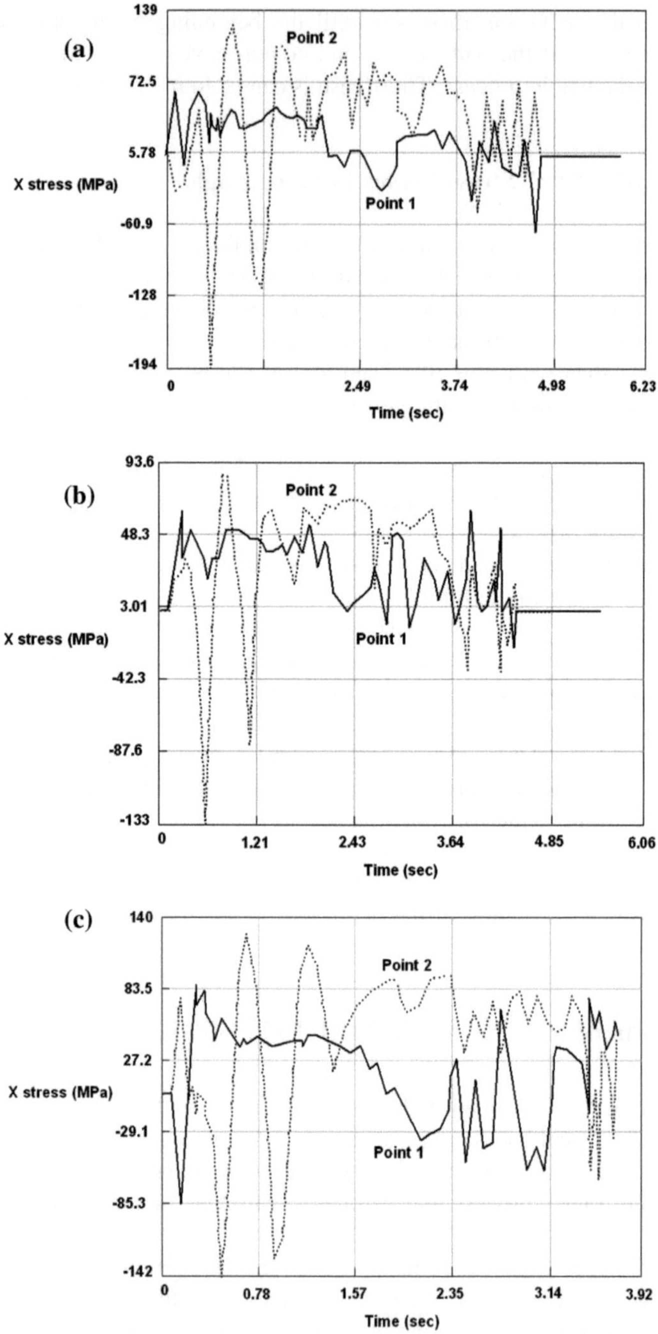

Fig. 10 "*X*" stress versus relative reduction ($\alpha = 25°$ and $v = 200$ mm/s): **a** $\delta = 1.41$, **b** $\delta = 1.51$, **c** $\delta = 1.61$

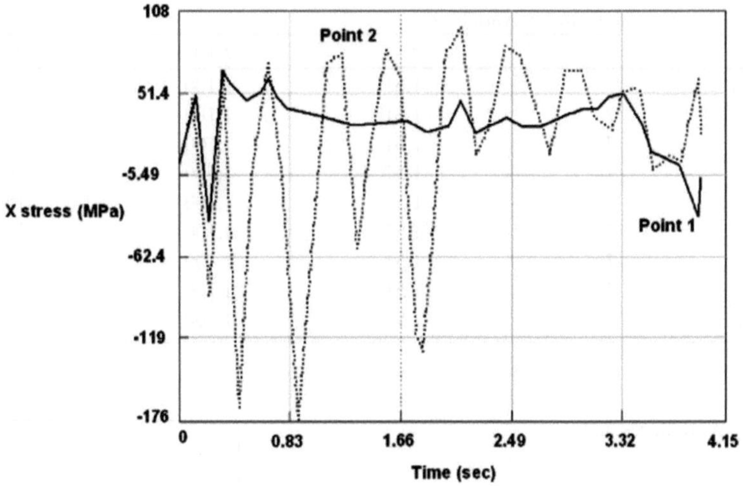

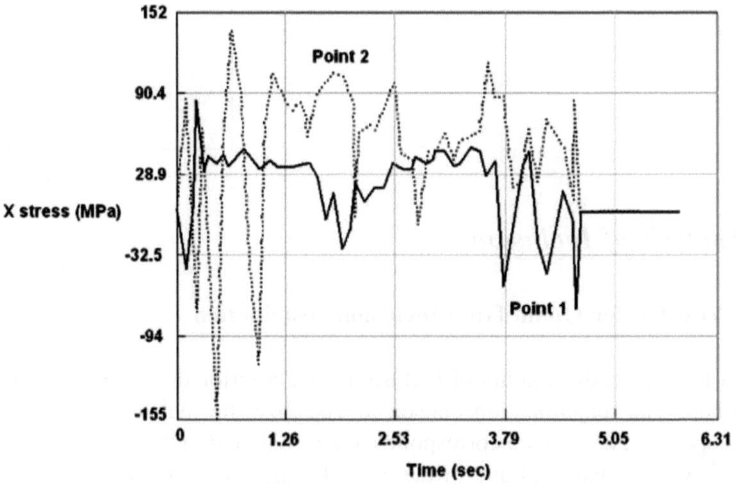

Fig. 11 "X" stress versus forming angle: **a** $\alpha = 10°$; $\delta = 1.53$; $v = 200$ mm/s, **b** $\alpha = 30°$; $\delta = 1.53$; $v = 200$ mm/s

Table 1 Chemical analysis of steel AISI 1045 (% in weight)

C	Mn	P (max)	S (max)
0.43/0.50	0.60/0.90	0.030	0.050

After the tests under the conditions shown in Table 2, the rolled parts were sectioned transversally in relation to the main axis in the central region for analysis of present defects.

Table 2 Process and geometric variables

α (°)	β (°)	δ	T (°C)	v (mm/s)
20	7	1.61	1,100	100
20	7	1.61	1,100	150
20	7	1.61	1,100	200
20	7	1.57	1,100	100
20	7	1.57	1,100	150
20	7	1.57	1,100	200
25	7	1.51	1,100	100
25	7	1.51	1,100	150
25	7	1.51	1,100	200

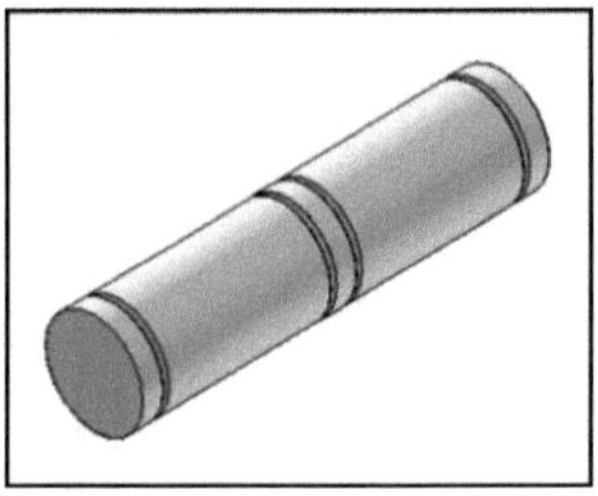

Fig. 12 Workpieces and slices

4.1 Analysis of Inclusions

4.1.1 Criterion for Quantifying Inclusion Distribution

In order to analyze the amount of inclusions, the criterion described as follow was established. The magnification chosen to visualize the inclusions at the optical microscope was 50×, which corresponds to a visual field in the workpiece equal to 7.4 mm^2. Within that field and in the central region of each slice, the amount of conclusions was then determined. A color code was also established, as can be seen in Fig. 13.

4.1.2 Analysis of Inclusions Distribution

The micrographic analysis of inclusions was performed with light optical microscopy and scanning electronic microscopy (SEM). A typical micrograph of a slice which had more than 100 inclusions can be seen in Fig. 14. The distribution of inclusions is homogeneous and their size varies.

Figures 15 and 16 present the micrographs obtained by SEM, as well as the chosen points for chemical analysis by energy dispersive spectrometer (EDS). The chemical composition of each point is presented in Table 3.

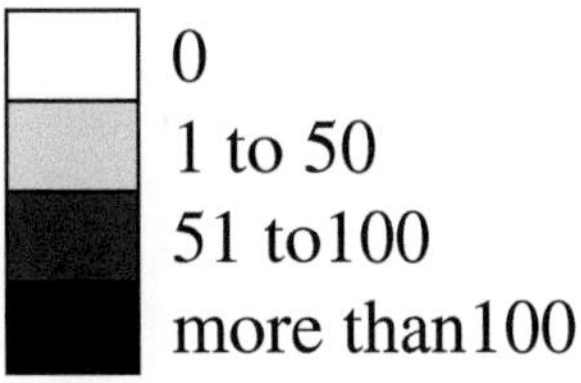

Fig. 13 Color code for the amount of inclusions

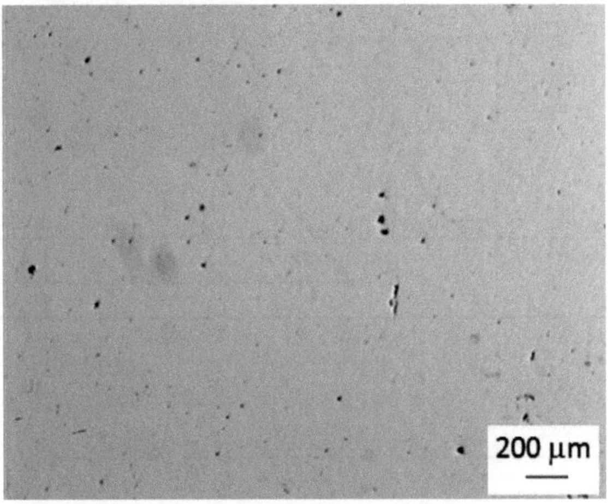

Fig. 14 Inclusions in steel AISI 1045

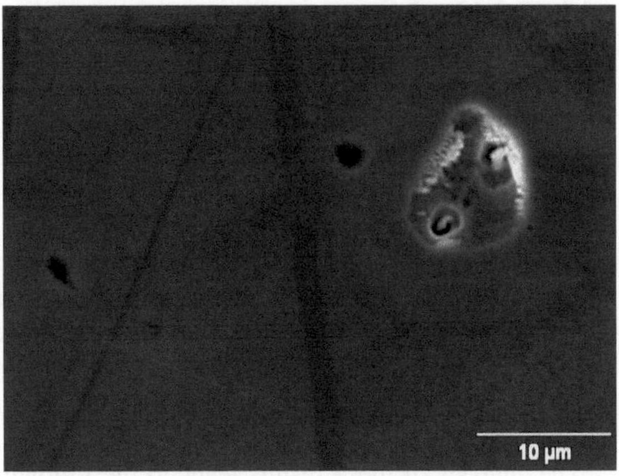

Fig. 15 Inclusions in steel AISI 1045 (SEM)

The chemical analysis of inclusions indicates high sulfur and manganese contents, which means they are manganese sulfides that are quite common for that commercial steel.

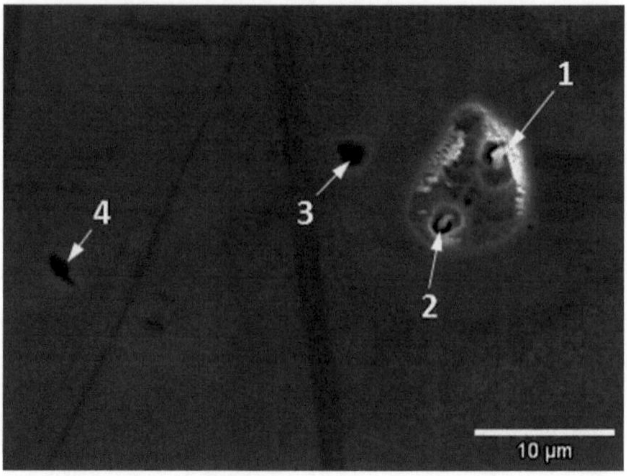

Fig. 16 Chosen points for chemical analysis of inclusions in steel AISI 1045

Table 3 Chemical composition of inclusions in steel AISI 1045

Point	% in weight			
	Si	S	Mn	Fe
1	0.38	13.71	19.43	66.48
2	0.17	9.32	14.86	75.66
3	0.21	8.89	15.54	75.36
4	0.17	14.27	27.41	58.15

4.2 Internal Defects

4.2.1 Classification

Internal defects found within rolled parts were classified according to their size. Those defects visible with the naked eye, with no magnification, were classified as "big" and those which needed a minimum magnification of 40× to be identified were classified as "small". A color code was established for the defects, similarly to the inclusions, as shown in Fig. 17.

Experimental results with the different conditions are shown in Fig. 18. At least two tests were performed for each condition. The largest rectangles represent the rolled parts and the smallest ones represent their respective slices. The conditions were grouped into columns (1, 2 and 3) representing the same forming angle and the same relative reduction and columns (A, B and C) representing the same rolling speed.

The analysis of these results leads to the following conclusions:

- The inclusion distribution along the bars was totally random. A correlation between inclusion content and internal defects generation could not be established. There were slices with high inclusion content adjacent to rolled parts with no defects as well as slices with no adjacent defects to rolled parts with big and small defects.

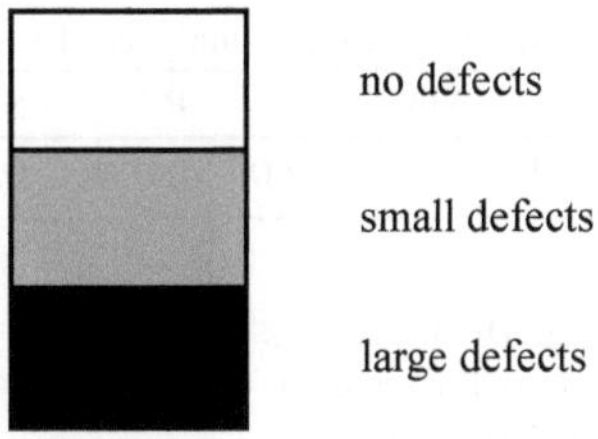

Fig. 17 Color code for the defect size classification

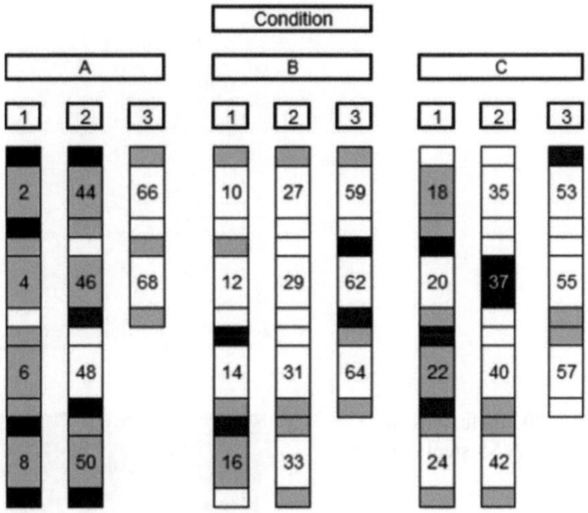

Fig. 18 Results of rolled parts with their respective slices

- For that steel, with that inclusion distribution, the effect of process and geometric variables is predominant over the effect of inclusions. The conditions with smaller forming angle and bigger relative reductions were more favorable for defect generation. This conclusion confirms results previously obtained by the author and by Idoyaga et al. [8].

5 Experimental Tests With Microalloyed Steel 38MnSiVS5

The microalloyed steel 38MnSiVS5, whose chemical composition is shown in Table 4, was used for another set of tests. The choice of that steel was due to the fact that microalloyed steels are increasingly used in the automotive industry, because they represent a great saving of time and energy since they do not require subsequent heat treatment reaching good mechanical properties when cooled from hot working temperatures.

A second reason that determined the use of that steel was the idea that it is a nobler material, produced in smaller scale with higher unit cost per kilo and thus, the inclusion control would be more effective, despite the results presented below

Table 4 Chemical analysis of microalloyed steel 38MnSiVS5 (% in weight)

C	Mn	Si	P	S	Cr	Ni	Mo	V	Cu	Al	N
0.37	1.41	0.60	0.014	0.055	0.11	0.10	0.02	0.09	0.04	0.011	0.0157

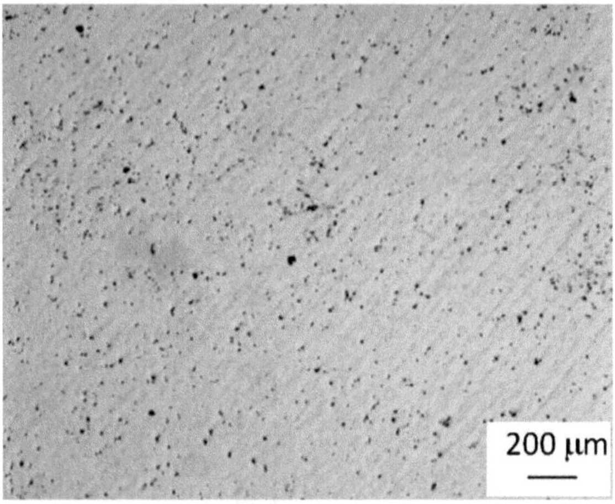

Fig. 19 Inclusions in microalloyed steel 38MnSiVS5

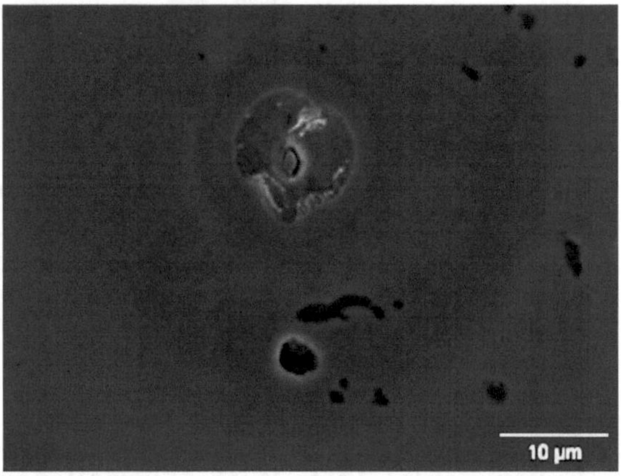

Fig. 20 Inclusions in 39MnSiVS5 steel (SEM)

show exactly the opposite. The analysis of inclusions made similarly to Steel AISI 1045 is presented next. The tests conditions were exactly the same as the Steel AISI 1045 ones.

5.1 Analysis of Inclusions

The same criterion for classification of the steel AISI 1045 was adopted. Micrographic analysis with light optical microscopy showed that all slices presented

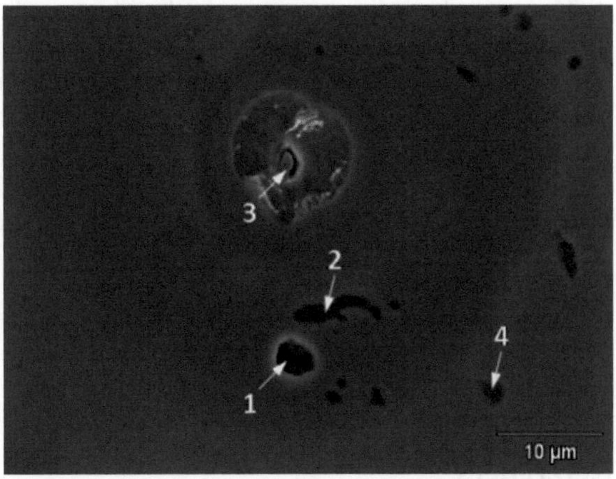

Fig. 21 Chosen points for chemical analysis of inclusions in 39MnSiVS5

Table 5 Chemical composition of the inclusions in 39MnSiVS5 steel

Point	% in weight					
	Al	Ti	Si	S	Mn	Fe
1	1.39	0.45	0.63	7.22	33.45	56.86
2	–	–	–	24.85	48.95	26.20
3	–	–	–	18.38	33.36	48.26
4	–	–	0.23	18.61	36.02	45.14

high inclusion content, significantly higher than steel AISI 1045 slices. A typical example of those micrographs is shown in Fig. 19.

Analyses of those steel inclusions are presented in Figs. 20 and 21. The high percentage of the elements shown in Table 5 indicates that these inclusions are manganese sulfides, similar to the steel AISI 1045 ones.

5.2 Internal Defects

All the rolled parts were sectioned and absolutely all had large internal defects. For this steel and with this inclusions content, the inclusion influence on the defects generation prevailed over the process and geometric variables influence.

6 Conclusion

Considering the several tests performed, results were not always as expected. Some tests produced parts without defects, while others under the same conditions produced parts with big and small defects.

That may be related to process limiting conditions or to the presence of inclusions in both steels which favors the crack formation and consequently the internal defects. Up to a certain inclusion content in the workpieces, as observed in steel AISI 1045, the geometric and process variables have a predominant influence on internal defects generation.

For materials like the microalloyed steel studied in this work, which contains high inclusion content, those inclusions are the most important factor to the internal crack formation.

Acknowledgments The authors wish to thank the financial support given by FAPESP (Foundation for Researchers of São Paulo State).

References

1. Wang, M., Li, X., Du, F.: Analysis of metal forming in two-roll cross wedge rolling process using finite element method. Int. J. Iron Steel Res. **16**(1), 38–43 (2009)
2. Silva, M.L.N., Regone, W., Button, S.T.: Microstructure and mechanical properties of microalloyed steel forgings manufactured from cross-wedge-rolled preforms. Scripta Materialia **54**(2), 213–217 (2006)
3. Dong, Y., Tagavi, K., Lovell, M.: Analysis of interfacial slip in cross-wedge rolling: a numerical and phenomenological investigation. J. Mater. Process. Technol. **97**, 44–53 (2000)
4. Qiang, L., Michael, R.L., William, S.: Experimental investigation of internal defects in the cross wedge rolling process. J. Mater. Process. Technol. **125–126**, 248–257 (2002)
5. Pater, Z.: Finite element analysis of cross wedge rolling. J. Mater. Process. Technol. **173**(2), 201–208 (2006)
6. Deform 3D version 3.1: Users' manual. Edited by Scientific Forming Technologies Corporation (2007)
7. Dieter, G.E.: Mechanical metallurgy. Guanabara, Rio de Janeiro (1981)
8. Idoyaga, Z., et al: Influence of tramp elements (P, Cu, S, Sn) on the mannesmann effect in the transversal hot rolling of engineering steels (Mannestramp). European Commission (2008)

Nonlinear Creep Model for Concrete in Analysis of Plates and Shells

Jure Radnić, Domagoj Matešan and Marija Smilović

Abstract A numerical model for analysis of reinforced and prestressed concrete plates and shells including creep, shrinkage and aging of concrete, already developed by the authors, has been updated with a nonlinear creep model for concrete. The model can be applied for all levels of concrete stresses, while its use for ultimate stress levels is still not fully tested. The presented nonlinear concrete creep model is simple, based on the well known linear model of concrete creep, and intended for simulation of practical concrete structures. For the verification of the presented model, an experimentally tested square concrete plate and cylindrical prestressed concrete shell were analysed numerically. The results of experimental tests at high stress levels and numerical results show good agreement.

Keywords Plate · Shell · Nonlinear concrete creep · Numerical model

1 Indroduction

As verified by experimental tests, a linear functional relationship between the creep strain of concrete in time and instantaneous elastic strain of concrete is valid only for low levels of concrete stresses. According to Bazant et al. [1–3], when the

J. Radnić (✉) · D. Matešan · M. Smilović
Faculty of Civil Engineering and Architecture, Matice Hrvatske 15,
21000 Split, Croatia
e-mail: jure.radnic@gradst.hr

D. Matešan
e-mail: domagoj.matesan@gradst.hr

M. Smilović
e-mail: marija.smilovic@gradst.hr

A. Öchsner et al. (eds.), *Materials with Complex Behaviour II*,
Advanced Structured Materials 16, DOI: 10.1007/978-3-642-22700-4_9,

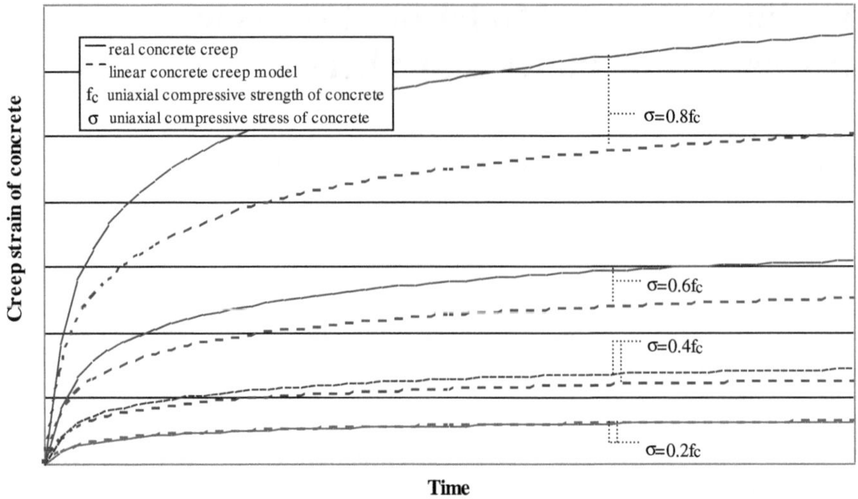

Fig. 1 Effect of concrete stress levels on accuracy of a linear concrete creep model

ratio between concrete stresses under a long-term static load and the average compressive strength of concrete is greater than 0.4, the aforementioned linear relationship between the instantaneous and time-dependent strains is no longer valid and there is a progressive non-linear increase of concrete creep strain (Fig. 1). When concrete stresses due to a long-term static load exceeds 80% of the average compressive strength of concrete, creep will cause failure of the concrete. There were numerous attempts to analytically describe the actual creep of concrete. Some nonlinear creep models for concrete, based on different rheological models and/or models of uniaxial experimental tests, can be found in [4–10].

Here, a nonlinear concrete creep model is presented. An empirical expression for nonlinear concrete creep is proposed that was used for numerical modelling of plates and shells under long-term load by finite element analysis. The model was verified in the numerical simulation of carried out experimental tests of concrete plates and prestressed concrete shells.

2 Linear Concrete Creep Model

A developed linear concrete creep model included in the numerical model for the analysis of concrete plates and shells under a long-term load [11, 12] will be described in short below.

The uniaxial creep strain has been calculated using the Glanville and Dischinger method [13], based on the assumption that the rate of creep is a function of the current uniaxial concrete stress and the time t elapsed after the loading, namely

$$\frac{d\varepsilon_t^c}{dt} = f(\sigma, t). \tag{1}$$

If time is divided into discrete time intervals Δt, when $t_n = t$ and $t_{n+1} = t + \Delta t$, an incremental version of Eq. (1) is

$$\Delta\varepsilon_{t_{n+1}}^c = \varepsilon_{t_{n+1}}^m \, \Delta\Phi_{t_n,t_{n+1}} = \varepsilon_{t_{n+1}}^m \left[\Phi_{t_{n+1}} - \Phi_{t_n}\right] \tag{2}$$

where $\Delta\varepsilon_{t_{n+1}}^c$ is the increase in creep strain between times t_n and t_{n+1}, ε_{n+1}^m is an instantaneous mechanical strain of concrete at t_{n+1}(it can be a nonlinear function of stress σ), $\Delta\Phi_{t_n,t_{n+1}}$ is the increase of creep coefficient between t_n and t_{n+1}, $\Phi_{t_{n+1}}$ is the creep coefficient at t_{n+1} and $\Phi_{t_n}$is the creep coefficient at t_n. The increment of the creep strain $\Delta\varepsilon_{t_{n+1}}^c$ is calculated based on the conditions at the beginning of the next time increment t_{n+1}. This method, in general, is very simple and describes the strain history very well even for sudden and irregular stress changes. Since the increment of the creep strain is based only on the current instantaneous strain (or stress) and time-dependent value, this method is very attractive in terms of calculations.

Creep coefficients values given in EUROCODE 2 [14] were used. Thus, Eq. (2) can be written in the following form

$$\Delta\varepsilon_{t_{n+1}}^c = \varepsilon_{t_{n+1}}^m \left[\Phi_{t_{n+1},t_0} - \Phi_{t_n,t_0}\right] \tag{3}$$

where

$$\Phi_{t_{n+1},t_0} = \Phi_0 \beta_{t_{n+1},t_0}^c \tag{4}$$

$$\Phi_{t_n,t_0} = \Phi_0 \beta_{t_n,t_0}^c. \tag{5}$$

If Eqs. 4 and 5 are inserted in (3), then

$$\Delta\varepsilon_{t_{n+1}}^c = \varepsilon_{t_{n+1}}^m \, \Phi_0 \left[\beta_{t_{n+1},t_0}^c - \beta_{t_n,t_0}^c\right]. \tag{6}$$

In the above equations, Φ_0 denotes the basic creep of concrete, while β_{t_{n+1},t_0}^c and β_{t_n,t_0}^c are the coefficients describing the creep development in time under loading.

Basic creep of concrete Φ_0 can be determined as

$$\Phi_0 = \Phi_{RH} \, \beta_{f_{cm}} \, \beta_{t_0} \tag{7}$$

where Φ_{RH} denotes the coefficient of relative humidity effect, β_{fcm} is the coefficient of concrete strength effect and β_{t_0} is the coefficient of concrete age effect on the basic creep at the beginning of loading. Then:

$$\Phi_{RH} = 1 + (1 - RH/100) \Big/ \left(0.1\, h_0^{1/3}\right) \tag{8}$$

$$\beta_{f_{cm}} = 16.8/f_{cm}^{0.5} \tag{9}$$

$$\beta_{t_0} = 1/(0.1 + t_0^{0.2}) \tag{10}$$

$$h_0 = 2A_C/u \tag{11}$$

In the above given equations RH is the relative humidity (%), h_0 is the nominal size of the cross-section (mm), f_{cm} is the mean compressive strength of concrete at the age of 28 days (N/mm^2), t_0 is the age of concrete at the time of first loading (days), A_c is the cross-sectional area (mm^2) and u is the perimeter of the part which is exposed to drying (mm).

Coefficients that describe the creep development in time can be calculated as

$$\beta^c_{t_{n+1},t_0} = [(t_{n+1} - t_0)/(\beta_H + t_{n+1} - t_0)]^{0.3} \tag{12}$$

$$\beta^c_{t_n,t_0} = [(t_n - t_0)/(\beta_H + t_n - t_0)]^{0.3}. \tag{13}$$

In the above given equations t_n and t_{n+1} are the time limits of the observed time increment (days), while β_H is the coefficient of the effect of the relative humidity RH (%) and the nominal size of the cross-section h_0(mm), given by

$$\beta_H = 1.5\left[1 + (0.012RH)^{18}\right] h_0 + 250 \leq 1500 \tag{14}$$

The effect of cement type on concrete creep can be taken into account if the concrete age at the time of first loading t_0 is adjusted as

$$t_0 = t_{0,T}\left\{9/\left[2 + \left(t_{0,T}\right)^{1.2}\right] + 1\right\}^{\alpha} \geq 0.5 \tag{15}$$

where $t_{0,T}$ is the adjusted age of concrete (days) at the moment of loading, also taking into account temperature effect, while α is the exponent dependant on the cement type:

$$\begin{aligned} &\alpha = -1 \text{ slowly hardening cement} \\ &\alpha = 0 \text{ normally hardening and rapidly hardening cement} \\ &\alpha = 1 \text{ rapidly hardening high strength cement} \end{aligned} \tag{16}$$

The effect of the temperature variation between 0 and 80°C on concrete hardening level can be taken into account by concrete age adjustment as follows

$$t_T = \sum_{i=1}^{n} \exp\{-[4{,}000/(273 + T_{\Delta t_i})] - 13.65\}\Delta t_i \tag{17}$$

where t_T is the age of concrete (days) adjusted according to the temperature effect, $T_{\Delta t_i}$ is the temperature (°C) in time interval Δt_i (number of days with the temperature T).

Since the analyzed spatial discretization refers to shell problems [11, 12], with strains ε_x, ε_y, γ_{xy}, γ_{xz}, γ_{yz}, the same creep coefficients were used for all strain components. Creep coefficients, namely, creep increments for different components of strain, were determined as for one-dimensional problem [11, 12].

3 Nonlinear Concrete Creep Models

Two empirical models of non-linear creep of concrete are briefly described hereinafter. Also, an original model for simulation of nonlinear concrete creep is presented. The model was verified in the numerical simulation of carried out experimental tests of concrete plate and concrete shell [15, 16] under relatively a high long-term static load.

3.1 Nonlinear Concrete Creep Model According to EUROCODE 2

According to the model given in Eurocode 2 [14], a nonlinear concrete creep at t_0 can be described accurately enough for average levels of the ratio between the concrete stresses due to long-term static load σ_c and the mean compressive strength of concrete f_{cm} in the range $0.4f_{cm}(t_0) < |\sigma_c| \leq 0.6f_{cm}(t_0)$. The coefficient of nonlinear concrete creep Φ_k is calculated as the coefficient of linear concrete creep Φ_0 multiplied by the creep factor$F(\sigma_c)$, namely,

$$\Phi_k = \Phi_0 \, F(\sigma_c). \tag{18}$$

A nonlinear creep factor $F(\sigma_c)$ is defined as a simple functional dependence between the instantaneous concrete stress due to static load σ_c and the mean compressive strength of concrete f_{cm} as

$$F(\sigma_c) = e^{1.5\left(\frac{|\sigma_c|}{f_{cm}(t_0)} - 0.4\right)} \tag{19}$$

The model is simple and provides satisfying results in practice. Its main disadvantage is a narrow stress level range that it covers (Fig. 2). Namely, it neglects the strong nonlinear concrete creep when the ratio between concrete stresses under a long-term static load and the average compressive strength of concrete is greater than 0.6.

3.2 Nonlinear Concrete Creep Model According to Bažant–Prasnnan and Bažant–Kim

An important researcher in the field of concrete creep modelling research is Bažant, who developed different concrete creep models (namely [1–8]). Two of those models [1–3] are presented hereinafter, which comprise high and very high concrete stress level ranges due to long-term static load.

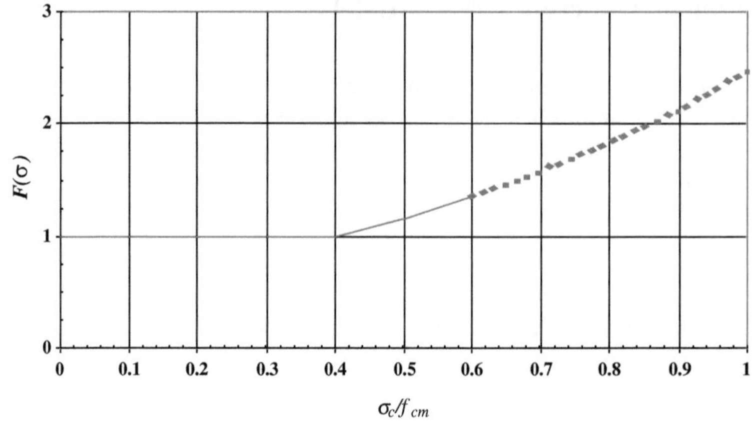

Fig. 2 Nonlinear concrete creep function according to EUROCODE 2 [14]

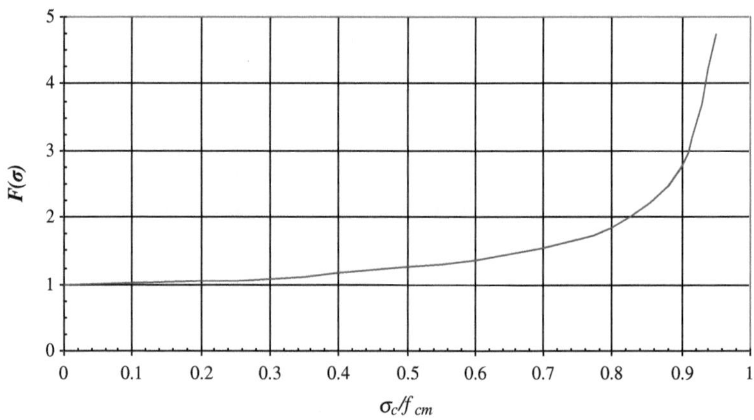

Fig. 3 Nonlinear concrete creep function according to Bažant–Prasannan [2, 3]

According to [2, 3], a nonlinear creep factor $F(\sigma_c)$ can be defined as an empirical expression for functional dependence between the instantaneous concrete stress due to static load σ_c and the compressive strength of concrete f_c as

$$F(\sigma_c) = \frac{1 + s^2}{1 - \Omega} \tag{20}$$

where $s = \sigma_c / f_c$, and $\Omega = s^{10}$(Fig. 3). Ω is a measure of the effect of concrete damage at high stresses, namely, when micro fractures develop. The expression remains valid in a wider range than expression (19), namely, at high stress levels. However, for concrete stresses exceeding $0.7 f_c$, expression (20) can be only used as provisional estimates.

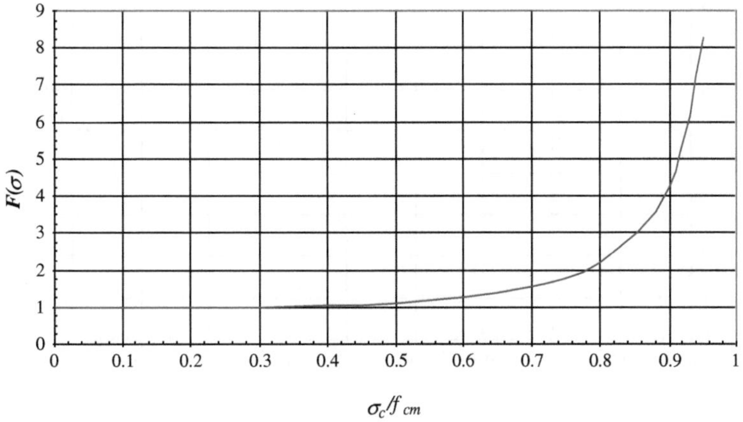

Fig. 4 Nonlinear concrete creep function according to Bažant–Kim [1]

For concrete stresses exceeding $0.7f_c$, when the ratio $(|\sigma_c|/f_c) \to 1$, namely, when $F(\sigma_c) \to \infty$, the following expression according to [1], is proposed (Fig. 4):

$$F(\sigma_c) = \frac{1 + 3s^5}{1 - \Omega}. \tag{21}$$

3.3 Proposed Nonlinear Concrete Creep Model

Similarly as for the previously described models, an empirical expression for nonlinear concrete creep is proposed that was used for numerical modelling of experimentally tested concrete plate and prestressed concrete shell under long-term load [15, 16]. Considering the need for a simple practical model for nonlinear creep of concrete, a simple function was adopted for a nonlinear creep factor $F(\sigma_c)$ as a function of the instantaneous concrete stress due to static load σ_c and the compressive strength of concrete f_c according to the expression

$$F(\sigma_c) = \frac{1}{1 - \left(\frac{|\sigma_c|}{f_c}\right)^3}. \tag{22}$$

This function covers all stress levels from $0 < |\sigma_c| \leq f_{cm}(t_0)$ (Fig. 5). It is continuous and simple for numerical applications. However, the proposed model is not completely valid for stress levels $|\sigma_c| > 0.8f_{cm}(t_0)$.

A proposed nonlinear concrete creep model is based on small-scale performed experimental tests, which are planned to be extended and intensified. Then, a more reliable nonlinear concrete creep model for all stress levels will be defined precisely.

Figure 6 shows the comparison of nonlinear concrete creep models according to Eqs. 19–22.

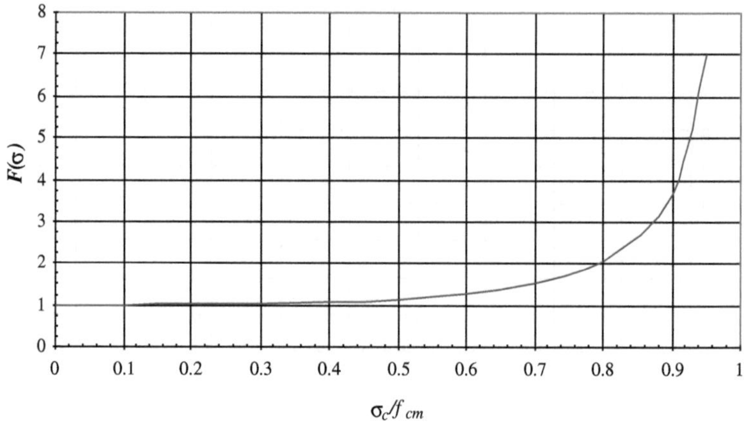

Fig. 5 Proposed nonlinear concrete creep function

4 Application of the Proposed Nonlinear Concrete Creep Model

A proposed model of nonlinear concrete creep according to Eq. (22) has been included in the numerical model for static analysis of prestressed concrete plates and shells under long-term load [11, 12]. The model was verified on the results of an experimentally tested reinforced concrete plate [15] and prestressed concrete shell [16] under relatively high levels of long-term static load. A detailed description of the performed experiments and developed numerical model can be found in previously listed references and will not be repeated here. Only a comparison between some experimentally determined and numerically obtained results will be given in short.

4.1 Numerical Analysis of the Concrete Plate Under Long-Term Load

First, a square reinforced concrete plate, supported at the edges and loaded by a uniform long-term load was experimentally tested [15]. Then, the behaviour of the same plate was analyzed by a previously developed numerical model [11, 12] with a linear creep model presented in Sect. 2 and the nonlinear creep model according to Eq. (22). The plate deflection as a function of time at mid-span is shown in Fig. 7.

As can be observed, there is a good correspondence between the results of experimental and numerical results, in particular those with the nonlinear concrete creep model.

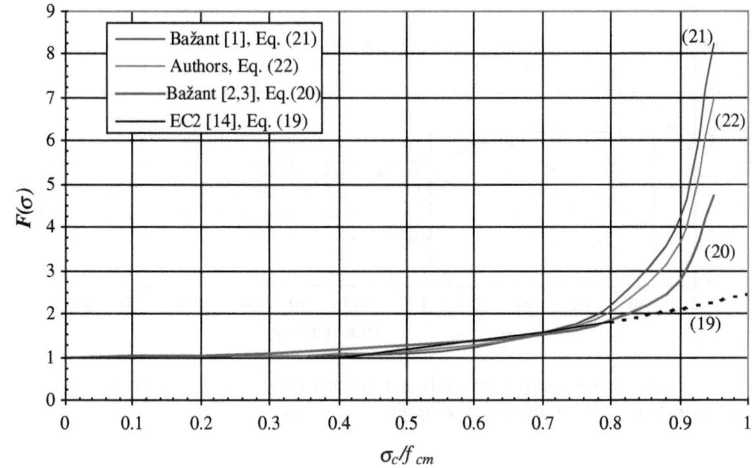

Fig. 6 Comparison of the described nonlinear concrete creep functions

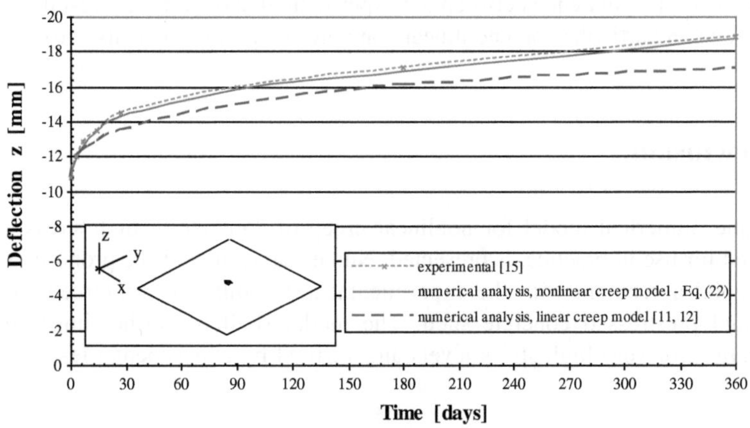

Fig. 7 Comparison between experimentally determined deflection at mid-span of reinforced concrete plate under long-term load and numerical results

4.2 Numerical Analysis of the Prestressed Concrete Shell Under Long-Term Load

Following the experimental test of the prestressed cylindrical concrete shell supported at the edges and loaded with uniform long-term load at its vertex [16], the shell was analyzed by a numerical model [11, 12]. A comparison between

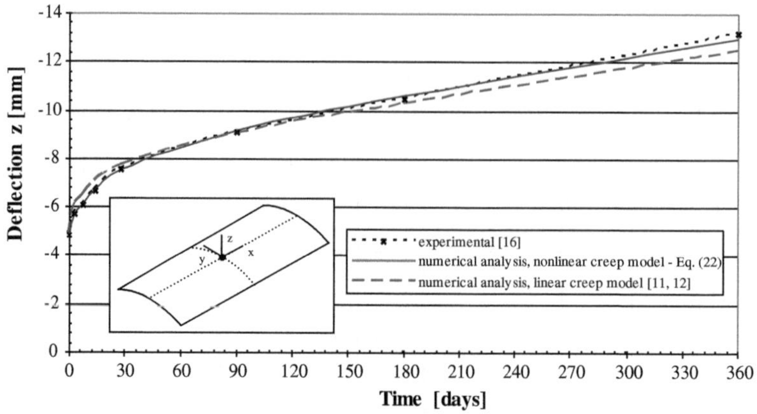

Fig. 8 Comparison between experimentally determined deflection of prestressed concrete shell under long term load at its midpoint and the numerical results

experimentally determined deflections at shell midpoint and the numerical results using the linear concrete creep model and nonlinear creep model according to Eq. 22 is shown in Fig. 8.

A good correspondence between the experimental and numerical results of shell deflections, in particular for a nonlinear concrete creep model can also be observed.

5 Conclusion

A simple numerical model for nonlinear creep of concrete at high stress levels, intended for use in practice, is presented. Numerical results of the proposed model show good agreement with the experimental test results for reinforced concrete plate and prestressed concrete shell. The model shall be further improved for application for very high stress levels and verified by experimental tests.

References

1. Bažant, Z.P., Kim, S.S.: Nonlinear creep of concrete—adaptation and flow. ASCE J. Eng. Mech. Div. **105**, 429–445 (1979)
2. Bažant, Z.P., Prasannan, S.: Solidification theory for concrete creep. Part I: formulation. J. Eng. Mech. **115**, 1691–1703 (1989)
3. Bažant, Z.P., Prasannan, S.: Solidification theory for concrete creep. Part II: verification and Application. J. Eng. Mech. **115**, 1704–1725 (1989)
4. Bažant, Z.P., Osman, E.: Double power law for basic creep of concrete. Mater. Struct. **9**, 3–11 (1976)
5. Bažant, Z.P., Asghari, A.A.: Constitutive law for nonlinear creep of concrete. J. Eng. Mech. Div. **103**, 113–124 (1977)

6. Bažant, Z.P., Panula, L.: Creep and shrinkage characterization for prestressed concrete structures. J. Prestress Concr. Inst. **25**, 86–122 (1980)
7. Bažant, Z.P., Chern, J.C.: Log-double-power law for concrete creep. ACI J. **82**, 675–685 (1985)
8. Bažant, Z.P., Kim, J.K.: Improved prediction model for time-dependent deformations of concrete: part 2–basic creep. Mater. Struct. **24**, 409–421 (1991)
9. Mazzotti, C., Savoia, M.: Nonlinear creep damage model for concrete under uniaxial compression. J. Eng. Mech. **129**, 1065–1075 (2003)
10. Ruiz, M.F., Muttoni, A., Gambarova, P.G.: Relationship between nonlinear creep and cracking of concrete under uniaxial compression. J. Adv. Concr. Tech. **5**, 383–393 (2007)
11. Radnić, J., Matešan, D.: Analysis of prestressed concrete shells subjected to long-term load. Gradjevinar (in Croatian) **62**, 183–196 (2010)
12. Radnić, J., Matešan, D.: Nonlinear time-dependent analysis of prestressed concrete shells. In: Öchsner, A., et al. (eds.) Materials with complex behaviour, pp. 165–179. Springer, Berlin (2010)
13. Glanville, W.H.: Studies in reinforced concrete—creep or flow of concrete under load. Department of Scientific and Industrial Research, Building research technical paper, **12**, p 111, London (1930)
14. EN 1992-1.: Eurocode 2: Design of concrete structures—Part 1: general rules and rules for buildings, European Standard, Brussels (2001)
15. Radnić, J., Matešan, D.: Experimental testing of RC slab behaviour under long-term load. MATWER **39**, 157–161 (2008)
16. Radnić, J., Matešan, D.: Testing of prestressed concrete shell under long-term loading and unloading. Exp. Mech. **50**, 575–588 (2010)

13. [illegible]
14. [illegible]
15. [illegible]
16. [illegible]

Numerical Model for Fluid–Structure Coupled Problems Under Seismic Load

Danijela Brzović, Goran Šunjić, Jure Radnić and Alen Harapin

Abstract This chapter briefly describes the numerical models for the simulation of fluid–structure coupled problems. The applied models are primarily intended to simulate the fluid–structure dynamic interaction in seismic conditions. The partition scheme of coupled (multi-field) problems is briefly described as the most common approach for the fluid–structure dynamic analysis. Models can simulate the most important effects of plane and spatial structures that are in direct contact with the fluid. Some of models' possibilities are illustrated in numerical analyses of the seismic behavior for four practical examples.

1 Introduction

Structures which are in direct contact with fluid, for example: dams, water tanks (reservoirs), off shore structures, pipelines and water towers etc., can often be encountered in engineering practice. Numerical models for real simulations of

D. Brzović · J. Radnić · A. Harapin (✉)
Faculty of Civil Engineering and Architecture, University of Split,
Matice hrvatske 15, Split, Croatia
e-mail: alen.harapin@gradst.hr

D. Brzović
e-mail: danijela.brzovic@gradst.hr

J. Radnić
e-mail: jure.radnic@gradst.hr

G. Šunjić
Faculty of Civil Engineering, University of Mostar,
Matice hrvatske bb, Mostar, Bosnia and Herzegovina
e-mail: goran.sunjic@tel.net.ba

A. Öchsner et al. (eds.), *Materials with Complex Behaviour II*,
Advanced Structured Materials 16, DOI: 10.1007/978-3-642-22700-4_10,

these structures have to include the simulation of the fluid–structure interaction to ascertain the real behavior of such a complex system. This problem is particularly emphasized under dynamic/seismic conditions and it is commonly referred to as a coupled (multi-field) problem.

A coupled multi-field problem involves two or more interacting fields, for example gravity dam with accumulation, water tower full of water etc. Such a problem is time dependent and the state of one field is continuously linked to the state of other fields and neither field can be solved independently from the other. Here, the coupling normally occurs through the differential equations representing the physical phenomena.

The most natural treatment for coupled problems is, as previously mentioned, partitioned analysis. In this approach the overall system is partitioned into zones or fields. Then the individual fields are solved independently by considering the interaction information transfer between them at every stage of the solution process.

The various advantages are: (i) the resulting model is very modular, (ii) it is easy to make any modifications, (iii) every modification in one field improves the whole model, (iv) the programmer/improver can have knowledge in (only) a single field.

This chapter briefly describes the partition approach in numerical modeling of the dynamic interaction of water-structure systems. The described model is suitable for problems with limited fluid motions, such as the response of offshore structures and dams to waves or earthquake.

2 Short Description of the Numerical Model's Basic Characteristics for the Simulation of the Dynamic Water-Structure System Interaction

In articles [1–3] the basic algorithms for fluid–structure interaction problems are given. Furthermore, articles [4, 5] present the development of non-linear numerical models for dynamic interactions of the fluid-soil-structure system for plane and spatial problems. Articles [6–11] present some recent works in this field.

All solutions shown here are based on the partitioned scheme where individual fields are solved independently by considering the interaction information transfer between them at every stage of the solution process. This approach allows the usage of ordinary approaches and appropriate mathematical/physical models for separate fields (structure and fluid) that include minor modifications for the influence of interactions.

Developed models and software are based on finite elements method for the spatial discretization and finite differences method for the time discretization of the system. For structure and soil the displacement formulation is used, and for fluid the displacement potential formulation is used. For spatial structure discretization

8 and 9 node elements can be used for plane problems, 20 and 27 node ("brick") elements can be used for 3D problems and for shell like structures 8 and 9 node shell elements can be used.

Very similar elements can be used for fluid: 8 and 9 node elements for plane problems and 20 and 27 node ("brick") elements for 3D problems.

Developed models include the most important non-linear effects of each field, such as:

- Material nonlinearity of the structure with the reinforced concrete model that can simulate:
  - Concrete yielding in compression,
  - Cracks occurrence and propagation in tension (opening and closing of cracks),
  - Tensile and shear stiffness of cracked concrete,
  - Yielding of steel or reinforcement in compression and tension,
  - Influence of strain rate effects on mechanical characteristic of concrete and reinforcing steel [12]
  - Influence of hydrostatic and hydrodynamic water pressures in structure cracks
- Material nonlinearity of water:
  - Cavitation,
  - Influence of suspensions in water;
- Geometrical nonlinearity of structure (large displacements).

2.1 *Equation for Coupled Fields Motions and Spatial Discretization*

Behaviors of the fluid–structure system (structure includes the structure itself as well as the surrounding soil) in dynamic load conditions, can be expressed with two second order differential equations [1–5]. If we use the displacement formulation for the structure and the displacement potential formulation for the fluid, dynamic equilibrium equations can be expressed in the following form:

$$\begin{aligned} \mathbf{M}_s\ddot{\mathbf{u}} + \mathbf{C}_s\dot{\mathbf{u}} + \mathbf{K}_s\mathbf{u} &= \mathbf{f}_s - \mathbf{M}_s\ddot{\mathbf{d}} + \mathbf{f}_{cs} \\ \mathbf{M}_f\ddot{\boldsymbol{\Psi}} + \mathbf{C}_f\dot{\boldsymbol{\Psi}} + \mathbf{K}_f\boldsymbol{\Psi} &= \mathbf{f}_f + \mathbf{f}_{cf} \end{aligned} \tag{1}$$

where

$$\begin{aligned} \mathbf{f}_{cs} &= \mathbf{Q}\boldsymbol{\Psi} \\ \mathbf{f}_{cf} &= -\rho_f\mathbf{Q}^T(\mathbf{u} + \mathbf{d}) \end{aligned} \tag{2}$$

In the above equations $\mathbf{M}_s$, $\mathbf{C}_s$ and $\mathbf{K}_s$ represent mass, damping and stiffness matrices for structure, and $\mathbf{M}_f$, $\mathbf{C}_f$ and $\mathbf{K}_f$ represent mass, damping and stiffness

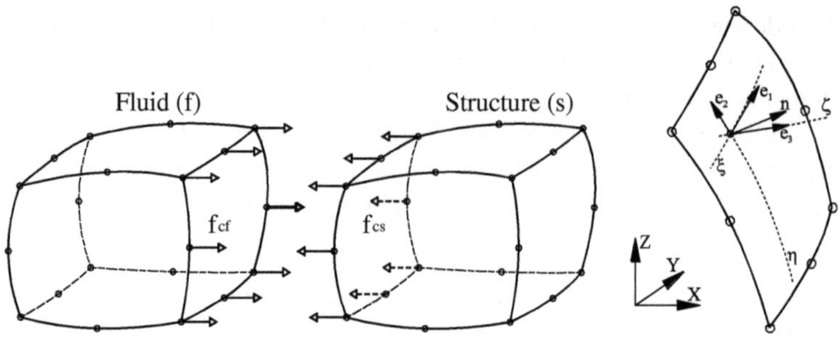

Fig. 1 Fluid–structure interaction surface and unit norm

matrices for fluid. Vectors $\mathbf{u}$, $\dot{\mathbf{u}}$, $\ddot{\mathbf{u}}$ represent structure's displacements and displacement's derivations (velocities and accelerations) and $\Psi, \dot{\Psi}, \ddot{\Psi}$ are the displacement potential and associated derivations. $\mathbf{Q}$ is the interaction matrix between structure and fluid.

Interaction between structure and base soil is modeled indirectly by contact elements in the connection surface. In fact, by applying the appropriate material model for contact elements, various effects in the contact surface can be simulated, such as: separating, embedment and sliding.

Fluid–structure interaction surface with fluid and structure elements is shown in Fig. 1. Interaction matrix $\mathbf{Q}$ includes only the surface integration and is defined as:

$$(\mathbf{Q})_{\mathrm{ij}} = \int_{\Gamma_{\mathrm{i}}} \mathbf{N}_{\mathrm{ui}}^{\mathrm{T}} \vec{n} \, \mathbf{N}_{\mathrm{pj}} \, \mathrm{d}\Gamma_{\mathrm{i}} \tag{3}$$

2.2 Solution Concept for the Dynamic Fluid–Structure Interaction Problem

Direct solution of the equation system (1) requires large computer capacity. So, the previously described partitioned scheme is ideal for this kind of problems. In that approach for every increment of the imposed load and every non-linear problem iteration step, each field is solved separately by including interaction forces on the contact surface between fluid and structure. Presentation of the solution scheme is given in Fig. 2.

In the presented approach, structure is solved first and fluid second. This approach allows the developed independent models to be used for each field, with additional calculations of the interaction forces only. Thus, in the fluid–structure interaction model, all non-linear effects of material and geometry, that are present in a particular field, can also be simulated in the coupled problem.

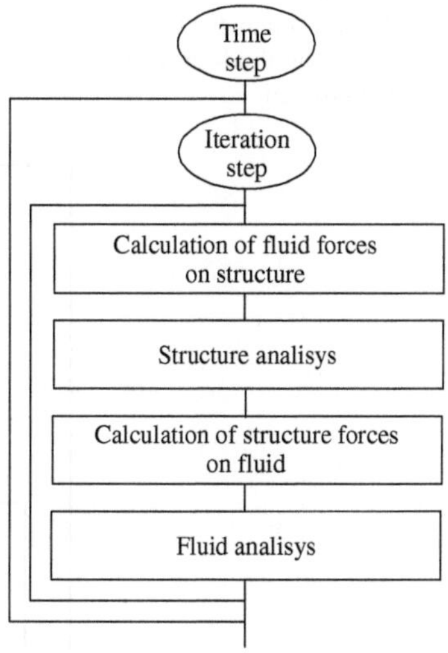

Fig. 2 Solution scheme for the fluid–structure coupled problem

2.3 Solution Algorithm

Solution algorithm for the fluid–structure coupled problem with non-linear fluid model (cavitation) is schematically presented in Fig. 3. For time integration, explicit-implicit algorithm developed by Hughes [13] is used.

Predicted values $\mathbf{u}$, $\dot{\mathbf{u}}$, $\ddot{\mathbf{u}}$ and Ψ, $\dot{\Psi}$, $\ddot{\Psi}$ at the beginning of every time step are corrected at the end of the same time step. For convergence control of the iterative procedure, the increase of the structure's displacements in comparison with current total displacements and the increase of the fluid's displacements potential in comparison with the current total displacements potential are simultaneously monitored. Various options of the Newton–Raphson method are used to solve the non-linear equations.

2.4 Finite Elements

For plane (2D) problems, 8-node and 9-node isoparametric elements are used for fluid and structure. For spatial (3D) problems, 20-node and 27-node ("brick") elements are used for fluid and structure. For thin curved structures, 8-node or 9-node degenerated shell elements can be used for structure and 20 or 27-node spatial element for fluid. Those shell elements are free of membrane and shear locking, according to [14].

For the simulation of connections between the foundation soil and the structure, 6-node contact elements can be used for plane and 16 or 18 nodes for spatial problems.

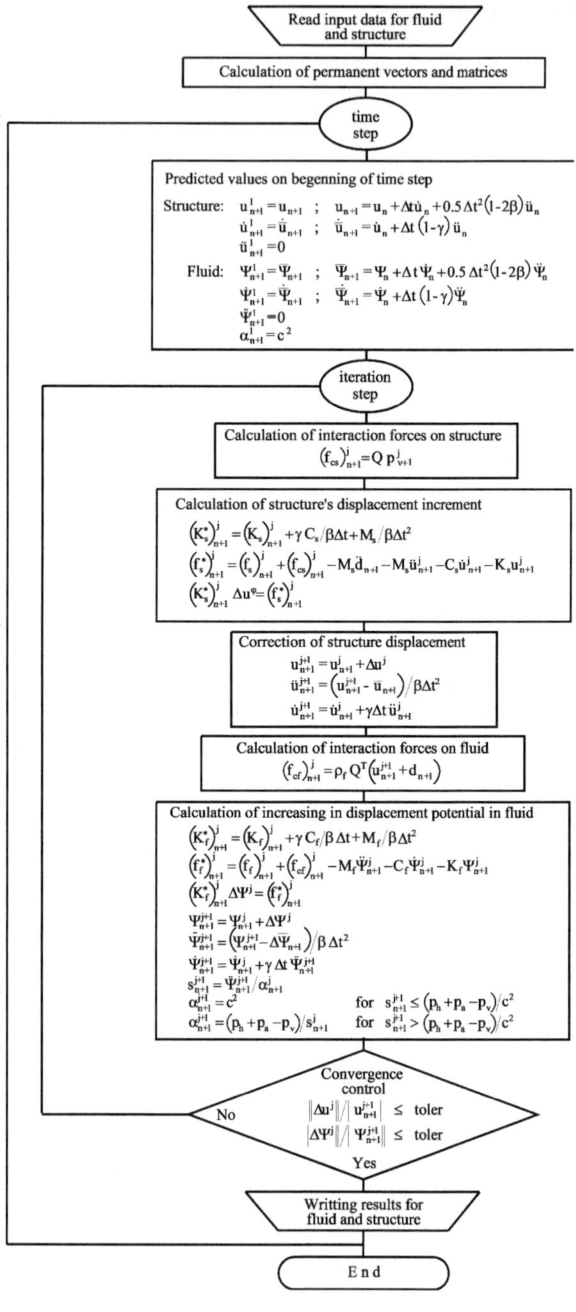

Fig. 3 Flow chart for the solution of the fluid–structure coupled problem

2.5 Material Model

For fluid simulations, two formulations are generally included: (i) pressure formulation, which presumes unlimited negative pressures in fluid and (ii) displacement potential formulation which can simulate cavitation effects in fluid.

In the fluid–structure interaction, nonlinearities are generally confined to the structural behavior where the fluid is considered linear. Pressure formulation, which presumes unlimited negative pressures in fluid, is very suitable for this approach.

The fluid can take some tension which depends upon the concentration and size of micro bubbles present in the fluid. However, if the absolute pressure in a subregion of fluid drops to a value close to vapor pressure of the fluid, bubbles are formed and this physical phenomenon is known as cavitation. Physically, cavitation occurs when the total absolute pressure is less then the vapor pressure of the fluid. Cavitation can cause significant damaging effects on solid surfaces.

Cavitation occurs when the total absolute pressure is less then the vapour pressure of the fluid i.e.:

$$p_{abs} = p + p_h + p_a \leq p_v \tag{4}$$

where p_{abs} is the total absolute pressure, p is hydrodynamic pressure, p_h is hydrostatic pressure, p_a is atmospheric pressure and p_v is vapour pressure. This implies that cavitation occurs when the hydrodynamic pressure drops below $(p_v - p_s)$. The vapour pressure of water, for all practical purposes, can be taken from 0.02 to 0.03 MPa.

The changes which the fluid may undergo under hydrodynamic excitation are a direct function of the mass dilatation s, defined as:

$$s = \nabla^T(\rho_f \mathbf{u}_f) = \mathrm{Div}(\rho_f \mathbf{u}_f) \tag{5}$$

where $\mathbf{u}_f$ is displacement of the fluid relative to the initial static state. As long as p_{abs} is greater then the vapour pressure p_v, a linear relation between s and p is assumed:

$$p = -\alpha s; \quad \alpha = c^2 \tag{6}$$

where c is the acoustic velocity of the fluid.

If Eq. 4 is true, cavitation occurs and the stage of linear fluid is no longer valid. A simple fluid model can be represented by the bilinear pressure-mass dilatation relation shown in Fig. 4. Cavitation, therefore, commences when the following condition is reached:

$$s \geq (p_h + p_a - p_v)/c^2 \tag{7}$$

If cavitation occurs, the iteration procedure, shown in Fig. 4, has to be performed to obtain the value of the coefficient α.

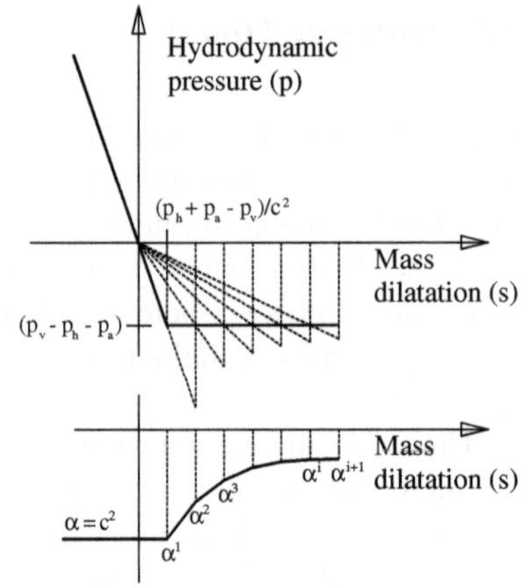

Fig. 4 Relation between mass dilatation and hydrodynamic pressure

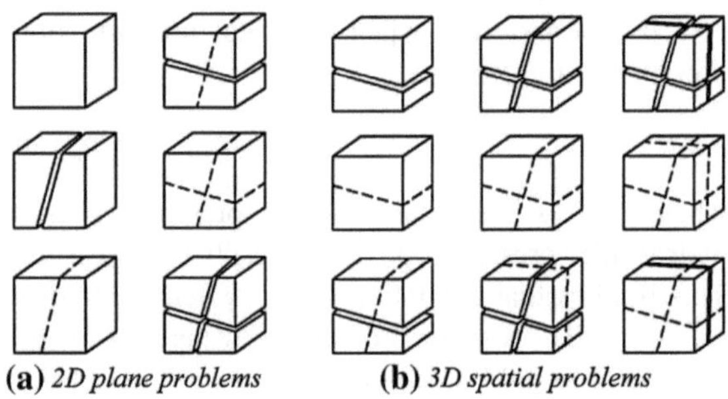

Fig. 5 Possible crack states at every integration point. **a** 2D plane problems. **b** 3D spatial problems

For the structure, the classic elastic, elasto-plastic and elasto-visco-plastic material model can be used.

A special material model was developed for the simulation of reinforced concrete structures [12, 15–20]. It includes the most important nonlinear effects of reinforced concrete behavior: yielding in compression and opening and propagation of cracks in tension, with tensile and shear stiffness of cracked concrete, as well as nonlinear behavior of reinforced steel. In every integration point of every element, simulation of cracks opening and closing is possible, according to Fig. 5.

These models will not be discussed here, but they can easily be found in quoted references.

2.6 Additional Model Characteristics

Solution of eigen value problem is also based on the partition solution scheme, with the Wilson–Yuan–Dickens (so-called WYD) method [21–24] as the solution procedure. In dynamic problems, as well as in the structures' response

Fig. 6 Koyna dam, photograph [25] and comparison with typical gravity dam [26]

calculations, eigen values and eigen vectors are needed to know the vibration characteristics (determination of time step length).

Radiation damping can be simulated on artificially formed fluid boundaries, as well as radiation and multi-axis structure damping for structure [3, 4, 13].

Simulation of fluid pressure in open cracks of a structure is included by additional nodal forces in finite elements that have cracks that fluid can get into.

As external dynamic forces, various time-dependant dynamic loads can be applied. Also, seismic base excitations can be applied to the model.

3 Examples

What follows are four complex practical examples which illustrate some possibilities of the developed models and the applied software.

3.1 Example 1: Koyna Dam

Koyna dam, built in 1963, is one of the largest dams in India (Fig. 6). It is an atypical gravitational dam, with a crest length of 853.44 m. It consists of 56 dilatation blocks of 17.07 m in thickness. Spillway length is 91.44 m. During construction, two accelerographs were embedded in the dam, and in one of them, in 1967, an earthquake that caused several significant damages was registered. Dominant damages of the dam manifested as horizontal cracks on the up-stream and down-stream sides on many blocks, especially on lines where the total thickness of the dam changes.

Figure 7 presents the main geometric data of the Koyna dam. Detailed information of the dam geometry, construction materials, damages (cracks) and earthquake characteristics can be found in [26, 27].

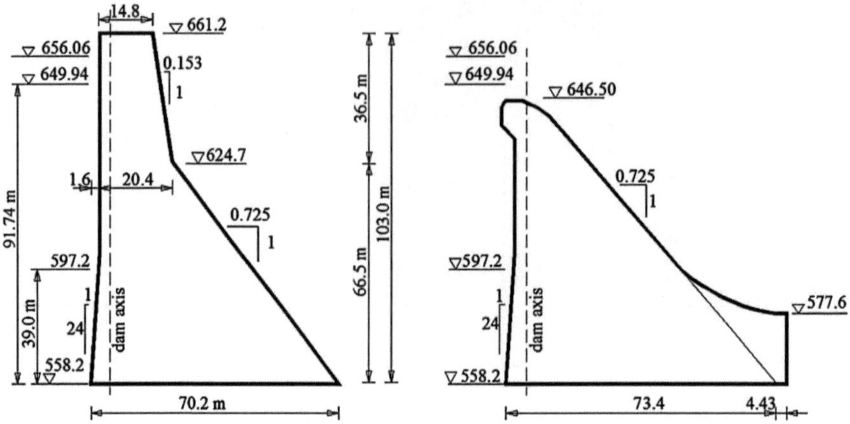

Fig. 7 Koyna dam—some geometrical data (all dimensions are in meters) [26]: **a** cross-section through dam body; **b** cross-section through spillway

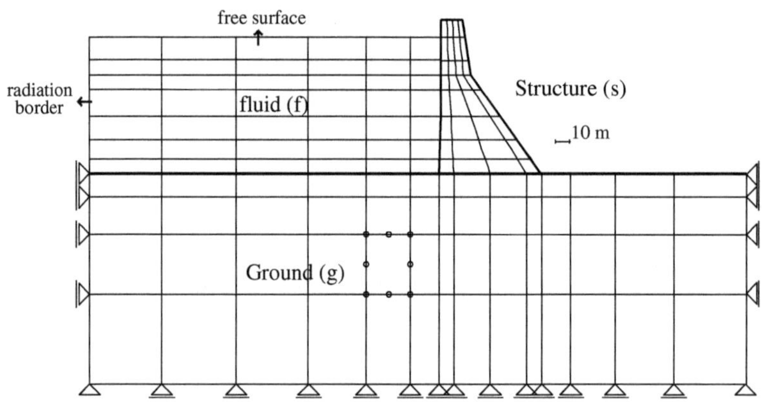

Fig. 8 Spatial discretization of Koyna dam

Spatial discretization of the system is presented in Fig. 8, and the used material characteristics are presented in Table 1. The behavior of the water-dam-soil system was analyzed for the previously mentioned registered earthquake. The system was analyzed separately for the linear and for the non-linear (cavitation) fluid model, with the following structure models: a) non-linear model without including the fluid pressure in open structure cracks (no FPC), b) non-linear model which includes the fluid pressure in open structure cracks (FPC).

Some numerical results are presented in Figs. 9 and 10. Other results can be found in [26, 27]. Dam damages calculated through numerical models match the real crack pattern very well.

Table 1 Material characteristics of the Koyna dam system

Fluid (water)	Structure (concrete dam)	Ground
$\rho_f = 1019.0\,\mathrm{kg/m^3}$ $c = 1439.0\,\mathrm{m/s}$ $p_a = 0.10\mathrm{MPa}$ $p_v = 0$	$E_s = 31640.0\mathrm{MPa}$ $v_s = 0.2$ $\rho_s = 2690.0\mathrm{kg/m^3}$ $(f'_c)_s = 24.6\mathrm{MPa}$ $(f'_t)_s = 2.46\mathrm{MPa}$ $(\varepsilon_{cu})_s = 0.003$ $(\max\varepsilon_t)_s = (\max\varepsilon_{sh})_s = 0.0012$	$E_g = 18000.0\mathrm{MPa}$ $v_g = 0.2$ $\rho_g = 1830.0\mathrm{kg/m^3}$ $(f'_c)_g = 20.0\mathrm{MPa}$; $(f'_t)_g = 2.0\mathrm{MPa}$ $(\varepsilon_{cu})_g = 0.003$; $(\max\varepsilon_t)_g = (\max\varepsilon_{sh})_g = 0.0017$

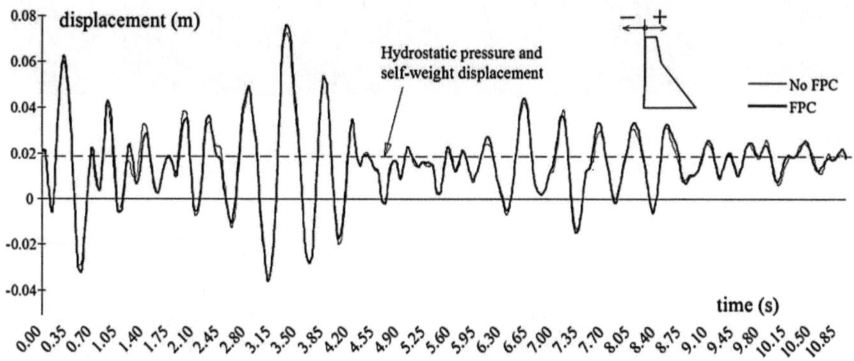

Fig. 9 Horizontal displacement of the Koyna dam crest for non-linear fluid model

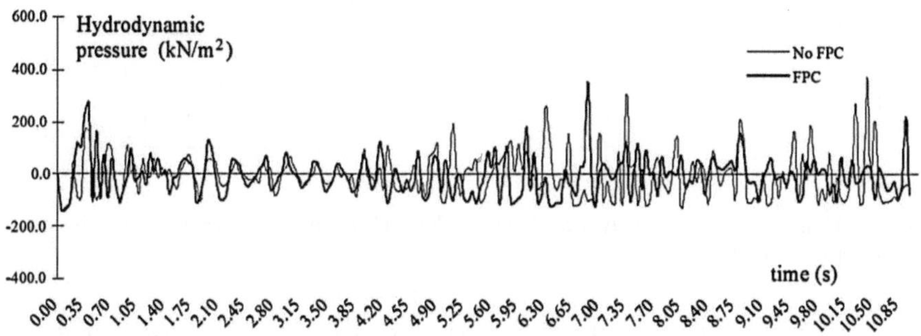

Fig. 10 Hydrodynamic pressure at the bottom of the Koyna dam

3.2 *Example 2: Grančarevo Dam*

The Grančarevo Arch Dam in Bosnia and Herzegovina (Figs. 11, 12) is a double-curvature concrete dam with a perimetral joint. The dam was constructed in 1968. The height of the dam is 123 m and the crest length is 439 m. Its bottom thickness

Fig. 11 Grančarevo arch dam

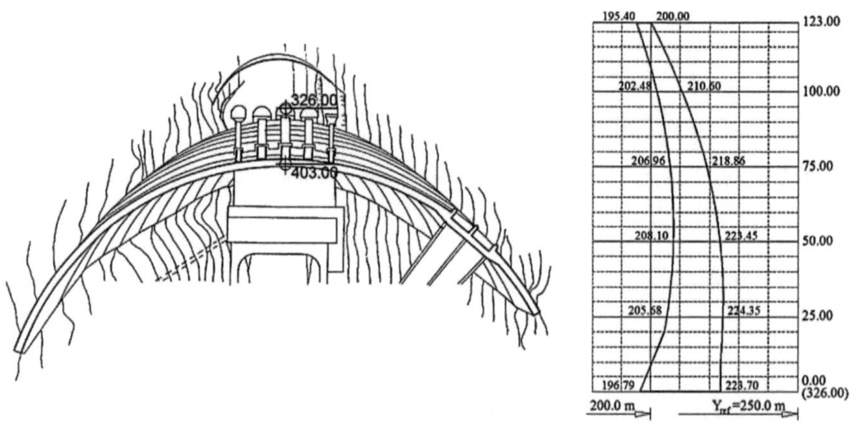

Fig. 12 Plan of the dam's body with land topology (*left*) and cross section through central cantilever (*right*) [30]

is 27 m and its top thickness 4.6 m. The dam's foundation dig was 230.000 m^3 and the volume of poured concrete was 376.000 m^3. The head of the dam is 100 m. The dam created the Bileća reservoir with a maximum water depth of 51 m and an available storage capacity of 1,100 million cubic meters. The Bileća reservoir is the largest storage lake in Balkan. Its dimensions are: total storage volume: 1,280 hm^3 and surface of the reservoir on normal top water level: 2,764 ha. Geometrical data tables (on Fig. 13) show basic geometrical characteristics for individual arches some of which are shown in Fig. 13. Other detailed information about dam can be found in [28–31].

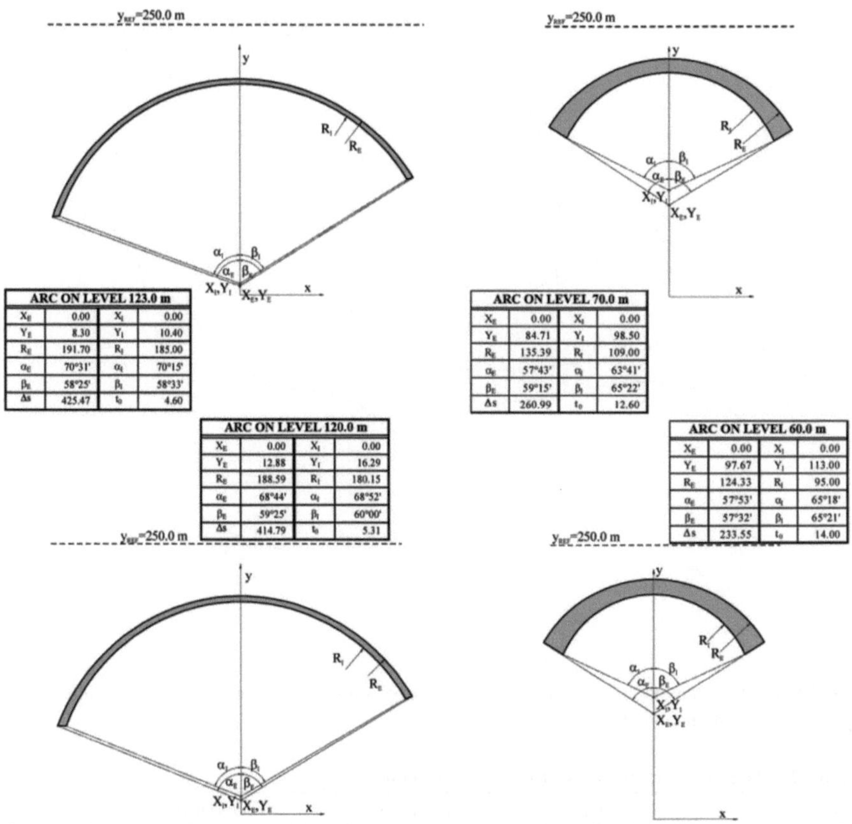

ARC ON LEVEL 123.0 m			
X_E	0.00	X_I	0.00
Y_E	8.30	Y_I	10.40
R_E	191.70	R_I	185.00
α_E	70°31'	α_I	70°15'
β_E	58°25'	β_I	58°33'
Δs	425.47	t_0	4.60

ARC ON LEVEL 70.0 m			
X_E	0.00	X_I	0.00
Y_E	84.71	Y_I	98.50
R_E	135.39	R_I	109.00
α_E	57°43'	α_I	63°41'
β_E	59°15'	β_I	65°22'
Δs	260.99	t_0	12.60

ARC ON LEVEL 120.0 m			
X_E	0.00	X_I	0.00
Y_E	12.88	Y_I	16.29
R_E	188.59	R_I	180.15
α_E	68°44'	α_I	68°52'
β_E	59°25'	β_I	60°00'
Δs	414.79	t_0	5.31

ARC ON LEVEL 60.0 m			
X_E	0.00	X_I	0.00
Y_E	97.67	Y_I	113.00
R_E	124.33	R_I	95.00
α_E	57°53'	α_I	65°18'
β_E	57°32'	β_I	65°21'
Δs	233.55	t_0	14.00

Fig. 13 Geometry of some arch elements of the Grančarevo dam [30]

The Institute of Earthquake Engineering and Engineering Seismology (IZIIS-Skopje, Macedonia) monitored the dam and performed several numerical simulations on different models, which were compared with results in situ [30]. All applied models included only the dam (structure), and water was treated as an additional mass on structure.

The complex model of the water-dam-foundation rock system is presented in Fig. 14. The behavior of this complex system was analyzed for the registered earthquake from 1986, [30, 31]. Material characteristics are given in Table 2.

The registered accelerations on the bottom of the dam (accelerograph 688, Fig. 15) were taken as imposed accelerations of the foundation's rock (excitation) along the canyon (perpendicular to the dam axis). The maximal registered imposed acceleration was 47.8 cm/s^2. The maximal registered acceleration on the dam was $a_{max,r}$ = 145.1 cm/s^2 (accelerograph 681, Fig. 15), and the maximal acceleration obtained through the numerical model was $a_{max,n}$ = 149.3 cm/s^2 (Fig. 16). Applied excitations cause hydrodynamic pressures that are always less than the hydrostatic pressure, so cavitation did not occur.

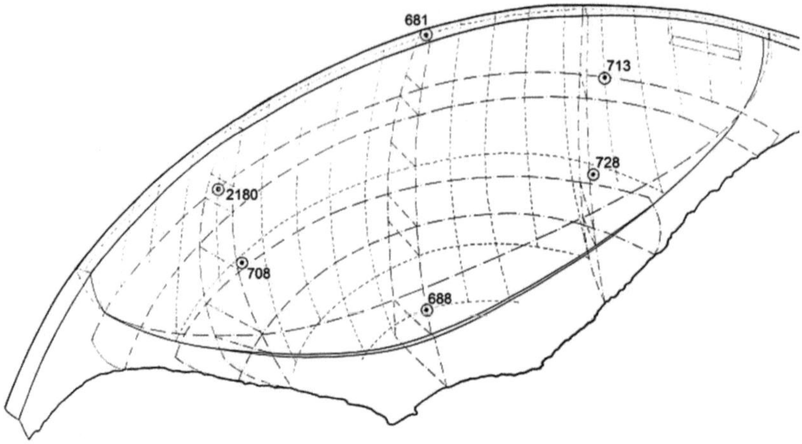

Fig. 14 Finite element mesh of the Grančarevo dam–water–foundation rock interaction system—axonometric view

Table 2 Material characteristics of Grančarevo dam system

Fluid (water)	Structure (concrete dam)	Foundation rock
$\rho_f = 981\,\mathrm{kg/m^3}$	$E_c = 33000.0\,\mathrm{MPa}$	$E_r = 80\,\mathrm{GN/m^2}$
$c = 1440.0\,\mathrm{m/s}$	$\nu_c = 0.15$	$\nu_r = 0.2$
	$\rho_c = 2400.0\,\mathrm{kg/m^3}$	$\rho_r = 2620.0\,\mathrm{kg/m^3}$
	$f_{ck} = 25\,\mathrm{MPa}$; $f_{ct} = 2.5\,\mathrm{MPa}$	$f_{rk} = 12.0\,\mathrm{MPa}$; $f_{rt} = 1.2\,\mathrm{MPa}$
	$\varepsilon_t = 0.083$; $\varepsilon_{t,max} = \varepsilon_{s,max} = 1.7$	

Some calculation results are presented in Figs. 16, 17 and 18. Figure 16 presents accelerations of the Grančarevo dam crest in time, Fig. 17 presents displacement of the Grančarevo dam crest in time and Fig. 18 presents hydrodynamic pressures on the bottom of the Grančarevo dam in time. Other results can be found in [30–32].

3.3 Example 3: Underwater Tank "Khazzan"

Khazzan (meaning: "To Store" in Arabic) was the name given to the tanks designed and built in late 1960s to store Dubai's Oil by Chicago Bridge and Iron Company. Dubai's Khazzans are unique in that they store Dubai's Oil under the Sea. Khazzan is a 500.000 barrel (80.000 m^3) oil storage tank (Fig. 19).

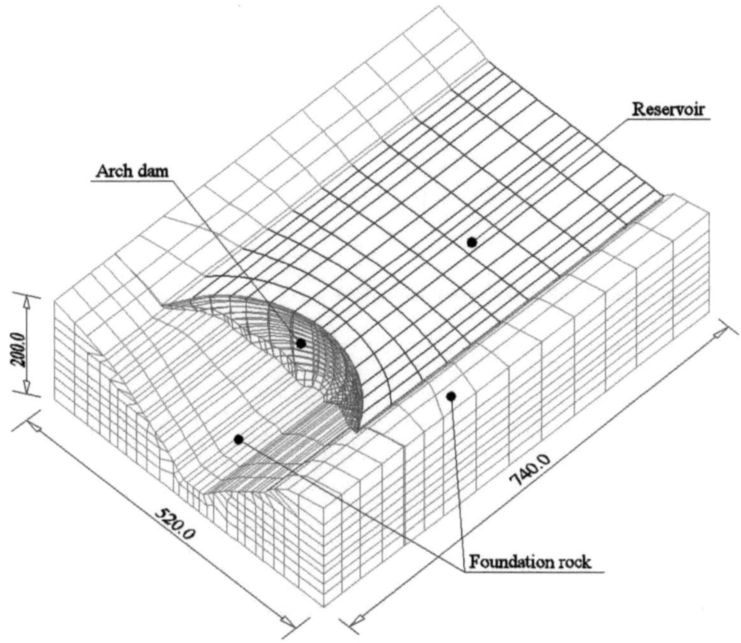

Fig. 15 Positions of accelerographs in the Grančarevo dam body [30]

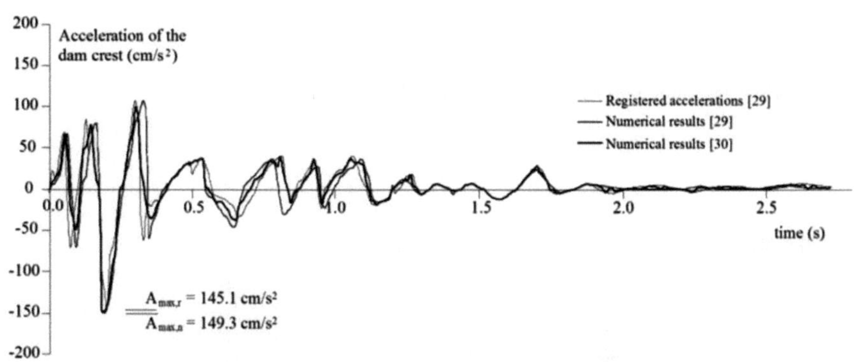

Fig. 16 Accelerations of the Grančarevo dam crest

The 15.000 ton structure is 80 m in diameter on bottom and 8 m diameter on top, and about 82.0 m in height. Sea depth is about 70 m, so tank crest is 12 m under sea level. It has no bottom and operates on the water displacement principle. It is filled by placing oil in the tank above water where the additional weight of the oil on the water creates an imbalance in pressure. This force pressures the water out of the tank through the openings in the wall at the bottom.

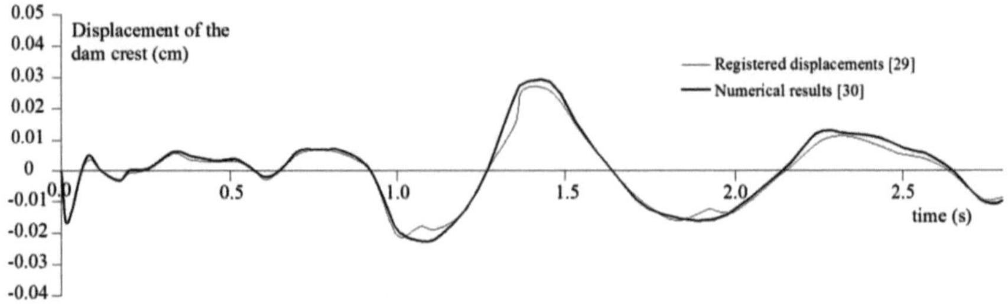

Fig. 17 Displacement of the Grančarevo dam crest

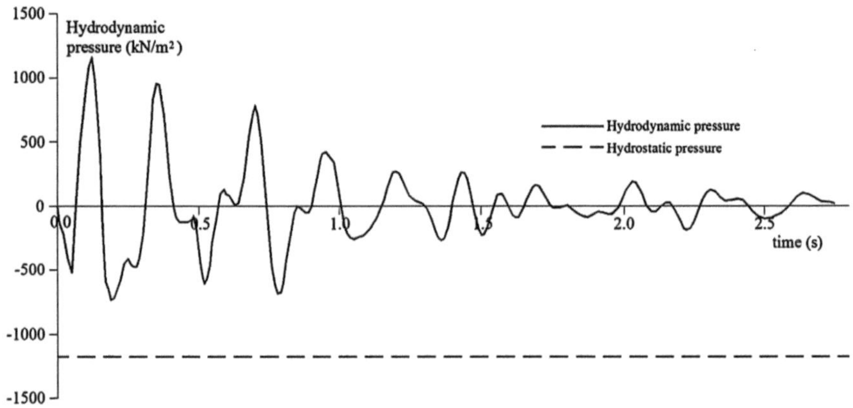

Fig. 18 Hydrodynamic pressures on the bottom of the Grančarevo dam

(a) – *Construction on shallow dewatered basin on shore* **(b)** – *Towing to the site*

Fig. 19 Oil-storage tank "Khazzan" [33, 34]. **a** Construction on shallow dewatered basin on shore. **b** Towing to the site

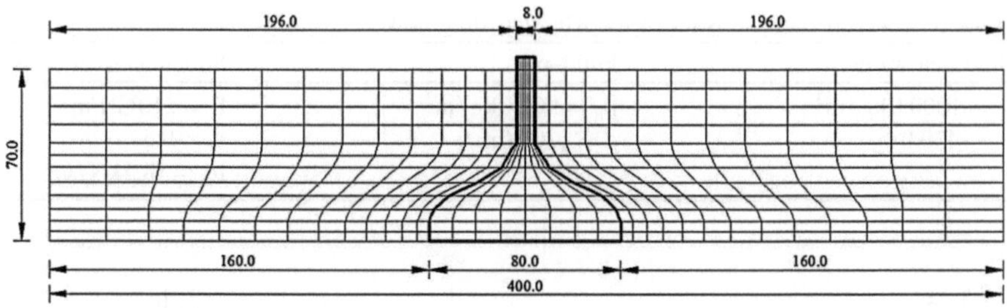

Fig. 20 Spatial discretization of the Kazzan tank, the oil in the tank and the surrounding sea water—longitudinal section of the finite element mesh (all dimensions in meters)

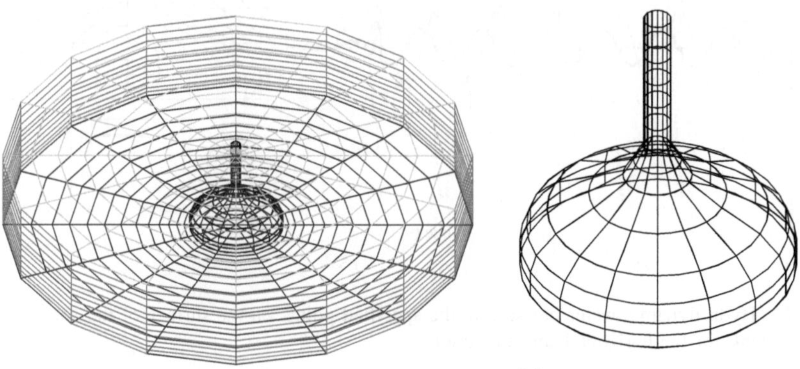

(a) – 3D view of finite element mesh

(b) 3D view of oil-storage tank finite element mesh

Fig. 21 Spatial discretization of the Kazzan tank, the oil in the tank and the surrounding see water—axonometric view. **a** 3D view of finite element mesh. **b** 3D view of oil-storage tank finite element mesh

Initial construction was in a shallow, dewatered basin. When the tank was sufficiently complete so that it could float as a single unit, using compressed air, the basin was flooded, and the tank, a bottomless hemisphere, was moved laterally into a deeper basin and seated on its floor by releasing the internal air pressure. The structure was then fully completed. Floated once again by filling the tank with compressed air, it was towed to the site and positioned by mooring lines, and the air was gradually released. It was allowed to slowly sink further and settle on the seafloor [33].

The geometrical characteristic of the model were taken from [33–35]. Figure 20 shows the vertical section of the oil tank with the adjacent part of the surrounding sea.

The sea-oil-tank system was modelled with the spatial 3D model, shown in Figs. 20 and 21. Spatial discretization of the liquid was done with 27-node 3D brick elements, and the structure with 9-node shell elements.

Table 3 Material characteristics of the Khazzan store tank

Fluid–sea water	Fluid–oil	Structure–steel (tank)
$\rho_f = 1000.0\,kg/m^3$ $c = 1430.0\,m/s$	$\rho_n = 900.0\,kg/m^3$ $c = 1300.0\,m/s$	$E = 210\,GN/m^2$ $\nu_t = 0.3$ $\rho_a = 78.5\,kN/m^3$

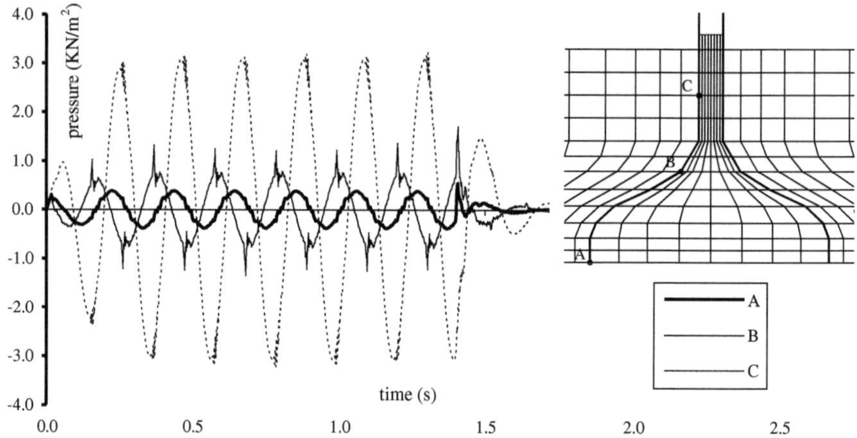

Fig. 22 Hydrodynamic water pressure in the specified points on the surface of the Kazzan oil-storage tank for the horizontal seismic action

The harmonic ground acceleration with the period of 0.207 s (which is in accordance with the first period of the sea-oil-tank system), and amplitudes of 0.3 g for the horizontal and 0.2 g for the vertical acceleration component is accepted. The material characteristics are shown in Table 3. Implicit time integration ($\Delta t = 0.002$ s) and diagonal mass matrix were used.

Some results are shown in Figs. 22, 23, 24 and 25, and a detailed description of the model and results can be found in [35].

Figure 22 shows the hydrodynamic water pressure for the horizontal seismic action in the specified points on the tank surface and Fig. 23 shows the horizontal displacements of the specified points of the tank for the horizontal seismic action. Figure 24 show the maximal displacements of the Kazzan oil-storage tank for the horizontal and the vertical seismic action.

3.4 Example 4: Underwater Tunnel "Høgsfjord"

The seismic behavior of the planned underwater tunnel "Høgsfjord" in Norway was analyzed. The tunnel is about 1,400 m long and 20 m immersed in the sea. It is connected to the sea-bed with cables of 200 m (axial distance) (Fig. 26).

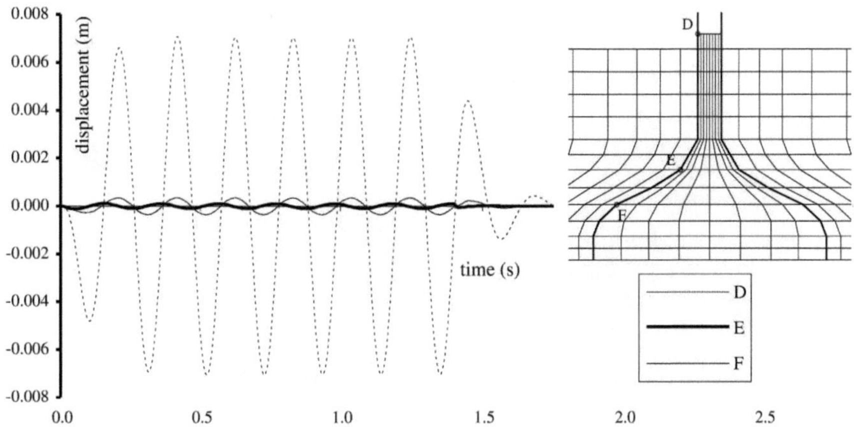

Fig. 23 Horizontal displacements of the specified points of the Kazzan oil-storage tank for the horizontal seismic action

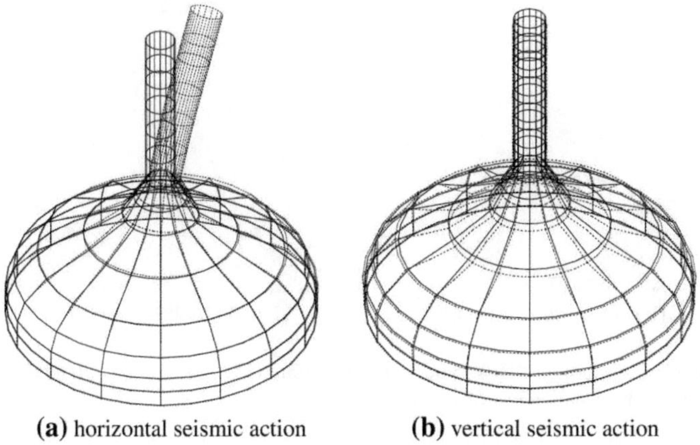

Fig. 24 The Kazzan oil-storage tank maximal displacements. **a** horizontal seismic action. **b** vertical seismic action

The tunnel has a circular cross-section with a 8.6 m inner diameter and 50–80 cm thick walls (Fig. 26). Intended construction material for the tunnel is prestressed concrete. Some other information about the planned structure can be found in [35, 36] (Fig. 27).

The seismic response to the vertical earthquake component was analyzed. A plane (2D) model was adopted with the discretization shown in Fig. 28. Some results are shown in Figs. 29, 30 and 31, and a detailed description of the model and results can be found in [35]. Displacements and stresses in the tunnel

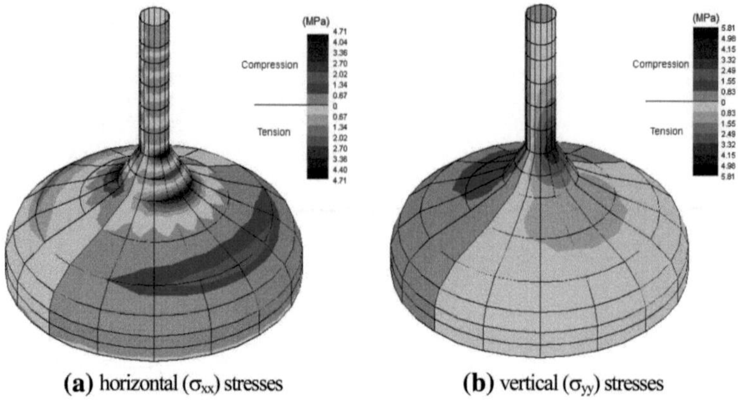

Fig. 25 Maximum stresses of the Kazzan oil-storage tank in t = 0.73 s, for the horizontal seismic action. **a** horizontal (σ_{xx}) stresses. **b** vertical (σ_{yy})

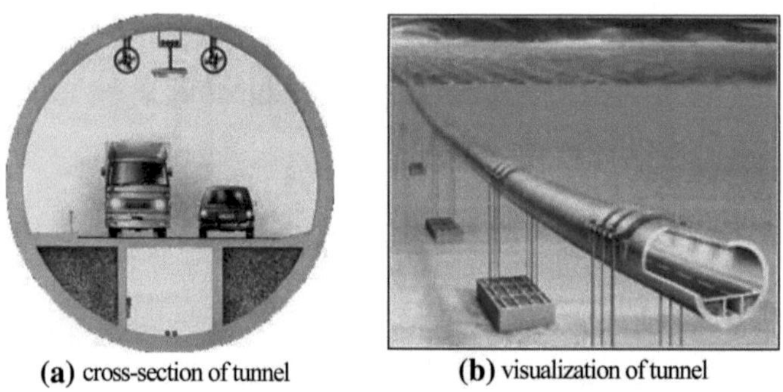

Fig. 26 Underwater tunnel "Høgsfjord", Photo: Statens Vegvesen Rogaland [36]. **a** cross-section of tunnel. **b** visualization of tunnel

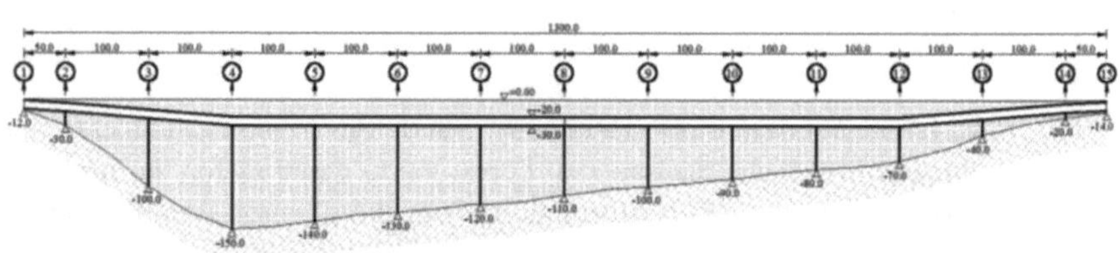

Fig. 27 Longitudinal section of "Høgsfjord" tunnel [35]

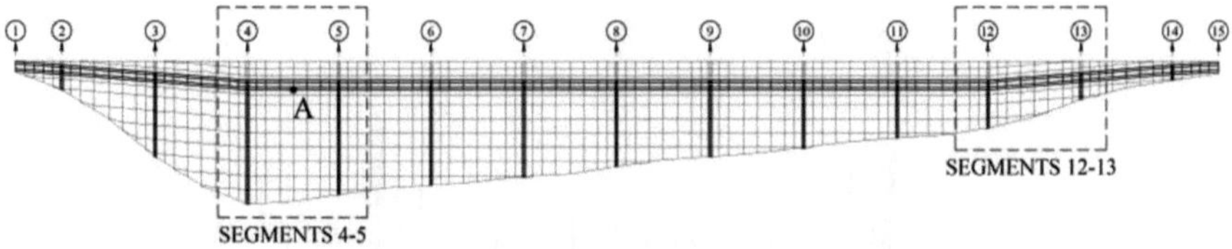

Fig. 28 Spatial discretization of "Høgsfjord" tunnel

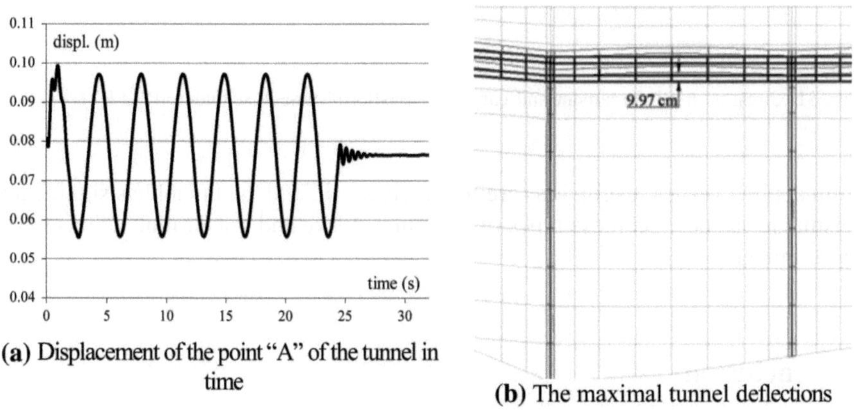

Fig. 29 Displacements of "Høgsfjord" tunnel, segments 4–5. **a** Displacement of the point "A" of the tunnel in time. **b** The maximal tunnel defelctions

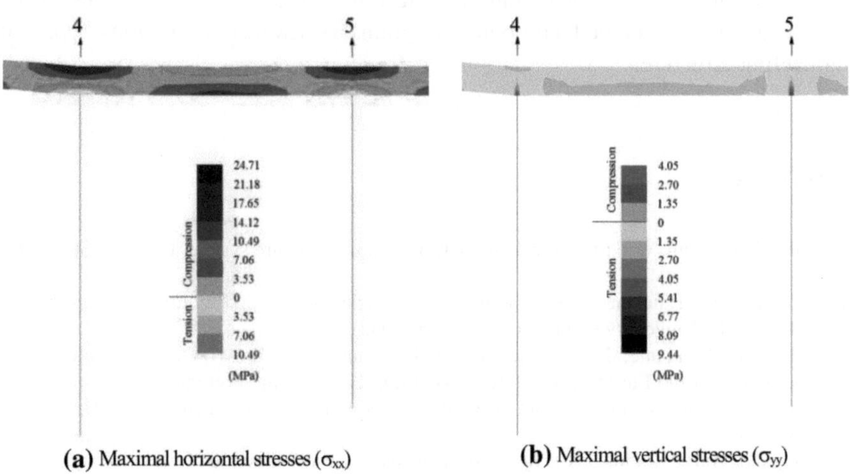

Fig. 30 Stresses of "Høgsfjord" tunnel, segments 4–5. **a** Maximal horizontal stresses (σ_{xx}). **b** Maximal vertical stresses (σ_{yy})

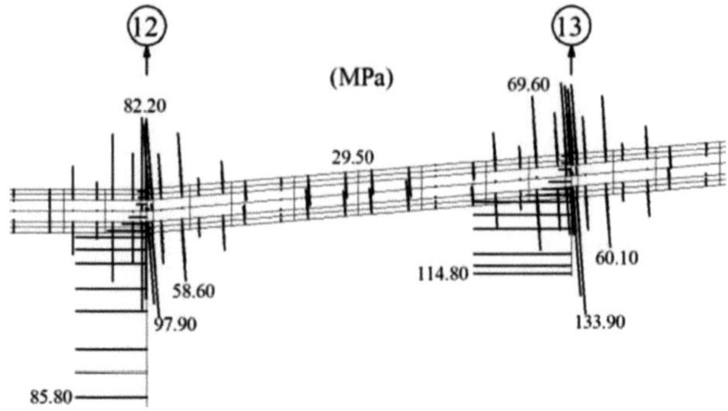

Fig. 31 Stresses in reinforcement and cables, "Høgsfjord" tunnel, segments 12–13

from applied vertical excitations are relatively small, and the tunnel has significant seismic resistance. On these types of structures, wave and sea current actions have more influence.

4 Conclusion

The presented models for the dynamic (seismic) analysis of various types of structures that are in contact with fluid can simulate some of the most important non-linear effects. The models are simple, reliable and can be used in a wide range of practical problems. Shown examples illustrate some of the possibilities of the models and the developed computer programs (software) for various types of engineering structures.

References

1. Bathe, K.J., Hahn, W.F.: On transient analysis of fluid-structure system. Comput. Struct. **10**, 383–391 (1979)
2. Paul Dilip, K.: Efficient dynamic solutions for single and coupled multiple field problems, Ph.D. Thesis, University College of Swansea (1982)
3. Radnić, J., Damjanić, F.B., Jović, V.: Hydrodynamic pressures on rigid structures. In: Proceedings of the European Conference on Earth Engineering, Portugal (1986)
4. Radnić, J.: Fluid-structure interaction with cavitation effect. Građevinar **7**, 269–275 (1987). (in Croatian)
5. Damjanić, F.B., Radnić, J.: Seismic analysis of fluid-structure interaction including cavitation. In: Proceedings of International Conference on Computer Modelling in Ocean Engineering, pp. 523–530. Balkema, Roterdam (1988)
6. Lofti, V.: Application of pseudo-symmetric technique in dynamic analysis of concrete gravity dams. Adv. Fluid Mech. **36**, 207–216 (2003)

7. Lofti, V.: Seismic analysis of concrete gravity dams by decoupled modal approach in time domain. Electron. J. Struct. Eng. **3** (2003) http://www.ejse.org/Archives/Fulltext/200301/06/20030106.pdf
8. Sekulović, M., Mrdak, R., Pejović, R., Mijušković, O.: Analysis of seismic response of high arch dam on basis of energy balance. 13th World Conference on Earthquake Engineering, Canada, Vancouver (2004)
9. Küçükarslan, S., Coşkun, S.B., Taşkin, B.: Transient analysis of dam-reservoir interaction including the reservoir bottom effects. J. Fluids Struct. **20**(8), 1073–1084 (2005)
10. Pin, F.D., Idelsohn, S., Oñate, E., Aubry, R.: The ALE/Lagrangian particle finite element method: a new approach to computation of free-surface flows and fluid-object interactions. Comput. Fluids **36**(1), 27–38 (2007)
11. Ortega, E., Oñate, E., Idelsohn, S.: An improved finite point method for three dimensional potential flows. Comput. Mech. **40**(6), 949–963 (2007)
12. Radnić, J.: Modelling of strain rate effects in dynamic analysis of R/C structures. Eng. Mod. **3**(1–2), 13–20 (1990)
13. Owen, D.R.J., Hinton, E.: Finite Elements in Plasticity. Pineridge Press, Swansea (1980)
14. Huang, H.C.: Static and Dynamic Analyses of Plates and Shells. Springer, Heilderberg (1989)
15. Bangash, M.J.H.: Concrete and Concrete Structures: Numerical Modelling and Applications. Elsevier Applied Science, New York (1989)
16. Radnić, J., Dešković, N.: Numerical model for dynamic analysis of RC structures including the strain rate effects. In: Proceedings of the 2nd International Conference on Computational Plasticity, Barcelona, pp. 65–71. Pineridge Press, Swansea (1989)
17. Phillips, D.V.: Numerical modelling of brittle materials; concrete and reinforced concrete. Lecture Notes on Nonlinear Engineering Computation, pp. C/1-78. TEMPUS-ACEM, Ljubljana, (1992)
18. Hofstetter, G., Mang, H.A.: Computational Mechanics of Reinforced Concrete Structures. Vieweg&Sohn, Weisbaden (1995)
19. Harapin, A., Radnić, J., Ćubela, D.: Numerical model for composite structures with experimental confirmation. Materialwissenschaft und Werkstofftechnik **39**(2), 143–156 (2008)
20. Galić, M., Marović, P., Nikolić, Ž., Harapin, A.: Numerical modelling of tension influences in 3D reinforced concrete structures. In: Onate E., Owen R., Suarez B. (eds.) Proceedings of the 10th International Conference on Computational Plasticity, pp. 539/1-539/4. CIMNE, Barcelona (2009)
21. Wilson, E.L., Yuan, M., Dickens, J.M.: Dynamic analysis by direct superposition of Ritz vectors. Earthq. Eng. Struct. Dyn **10**, 813–832 (1982)
22. Yuan, M., Chen, P., Xiong, S., Li, Y., Wilson, E.L.: The WYD method in large eigenvalue problems. Eng. Comput. **6**, 49–57 (1989)
23. Mihanović, A., Schönauer, M.: Modified WYD method in large dynamics eigen problems. In: Proceedings of the 19th Symposium of Yugoslav Society of Mechanics, Bled (1989) (in Croatian)
24. Harapin, A., Radnić, J., Brzović, D.: WYD method for an eigen solution of coupled problems. Int. J. Multiphysics **3**(2), 167–176 (2009)
25. www.indianetzone.com/34/koyna_dam_maharashtra.htm
26. Chopra, A.K., Chakrabarti, P.: The Koyna earthquake and the damage of Koyna dam. Bull. Seismol. Soc. Am. **63**, 381–397 (1973)
27. Krishna, J., Chandrasekaran, A.R., Saini, S.S.: Analysis of Koyna accelerogram of December 11, 1967. Bull. Seismol. Soc. Am. **59**(4), 1719–1731 (1969)
28. "Esperienze Statiche su Modello Della Diga di Grancarevo", I.S.M.E.S. Instituto Sperimentale Modelli e Strutture, Bergamo, , pratica no. 271 Settembre (1960) (in Italian)
29. "Sulla Stabilita' Della Roccia di Fondazione Della Diga di Grancarevo Verificata Anche a Mezzo Modello Geomeccanico", I.S.M.E.S. Instituto Sperimentale Modelli e Strutture, Bergamo, Settembre (1963) (in Italian)

30. Bičkovski, V., Bojadžiev, M.: Studies of static and seismic analysis of Grančarevo dam, The Institute of Earthquake Engineering and Engineering Seismology University "Ss. Cyril and Methodius" (IZIIS) Skopje, Macedonia, Report IZIIS 88-30, (1988) (in Serbian)
31. Harapin, A.: Numerical model of fluid-structure dynamic interaction. Ph.D. Thesis, University of Split, Faculty of civil engineering (2000)
32. Pejović, R., Mrdak, R., Živaljević, R., Mijušković, O.: An analysis of seismic resistance of the Grančarevo concrete dam. Građevinar **58**, 447–453 (2006). (in Croatian)
33. www.dubaiasitusedtobe.com
34. www.coastal.udel.edu
35. Šunjić, G.: Numerical model of seismic response of submerged structures. MD Thesis, University of Split, Faculty of civil engineering and architecture (2003) (in Croatian)
36. www.ntnu.no/gemini/1998-01E/36.html

Numerical Model for the Determination of the Soil Retention Curve from Global Characteristics Obtained via a Centrifuge

B. Malengier, J. Kačur and P. Kišoň

Abstract A novel centrifuge set-up for the study of unsaturated flow characteristics in porous media is examined. In this set-up, simple boundary conditions can be used, but a free moving boundary between unsaturated-saturated flow arises. A precise and numerically efficient approximation is presented for the mathematical model based on Richards' nonlinear and degenerate equation expressed in terms of effective saturation using the Van Genuchten–Mualem approach for the soil parameters in the unsaturated zone. Sensitivity of the measurable quantities (rotational moment, center of gravity and time period to achieve quasi steady state) on the soil parameters is investigated in several numerical experiments. They show that the set-up is suitable for the determination of the soil parameters via the solution of an inverse problem in an iterative way.

B. Malengier (✉)
Research Group NAM2, Department of Mathematical Analysis, Ghent University, Galglaan 2, 9000 Ghent, Belgium
e-mail: bm@cage.UGent.be

J. Kačur · P. Kišoň
Faculty of Mathematics, Physics and Informatics, Comenius University, Bratislava, Slovakia
e-mail: kacur@fmph.uniba.sk

P. Kišoň
e-mail: kisson@fmph.uniba.sk

A. Öchsner et al. (eds.), *Materials with Complex Behaviour II*,
Advanced Structured Materials 16, DOI: 10.1007/978-3-642-22700-4_11,

1 Introduction

To predict the flow and solute transport in soils, one needs the soil hydraulic properties in terms of soil parameters. Once determined, these parameters can be used as input data in the governing mathematical model. For unsaturated flow, this model is given in terms of the saturation and the pressure head in Richards' equation (see below), which is a nonlinear and degenerate parabolic equation. Furthermore, when part of the sample is saturated, free boundaries between the saturated zone and the partially saturated zone arise, as well as between the dry and the partially saturated zone. This is a major problem for many modeling approaches, leading to experimental set-ups that avoid the formation of these boundaries.

The soil retention and hydraulic permeability functions linking the saturation and pressure head for unsaturated flow are expressed using the Van Genuchten–Mualem ansatz by means of soil parameters. Measuring these soil parameters is usually time consuming and tedious, especially for low conductive porous media. Several set-ups based on centrifugation have been proposed to obtain a large acceleration of the processes involved, see [2–6] and citations therein. These techniques have several disadvantages. Aiming for a steady-state flow regime inside the centrifuge [2, 5] requires expensive and/or complex apparatus, and obtains only a few water content versus conductivity measurements per run. Also, transient set-ups based on keeping a top boundary at a fixed prescribed setting [3] are expensive. The quasi-steady centrifuge (QSC) method [1] is a simpler technique (a slowly emptying reservoir at the top that is refilled when needed), but requires that the criterion for steadiness of flow through the sample is relaxed, leading to higher uncertainty in the obtained results.

The alternatives for determining conductivity with a steady-state flow, combine transient flow with parameter estimation techniques, see e.g. [3, 6]. In this way, the conductivity and retention curve can be determined inversely over a large saturation domain. These methods require experiments of some state variables which relate to the conductivity. One-step or multi-step outflow methods are common in column experiments. The measurements are then used to estimate the hydraulic parameters. This technique is transferred to the centrifuge device in [6]. Good results are obtained, but there remain some disadvantages to this technique: there are few measurements close to saturation, leading to a high error in the prediction of the conductivity close to saturation, the sample needs to be disturbed to introduce electrodes, and there is a very long waiting time in order to achieve equilibrium when the equilibrium analysis approach is used.

The main goal of this manuscript is to develop a precise numerical method enabling to determine the soil parameters (via solution of inverse problem) in a very simple way requiring very cheap measurements.

In this chapter we focus on a partially saturated sample which is sealed at the right boundary (from the center of centrifuge) and has no inflow at the left boundary. The only measurements required are the rotational momentum and the center of gravity of the sample at several time values, preferably also at the equilibria corresponding to predetermined rotational speeds. These measurements are sufficient due to the fact that the saturation profile at the equilibria do not depend on the initial distribution of water in the specimen, but only on its amount, which, when the right boundary of the sample is sealed, is identical in all equilibria.

To use this procedure, we have to face serious difficulties in the numerical modeling. The main one is that if the right side of the sample reaches effective saturation, an interface between partially saturated zone and saturated zone appears. This boundary is very difficult to control numerically, causing problems with the mass balance conservation which is very important in this set-up.

To reach the equilibrium is an infinite asymptotic process, but after some time (e.g. 1–3 days for low conductive material) the change of the rotational momentum and of the center of gravity can no longer be measured. At that moment, the rotational speed is increased, and the system moves towards a new corresponding equilibrium. Note that even when equilibrium was not reached and a small error is present in the measurements of the rotational momentum and the center of gravity, this will not influence the error at the higher equilibrium level. This error depends only on the running time of centrifugation at the actual rotational speed. The differences between applied rotational speeds are chosen in such a way that that the differences in outputs (rotational momentum and center of gravity) are technically well distinguishable.

Next, the soil parameters and eventually the amount of originally infiltrated water, can be determined by minimizing a cost functional expressing the distance between the measured and the computed output, e.g., with the Levenberg–Marquard method. The advantage of this approach is that the full range of saturation values are present in the setup, while preventing outflow means equilibrium can be obtained faster. However, due to the set-up, it is clear that the water flows from the unsaturated zone to the saturated zone, with no flow occurring in the saturated zone. Indeed, we notice that the rotational momentum and center of gravity are not sufficiently sensitive on the "saturated hydraulic conductivity". This parameter is hence better determined from saturated flow experiments, see e.g. [4].

In the numerical method, we reduce the mathematical model to a system of ordinary differential equations (ODE) using the method of lines (MOL), which has already been successfully applied to Richards' equation in e.g. [7]. As a variation, a reduction to a system of ODE and algebraic equations (DAE) is considered. Our main contribution is in correctly handling the moving free boundary. The obtained system can be solved with ODE/DAE solvers for stiff systems. The numerical method can be successfully applied in other centrifugation settings (concerning control of the inflow, or control of the outflow) as, e.g., in [3, 6].

In Sect. 2, we present the mathematical model, giving specific attention to the movement of the free boundary. In Sect. 3 the numerical method based on the MOL approach is given, while in Sect. 4 the approach to determine the saturated hydraulic conductivity is explained. We finish in Sect. 5 with several numerical experiments showing the sensitivity of the output parameters on the soil parameters.

2 Mathematical Model

We consider a one dimensional model for a partially saturated sample in the form of a tube. The tube starts (top or left boundary) at the distance $r = r_0$ from the center of the centrifuge and ends at the distance $r = r_0 + L$. The right boundary of the specimen is isolated. Flow in porous media under centrifugation is modeled by Darcy's equation in the saturated region and by Richards' equation in the unsaturated region (see, e.g., [3, 6]). So

$$\partial_r \left[K_s \left(\partial_r h - \frac{\omega^2}{g} r \right) \right] = 0, \tag{1}$$

in the saturated region, and

$$\partial_t \theta = \partial_r \left[k(\theta) \left(\partial_r h - \frac{\omega^2}{g} r \right) \right], \tag{2}$$

in the unsaturated region. Here, h is the piesometric head, θ the saturation of the porous medium, ω the angular speed of rotation (in radians per second), K_s the hydraulic conductivity in the saturated region, g the gravitational constant and the function $k(\theta)$ describes the hydraulic conductivity in the unsaturated region. Denote by $u = \frac{\theta - \theta_r}{\theta_s - \theta_r}$ the effective saturation, where θ_s is the volumetric water content at saturation and θ_r is the residual volumetric water content. We have $u \in (0, 1)$, since $\theta \in (\theta_s, \theta_r)$. The soil hydraulic properties are represented by empirical expressions (see [8]),

$$u = \frac{1}{(1 + (\gamma h)^n)^m}, \quad h \in (-\infty, 0), \quad k(u) = K_s u^{1/2} [1 - (1 - u^{1/m})^m]^2, \tag{3}$$

where $m = 1 - 1/n$, $n > 1$ and γ are empirical soil parameters. Determining γ, n and K_s from the experiments means the soil retention curve has been determined. Note that we do not take hysteresis of the retention curve in consideration with this model.

It is possible to rewrite the flow in unsaturated form as

$$\partial_t u = \partial_r \left(D(u) \partial_r u - \frac{\omega^2}{g} k(u) r \right), \tag{4}$$

where

$$D(u) = -\frac{K_s}{(n-1)\gamma(\theta_s - \theta_r)} u^{1/2 - 1/m} (1 - u^{1/m})^{-m} \times [1 - (1 - u^{1/m})^m]^2. \tag{5}$$

Equation 4 is strongly nonlinear and degenerate. We note that $D(0) = 0$, $D(1) = \infty$. Equilibria at the high rotational speed can be expected to have a fully saturated zone (supposing the initial amount of infiltrated water is sufficiently large), which appears at the right boundary and of which the front evolves to the left of the specimen (under non-decreasing rotational speed). We denote the position of this interface by $s(t)$. This saturated zone is governed by Darcy's equation, but $s(t)$ is unknown and time dependent. The time evolution of $s(t)$ is difficult to compute. The dynamics of this region is linked with the (finite) interface flux q_i

$$q_i = -\left(D(u)\partial_r u - \frac{\omega^2}{g}k(u)r\right)\Bigg|_{r=s(t)},$$

and based on a mass balance argument we can expect $\dot{s}(t) = -q_i$. Unfortunately, we cannot use this model for the determination of the time evolution of $s(t)$, since at $r = s(t)$ it holds $u = 1$ and $D(1) = \infty$. Consequently, $\partial_r u|_{r=s(t)} = 0$.

If we transform Richards' equation in terms of the piesometric head using (3), we obtain

$$d_s(h)\partial_t h = k_0 \partial_r \left[\bar{k}(h)\partial_r h - \frac{\omega^2}{g}\bar{k}(h)r\right], \tag{6}$$

with $k_0 = \frac{K_s}{\theta_s - \theta_r}$, where $k_0\bar{k}(h)$ is the hydraulic conductivity function,

$$\bar{k}(h) = \frac{1}{(1 + (\gamma h)^n)^{m/2}}\left(1 - \frac{(\gamma h)^{n-1}}{(1 + (\gamma h)^n)^m}\right)^2,$$

and the specific moisture capacity function $d_s(h) = \mathrm{d}u/\mathrm{d}h$ is given by

$$d_s(h) = -\gamma(n-1)\frac{(\gamma h)^{n-1}}{(1 + (\gamma h)^n)^{1+m}}.$$

We can see that $\bar{k}(h) \to 1$ for $h \to 0$. In Fig. 1 we present the graph of the functions $\bar{k}(h)$ and $100d_s(h)$ for $h \in (-200, 0)$, and parameter values $K_s = 2.4 \times 10^{-5}$, $n = 2.81$, $\gamma = -0.0189$. As we can see, Eq. 6 also degenerates at $h = 0$. This has to be taken into account when saturation becomes 1 at the right boundary of specimen. After this moment, $t = t_1$, the mathematical model must be changed to reflect the physical phenomenon. At the right hand side of the (isolated) specimen appears a saturated zone with an interface $s(t)$ moving from the right boundary to the left. The flux at the interface $s(t)$ is equal to $-\dot{s}(t)$, but also in this pressure-head form of Richards' equation it is difficult to approximate correctly $\partial_r h|_{x=s(t)}$, which leads to a significant error in the mass balance.

Therefore, to determine the interface $s(t)$, we will consider the algebraic equation

$$\int_{r_0}^{r_0+L} u(h(x,t))\mathrm{d}t + L - s(t) = M_w, \quad s(0) = L, \tag{7}$$

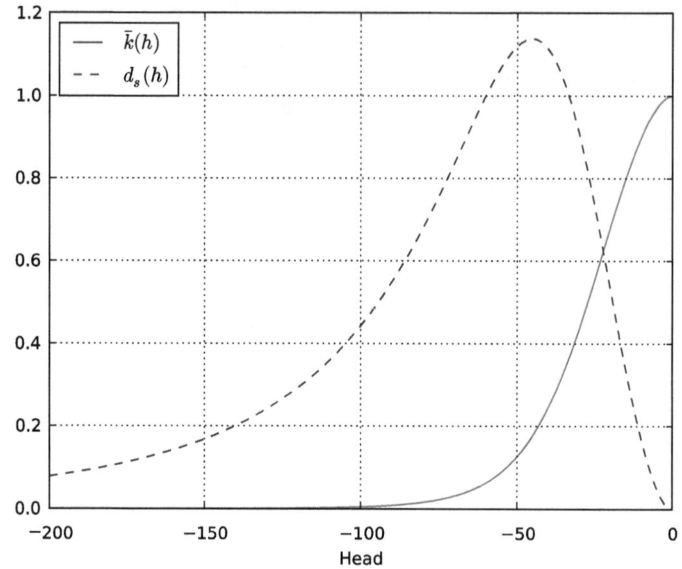

Fig. 1 $\bar{k}(h)$ and $100 \times d_s(h)$ for $n = 2.81$, $\gamma = -0.0189$

where M_w is the amount of infiltrated water (which remains constant during the centrifugation). This condition reflects the global mass balance in the specimen and does not suffer from a flux approximation at $r = r_0 + s(t)$.

Then, mathematical model (6) only needs to be solved over the interval $r \in (r_0, r_0 + s(t))$ with right boundary condition $h(r_0 + s(t)) = 0$ for all t. We approximate this mathematical model in the next section.

3 Numerical Method

For the output parameters that will be measured (gravity center and rotational momentum), there is no need to model the head in the saturated zone, as we consider the compressibility of water to be negligible. The numerical approximation of (6–7) results in a coupled system of a partial differential equation (PDE) and an algebraic equation. Moreover, the solution domain is a moving region, with unknown interface $s(t)$, which has to be determined.

We shift (6) to the domain $r \in (0, s(t))$ and use the fixed domain transformation $y = \frac{r}{s(t)}$. This gives

$$d_s(h)\left(\mathrm{d}_t h(y,t) - y\frac{\dot{s}(t)}{s(t)}\partial_y h\right) = k_0 \frac{1}{s(t)^2}\partial_y\left(\bar{k}(h)\partial_y h - \bar{k}(h)\frac{\omega^2 s}{g}(r_0 + ys(t))\right). \quad (8)$$

Consider the space discretization $0 = y_0$ and $\alpha_0 = 0$, $\alpha_i := y_i - y_{i-1}$, $i = 1, \ldots, N$ and integrate (8) over $I_i := (y_{i-1/2}, y_{i+1/2})$ for $i = 1, \ldots, N-1$ where $y_{i-1/2} := (y_i + y_{i-1})/2$, $y_{i+1/2} := (y_i + y_{i+1})/2$.

We denote by $h_i(t) \approx h(y_i, t)$, $\forall i = 1, \ldots, N-1$, and approximate $d_t h(y,t) \approx \dot{h}_i(t)$ in the interval I_i. We approximate

$$\partial_y h\big|_{y=y_{i+1/2}} \approx \frac{h_{i+1}(t) - h_i(t)}{\alpha_{i+1}} =: \partial^+ h_i$$

and similarly we approximate $\partial_y h\big|_{y=y_{i-1/2}}$ and denote it by : $\partial^- h_i$. Let $\mathscr{L}(z; y_i)$ be the second order Lagrange polynomial crossing the points (y_{i-1}, h_{i-1}), (y_i, h_i) and (y_{i+1}, h_{i+1}). We use the abbreviation $k_{i+1/2} := \bar{k}(h_{y_{i+1/2}})$. Then, the approximation of (8) (based on finite volume type approximation) at the point $y = y_i$ reads as follows

$$d_s(h_i)\left(\dot{h}_i - \frac{\dot{s}y_i d\mathscr{L}(z;y_i)}{s \quad dz}\bigg|_{z=y_i}\right) = k_0 \frac{2}{\alpha_i + \alpha_{i+1}} \frac{1}{s^2}\Big[k_{i+1/2}\partial^+ h_i - k_{i-1/2}\partial^- h_i$$
$$- \frac{\omega^2 s}{g}\left(k_{i+1/2}(r_0 + s y_{i+1/2}) - k_{i-1/2}(r_0 + s y_{i-1/2})\right)\Big] \tag{9}$$

for $i = 1, \ldots, N-1$. We add the corresponding equation at point y_0 taking into account that the flux is zero there. In a similar way as in (9) (following the finite volume type of approximation) we obtain

$$d_s(h_0)\dot{h}_0 = k_0 \frac{2 \quad 1}{\alpha(1) s^2} \times \left[k_{1/2}\partial^+ h_1 - \frac{\omega^2 s}{g}\left(k_{1/2}(r_0 + s y_{1/2})\right)\right]. \tag{10}$$

At the point $y_N = 1$ we have $h_N(t) = 0$, so no additional equation is considered. We approximate the amount of water M_w using the trapezoidal rule for the integration. Define

$$Q(t) \approx u_0 \alpha_1/2 + \alpha_N/2 + \sum_1^{N-1} \frac{\alpha_i + \alpha_{i+1}}{2} u_i,$$

where $u_i = \frac{1}{(1+(\gamma h_i)^n)^m}$. Then, system (9–10) will be completed by the algebraic equation

$$0 = L - s(t)[1 - Q(t)] - M_w. \tag{11}$$

This algebraic equation is used instead of an ODE equation that models $\dot{s}(t)$. System (9–11) is degenerate and is of the form

$$M(t,z)\dot{z}(t) = f(t,z) \tag{12}$$

where $z = (h_0, h_1, \ldots, h_{N-1}, s)$. The last equation of this system is just (11). This system can be readily solved, e.g., by the solver "ode15s" in MATLAB® or the "ida" solver of the Sundials package.

As is usual with these solvers, some regularization in (11) is needed as well as a tuning of the space discretization. Most important is to have a "good" starting point.

If the equilibria have the property $h_N < 0$, then no interface appears. It is then needed to set $s(t) := L$ in the previous mathematical model and replace algebraic equation (11) by an ODE equation for $\dot{h}_N$ which will be similar to (10). Successively increasing the rotational speed of the centrifuge increases the head at the right boundary. The model remains in the state where $s(t) := L$ up to the point when $h(N) = 0$, at which point the computation is automatically halted. The full model (9–11) is used onwards to compute the equilibrium states.

In numerical equilibrium experiments it is observed, as expected, that the values of the rotational moment M_r and the center of gravity G_c are not very sensitive to the K_s parameter. Also, the transient experiments where the time sections between different equilibria are measured, are not very sensitive. The saturated conductivity K_s can only be determined from measurements of M_r, G_c that are accurate up to three digits. Therefore, another method must be used for the determination of K_s.

4 Alternative Experiments

4.1 Saturated Flow

For the determination of the saturated conductivity, we propose to use the method put forward in [4], a water reservoir put to the left of a saturated sample and collection of the water in an outflow reservoir, with the addition of allowing for transient measurements. We specifically use the ability to measure when a reservoir has completely drained out, combined with the measurements of the rotational moment.

This leads to the following equation for the dropping water level $\ell(t)$ in the reservoir,

$$\dot{\ell}(t) = -K_s \frac{\omega^2}{2gL}\Big[L^2 - \ell(t)^2 + 2r_0(L + \ell(t))\Big] \equiv -q_F(t), \tag{13}$$

with $\ell(0) = l_0$ and $\ell(T_e) = 0$. Solving this ODE, we obtain the relation between T_e and K_s, whereas $\ell(t)$ fully determines the change of the rotational moment $M_r(t)$ over time.

4.2 Water Reservoir and Outflow Reservoir

The mathematical and numerical model presented can be extended also to allow for a water reservoir to the left of an unsaturated sample, and an outflow reservoir to the right. This allows several different centrifugation experiments to be performed, and allows to change the set-up during an experiment. For example, the following scenario is possible: 1. Start from a saturated sample and a water reservoir to the left. This makes it possible to determine K_s. 2. Continue with outflow of the water content, making the sample unsaturated. 3. Isolate the right boundary (that is, close it), which means we have the problem as described in the previous two sections. 4. Continue step 2 and 3 of above so as to change the global water content.

The advantage of the above centrifugation scenario is that all parameters can be determined with one ground sample, and that more saturation levels are sampled during the entire experiment. The main point to arrive at an accurate solution of the model doesn't change: an algebraic equation for mass balance determines the difficulty to control unknowns. With a closed right boundary, this is the moving interface, with an open boundary, this will be the outflow flux.

5 Numerical Experiments

For the first experiments we use as data $r_0 = 10$, $L = 10$, $\omega = 30$, $K_s = 2.4 \times 10^{-5}$, $\theta_r = 0.02$, $\theta_s = 0.4$, $\gamma = -0.0189$, $n = 2.81$, except where sequences are compared to investigate the sensitivity of the set-up on the parameters. A uniformly distributed space discretization with $N = 40$ grid points is used.

The formulas for M_r, G_c and M_w at time t are:

$$M_r = \frac{s(t)}{2} \int_0^1 (r_0 + s(t)z)^2 u(t,z)dz + \frac{1}{6}(L^3 - s(t)^3),$$

$$M_w = s(t) \int_0^1 u(t,z)dz + \frac{1}{2}(L^2 - s(t)^2), \quad G_c = s(t) \int_0^1 yu(t,z)dz/M_w,$$

and are all evaluated numerically using the trapezoidal rule. Note that if $u(t,1)$ then $s(t) = L$. The sensitivity of the measured quantities on the changing water content is very good. The following experiments allow to determine the contribution a change in the different soil parameters has on the measurements.

Table 1 Rotational momentum, center of gravity, water amount for Exp. 5.1

$\frac{\text{time}}{2000}$ s	$M_{r,e} \times 10^{-6}$	G_c	M_w
0	1.5201	7.2512	4.0141
1	1.5248	7.2813	4.0128
3	1.5299	7.3119	4.0131
7	1.5345	7.3413	4.0133
15	1.5389	7.3697	4.0134
31	1.5430	7.3972	4.0134
63	1.5469	7.4234	4.0135
127	1.5505	7.4478	4.0135
255	1.5537	7.4699	4.0136
511	1.5565	7.4893	4.0136
770	1.5588	7.5056	4.0136
1800	1.5645	7.5462	4.0133
2300	1.5649	7.5494	4.0133

5.1 Reaching Equilibrium

To investigate the head profiles we start this experiment from the equilibrium corresponding to $\omega = 40$ and a rotational speed $\omega = 50$. The centrifuge normally operates up to $T_e = 1.540.000$ seconds. At that time, equilibrium for $\omega = 50$ is almost reached. We compare 13 values, the starting value, nine increasing time steps (with $\Delta t_j = t_{j+1} - t_j = 2000 \times 2^j, j = 1, \ldots, 9$), the sensible end time step $T_e = 770 \times 2.000$s, and two extra time steps to investigate the very long time behavior. The measured values for the rotational momentum, gravity center, and water amount, are given in Table 1. The small change between the last two values in Table 1 demonstrates that equilibrium is eventually reached.

We can conclude that reaching equilibrium is a very slow process. The reason for this is that the hydraulic permeability at low head is negligibly small, so it takes a very long time to reach the equilibrium. If the centrifugation is continued, also the section with low head obtains the required parabolic shape associated with the equilibrium. Note however, that the other part of the head profile (for higher head values) is changing insignificantly. Therefore, we arrive at the conclusion that it makes sense to increase the rotational speed and not wait for these lower head values to stabilize.

5.2 Dependence on n

In this experiment, we demonstrate the sensitivity of M_r and G_c to the model parameter n. We start with a constant saturation $u = 0.4$ and apply the rotational speed $\omega = 20$. The centrifuge is operated for 800.000s. In Fig. 2 the obtained

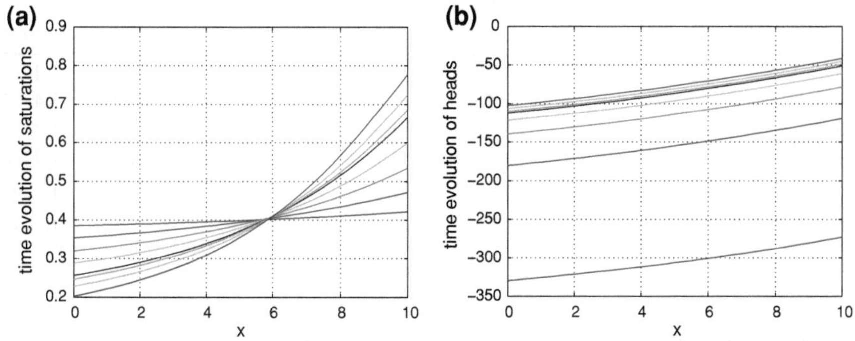

Fig. 2 Equilibrium profiles at $\omega = 20$ for n = 1.51; 1.81; 2.11; 2.41; 2.71; 2.81; 3.01 and 3.31. **a** Saturation profiles, **b** head profiles

Table 2 Rotational momentum, center of gravity and water amount for Exp. 5.2

n	$M_r e \times 10^{-6}$	G_c	M_w
1.51	0.1887	5.0736	4.0043
1.81	0.1927	5.2391	4.0052
2.11	0.2020	5.6188	4.0043
2.41	0.2068	5.8096	4.0062
2.71	0.2083	5.8701	4.0054
2.81	0.2112	5.9870	4.052
3.01	0.2153	6.1524	4.0060
3.31	1.5505	7.4478	4.0135

equilibrium profiles are depicted for successively n = 1.51; 1.81; 2.11; 2.41; 2.71; 2.81; 3.01 and 3.31.

The resulting values for M_r, G_c and M_w are given in Table 2, and indicate a good sensitivity.

5.3 Dependence on γ

We now investigate the sensitivity of M_r and G_c to the γ soil retention curve parameter. We again use a rotational speed of $\omega = 50$, starting from the equilibrium position at $\omega = 35$. As values for γ we consider $\gamma = -\gamma_0 \times 10^2$ with $\gamma_0 \in (1.59; 2.19)$ where increments of size 0.1 are used. The values of M_r and G_c are listed in Tables 3 and 4, respectively. The corresponding saturation and head profiles at time section $t = 10^5$ are given in Fig. 3. In Tables 3 and 4 the water amount is 4.05. The sensitivity on γ is less than that of n, but is sufficient.

Table 3 Rotational momentum $M_r \times 10^{-6}$ for Exp. 5.3

time\γ_0	1.59	1.69	1.79	1.89	1.99	2.09	2.19
1000	1.5189	1.5093	1.5013	1.4949	1.4896	1.4854	1.4819
3000	1.5352	1.5213	1.5097	1.5000	1.4919	1.4852	1.4797
5000	1.5438	1.5278	1.5144	1.5030	1.4935	1.4855	1.4788
10^4	1.5565	1.5376	1.5216	1.5079	1.4963	1.4864	1.4780
5×10^4	1.5901	1.5645	1.5423	1.5231	1.5063	1.4917	1.4791
10^5	1.6058	1.5777	1.5530	1.5313	1.5124	1.4958	1.4812

Table 4 Center of gravity for Exp. 5.3

time\γ_0	1.59	1.69	1.79	1.89	1.99	2.09	2.19
1000	7.1117	7.1031	7.0948	7.0872	7.0805	7.0744	7.0693
3000	7.1487	7.1379	7.1271	7.1170	7.1076	7.0990	7.0912
5000	7.1697	7.1581	7.1464	7.1351	7.1245	7.1146	7.1056
10^4	7.2026	7.1903	7.1775	7.1649	7.1527	7.1411	7.1303
5×10^4	7.3012	7.2898	7.2763	7.2617	7.2466	7.2314	7.2167
10^5	7.3512	7.3429	7.3309	7.3167	7.3012	7.2851	7.2689

Nevertheless, taking transient information into account, as given in the rows of Tables 3 and 4, will benefit the determination of γ via this experimental set-up.

5.4 Inverse Determination of γ, **n** *and* K_s

In this numerical experiment, we use $r_0 = 30\, L = 10, \omega = 20$, $K_s = 2.4 \times 10^{-5}$, $\theta_r = 0.02, \theta_s = 0.4$, $\gamma = -0.0189$ and $n = 2.81$. The space discretization for $T \in (0, T_1)$ is not equidistant. Here T_1 is the time needed to empty the left water reservoir. We shall consider $N = 40$ grid points with geometrical distribution as follows. The first space interval is $d_1 = 1/20$ and then $d_{i+1} = qd_i$ with q Once the water reservoir is empty, a uniform space discretization with $N = 40$ is used.

In this experiment, we restore the soil parameters applying the following centrifugation scenario. First, we centrifugate the fully saturated sample along the time 10^4 and collect $data_1 = \{M_1, G_1, M_{w,1}\}$. Then, we isolate the right boundary of the sample and centrifugate it tor $t = 5000$ s with rotational speed $\omega = 15$. Then, we obtain $data_2 = \{M_2, G_2, M_{w,2}\}$ (where $M_{w,2} = M_{w,1}$, since we have zero output). After this, we repeat these two steps with the same sample at the same running time $t = 5000$ s and $\omega = 20$. Successively we obtain $data_3 = \{M_3, G_3, M_{w,3}\}$, $data_4 = \{M_4, G_4, M_{w,4}\}$ where $(M_{w,4} = M_{w,3})$ and continue up to $data_7$. Then, the total measurement data is represented by the vector $data = \{data_1, data_2, \ldots, data_7\}$. To imitate a realistic situation, we perturb every

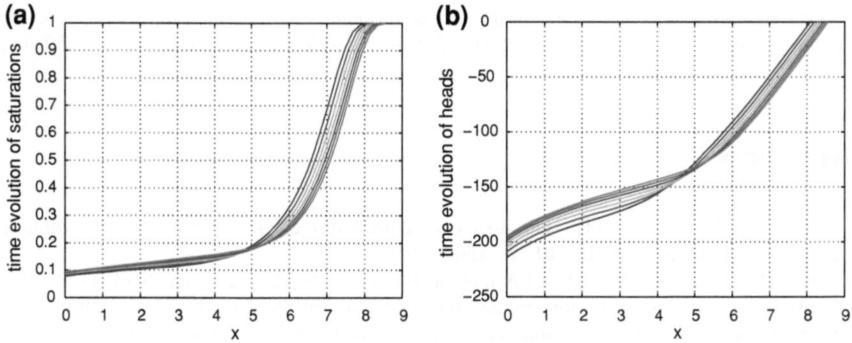

Fig. 3 Saturation profiles at equilibrium for $\omega = 50, \gamma_0 = 1.59; 1.69; 1.79; 1.89; 1.99; 2.09; 2.19$. **a** Saturation profiles, **b** head profiles

Table 5 LM-iterations for determination of γ, n, K_s

Iteration	$-100 \times \gamma$	n	$K_s \times 10^5$	*RMS*
0	1	2	1.6	3.3977
1	1.7643	3.4612	1.6149	5.057×10^{-2}
3	1.8555	2.8423	2.2320	2.878×10^{-4}
5	1.8469	2.8519	2.2185	1.506×10^{-4}

component of *data* by $0.01 \cdot (\text{rand} - 0.5)$, where rand is a generator of random numbers from (0,1). This corresponds to 0.5% noise. Next, we apply the Levenberg–Marquardt method to restore the soil parameters, starting from initial parameters $\gamma = -0.01, n = 2, K_s = 1.6 \times 10^{-5}$. The corresponding iterations of the LM method are presented in Table 5.

6 Conclusion

In this chapter, it is shown that global characteristics measured with a centrifuge can be used to determine the soil retention curve of ground samples. In order for this to work, transient data must be used, different centrifugation scenario's must be coupled to obtain sufficient information, and a very precise numerical model must be used. Specifically, this model must be able to accurately track the moving interface. We further draw attention that in the alternative scenario using outflow, no outflow boundary condition is imposed. This gives more freedom to the experimentator. Instead, for all simulations, an algebraic equation based on mass balance is used to obtain a solution.

Acknowledgments The second and third author confirm financial support by the Slovak Research and Development Agency under contract APVV-0184-10, APVV-0743-10 and VEGA-1/0502/09.

References

1. Caputo, M.C., Nimmo, J.R.: Quasi-steady centrifuge method for unsaturated hydraulic properties. Water Resour. Res. **41**, W11504 (2005)
2. Conca, J.L., Wright, J.V.: The UFA method for rapid, direct measurements of unsaturated transport properties in soil, sediment and rock. Aust. J. Soil Res. **36**, 291–315 (1998)
3. Constales, D., Kačur, J.: Determination of soil parameters via the solution of inverse problems in infiltration. Comput. Geosci. **5**, 25–46 (2004)
4. Nimmo, R.J., Mello, K.A.: Centrifugal techniques for measuring saturated hydraulic conductivity. Water Resour. Res. **27**(6), 1263–1269 (1991)
5. Nimmo, R.J., Akstin, K.C., Mello, K.A.: Improved apparatus for measuring hydraulic conductivity at low water content. Soil Sci. Soc. Am. J. **56**, 1758–1761 (1992)
6. Šimůnek, J., Nimmo, J.R.: Estimating soil hydraulic parameters from transient flow experiments in a centrifuge using parameter optimization technique. Water Resour. Res. **41**, W04015 (2005)
7. Tocci, M.D., Kelley, C.T.: Accurate and economical solution of the pressure-head form of Richards' equation by the method of lines. Adv. Water Resour. **20**(1), 1–14 (1997)
8. van Genuchten, M.Th.: A closed-form equation for predicting the hydraulic conductivity of unsaturated soils. Soil Sci. Soc. Am. J. **44**, 892–898 (1980)

Numerical Investigation of Chatter in Cold Rolling Mills

S. K. Dwivedy, S. S. Dhutekar and P. Eberhard

Abstract In this work, considering a four high cold rolling mill and using a dynamic friction model, expressions for the variation of pressure in the roll bite have been developed. The effects of parameters used in the dynamic friction model on the variation of pressure and shear stress are investigated. The numerically obtained horizontal and vertical work roll deflections using the dynamic friction model have been compared with those obtained by the conventionally used constant friction model. The effects of rolling parameters like strip thickness; periodic back tension and strip velocity on the work roll deflections have been studied. This work will find applications in predicting the critical system parameters in cold rolling to avoid chatter.

Keywords Chatter · Cold rolling mill · Dynamic friction model · Constant friction model · Roll bite

Nomenclature

M	Mass per unit length of the work roll (kg/m)
y	Vertical displacement of the work roll (m)
f^s	Reaction force from metal sheet (N/m)
D_w	Diameter of work roll (m)
D_b	Diameter of backup roll (m)
E	Young's modulus of the material (GPa)

S. K. Dwivedy (✉) · S. S. Dhutekar
Mechanical Engineering Department, Indian Institute of Technology Guwahati, Guwahati 781309, India
e-mail: dwivedy@iitg.ernet.in

P. Eberhard
Institute of Engineering and Computational Mechanics, University of Stuttgart, Pfaffenwaldring 9, 70550 Stuttgart, Germany
e-mail: eberhard@itm.uni-stuttgart.de

A. Öchsner et al. (eds.), *Materials with Complex Behaviour II*, Advanced Structured Materials 16, DOI: 10.1007/978-3-642-22700-4_12,

μ	Poisson's ratio of the material
f_s^s	Steady sheet force (N/m)
f_d^s	Dynamic sheet force (N/m)
y_s	Work roll displacement due to the steady sheet force (m)
y_d	Work roll displacement due to the dynamic part of sheet force (m)
$\dot{y}_d$	Rate of change of dynamic roll gap displacement (m/s)
h_c	Gap between two work rolls (m)
h_{c0}	Gap between two work rolls at $t = 0$ (m)
$\dot{h}_c$	Rate of change of roll gap (m/s)
ω_n	Natural frequency of the system not considering f_d^s (Hz)
h_1	Strip thickness at entry (m)
h_2	Strip thickness at exit (m)
R	Radius of work roll (m)
u_1	Strip velocity at entry (m/s)
τ_y	Strip shear yield strength (MPa)
σ_{XX}	Normal stress in X-direction (MPa)
σ_{XY}	Normal stress in Y-direction (MPa)
τ_{XY}	Shear stress (MPa)
m	Contact friction coefficient between the work roll and the strip
τ_s	Shear stress at the surface of strip (MPa)
x_1	Distance measured from strip entry to the centerline of rolls (m)
x_2	Strip exit position (m)
x_n	Distance of neutral plane from the centerline of rolls (m)
m_1	Friction factor between x_n and x_1 (considered positive)
m_2	Friction factor between x_n and x_2 (considered negative)
p	Roll pressure (MPa)

1 Introduction

Chatter in rolling mills is the undesirable mechanical vibration observed in most of the rolling mills operating at high speed and rolling thin strips. It results in unacceptable gauge variations in the rolled strip, affects the surface quality, damages mill components and produces undesirable noise. Hence, it may lead to loss in productivity if sufficient care hasn't been taken to prevent it. Chatter is a type of self excited vibration which is believed to arise in rolling operations as a consequence of the interaction between structural dynamics of the rolling mill stand and the dynamics of the rolling operation itself. Generally three types, i.e., torsional, third octave and fifth octave chatters are observed in rolling mills. Torsional chatter is related to torsional vibrations and is generally observed in 5–15 Hz range. The other two types are related to vertical vibrations of the roll system. While third octave chatter falls in the range of 125–240 Hz, fifth octave chatter falls in the range of 550–650 Hz. These are characterized by the frequency range of musical octaves. It has been observed that third octave chatter is responsible for gauge variations and fifth octave chatter causes damage to the backup roll and in turn parallel strip marks are found in the rolling sheet. Over the years there has been lot of research on chatter in rolling mills. A brief review of the dynamics and chatter model developed by different researchers is given below.

Roberts [1] considered the rolling mill as a single spring-mass system and analytically predicted the natural frequency to show the origin of fifth octave chatter. Chefneux et al. [2] developed a numerical model to show that chatter tends to occur when there is a sudden change in rolling force which may be due to a weld line or sudden change in lubrication. Johnson and Qi [3] and Johnson [4] studied the effect of friction and inelastic deformation on chatter in sheet rolling and developed analytical expressions for rolling force and corresponding spring and damping coefficients. They have shown that inter stand tension plays a great role in the chatter of rolling mill. They modeled the mill as two degrees and four degrees of freedom systems and explained fifth octave chatter. Yun et al. [5] reviewed different chatter models and in a series of papers [6–8] developed correlation between different rolling parameters such as strip speed, tension at entry and exit, rolling force and rolling torque. Performing experiments, they tried to understand the conditions which lead to the dynamic instability and proposed that negative damping, mode coupling and regeneration are the basic mechanisms which lead to chatter in rolling. To carry out the investigation of chatter due to negative damping, a unimodal structural model (the structure is allowed to oscillate only in one direction perpendicular to the flow of strip) was formulated. There they coupled a dynamic rolling model with an unimodal chatter model and simulated the results to show roll force, roll gap, back tension variations leads or lags in phase to produce negative damping. In mode coupling, they took roll vibrations in more than one direction and attempted to show this as one of the causes of chatter [8].

Hu and Ehmann [9] proposed a dynamic rolling model considering homogenous material and the movement of the roll in both directions has been considered. Using a linearised model they validated their results with experiments. They also developed a dynamic rolling model considering non-homogenous material [10]. It may be noted that in these works, chatter studies have not been carried out. A constant friction model has been considered in these works.

Kimura et al. [11] studied the chatter problem in a five stand continuous rolling mill and showed the influence of rolling speed and friction coefficient on vibration in rolling mill. They deduced an optimal range of friction coefficients and proposed a stability index to damp the vibration and make the mill stable against any disturbance. Meehan [12] also developed a comprehensive stability criterion for the third octave rolling chatter and suggested a critical rolling speed below which chatter does not occur. He used a spring-mass-damper model and with the help of simplified block diagrams illustrated the roll stack vibrations and inter stand tension interaction. Lin et al. [13] presented a nonlinear dynamic model to describe the dynamic interaction between work rolls and metal sheets and obtained critical speed for initiation of fifth octave chatter. They coupled a work roll sub model with a roll bite sub model to produce simulation results and concluded that even though the governing dynamics is highly nonlinear, the rolling chatter instability is nothing else than mode excitation or beating and thus linear. Their results correspond well with a physical rolling mill and the frequency predicted for a 4-H rolling mill falls within 550–650 Hz (fifth octave chatter range). Niziol and Swiatoniowski [14] and Hu et al. in a series of paper [15, 16] extended their previous work [9, 10] to study third octave chatter in a single stand [15] and a multi-stand mill [16]. Their model combined a homogeneous process model with suitable mill structure models to obtain a linear characteristic equation which was investigated to study the stability of the system.

In all of the mentioned cases, either constant friction or Coulomb friction models have been used to obtain the expression for roll force and the system stiffness and damping properties of the roll. However, these models do not correctly predict distribution of contact stresses in flat rolling, particularly in the neutral plane. Tan et al. [17] proposed a dynamic friction model considering both friction and viscosity effects in flat rolling. They have validated their model using the published experimental data.

In the present work, an attempt has been made to investigate the chatter phenomena by combining the roll-bite sub-model of Hu et al. [9] and the dynamic friction model of Tan et al. [17] for a four-high cold rolling mill. It may be noted that unlike constant friction or Coulomb friction models where only the coefficient of friction (μ) is required to obtain the shear stress during rolling, in the case of a dynamic friction model one requires four parameters $(\beta_p, k_p, \beta_f, k_f)$ for doing the computation. Initially, numerical investigations have been carried out to study the effect of these parameters on the variation of pressure and shear stress in the roll bite. Then, a comparative study has been made between the deflection obtained using dynamic friction model and the constant friction model for different strip velocity, strip thickness and back tension.

2 Mathematical Model

Figure 1a shows the schematic diagram of a four high stand rolling mill with the metal strip passing through the work rolls which are supported by backup rolls. It is assumed that the physical contact between work roll and backup roll can be modeled as springs, as shown in Fig. 1b. The damping between them is assumed to be negligible. Further, it is assumed that the work roll and backup roll are always in contact during rolling process.

Considering both upper and lower work rolls, a chatter model of work rolls has four degrees of freedom. These are vertical and horizontal displacements of both the work rolls. Considering the symmetry of rolling process, the degrees of freedom can be reduced to the horizontal and vertical displacement of any of the two

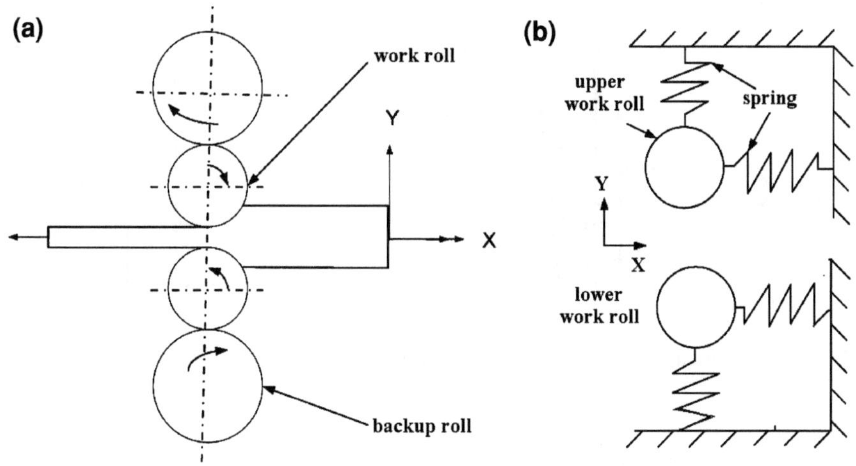

Fig. 1 **a** Schematic representation of a four-high rolling mill, **b** representation of rolling mill by a spring-mass system

work rolls. Now, considering the vibration of the upper roll only, see Fig. 1b, and applying Newton's second law, one can write the governing differential equation of motion for a spring mass system in vertical and horizontal directions as follows.

$$M\frac{d^2y}{dt^2} + K_1 y = f_y^s, \frac{d^2x}{dt^2} + K_2 x = f_x^s \tag{1}$$

Here M is the mass per unit length of work roll, K_1 is the spring constant which represents the contact between the work roll and the backup roll, see Fig. 1, K_2 represents the stiffness of the equivalent spring in the horizontal direction, x and y are the deflections of the work roll in horizontal and vertical direction, respectively. Also, f_x^s and f_y^s, respectively, represent force per unit width on work rolls in horizontal and vertical directions while rolling. The expression for these forces can be obtained by using different friction models such as a constant friction model, Coulomb's friction model or a dynamic friction model. While the expression for forces for a constant friction model has been obtained by Hu and Ehmann [9], here these expressions for a dynamic friction model are derived.

Following the procedure of Lin et al. [13], the total sheet force and total displacement can be divided into steady and dynamic parts. Hence, they can be written as

$$f^s = f_s^s + f_d^s \text{ and } y = y_s + y_d \tag{2}$$

Considering homogeneous deformation of the sheet, the different quantities used in the roll-bite are shown in Fig. 2a. The roll gap in vertical direction h_c and in the horizontal direction x_c and their rate of change of roll gap $\dot{h}_c$ and $\dot{x}_c$in respective directions are given by

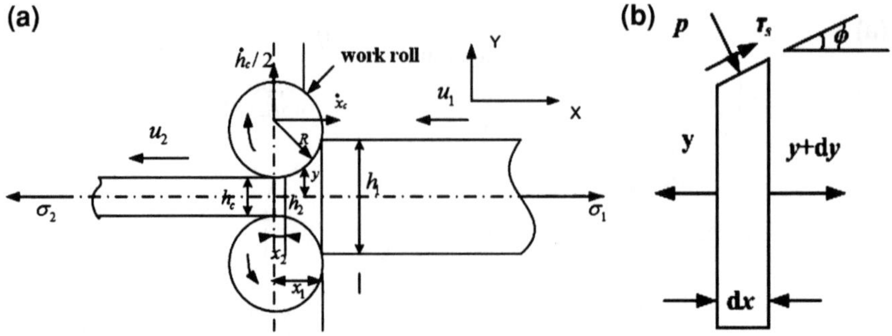

Fig. 2 **a** Roll bite model, **b** Half slice of the roll bite

$$h_c = h_{c0} + 2y_d,\ \dot{h}_c = 2\dot{y}_d \text{ and } x_c = x_{co} + x_d,\ \dot{x}_c = \dot{x}_d \tag{3}$$

The quantities h_{c0} and x_{c0} are the initial roll gap in vertical and horizontal direction i.e., when time t is equal to 0. From Fig. 2a, the expression for strip thickness h varying in the roll bite in terms of the vertical roll gap h_c, horizontal roll gap x_c, radius of work roll R is given by

$$h = h_c + (x - x_c)^2 \Big/ R \tag{4}$$

The half varying thickness is $y = h/2$. It is defined so as to properly apply the boundary conditions. Applying material volume preservation, one can obtain

$$uh = u_1 h_1 - (x_1 - x)\dot{h}_c + (h_1 - h)\dot{x}_c \tag{5}$$

where u is the strip velocity, u_1 and $h_1$are the strip velocity and thickness, respectively, at the inlet. Differentiating u with respect to y we get

$$\frac{\mathrm{d}u}{\mathrm{d}y} = \frac{-(\dot{x}_c + u)}{y} \tag{6}$$

The contact length at inlet x_1 is given as,

$$x_1 = x_c + \sqrt{R(h_1 - h_c)} \tag{7}$$

The position of the strip at the exit of roll bite x_2 is given as

$$x_2 = x_c + \frac{R h_c \dot{h}_c}{2\left(u_1 h_1 - (x_1 - x)\dot{h}_c + h_1 \dot{x}_c\right)} \tag{8}$$

The location of the neutral point x_n can be determined by equating the strip velocity with the roll velocity. Then x_n can be given by

$$x_n = \frac{2x_c(v_r + \dot{x}_c) + \dot{h}_c R + \sqrt{\left(2x_c(v_r + \dot{x}_c) + \dot{h}_c R\right)^2 - 4(v_r + \dot{x}_c)x_c^2(v_r + \dot{x}_c)}}{2(v_r + \dot{x}_c)} \tag{9}$$

The exit strip velocity u_2 can be determined from Eq. 5 as

$$u_2 = \frac{u_1 h_1 - (x_1 - x_2)\dot{h}_c + (h_1 - h_2)\dot{x}_c}{h_2}. \tag{10}$$

From Fig. 2b, the equilibrium equation can be determined from the slice of roll bite.

$$-\sigma_x y + (\sigma_x + \mathrm{d}\sigma_x)(y + \mathrm{d}y) + p \tan\phi \mathrm{d}x \pm \tau_s \mathrm{d}x = 0. \tag{11}$$

The shear stress τ_s with a dynamic friction model can be given according to Tan et al. [17] as

$$\tau_s = \beta \frac{\mathrm{d}u}{\mathrm{d}y} p. \tag{12}$$

Here β is the dynamic coefficient of friction, du/dy is the strip velocity across the rolling direction and p is the rolling pressure. Then, applying the von-Misses criterion for the plane strain condition using σ_x and τ_y as the horizontal tensile stress and yield strength in shear for the strip material, one obtains

$$\sigma_x + p = 2\tau_y. \tag{13}$$

Differentiating Eq. 13 with respect to x, one gets

$$\frac{d\sigma_x}{dx} = -\frac{dp}{dx}. \tag{14}$$

Inserting Eq. 14 in Eq. 11 and considering only positive sign of β, one obtains after rearranging

$$2\tau_y \frac{\mathrm{d}y}{\mathrm{d}x} - y\frac{\mathrm{d}p}{\mathrm{d}x} + \beta\frac{\mathrm{d}u}{\mathrm{d}y}p = 0,$$

$$\text{or,} \quad \frac{\mathrm{d}u}{\mathrm{d}y} = \frac{-2\tau_y}{\beta p}\frac{\mathrm{d}y}{\mathrm{d}x} + \frac{y}{\beta p}\frac{\mathrm{d}p}{\mathrm{d}x}. \tag{15}$$

Integrating Eq. 15 with respect to y gives

$$u = \frac{-2\tau_y y}{\beta p}\frac{x - x_c}{R} + \frac{y^2}{2\beta p}\frac{\mathrm{d}p}{\mathrm{d}x} + c_1 \tag{16}$$

Applying the boundary conditions at the exit of the strip from the roll bite yields the constant c_1,

$$y = 0,\ x = x_c,\ u = u_2,\ c_1 = u_2. \tag{17}$$

Then, the differential equation in p follows

$$\frac{\mathrm{d}p}{\mathrm{d}x} - \frac{2\beta}{y^2}(u - u_2)p = \frac{4\tau_y(x - x_c)}{Ry}. \tag{18}$$

Taking $A(x) = (2\beta/y^2)(u_2 - u)$ and $B(x) = 4k(x - x_c)/Ry$, Eq. 20 can be rewritten as

$$\frac{\mathrm{d}p}{\mathrm{d}x} - A(x)p = B(x). \tag{19}$$

Solving Eq. 19 gives the expression for the variation of pressure in the roll bite as

$$p(x) = \frac{(2\tau_y - \sigma_1)\left(1 - 2\beta(A_1 + B_1 + C_1) + 2\tau_y \ln\left(\frac{h}{h_1}\right)\right)}{1 - 2\beta(D_1 + E_1 + F_1)}, \tag{20}$$

where the abbreviations are,

$$A_1 = \left(\frac{R}{4(h_c h_1^2)}\left((x_c - x_1)(u_1 h_1 - \dot{h}_c x_1 + \dot{x}_c h_1 + \dot{h}_c x_c)\right)\right)$$
$$B_1 = (3/(8h_c^2 h_1))\left((x_c - x_1)(u_1 h_1 - \dot{h}_c x_1 + \dot{x}_c h_1 + \dot{h}_c x_c)\right)$$
$$C_1 = \left(3R\tan^{-1}\left((x_1 - x_c)/\sqrt{Rh_c}\right)/\left(8h_c^2\sqrt{Rh_c}\right)\right)\left(h_1(-u_1 - \dot{x}_c) + \dot{h}_c(x_1 - x_c)\right)$$

$$D_1 = \left(R^2\left((x_c - x)(u_1 h_1 - \dot{h}_c x_1 + \dot{x}_c h_1 + \dot{h}_c x_c)\right)/4\left(h_c\left(Rh_c + (x - x_c)^2\right)^2\right)\right)$$
$$E_1 = \left(3R/8\left(h_c^2\left(Rh_c + (x - x_c)^2\right)\right)\right)\left((x_c - x_1)(u_1 h_1 - \dot{h}_c x_1 + \dot{x}_c h_1 + \dot{h}_c x_c)\right)$$
$$F_1 = 3R\tan^{-1}\left((x - x_c)/\sqrt{Rh_c}\right)\left(h_1(-u_1 - \dot{x}_c) + \dot{h}_c(x_1 - x_c)\right)/\left(8h_c^2\sqrt{Rh_c}\right)$$

Then, the roll force can be computed as

$$f_x^s = \int_{x_1}^{x_2} p\tan\phi\,\mathrm{d}x + \int_{x_1}^{x_2} (\pm)\tau_s\,\mathrm{d}x, \quad f_y^s = \int_{x_1}^{x_2} p\,\mathrm{d}x + \int_{x_1}^{x_2} (\pm)\tau_s \tan\phi\,\mathrm{d}x. \tag{21}$$

These forces are used in Eq. 1 to numerically compute the vertical and horizontal work roll deflections. As proposed by Tan et al. [17], while computing pressure p in the roll bite, β should be replaced by β_p and the strip velocity term should be multiplied by a term k_p. Similarly, in the shear stress expression β should be replaced by β_f and the strip velocity term should be multiplied by a term k_f. In their work, they obtained these parameters by comparing their result with those obtained from experiments. In the following section, the effect of k_p and β_p on the variation of pressure and shear stress in the roll bite has been investigated. Also, incorporating the expression for forces as given in Eq. 21, the time responses

Table 1 System parameters used in numerical analysis

M	Mass per unit length of work roll	2299.98 kg/m
D_w	Diameter of work roll	0.61 m
D_b	Diameter of backup roll	1.52 m
E	Young's modulus of steel	207 GPa
μ	Poisson's ratio	0.3
f_s^s	Steady sheet force	2.6 MN/m
τ_y	Strip shear yield strength (for 3004 Al alloy)	110 MPa

Table 2 Input of roll bite sub-model

σ_1	Horizontal tensile stress at entry (<170 MPa: 3004 Aluminum tensile yield strength)
σ_2	Horizontal tensile stress at exit (<170 MPa: 3004 Aluminum tensile yield strength)
u_1	Strip velocity at entry (typical value 0–25.4 m/s)
h_1	Strip thickness at entry (typical value 2.54–0.001 mm)

for horizontal and vertical work roll displacements are determined by numerically solving Eq. 1.

3 Results and Discussions

In this section, initially the effects of the parameters used in the dynamic friction model on the variation of pressure and shear stress in the roll bite are considered. The parameters used in this analysis are given in Tables 1 and 2.

Figure 3 shows the flow chart used for computation of the work roll deflection and velocity in both horizontal and vertical direction. Considering the system parameters such as mass, stiffness in horizontal and vertical direction, thickness and velocity at the entry and the roll radius as the input parameters, in the first step, the strip velocity, pressure and shear stresses are calculated in the roll bite. Using the calculated values of exit thickness and velocity and the position of the neutral plane and thickness at the neutral plane, the horizontal and vertical components of the roll forces are computed. These forces are used in the dynamic equation of motion to compute the work roll deflection in horizontal and vertical direction.

Figure 4 shows the variation of pressure and shear stress considering different values of the coefficient of the dynamic friction model. It may be observed that for constant k_f and β_f, and for all values of β_p, the zero shear stress occurs at a distance from the exit point. Though Tan et al. [17] claimed that the dynamic friction model is better than thother two friction models, its applicability highly depends on the correct prediction of these four constants. It is shown that the pressure variation is very sensitive to the change in the parameter β_p. While a

Fig. 3 Flow chart showing the calculation of the horizontal and vertical work roll deflection

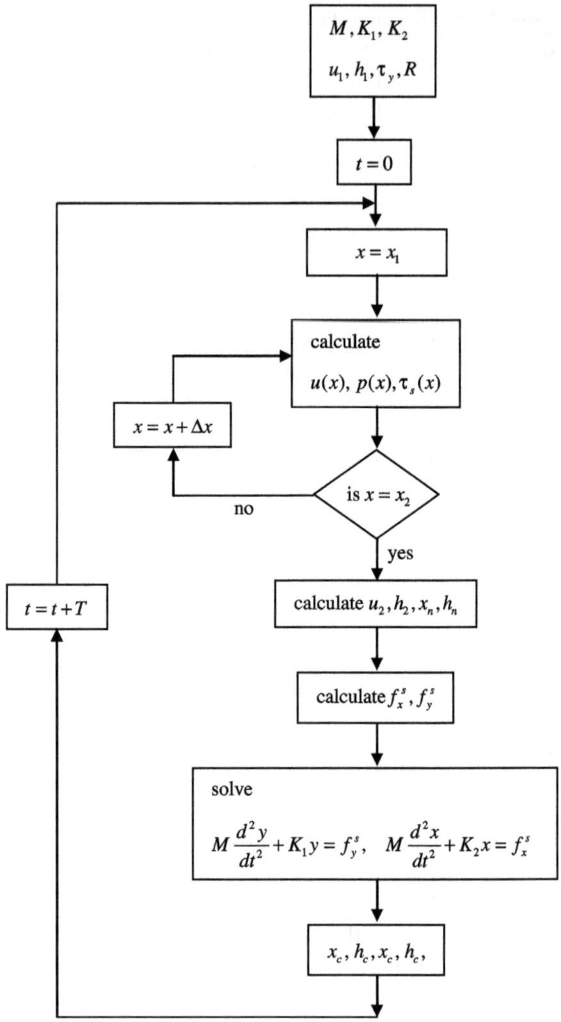

lower value of β_p predicts more accurately the pressure distribution in the forward slip zone, it fails to predict the pressure distribution in the backward slip zone. Also, as pointed out in [17], the pressure hill does not coincide with the neutral plane. Hence, if one calculates the pressure distribution for both the forward slip zone and for the backward slip zone, a jump in the pressure can be observed at the neutral point, as shown in Fig. 5. This jump is most probably the driving factor for chatter at higher speed while rolling a thin strip.

Figure 6 shows the horizontal and vertical work roll deflections obtained by using both dynamic friction model and constant friction model. Here the rolling speed is taken as 20 m/s. With an increase in rolling speed, the dynamic friction model clearly predicts unstable response in vertical direction, while the constant

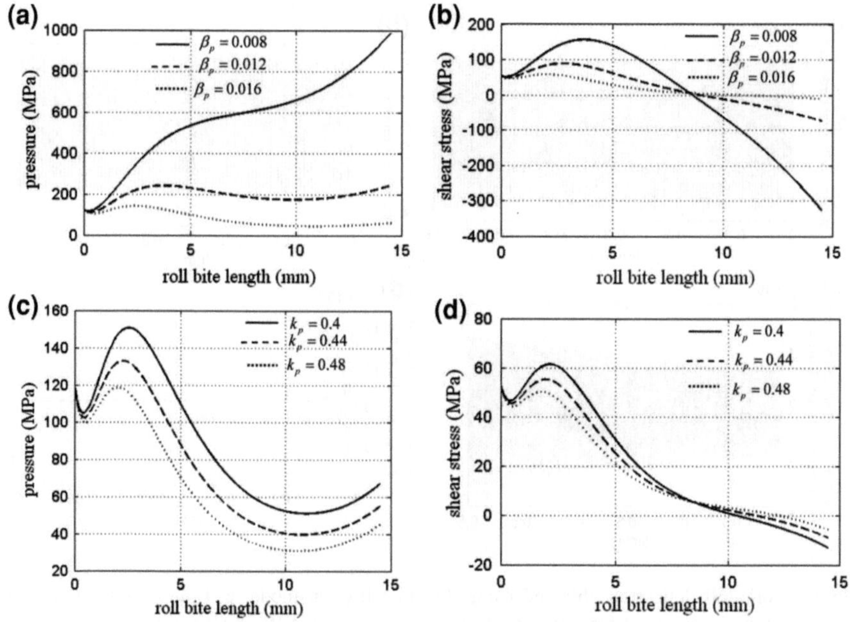

Fig. 4 Variation of pressure and shear stress in the roll bite with β_p and k_p. System parameters as in Tan et al. [17]

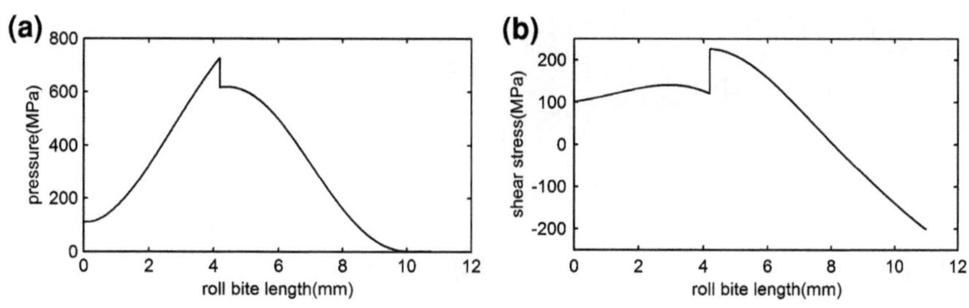

Fig. 5 **a** pressure variation in roll bite, **b** shear stress variation in roll bite for 3.17 mm strip thickness at entry, with reduction of 27.44%, strip shear yield strength 110 MPa and roll velocity 40 mm/s

friction model predicts constant amplitude (82 μm) of vertical deflection. Both of these methods predict stable horizontal work roll deflection. The dynamic friction model gives a higher value of deflection than the constant friction model. It should be noted that in this investigation no front or back tension was considered.

To study the effect of inter stand tension, which is a very important parameter in tandem rolling mills, in the present case a time varying back tension with

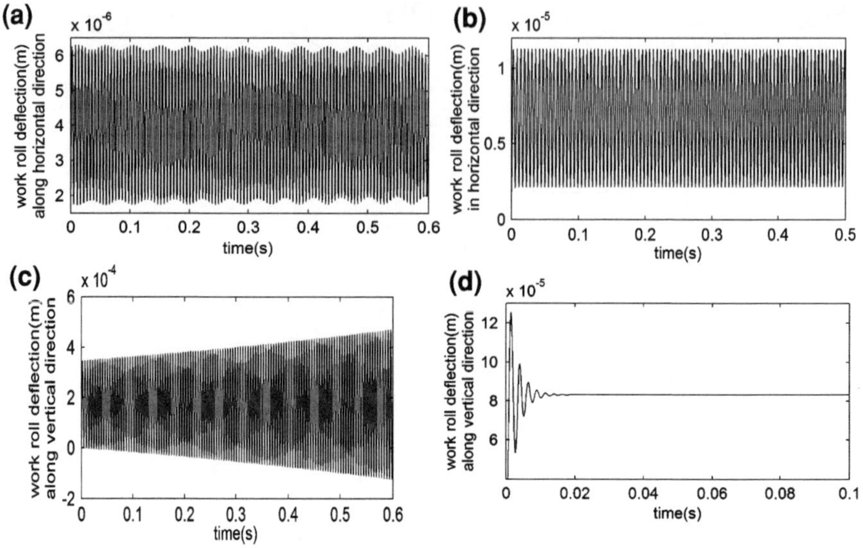

Fig. 6 Work roll deflection obtained using dynamic friction model **a**, **b** and constant friction model **c**, **d**. **a**, **c** horizontal direction, **b**, **d** vertical direction for 1 mm strip thickness

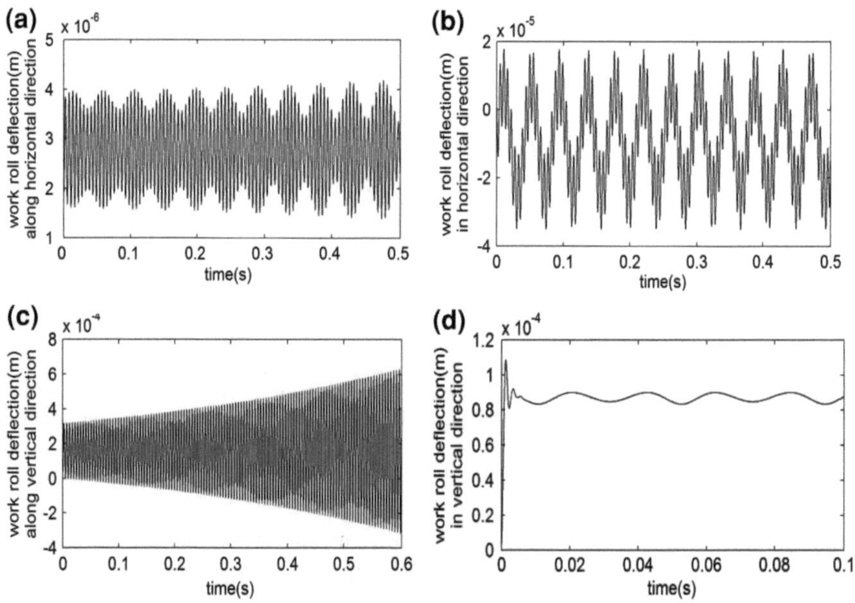

Fig. 7 Effect of variation of tension using dynamic friction model **a**, **c** and constant friction model **b**, **d** on work roll deflection **a**, **b** horizontal direction, **c**, **d** vertical direction. Work roll vibrations at 10 m/s, 1 mm strip thickness and back tension varying with sine function at 150 Hz

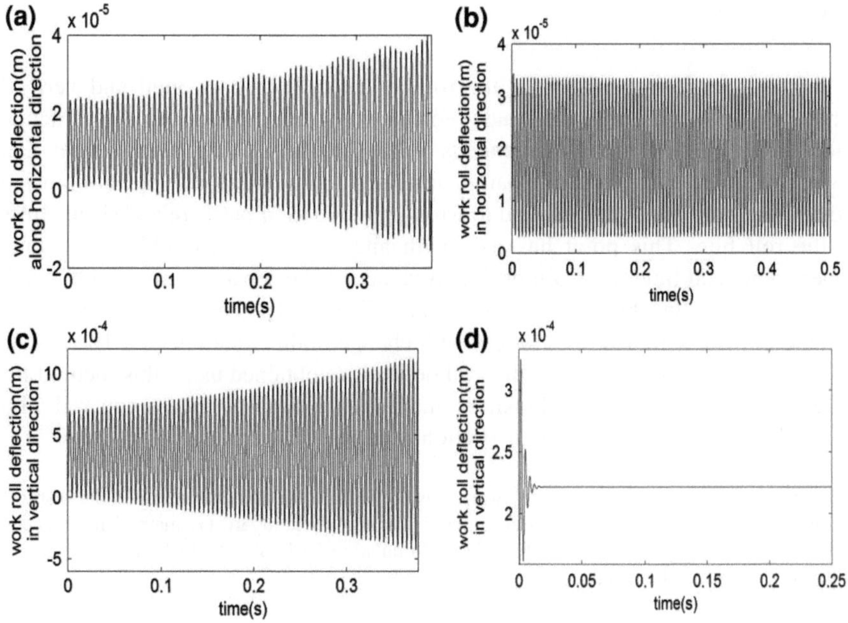

Fig. 8 Deflection of work roll for strip thickness of 2.5 mm and at 20 m/s using dynamic friction model **a**, **c** and constant friction model **b**, **d**; **a**, **b** horizontal direction, **c**, **d** vertical direction

frequency 150 Hz has been introduced. This causes large variations in the horizontal component of the roll force. Figure 7 shows the work roll deflections obtained using dynamic friction model and constant friction model. Here, while both methods show a clear increase in vibration, constant friction model predicts a higher value of horizontal roll deflection. This back tension also affects the deflection in the vertical direction. This may be due to the coupling between the forces in the horizontal and vertical direction. This observation is similar to those observed in linear parametrically excited systems where a time varying force applied in horizontal direction causes a deflection in the vertical direction. Here, while constant friction model predicts periodic stable solutions, dynamic friction model clearly shows the vibration of the system growing with time.

Figure 8 shows the time response obtained for the system when the strip thickness is increased from 1 mm to 2.5 mm. While clearly the horizontal and vertical components of the work roll deflections obtained by dynamic friction model are increasing with time showing an unstable system, the constant friction model shows stable steady state vibration.

4 Conclusion

In this work, the expression for the roll forces in both horizontal and vertical directions of a four high mill has been developed incorporating the dynamic friction model with the homogeneous rolling model. It has been shown that improper choice of the dynamic parameters used for the pressure calculation using a dynamic friction model may lead to wrong prediction of pressure and shear stress in the roll bite. This point has its significance as it is impossible to conduct experiment each time to obtain optimum values of these parameters which match the experimentally obtained pressure curve. This study gives an idea about the variation of pressure and shear stress with change in these parameters. The results for the horizontal and vertical work roll deflections obtained using this method for different strip velocity, strip thickness and back tension has been compared with those obtained using the constant friction model.

Acknowledgments Part of the present work was done while the first author visited the Institute of Engineering and Computational Mechanics, University of Stuttgart, Germany. The financial assistance provided by the Cluster of Excellence Simulation Technology (SimTech) University of Stuttgart is greatly acknowledged.

References

1. Roberts, W.L.: Four high mill stand chatter of the fifth octave-mode. Iron Steel Eng. **55**, 41–47 (1978)
2. Chefneux, L., Fischbach, J.P., Gouzou, J.: Study and industrial control of chatter in cold rolling. Iron Steel Eng. **61**, 17–26 (1984)
3. Johnson, R.E., Qi, Q.: Chatter dynamics in sheet rolling. Int. J. Mech. Sc. **36**, 617–630 (1993)
4. Johnson, R.E.: The effect of friction and inelastic deformation on chatter in sheet rolling. Proc. R. Soc. Lond. A **445**, 479–499 (1994)
5. Yun, I.S., Wilson, W.R.D., Ehmann, K.F.: Review of chatter studies in cold rolling. J. Machine Tool Manuf. **38**, 1499–1530 (1998)
6. Yun, I.S., Wilson, W.R.D., Ehmann, K.F.: Chatter in strip rolling process—Part I: dynamic rolling modeling, ASME. J. Manuf. Sci. Eng. **120**, 330–336 (1998)
7. Yun, I.S., Wilson, W.R.D., Ehmann, K.F.: Chatter in strip rolling process—Part II: dynamic rolling experiments ASME. J. Manuf. Sci. Eng. **120**, 337–342 (1998)
8. Yun, I.S., Wilson, W.R.D., Ehmann, K.F.: Chatter in strip rolling process—Part III: chatter model ASME. J. Manuf. Sci. Eng. **120**, 343–348 (1998)
9. Hu, P.H., Ehmann, K.F.: A dynamic model of rolling process. Part-I: homogenous model. J. Machine Tool Manuf. **40**, 1–19 (1999)
10. Hu, P.H., Ehmann, K.F.: A dynamic model of rolling process. Part-II: inhomogeneous model. J. Machine Tool Manuf. **40**, 21–31 (1999)
11. Kimura, Y., Sodani, Y., Nishiura, N., Ikeuchi, N., Mihara, Y.: Chatter analysis in tandem cold rolling mills. Iron Steel Inst. Jpn **43**, 77–84 (2002)
12. Meehan, A.P.: Vibration instability in rolling mills modeling and experimental results ASME. J. Vib. Acoust. **124**, 221–228 (2002)
13. Lin, Y.J., Suh, C.S., Langari, R., Naoh, S.T.: On characteristics and mechanism of rolling instability and chatter, ASME. J. Manuf. Sci. Eng. **125**, 778–786 (2003)

14. Niziol, J., Swiatoniowski, A.: Numerical analysis of the vertical vibrations of rolling mills and their negative effect on the sheet quality. J. Mat. Proc. Tech. **162–163**, 546–550 (2005)
15. Hu, P.H., Zhao, H., Ehmann, K.F.: Third octave mode chatter in rolling. Part1: chatter model. Proc. IMechE Part B: J. Eng. Manf. **220**, 1267–1277 (2006)
16. Hu, P.H., Zhao, H., Ehmann, K.F.: Third octave mode chatter in rolling. Part2: stability of a single stand mill. Proc. IMechE Part B: J. Eng. Manuf. **220**, 1279–1292 (2005)
17. Tan, X., Yan, X.T., Juster, N.P., Raghunathan, S., Wang, J.: Dynamic friction model and its application in flat rolling. J. Mat. Proc. Tech. **207**, 222–234 (2008)

Heavy Oils Transportation in Catenary Pipeline Riser: Modeling and Simulation

Severino Rodrigues de Farias Neto, Jobsan Sueny de Souza Santos, Kelen Cristina de Oliveira Crivelaro, Fabiana Pimentel Macêdo Farias and Antonio Gilson Barbosa de Lima

Abstract This chapter presents information about multiphase flows such as definition, flow pattern and modeling. Application to petroleum industry has been given to water-heavy oil flow in catenary riser. In offshore platforms a catenary riser is often used to carry heavy ultraviscous oils. However, the high viscosity of these oils provides an elevated pressure drop in the flow. Several studies report the use of the core-flow technique in vertical and horizontal pipes to reduce the pressure drop in the transport of heavy oils. Nevertheless, so far no record of studies using catenary riser was found. Results of velocity, pressure, temperature and volumetric fraction distribution were presented and analyzed. The pressure drop in the catenary riser decreased 3.28 times compared with the single-phase oil

S. R. de Farias Neto (✉) · J. S. de Souza Santos
Department of Chemical Engineering, Center of Science and Technology,
Federal University of Campina Grande (UFCG), PO Box 10057,
Campina Grande PB, 58429-900, Brazil
e-mail: fariasn@deq.ufcg.edu.brs.fariasn@gmail.com

J. S. de Souza Santos
e-mail: sueny_santos@yahoo.com.br

F. P. M. Farias
Department of Technology, Semi-Arid Sustainable Development Center,
Federal University of Campina Grande (UFCG), Sumé PB,
58540-000, Brazil
e-mail: fabianapimentel@ufcg.edu.br

K. C. de Oliveira Crivelaro · A. G. B. de Lima
Department of Mechanical Engineering, Center of Science and Technology,
Federal University of Campina Grande (UFCG), PO Box 10057,
Campina Grande PB, 58429-900, Brazil
e-mail: kelencrivelaro@yahoo.com.br

A. G. B. de Lima
e-mail: gilson@dem.ufcg.edu.br

A. Öchsner et al. (eds.), *Materials with Complex Behaviour II*,
Advanced Structured Materials 16, DOI: 10.1007/978-3-642-22700-4_13,

flow. This low value compared with the available literature was attributed to the presence of regions of adhesion along the surface of the overhead line and the high viscosity of the produced water.

Keywords Core-flow · Riser · Catenary · Heavy oil · Numerical simulation

Nomenclature

$A_{\alpha\beta}$	Interfacial area density (1/m)
C_D	Drag coefficient (-)
$\mathbf{D}_{\alpha\beta}$	Drag force (N)
$d_{\alpha\beta}$	Mixture length scale (m)
d_β	Mean diameter (m)
Eo	Eötvös number (-)
f_α	Volume fraction (-)
g	Gravitational acceleration (m/s^2)
h_α	Specific enthalphy (J/kg)
$h_{\alpha\beta}$	Heat transfer coefficient W/m^2 K
k	Turbulent kinetic energy (m^2/s^2)
$\mathbf{LB}_{\alpha\beta}$	Lubrication force (N)
$\mathbf{L}_{\alpha\beta}$	Lift force (N)
$\mathbf{M}_\alpha$	Total force on phase α (N)
$\mathbf{M}_{\alpha\beta}$	Interphase momentum transfer (N)
Nu	Nusselt number (-)
Pr	Prandtl number (-)
Q_m	Heavy oil and water volumetric flow (m^3/s)
Q_α	Interphase heat transfer (W/m^2)
Re	Reynolds number (-)
$\mathbf{S}_{\mathrm{M}\alpha}$	Mass source (kg/m^{3}s)
$S_{q\alpha}$	External heat source (kg/m s^3)
$\mathbf{S}_\alpha$	Momentum sources (kg//m^2 s^2)
t	Time (s)
$\mathbf{TD}_{\alpha\beta}$	Turbulent dispersion force (N)
$\mathbf{U}_\alpha$	Velocity vector (m/s)
$\mathbf{VM}_{\alpha\beta}$	Virtual mass force (N)
$\Gamma_{\alpha\beta}$	Mass flow rate per unit volume (kg/m^3 s)
ε	Turbulence dissipation rate (m^2/s^3)
λ_α	Thermal conductivity (W/m K)
$\lambda_{\alpha\beta}$	Misture conductivity scale (m)
μ_t	Turbulent viscosity (kg/m s)
μ_α	Dynamic viscosity (kg/m s)
ρ_α	Density (kg/m^3)
$\rho_{\alpha\beta}$	Mixture density (kg/m^3)
σ	Surface tension coefficien t (N/m^2)
σ_ρ, σ_ε, σ_K,	Empirical constants to k-ε turbulence model (-)

1 General Features of Multiphase Flow

The incidence of multiphase flows is involved in many industrial processes. Examples are fluidized beds, fluid catalytic cracking reactors, bubble column reactors, combustion, treatment of environmental pollution, and many others operations. Multiphase flows refer to situations in which two or more fluids are present and each of the fluids are immiscible or when there are different thermodynamic phases (gas–liquid, liquid–liquid, gas–liquid–solid and so on). One of the most challenging aspects of dealing with multiphase flow is the fact that it can take many different forms (flow patterns). Therefore, it can be classified broadly into dispersed flow (continuous–dispersed flows) and stratified flow (continuous–continuous flow). A dispersed flow pattern occurs when one or more phases are dispersed (bubbles, droplets or solid particles) within another continuous phase, while in a stratified flow pattern the two phases are separated by a continuous interface. In most of the reported studies, the identification and classification of the flow pattern is based on visual observations.

In the petroleum industry, a complex mixture of hydrocarbon can exist as a single-phase (liquid or gas) or as a two-phase mixture or multiphase mixture, depending on temperature, pressure, and the composition of mixture. The production and transportation of this mixture leads to a stream with different multiphase variations (oil–gas, oil–water, oil–water–gas, and so on). For many energy plants, annular two-phase flow is the most common flow pattern in boiling heat transfer systems, such as boiler, heat exchanger and in pipes, often used for the production and transport of gas.

In the specific case of two-phase immiscible flow (liquid–liquid flow) the annular flow pattern depend on the flow rates of the phases, geometry and roughness of pipe, the flow properties of the phases, and interfacial tension between the phases. The knowledge about the hydrodynamic properties associated with these flows is extremely important to ensure safe design and efficient operation of an offshore transportation pipeline. The annular flow pattern stands out among the other flow regimes, for their important application in the transport of heavy ultra-viscous oil. Its very interesting feature is that the frictional pressure loss is comparable to that of single-phase flow of a thinner fluid in the same pipe at mixture flow rate. To obtain this flow pattern, water is injected in the oil such that it flows as an annular film along the pipe wall while oil flows in the core region. Many oil companies have shown an interest in this flow pattern, i.e., in the technology of water-lubricated transport of heavy oils. These oils frequently are very viscous and somewhat lighter than water. Typical oils might have a viscosity of 100 Pa.s and a density of 990 kg/m^3 at 25°C.

Most flow models have been published to predict the flow characteristics of the core-annular flow in horizontal and vertical pipelines to heavy oil transportation and artificial lifting in heavy oil wells. These models require theoretical and experimental information about the pressure-gradient and flow patterns, where a thin fluid surrounds a viscous one [4, 5, 7, 11, 13–15, 19, 20, 22, 23]. Application of a computational fluid dynamics (CFD) based approach is being increasingly adopted to predict the detailed thermo fluid mechanics of core-annular flow [1, 8, 9, 17, 26].

Core-annular flow of highly viscous oil and water in a horizontal pipe was analyzed by Ooms et al. [15]. They assumed that the oil viscosity is so high that the oil–water interface can be treated as a solid–liquid interface and, the buoyancy forces generated due to the density difference of oil and water are counterbalanced by the lubrication forces acting on the core. They showed that the theoretical predictions agreed with their experimental data.

Bai et al. [4] identified a new flow type namely bamboo waves in upflow and corkscrew waves in downflow. They observed that the single oil pressure drop was about 200 times larger than in case of water lubricated flow for the same oil superficial velocity.

Different strategies for preventing oil from fouling the walls of core-annular flow pipelines and also for restart from an unexpected pipeline shut-down was present by Arney et al. [3]. Experiments in a pilot scale cement-lined core-annular flow pipeline showed that cement-lined pipes can resist fouling by oil and facilitate the restart of transport operation.

Rovinsky et al. [25] in another annular flow study concluded that the velocity profiles, pressure drop reduction factor and power saving factor depend on the viscosity ratio of two phases. They also report that the power saving factor does increase with increase of viscosity ratio.

Frictional pressure drop measurements for upward vertical core-flow in a 1 in. pipe, using a 17.6 Pa.s, 963 kg/m^3 oil and water at room temperature was measured by Prada and Bannwart [18]. The authors reported a decrease by more than 1,000 times with respect to single phase oil flow, being comparable to the flow of water alone in the pipe at mixture flow rate. Similar results were obtained by Rodriguez et al. [24] using this technique to heavy ultraviscous oil transport.

The aforesaid survey brings out the importance of core flow to heavy oil transportation, and the different aspects of this phenomenon were reviewed by Ghosh et al. [28], like nozzle design, wettability characteristics, restart procedure, etc., and allow us to say which dynamics of core-annular flow is yet not completely understood. It should be emphasized that, the two-phase annular flow patterns with heat transfer are customary in many industrial processes. The reliable prediction of pressure drop and heat transfer rates associated with these processes are essential to developing more reliability to heavy oil transportation, reducing or eliminating problems and cost. Even less information is available on non-isothermal two-phase core-annular flow in the catenary riser. Some studies have been initiated by the present authors in this direction that was conducted at the Departments of Chemical Engineering and Mechanical Engineering at Federal University of Campina Grande (UFCG), Brazil.

2 Multiphase Transport Modeling

For the studies of multiphase flows models are needed that are able to predict the detailed hydrodynamic behavior of those flows. Two different ways present highlights: experimental and theoretical. In this last case, the power of modern computers can be employed to address the complexity of the flow. However, the predictive capability must rely on complex theoretical and computational models for each phases or components. According to Ranade [21] there are three main approaches for modeling multiphase flows:

- Volume of fluid (VOF) approach (Eulerian framework for both the phases with reformulation of interface forces on volumetric basis).
- Eulerian framework for the continuous phase and Lagrangian framework for all the dispersed phases.
- Eulerian framework for all phases (without explicitly accounting for the interface between phases).

If Eulerian framework is assumed, two distinct models can be distinguished: homogeneous and inhomogeneous models. Inhomogeneous multiphase flow occurs when separate velocity fields and other fields (temperature, turbulence, etc.) exist for each phase. The pressure field is shared by all fluids. The fluids interact via interphase transfer terms; while in the homogeneous multiphase flow has the same velocity and pressure fields, and other relevant fields.

In the inhomogeneous multiphase model there are different sub-models how differ in the way which these models are compared. The differences are related mainly to interfacial area density and the interphase transfer terms. These are:

- The particle model—this model is applied when one of the phases is continuous and the other is dispersed (solid particles, liquid droplets or gas bubble).
- The mixture model—this model treats both phases symmetrically and requires both phases to be continuous.
- Free surface model—this model is applicable to free surface flows.

The set of governing equations that describe the three-dimensional two-phase flow, transient and non-isothermal in a catenary riser shall consist of conservation laws of mass, momentum and energy as described in the following.

2.1 Continuity Equation

A continuity equation corresponds to a partial differential equation that describes the transport of some type of conserved quantity. For a multiphase flow this equation is defined by:

$$\frac{\partial}{\partial t}(f_\alpha \rho_\alpha) + \nabla \cdot (f_\alpha \rho_\alpha \mathbf{U}_\alpha) = S_{M\alpha} + \sum_{\beta=1}^{N_P} \Gamma_{\alpha\beta} \quad (1)$$

where the subscript α represent the phases involved (water and heavy oil); f is the volume fraction; ρ is density and $\mathbf{U}$ is the velocity vector; $\mathbf{S}_{M\alpha}$ describes user specified mass sources, and $\Gamma_{\alpha\beta}$ is the mass flow rate per unit volume from phase α to phase β. This term only occurs if interphase mass transfer takes place. For details see ANSYS [2].

2.2 Momentum Equation

The linear momentum equation can be developed by Newton's Second Law and relates that the sum of all forces applied on the control volume is equal to the sum of the rate of change of momentum inside the control volume and the net flux of momentum through the control surface. In the case of multiphase flow this equation is given by:

$$\frac{\partial}{\partial t}(f_\alpha \rho_\alpha \mathbf{U}_\alpha) + \nabla \cdot [f_\alpha(\rho_\alpha \mathbf{U}_\alpha \otimes \mathbf{U}_\alpha)] = -f_\alpha \nabla p_\alpha + \nabla \cdot \{f_\alpha \mu_\alpha [\nabla \mathbf{U}_\alpha + (\nabla \mathbf{U}_\alpha)^T]\} + \sum_{\beta=1}^{N_P} \left(\Gamma_{\alpha\beta}\mathbf{U}_\beta - \Gamma_{\beta\alpha}\mathbf{U}_\alpha\right) + \mathbf{S}_\alpha + \mathbf{M}_\alpha \tag{2}$$

where μ is dynamic viscosity; $\mathbf{S}_\alpha$ describes momentum sources due to external body (buoyancy force and rotational force); the term $\Gamma_{\alpha\beta}\mathbf{U}_\beta - \Gamma_{\beta\alpha}\mathbf{U}_\alpha$ represents momentum transfer induced by interphase mass transfer and occurs when mass is carried from one phase into another. $\Gamma_{\alpha\beta}$ is a positive mass flow rate per unit volume from phase β to phase α. $\mathbf{M}_\alpha$ describes the total force on phase α due to interaction with other phases, such as drag force, lift force, virtual mass force, etc., and is given by:

$$\mathbf{M}_\alpha = \sum_{\beta \neq \alpha} \mathbf{M}_{\alpha\beta} \tag{3}$$

Here, $\mathbf{M}_{\alpha\beta}$ is the interphase momentum transfer, which occurs due to interfacial forces acting on each phase α, due to interaction with another phase β.

The total interfacial force acting between two phases may arise from several independent physical effects:

$$\mathbf{M}_\alpha = \mathbf{D}_{\alpha\beta} + \mathbf{L}_{\alpha\beta} + \mathbf{LB}_{\alpha\beta} + \mathbf{VM}_{\alpha\beta} + \mathbf{TD}_{\alpha\beta} \tag{4}$$

The lift force ($\mathbf{L}_{\alpha\beta}$) acts perpendicular to the main relative velocity and thus contributes to the lateral void distribution. The wall lubrication force ($\mathbf{LB}\alpha_\beta$) occurs when the dispersed phase is observed to concentrate in a region close to the wall, but not immediately adjacent to the wall. The virtual mass force ($\mathbf{VM}_{\alpha\beta}$) accounts for transient behavior in the flow field. It represents the force required to accelerate the apparent mass of the surrounding continuous phase, when the relative velocity changes. The interphase turbulent dispersion force ($\mathbf{TD}_{\alpha\beta}$) corresponds to an additional dispersion of the phases of a region of high volume fraction to another of low volume fraction due to turbulent fluctuations. This is caused by the combined action of eddies and interfacial drag [12].

Drag forces ($\mathbf{D}_{\alpha\beta}$) act in the direction opposing the relative flow between the phases, which is defined by the following equations:

- For the particle model

$$\mathbf{D}_{\alpha\beta} = \frac{C_D}{8}\rho_\alpha A_{\alpha\beta}\left|\mathbf{U}_\beta - \mathbf{U}_\alpha\right|\left(\mathbf{U}_\beta - \mathbf{U}_\alpha\right) \tag{5}$$

  Here, α represents the continuous phase and β the dispersed phase and $A_{\alpha\beta}$ corresponds to the interfacial area density, which is characterized by the interfacial area per unit volume between phase α and phase β given by:

$$A_{\alpha\beta} = \frac{6f_\beta}{d_\beta} \tag{6}$$

  where d_β is the mean diameter of the spherical particles and f_β is the volume fraction of the dispersed phase.
- For the mixture model

$$\mathbf{D}_{\alpha\beta} = C_D\rho_{\alpha\beta}A_{\alpha\beta}\left|\mathbf{U}_\beta - \mathbf{U}_\alpha\right|\left(\mathbf{U}_\beta - \mathbf{U}_\alpha\right) \tag{7}$$

  Here, $\rho_{\alpha\beta}$ is the mixture density and the interfacial area density, $A_{\alpha\beta}$, defined respectively by:

$$\rho_{\alpha\beta} = f_\alpha\rho_\alpha + f_\beta\rho_\beta \tag{8}$$

$$A_{\alpha\beta} = \frac{f_\alpha f_\beta}{d_{\alpha\beta}} \tag{9}$$

  where $d\alpha_\beta$ is a mixture length scale.
- For the free surface model.
  In this model the interphase drag is calculated in the same way as for the mixture model, except that the interfacial area density is given by:

$$A_{\alpha\beta} = |\nabla f_\alpha| \tag{10}$$

  When more than two phases are present, these parameters can be written as follows:

$$A_{\alpha\beta} = \frac{2|\nabla f_\alpha||\nabla f_\beta|}{|\nabla f_\alpha| + |\nabla f_\beta|} \tag{11}$$

  The drag coefficient (C_D) is the obtained by empirical or theoretical correlation. As an example, we can write:

- Schiller-Neumann Drag Model—this model is only used for solid spherical particles, or for fluid particles that are sufficiently small and considered like spherical. This parameter depends on the particle Reynolds number (Re) as follows:

$$C_D = \frac{24}{\mathrm{Re}}\left(1 + 0.15\mathrm{Re}^{0.687}\right) \tag{12}$$

with $0.2 < \mathrm{Re} < 500$. The particle Reynolds number is given by:

$$\mathrm{Re} = \frac{\rho_\alpha d_\beta \left(U_\alpha - U_\beta\right)}{\mu_\alpha} \tag{13}$$

where $(U\alpha - U_\beta)$ is the relative velocity.

- Ishii-Zuber Drag Model—this model is applied to distorted particle regime. In this case the drag coefficient is independent of the Reynolds number, but dependent on the particle shape through the dimensionless group known as the Eötvös number, *Eo*. For an elliptical particle we can use the following equation:

$$C_D = \frac{2}{3} Eo^{1/2} \tag{14}$$

where the Eötvös number (*Eo*) represents the ratio between gravitational and surface tension forces, defined by:

$$Eo = \frac{g \Delta \rho d_\beta^2}{\sigma} \tag{15}$$

where, $\Delta\rho$ is the density difference between the phases, g is the gravitational acceleration, and σ is the surface tension coefficient.

2.3 Turbulence Model

Turbulence occurs when the inertia forces in the fluid becomes significant comparable to viscous forces. This one consists basically of fluctuations in the flow in time and space. To enable the effects of turbulence without recourse to a prohibitively fine mesh it is necessary to use a turbulence model that offers a good compromise between numerical effort and computational accuracy. Several turbulence models are reported in the literature, the k–ε two-equation model is one these models. This one uses the gradient diffusion hypothesis to relate the Reynolds stress to the mean velocity gradients and turbulent viscosity. In this model, the turbulence kinetic energy (k) is defined as the variance of the fluctuations in velocity and turbulence eddy dissipation (ε) corresponds to the dissipation rate of the velocity fluctuations.

In multiphase flow, the values of k and ε come directly from the differential transport equations for the turbulent kinetic energy and the turbulence dissipation rate as follows:

$$\frac{\partial}{\partial t}(\rho_\alpha f_\alpha k_\alpha) + \nabla \cdot \left\{ f_\alpha \left[\rho_\alpha \mathbf{U}_\alpha k_\alpha - \left(\mu + \frac{\mu_{t\alpha}}{\sigma_k} \right) \nabla k_\alpha \right] \right\} = f_\alpha (G_\alpha - \rho_\alpha \varepsilon_\alpha) \tag{16}$$

$$\begin{aligned} &\frac{\partial}{\partial t}(\rho_\alpha f_\alpha \varepsilon_\alpha) + \nabla \cdot \left\{ f_\alpha \rho_\alpha \mathbf{U}_\alpha \varepsilon_\alpha - \left(\mu + \frac{\mu_{t\alpha}}{\sigma_\varepsilon} \right) \nabla \varepsilon_\alpha \right\} \\ &= f_\alpha \frac{\varepsilon_\alpha}{k_\alpha} (C_1 G_\alpha - C_2 \rho_\alpha \varepsilon_\alpha) \end{aligned} \tag{17}$$

where the empirical constants $C_1 = 1.44$; $C_2 = 1.92$; $\sigma_\varepsilon = 1.2$ and $\sigma_k = 1.0$. G_α is the turbulence production due to viscous and buoyancy forces, which is modeled as follows:

$$G_\alpha = \mu_{t\alpha} \nabla \mathbf{U}_\alpha \cdot \left(\nabla \mathbf{U}_\alpha + \nabla \mathbf{U}_\alpha^T \right) - \frac{\mu_{t\alpha}}{\rho \sigma_\rho} g \cdot \nabla \rho \tag{18}$$

where $\sigma_\rho = 1$; μ_t correspond to the turbulent viscosity defined by:

$$\mu_{t\alpha} = c_\mu \rho_\alpha \left(\frac{k_\alpha^2}{\varepsilon_\alpha} \right) \tag{19}$$

where c_μ is a constant equal 0.99.

2.4 Thermal Energy Equation

The thermal energy equation for the multiphase flow is given by:

$$\frac{\partial}{\partial t}(f_\alpha \rho h_\alpha) + \nabla \cdot [f_\alpha (\rho_\alpha \mathbf{U}_\alpha h_\alpha - \lambda_\alpha \nabla T_\alpha)] = Q_\alpha + S_{q\alpha} \tag{20}$$

where h_α, T_α, λ_α denote the static enthalpy, the temperature, and the thermal conductivity of phase α; $S_{q\alpha}$ describes external heat sources; Q_α denotes interphase heat transfer across interfaces with other phases, and is given by:

$$Q_\alpha = \sum_{\beta \neq \alpha} h_{\alpha\beta} A_{\alpha\beta} (T_\beta - T_\alpha) \tag{21}$$

here, $h_{\alpha\beta}$ is the heat transfer coefficient, which is the amount of heat energy crossing a unit area per unit time per unit temperature difference between the phases.

The Nusselt number ($\mathrm{Nu}_{\alpha\beta}$) is the ratio of convective to conductive heat transfer across (normal to) the boundary, and is given as follows:

- For the particle model

$$h_{\alpha\beta} = \frac{\lambda_{\alpha}\mathrm{Nu}_{\alpha\beta}}{d_{\beta}} \quad (22)$$

where d_{β} is the diameter of dispersed phase. The Nusselt number can be obtained by empirical correlation, for example, the Ranz–Marshall correlation for spherical particles:

$$\mathrm{Nu}_{\alpha\beta} = 2 + 0.6\mathrm{Re}^{0.5}\mathrm{Pr}^{0.3} \quad (23)$$

with $0 < \mathrm{Re} < 200$ and $0 < \mathrm{Pr} < 250$. Here, Pr is the Prandtl number given by:

$$\mathrm{Pr} = \frac{\mu_{\alpha}C_{p\alpha}}{\lambda_{\alpha}} \quad (24)$$

where $C_{p\alpha}$ represents the specific heat of the phase α.

- For the mixture model

$$h_{\alpha\beta} = \frac{\lambda_{\alpha\beta}\mathrm{Nu}_{\alpha\beta}}{d_{\alpha\beta}} \quad (25)$$

where $d_{\alpha\beta}$ is the mixture length scale and $\lambda_{\alpha\beta}$ is a mixture conductivity scale, given by:

$$\lambda_{\alpha\beta} = f_{\alpha}\lambda_{\alpha} + f_{\beta}\lambda_{\beta} \quad (26)$$

It is noticed that Eq. 26 corresponds to the weighted equations that account the effect of the volumetric fraction of the phases.

3 Heat Transfer and Fluid Flow in Catenary Riser

On offshore structures are usually flexible pipes connecting the platform with the Christmas tree at points further away, which give the riser configuration in the form of a catenary. Often these pipes are surrounded by buoyancy modules to reduce their weight, especially when it is used in deep water. In many circumstances, the mixture flowing through curved pipelines can be greatly affected by the maldistribution of the phases. Situations more dramatic occur when the fluid flow is accompanied by heat transfer. In this case we have paraffin deposition and higher oil viscosity into the pipe which provokes increase of the pumping power and until flow obstruction for severe situations.

The complexity of the phenomena associated with the maldistribution of phases has been addressed by a particular flow pattern during the transport of produced fluids and sediments from the well on the seabed to the production platform. In this sense, a numerical study was developed for the annular heavy oil–water flow at a

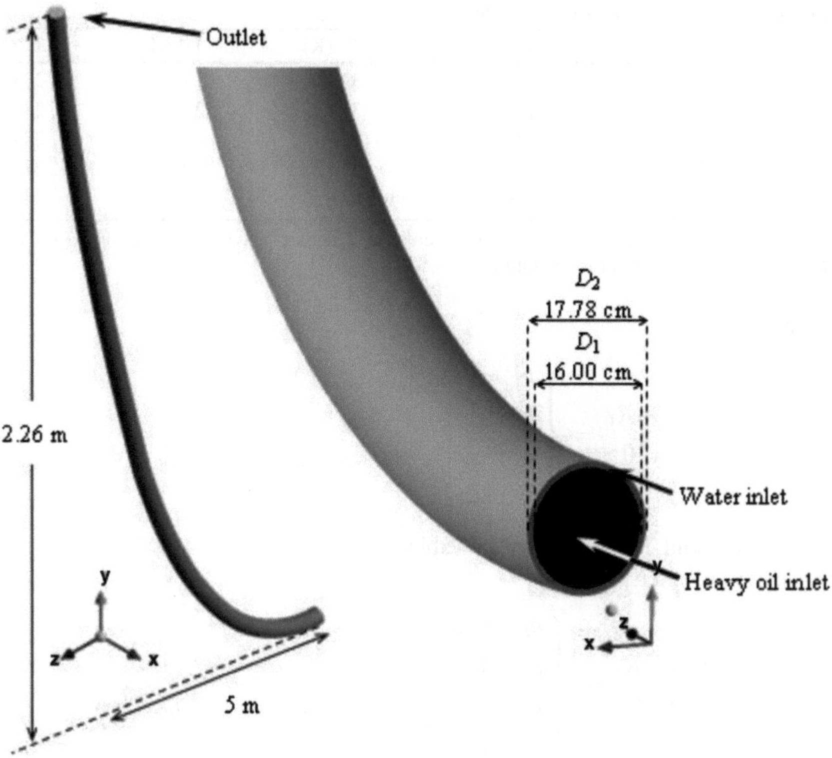

Fig. 1 Illustration of catenary riser and dimensions

submerged catenary riser illustrated in Fig. 1 considering a non-isothermal flow and unsteady state regime.

3.1 Boundary Layers and Physical Properties

A numerical solution for conservation equations was developed using ANSYS CFX Commercial code which the following initial conditions and boundary conditions applied to the catenary riser illustrated in Fig. 1.

(a) For the initial conditions, it was considered that the catenary riser is full with water, and it has a null velocity vector ($u_{x,w} = u_{y,w} = u_{z,w} = 0$).
(b) Uniform velocity and volume fraction of water and heavy oil at the inlet surfaces are given as follows:

Table 1 Simulations data using core-flow

Case	u_o (m/s)	u_w (m/s)	T_o (K)	T_w (K)	T_p (K)
1	1.20	1.90	323	298	283
2	1.20	1.90	373	298	283
3	1.20	1.90	423	298	283
4	1.20	1.90	473	298	283
5	1.20	1.90	523	298	283

u_o, inlet heavy oil velocity; u_w, inlet water velocity; T_o, inlet heavy oil temperature; T_w, inlet water temperature; T_p, wall temperature

- Water inlet:

$$\begin{matrix} R_2 < r < R_1 \\ z = 0 \end{matrix} \Rightarrow \begin{cases} u_{z,w} = u_w \\ f_w = 1 \\ u_{z,o} = u_{x,o} = u_{x,w} = u_{y,o} = u_{y,w} = f_o = 0 \\ T = T_w \end{cases}$$

values of uw and Tw are shown in Table 1

- Heavy oil inlet:

$$\begin{matrix} 0 < r < R_1 \\ z = 0 \end{matrix} \Rightarrow \begin{cases} u_{z,o} = u_o \\ f_o = 1 \\ u_{z,w} = u_{x,o} = u_{x,w} = u_{y,o} = u_{y,w} = f_o = 0 \\ T = T_o \end{cases}$$

values of uo and To are shown in Table 1

(c) No slip conditions at the wall catenary riser/fluid interface:

$$\begin{matrix} r = R_2 \\ 0 \leq z \leq L \end{matrix} \Rightarrow \begin{cases} u_{x,w} = u_{y,w} = u_{z,w} = 0 \\ u_{x,o} = u_{y,o} = u_{z,o} = 0 \\ T = T_p \end{cases}$$

values of Tp are shown in Table 1

(d) Static pressure on the outlet catenary riser is equal to 98 kPa.

(e) The drag coefficient $C_D = 0.44$.

Aiming to compare the results of the transport of heavy oils using core-annular flow with oil and water single-phase flow the new values of inlet velocities of water (u_{ws}) and heavy oil (u_{os}) was determined. The values of these velocities were determined by making the sum of heavy oil and water volumetric flow rate equal to the mixture, Q_m. Then, we can write:

$$u_{ws} = u_{os} = \frac{4 \cdot Q_m}{\pi D^2} \tag{27}$$

Here D corresponds to the catenary riser diameter.

Table 2 Simulations data using single-phase flow to water and heavy oil

Case	u_{oas} (m/s)	u_{ocs} (m/s)	T_{as} (K)	T_{cs} (K)	T_p (K)
14	1.33	1.33	423	298	283
Case	u_{oas} (m/s)	u_{ocs} (m/s)	T_{as} (K)	T_{cs} (K)	T_p (K)
15	1.33	1.33	423	298	283

u_{oas}, inlet heavy oil velocity in the annular section; u_{was}, inlet water velocity in the annular section; u_{ocs}, inlet heavy oil velocity on the circular section; u_{wcs}, inlet water velocity on the circular section; T_{as}, inlet temperature on the annular section; T_{cs}, inlet water temperature on the cylindrical section; T_p, wall temperature

Table 3 Thermal–physical properties of the fluid phases

	Produced water	Heavy oil
Density (kg/m^3)	997	989
Viscosity (Pa s)[a]	$\mu_w = \frac{2185}{(40.12-0.0051547 \cdot T) \cdot T-1000}$	$\mu_o = 51.87 \cdot e^{\left[-2.3935 \cdot \left(\frac{T-T_{min}}{T_{max}-T_{min}}\right)\right]}$
Heat capacity (J/kg K)	4,181.7	1,800
Thermal conductivity (W/m K)	0.6069	0.147
Surface tension (N/m)	0.062	

[a] Produced water viscosity correlation was based in the work of Babadagli and Al-Bemami [29] with $72 < T < 440$°F (Fahrenheit)

Parameters T_{min} and T_{max} of heavy oil viscosity correlation is 273 K and 573 K, respectively (Perry [30])

In Table 2 we illustrate the inlet velocities to water and heavy oil used to simulated the water and heavy oil single-phase flows.

The thermal–physical properties of water and heavy oil used in all simulations are shown in Table 3.

3.2 Numerical Grid

All simulations were developed using the numerical grid illustrated in Fig. 2. By using the ANSYS CFX-Build 5.5 an unstructured mesh was obtained after several refinements contained 346,075 elements. This mesh was optimal for good predictions and reasonable computational time for simulations.

3.3 Hydrodynamics Effects

In order to illustrate the formation of core-annular flow on the catenary riser we present the volume fraction fields in the yz plane passing through the axis in Fig. 3.

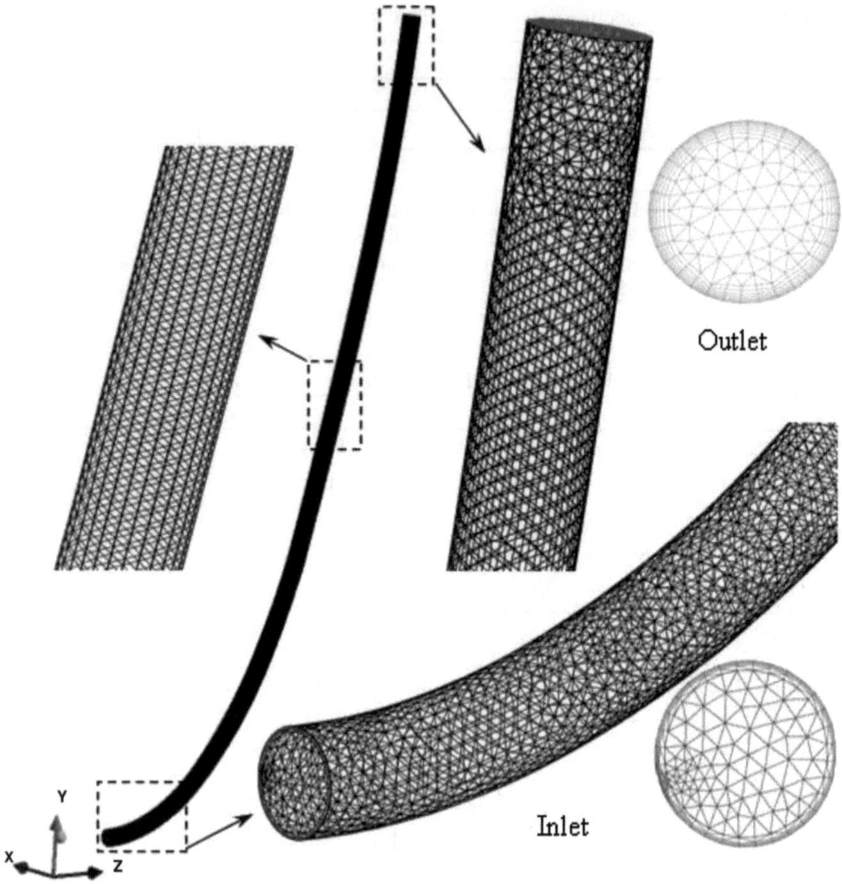

Fig. 2 Numerical mesh used in all simulations

Note that there is a formation of an annular pattern that moves along the catenary riser. We can observe near the inlet section that the density difference between the phases tends to lift the core in the upper part of the pipe (Fig. 4). In Fig. 3b, c, d it can be observed the influence of pipe curvature about phases distributions. It is verified that the heavy oil tends to collide with the wall in the lower region of the catenary (Fig. 3b), and the lubricating forces tend to move the core to the centre. A competition between these effects gives a variation in the position of the core inside the catenary (Fig. 3c, d). When lubricating forces are small (small water velocity) the core approaches of the pipe touching the wall as shown in Fig. 5. This figure displays that oil touches the inner wall of the catenary in different positions, which can be attributed to the imbalance between the forces acting on the annular flow. The difference of density makes the oil-core to move closer to the upper wall of the catenary, while the lubrication forces tend to repel the core to

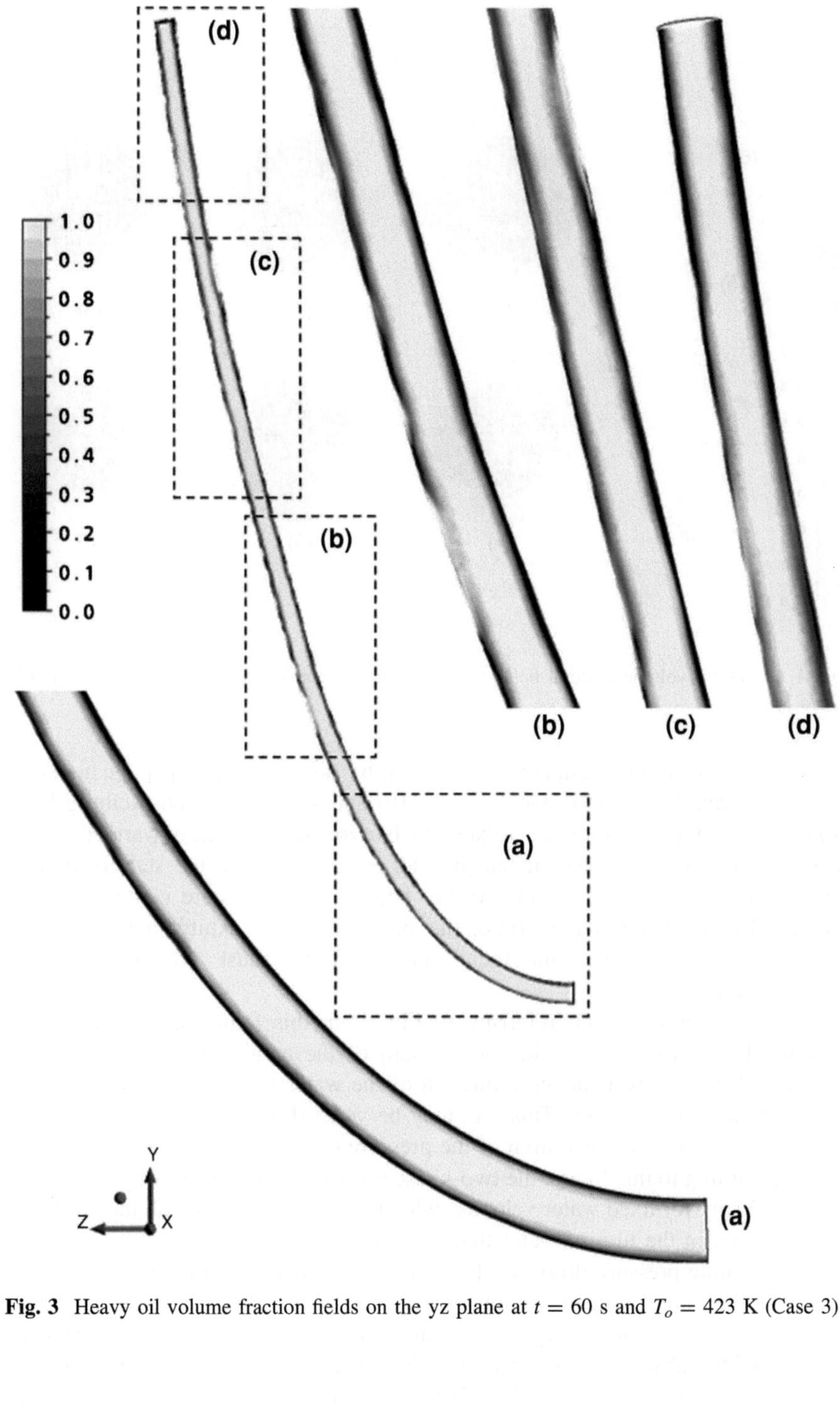

Fig. 3 Heavy oil volume fraction fields on the yz plane at $t = 60$ s and $T_o = 423$ K (Case 3)

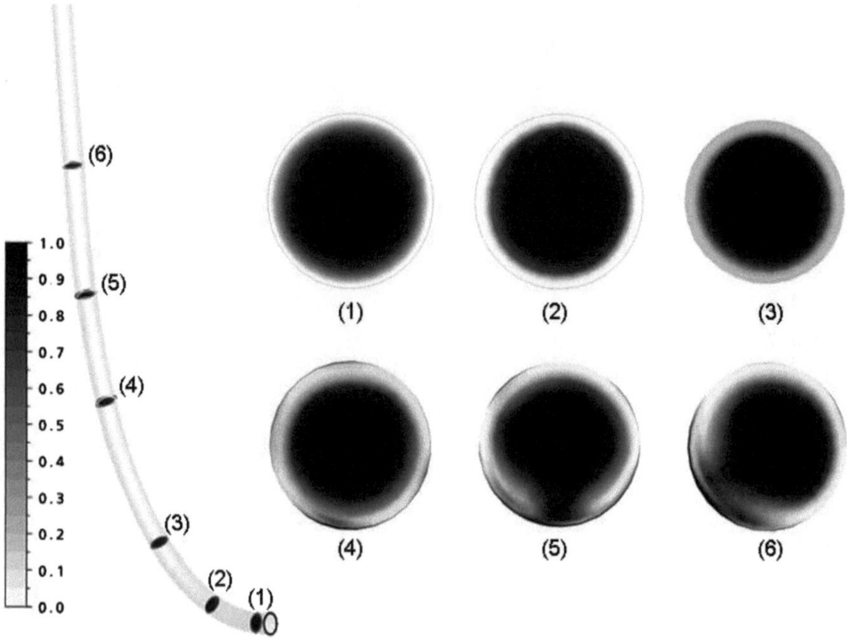

Fig. 4 Heavy oil volume fraction fields on several cross sections at $t = 60$ s and $T_o = 423$ K (Case 3)

catenary center. Similar results were obtained by Bensakria et al. [7], Huang et al. [10], Ooms and Poesio [14], Ooms et al. [16] and Vanaparthy and Meiburg [27]. However, all these studies are related to horizontal tubes. Bannwart et al. [6] reports that wettability phenomena play an essential role in the stabilization of horizontal core-annular flow. In order to keep the viscous core completely surrounded by water for a long time, the pipe wall must exhibit a hydrophilic-oleophobic behavior, i.e., interfacial tension forces must overcome oil-wall adhesion forces.

Figure 6 shows a detail referring to Fig. 3. In this figure are displayed the superficial velocity vector field and a zoom of the area highlighted. It can be observed that there is a sudden reduction of the water velocity close to the wall, leading to near-zero values. This fact may be verified in Fig. 6b, c.

Figure 7 presents the variation of the pressure drop versus oil to water velocity ratio. According to this figure, the two-phase pressure drop increases by increasing the oil velocity for fixed water velocity, which can be associated with increased oil concentration in the film of water that surrounds the core (Fig. 3, 4).

The absolute pressure drop as a function of the operation time in the catenary riser was measured with or without water lubrication. Results are presented in Fig. 9 (Case 3—Table 1). By using water lubrication, it can be observed an increase of the absolute pressure drop, which becomes almost constant after the first 20 s (approximately 150 kPa). Without lubrication, it can be observed an

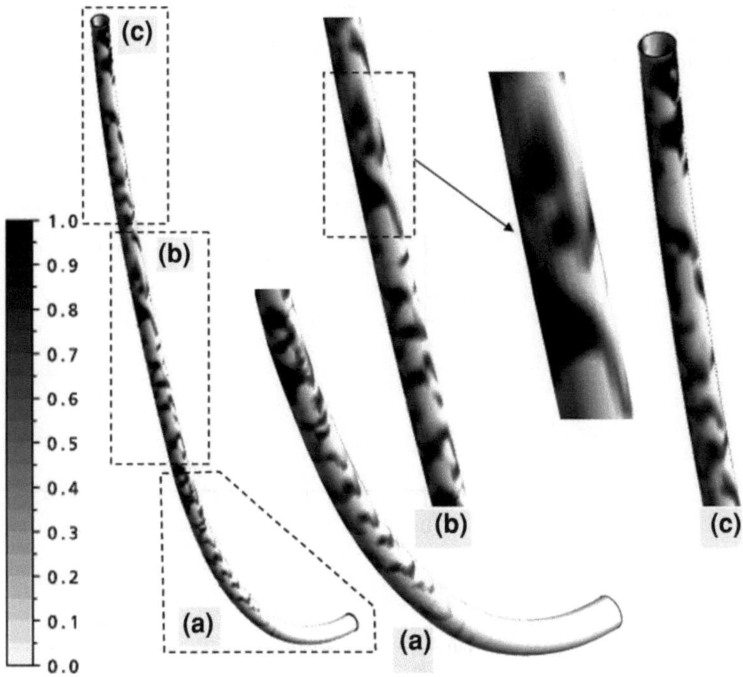

Fig. 5 Heavy oil volume fraction fields on the wall of the catenary at $t = 60$ s and $T_o = 423$ K (Case 3)

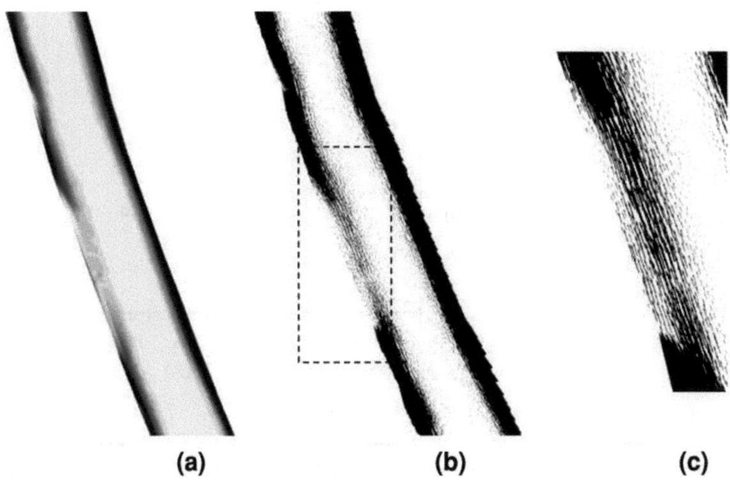

Fig. 6 Details referring to Fig. 3 **a** oil volume fraction distribution; **b** water superficial velocity vector field; **c** zoom of the vector field highlighted

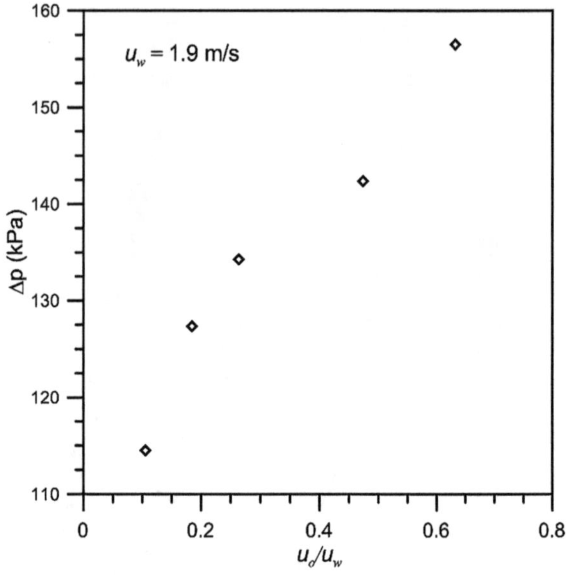

Fig. 7 Pressure drop in the catenary as a function of the oil–water velocity ratio

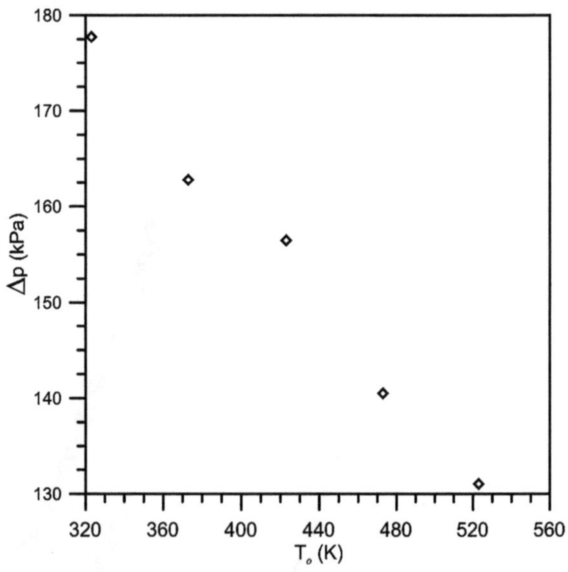

Fig. 8 Pressure drop in the catenary as a function of the oil inlet temperature

adverse behavior and it is verified a decrease of absolute pressure drop with the time.

Comparisons between absolute pressure drop with and without lubrication show a large discrepancy as one can see in Fig. 9. The absolute pressure drop for heavy oil transportation without lubrication was 3.28 times higher than with lubrication

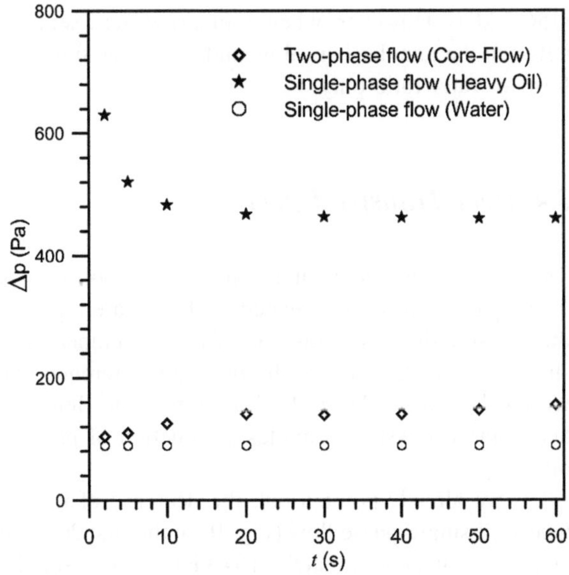

Fig. 9 Comparison of pressure drop between the water and oil single-phase flow, and annular-core flow (Case 3)

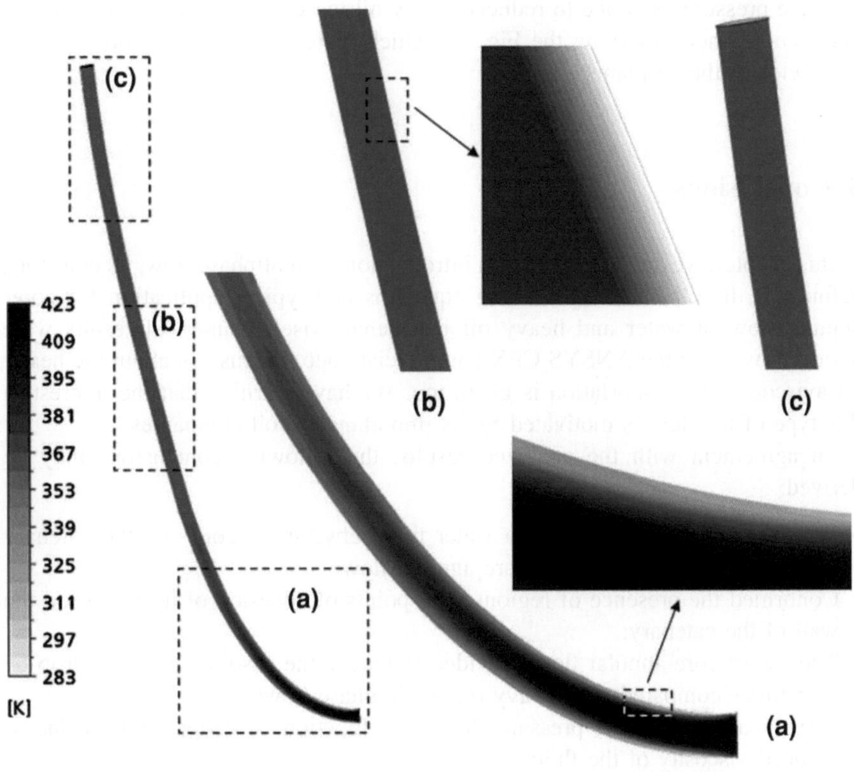

Fig. 10 Heavy oil temperature fields on the yz plane at $t = 60$ s and $T_o = 423$ K (Case 3)

(156.5 kPa). However, when comparing the results of the absolute pressure drop between the annular-core flow and the water single-phase flow, we get 1.59 core-flow to water ratio only.

3.4 Heat Transfer Effects

The effect of the temperature on the heavy oil transportation using the core-flow technique was also investigated. In Fig. 8 are represented the numerical results of the pressure drop as a function of the oil temperature in the entrance. The results show a linear decrease of the pressure difference with temperature, as expected, with a slope of −231.58 Pa/K. According to Bensakhria et al. [7], the addition of heat transport system provides a reduction of pressure drop in the flow of heavy oil.

As we already pointed out, the decrease of the absolute pressure drop for the heavy oil single-phase flow (Fig. 9) in the first 20 s, giving a difference (between 2 and 20 s) of approximately 164 kPa. Such result is due to the boundary layer imposed in the entrance section ($T_o = 423$ K), which provides a reduction in absolute pressure drop due to reduced heavy oil viscosity. This fact can be better understood when observing the Fig. 10, which represents the heavy oil temperature fields in the yz plane.

4 Conclusions

In this chapter, we have presented an introduction to multiphase flow, its concepts, definitions, flow patterns, governing equations and typical application for core-annular flow of water and heavy oil in catenary riser. Numerical results were obtained by using the ANSYS CFX commercial code. Discussion about the heavy ultraviscous oil transportation is given too. We have clarified that the interest in this type of problem is motivated by its importance to oil companies.

In agreement with the analyzed results, the following conclusions may be derived:

- We observed the presence of a water film between the core and the catenary riser wall, characterizing the core-annular flow;
- Confirmed the presence of regions with points of adhesion of heavy oil on the wall of the catenary;
- The use of core-annular flow provided to reduce the absolute pressure drop by 3.28 times compared with heavy oil single-phase flow;
- The linear decrease of pressure drop as a function of temperature is due to reduced viscosity of the fluids.

Acknowledgments The authors thank to Brazilian offices CNPq, ANP/UFCG-PRH-25, FINEP, and CAPES, to Brazilian private enterprises PETROBRAS and JBR Engenharia LTDa, for the granted financial support and to researcher reported in the text for their contributions for improvement of this work. The authors acknowledge also the opportunity given by the Editors to present our researches in the chapter.

References

1. Andrade, T.H.F.: Numerical study of heavy oil transport on pipe lubrificated by water. Master Sci. Chem. Eng. Federal University of Campina Grande, PB-Brasil (2008) (In Portuguese)
2. ANSYS: Solver theory guide of ANSYS CFX 11.0. ANSYS Europe Ltd, USA (2006)
3. Arney, M.S., Ribeiro, G.S., Guevara, E., Bai, R., Joseph, D.D.: Cement-lined pipes for water lubricated transport of heavy oil. Int. J. Multiph. Flow **22**, 207–221 (1996)
4. Bai, R., Chen, K., Joseph, D.D.: Lubricated pipelining: stability of core-annular flow. Part. 5, experiments and comparison with theory. J. Fluid Mech. **240**, 97–132 (1992)
5. Bannwart, A.C.: A simple model for pressure drop in horizontal core annular flow. J. Braz. Soc. Mech. Sci. **21**, 233–244 (1999)
6. Bannwart, A.C., Rodriguez, O.M.H., Trevisan, F.E., Vieira, F.F., de Carvalho, C.H.M.: Experimental investigation on liquid–liquid–gas flow: flow patterns and pressure-gradient. J Pet. Sci. Eng. **65**, 1–13 (2009)
7. Bensakhria, A., Peysson, Y., Antonini, G.: Experimental study of the pipeline lubrication for heavy oil transport. Oil Gas Sci. Technol. **59**, 523–533 (2004)
8. Crivelaro, K.C.O., Damacena, Y.T., Andrade, T.H.F., Lima, A.G.B., Farias Neto, S.R.: Numerical simulation of heavy oil flows in pipes using the core-annular flow technique. WIT Trans. Eng. Sci. **63**, 193–203 (2009)
9. Damacena, Y.T.: Reduction of friction during the transport of heavy oils in pipelines. Monogr. in Mech. Eng., Federal University of Campina Grande, p. 111 (2009) (In Portuguese)
10. Huang, A., Christodoulou, C., Joseph, D.D.: Friction factor and holdup studies for lubricated pipelining-ii laminar and k-e models for eccentric core-flow. Int. J. Multiph. Flow **20**, 481–491 (1994)
11. Joseph, D.D., Bai, R., Chen, K.P., Renardy, Y.Y.: Core-annular flows. Annu. Rev. Fluid Mech. **29**, 65–90 (1997)
12. Kleinstreuer, C.: Two-phase flow: theory and applications. Taylor & Francis, New York (2003)
13. Oliemans, R.V.A., Ooms, G., Wu, H.L., Duijvestijn, A.: The core-annular oil/water flow turbulent-lubricating-film model and measurements in a 5 cm pipe loop. Int. J. Multiph. Flow **13**, 23–31 (1987)
14. Ooms, G., Poesio, P.: Stationary core-annular flow through a horizontal pipe. Phys. Rev. 68 (2003)
15. Ooms, G., Segal, A., Van Der Wees, A.J., Meerhoff, R., Oliemans, R.V.A.A.: Theoretical model for core-annular flow of a very viscous oil core and a water annulus through a horizontal pipe. Int. J. Multiph. Flow **10**, 41–60 (1984)
16. Ooms, G., Vuik, C., Poesio, P.: Core-annular flow through a horizontal pipe: hydrodynamic counterbalancing of buoyancy force on core. Phys. Fluids, (2007) 19:092103-092103-17
17. Pereira Filho, G.H.S.: Non-isothermal heavy oils transportation in submerged risers Monogr. Mech. Eng. Federal University of Campina Grande, p. 111 (2010) (in Portuguese)
18. Prada, J.W.V., Bannwart, A.C.: Modeling of vertical core annular flows and application to heavy oil production.In: Proc. of ETCE/OMAE: energy for the new millen, February 14–17, New Orleans, LA (2000)

19. Prada, J.W., Bannwart, A.C. Pressure drop in vertical core annular flow. J. Fr. Soc. Mech. Sci. 23 (2001)
20. Preziosi, L., Chen, K., Joseph, D.D.: Lubricated pipelining: stability of core-annular flow. J. Fluid. Mech. **201**, 323–356 (1989)
21. Ranade, V.V.: Computational flow modeling for chemical reactor engineering. Academic Press, San Diego (2002)
22. Rodriguez, O.M.H., Bannwart, A.C.: Analytical model for interfacial waves in vertical core flow. J. Pet. Sci. Eng **54**, 173–182 (2006)
23. Rodriguez, O.M.H., Bannwart, A.C.: Experimental study on interfacial waves in vertical core flow. J. Pet. Sci. Eng. **54**, 140–148 (2006)
24. Rodriguez, O.M.H., Bannwart, A.C., Carvalho, C.H.M.: Pressure loss in core-annular flow: modeling, experimental investigation and full-scale experiments. J. Pet. Sci. Eng. **65**, 67–75 (2009)
25. Rovinsky, J., Brauner, N., Moalem Maron, D.: Analytical solution of laminar two-phase flow in the limit of fully eccentric core-annular configuration. Int. J. Multiph. Flow **23**, 523–543 (1997)
26. Santos, J.S.S.: Numerical study of the lubrification of submerged risers for the heavy oil transportation, Master Sci. Chem. Eng. Federal University of Campina Grande, PB- Brasil (2009) (In Portuguese)
27. Vanaparthy, S.H., Meiburg, E.: Variable density and viscosity, miscible displacements in capillary tubes. Eur. J. Mech B/Fluids **27**, 268–289 (2008)
28. Ghosh, S., Mandal, T.K., Das, G., Das, P.K.: Review of oil water core annular flow. Renew. Sustain. Energy Rev. **13**(8), 1957–1965 (2009)
29. Babadagli, T., Al-Bemani, A.: Investigations on matrix recovery during steam injection into heavy-oil containing carbonate rocks. J. Petroleum Sci. Eng. **58**, 259–274 (2007)
30. Perry, J.H. (ed.): Chemical Engineer's Handbook, 4th edn. McGraw-Hill, New York (1963)

First Principle Study on the Lead-Free Perovskite Structure of $SnTiO_3$

M. F. M. Taib, K. H. K. Arifin, M. K. Yaakob, A. Chandra, A. K. Arof and M. Z. A. Yahya

Abstract The electronic band structure, density of states, dielectric function, born effective charges, and phonon dispersion of perovskite $SnTiO_3$ (ST) are investigated from the first principles calculation using Density Functional Theory within local density approximation. Calculated along the high symmetry direction in the Brillouin zone, the resulting band gap of ST is 0.967 eV. The dielectric function in this work shows the details of the absorptive transitions from the valence bands to the conduction bands in the ST compound. In addition, an analysis of the born effectives charges and phonon dispersion of ST shows that this compound has a

M. F. M. Taib · K. H. K. Arifin · M. K. Yaakob · M. Z. A. Yahya (✉)
Faculty of Applied Sciences, Universiti Teknologi MARA,
40450 Shah Alam, Malaysia
e-mail: mzay@salam.uitm.edu.my

M. F. M. Taib
e-mail: mfm_taib@yahoo.com

K. H. K. Arifin
e-mail: lme_khka@yahoo.com

M. K. Yaakob
e-mail: muskamil20@yahoo.com

M. Z. A. Yahya
Ionics Materials and Devices (iMADE) Research Laboratory, Institute of Science
Universiti Teknologi MARA, 40450 Shah Alam, Malaysia

A. Chandra
Department of Physics and Meteorology, Indian Institute of Technology,
Kharagpur, 721302 India
e-mail: rishu_chandra@yahoo.com

A. K. Arof
Physics Department, University of Malaya, 50603 Kuala Lumpur, Malaysia
e-mail: akarof@um.edu.my

A. Öchsner et al. (eds.), *Materials with Complex Behaviour II*,
Advanced Structured Materials 16, DOI: 10.1007/978-3-642-22700-4_14,

covalent band between Ti–O and Sn–O that provides a stable tetragonal structure, P4mm, which refers to the imaginary value of TO1($-213.222\ cm^{-1}$) in phonon calculation. All results are compared, and they showed good agreement with other calculated values using different methods.

Keywords Band structure • Density of state • Dielectric function • Phonon dispersion • Born effective charges • Density Functional Theory • Local density approximation

1 Introduction

Numerous studies on the crystal in the perovskite family have been conducted to meet the high demand brought about by the rapid development of ferroelectric devices such as non-volatile memory, optical waveguides, laser frequency modulations, and piezoelectric transducers [1–3]. Lead-based ferroelectric ceramics such as $PbTiO_3$ and $PbZrO_3$ are important members of the perovskite family that have widely used in industrial applications. However, PZT ($Pb(Ti,Zr)O_3$) materials contain toxic Pb, which can contribute to serious environment pollution [4]. To overcome this problem, the Pb atom in A-site perovskite ceramic must be replaced with an alternative material such as Tin (Sn), which is expected to give good performance. $SnTiO_3$ (ST) is a new potential ferroelectric material because of its high dielectric constant and polarization.

Yahong et al. [4] showed that the disproportion and oxidation of Sn^{2+} can be avoided using a spark plasma sintering method. The results of their experiments showed that perovskite oxide $BaTiO_3$ had excellent ferroelectric properties, such as visible light absorption ability. Suzuki et al. [5] also investigated the characteristics of Sn^{2+} in $(Ba_{1-x}\ Ca_x)TiO_3$ compound and found that Sn^{2+} could change or enhance the properties of the material when it was doped in the system. They reported that the results of the tetragonality increased directly with increased Sn^{2+} in the $(Ba_{1-x}\ Ca_x)TiO_3$ compound as confirmed by X-ray diffraction profiles. In addition, the phase transition temperature of $(Ba_{1-x}\ Ca_x)TiO_3$ increased from 130 to 155°C when doped with Sn^{2+}.

Other investigations delved into the use of the first principles calculation on ST. This study is important in predicting the properties of ST without having to synthesize the compound. Konishi et al. [6] calculated the electronic structure of ST using plane-wave pseudopotential (PWPP). Lebedev [7] calculated the phonon spectra of ST using ABINIT computer code, while Uratani et al. [8] used ST in the tetragonal structure (P4mm and P4/mm) and Matar et al. [9] reported on the properties of ST in tetragonal structure using the VASP computer code. Matar also reported on the moment of spontaneous polarization for an ST value of $P_{ST} = 1.1\ cm^{-2}$, as large as that of $PbTiO_3$ $P_{ST} = 0.72\ cm^{-2}$. These studies indicated the excellent properties of ferroelectric oxide, which could be potentially used in ferroelectric devices.

Fig. 1 Cubic structure of ST

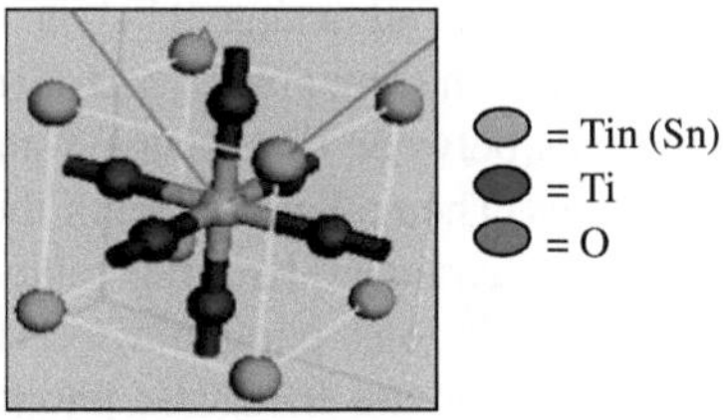

However, no other detailed analysis of the electronic, optical, and structural properties in Pm3m (cubic) structure of ST using the Cambridge Serial Total Energy (CASTEP) computer code has been reported. In this work, the first principles studies on the properties of ST were performed using CASTEP computer code. The behavior and characteristics of ferroelectricity, such as energy band gap, density of state (DOS), dielectric function, born effective charges (BEC), and phonon dispersion are studied and are compared with other computational methods such as ABINIT and VASP.

2 Computational Method

In this study, the first principle calculations were performed using the the Cambridge Serial Total Energy (CASTEP) computer code [10]. In all the calculations of ST compound in the cubic ABO_3 perovskite structure (Pm3m space group), the A (Sn) occupied the corner of cubic structure (0, 0, 0), B(Ti) in the body-centered cubic at (0.5, 0.5, 0.5) and O atoms at the face-centered cubic (0, 0.5, 0.5) and (0.5, 0, 0.5), as shown in Fig. 1. The electrons in Sn (5s, 5p), Ti (3d 4s), and O (2s 2p) were treated as valence states. The exchange–correlation energy functional was evaluated within the local density approximation (LDA) using the Caperly–Alders parameterized by the Perdew–Zunger scheme (CA–PZ) [11, 12] as electron–ion interactions as the Vanderbilt ultrasoft pseudo-potential [13].

To determine the structural parameters of ST, CASTEP was used based on the Broyden Goldfarb Shenno (BFGS) minimization technique. This method usually provides the fastest way to find the lowest energy structure. It is the only scheme that supports cell optimization in CASTEP. This geometry optimization was performed with convergence of energy change per atom of less than 5×10^{-6} eV, residual force of less than 0.01 eV/Å, stress below 0.02 GPa, and displacement of atoms during the geometry optimization of less than 0.0005 Å.

The structural, band structure, and DOS of ST were calculated with the kinetic cut-off energy for the plane wave expansion taken as 380 eV and $6 \times 6 \times 6$ *k*-points according to the Monkhorst–Pack scheme in the Brillouin zone integration, which was employed to obtain good convergence.

The optical properties may be acquired from the knowledge of the complex dielectric function $\varepsilon(\omega) = \varepsilon_1(\omega) + i\varepsilon_2(\omega)$. In this work, $\varepsilon(\omega)$ was performed using

LDA, whose advantage [14] is the largest underestimation of $\varepsilon_2(\omega)$ peaks of materials. The imaginary part $\varepsilon_2(\omega)$ was calculated from the momentum matrix elements between the occupied and unoccupied wave functions within the selection rules. The real part $\varepsilon_1(\omega)$ of the dielectric function can be extracted from $\varepsilon_2(\omega)$ using the Kramers–Kroning relation. Similar to those in the band structure and DOS calculations, phonon dispersion curves were calculated within LDA for ST. However, for the BEC and phonon dispersion calculation, the norm-conserving pseudo-potential was used with 550 eV cut-off energy.

3 Results and Discussions

(a) *Band structure and DOS of ST*

The calculation of ST was performed at parameter $a = 4.01\text{Å}$ at cubic structure. Figure 2a shows the calculated electronic-band structure of ST along the various symmetry lines G, M, R, and X in the Brillouin Zone using LDA approach. In this work, the highest valence bands (VB), which lie close to the Fermi level (E_F), are dominated by the O 2p at X point. The conduction bands (CB) have compounds primarily from the Ti d-state and the bands occur at G point for ST. The result for ST shows an indirect band gap with 0.967 eV. The value of energy band gap calculated in this work is significantly less than the 1.19 eV obtained by Rozo [15], which was obtained using DFT with the generalized gradient approximation (GGA).

The LDA method underestimates the band gap; thus, to adjust the theoretical band gap to the experimental result, we estimated a multiplicative correction factor of 1.66 to the calculated band gaps [16]. We estimated the experimental value of the band gap for ST to be 1.605 eV. The value of band gap of ST was small and needed some modification to enhance its value and become useful for ferroelectric application.

The total and partial DOS of ST were calculated to understand the chemical bonding of the materials as shown in Fig. 2b. The lowest state for ST was O 2s and was located at around -16 eV. In our calculation, the O 2s states were separated from the Sn 5s state approximately 8.5 eV in ST. In these compounds, the upper valence parts were dominated by O 2p states and the CB had compounds primarily formed from the Ti d-state and Sn 5p. In addition, the CB of ST had small contributions from Sn 5p. From ST, the lowest CB comprised Ti 3d and Sn 5p formed at 0.97–9.5 eV. The top VB was dominated by O 2p and consisted of Sn 5s at -7.5–0 eV. The energy gap existing in the structure was separated between Ti 3d treated as conduction state and maximum O 2p.

(b) *Dielectric function of ST*

The calculated optical properties for dielectric function for ST for the energy range of up to 50 eV are presented in Fig. 3. To explain the peak observed in the

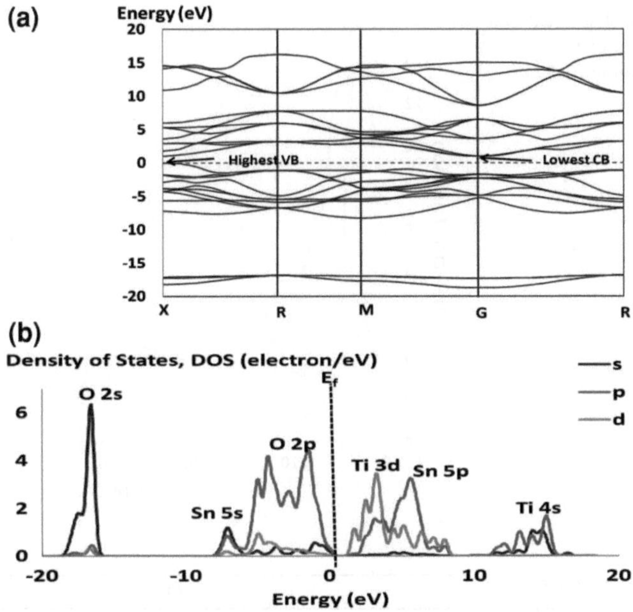

Fig. 2 Calculated **a** energy band structure and **b** total DOS of ST

optical spectra, it is customary to consider transitions from occupied to unoccupied bands in the electronic energy band structure, particularly at high symmetry points in the Brillouin zone [17]. These calculated optical properties (dielectric function) could be helpful in understanding the electronic structure in the cubic phase of ST. Figure 3 shows the real and imaginary parts of the dielectric function for cubic ST compound. The imaginary part can be obtained directly from the band structure, while whereas the real part can be derived according to the Kramers–Kronig relations explained in the computational method.

Figure 3 illustrates the four peaks of the imaginary part, which are at 4–37 eV. These peaks are the absorptive transitions from the VB to the CB. These peaks are labeled A, B, C, and D and are located at 4.37, 13.6, 20.1, and 36.3 eV, respectively. According to the analysis of the electronic structure of cubic ST in Fig. 2, peak A originated from the transitions of O 2p into Ti 3d CB, peak B originated from the transitions of O 2p into Ti 3d CB as well as O 2p into Sn 5p and Ti 4s CB, and peak C originated from the transitions of O 2s and Sn 5s into Ti 3d CB. Peak D had no transition because the DOS in Fig. 2 showed no peak. For the real part, there were three peaks located at 2.8, 12.7, and 34.8 eV, respectively. Additionally, the calculated static dielectric constant of Pm3m cubic ST was about 9.75.

(c) *BEC and phonon dispersion of ST*

Table 1 illustrates the BEC of ST in the cubic structure, which was compared with the work of Lebedev [7]. However, the work of Lebedev was performed using

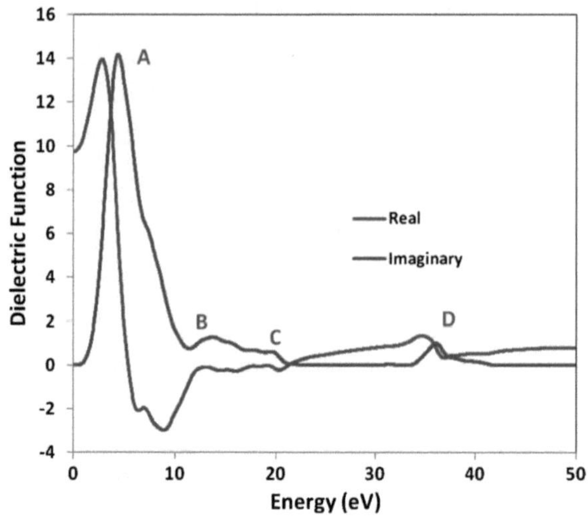

Fig. 3 Graph of the real and imaginary parts of the dielectric function of ST

Table 1 Bond effective charges of ST compounds

	ABO_3	
	ST	[7]
Z*Sn	4.605	4.255
Z*Ti	6.548	7.529
Z*O⊥	−2.882	-2.745
Z*O‖	−5.389	-6.294

the ABINIT computer code. BEC is important to identifying the long-range part of the interatomic force constants and making the interpolation of phonon frequency tractable. For the case of ST, Z*Ti (6.548) and Z*O‖ (−5.389) were anomalously large with respect to the nominal ionic charges (+4 for Ti and −2 for O). This surprising phenomenon was explained recently in connection with dynamic charges of hybridization between O 2p with Ti 3d [18–20]. The Z*Sn is significantly greater, indicating that the bonding Sn–O becomes more covalent in character and the value of Z*Sn for ST is 4.605 in Table 1.

Phonon dispersion is important in determining the properties of solids, particularly when referring to the frequencies of vibration atoms in a solid structure. The unstable modes, which determine the nature of the transition of the compounds, have imaginary frequency (below the zero frequency line). In this work, the unstable modes of ST exhibited at X, M, R, and G points are illustrated in Fig. 4. The longitudinal optical existed at 6, 12 and 15th mods with frequencies of 73.639 (LO1), 351.475 (LO2), and 780.969 cm^{-1} (LO3) at G point, respectively. In contrast, the first transverse optical for ST existed in the imaginary part with a value of −213.222 cm^{-1} (TO1) and remained unstable along the X, R, M, and G points, while TO2 and TO3 existed at the 10–11 and 13–14 modes with

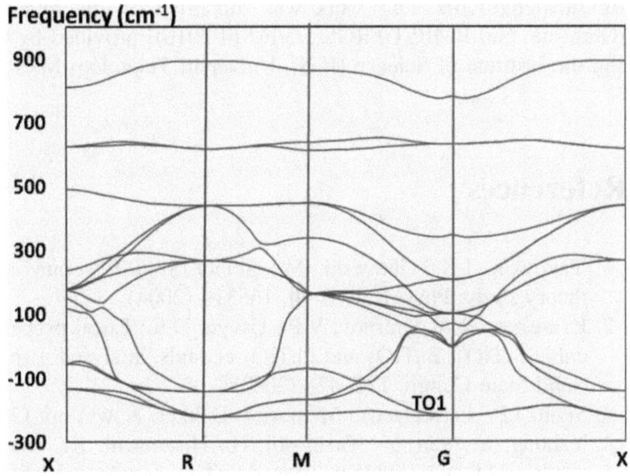

Fig. 4 Calculated phonon dispersion for ST

208.887 and 631.475 cm^{-1}, respectively. The instability of ferroelectric oxide of a transverse optic (TO1) in the phonon dispersion is an important key to explaining structural instability and lattice dynamic.

In this work, we expected that the nature of the ferroelectric instability of the ST, would most favorably be the tetragonal P4mm phase due to the imaginary phonon energies associated with structural instability of the crystals represented at TO1 (G point). The analysis emphasized the correlation of the Ti displacement and showed that the Ti and O atomic displacements were only weakly coupled. However, this small coupling remains important in reproducing ferroelectric instability. Hence, ST is a good candidate to replace $PbTiO_3$ because both compounds have similar properties especially in the transition structure. Based on the analysis of phonon dispersion in this work, ST can change to tetragonal structure (P4mm) and can be expected to have excellent properties for application to ferroelectric devices.

4 Conclusion

In this paper, we described the first principles band structure, DOS, dielectric function, BEC, and phonon dispersion of cubic ST. All calculations show good agreement with other reported calculations. The results showed that the energy band gap of ST was 0.967 eV (indirect band gap at X–G point). In addition, the transition electron from the VB into CB was explained by the peak occurring in the dielectric function of ST. The BEC results proved that Ti–O and Sn–O bonds in the ST compound have covalent characteristics. The phonon dispersion calculation shows that the ST compound provides a stable tetragonal structure P4mm relative to cubic. However, the single-component ferroelectric material in bulk of ST requires some modifications to display superior functional properties, which would be beneficial for future applications, particularly in enhancing the energy gap using the first principle calculation.

Acknowledgments This work was supported by the Fundamental Research Grant Scheme [Grant no. 600-RMI/ST/FRGS 5/3/Fst (4/2010)] provided by the Ministry of Higher Education and the Institute of Science (IOS), Universiti Teknologi MARA Malaysia.

References

1. Yakovkin, I.N., Gutowski, M.: $SrTiO_3$/Si (001) epitaxial interface: a density functional theory study. Phys. Rev. B **70**, 165319 (2004)
2. Evarestov, R.A, Smirnov, V.P., Usvyat D.E.: Local properties of the electronic structure of cubic $SrTiO_3$, $BaTiO_3$ and $PbTiO_3$ crystals, analysed using Wannier-type atomic functions. Solid State Comm. **127**, 423 (2003)
3. Scott, J.F.: Ferroelectric Memories. Dekker, New York (2000)
4. Yahong, X., Shu, Y., Takatoshi, H., Hisamachi, K., Tsugio, S.: Microwave-hydrothermal synthesis of nano-sized Sn^{2+}-doped $BaTiO_3$ powders and dielectric properties of corresponding ceramics obtained by spark plasma sintering method. J. Mater. Sci. **44**, 4834–4839 (2009)
5. Suzuki, S., Takeda, T., Ando, A., Takagi, H.: Ferroelectric phase transition in Sn^{2+} ions doped Ba,Ca..TiO_3 ceramics. Appl. Phys. Lett. **96**, 132903 (2010)
6. Konishi, Y., Ohsawa, M., Yonezawa, Y., Tanimura, Y., Chikyow, T, Wakisaka, T., Koinuma, H., Miyamoto, A., Kubo, M., Sasata, K.: Possible ferroelectricity in $SnTiO_3$ by first-principles calculation. In: Materials research society symposia proceedings. **748**, U3.13.1 (2003)
7. Lebedev, L.I.: Ab initio calculations of phonon spectra in $ATiO_3$ perovskite crystals (A= Ca, Sr, Ba, Ra, Cd, Zn, Mg, Ge, Sn, Pb). Phys. Solid State **51**(20), 362–372 (2009)
8. Uratani, Y., Shishidou, T., Oguchi, T.: First-principles study of lead-free piezoelectric $SnTiO_3$. Jpn. J. Appl. Phys. **47**, 7735–7739 (2008)
9. Matar, S.F., Baraille, I., Subramaniam, M.A.: First principles studies of $SnTiO_3$ perovskite as potential environmentally benign ferroelectric material. Chem. Phys. **355**, 43 (2009)
10. Segall, M.D., Lindan, P.J.D., Probert, M.J., Pickard, C.J., Hasnip, P.J., Clark, S.J., Payne, M.C.: First-principles simulation: ideas, illustrations and the CASTEP code. J. Phys. Condens. Matter **2717**, 14 (2002)
11. Ceperley, D.M., Alder, B.J.: Ground State of the electron Gas by a Stochastic method. Phys. Rev. Lett. **45**, 566 (1980)
12. Perdew, J.P., Zunger, A.: Self interaction correction to density-functional approximation for many-electron system. Phys Rev. B **23**, 5048 (1981)
13. Venderbilt, D.: Soft self-consistent pseudopotentials is a generalized eigenvalue formalism. Phys. Rev. B **41**, 7892–7895 (1990)
14. Cherrad, D., Maouche, D.: Structural, electronic and optical properties of $SrHfO_3$ (I4/mcm, Imma, Cmcm, P4/mbm and P4mm) phases. Phys. B **405**, 3862–3868 (2010)
15. Rozo, E.E., Fajardo, F., Landinez Tellez, D.A., Rodrıguez Arbey, J., Roa–Rojas, J.: Electronic structure and band of perovskite $SnTiO_3$. Int. J. Phys. 40 (2008)
16. Tang, S., Wallace, R.M., Seabaugh, A., King–Smith, D.: Evaluating the minimum thickness of gate oxide on silicon using first-principles method. Appl. Surf. Sci. **135**, 137 (1998)
17. Bouhemadou, A., Djabi, F., Khenata, R.: First principles study of structural, elastic, electronic and optical properties of the cubic perovskite $BaHfO_3$. Phys. Lett. A **372**, 4527–4531 (2008)
18. Ghosez, Ph., Gonze, X., Lambin, Ph., Michenaud, J.-P.: Born effective charges of barium titanate: band by band decomposition and sensitivity to structural features. Phys. Rev. B **51**, 6765 (1995)
19. Ghosez, Ph., Gonze, X., Michenaud, J.-P.: A microscopic study of barium titanate. Ferroelectrics. **164**, 113 (1995)
20. Posternak, M., Resta, R., Baldereschi, A.: Role of covalent bonding in the polarization of pervoskite oxide: the case of $KNbO_3$. Phys. Rev. B **50**, 8911 (1994)

Strength Anisotropy in Prestressing Steel Wires

Jesús Toribio, Beatriz González and Juan-Carlos Matos

Abstract Cold-drawn prestressing steel wires exhibit strength anisotropy in the form of fracture path deflection towards a direction approaching the wire axis, or cold drawing line, as a consequence of the pearlitic microstructure orientation induced by the manufacturing procedure. Such a crack path deflection is initiated at certain nuclei (fracture origins) at which axial cracking appears in the cold drawing direction (or wire axis) in the form of micro-cleavage units producing a macroscopic phenomenon of *pop-in* in the load–displacement curve. This chapter shows that such fracture initiators appear at a certain distance from the fatigue pre-crack tip at which a local maximum of the cleavage stress is located.

Keywords Cold drawn steel · Cleavage stress · Fracture path deflection · Anisotropy

J. Toribio (✉) · B. González
Department of Materials Engineering, University of Salamanca,
E.P.S.,Campus Viriato, Avda. Requejo 33, 49022 Zamora, Spain
e-mail: toribio@usal.es

B. González
e-mail: bgonzalez@usal.es

J.-C. Matos
Department of Computing Engineering, University of Salamanca,
E.P.S., Campus Viriato, Avda. Requejo 33, 49022 Zamora, Spain
e-mail: jcmatos@usal.es

A. Öchsner et al. (eds.), *Materials with Complex Behaviour II*,
Advanced Structured Materials 16, DOI: 10.1007/978-3-642-22700-4_15,

1 Introduction

It is well known that macroscopic fracture behaviour of materials is strongly dependent on microstructural features controlling the specific micromechanisms of fracture in the process zone [3, 6]. In the case of randomly oriented pearlitic microstructures, the prior austenite grain size was shown to be the microstructural parameter governing the fracture process [1, 5]. However, when pearlitic steels are heavily cold drawn to produce prestressing steel used in prestressed concrete, such a manufacturing procedure affects the microstructural arrangement in the form of a progressive orientation and slenderising of the pearlitic colonies in the drawing direction [8, 9], together with an also progressive orientation of the ferrite/cementite lamellae in the drawing direction and an increase of packing density, i.e., a drawing-induced decrease of interlamellar spacing of pearlite [10, 11].

Previous research work [12] established the microstructural bases of the anisotropic fracture behaviour of heavily drawn steel, assuming that such a particular behaviour was due to the existence of the so called pearlitic pseudocolonies (cf. [12]), i.e., extremely slender colonies, aligned quasi-parallel to the wire axis or drawing direction, and with specially high local interlamellar spacing due to the fact that the cementite plates are not oriented along the wire axis direction and in some cases are pre-fractured by shear during the manufacturing process, according to the model proposed by [4]. Thus, the pearlitic pseudocolonies are preferential fracture paths with minimum local fracture resistance.

This chapter goes further in the analysis of anisotropic fracture behaviour of heavily drawn steels and associates fracture path deflection in a direction close to the wire axis with a local maximum of a cleavage stress perpendicular to the axial crack, i.e., parallel to the initial macroscopic crack, such a stress being computed at the fracture instant.

2 Experimental Procedure

Material was an eutectoid steel whose chemical composition is given in Table 1. It comes from a previously hot rolled bar that was heavily cold drawn in seven passes (plus a stress relieving procedure to eliminate or, at least reduce, residual stresses) finally producing heavily drawn pearlitic steel in the form of prestressing steel wires with an oriented microstructure, as seen in Fig. 1 in which the longitudinal section shows an oriented arrangement, whereas the transverse section shows a randomly-oriented microstructure.

The drawing process produces important microstructural changes in the steel at the two basic microstructural levels of pearlitic colonies and lamellae. The colonies become progressively enlarged and oriented in axial direction with

Table 1 Chemical composition of the steel

% C	% Mn	% Si	% P	% Cr	% V
0.789	0.698	0.226	0.011	0.271	0.078

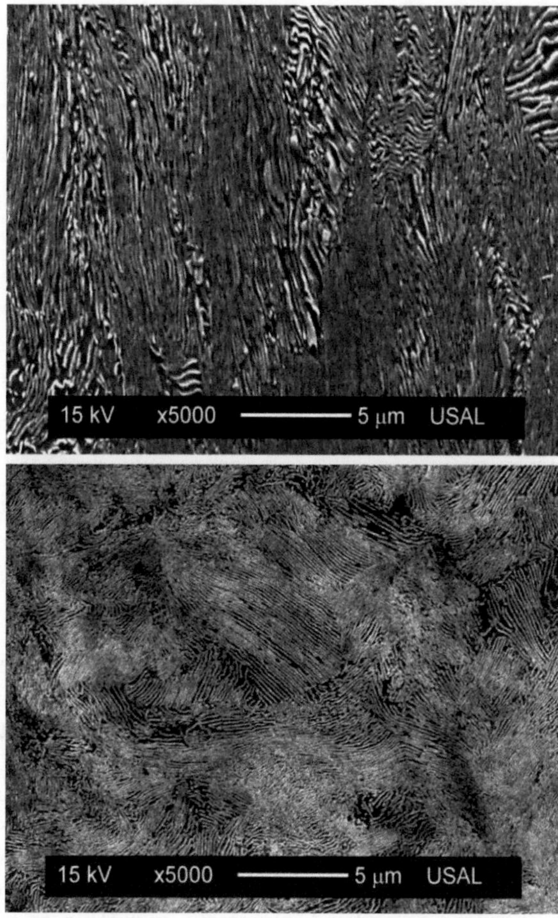

Fig. 1 Longitudinal (*top*) and transverse (*bottom*) metallographic sections of the cold drawn prestressing steel wire. In the longitudinal cut, the wire axis or cold drawing direction corresponds to the vertical side of the micrograph, whereas the transverse cut is oriented in radial direction

cold drawing [8, 9]. With regard to the lamellae, they are also axially oriented after drawing and at the same time the pearlite interlamellar spacing decreases with the level of cumulative plastic strain [10, 11]. Therefore, the microstructure becomes progressively packed and oriented with cold drawing, and such an orientation is the cause of the anisotropic fracture behaviour of the cold drawn steel.

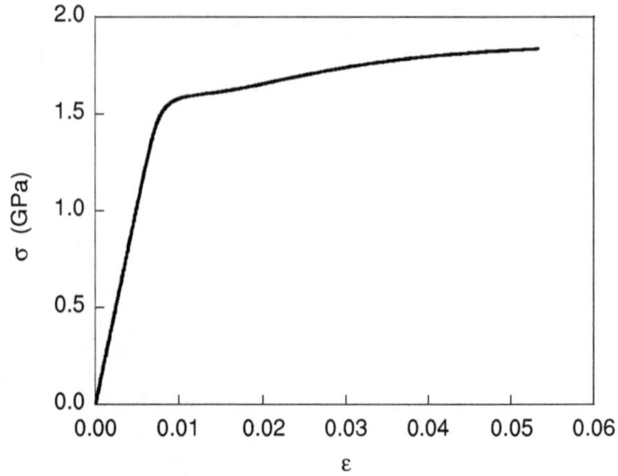

Fig. 2 Curve σ–ε of the considered steel

Mechanical properties of the material, obtained from standard tension tests (Fig. 2) were as follows: Young's modulus $E = 205$ GPa, yield strength $\sigma_Y = 1.57$ GPa, ultimate tensile strength (UTS) $\sigma_R = 1.84$ GPa and strain at UTS $\varepsilon_R = 0.053$. This implies a significant improvement of conventional mechanical properties in relation to plain hot-rolled pearlitic steel (not cold drawn). In particular, both yield strength and UTS clearly increase with cold drawing.

In addition, fracture test on prestressing steel wire allowed the evaluation of the material fracture toughness in transverse direction K_{IC} (0°) = 152.1 MPa $m^{1/2}$ and the same critical parameter in axial direction K_{IC} (90°) = 54.2 MPa $m^{1/2}$. The cold drawing procedure generates microstructural orientation in the pearlitic arrangement of the steel, cf. Fig. 1, thus producing two different values of the *directional* fracture toughness in axial and transverse directions, the axial toughness (associated with longitudinal splitting or delamination) being clearly lower than the transverse one (linked to fracture of the strongest units), as discussed elsewhere [2, 13, 14].

Samples for testing were cylindrical rods with a length of 300 mm and a diameter of 7 mm taken from the commercial wires. After axial tensile fatigue with a sinusoidal wave (at a frequency of 10 Hz and *R*-ratio equal to 0) under load control and decreasing loading steps, specimens were subjected to monotonic tensile loading under displacement control up to fracture, the crosshead speed being 2 mm/min. An extensometer with a gage length of 25 mm was placed in front of the crack mouth (symmetrically in relation to the crack faces), so that both the load applied on the sample (*F*) and the relative displacement by the extensometer (*u*) were recorded to plot the load–displacement curve (*F*–*u*). Some tests were interrupted before final fracture in order to perform a fracto-metallographic analysis on the unloaded sample (evaluating both the fracture path and the microstructure of the material). To this end, specimens were cut, mounted, grounded, polished and attacked with 4% Nital.

Fig. 3 Fracture surface in a sample taking from a commercial prestressing steel wire; *top* view (*left*) and front view (*right*)

3 Experimental Results

Fracture surface in heavily drawn steels shows cracking path deflections, so that the fracture path is not contained in the transverse section of the wire (Fig. 3), with frequent secondary cracking in axial direction, a signal of anisotropic fracture behaviour. The fracture angle, in relation to the transverse axis of the wire (i.e., the fatigue propagation direction), is 39° in heavily drawn steels with a high level of cumulative plastic strain (1.09) as a consequence of seven steps of cold drawing undergone by the steel wire.

The *F*–*u* plot in the fracture tests on cold drawn steel always consisted of three stages, as shown in Fig. 4. After an initial elastic period (linear behaviour), the load F_e may be defined as the end of the elastic stage. Later the plot becomes curved up to a load F_Y at which a *pop-in* appears in the form of sudden (and small) decrease of load. Finally the increase of load continues up to final fracture at a load F_{max}.

The fracto-metallographic analysis of a test interrupted between F_Y and F_{max} (Fig. 5) shows vertical cracking in axial direction, i.e., parallel to the drawing axis. The *pop-in* associated with the load F_Y is not produced by plastic yielding but by a kind of microstructural yielding due to the appearance of the local axial cracking (cf. Fig. 5). This is consistent with the ideas presented by [7], according to which the *pop-in* in the load–displacement curve is produced by a small amount of abrupt crack extension and could be related to the presence of heterogeneities in the material in the form of large inclusions, carbides or, in the case of the cold-drawn pearlitic steel under study, the afore-said pearlitic pseudocolonies or any other microstructural pre-defect (pre-damage) created in the steel during manufacture by heavy drawing.

During the critical phase of fracture, the initial fatigue crack tip (left hand side of Fig. 5) exhibits an increase of crack tip opening displacement (CTOD) in relation to the typical value associated with subcritical regime of fatigue.

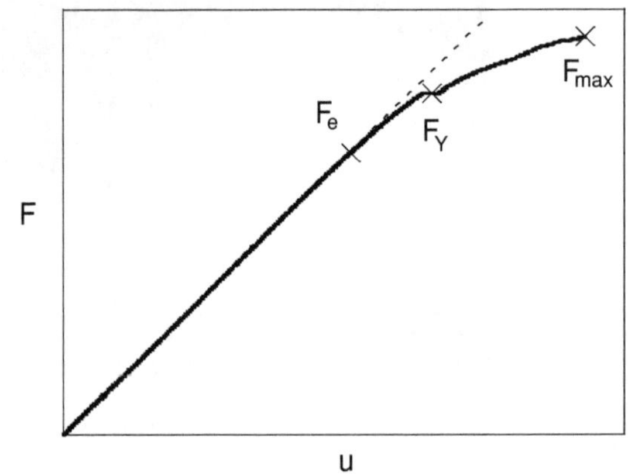

Fig. 4 Load–displacement curve in the fracture test

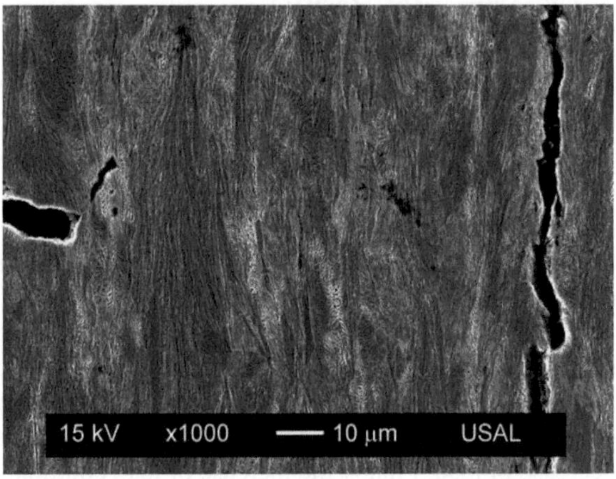

Fig. 5 Metallographic section perpendicular to the crack front, in a fracture test interrupted at a load level after the *pop-in*, i.e., between F_Y and F_{max}, cf. Fig. 4, showing vertical cracking in axial direction, i.e., parallel to the cold drawing axis

In addition, a cornered crack tip shape appears in the material and a small micro-crack initially inclined 60° in relation to the fatigue crack propagation direction. The vertical cracking (right hand side of Fig. 5) exhibits irregularities in the form of micro-discontinuities, differences in the size of the crack opening displacement and small portions of the cracking path oriented in transverse directions. These vertical cracks are generated by delamination in axial (drawing) direction accompanied by secondary delamination (also in axial direction) in their vicinity.

Figure 6 shows a metallographic section of the cracking path just before the fracture instant and the fractographic aspect of the vertical cracking path showing cleavage appearance (Fig. 7), i.e., unstable (brittle) fracture. This is again consistent with a sudden drop (*pop-in*) in the load–displacement plot. It is not conventional cleavage, but a sort of *oriented and enlarged cleavage*, its enlargement and orientation being in the cold drawing direction (wire axis). Considering the cleavage facet as associated with a common crystallographic orientation of ferrite

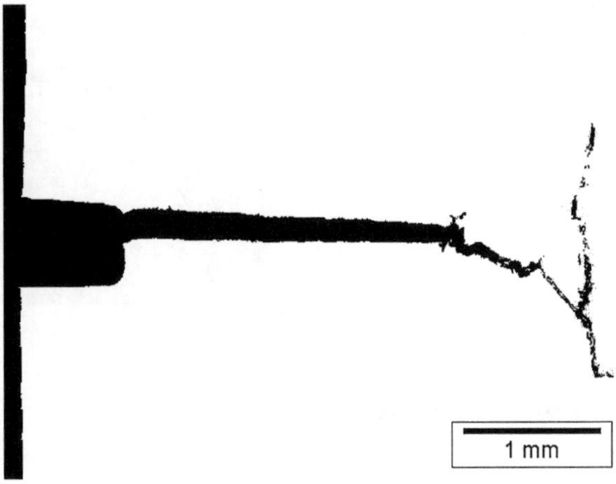

Fig. 6 Metallographic section of the cracking path just before the failure instant

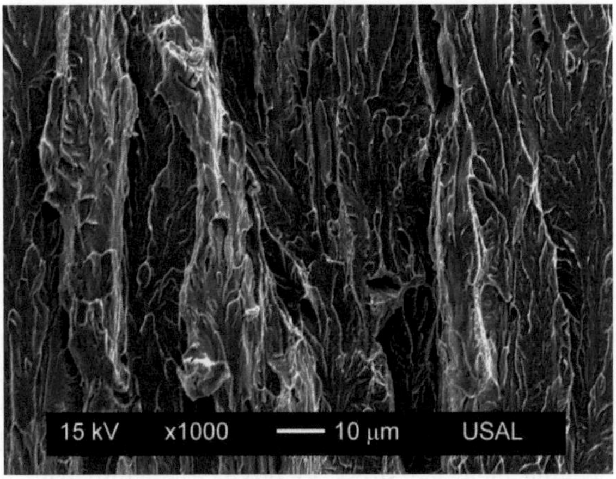

Fig. 7 Fractographic aspect of the vertical cracking path showing enlarged-oriented cleavage

[5], the enlarged cleavage facet unit could be linked with an oriented and elongated colony of pearlite that undoubtedly maintains a shared crystallographic alignment during the drawing procedure.

With regard to the appearance of axial cracking not at the crack tip, but a certain distance ahead of it, the mechanical stress–strain field in the *K*-dominance region could be responsible for the described fact, in addition to the clear microstructural weakness in axial direction (pseudocolonies, micro-cracks, damage, pre-defects, etc.) as a consequence of drawing. Next section of the chapter provides a numerical analysis to clarify this important item.

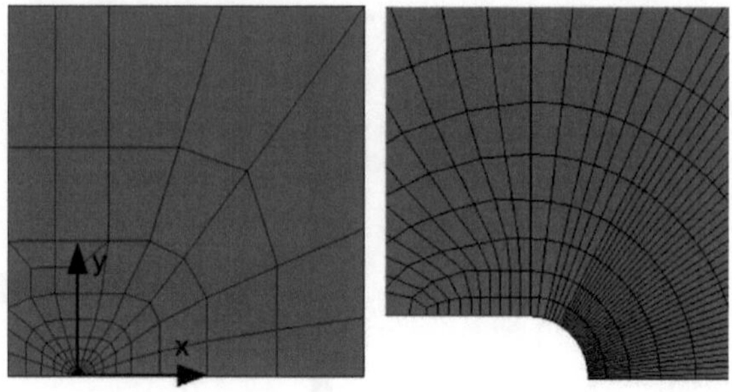

Fig. 8 Full mesh (*left*) and detail of the same in the vicinity of the crack tip (*right*)

4 Finite Element Analysis

A large-displacement elastoplastic finite element analysis was performed within updated lagrangian formulation with additive decomposition of strain rates. The material properties were those associated with the prestressing steel analyzed in this chapter (cf. Fig. 2) with elastic–plastic isotropic behaviour and von Mises yield criterion, and the analysis was performed under the control of the stress intensity factor K in the K-dominance region, so that a standard geometry with an edge crack could be used to reproduce the stress–strain state in the vicinity of the crack tip for any loading level up to final fracture in the tests.

Figure 8 shows the full finite element mesh (left) and a detail of the same (right) in the close vicinity of the crack tip where mesh refinement is more defined to properly reproduce the stress gradient in the area. The element type is a four-node, isoparametric, arbitrary quadrilateral specially design for axisymmetric applications.

Considering the axial cracking (fracture path deflection) described in the previous section and associated with cleavage appearance, it is interesting to compute the values of a cleavage stress which could act as a driving force for cracking (in addition to the microstructural weakness in that axial direction). To this end, the opening cleavage stress to produce fracture path deflection in axial (vertical) direction (perpendicular to the crack plane) should be the horizontal stress in the crack direction, i.e., the σ_{xx} component of the stress tensor, whose distributions for different loading levels are given in Fig. 9. The loading levels represent $\sigma_c/100$, $\sigma_c/10$, $\sigma_c/2$ and σ_c, where σ_c is the critical remote stress at the fracture instant, i.e., that applied far from the crack in axial (vertical) direction.

5 Discussion

Numerical results of Fig. 9 show, firstly, that a local maximum of such a cleavage stress appears at a certain distance form the crack tip (but not exactly at the tip) and, secondly, that the location of such a maximum moves far from the crack tip as

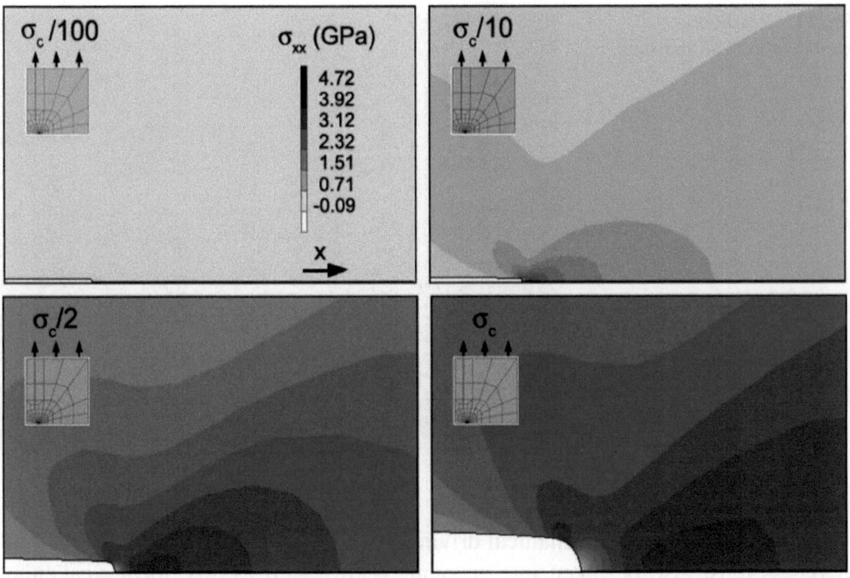

Fig. 9 Distributions of cleavage stress σ_{xx} (x is the crack direction) at increasing loading levels $\sigma_c/100$, $\sigma_c/10$, $\sigma_c/2$ and σ_c, where σ_c is the critical remote stress at failure. Large deformations at the crack tip shape are also shown

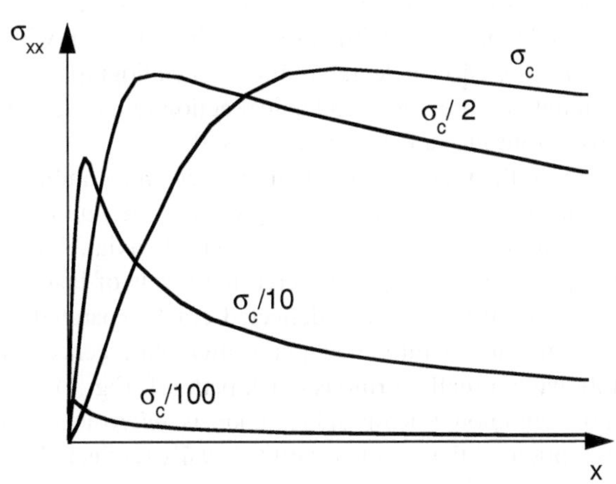

Fig. 10 Evolution of the cleavage stress profile σ_{xx} at increasing loading levels $\sigma_c/100$, $\sigma_c/10$, $\sigma_c/2$ and σ_c

the loading process goes on, as shown in Fig. 10 by means of the cleavage stress profiles at increasing loading levels, so that the probability of appearance of cleavage crack deflection increases with the level of loading, the horizontal cleavage stress being the mechanical responsible for cracking.

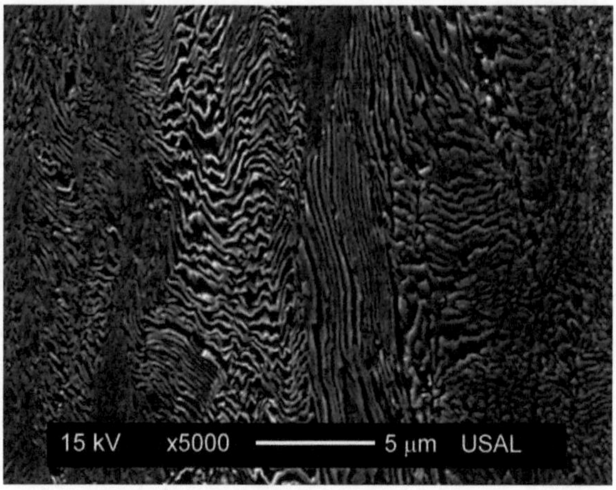

Fig. 11 Pearlitic pseudocolonies in the steel

In addition to such a mechanical driving force for cracking, the microstructural weakness must also be taken into account. With regard to this, local weakening can be observed in heavily drawn steel in the form of *pearlitic pseudocolonies* (Fig. 11), special microfracture units oriented in the direction of the drawing axis where the lamellae are not properly oriented in axial direction and, as a consequence of such a disorientation and of the drawing force, they have anomalous (very high) local interlamellar spacing (quite higher than the average pearlite interlamellar spacing). In addition, severe pre-damage appears in the form of cracking (always in axial direction) as a consequence of the manufacturing procedure by heavy drawing, as shown in Fig. 12 in which an array of axial (vertical) cracks is clearly detectable after a metallographic analysis of the supplied wire, without any previous mechanical action (either fatigue or fracture) on the sample, apart from the drawing process itself.

From the macroscopic point of view, the combination of axial pre-damage and the increasing cleavage stress produces the *pop-in* in the load–displacement plot (cf. Fig. 4). After this sudden, abrupt, axial, local extension of cracking, the sample is able to support increasing levels of loading with the main macroscopic crack and the secondary deflected crack separated by a certain distance. Finally catastrophic fracture takes place after coalescence between the two cracked areas following a really tortuous crack path (cf. Fig. 6). In this new fracture surface the most common fractographic mode is *micro-void coalescence* (MVC), with the exception of the frequent vertical walls (deflected crack paths) consistent of oriented and enlarged cleavage.

As a summary, the marked microstructural orientation in pearlitic steel after cold drawing (at the two levels of pearlitic colonies and lamellae) produces strength anisotropy in the material. At the macroscopic level, it manifests itself through the presence of deflections in the cracking path and pronounced initial fracture angle in relation to the transverse direction, due to the elevated cleavage

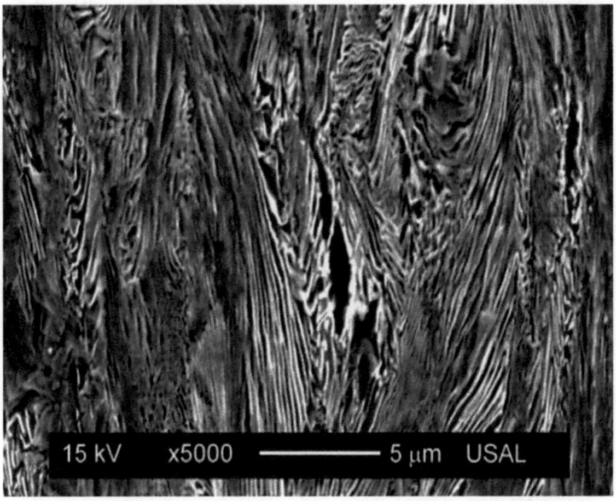

Fig. 12 Axial cracks (aligned in the drawing direction) produced by cold drawing

stress next to the crack tip and the microstructure weakness in axial direction producing vertical microcracking. At the microscopic level, it manifests itself through the existence of zones with enlarged cleavage oriented in the drawing axis direction (over the vertical walls) and fracture regions where the predominant fractographic mode is MVC (over the transversal and inclined surfaces).

6 Conclusions

The following conclusions may be drawn on the basis of the experimental, fractographic and numerical analysis performed in this chapter on commercial prestressing steel:

(I) Cold-drawn prestressing steel wires exhibit strength anisotropy in the form of fracture path deflection towards a direction approaching the wire axis, or cold drawing direction.
(II) Such an anisotropic behaviour is a clear macroscopic consequence of the pearlitic microstructure orientation induced by the manufacturing procedure at the two levels of the pearlitic colonies and lamellae.
(III) Crack path deflection is initiated at certain nuclei (fracture origins) at which axial cracking appear in the cold drawing direction (or wire axis) in the form of micro-cleavage units.
(IV) This cracking produces a sudden and small decrease of load in the load–displacement curve (*pop-in* phenomenon) as a consequence of the slight loss of bearing capacity of the sample.
(V) The typical fractographic mode associated with the frequent vertical walls appearing in the fracture surface is enlarged cleavage, oriented in the drawing direction. In other fracture areas the failure mode is MVC.

(VI) The cleavage stress is responsible for this locally deflected cracking appearing at a certain distance from the crack tip, in addition to the pre-damage (at the micro-level) after cold drawing.

Acknowledgments The authors wish to acknowledge the financial support provided by the following Spanish Institutions: Ministry for Science and Technology (MCYT; Grant MAT2002-01831), Ministry for Education and Science (MEC; Grant BIA2005-08965), Ministry for Science and Innovation (MICINN; Grant BIA2008-06810), *Junta de Castilla y León* (JCyL; Grants SA067A05, SA111A07 and SA039A08), and the steel supplied by EMESA TREFILERÍA.

References

1. Alexander, D., Bernstein, I.: The cleavage plane of pearlite. Metall. Mater. Trans. **13A**, 1865–1868 (1982)
2. Astiz, M.A., Valiente, A., Elices, M., Bui, H.D.: Anisotropic fracture behaviour of prestressing steels. In: Life assessment of dynamically loaded materials and structures—ECF5 (ed.) Faria LO, pp. 385–396. EMAS, Lisbon, Portugal (1985)
3. Fernández-Vicente, A., Carsí, M., Peñalba, F., Taleff, E., Ruano, O.A.: Toughness dependence on the microstructural parameters for an ultrahigh carbon steel (1.3 wt% C). Mater. Sci. Eng. **A335**, 175–185 (2002)
4. Miller, L.E., Smith, G.C.: Tensile fracture in carbon steels. J. Iron Steel Inst. **208**(11), 998–1005 (1970)
5. Park, Y., Bernstein, I.: The process of crack initiation and effective grain size for cleavage fracture in pearlitic eutectoid steel. Metall. Mater. Trans. **10A**, 1653–1664 (1979)
6. Rosenfield, A.R., Majumdar, B.S.: Micromechanisms and toughness for cleavage fracture of steel. Nucl. Eng. Des. **105**, 51–57 (1987)
7. Singh, U.P., Banerjee, S.: On the origin of pop-in crack extension. Acta. Metall. Mater. **39**, 1073–1084 (1991)
8. Toribio, J., Ovejero, E.: Microstructure evolution in a pearlitic steel subjected to progressive plastic deformation. Mater. Sci. Eng. **A234–236**, 579–582 (1997)
9. Toribio, J., Ovejero, E.: Microstructure orientation in a pearlitic steel subjected to progressive plastic deformation. J. Mater. Sci. Lett. **17**, 1045–1048 (1998)
10. Toribio, J., Ovejero, E.: Effect of cumulative cold drawing on the pearlite interlamellar spacing in eutectoid steel. Scripta Mater. **39**, 323–328 (1998)
11. Toribio, J., Ovejero, E.: Effect of cold drawing on microstructure and corrosion performance of high-strength steel. Mech. Time-Depend. Mater. **1**, 307–319 (1998)
12. Toribio, J., Ovejero, E., Toledano, M.: Microstructural bases of anisotropic fracture behaviour of heavily drawn steel. Int. J. Fract. **87**, L83–L88 (1997)
13. Toribio, J., Valiente, A.: Approximate evaluation of directional toughness in heavily drawn pearlitic steels. Mater. Lett. **58**, 3514–3517 (2004)
14. Toribio, J., Valiente, A.: Failure analysis of cold drawn eutectoid steel wires for prestressed concrete. Eng. Fail. Anal. **13**, 301–311 (2006)

Tensile Characteristics of High-Toughness Steel at High Temperatures

S. H. Park, Y. K. Yoon and J. H. Kim

Abstract In this study, material properties of a high toughness steel under high temperatures were investigated through the tensile test. A-basis and B-basis strength are taken from sampling based calculation by using noncentral t-distribution. The sampling-based basis strength gives a more conservative value than that taken by using a normal distribution. From material properties as a function of temperature, shear-lip size and morphology of fractured surface, it is confirmed that there is a ductility minimum temperature at 500°C.

Keywords Tensile test · Sampling-based basis strength · Temperature · Noncentral t-distribution · Ductility minimum temperature

1 Introduction

High-strength and high-toughness steels have traditionally been used in the case of earth-penetrating munitions or gravity-dropped weapons, which requires an understanding of the dynamic mechanical properties at relevant strain-rates.

S. H. Park
Defense Systems Test Center, Agency for Defense Development, Taeahn P.O. Box 1Taeahn, 357-900, Korea
e-mail: redfox89@dreamwiz.com

Y. K. Yoon · J. H. Kim (✉)
Department of Mechanical Design Engineering, Chungnam National University, 220 Gung-dong Yuseong-gu, Daejeon, 305-764, Korea
e-mail: kimjhoon@cnu.ac.kr

Y. K. Yoon
e-mail: genius0247@naver.com

A. Öchsner et al. (eds.), *Materials with Complex Behaviour II*,
Advanced Structured Materials 16, DOI: 10.1007/978-3-642-22700-4_16,

A structural steel shows different mechanical and thermal properties at room and high temperatures. So, material characteristics at high temperatures must be considered in structural design such as aircraft parts.

Aircraft structural design still relies on the deterministic design code such as federal aviation administration (FAA) regulations. In deterministic design, conservative material properties such as A-basis or B-basis strength are used and safety factors are introduced to protect against uncertainties. Generally, the material properties are acquired from tensile tests and the samples are taken from material's population. Moreover, we can only acquire limited number of test data because of the cost to get test material, the operation time of equipment, etc. So, the true material's properties cannot help being estimated from test samples. Because the nature of mechanical behavior of materials and failure are probabilistic, the strength of materials has variations from size effect, surface finish, notch effect, etc. and the stress varies due to stress concentration, temperature factor, stress combinations, etc. Therefore, it is compulsory to get A-basis or B-basis strength from sampling-based method.

Fractography is critical to failure analysis of metals and plastics. This shows us the mode of failure. It is well known that the ductility minimum occurs in an intermediate temperature range for all ductile metals and alloys. Fractography is a useful tool to investigate this ductility minimum temperature.

The purpose of this study is to investigate the material's tensile properties at high temperatures. Some methods to calculate the A-basis and B-basis strengths are reviewed. By detailed investigation on the fractured surface and material property at elevated temperatures, it will be confirmed that there is a ductility minimum temperature (DMT).

2 Theory

2.1 Basis of Strength of Material

The allowable stress should be less than the material's strength to prevent failure. The material's strength is determined by standard test methods and specimens. But it has a statistical distribution from the deviation of manufacture, microstructure, test environment, etc. Moreover, there is variation of loads acting on the structural materials from load and environmental conditions. So, the test result of the material's strength is non-deterministic but statistical. It is important to investigate the distribution of the material's strength to analyze its characteristics statistically. It is well known that the strength data has a normal distribution [1].

In the statistically-based design approach, the material property of interest must be regarded as a random variable which is a quantity that varies from specimen to specimen according to some probability distribution. A designated value for a material is the minimum value of a material property expected to be used in the fabrication of the structure. Basis values are 95% lower confidence limits on prescribed percentiles, which are also sometimes referred to as tolerance limits.

Aircraft structural design still relies on the FAA deterministic design code. The FAA regulations (FAR-25.613) [2] state that in deterministic design, conservative material properties are characterized as A-basis or B-basis values. In U.S. Military Handbooks, MIL-HDBK-5 J [3] for metallic materials and elements for aerospace vehicle structures and MIL-HDBK-17-1F [4] for polymer matrix composites, the detailed methods to determine the allowable stress from A-basis or B-basis for military planes, light alloy materials and composite materials are described.

In the definition of conservative material properties by the FAA regulation (FAR 25.613), A-basis value is defined as a 95% lower confidence bound on the first percentile of a specified population of measurements. On the other hand, B-basis value is defined as 95% lower confidence bound on the tenth percentile of a specified population of measurements. In other words, A-basis value is a 95% lower tolerance bound for the upper 99% of a specified population and B-basis value is a 95% lower tolerance bound for the upper 90% of a specified population. A-basis or B-basis material property values depend on the failure path in the structure. A-basis values are used when there is a single failure path in the structure, while B-basis values are used when there are multiple failure paths in the structure.

2.2 Methods to Acquire Basis-Values

According to the military handbook of MIL-HDBK-17-1F, the basis values are calculated by

$$Basis = \bar{X} - ks \tag{1}$$

where $\bar{X}$, sample mean; s, sample standard deviation; k, tolerance coefficient.

One-sided A-basis tolerance factor, k_A, for the normal distribution can be acquired as $1/\sqrt{n}$ times the 0.95th quantile of the noncentral t-distribution with noncentrality parameter $2.326\sqrt{n}$ and $n-1$ degrees of freedom. An approximation to the k_A value is

$$k_A \approx 2.326 + \exp\{1.34 - 0.522\ \ln(n) + 3.87/n\} \tag{2}$$

Similarly, one-sided B-basis tolerance factor, k_B, for the normal distribution can be calculated as $1/\sqrt{n}$ times the 0.95th quantile of the noncentral t-distribution with noncentrality parameter $2.326\sqrt{n}$ and $n-1$ degrees of freedom. An approximation to the k_B values is

$$k_B \approx 1.282 + \exp\{0.958 - 0.520\ \ln(n) + 3.19/n\} \tag{3}$$

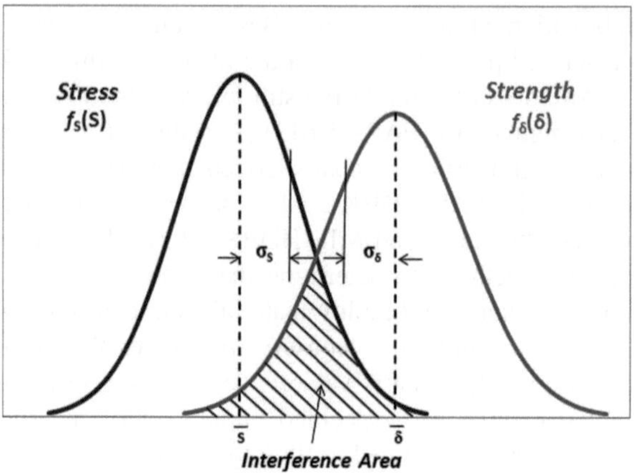

Fig. 1 Stress-strength interference model

Acar [5] suggested the equation to calculate the tolerance coefficient, k, for normal distribution given by

$$k = \frac{z_{1-p} + \sqrt{z_{1-p}^2 - ab}}{a}, \; a = 1 - \frac{z_{1-\gamma}^2}{2(N-1)}, \; b = z_{1-p}^2 - \frac{z_{1-\gamma}^2}{N} \tag{4}$$

where N, sample size; z_{1-p}, critical value of normal distribution that is exceeded with a probability of 1-p (for A-basis value p = 0.99 while for B-basis value p = 0.90); $z_{1-\gamma}$, critical value of normal distribution that is exceeded with a probability of 1-γ ($\gamma = 0395$ for both A-basis and B-basis values).

For the B-basis value, $k_B = z_{1-p} = z_{0.1} = \Phi(0.1) = 1.282$, where Φ is the cumulative distribution function (c.d.f.) of the standard normal distribution. $z_{0.1} = \Phi(0.1) = 1.282$ is the critical value of normal distribution that is exceeded with a probability of 10%. Of course, $z_{1-\gamma}$ corresponds to $z_{0.05}$.

One-sided A-basis and B-basis tolerance limit factors, k_A and k_B respectively, for the normal distribution for sample size, n are presented in MIL-HDBK-17-1F as Tables. For infinite number of samples, $n = \infty$, $k_A = 2.326$ and $k_B = 1.282$.

2.3 Determination of Basis Values from Interference Model

The allowable stress and the strength of a material are statistical quantities. A component, a subsystem, or a system fails when the stress, in general, exceeds the strength. The reliability of the system or the probability of failure may be estimated from stress-strength interference model [6], as shown in Fig. 1. In this model, the shaded portion is an interference area, which is indicative of the probability of failure.

Let the probability density function (p.d.f.) for the stress (s) be denoted by $f_s(s)$, and that for the strength (δ) by $f_\delta(\delta)$, as shown in Fig. 1. Then, the reliability R is

$$R = \int_{-\infty}^{\infty} f_s(s) \left[\int_{s}^{\infty} f_\delta(\delta) d\delta \right] ds \tag{5}$$

In general, the strength of a material has a normal distribution. If the applied stress distributes normally, the p.d.f. for the stress f_s (s) and that for the strength f_δ (δ) are expressed as follows

$$f_s(s) = \frac{1}{\sqrt{2\pi}\sigma_s} e^{-\frac{1}{2}\left(\frac{s-\mu_s}{\sigma_s}\right)^2}, \ -\infty < s < \infty \tag{6}$$

$$f_\delta(\delta) = \frac{1}{\sqrt{2\pi}\sigma_\delta} e^{-\frac{1}{2}\left(\frac{s-\mu_\delta}{\sigma_\delta}\right)^2}, \ -\infty < \delta < \infty \tag{7}$$

Let us define $y = \delta$ - s. It is well known that the random variable y is normally distributed with a mean of $\mu_y = \mu_\delta - \mu_s$ and a standard deviation of $\sigma_y = \sqrt{\sigma_\delta^2 + \sigma_s^2}$. Then, the reliability R which is the probability to overcome the applied stress can be expressed in terms of y as

$$\begin{aligned} R &= P(\delta > s) = P(y > 0) \\ &= \int_0^{\infty} \frac{1}{\sigma_y \sqrt{2\pi}} e^{-\frac{1}{2}\left(\frac{y-\mu_y}{\sigma_y}\right)^2} dy \end{aligned} \tag{8}$$

If we let $z = (y - \mu_y)/\sigma_y$, then

$$R = \frac{1}{\sqrt{2\pi}} \int_{-\frac{\mu_\delta - \mu_s}{\sqrt{\sigma_\delta^2 + \sigma_\delta^2}}}^{\infty} e^{-\frac{1}{2}z^2} dz \tag{9}$$

Because the random variable $z = (y - \mu_y)/\sigma_y$ is clearly the standard normal variable, the reliability can be found by merely referring to the normal tables. Equation 9 may be expressed using standard normal distribution function, F( ), as

$$R = 1 - F\left(-\frac{\mu_\delta - \mu_s}{\sqrt{\sigma_\delta^2 + \sigma_\delta^2}} \right) \tag{10}$$

Because the allowable stress is a maximum stress that may be safely applied to a component, S_a, the probability of failure, P_f, is expressed as

$$P_f = F\left(-\frac{\mu_\delta - \mu_s}{\sqrt{\sigma_s^2 + \sigma_\delta^2}}\right) = 1 - F\left(\frac{\mu_s - \mu_\delta}{\sqrt{\sigma_s^2 + \sigma_\delta^2}}\right) \tag{11}$$

If we assume that the strength, S, is normally distributed with known mean and standard deviation of population and if we let $\mu_s = \mu, \sigma_s = \sigma, \mu_\delta = S_a, \sigma_\delta = 0$, Eq. 11 is

$$P_f = F\left(\frac{S_a - \mu}{\sigma}\right) \tag{12}$$

The allowable stress, S_a, must be determined as the probability of failure, P_f, is below a certain value of P. If we let $P_f = P$ and rearrange Eq. 9, then

$$F^{-1}(P) = -F^{-1}(1-P) \tag{13}$$

$$S_a = \mu - F^{-1}(1-P)\sigma = \mu - u_p\sigma \tag{14}$$

Here, $u_p = F^{-1}(1-P)$ is the upper probability of standard normal distribution. For example, when $P = 0.01$, then $u_p = 2.326$ and when $P = 0.05$, then $u_p = 1.282$.

2.4 Determination of Sampling-Based Basis Strength

Unfortunately, in general we cannot know the mean and the standard deviation of a population. So, we can merely estimate them from the mean and the standard deviation of samples. If we assume that the strength, S, is normally distributed with unknown mean and standard deviation of population and the mean and the standard deviation of n samples are $\hat{\mu}$ and $\hat{\sigma}$ respectively then the allowable stress $\hat{\mu} - k\hat{\sigma}$ should satisfy Eq. 15, as shown Fig. 2.

$$P[\hat{\mu} - k\hat{\sigma} > \mu - u_p\sigma] = \gamma \tag{15}$$

If we manipulate and rearrange Eq. 15, Eq. 16 is obtained.

$$P[(\hat{\mu} - \mu) + u_p\sigma > k\hat{\sigma}] = \gamma$$

$$P\left[\frac{\sqrt{n}(\hat{\mu} - \mu) + \sigma\sqrt{n}u_p}{\hat{\sigma}} > \sqrt{n}k\right] = \gamma$$

$$P\left[\frac{(\hat{\mu} - \mu)\sqrt{n}/\sigma + \sqrt{n}u_p}{\hat{\sigma}/\sigma} > \sqrt{n}k\right] = \gamma$$

$$P\left[\frac{\frac{\hat{\mu}-\mu}{\sigma/\sqrt{n}} + \sqrt{n}u_p}{\sqrt{\frac{(n-1)\,\hat{\sigma}^2/\sigma^2}{n-1}}} > \sqrt{n}k\right] = \gamma \tag{16}$$

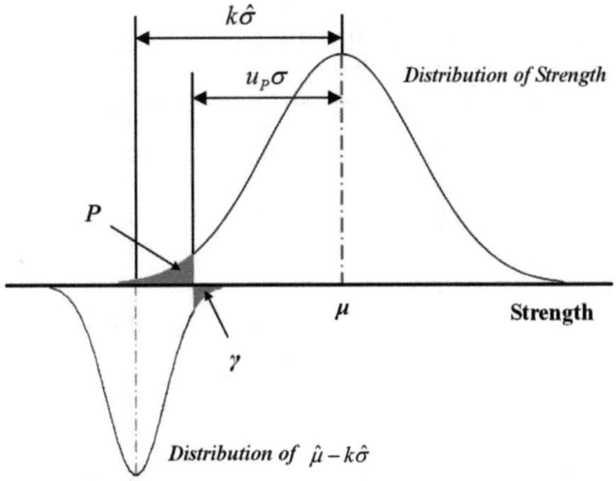

Fig. 2 Distribution of $\hat{\mu} - k\hat{\sigma}$

If Z is a normally distributed random variable with unit variance and zero mean, and V is a χ^2distributed random variable with v degrees of freedom that is statistically independent of Z, then $T = \frac{Z+\mu}{\sqrt{V/v}}$ is a noncentral t-distributed random variable with v degrees of freedom and noncentrality parameter μ. Because the χ^2-distribution with k degrees of freedom is defined as the distribution of a sum of the squares of k independent standard normal variables, $(n-1)\,\hat{\sigma}^2/\sigma^2$ will be distributed as χ^2-distribution with $v = n - 1$ degrees of freedom in Eq. 16 and apparently $(\hat{\mu} - \mu)/(\sigma/\sqrt{n})$ will be distributed as standard normal distribution, $N(0,1)$. So, the left side of inequality in Eq. 16 distributes as noncentral t-distribution with $v = n - 1$ degrees of freedom and noncentrality parameter $\mu = \sqrt{n}u_p$. So, if the upper probability of noncentral t-distribution, γ, is $t'_\gamma(n-1, \sqrt{n}u_p)$,, the value of k should satisfy Eq. 17

$$\sqrt{n}k = t_\gamma{}'(n-1, \sqrt{n}u_p) \tag{17}$$

The risk of $S_a > \mu - u_p\sigma$ can be restricted within γ if we determine the allowable stress as $S_a = \hat{\mu} - k\hat{\sigma}$ by selecting the appropriate value of k. In this way, the tolerance coefficients, k_n, for the number of samples are calculated and shown as Tables in MIL-HDBK-17-1F. In Table 1, three methods to get the tolerance coefficient are compared. The values listed in the MIL-HDBK-17-1F are probably accurate the most but the values calculated by approximate equations may be accurate to within 0.2% of the tabulated values for n greater than or equal to 16. The values from Acar's equation are also acceptable but less accurate than approximate equations of Eqs. 2 and 3. The tolerance coefficients for the infinite number of samples are the same as the case of which we know the mean and the standard deviation of population as shown in Eq. 14.

A test of specimens obtained in a material can be regarded as sampling from a population. If the number of test samples is infinite then the mean and the standard deviation of samples and population will be the same. Unfortunately, we cannot

Table 1 Tolerance coefficients, k_n, for the number of samples between methods

Number of samples (n)	Tables in MIL-HDBK-17-1F		Approximate Eqs. 2 and 3		Acar's Eq. 4	
	k_A	k_B	k_A	k_B	k_A	k_B
2	37.094	20.581	20.741	10.240	−13.999	−7.375
3	10.553	6.157	10.145	5.545	13.335	7.625
5	5.741	3.408	5.901	3.418	5.750	3.382
10	3.981	2.355	4.017	2.365	3.940	2.322
16	3.464	2.034	3.470	2.035	3.437	2.013
20	3.295	1.927	3.296	1.926	3.274	1.911
30	3.064	1.778	3.062	1.777	3.050	1.768
∞	**2.326**	**1.282**				

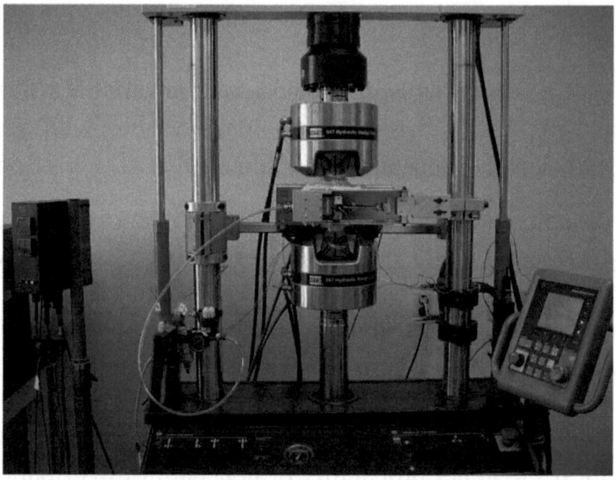

Fig. 3 Experimental setup for tensile tests under various temperatures

know the mean and the standard deviation of a population and cannot increase the number of samples infinitely. As the sample size increases the allowable stress also increase because the certainty of estimate of $\hat{\mu}$ increase. When we can test infinite number of samples, the allowable stress is the same as the case of which we know the mean and the standard deviation of population. As shown in Table 1, more than 16 samples are necessary to estimate the population's mean and deviation more accurately. Considering the cost of test and the accuracy of estimation, at least 10 samples are essential.

3 Experimental Procedure

Tensile tests were conducted in accordance with ASTM E8 [7] and E21 [8] using an MTS 810 servo-hydraulic test machine, as shown in Fig. 3. Figure 4 is the geometry of the test specimen. Thermocouples (K type) were attached at the center

Fig. 4 Test specimen geometry (dimensions in mm)

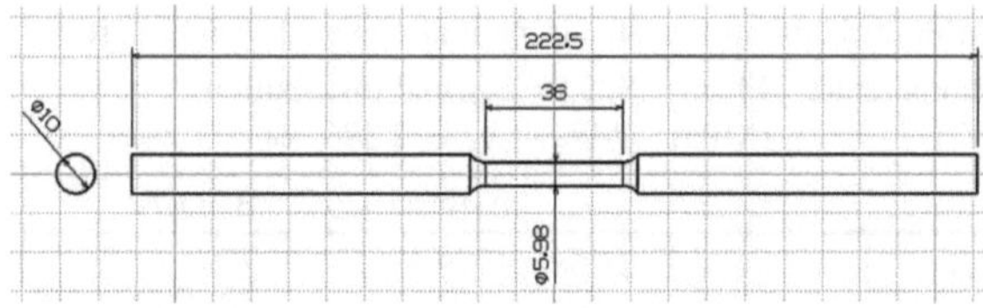

Table 2 Tensile characteristics at elevated temperatures

Temperature (°C)	Young's modulus (GPa)	Elongation (%)	Reduction in area (%)
R.T.	184.0	57.8	18.6
100	170.2	60.5	15.3
200	171.1	61.3	13.9
300	160.7	58.4	12.6
400	155.9	58.7	12.3
500	132.5	45.2	10.9
550	108.7	59.7	15.9
600	92.7	76.7	30.7
700	67.1	91.5	40.6
800	56.2	97.2	42.2
900	36.4	97.8	55.4
1000	54.9	73.2	49.4

of it. The test was run in strain control mode at a constant strain rate of 2 mm/min. The tensile tests were conducted at room temperature (R.T.), 100, 200, 300, 400, 500, 550, 600, 700, 800, 900 and 1000°C. At each temperature, for ten specimens a steady state tensile test was conducted. The tensile load was applied to specimen after the specimen was heated up to a specified temperature and maintained for 30 min. A tensile loading rate of 2 mm/min was used, leading to an initial strain rate of 10^{-3} s^{-1}, and the heating rate of the furnace is 1.9–2.0°C/sec. An extensometer with a total gauge length of 25 mm was used.

4 Results and Discussion

4.1 Tensile Characteristics at Elevated Temperatures

The mean value of Young's modulus, elongation and reduction in area at elevated temperatures are shown in Table 2. The mean, standard deviation, A-basis and B-basis of ultimate tensile strength and yield strength at elevated temperatures are shown in Tables 3 and 4 respectively.

Young's modulus as a function of temperature is shown in Fig. 5. Young's modulus decreases gradually as the temperature increase. There is an abrupt

Table 3 Ultimate tensile strength at elevated temperatures

Temperature (°C)	Mean (MPa)	S.D. (MPa)	Normal distribution ($k_A = 2.326$, $k_B = 1.282$)		Noncentral t-distribution ($k_A = 3.981$, $k_B = 2.355$)	
			A-Basis	B-Basis	A-Basis	B-Basis
R.T.	2034.7	23.68	1979.6	2004.3	1940.4	1978.9
100	1935.4	31.95	1861.1	1894.4	1808.2	1860.2
200	1824.9	6.58	1809.6	1816.5	1798.7	1809.4
300	1764.5	28.36	1698.5	1728.1	1651.6	1697.7
400	1656.2	14.63	1622.2	1637.4	1598.0	1621.7
500	1407.1	37.31	1320.3	1359.3	1258.6	1319.2
550	981.3	16.38	943.2	960.3	916.1	942.7
600	619.0	3.22	611.5	614.9	606.2	611.4
700	425.4	5.21	413.3	418.7	404.7	413.1
800	178.5	3.70	169.9	173.8	163.8	169.8
900	100.1	12.27	71.6	84.4	51.3	71.2
1000	57.7	2.20	52.5	54.8	48.9	52.5

Table 4 Yield strength at elevated temperatures

Temperature (°C)	Mean (MPa)	S.D. (MPa)	Normal distribution ($k_A = 2.326$, $k_B = 1.282$)		Noncentral t-distribution ($k_A = 3.981$, $k_B = 2.355$)	
			A-Basis	B-Basis	A-Basis	B-Basis
R.T.	1764.3	24.10	1708.2	1733.4	1668.4	1707.5
100	1685.4	32.40	1610.0	1643.9	1556.4	1609.1
200	1604.2	13.60	1572.6	1586.8	1550.1	1572.2
300	1535.5	13.80	1503.4	1517.8	1480.6	1503.0
400	1425.4	13.50	1394.0	1408.1	1371.7	1393.6
500	1242.7	22.60	1190.1	1213.7	1152.7	1189.5
550	926.9	8.10	908.1	916.5	894.7	907.8
600	559.1	2.60	553.1	555.8	548.7	553.0
700	372.1	8.40	352.6	361.3	338.7	352.3
800	152.5	5.30	140.2	145.7	131.4	140.0
900	74.7	12.40	45.9	58.8	25.3	45.5
1000	41.3	4.30	31.3	35.8	24.2	31.2

change of decreasing rate in Young's modulus at 500°C. The reduction in area and elongation gradually fall or are constant till 500°C but rapidly rise to 900°C and then fall, as shown in Fig. 6. The stress–strain curve at room and elevated temperatures is shown in Fig. 7 which shows a smooth elastic–plastic transition. As shown in Fig. 8, the ultimate tensile strength and the yield strength decrease gradually to the temperature of 500°C but rapidly fall over at 500°C and gradually decrease over the temperature of 800°C.

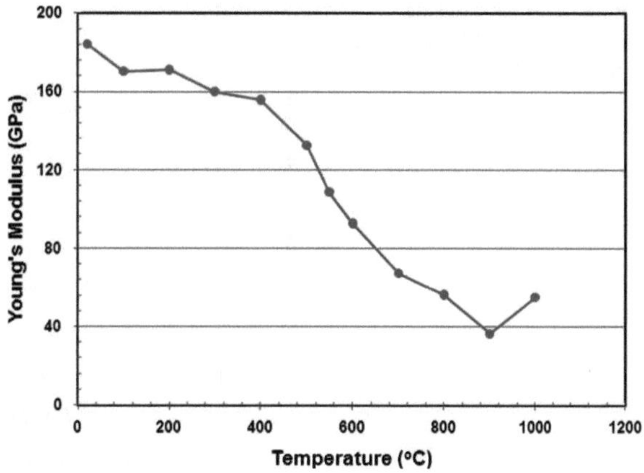

Fig. 5 Young's modulus as a function of temperature

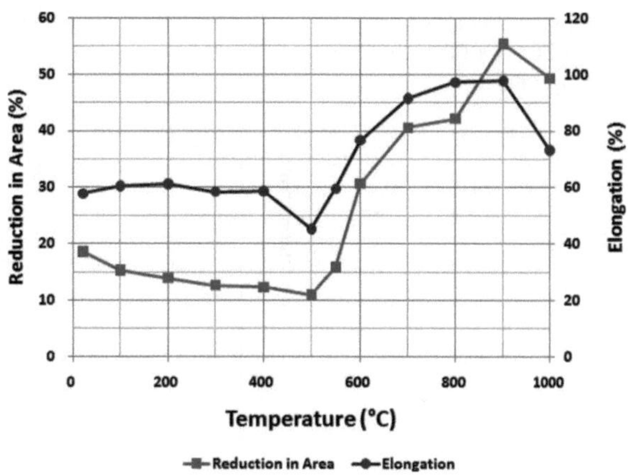

Fig. 6 Reduction in area and elongation as a function of temperature

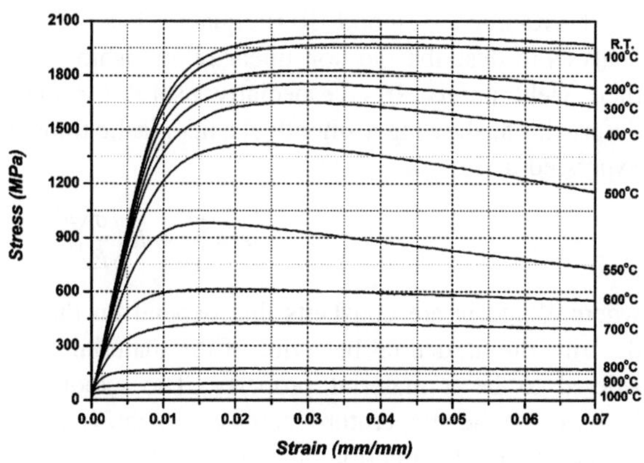

Fig. 7 Stress-strain curves at elevated temperatures

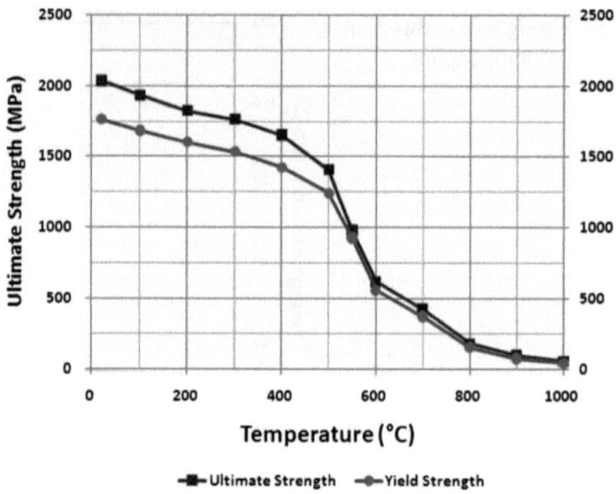

Fig. 8 Ultimate tensile strength and yield strength as a function of temperature

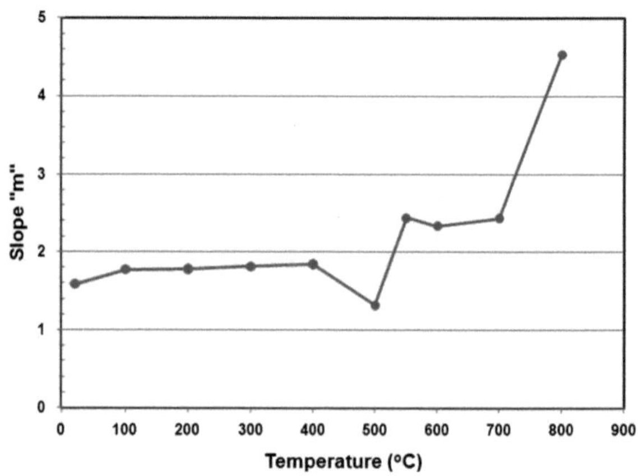

Fig. 9 Exponent parameter of RO Eq. as a function of temperatures

The Ramberg–Osgood (RO) equation has been used in most modern strength analysis to describe the non linear relationship between stress and strain in near their yield points. It is especially useful for metals that harden with plastic deformation, showing a smooth elastic–plastic transition. The equation for strain is expressed as Eq. 18.

$$\varepsilon = \frac{\sigma}{E} + K\left(\frac{\sigma}{E}\right)^m \tag{18}$$

where K and m are constants that depend on the material being considered and describe the hardening behavior of the material. So, these values have been considered to be a material property. The exponent parameter of RO equation increases as the temperature increases, as shown in Fig. 9. In Fig. 10, the exponent parameter of RO equation decreases as Young's modulus increase.

Fig. 10 Exponent parameter of RO Eq. as a function of Young's modulus

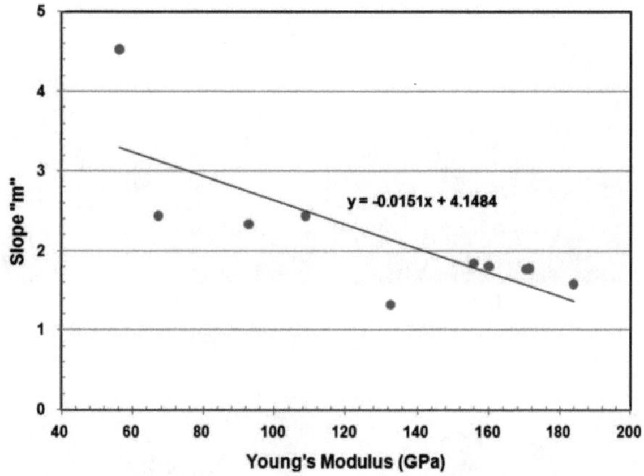

4.2 Ductility Minimum Temperature and Fractured Surface

Rhines and Wray [9] have pointed out that the minimum ductility occurs at an intermediate temperature range for all ductile metals and alloys. At low temperatures, fracture occurs by the usual transgranular crack propagation mechanism and ductility is high. At temperatures near minimum, deformation occurs by grain boundary sliding, causing the formation of intergranular cavities at triple junctions. At high temperatures, recrystallization occurs simultaneously with intergranular cavity formation as a result of which intergranular crack propagation is retarded. The ductility, therefore, increases at high temperatures. Sikka et al. [10] have reported elevated temperature tensile ductility minima in AISI 304 and 316 stainless steels and their metallographic findings were found to be consistent with the model proposed by Rhines and Wray. That is, the ductility minimum was associated with the temperature, the strain rate and the metallurgical condition under which intergranular crack propagation was not inhibited [11]. Boyce et al. [12] suggested that at slower strain rates or higher temperatures, flow is controlled by long-range obstacles to dislocation motion and is largely strain-rate insensitive. At lower temperature or higher strain-rate, weaker short range obstacles become controlling due to the time-dependent diffusion-limited mechanisms such as climb which are necessary to overcome these short range obstacles, leading to stronger strain-rate dependence. The variation of ductility from different fracture mechanism is the cause of variation of the elongation with temperature.

In the last study of the present analysis for the AISI 304 austenitic stainless steel [13], there was also ductility minimum at the temperature of 500°C. From the result of EDX (Energy Dispersive X-ray) microanalysis, it has been confirmed that AISI 304 austenitic stainless steel has a characteristic of being harder and more brittle at temperatures between 500 and 800°C and this is due to a precipitation of α phase chrome carbide ($Cr_{23}C_6$) in this temperature range. The density of chrome

Fig. 11 Low-magnification images of cup-and-cone fracture morphology for the test specimen

in the material decreases as a result of the precipitation and the weight % of carbon and chrome decreased above 500°C from EDX analysis. In this study, Figs. 6 and 8 show that there is a ductility minimum at 500°C and this is the same as for AISI 304 stainless steel.

The fractured surfaces were inspected to observe temperature dependent changes in fractography. In all cases, the cylindrical test specimen resulted in a cup-and-cone fracture morphology. According to Metals Handbook [14], the cup-and-cone morphology consists of three zones: fibrous zone, radial zone and shear-lip zone. At the center of the tensile specimen, there is a fibrous zone associated with early coalescence and slow, stable crack growth. In the radial zone, fast and unstable cracks propagate away from the fibrous zone. Around the rim of fracture, there is a shear-lip zone from planes of maximum shear ahead of the growing crack tip.

As shown in Fig. 11, the relative size of the shear-lip diminishes as the temperature increases and has an abrupt change after the ductility minimum temperature of 500°C. The morphology of the fibrous zones at each temperature is shown in Fig. 12. High-magnification images from scanning electron microscope (SEM) show that there is a typical microvoid coalescence morphology expected in ductile metal fracture. The dimple size grows as the temperature rises and has abrupt change after the ductility minimum temperature of 500°C.

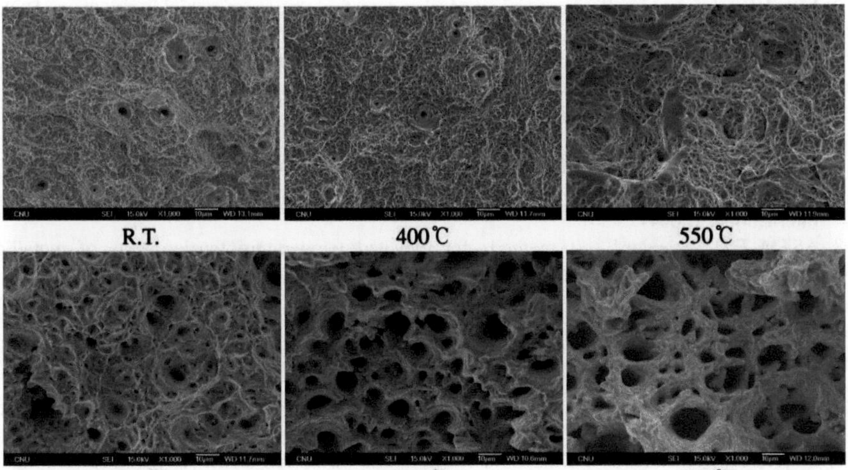

Fig. 12 High-magnification of the ductile dimples in the fibrous zone of the tensile fracture surfaces

5 Conclusions

This chapter aimed to get sampling-based basis strength of high toughness steel at room and elevated temperatures and investigate the ductility minimum temperature from inspecting the material property and fractured surface. The following conclusions can be drawn.

(1) The tensile strength, yield strength and Young's modulus decrease as the temperature increases. But the elongation and the reduction in area decrease as the temperature increases below 500°C which corresponds to a ductility minimum temperature. From low and high magnification of the fractured surface, typical evidence of ductile fracture is shown. The abrupt change in the size of shear-lip and dimple size showed that there is a ductility minimum temperature at 500°C.
(2) The sampling-based basis strength values were taken from mean, standard deviation of tensile test results and tolerance coefficient from noncentral t-distribution. The sampling-based basis strength gives us more conservative strength values than from normal distribution.

References

1. Song, J.H., Park, J.H.: Introduction to reliability engineering, Intervision, Seoul, 217-219 (2007)
2. FAR-25.613: Material strength properties and material design values. Federal Aviation Administration (USA) (2003)

3. MIL-HDBK-5 J: Metallic materials and elements for aerospace vehicle structures. DOD (USA) (2003)
4. MIL-HDBK-17-1F: Polymer matrix composites guidelines for characterization of structural materials. DOD (USA) (2002)
5. Acar, E.: Aircraft structural safety: Effects of explicit and implicit safety measures and uncertainty reduction mechanics. Ph.D Dissertation. University of Florida (2006)
6. Kapur, K.C., Lamberson, L.R.: Reliability in Engineering Design. Wiley, New York (1977)
7. ASTM E8: Standard test methods for tension testing of metallic materials. ASTM International (2004)
8. ASTM E21: Standard test methods for elevated temperature tension tests of metallic materials. ASTM International (2005)
9. Rhines, F.N., Wray, P.J.: Investigation of the intermediate temperature ductility minimum in metals. Trans. ASM **54**, 117 (1961)
10. Sikka, V.K., Swindeman, R.W., Brinkman, C.R.: Proceedings of 4th International Conference on Fracture, vol. 2, p. 561. (1977)
11. Mannan, S.L., Samuel, K.G., Rodriguez, P.: Influence of temperature and grain size on the tensile ductility of AISI 316 stainless steel. Mater. Sci. Eng. **68**, 143–149 (1984)
12. Boyce, B.L., Crenshaw, T.B., Dilmore, M.F.: The Strain-rate Sensitivity of High-Strength High-toughness Steels. Sandia National Laboratories, Albuquerque (2007)
13. Park, S.H., Kim, J.K., Kim, J.H.: A statistical study on stress-strain relation of AISI 304 stainless steel under elevated temperatures. Mater. Complex Behav. Adv. Struct. Mater. **3**, 3–18 (2010)
14. Metals Handbook: Fractography and Atlas of Fractographs, vol. 9, 8th edn. American Society for Metals, Metals Park (1974)

Part II
Non-classical Materials and Structures

Damage and Fracture Analysis of Rubber Component

Chang-Su Woo, Wan-Doo Kim, Hyun-Sung Park and Wae-Gi Shin

Abstract Rubber components have been widely used in automotive industry as anti-vibration components for many years. These subjected to fluctuating loads, often fail due to the nucleation and growth of defects or cracks. To prevent such failures, it is necessary to understand the fatigue failure mechanism for rubber materials and evaluate the fatigue life for rubber components. Fatigue lifetime prediction and evaluation are the key technologies to assure the safety and reliability of automotive rubber components. The objective of this study is to develop the durability analysis process for vulcanized rubber components, which is applicable to predict fatigue lifetime at initial product design step. Fatigue lifetime prediction methodology of vulcanized natural rubber was proposed by incorporating the finite element analysis and fatigue damage parameter of maximum Green-Lagrange strains appearing at the critical location determined from fatigue test. In order to develop an appropriate fatigue damage parameter of the rubber material, a series of displacement controlled fatigue tests was conducted using 3-dimensional dumbbell specimens with different levels of mean displacement. It was shown that the maximum Green-Lagrange strain was a proper damage parameter, taking the mean displacement effects into account. Nonlinear finite

C.-S. Woo (✉) · W.-D. Kim · H.-S. Park
Department of Nano Mechanics, Korea Institute of Machinery & Materials, 171 Jang-dong, Yusung-gu, Daejeon 305-343, Korea
e-mail: cswoo@kimm.re.kr

W.-D. Kim
e-mail: wdkim@kimm.re.kr

H.-S. Park
e-mail: hspark@kimm.re.kr

W.-G. Shin
PAVCO Ltd., 291-7 Bonri-ri, Nongong-eup, Dalseong-gun, Daegu 711-855, Korea
e-mail: wgshin@ph.co.kr

A. Öchsner et al. (eds.), *Materials with Complex Behaviour II*,
Advanced Structured Materials 16, DOI: 10.1007/978-3-642-22700-4_17,

element analyses of the engine mount insulator and 3-dimensional dumbbell specimens were performed based on a hyper-elastic material model determined from the simple tension, equi-biaxial tension and planar test. Fatigue lifetime prediction of engine mount insulator was made by incorporating the maximum Green-Lagrange strain values, which was evaluated from the finite element analysis and fatigue tests, respectively. Predicted fatigue lives of the rubber component showed a fairly good agreement with the experimental fatigue lives. Fatigue analysis procedure employed in this study could be used approximately for the fatigue design.

Keywords Rubber component · Strain energy function · Finite element analysis · Fatigue test · Damage parameter · Lifetime prediction

1 Introduction

Rubber's ability to withstand very large strains without permanent deformation or fracture makes it ideal for many applications including tires, vibration isolators, seals, hoses, belts, impact bumper, medical devices and structural bearing to name a few [1, 2]. These rubber components subjected to fluctuating loads often fail due to the nucleation and growth of defects or cracks. To prevent such failures, it is necessary to understand the fatigue failure mechanism for rubber materials and to evaluate the fatigue life for rubber components. For these reasons, not only the rubber component manufacturers but also their customers like automotive makers perform a series of strict fatigue test on the components such as component fatigue tests and driving fatigue tests.

Currently, designers rely on their own trial-error based experiences for the fatigue design. Thus, those designs depending on only experience may result in disqualification from the fatigue test during final product evaluation. Those fatigue failures of any new designs are prohibitive for automotive manufacturers. In order to avoid this problem, many researchers [3–5] are focusing on evaluation of fatigue life using computer added engineering techniques that could supplement drawbacks of evaluation through tests and could significantly reduce the time for fatigue-proof design. However, there are the following problems. First, the rubber materials show particular mechanical properties according to compounding ingredients and manufacturing conditions [6, 7]. Therefore, in order to evaluate the fatigue life of designed rubber components, the material properties of the components should be obtained. It is practically impossible to measure the material properties for the whole component. Second, some parameters like stress, strain, Strain Energy Density (SED) and so on are .generally used to estimate fatigue life of rubber components [8–10] but the question remains how we should use these parameters to estimate component life and what the limitation of the parameters is.

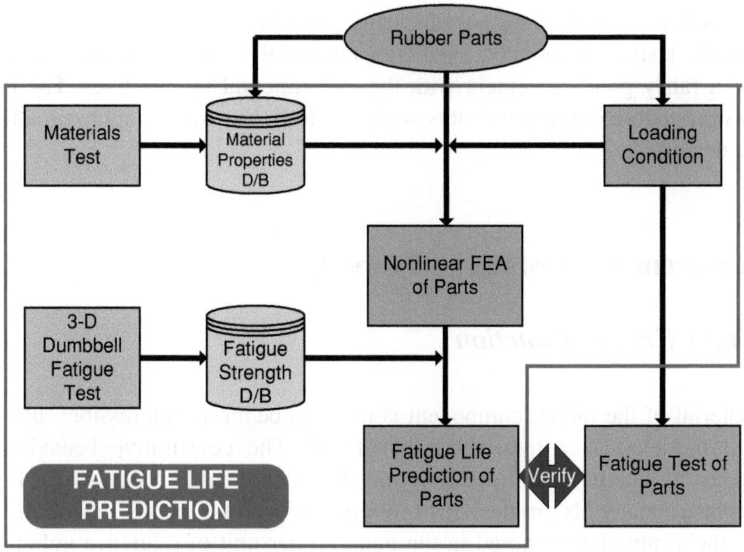

Fig. 1 Procedure to fatigue life prediction system

The fatigue lifetime prediction on the rubber components has been increasing according to the extension of warranty period of the automotive components. A design of rubber components against fatigue failure is one of the critical issues to prevent the failures during the operation. Therefore, fatigue lifetime prediction and evaluation are the key technologies to assure the safety and reliability of mechanical rubber components [11, 12]. Fatigue lifetime evaluation of rubber components has hitherto relied mainly on a real load test, road simulator test or bench fatigue test. Although the above methods have advantages in accuracy of fatigue life, they cannot be used before the first prototype is made and the fatigue test should be always conducted whenever material or geometry changes are made [13]. In order to predict the fatigue life of the rubber components at the design stage, a simple procedure of life prediction is suggested in Fig. 1.

In this chapter, engine mount insulator for automobile, which is damaged by repeated loading during operation, is selected for a typical application of fatigue life prediction methodology developed in the research. Uniaxial tension, equi-biaxial tension and planar tests were conducted to determine the nonlinear material constants of the rubber components. The maximum Green-Lagrange strain of 3 dimensional dumbbell specimens and engine mount insulator were obtained from a nonlinear finite element analysis (FEA) using the hyper-elastic material model determined from the material tests. We used the 3-dimensional dumbbell specimens to evaluate the fatigue life of rubber materials. Fatigue tests of 3-dimensional dumbbell specimens with various mean strains were performed, and a fatigue life curve equation represented by the maximum Green-Lagrange strain was obtained. Fatigue lifetime prediction of engine mount insulator was made by incorporating the maximum Green-Lagrange strain values, which was evaluated from the FEA and fatigue tests, respectively. Predicted fatigue lives of the rubber component showed a fairly good agreement with the experimental fatigue lives. The fatigue analysis procedure employed in this study could be used as an approximation for fatigue design.

2 Measurement of Material Property

2.1 Stain Energy Function

The material of the rubber component is taken to be an incompressible rubberlike material modeled as a hyper-elastic material. The constitutive behavior of a hyper-elastic material is defined as a total stress–total strain relationship [14, 15]. Hyper-elastic materials are described in terms of a strain energy potential, which defines the strain energy stored in the material per unit of reference volume as a function of the strain at that point in the material. The strain energy functions have been represented either in terms of the strain invariants that are functions of the stretch ratios, or directly in terms of the principal stretch. Successful modeling and design of rubber components relies on both the selection of an appropriate strain energy function and an accurate determination of material coefficient in the function. Material coefficient in the strain energy functions can be determined from the curve fitting of experimental stress–strain data. There are several different types of experiments, including simple tension, equi-biaxial tension and pure shear tests. In general, a combination of simple tension, equi-biaxial tension and pure shear tests are used to determine the material coefficient. The classical Mooney-Rivlin and Ogden model are an example of a Hyper-elastic model that is implemented in FEA [16].

In order to explain the deformation of the rubber materials, it is assumed that the material has elastic behavior and is isotropic. Then, strain energy function (W) can be written as Eq. 1, with strain invariant functions (I_1, I_2, I_3) and principal stretch functions $(\lambda_1, \lambda_2, \lambda_3)$.

$$W = W(I_1, I_2, I_3), \quad W = W(\lambda_1, \lambda_2, \lambda_3) \tag{1}$$

When the material is isotropic, I_1, I_2, I_3 can be expressed as follows;

$$\begin{aligned} I_1 &= \lambda_1^2 + \lambda_2^2 + \lambda_3^2 \\ I_2 &= \lambda_1^2\lambda_2^2 + \lambda_2^2\lambda_3^2 + \lambda_3^2\lambda_1^2 \\ I_3 &= \lambda_1^2\lambda_2^2\lambda_3^2 \end{aligned} \tag{2}$$

Most rubber materials are incompressible and its bulk modulus is much greater than its shear modulus. Thus, it is widely accepted to presume the materials to be incompressible when they are under less restriction. When the materials are incompressible in Eq. 2, $\lambda_1, \lambda_2, \lambda_3 = 1$ and $I_3 = 1$. Since, Eq. 1 can be rewritten as follows,

$$W = W(I_1, I_2) \tag{3}$$

Strain energy function, which is widely used to analyze deformation of incompressible materials, can be described with Mooney-Rivlin's function and Ogden's function.

$$\text{Mooney-Rivlin's function} : W = \sum_{n=1}^{N} C_{ij}(I_1 - 3)^i (I_2 - 3)^j \tag{4}$$

$$\text{Ogden's function} : W = \sum_{n=1}^{N} \frac{\mu_n}{\alpha_n}(\lambda_1^{\alpha_n} + \lambda_2^{\alpha_n} + \lambda_3^{\alpha_n} - 3) \tag{5}$$

where C_{ij}, μ_n, α_n are material constants determined experimentally from the stress–strain relationship.

2.2 Uniaxial Tension Test

The rubber material property, which is essential in FEA, is expressed with the coefficient values of strain energy function and these values are determined by fitting stress–strain data obtained from the material tests under various load conditions into the stress–strain curve induced from strain energy function. And it is determined to minimize the differences between the test values and calculated values. Therefore, we analyzed the property of the material and determined the nonlinear material coefficient, which is necessary in finite element analysis, by conducting uniaxial, equi-biaxial tension and pure shear tests [17].

Figure 2a shows the uniaxial tension test by using non contacting strain measurement (laser extensometer). When rubber materials are deformed, their network structure lose their stiffness due to modification and reformation, and damping properties change. Mullins suggested this is due to the stress–strain response, called the Mullins effect [18]. This is more prevalent in carbon black-filled rubbers. In other words, the stiffness of rubber depends on its history and strain range. In addition, the stress–strain curve in initial stage is not repeated anywhere and the curve stabilize after receiving approximately five repetitive loads within the same strain range. The stress–strain curve exhibits yet another change when the rubber material is subjected to a larger strain than the previous one. Lastly, the rubber material possesses properties in which fixed permanent deformation occurs when the material returns to the initial strain value; the strain is not equal to 0 even if the stress is equal 0. Figure 2b shows the Mullins effect in uniaxial tension test.

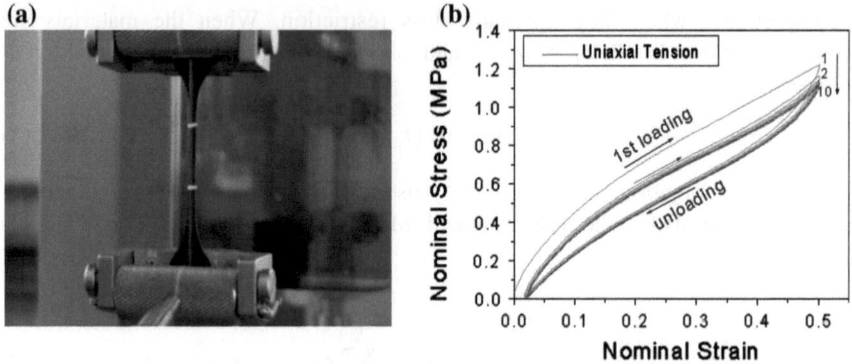

Fig. 2 Uniaxial tension test and Mullins effect **a** Laser extensometer **b** Mullins effect

2.3 Equi-Biaxial Tension Test

In uniaxial compression test, it is very difficult to obtain the pure compressed stress–strain relationship because of the frictions on the grip and the contact plane of rubber test specimen. Also, there is some bubbling phenomenon in the middle part of the test sample due to this friction. Therefore, it is hard to say that the property values of materials obtained from uniaxial compression tests are accurate. Thus, Woo et al. [5] suggested equi-biaxial tension tests, in which the pure strain values can be obtained, in order to resolve such issues in uniaxial compression test. For equi-biaxial tension tests, we prepared round shaped test specimen (Fig. 3) with 16 grips placed on the outer edges of the test specimen in order to apply evenly distributed loads in the direction of the circumference.

FEA of the specimen is required to determine the appropriate geometry of the clamping point in Fig. 3a. The equi-biaxial strain state may be achieved by radial stretching a circular disc in Fig. 3b. Once again, a non-contacting strain measuring device must be used such that strain is measured away from the clamp edges in Fig. 3c.

2.4 Pure Shear Test

A shear strain state is a more important mode of deformation for engineering applications than tension. The quad lap simple shear test piece is standardized [19]. But, the pure shear test is not yet standardized. There are two difficulties in the simple shear test. The first difficulty is making the specimen. This may require either bonding to rigid supports during vulcanization or molding blocks bonded with a high modulus adhesive. Secondly, the low shear strain range is limited because the rigid plates are bent on straining. Alternatively, the pure shear test can

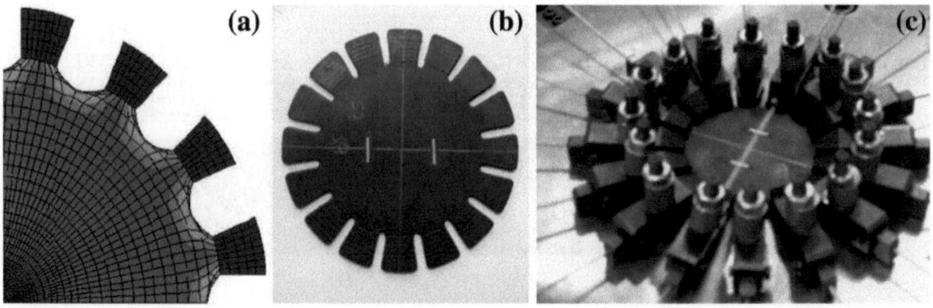

Fig. 3 Specimen and grips for equi-biaxial tension test **a** Finite element model **b** Specimen **c** Equi-biaxial tension test

be developed high strain range than simple shear test. If the material is incompressible and the width of the specimen is longer than the height, a pure shear state exists in the specimen at 45° angle to the stretching direction. Aspect ratio of the specimen is most significant in the pure shear test because the specimen is perfectly constrained in the horizontal direction.

Figure 4 shows the deformed shapes obtained by FEA at 100% stretching for aspect ratios of 5:1 and 10:1. Stress–strain curves obtained from the tests are shown Fig. 4c, compared to those predicted by FEA. Even though there exist some differences in the stress–strain responses between the experiment and the analysis, fairly good correlations are observed. A better agreement can be seen for the aspect ratio of 10:1, compared to 5:1. The differences are attributed to the specimen slippage from the clamp edges, leading to the inadequate states of pure shear strain. Therefore, it is necessary to design a gripping device to prevent specimen slippage, in order to improve the test accuracy.

2.5 Non-Linear Material Component for Hyper-Elastic Material

Figure 5 shows the stress–strain curves obtained from the uniaxial tension, equi-biaxial tension and pure shear tests in which we applied five repetitive loads in each of the vertical and horizontal direction with 25, 50 and 100% of the strain range for natural rubber. According to Fig. 5, the stress–strain curves during the second repetition showed a greater decline than in the first repetition. The stress–strain curve gradually decreased as the number of repetitions increased, and ultimately stabilized to a fixed stress–strain value.

In order to predict the behavior of the rubber components using the finite element analysis, the rubber material constants must be determined from the stabilized cyclic stress–strain curve. The stress–strain curve varies significantly depending on the cyclic strain levels. A 5th loading cycle was selected as the stabilized stress–strain relationship in this study. But this stabilized relation should

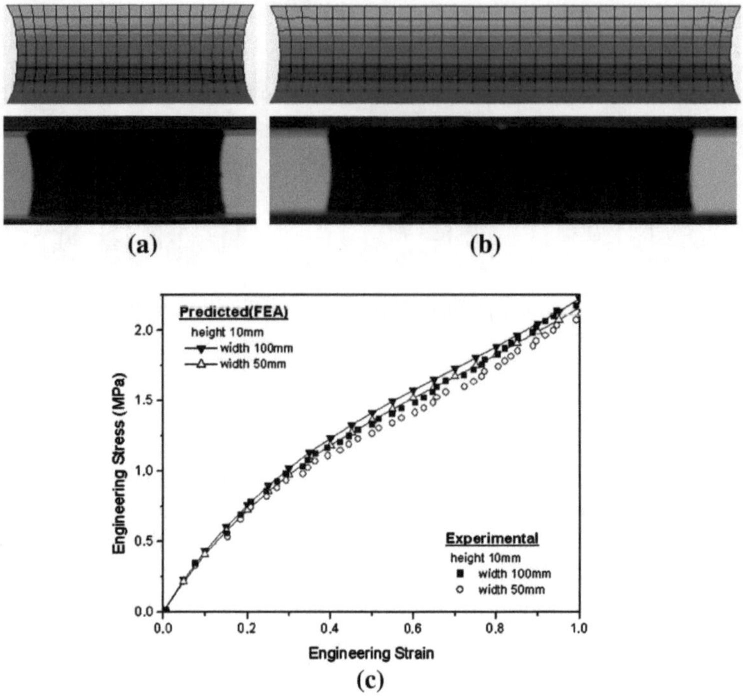

Fig. 4 Deformed shape of pure shear test **a** Aspect ratio 5:1 **b** Aspect ratio 10:1 **c** Stress–strain curves

be shifted to pass through the origin of the curve, to satisfy the hyper-elastic nature of rubber. Figure 6 shows the stress–strain relation of rubber material for various physical tests. The shift of curve meant that the gage length and initial cross sectional area were changed as shown in Eq. 6.

$$\varepsilon = \frac{\varepsilon' - \varepsilon_p}{1 + \varepsilon_p} \qquad \sigma = \sigma'(1 + \varepsilon_p) \tag{6}$$

We performed the curve fitting with uniaxial tension, equi-biaxial tension and pure shear test data. Table 1 contains the values of rubber material coefficient calculated in each case.

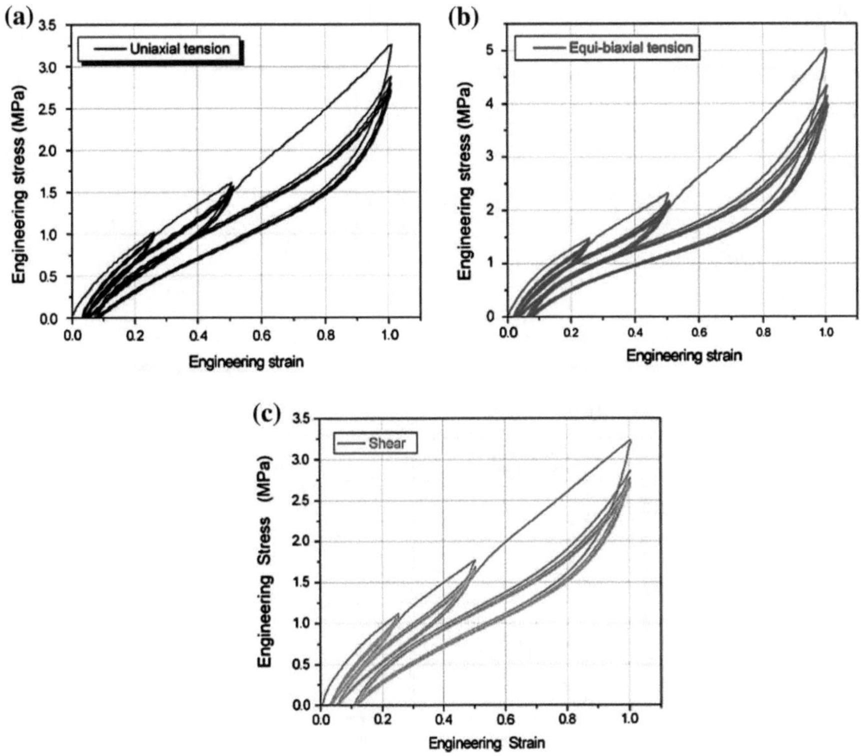

Fig. 5 Stress–strain curves at various loading and strain range **a** Uniaxial tension **b** Equi-biaxial tension **c** Pure shear

3 Fatigue Lifetime Prediction and Evaluation

3.1 Fatigue Test Specimen

The fatigue test piece has the basic shape of the 3-dimensional dumbbell specimen with a metal fitting cure bonded to each end. The geometry of the central part of the cylinder was designed to meet the following criteria in relation to fatigue test data for rubber components and strain distribution profile. The test piece should be capable of compression and tensile deformation without developing slackness under cyclic deformation. It should have a smooth strain distribution and the position at which maximum tensile strain develops should be the same for any deformation. The 3-dimensional dumbbell specimen has an elliptical cross-section and parting lines are located on the minor axis of specimen to avoid undesirable failure at the surface discontinuities [20]. The finite element model of the fatigue test specimen is shown in Fig. 7a. Figure 7b shows the strain distribution according

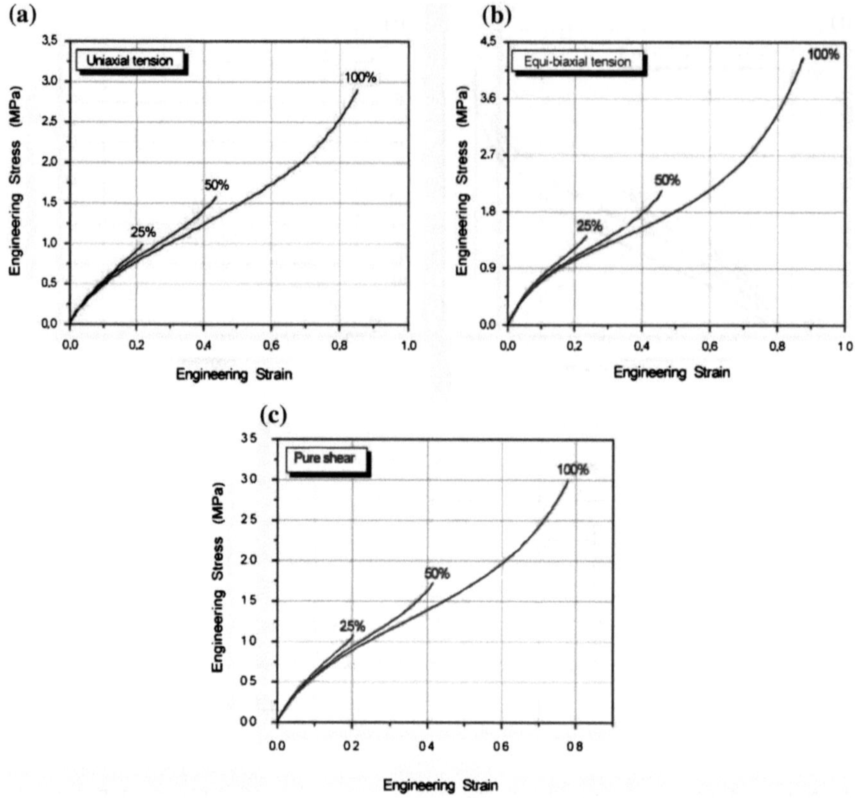

Fig. 6 Stabilized stress–strain curve and shift curve to zero **a** Uniaxial tension **b** Equi-biaxial tension **c** Pure shear

Table 1 Mooney-Rivlin and Ogden function of rubber material

Strain	Mooney-Rivlin 2-terms			Ogden 3-terms						
	C_{10}	C_{01}	GM	μ_1	α_1	μ_2	α_2	μ_3	α_3	Go
25%	0.889	0	1.779	1.1e−4	2.291	0.968	3.710	4.5e-4	3.081	1.796
50%	0.772	0	1.543	2.0e−5	1.030	2.140	1.3e−6	1.182	2.602	1.532
100%	0.723	0.009	1.463	0.028	4.391	5.813	0.031	0.953	2.593	1.390

to FEM analysis from 3-dimensional dumbbell specimen in compression and tension. Maximum Green-Lagrange strain was found to develop at a constant position in the surface at the centre of the rubber part of the test piece in both compression and tension.

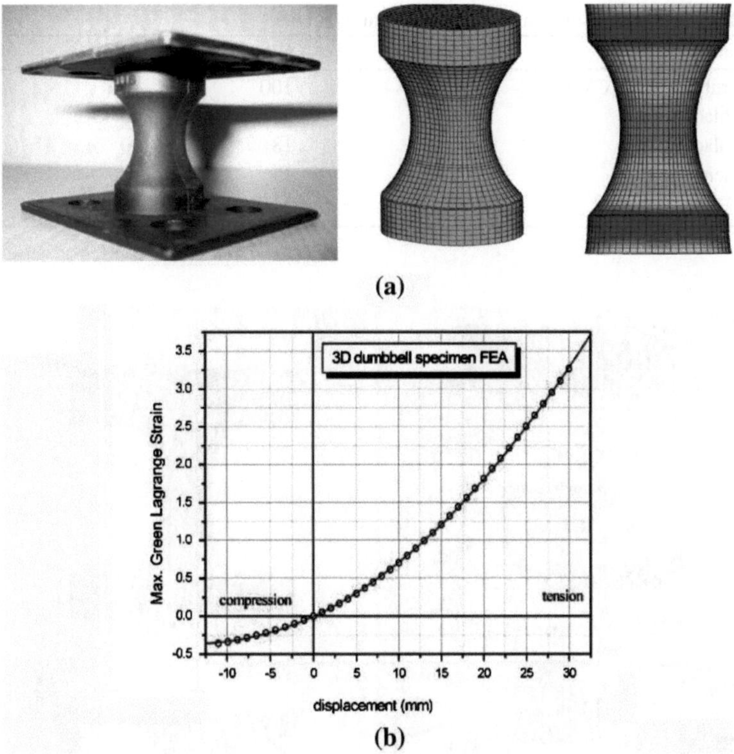

Fig. 7 3-dimensional fatigue test specimen **a** Finite element model **b** Displacement–strain curve

3.2 Fatigue Damage Parameter

In order to evaluate a fatigue damage parameter of the natural rubber material and to determine the experimental fatigue life, fatigue tests of 3-dimensional dumbbell specimens were performed using the fatigue testing system. The material used in this study is a carbon-filled vulcanized natural rubber, which have the hardness of the International Rubber Hardness Degree 50, 55, 60, 65(NR50, NR55, NR60, NR65). Compound recipes, including applied cure conditions, are summarized in Table 2.

Fatigue tests were conducted in an ambient temperature under the stroke-controlled condition with a sine waveform of 5 Hz and the mean displacement was 0, 3, 5, 8, 10 mm at the various displacement range. The fatigue failure was defined as a number of cycles at which the maximum load dropped by 20%. As increasing the fatigue cycles, the maximum load decreased little by little. When the crack grew over the critical size, the maximum load decreased suddenly corresponding to final failure. Figure 8 shows the fatigue testing system and fatigue life curve. Figure 9 shows the relationship between the displacement

Table 2 Compound recipes investigated of natural rubber

Ingredient (grade)	NR50	NR55	NR60	NR65
Natural rubber (SMRCV 60)	100	100	100	100
Carbon black (FEF)	22	27	40	40
Carbon black (SRF)	15	18	20	32
Stearic acid	1	1	1	1
ZnO	5	5	5	5

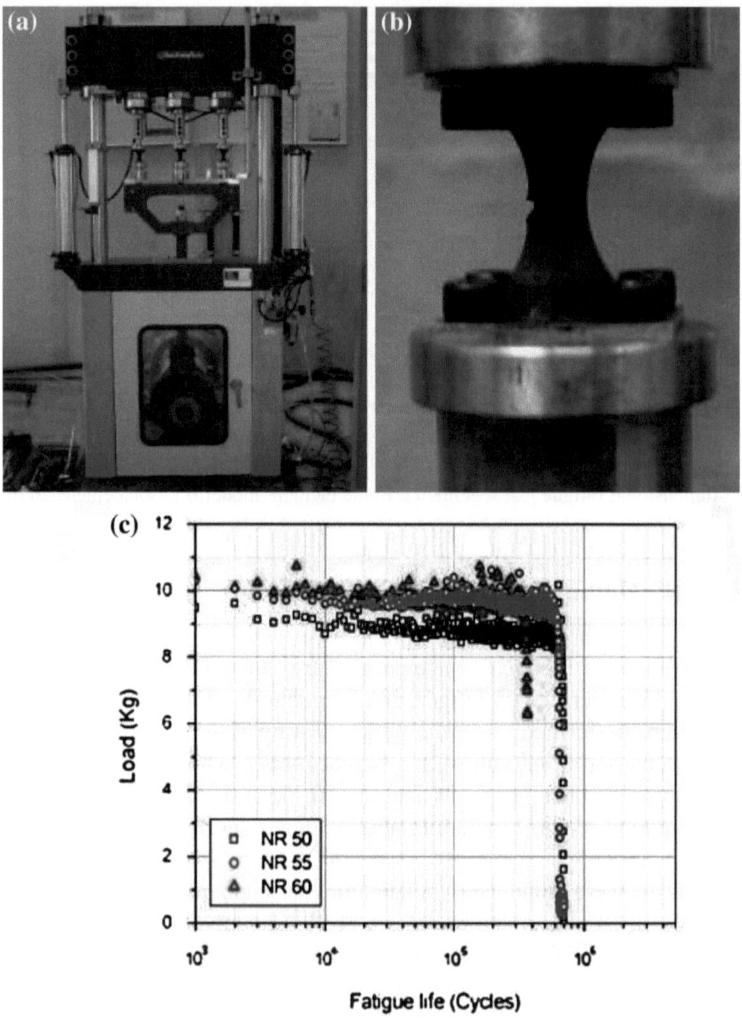

Fig. 8 Fatigue test of 3-dimensional dumbbell specimen **a** Fatigue test **b** Crack initiation **c** Crack initiation

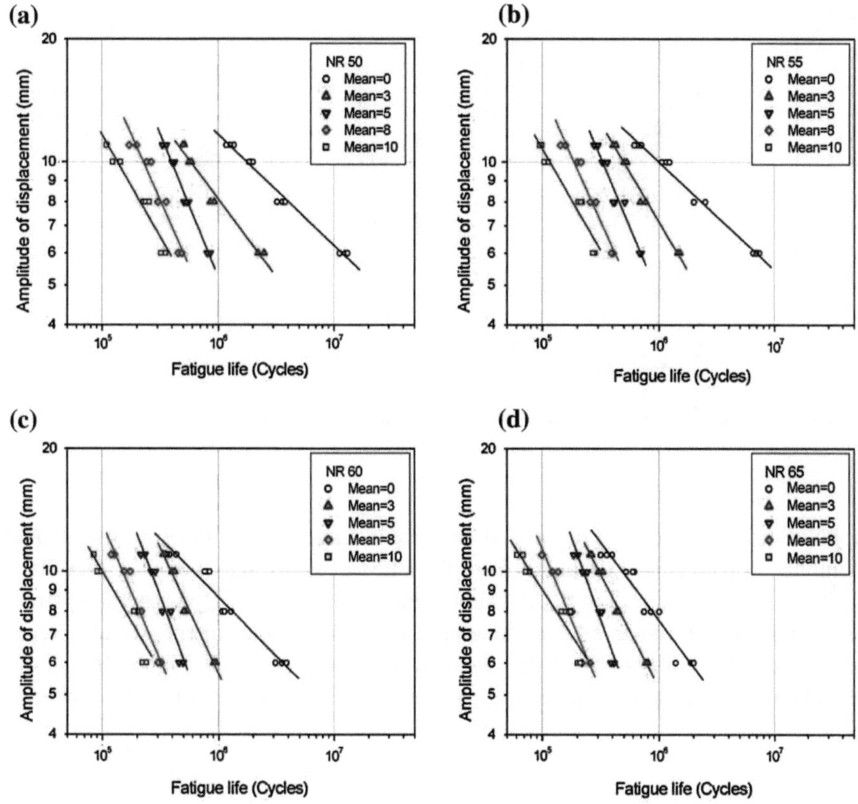

Fig. 9 Fatigue lives of 3-dimensional dumbbell specimen for natural rubber **a** NR50 **b** NR55 **c** NR60 **d** NR65

amplitude and the fatigue life of 3-dimensional dumbbell specimens. The fatigue life at the same displacement amplitude decreases as the mean displacement and hardness increase.

The fatigue process begins with the accumulation of damage at a localized region due to alternating load and displacement, leading to crack nucleation, growth, and final fracture [21, 22]. The crack nucleation life of the component may be defined as the number of cycles required for the appearance of a macro-crack. Therefore, the crack nucleation life of the component can be related to the life of a smooth specimen that is cycled to the same stresses or strains as the material at the critical region of the component. In this study, the fatigue damage of the natural rubber was evaluated from smooth dumbbell specimens.

Figure 10a shows the relationship between the maximum tension displacement and fatigue life of 3-dimensional dumbbell specimens. The fatigue life decreases as the maximum tension displacement increase. It is possible to express the fatigue life with the maximum tension displacement fairly good. Relationship between the

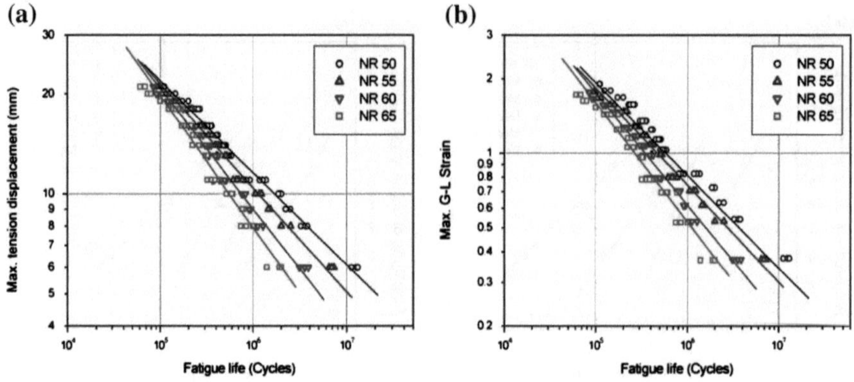

Fig. 10 Fatigue life curve for natural rubber **a** Displacement and fatigue life **b** Strain and fatigue life

applied displacement and corresponding Green-Lagrange strain of 3-dimensional dumbbell specimen are obtained from the FEM analysis and shown in Fig. 7b.

The fatigue life curve of natural rubber expressed by the Green-Lagrange strain as a damage parameter can be generated from the displacement versus Green-Lagrange strain curve of the 3-dimensional dumbbell specimen. Figure 10b shows the maximum Green-Lagrange strain versus cycles to failure of natural rubber.

3.3 Fatigue Lifetime Prediction

The fatigue life of the 3-dimensional dumbbell specimens represented by the maximum Green-Lagrange strain parameter ($\varepsilon_{G\text{-}L}$) are shown in Fig. 10b, where the Green-Lagrange strain for each dumbbell specimen is calculated from the displacement versus Green-Lagrange strain curve in Fig. 7b. It can be seen from Fig. 10b that the fatigue lives with different hardness can be effectively represented by a function using the maximum Green-Lagrange strain, thus taking into account the mean displacement and amplitude.

By using the fatigue test and FEM analysis, the normalized maximum strain defined as dividing by elongation at break (ε_{EB}) for the maximum Green-Lagrange strain ($\varepsilon_{G\text{-}L}$) for each hardness. Figure 11 shows the relation between normalized maximum strain and fatigue life. It was observed that the maximum Green-Lagrange strain was a good parameter to account for the hardness, mean displacement and amplitude effects. Fatigue lives of the 3-dimensional dumbbell specimen represented by the maximum Green-Lagrange strain and elongation at break are shown in Eq. 7. The fatigue lives are effectively represented by a single function using the maximum Green-Lagrange strain and elongation at break for each natural rubber materials.

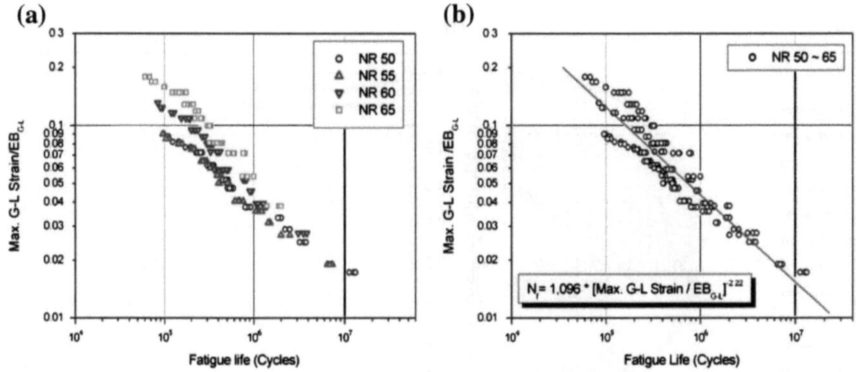

Fig. 11 Fatigue lifetime prediction curve of natural rubber **a** G-L strain versus fatigue life **b** Fatigue life prediction curve

$$N_f = 1096 \cdot \left[\varepsilon_{G-L}/\varepsilon_{EB}\right]^{-2.22} \tag{7}$$

3.4 Fatigue Lifetime Evaluation of Rubber Component

The methodology was applied to the fatigue life estimation of a rubber engine mount for automotive component. FEA is very important in design procedure to assure the safety and reliability of automotive rubber components. A nonlinear FEA was executed to evaluate the strain distribution by using a commercial finite element code. Material constants representing the Ogden strain energy potential of order 3 were used for defining the constitutive relation of the natural rubber.

Figure 12 shows the Green-Lagrange strain distribution of a rubber engine mount under a tensile displacement. The maximum Green-Lagrange strain at the critical location was used for evaluating the fatigue damage parameter of the rubber engine mount.

Static and fatigue tests were performed using a servo-hydraulic fatigue testing system shown in Fig. 13a. The experimental and FEA load–displacement curves compare very well (Fig. 13b). Maximum Green-Lagrange strain occurred at the location indicated in Fig. 14, and fatigue cracking at the critical location was observed during the fatigue test of the rubber engine mount. Fatigue failure was initiated at the critical location during the fatigue test of the rubber component, and was predicted by FEA. Correlation between the fatigue test life and the predicted life of the rubber engine mount using the normalized maximum strain is shown in Fig. 15. The predicted fatigue lives of the rubber engine mount agreed fairly well with the experimental fatigue lives.

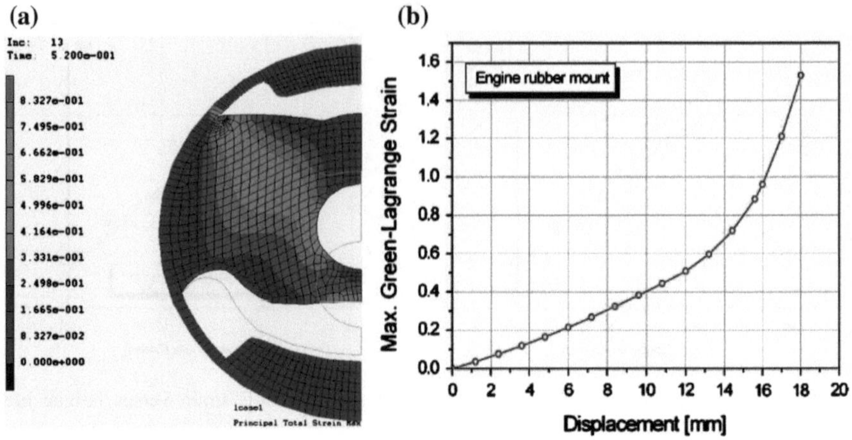

Fig. 12 Green-Lagrange strain distribution of the engine rubber mount **a** Finite element analysis **b** Max. G-L strain versus displacement

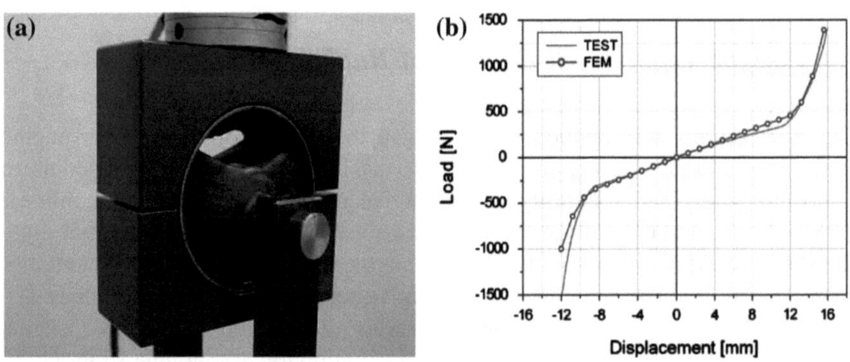

Fig. 13 Load–displacement curves of experimental and finite element analysis **a** Static test **b** Load and displacement curves

Fig. 14 Failure of the engine rubber mount **a** Finite element analysis **b** Experimental

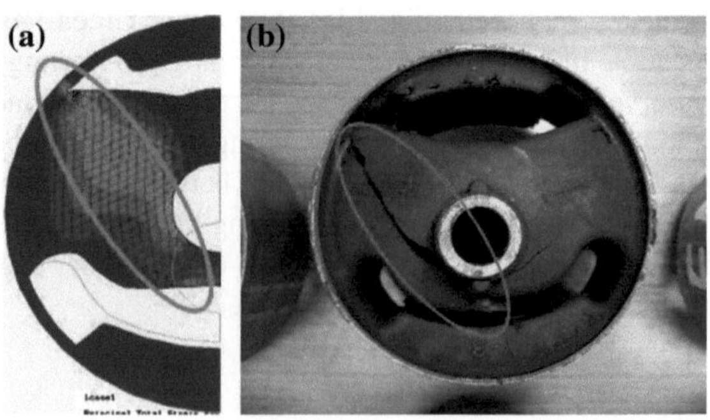

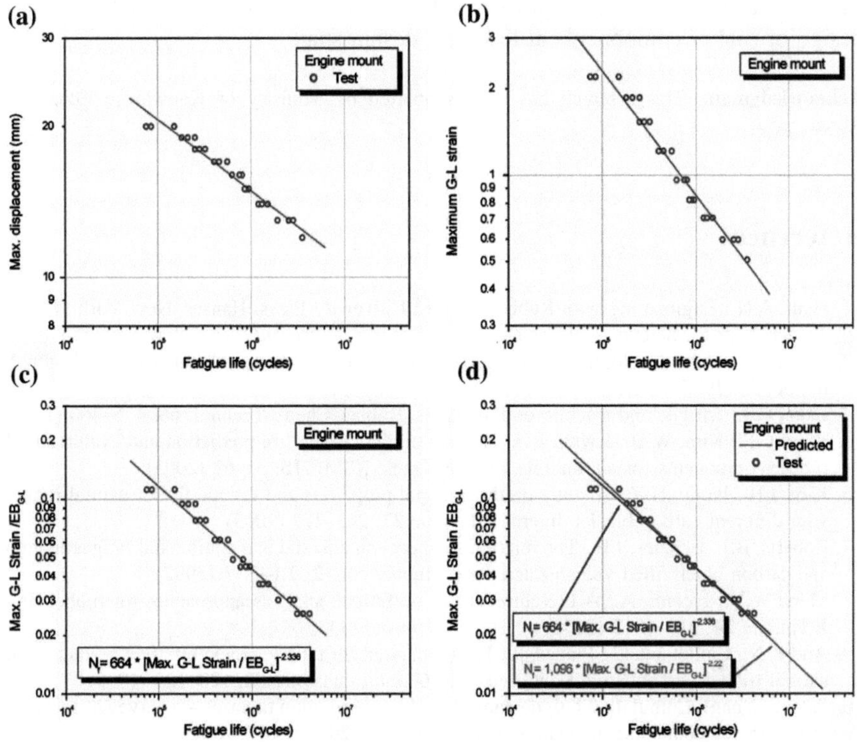

Fig. 15 Fatigue life for rubber engine mount **a** Max. displacement—fatigue life **b** Max. G-L strain-fatigue life **c** Fatigue life of engine mount **d** Comparison between predicted and experimental fatigue test

4 Conclusion

The fatigue lifetime prediction on the rubber components has been increasing according to the extension of warranty period of the automotive components. A design of rubber components against fatigue failure is one of the critical issues to prevent the failures during the operation. Therefore, the fatigue lifetime prediction and evaluation are the key technologies to assure the safety and reliability of mechanical rubber components.

In this chapter, to develop the durability analysis process for vulcanized rubber components, a fatigue lifetime prediction methodology was proposed by incorporating FEA and a fatigue damage parameter. In order to investigate the applicability of commonly used fatigue damage parameters, fatigue tests and corresponding FEA were carried out and optimum fatigue damage parameter were selected.

The fatigue lifetime of the rubber component was effectively represented by the maximum Green-Lagrange strain. Predicted fatigue lives of the rubber component were in fairly good agreements with the experimental lives. Therefore, the fatigue life estimation procedure employed in this study could be used for the fatigue design of rubber components at the early design stage.

Acknowledgment This research has been supported by Ministry of Knowledge Economy, Korea.

References

1. Gent, A.N.: Engineering with Rubber. Oxford University Press, Hanser, New York (1992)
2. Trederick, R.E.: Science and Technology of Rubber. Accdemic Press, New York (1978)
3. Tomas, A.J.: The development of fracture mechanics for elastomers. Rubber Chem. Technol. **67**, 50–60 (1994)
4. Lake, G.J.: Fatigue and fracture of elastomers. Rubber Chem. Technol. **68**, 435–460 (1995)
5. Woo, C.S., Kim, W.D., Kwon, J.D.: A study on the fatigue life prediction and evaluation of rubber components for automobile vehicle. Trans. KSME **13**, 56–62 (2005)
6. Kim, J.H., Jeong, H.Y.: A study on the material properties and fatigue life of natural rubber with different carbon blacks. Intern. J. Fatigue **27**, 263–272 (2005)
7. Roberts, B.J., Benzies, J.B.: The relationship between uniaxial and equibiaxial fatigue in gum and carbon black filled vulcanizates. Proc. Rubber con. **2**, 2.1–2.13 (1997)
8. Mars, W.V., Fatemi, A.: A literature survey on fatigue analysis approaches for rubber. Int. J. Fatigue **24**, 913–1012 (2002)
9. Andre, N., Cailletaud, G., Piques, R.: Haigh diagram for fatigue crack initiation prediction of natural rubber components. Kautschuk Und Gummi dunstoffe **52**, 120–123 (1999)
10. Fatemi, Yang.: Cumulative fatigue damage and life prediction theories—a survey of the state of the art for homogeneous materials. Int. J. Fatigue **20**, 9–34 (1998)
11. Hamed, G.R.: Molecular aspects of fatigue and fracture of rubber. Rubber Chem. Technol. **67**, 529 (1994)
12. Klenke, D., Beste, A.: Ensurance of the fatigue life of metal-rubber components. Gummi Kunsftoffe **40**, 1067–1071 (1987)
13. Thomas, A.G.: Factors influencing the strength of rubbers. J. Polym. Sci. **48**, 145–157 (1974)
14. Agyris, J.H., Dunne, P.C., Angelpoulos, T., Bichat, B.: Large nature strains and some special difficulties due to nonlinearity and incompressibility in finite elements. Comp. Meth. Appl. Eng. **4**, 219–278 (1974)
15. Swanson, S.H., Christensen, L.W.: Large deformation finite element calculations for slightly compressible hyperelasticitic materials. Comput. Struct. **21**, 81–88 (1985)
16. Norman, K.N., Kao, B.G.: Finite element analysis of viscoelastic elastomeric structures vibrating about non-linear statically stressed configurations, SAE paper 811309 (1981)
17. Day, J., Miller, K.: Equibiaxial stretching of elastometric sheets, an analytical verification of an experiment technique. ABAQUS Users Conf. Proc. **35**, 205–219 (2000)
18. Mullins, L.: Softening of rubber by deformation. Rubber Chem. Techno. **42**, 339–362 (1969)
19. ISO 1827.: Rubber, vulcanized or thermoplastic—Determination of shear modulus and adhesion to rigid plates—Quadruple shear methods. TC 45, Rubber and rubber products (2007)
20. Yamaguchi, H., Nakagawa, M.: Fatigue test technique for rubber materials of vibration insulator. Int. polym. Sci. Technol. **20**, 64–69 (1993)
21. Oh, H.L.: The fatigue life model of rubber bushing. Rubber Chem. Technol. **53**, 1226–1238 (1980)
22. Alshuth, T., Abraham, F.: Parameter dependence and prediction of fatigue life of elastomer products. Rubber Chem. Technol. **75**, 635–642 (2002)

The Thermal Shock Resistance and Mechanical Properties at Elevated Temperature of Transparent Ceramics

Marek Boniecki, Zdzisław Librant, Tomasz Sadowski and Władysław Wesołowski

Abstract In this work, the thermal shock resistance and mechanical properties at elevated temperature of transparent ceramics (spinel ($MgAl_2O_4$) and yttria (Y_2O_3)) were studied. The thermal shocks were done by fast inserting ceramic samples (disk shape) into a hot furnace (1000°C). Vickers indentations were made on the polished sample surfaces. Before and after shocks, the measurements of crack lengths were made and next a parameter R_m (an indicator of thermal stress resistance) was obtained. Hence the maximum thermal stresses was calculated using fracture toughness K_{Ic}. The measurements of bending strength σ_c and K_{Ic} as a function of temperature were carried out. Young's modulus and Vickers hardness were measured at room temperature. For spinel, fracture toughness K_{Ic} reached the maximum value at room temperature and minimum at 800°C. Above this temperature, K_{Ic} increased up to 1400°C. Bending strength σ_c attained the minimum value at 800 and 1000°C. At room temperature and at 1200°C it has almost the same value. For yttria, K_{Ic} and σ_c are higher at temperature above 600°C than at room temperature and remains almost constant up to 1500°C. In order to explain these observations, some hypotheses were proposed.

M. Boniecki (✉) · Z. Librant · W. Wesołowski
Institute of Electronic Materials Technology, 133 Wólczyńska Str, 01-919 Warsaw, Poland
e-mail: marek.boniecki@itme.edu.pl

Z. Librant
e-mail: zdzislaw.librant@itme.edu.pl

T. Sadowski
Lublin University of Technology, 40 Nadbystrzycka Str, 20-618 Lublin, Poland
e-mail: sadowski.t@gmail.com

W. Wesołowski
e-mail: wlawes@wp.pl

A. Öchsner et al. (eds.), *Materials with Complex Behaviour II*, Advanced Structured Materials 16, DOI: 10.1007/978-3-642-22700-4_18,

Keywords Spinel ($MgAl_2O_4$) · Yttria (Y_2O_3) · Thermal shock resistance · Bending strength · Fracture toughness

1 Introduction

Transparent ceramics with optical qualities comparable to single crystals of similar compositions have been developed in recent years [1]. This is a result of improvement in fabrication technology mainly by a significant progress in pure powder preparation and in sintering methods. These high-temperature materials with good thermal and mechanical properties can replace traditionally used materials as glass and quartz in some armour applications or in lamp envelopes and infrared emitter and detector covers production. The most known and common ceramics as alumina (α-Al_2O_3) are rather translucent than transparent and are birefringent as a result of their hexagonal structure. The birefringence leads to the light scattering in a polycrystalline material [2]. In order to avoid this disadvantageous phenomenon, ceramics with cubic crystal lattice should be used. At Institute of Electronic Materials Technology in Warsaw (Poland), two cubic transparent ceramics: spinel $MgAl_2O_4$ and yttria Y_2O_3 were prepared. For potential applications of these ceramics, information about their thermal and mechanical properties is needed. The object of this work was to determine the thermal shock resistance and mechanical properties at elevated temperature of these transparent ceramics.

2 Theoretical Background

A sample subjected to a thermal shock is under unsteady or transient state. In this case, the temperature at any point changes with time. The individual parts of the sample expand during heating and shrink during cooling in a different degree. Because they are parts of one sample, then the whole sample extends or shrinks to average dimensions which leads to a state, where in hotter parts of the sample compressive stresses appear but in the cooler parts tensile stresses occur. There are flaws in the sample which have been created during the fabrication process. These flaws could be treated as cracks. The stress acting over the longest crack of length c_0 can cause rapid fracture if the critical stress intensity factor K_{Ic} is reached. Then, the critical stress σ_c can be calculated as [3]:

$$\sigma_c = \frac{K_{Ic}}{Y\sqrt{c_0}} \quad (1)$$

where Y is a geometric factor.

The maximum stress σ_T caused by a thermal shock is given by [3]:

$$\sigma_T = \frac{E\alpha\Delta T}{1-v} f(\beta) \tag{2}$$

where E is Young's modulus, α is the linear thermal expansion coefficient, v is Poisson's ratio and β is the Biot modulus, defined as

$$\beta = \frac{rh}{k} \tag{3}$$

where r is a specimen dimension, h is the heat transfer coefficient and k is the thermal conductivity of the specimen. The function $f(\beta)$ depends on the specimen configuration.

However, the thermal stresses determination and their influence on brittle fracture of ceramics is not easy because the heat transfer h which determines β is not known with any significant precision as well as flaw size distribution in the sample. Therefore, Vickers indentation technique was applied [4, 5]. It creates a median-radial crack system which results in surface cracks (seen on the polished surfaces). These cracks, of length c_0, are in an equilibrium between a central residual opening force and the fracture toughness of the material K_{Ic}. For a given indentation load P, this gives [6]:

$$K_{Ic} = \chi P c_0^{-1.5} \tag{4}$$

where χ is a proportionality factor which is given as:

$$\chi = \xi\sqrt{E/H} \tag{5}$$

where ξ is a constant for Vickers produced radial cracks, H is hardness. In [6], the constant ξ was experimentally evaluated (0.016 ± 0.004) using results of toughness obtained by double cantilever beam technique for several materials.

If an external stress σ is now applied, the median-radial cracks will increase their size c under the superimposition of the decreasing residual and increasing applied stress intensity fields. This is written as [7]:

$$K_{Ic} = \chi P c^{-1.5} + Y\sigma c^{0.5} \tag{6}$$

Equation (6) indicates the possible existence of stable crack in equilibrium with the toughness of the material. Differentiation of σ with respect to the crack length c shows that stable crack growth occurs between the initial crack length c_0 and the maximum crack length c_m obtained as [7]:

$$c_m = (4\chi P/K_{Ic})^{2/3} \tag{7}$$

But the maximum value attained by the applied stress during stable crack extension is:

$$\sigma_m = 3K_{Ic}/4Yc_m^{0.5} \tag{8}$$

Because $c_0 = (\chi P/K_{Ic})^{2/3}$ from Eq. (4), then $c_m/c_O = 4^{2/3} = 2.52$.

Transient thermal stresses act like other applied mechanical stresses. Thus, if an indented sample is thermally shocked, the introduced, by an indenter, cracks will extend. This extension is limited by the maximum value of the thermal stress.

In [5], a parameter R_m was proposed defined as:

$$R_m = \left(\frac{K_{Ic}}{\sigma_T}\right)^2 \tag{9}$$

This parameter has the dimension of a length and is proportional to the maximum surface flaw size that can be withstood by the material for a given shock condition. R_m can be calculated using expressions (4) and (6):

$$R_m = \frac{Y^2 c}{\left[1 - \left(\frac{c}{c_0}\right)^{-1.5}\right]^2} \tag{10}$$

R_m is the slope of the Y^2c versus $[1-(c/c_0)^{-1.5}]^2$ curve. It should be underlined that R_m can be determined without knowledge of any material property or thermal shock conditions. It is derived only from direct measurement of cracks lengths before and after thermal shock. The definition of R_m could be compared to that of the fourth thermal shock resistance parameter (defined by Hasselman in [8]) $R'''' = (K_{Ic}/\sigma_c)^2(1 + \nu)$ (where σ_c is the strength of the material). R'''' is thus expected to be the lower limit of $(1 + \nu)R_m$; it depends on the distribution of initial flaws and it is independent of the thermal shock conditions. Parameter R_m does not depend on the distribution of flaws in the material, but since it depends on σ_T, it is affected by thermal stress conditions. Thus, it is an indicator of the severity of the thermal shock conditions. The higher it is, the less severe these conditions are. Having K_{Ic} and R_m, one can evaluate the maximum thermal stress σ_T for a given thermal shock condition.

3 Experimental Procedure

Figure 1 presents the method of the sample preparation of spinel and yttria.

The following powders were used:

- $MgAl_2O_4$ (spinel S30CR type) from Baikalox France of impurity content about 45 ppm and crystallite size about 0.2 μm.
- Y_2O_3 from Metall USA, purity 99.999%, crystallite size of about 2 μm.

For the thermal shock resistance study, disks about 30 mm in diameter and 1 mm thick were made. Spinel disks were sintered at 1750°C for 6 h in vacuum

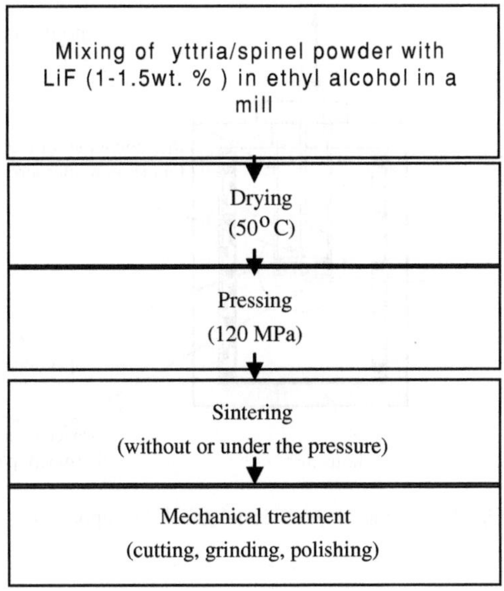

Fig. 1 Scheme of the sample preparation

and yttria disks were sintered at 1450°C for 2 h under a pressure of 30 MPa in argon. After grinding to the final dimensions, both disk surfaces were polished.

For bend strength, fracture toughness and Young's modulus measurements, beams of 2.5 × 4 × 40 mm^3, 1.5 × 4 × 40 mm^3 and 1 × 4 × 40 mm^3 respectively were made. Spinel beams were sintered at 1550°C for 2 h under a pressure of 30 MPa in argon, but yttria beams were sintered in the same conditions as yttria disks. Density of the obtained samples should be close to theoretical because they have a good transparency.

Thermal shocks were performed in air by quick inserting the disks at room temperature into the hot furnace heated to 1000°C (Fig. 2). Disks were hold in the furnace for several minutes in order to reach the furnace temperature and next were cooled slowly to room temperature (in the furnace).

One of the disk surface was sheltered by 5 mm thick alumina disk, before the shock Vickers indentations were introduced on the sheltered surfaces. The length of surface cracks before and after thermal shocks were measured with an optical microscope.

Strength tests were carried out at room and elevated temperature using a four-point bending device (20–10 mm, outer–inner spans) made of steel/alumina at room/high temperature with a loading rate of 1 mm/min. Strength σ_c was calculated from Eq. (11):

$$\sigma_c = \frac{1.5F_c(L-l)}{bw^2} \tag{11}$$

where F_c is the failure load, L and l are the outer and inner span respectively, b is the sample width = 4 mm and w is the sample thickness = 2.5 mm.

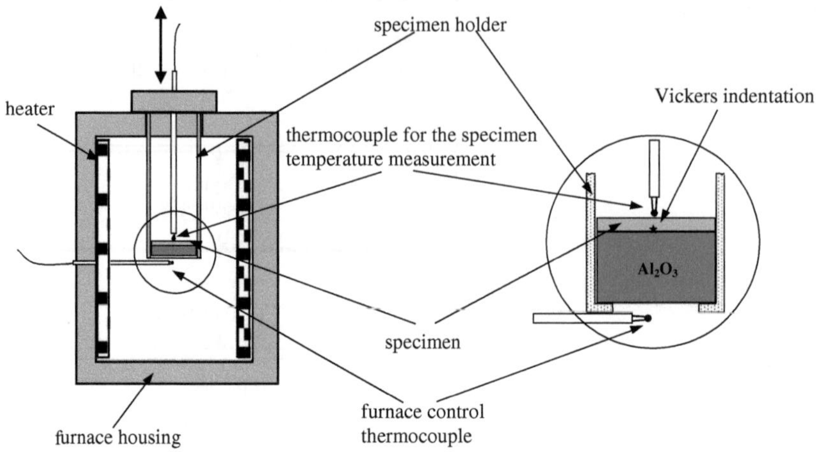

Fig. 2 Scheme of the thermal shock test apparatus

The fracture toughness was determined by two methods. In the first method bars for bending with notches of 1.1 mm deep were used (measurements were made at room and elevated temperature). The notches were made by a two-steps. At first a 0.9 mm deep notch was made using a 0.2 mm thick diamond saw and next was deepened for 0.2 mm using a 0.025 mm thick saw. Tests were carried out using three-point bending device with $L = 37$ mm, made of steel (at room temperature) and with $L = 20$ mm, made of alumina (at high temperature) with a loading rate of 1 mm/min. Fracture toughness K_{Ic} was calculated from Eq. (12):

$$K_{Ic} = Y_n \frac{1.5 F_c L}{b w^2} c_n^{0.5} \tag{12}$$

where Y_n is a geometric factor for the notched beam calculated accordingly to [9], $b = 1.5$ mm, $w = 4$ mm, c_n is the notch length $= 1.1$ mm.

In the second method fracture toughness was determined from Vickers cracks made on disks for thermal shock testing. Values of K_{Ic} were evaluated using Eq. (4, 5). Measurements were made only at room temperature.

Young's modulus was determined only at room temperature using a three-point bending device equipped with a deflection gauge. Young's modulus E was calculated from Eq. (13) [9]:

$$E = \frac{L^2}{b w^2 C}\left[\frac{L}{4w} + \frac{(1+\nu)w}{2L}\right] \tag{13}$$

where $L = 37$ mm, $b = 4$ mm, $w = 1$ mm, C is the compliance $= \Delta y/\Delta F$ (Δy is the increase of deflection, ΔF is the increase of load), $\nu = 0.26$ and 0.3 for spinel and yttria respectively [10].

Spinel, $d = 3.2 \pm 2.1$ μm Y_2O_3, $d = 42.2 \pm 20.9$ μm

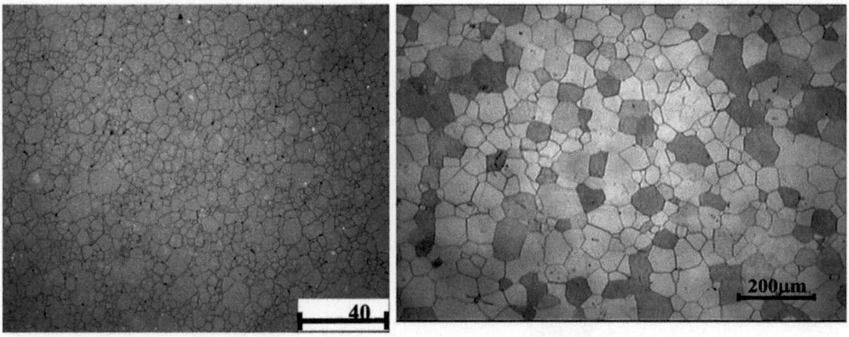

Fig. 3 Optical microscope pictures of the microstructures of studied ceramics. Values of average grain size *d* are given above

Some of the polished spinel samples were thermally etched at 1450°C for 1 h in air but Y_2O_3 samples were chemically etched in boiling water solution of HCl for 3 min in order to reveal the microstructure. Next, the microstructure seen under the optical microscope (Zeiss) were analyzed with the computer programme for Feret's diameter method in order to obtain grain size distribution.

4 Results and Discussion

The microstructures of the ceramics are shown in Fig. 3.

The shown spinel microstructure concerns the disks. For the spinel beams $d = 36.1 \pm 19.7$ μm.

In Fig. 4, a Vickers indentation made on the polished surface of Y_2O_3 disk (as an example) is presented.

The results of thermal shocks made on Vickers indented disks are shown in Fig. 5.

Mechanical properties as a function of temperature for the studied ceramics are presented in Fig. 6 and 7. In Fig. 6, data from [11] are also shown for comparison. In [11], hot-pressed polycrystalline transparent spinel was studied (with a theoretical density of 3.58 g/cm^3) with the an average grain size of about 35 μm. Fracture testing was conducted using three-point fixture at a crosshead speed 0.01 mm/min. In [11], three test specimens were used in order to obtain K_{Ic}. The results obtained for straight-notched specimens were chosen because similar samples were used in this work.

From the line slopes of Fig. 5, the values of parameter R_m and hence maximum thermal stresses σ_T (using K_{Ic} values) were estimated (Tables 1 and 2). For spinel the R_m value with the smallest standard deviation *sd* was chosen (the case without

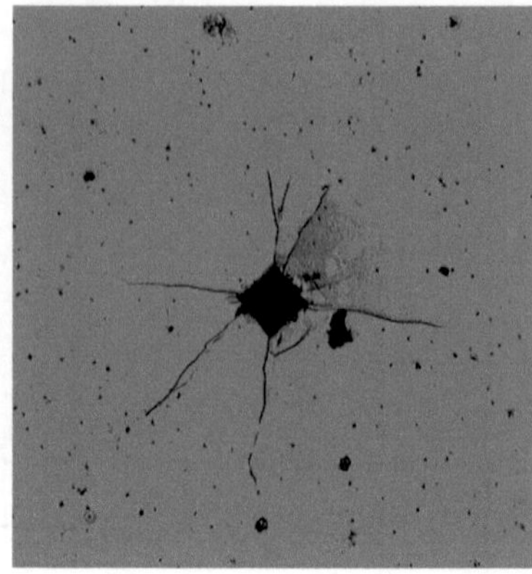

Fig. 4 Vickers indentation made on polished surface of Y_2O_3 disk under the load $P = 29.4$ N

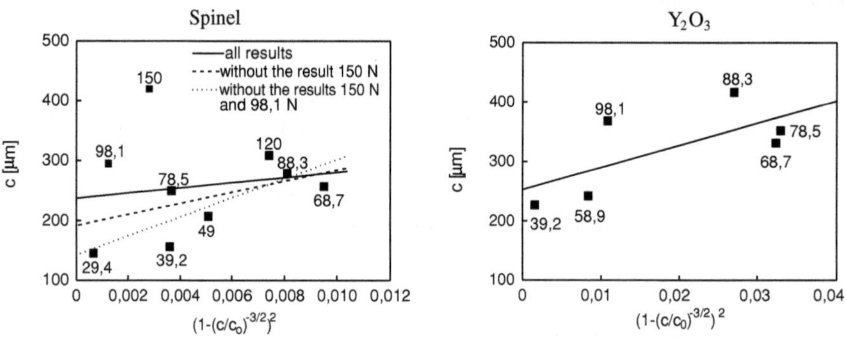

Fig. 5 Vickers indentation crack growth in spinel and Y_2O_3 under the thermal shocks. On the ordinate is a crack length *c* after thermal shocks; in turn on the abscissa is an expression containing c/c_0 (see Eq. (10)). Numbers next to the experimental points mean the Vickers indenter *P* load in N. Position of the lines were calculated by least square method. Slopes of the lines were used next for R_m determination

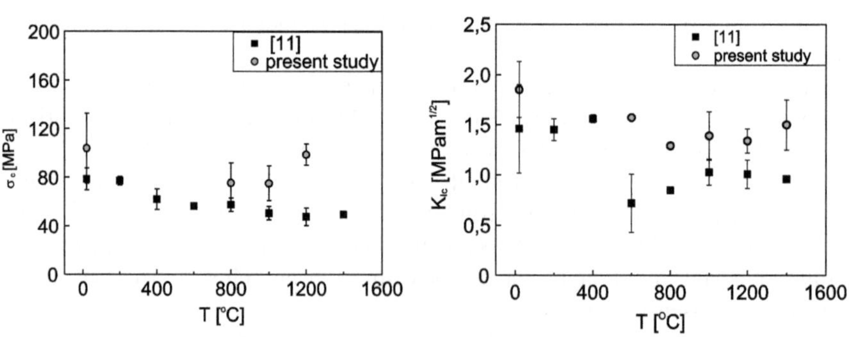

Fig. 6 Strength σ_c and fracture toughness K_{Ic} as a function of temperature *T* for spinel. For comparison, data from [11] are shown. Mean values and standard deviations (marked as bars) were calculated for five samples

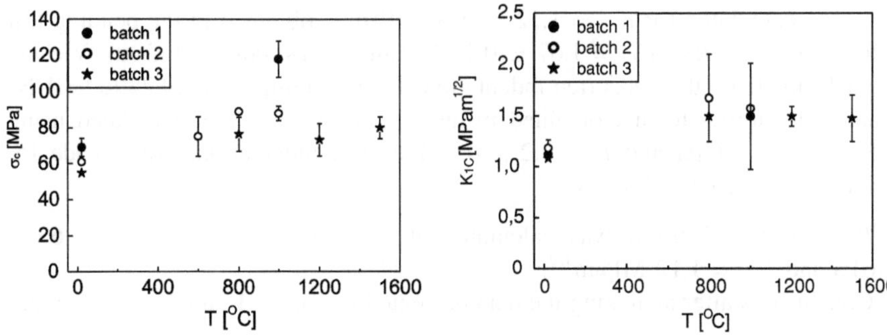

Fig. 7 Strength σ_c and fracture toughness K_{Ic} as a function of temperature T for Y_2O_3. Measurements were made for three batches. These batches had almost the same microstructure (grain sizes). Mean values and standard deviations (marked as bars) were calculated for five samples for each batch. For the sake of clarity only the biggest standard deviations are shown

Table 1 Parameters R_m and maximum thermal stress σ_T for spinel

R_m (μm)	$\overset{*}{\sigma_T}$ (MPa)	$\overset{**}{\sigma_T}$ (MPa)
26571	8.5	11.3
17507	10.4	14.0
35635	7.3	9.8

Table 2 Parameter R_m and maximum thermal stress σ_T for Y_2O_3

R_m(μm)	σ_T (MPa)
6,195	14.2
2,862	20.9
9,528	11.5

experimental points for $P = 98.1$ and 150 N). Geometric factor Y was taken as 1.29 according to [12]. The underlined numbers are the average value of R_m M. The next numbers in Tables 1 and 2 are $M-sd$ and $M + sd$. Values of K_{Ic} at room temperature were used here. We assumed that just after putting the sample into the hot furnace the temperature of the sheltered surface should be close to the room temperature. For comparison the fourth thermal shock resistance parameter R'''' [8] was calculated.

- $\overset{*}{\sigma_T}$ —was calculated from Eq. 9 in which $K_{Ic} = 1.38 \pm 0.18$ MPam$^{1/2}$ was in turn obtained from Vickers crack using Eqs. 4 and 5, where $E = 264 \pm 5$ GPa (measured for five samples) and $H = 14.3 \pm 0.7$ GPa. Ten indentations were made for load P ranging from 29.4 to 150 N
- $\overset{**}{\sigma_T}$ —was calculated from Eq. (9) in which $K_{Ic} = 1.85 \pm 0.28$ MPam$^{1/2}$ was obtained in bending test for 5 notched beams at room temperature.
- $R''''/(1 + v) = (K_{Ic}/\sigma_c)^2$ was calculated at room temperature for $\sigma_c = 104 \pm 29$ MPa. It equals 176 and 316 μm respectively for $K_{Ic} = 1.38$ and 1.85 MPam$^{1/2}$

- σ_T was calculated for $K_{Ic} = 1.12 \pm 0.06$ MPam$^{1/2}$ obtained in the bending test for three batches. $K_{Ic} = 1.07 \pm 0.15$ MPam$^{0.5}$ was obtained from Vickers cracks made on the disks (ten indentations for P ranging from 29.24 to 98.1 N) and it is similar to that obtained in the bending test. It was calculated using $E = 172 \pm 6$ GPa and $H = 7.2 \pm 0.3$ GPa. Ten indentations were made for load P from 29.4 to 98.1 N.

$R''''(1 + \nu) = 316$ μm was calculated at room temperature for $\sigma_c = 63 \pm 7$ MPa and $K_{Ic} = 1.12$ MPam$^{1/2}$.

Conclusions after analyzing the data presented in Tables 1 and 2 are following:

- Spinel is more resistant to the thermal shocks than yttria (R_m for spinel is higher than for yttria, but $R''''/(1 + \nu)$ is similar for both of them),
- The chosen thermal shock conditions seem not to be severe for the studied ceramics because R_m is about two orders of magnitude for spinel and one order of magnitude for yttria higher than $R''''/(1 + \nu)$,
- Some higher σ_T values for yttria than for spinel were probably caused by two factors: higher linear thermal expansion coefficient α for yttria and higher Biot modulus β because of smaller thermal conductivity for Y_2O_3 than for spinel [10]. The function $f(\beta)$ and hence σ_T (Eq. 2) increases as β increases [3]. However, it should be noticed that Young's modulus E is bigger for spinel.

The origin of the tensile thermal stresses generated on the sheltered surface of samples seems to be obvious. This surface should be cooler than the exposed one just after putting the sample into the hot furnace (Fig. 2). The big scatter of results seen in Fig. 5 is a consequence of a complicated crack system arising around Vickers indentations (Fig. 4). Cracks are treated as straight lines during length measurements, but they are not and additively some of them are bifurcated. Hence there are big errors in crack length measurements and big standard deviations in the obtained parameter R_m.

As indicated previously, we obtained two kinds of spinel ceramics. The first used for the disk production had a grain size $d = 3.2 \pm 2.1$ μm, but the second for the beam production had a grain size $d = 36.1 \pm 19.7$ μm. The thermal shocks were on disks but the maximum thermal stress σ_T and Hasselman parametr R'''' were calculated using K_{Ic} and σ_c obtained from bending tests of beams. The difference between results of K_{Ic} obtained by the indentation method using disks and by the bending test using beams (Table 1) could point out that the microstructure significantly influenced the mechanical properties of the ceramics. In order to explain this problem, authors carefully analyzed the literature reports about it. Stewart and Bradt in [13] presented fracture toughness data for spinel with grain size from 5 to 38 μm and they determined that K_{Ic} is independent of grain size and equals about 1.93 MPam$^{0.5}$ at room temperature. In Table 3, some mechanical data for spinel with various grain sizes from the literature were collected.

The review of data in Table 3 shows that spinel strength σ_c rather increases as grain size decreases (according to Hall–Petch relationship [17]). A similar result is

Table 3 Strength σ_c, fracture toughness K_{Ic} and Young's modulus E as a function of grain size d for spinel at room temperature

Reference	d (μm)	σ_c (MPa)	K_{Ic} (MPam$^{0.5}$)	E (GPa)
This work	36 ± 20	104 ± 29	1.85 ± 0.28	264 ± 5
This work	3.2 ± 2.1		1.38 ± 0.18	
[11]	35 (10–100)	80	1.46 ± 0.44	258
[14]	50–200		2.2	275
[15]	1.5 ± 0.8	129 ± 20	3.0 ± 0.1	258 ± 2
		194 ± 20		
[16]	1.9 ± 1.7	120 ± 20	2.4 ± 0.24	258 ± 4

obtained for K_{Ic} (the value 1.38 MPam$^{0.5}$ obtained by Vickers indentation method for $d = 3.2$ m for disks seems to be to low) but Young's modulus E is independent of grain size. This conclusion is not consistent with the results presented in [13]. However, it was established that errors resulting from using K_{Ic} from beams for σ_T calculations for disks in Table 1 are insignificant in comparison with errors of R_m determination. One can say the same about the error in the calculation of parameter R'''' in Table 1 because K_{Ic} and σ_c simultaneously rise as grain size decreases.

An unexpected feature is observed in Fig. 6—σ_c for spinel decreases as a function of temperature up to 800°C, remains constant up to 1000°C and next increases up to 1200°C to a value close to the measured one at room temperature. It could be connected with the appearance of crack tip plasticity at elevated temperatures [15]. In this situation, the following relationship proposed in [18] between σ_c and plastic zone extension s at the crack tip can be applied:

$$\sigma_c = \sigma_{c0}\sqrt{\frac{2s}{2s + d}} \tag{14}$$

where $\sigma_{cO} = 104$ MPa is the strength of spinel at room temperature, $d = 36.1$ μm is a median grain size of spinel (for beams).

It was evaluated from Eq. 14 that for $\sigma_c = 75$ MPa at $T = 800°$ and 1000°C (Fig. 6), $s \approx 20$ μm, but for $\sigma_c = 99$ MPa at $T = 1200$°C $s \approx 175$ μm. It means that the plastic zone extension increases as a function of temperature, which seems to be obvious. A similar effect as for σ_c is observed for K_{Ic} (Fig. 6). K_{Ic} decreases up to $T = 800$°C, remains constant up to 1200°C and next increases up to 1400°C. The increase of K_{Ic} for spinel at temperatures higher than 800°C is also registered by other authors [11, 14, 15] (data from [11] are shown in Fig. 6). This phenomenon was explained by plasticity of crack tip due to cation diffusion [15]. Our results differ from others because the strength growth is also observed at temperatures higher than 800°C but in the literature [11, 15] strength of spinel decreases monotonically as a function of temperature (Fig. 6). In order to reveal the mode of fracture as a function of temperature, the fracture surfaces were observed by scanning electron microscopy (Fig. 8).

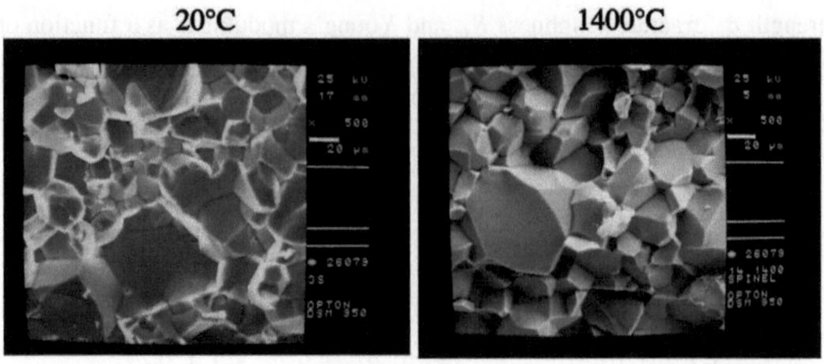

Fig. 8 Scanning electron micrographs of fracture surfaces of spinel samples at room and at the highest temperature

The micrographs analysis indicates that there is a mixed transgranular and intergranular mode of fracture but it seems that at high temperature prevails rather the intergranular one. However, the largest grains failed transgranularly even at the highest temperature. These observations are consistent with earlier studies [13, 14]. The increased fraction of intergranular fracture and probably arising of plastic zone ahead of crack tip can cause the increase of σ_c and K_{Ic} at temperatures higher than 800°C for spinel.

Concerning the mechanical properties of yttria, it needs to be said that only limited work has been published. Therefore, the authors studied three batches of the yttria ceramics (Fig. 7) in order to obtain the verified results. Except one result concerning unexpected high value of σ_c at 1000°C for batch 1, the results quite consistent. The tendency is that for T $\geq$ 600°C, σ_c and K_{Ic} are higher than at room temperature and they are almost constant in the range of temperature : 600°–1500°C for σ_c and 800°–1500°C for K_{Ic}. In Fig. 9, fracture surfaces after strength tests are shown for room and the highest temperature.

At room temperature, a transgranular mode of fracture is dominant but at 1500°C there is rather mixed (transgranular and intergranular) mode of fracture.

In [19] it was revealed that up to 1000°C, brittle fracture during compressive test took place in yttria and it was experimentally showed there that more cracks were generated in the specimens at 1000°C than at room temperature. This phenomenon led to the decreasing of compression strength from 1100 MPa at room temperature to 600 MPa at 1000°C.

Plastic deformation of yttria became prominent at 1200°C [19], meaning that brittle-to ductile temperature is somewhere between 1000 and 1200°C. Plastic deformation of Y_2O_3 by dislocation motion at 1200°C was confirmed in [19] by transmision electron microscopy. At and above this temperature, yttria samples had residual deformation in response to compressive loading, and catastrophic brittle fracture was not observed. With an increase in temperature, compression strength of yttria decreased to about 100 MPa at 1600°C. In this work a rather opposite behaviour of bending strength and fracture toughness of yttria with increasing temperature was observed. It could be explained by the fact that at

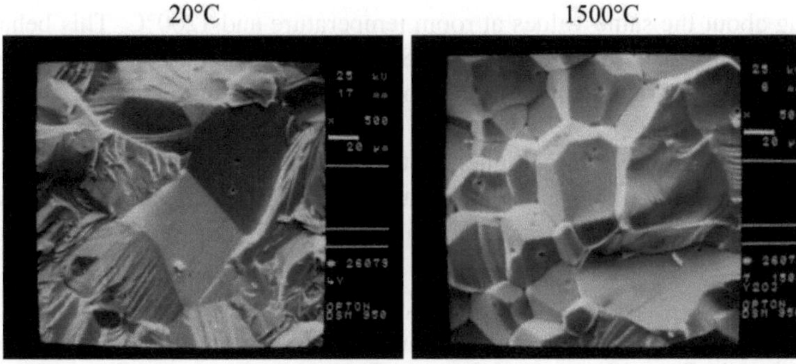

Fig. 9 Scanning electron micrographs of fracture surfaces of yttria samples at room and at the highest temperature

elevated temperature (above 600°C), grain boundary strength of the ceramics weakens and tensile stress acting on the notch tip or the flaw on the tension surface of the sample during loading could cause a crack deflection (in this case—a crack propagation through grain boundary interfaces) which would lead to the toughness increase [20]. This fracture increase is accompanied by the same bending strength increase ($K_{Ic}/\sigma_c \approx 0.02\ m^{0.5}$ for all data from room temperature to 1500°C). During loading of the samples at and above 1200°C, plastic deformation was not visible due to the high loading rate (1 mm/min).

5 Conclusions

Thermal shock resistance and mechanical properties of two transparent ceramics, $MgAl_2O_4$ (spinel) and Y_2O_3 (yttria), up to 1500°C were measured. The following conclusions can be drawn:

- Spinel is more resistant to the thermal shocks than yttria (parameter R_m, an indicator of severity of thermal shock conditions is higher for spinel than for yttria, but the fourth thermal shock resistance parameter defined by Hasselman R'''' divided by $(1 + v)$ is similar for both of them),
- The chosen thermal shock conditions seem not to be severe for the studied ceramics because R_m is about two orders of magnitude for spinel and one order of magnitude for yttria higher than $R''''/(1 + v)$,
- Some higher maximum thermal stress σ_T values for yttria than for spinel were probably caused by higher linear thermal expansion coefficient and Biot modulus for yttria than for spinel,
- For spinel, fracture toughness K_{Ic} reached the maximum value at room temperature and minimum at 800°C. Above this temperature, K_{Ic} increased up to 1400°C. Bending strength σ_c reached the minimum value at 800 and 1000°C,

having about the same values at room temperature and 1200°C. This behaviour of K_{Ic} and σ_c at temperatures above 800°C could be caused by the increased fraction of intergranular fracture and probably by arising of plastic zone ahead of the crack tip,

- For yttria, K_{Ic} and σ_c are higher at temperature above 600°C than at room temperature and remains almost constant up to 1500°C. It was probably caused by crack deflection. The explanation of this phenomenon needs further studies.

Acknowledgments This work was supported by the Polish Ministry of Education and Science (Project No. 438/E-241/S/2009). The authors thank Dr Anna Wajler and Mgr Helena Węglarz for preparing samples.Financial support of Structural Funds in the Operational Programme - Innovative Economy (IE OP) financed from the European Regional Development Fund - Project “Modern material technologies in aerospace industry”, No POIG.0101.02-00-015/08 is gratefully acknowledged. (task: ZB-10**:** Modern thermal barrier coatings for critical parts of engines).

References

1. Krell, A., Hutzler, T., Klimke, J.: Transmission physics and consequences for materials selection, manufacturing and applications. J. Eur. Ceram. Soc. **29**, 207–221 (2009)
2. Apetz, R., van Bruggen, M.P.B.: Transparent alumina: A light-scattering model. J. Am. Ceram. Soc. **86**, 480–486 (2003)
3. Evans, A.G., Linzer, M., Jonson, H., Hasselman, D.P.H., Kipp, M.E.: Thermal fracture studies in ceramic systems using an acoustic emission technique. J. Mater. Sci. **10**, 1608–1615 (1975)
4. Osterstock, F.: Contact damage submitted to thermal shock: a method to evaluate and simulate thermal shock resistance of brittle materials. Mat. Sci. Eng. A. **168**, 41–44 (1993)
5. Tancret, F., Osterstock, F.: The Vickers indentation technique used to evaluate thermal shock resistance of brittle materials. Scripta Mater. **37**, 443–447 (1997)
6. Anstis, G.R., Chantikul, P., Lawn, B.R., Marshall, D.B.: A critical evaluation of indentation techniques for measuring fracture toughness: I, Direct crack measurements. J. Am. Ceram. Soc. **64**, 533–538 (1981)
7. Chantikul, P., Anstis, G.R., Lawn, B.R., Marshall, D.B.D.B.: A critical evaluation of indentation techniques for measuring fracture toughness: II, Strength method. J. Am. Ceram. Soc. **64**, 539–543 (1981)
8. Hasselman, D.P.H.: Unified theory of thermal shock fracture initiation and crack propagation in brittle ceramics. J. Am. Ceram. Soc. **52**, 600–604 (1969)
9. Fett, T., Munz, D.: Subcritical crack growth of macrocracks in alumina with R-curve behavior. J. Am. Ceram. Soc. **75**, 958–963 (1992)
10. Harris, D.C.: Durable 3–5 μm transmitting infrared window material. Infrared. Phys. Technol. **39**, 185–201 (1998)
11. Ghosh, A., White, K.W., Jenkins, M.G., Kobayashi, A.S., Bradt, R.C.: Fracture resistance of a transparent magnesium aluminate spinel. J. Am. Ceram. Soc. **74**, 1624–1630 (1991)
12. Smith, S.M., Scattergood, R.O.: Crack–shape effects for indentation fracture toughness measurements. J. Am. Ceram. Soc. **75**, 305–315 (1992)
13. Stewart, R.L., Bradt, R.C.: Fracture of polycrystalline $MgAl_2O_4$. J. Am. Ceram. Soc. **63**, 619–623 (1980)
14. White, K.W., Kelkar, G.P.: Fracture mechanisms of a coarse-grained, transparent $MgAl_2O_4$ at elevated temperatures. J. Am. Ceram. Soc. **75**, 3440–3444 (1992)

15. Baudin, C., Martinez, R., Pena, P.: High-temperature mechanical behavior of stoichiometric magnesium spinel. J. Am. Ceram. Soc. **78**, 1857–1862 (1995)
16. Boniecki, M.: Superplasticity phenomenon in selected oxide ceramics. Prace. ITME. **58**: 112 pages (in Polish) (2008)
17. Masumura, R.A., Hazzledine, P.M., Pande, C.S.: Yield stress of fine grained materials. Acta. Mater. **46**, 4527–4534 (1998)
18. Armstrong, R.W.: Grain size dependent alumina fracture mechanics stress intensity. Int. J. Refr. Met. Hard. Mater. **19**, 251–255 (2001)
19. Ünal, Ö., Akinc, M.: Compressive properties of yttrium oxide. J. Am. Ceram. Soc. **79**, 805–808 (1996)
20. Clarke, D.R., Faber, K.T.: Fracture of ceramics and glasses. J. Phys. Chem. Solids. **48**, 1115–1157 (1987)

Analysis of the Effect of the Elliptical Ratio in Tubular Energy Absorbers Under Quasi-Static Conditions

Ahmad Baroutaji and Abdul Ghani Olabi

Abstract Tubular systems are proposed to use as energy absorber because they are cheap and they are easy to manufacture; recently some researchers use elliptical tubes as energy absorbers. In this work, the influence of elliptical ratio on energy absorption capability and load carrying capacity and stresses of mild steel elliptical tubes has been investigated both experimentally and numerically. The experimental analysis was conducted by using a Zwick Type BT1-FB050TN testing machine. In addition to the experimental work, a computational analysis using ANSYS is made to predict the loading and response of such tubes where series of models were performed with elliptical ratios ranging from 0.5–1.5. Comparison of experimental and numerical force- and energy-displacement responses are presented.

Keywords Energy absorber · ANSYS · Quasi-static analysis

1 Introduction

An energy absorber can be identified as a mechanism that converts, totally or partially, kinetic energy of a mass moving at high speed into another form of energy during the collision. The energy transformed is either reversible, like elastic strain energy in solids, or irreversible, like plastic deformation energy.

A. Baroutaji (✉) · A. G. Olabi
School of Mechanical and Manufacturing Engineering, Dublin City University, Dublin, Ireland
e-mail: ahmad.baroutaji2@mail.dcu.ie

A. G. Olabi
e-mail: abdul.olabi@dcu.ie

A. Öchsner et al. (eds.), *Materials with Complex Behaviour II*,
Advanced Structured Materials 16, DOI: 10.1007/978-3-642-22700-4_19,

Energy absorption through material deformation has been investigated over the last three decades, particularly in the form of tubular systems.

Circular or square sectioned tubes are widely used as structural elements due to their prevalent occurrence and because they are readily available for selection in the design process.

A number of authors studied the lateral compression of a single circular tube and its strain hardening phenomena [1]. These researchers were among the first to study such problems and each one of them gave a slightly different deformation mechanism for the compression of a tube between rigid flat platens.

A rigid linear strain hardening material model was used to investigate the effect of strain hardening phenomena in order to predict the force-deflection response of tubes under lateral compression more accurately [2]. The collapse behavior of elliptical tubes under quasi-static axial or lateral compression has been studied by Wu and Carne [3, 4]. They carried out a study on metallic elliptical cross-section tubes and the crushing behavior of braced elliptical tube was investigated. Their treatment was used in the field of crashworthiness to decrease the effects of collision in roadside safety applications. New systems of elliptical energy absorbers were proposed by Olabi et al. [5]. They used nested elliptical tubes of varying diameter to build their design. They investigated the quasi-static lateral compression of these systems between rigid platens travelling with a velocity of 3 mm/min to ensure that no dynamic effects were affected. Experimental and numerical study were conducted on these systems to see their responses and it was found that these systems are well suited to applications where space or volume restrictions are an important design consideration without compromising energy absorbing requirements.

The importance of conducting quasi-static analysis for energy absorption systems is for predicting the behavior of these systems under dynamic load since the same pre-dominant geometrical effects will also occur under dynamic loading conditions and also for investigating the effects of strain rate on the system when applying dynamic load.

The primary aim of this study is to examine the effects of changing the geometry shape of metallic circular tubes into an ellipse on the energy absorbed by this tube.

2 Numerical Procedure

Computational analyses and finite element models of these energy absorbers were developed by using the implicit finite element code ANSYS.

Figure 1 displays the finite element model used to simulate the responses of these absorbers. In this model, 3D-structural solid element (solid 45) which has eight nodes with large strain, large deflection and plasticity capabilities was used to model the circular and elliptical tube profile. The moving mass was modeled as

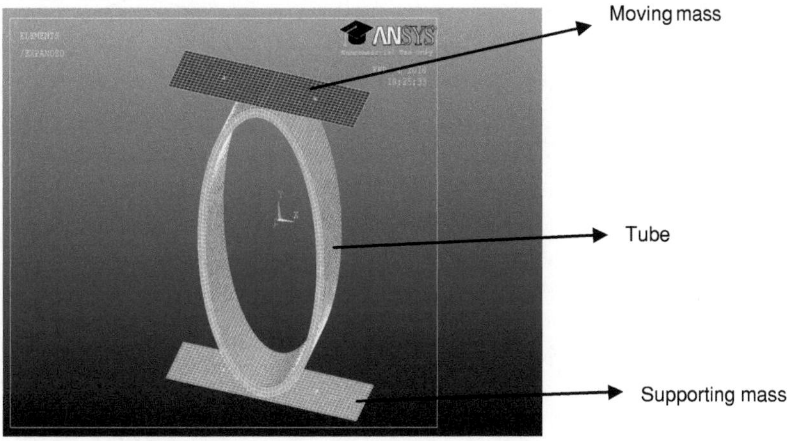

Fig. 1 Finite element model for elliptical tube with moving and supporting mass

a rigid body and constrained to move vertically over the y axis. The supporting mass was defined as a rigid body as well and constrained to be fix in its place.

The numerical models have three nonlinear features.

- Nonlinear material model where a bilinear isotropic hardening material model was employed to define the material of the elliptical and circular tubes. A value of 500 MPa was assigned to the yield stress of the elliptical and circular tubes and a non-zero value of 1500 MPa for the plastic modulus was selected to depict the plastic portion of the stress–strain curve. The same value of plastic modulus was used by [6].
- An augmented Lagrangian penalty option describes the contact algorithm used to capture the models changing contact status throughout the deformation stroke.
- Finally, the third non-linearity being large deflection large strain deformation, this feature should be included since changes in volume will happen due to applying large displacements.

Series of models were performed with elliptical ratios ranging from 0.5–1.5. The elliptical ratio is given by

$$r = \frac{D_1}{D_2} \tag{1}$$

where D_1, D_2 are horizontal and vertical semi-axes of the elliptical tube and r is the elliptical ratio.

Figure 2 displays the horizontal and vertical semi-axes of the elliptical tube.

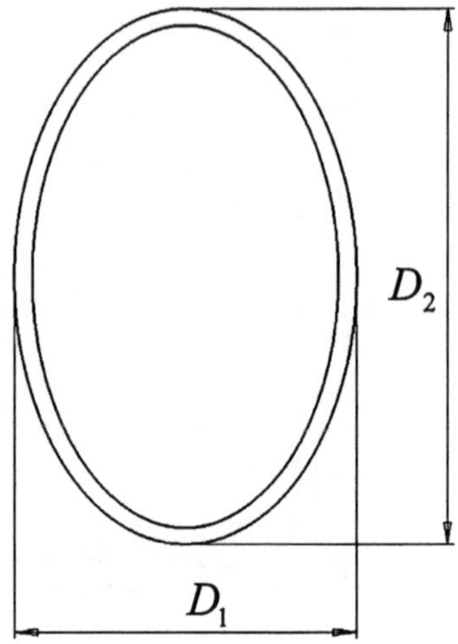

Fig. 2 Elliptical tube with its dimensions

3 Numerical Results

3.1 Analysis the Effect of Elliptical Ratio on the Load-Displacement Curve

Figure 3 displays the force-displacement curve for elliptical tubes with different elliptical ratios. It can be seen from Fig. 3 that the tubes resistance increase linearly at the beginning of the compression process until reaching the yielding point when it change in a non-linear manner until the end of the process.

It was observed that the elliptical ratio has a great influence on the load-displacement curve where by reducing the elliptical ratio the load-displacement curve will has a peak at the beginning of the compression process and with increasing the elliptical ratio this peak load moves to the end of compression process.

3.2 Analysis of the Effect of Elliptical Ratio on the Energy-Displacement Curve

Figure 4 shows the energy-displacement curve for each of the elliptical tubes. It can be seen from the figure that all tubes have a non-linear relation in terms of energy with displacement. The energy response curve is more stable for the tubes with an elliptical ratio higher than 1.

On the other hand the energy absorbed by tubes with a higher elliptical ratio is smaller at any value of displacement than the energy absorbed by the tube with a lower elliptical ratio, but at the end of the compression process the value of energy is the same for all tubes.

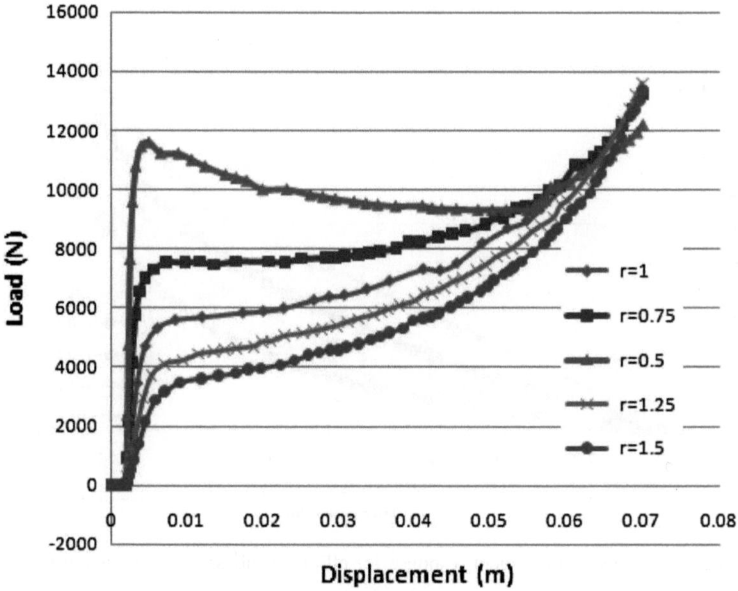

Fig. 3 Load-displacement response for elliptical tubes with various elliptical ratios

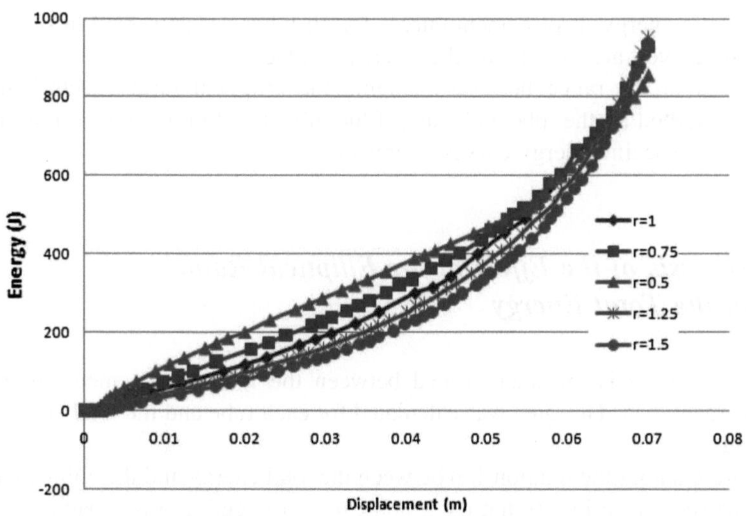

Fig. 4 Energy-displacement response for elliptical tubes with various elliptical ratios

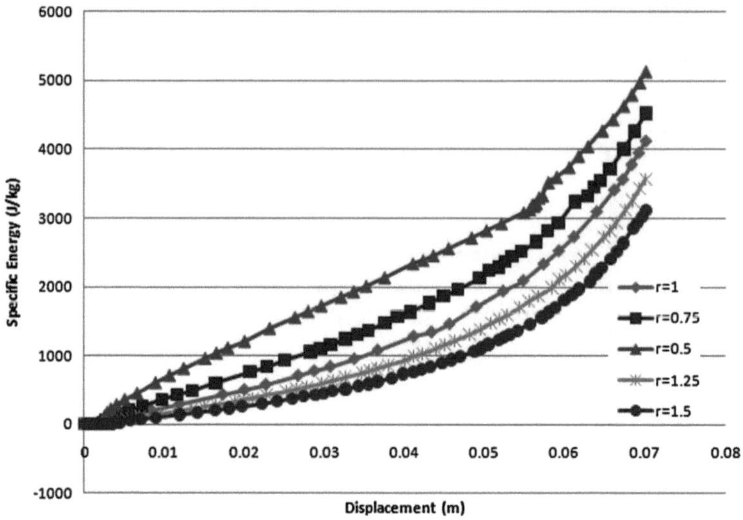

Fig. 5 Specific energy-displacement response for elliptical tubes with various elliptical ratios

3.3 Analysis of the Effect of the Elliptical Ratio on the Specific Energy-Displacement Curve

The specific energy-displacement curves for all tubes are presented in Fig. 5.

These curves are similar to the energy-displacement curves but with some differences in the range because changing the elliptical ratio affects both the energy absorbed by the tube and mass of the tube therefore the effect of elliptical ratio on the specific energy curves is greater.

3.4 Analysis of the Effect of the Elliptical Ratio on the Total Energy

The total energy is the area located between the load-displacement curve and displacement axis. This area was calculated for each tube and the total energy was obtained.

Figure 6 shows the relationship between the total energy and the elliptical ratio. It can be seen from Fig. 6 that the graph can be divided into two parts. The first one is for tubes with an elliptical ratio lower than ($r = 1$). This part has a nearly linear relationship between energy and elliptical ratio where the energy decreases when the elliptical ratio increases. The second part is for tubes with elliptical ratio higher than ($r = 1$). In this part the energy also decreases when the elliptical ratio increases but the slope is different from the first range.

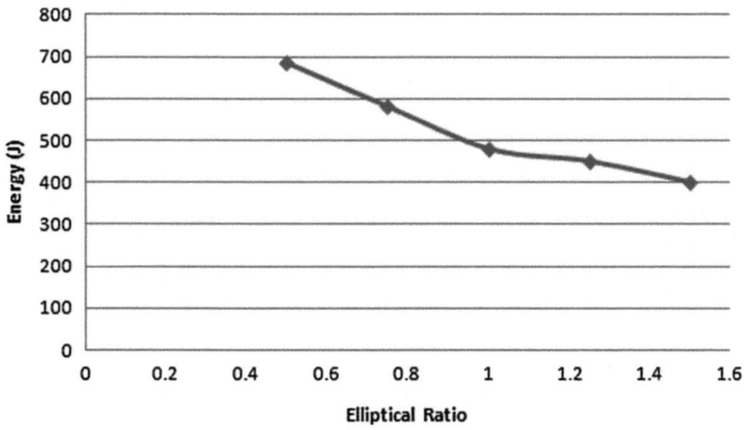

Fig. 6 Energy—elliptical ratio curve

3.5 Analysis of the Effect of the Elliptical Ratio on the Total Specific Energy

The total specific energy is the total energy absorbed by the tube divided by the mass of this tube.

$$e = \frac{E}{m} \tag{2}$$

where e is specific energy absorbed by tube, E is the total energy absorbed by tube and m is the mass of the tube.

Figure 7 displays the relationship between the specific energy absorption and the elliptical ratio.

It can be seen from Fig. 7 that the specific energy also decreases when elliptical ratio increases.

3.6 Analysis of the Effect of Elliptical Ratio on the Stress and Stress Distribution

The value of stresses at the critical point (the point which has a maximum stress) has been affected by changing the elliptical ratio. Figure 8 shows the relationship between the stress at the critical point and displacement of the rigid body for all tubes.

It is clear that the tubes which have lower elliptical ratio have a higher value of stresses at the critical point. It is clear also that the elastic range (the range before the yield stress) is not affected by changing the elliptical ratio.

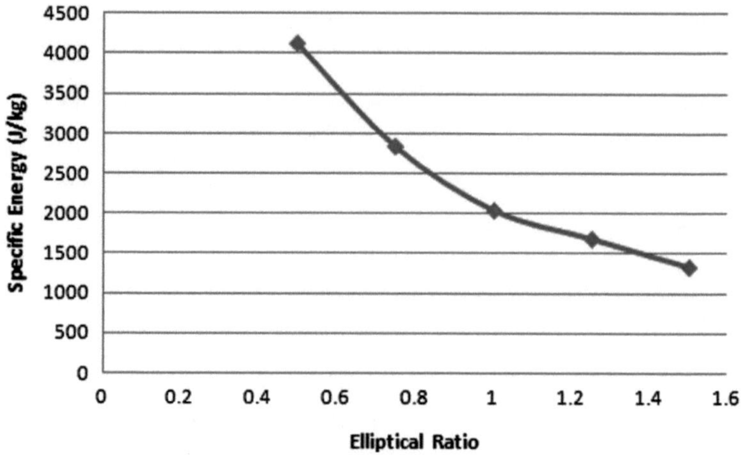

Fig. 7 Specific energy—elliptical ratio curve

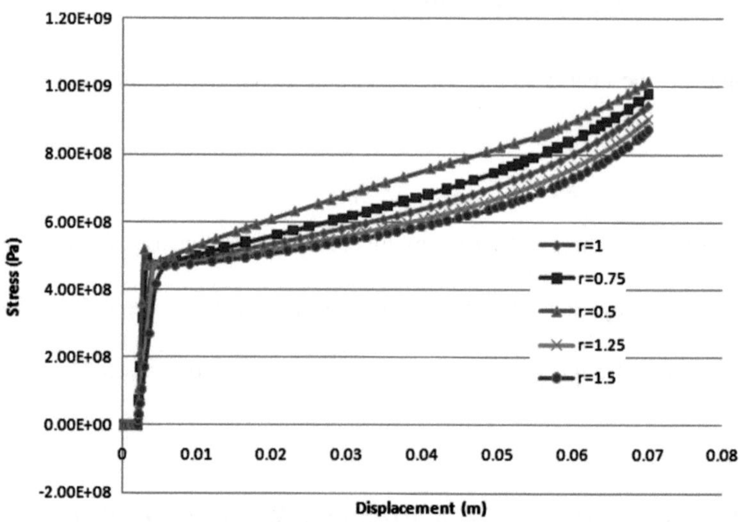

Fig. 8 Stress at critical point-displacement response for elliptical tubes with various elliptical ratios

Figure 9 illustrates the stress distribution of an elliptical tube with an elliptical ratio $r = 1.25$ and r = 1.5. It can be seen that the elliptical ratio does not has any effect on the stress distribution.

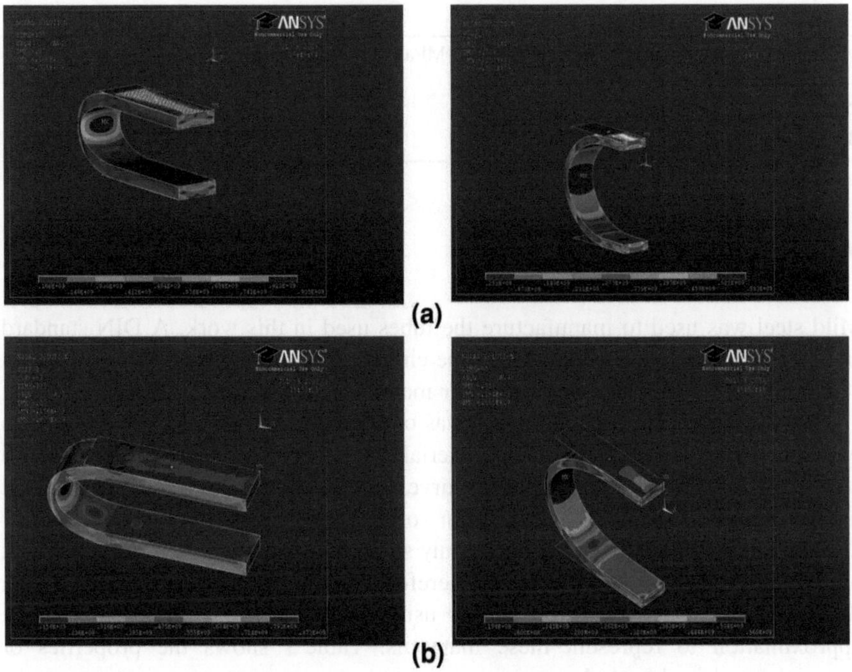

Fig. 9 Graphic display of stress distribution for two elliptical tubes **a** elliptical ratio (r = 1.25) **b** elliptical ratio (r = 1.5)

4 Experimental Work

4.1 Experimental Set up

The quasi-static tests of the elliptical and circular samples were conducted by using the Zwick Type BT1-FB050TN testing machine. The loading frame has a maximum capacity of 50 kN.

To simulate the quasi-static conditions a displacement rate of 10 mm/min was applied to the moving head whereas velocities between 0.5 and 15 mm/min have been used by many researchers to represent the quasi-static lateral compression of tubes between various indenters [7, 8]. Due to unavailability of elliptical tube in the market the elliptical tubes were prepared in the workshops where many samples of elliptical tubes with elliptical ratio (r = 0.5) were obtained by cutting metal sheets of mild steel.

Table 1 Material properties

Material type	Tube shape	Yield stress (MPa)	Young's modulus (GPa)	Tangent modulus
DIN 2393 ST 37.2	Circular	725	156	1500
BS 970 EN3B	Elliptical	500	216	1500

4.2 Material Properties

Mild steel was used to manufacture the tubes used in this work. A DIN standard (DIN 2393 ST 37.2) was selected for the circular tubes and metal sheets of British standard (BS 970 EN3B) were used for manufacturing elliptical tubes.

The true static stress–strain curve was obtained from a tensile test. More than two dog bone samples for each material were prepared in order to confirm consistent results. The stress–strain curves obtained from the samples showed highly rare phenomena in which strain softening happened almost immediately after yielding with no apparent sign of any strain hardening which is not consistent with normal behavior of mild steel. Therefore, the material properties of tubes in the numerical study were simulated by using a bilinear stress–strain curve as an approximation to represent these materials. Table 1 shows the properties of materials used in this study.

5 Experimental Results and Validation

The Fig. 10 illustrates the comparison between the numerical results obtained with ANSYS and experimental results. The finite element models used in this section have different material properties in order to fit the materials used in the experiments whereas the numerical section assumed that all models have the same material properties.

It can be seen that numerical curve gives an under prediction for the crash force followed by over prediction for the rest of the process. It can be seen also that the results compared well in terms of energy so it is possible to consider the results of the numerical work validated.

6 Analysis of the Energy Absorption Characteristics

The effectiveness of energy absorber systems can be described by many indictors such as crush efficiency, energy efficiency, specific energy absorption capacity and weight effectiveness.

Table 2 explains the various indictors mentioned above.

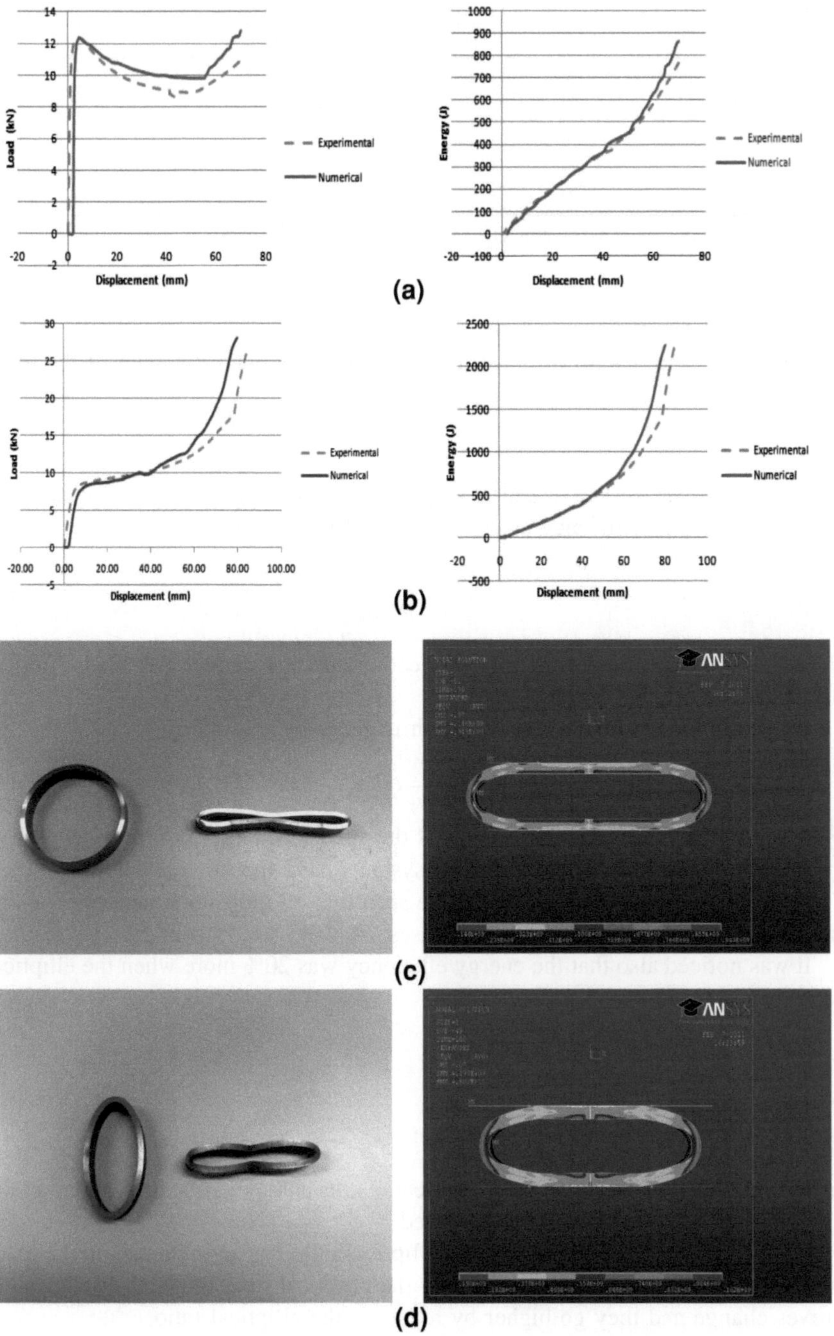

Fig. 10 **a** Experimental results and numerical solutions for a circular tube **b** Experimental results and numerical solutions for an elliptical tube **c** Initial and final stages of circular tube **d** Initial and final stages of elliptical tube

Table 2 Indicators for describing the effectiveness of energy absorber systems

r	e_g (%)	e_g (%)	W_{eff}
0.5	70	56	2885
0.75	70	44	1985
1	70	35	1427
1.25	70	33	314
1.5	70	30	280

e_g: Crush efficiency e_E: Energy efficiency W_{eff}: Weight effectiveness

$e_g$is the crush efficiency which was defined by [9] as the stroke length divided by a characteristic length of the structure such as the outer diameter of the tube.

$$e_g = \frac{L}{D} \tag{3}$$

Where $L$is the stroke length and D is the major axis length of the elliptical tube.

$e_E$is the energy efficiency as described by [9] and given by

$$e_E = \frac{A}{P_d * L_0} \tag{4}$$

Where A is the area under the force deflection curve, $P_d$is the peak load observed and L_0 is the original length.

$W_{eff}$is the weight effectiveness, which is given by

$$W_{eff} = e_g * S_s^c \tag{5}$$

Where $e_g$is the crush efficiency and $S_s^c$is the specific energy absorption capacity.

It can be seen from Table 2 that all systems have the same crush efficiency.

Upon inspection of Table 2, it can be seen that all indicators mentioned above increase when the elliptical ratio decreases.

It was noticed also that the energy efficiency was 20% more when the elliptical ratio was reduced to 0.5.

7 Discussion

In this section, depending on the numerical results and experimental validation the effects of elliptical ratio will be discussed.

It was noticed in the Fig. 3 that the elliptical ratio has an influence on the load-displacement curves where by changing the elliptical ratio the load-displacement curves change and they go higher by reducing the elliptical ratio.

The increase in force observed in the elliptical tubes which have elliptical ratio lower than 1 is because of the increase in the initial collapse load where the value of initial collapse load increases when the horizontal dimension D_1 of the elliptical is decreased. Once the load increases the energy absorbed by the tube also increases and that is clear in Fig. 4 where the elliptical tubes with elliptical ratio lower than 1 absorb more in terms of energy and specific energy. It was noticed that all tubes have non-linear responses of energy that is because the loads also have non-linear responses. Figures 5 and 6 show the effect of elliptical ratio on the total energy and total specific energy respectively. The increase in force in the elliptical tubes which have an elliptical ratio lower than 1 make the area located under the load-displacement curve greater and this leads to an increase in the total and specific energy absorbed by these tubes. When the elliptical ratio increases the energy and specific energy absorbed by the tubes decrease where the area located under the load-displacement curve decreases. The effect of elliptical ratio on the stress and stress distribution were presented in Sect. 3.6. It was discovered that changing of elliptical ratio does not affect on the stress distribution over the elliptical tubes so it does not affect on the location of fracture point, but it affect on the value of the stress at critical point where at any displacement the stress is greater in the tubes which have lower elliptical ratio so the elliptical tubes which have lower elliptical ratio need lower displacement to fracture. In Sect. 6 an analysis on the characteristics of energy absorption systems was presented, it was found that all tubes have the same value of crush efficiency because of applying the same displacement to all tubes. It was found also that both energy efficiency and weight effectiveness increase when elliptical ratio decreases that is because of increasing the energy absorbed by these tubes.

It was found in Sect. 5 that the numerical analysis gave an under prediction for the crash force followed by over prediction for the rest of the process. This difference between the results is because of using the bilinear material model which considers the flow of stress increase as strain increase but this does not happen in real condition. It was found also that the numerical code could not predict exactly where and when the plastic hinges should generate. Figures 10c and d show the inability of numerical code to capture hinges at the contact points A, B, C and D.

8 Conclusion

A number of experiments and numerical simulations were done to determine the effects of changing the geometry of a circular tube into an elliptical tube.

The numerical method gave a little over predictions in the collapse of the tubes because of using the bilinear material model and inability of finite element code (ANSYS) to predict exactly when and where plastic hinges should generate.

The numerical compared well with experimental results especially in terms of specific energy.

It has been found also that changing the elliptical ratio of the tube alter the load-deflection curve and this leads to a change of the energy absorbed by the tube. Changing the geometrical shape of the tube leads to change the volume of this tube and thus the mass of the tube change also and this leads to a change of the specific energy absorbed by the tube.

By reducing the elliptical ratio to 0.5 the tube absorbs 43.3% more energy and the system will gain 102% more in terms of specific energy.

The indictors which express the performance of energy absorber systems were calculated, it was found that all these indicators increase when the elliptical ratio decreases.

References

1. Burton, R., Craig, J.: An Investigation into the Energy Absorbing Properties of Metal Tubes Loaded in the Transverse Direction. BSc (Eng) Report. University of Bristol, England (1963)
2. Reid, S., Reddy, T.Y.: Effect of strain hardening on the lateral compression of tubes between rigid plates. Int. J. Solids Struct. **14**(3), 213–225 (1978)
3. Wu, L., Carney, J.F.: Experimental analyses of collapse behaviors of braced elliptical tubes under lateral compression. Int. J. Mech. Sci. **40**(8), 761–777 (1998)
4. Wu, L., Carney, J.F.: Initial collapse of braced elliptical tubes under lateral compression. Int. J. Mech. Sci. **39**(9), 1023–1036 (1997)
5. Olabi, A., Morris, E., Hashmi, M., Gilchrist, M.: Optimised design of nested oblong tube energy absorbers under lateral impact loading. Int. J. Impact Eng. **35**(1), 10–26 (2008)
6. Reddy, T.Y., Reid, S.: Phenomena associated with the crushing of metal tubes between rigid plates. Int. J. Solids Struct. **16**(6), 545–562 (1980)
7. Reddy, Y.: Lateral compression of tubes and tube-systems with side constraints. Int. J. Mech. Sci. **21**(3), 187–199 (1979)
8. Morris, E., Olabi, A., Hashmi, M.: Analysis of nested tube type energy absorbers with different indenters and exterior constraints. Thin-Walled Struct. **44**(8), 872–885 (2006)
9. Thornton, P., Mahmood, H., Magee, C.: Energy absorption by structural collapse. Struct. Crashworthiness 96–117 (1983)

Dynamic Response Analysis of Thermoplastic Polyurethane

V. Fontanari, M. Avalle, C. Migliaresi, L. Peroni and B. D. Monelli

Abstract The proper modeling of the strain-rate dependence of rubber-like polymers is of high importance in energy-absorbing systems design. There are several possibilities for modeling the material dynamic response of such materials, but the field of applicability, drawbacks and critical aspects of each developed approach are not definitely assessed yet. The present chapter discusses these topics for Thermoplastic PolyUrethane (TPU), being TPU one of the most used shock-absorber materials. The qualification of the most promising approaches is numerically established by performing Finite Element Analysis (FEA) of tensile and compressive tests at different strain-rates. For the selected constitutive models, the corresponding constitutive parameters are experimentally determined and then implemented in FE models. The assessment of each material model is then carried out by comparing the predicted response with the experimental one.

V. Fontanari (✉) · C. Migliaresi · B. D. Monelli
Department of Materials Engineering and Industrial Technologies,
University of Trento, Via Mesiano, 77–38121 Trento TN, Italy
e-mail: vigilio.fontanari@unitn.it

C. Migliaresi
e-mail: claudio.migliaresi@unitn.it

B. D. Monelli
e-mail: bernardo.monelli@unitn.it

M. Avalle · L. Peroni
Department of Mechanical Engineering, Politecnico di Torino,
Piazza Sant'Eusebio, 5–13100 Vercelli VC, Italy
e-mail: massimiliano.avalle@polito.it

L. Peroni
e-mail: lorenzo.peroni@polito.it

A. Öchsner et al. (eds.), *Materials with Complex Behaviour II*,
Advanced Structured Materials 16, DOI: 10.1007/978-3-642-22700-4_20,

Keywords Impact · Nonlinear hyperelasticity · Nonlinear viscoelasticity · TPU · FEA

1 Introduction

Nowadays, polymeric materials are commonly employed in shock-absorbing systems. This class of materials presents peculiar constitutive properties, mainly linked to the strain-rate dependence of their mechanical response. The design of such energy-absorbing devices must be carried out after thorough determination of the energy absorbed during the impact. This quantity is tightly related to the material's behaviour. The knowledge of the tensile and compressive material response as well as their strain-rate sensitivity is of paramount importance: during impact indeed, the material may experience both tensile and compressive stress states characterized by time-varying strain rates. Currently, tensile impact and weight-drop tests constitute the most common ways to assess the material's shock-absorbing capability under tensile and compressive conditions, respectively. However, the experimental approach is often expensive and time-consuming, while the use of predictive analytical/numerical models undoubtedly represents a valid alternative for estimating this quantity.

In rubber-like Thermoplastic PolyUrethane (TPU), the strain rate produces significant effects in all the three stages of the material deformation process (Fig. 1), depending upon the loading condition (i.e. tensile or compressive) [1–6]. Strain rate effects are particularly evident on the initial slope of the curve and on the extension of the characteristic plateau. This mechanical behavior arise as a consequence of the strain-rate dependence of the polymeric chains rearrangement phenomena necessary for bearing the applied loads [7].

Several constitutive models accounting for strain-rate effects can be found in literature [2, 8–17]. However, despite the large amount of proposed models, the application range and limits as well as the critical issues of each approach have not been fully explained yet. The present work addresses these aspects by comparing the predictive capabilities of different constitutive approaches, considered as the most promising for studying hyperelastic behavior of a TPU widely used for producing shock-absorber devices. To this purpose, tensile and compressive tests have been carried out at different strain rates and the experimental results have been compared with the outcomes of computational models ad hoc developed to simulate the performed mechanical tests.

2 Theoretical Background: Nonlinear Hyper-Viscoelastic Constitutive Model

The constitutive models, up to now proposed in the scientific literature, can be classified into three macro-groups: empirical [8, 9], phenomenological [10–13],

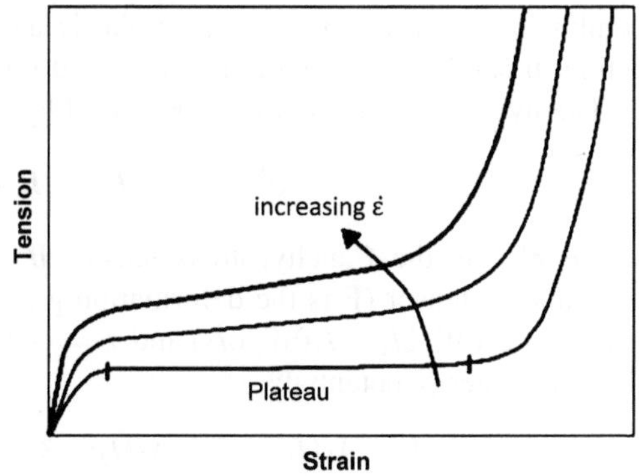

Fig. 1 Schematic representation of the strain-rate effects onto the stress–strain curves ($\sigma - \varepsilon$) of rubber-like polymeric materials

and micro-mechanical [14–17]. Even though the models belonging to the first two groups explicitly represent the material's response and its energy absorption capability, only the micro-mechanical models can be used to establish the stress strain evolution produced by the loading conditions into the material. From this point of view, the micro-mechanical models offer the possibility, at least theoretically, of directly operating on the tuning of material properties in order to optimize its mechanical response. Anyway, the second and the third of the three aforementioned groups offer the largest amount of information on the material response towards the comprehension of the phenomena observed during dynamic loading; for these reasons, the attention will be focused on these two latter groups, while the purely empirical models have been discarded, since their usefulness is strictly confined to the specific material under examination.

Even though elastomeric TPU only recently found wide use in many engineering applications, its constitutive properties and the role played by the strain-rate in the definition of the constitutive law have been extensively studied [1–6]. The peculiar characteristic of this class of materials is the concomitant occurrence in their mechanical response of a hyperelastic and a viscoelastic component, both of them nonlinear and strain-rate sensitive. Therefore, only constitutive models accounting for these two components are reputed to be able to correctly reproduce the mechanical behaviour of these materials. In the following, two of the most promising models developed in the literature: the hyper-viscoelastic models proposed by Yang et al. [10] and by Bergstrom and Boyce [14] will be presented. The study has not been extended to the remaining cited references, since these are derived from the two aforementioned constitutive models.

2.1 Yang Nonlinear Hyper-Viscoelastic Constitutive Model

The Yang nonlinear hyper-viscoelastic model assumes the material as homogeneous, isotropic and incompressible and the macroscopic behavior to be a combination of a quasi-static hyperelastic and a nonlinear viscoelastic response acting in parallel. The model is thus based on a phenomenological approach.

The hyperelastic response is expressed by the Mooney–Rivlin model [18, 19]

$$\boldsymbol{\sigma}^{hyp} = -p_e \boldsymbol{I} + \alpha_1 \boldsymbol{B} + \alpha_2 \boldsymbol{B} \cdot \boldsymbol{B} \tag{1}$$

where $\boldsymbol{\sigma}^{hyp}$ is the Cauchy stress tensor, $\boldsymbol{B} = \boldsymbol{F} \cdot \boldsymbol{F}^T$ is the left Cauchy-Green deformation tensor (**F** is the deformation gradient), p_e is the hydrostatic pressure and $\alpha_1 = 2(\partial W/\partial I_1 + I_1 \partial W/\partial I_2)$ and $\alpha_2 = -2\partial W/\partial I_2$ are the terms deriving from the strain energy potential:

$$W = A_1(I_1 - 3) + A_2(I_2 - 3) + A_3(I_1 - 3)(I_2 - 3) \tag{2}$$

where $I_1 = tr(\boldsymbol{B})$, $I_2 = \left[I_1^2 - tr\left(\boldsymbol{B}^2\right)\right]$ and $I_3 = det(\boldsymbol{B})$ are the three invariants respectively. The parameters A_1, A_2 and A_3 are material constants that can be determined by fitting the $\sigma - \varepsilon$ curve of the material (tension/compression) under quasi-static loading conditions.

The viscoelastic component $(\boldsymbol{\sigma}^v)$ is given by the T-BKZ model [20–22], expressing the Cauchy stress tensor $\boldsymbol{\sigma}^v$ as follows:

$$\boldsymbol{\sigma}^v = -p^v + \boldsymbol{F}(t) \cdot \boldsymbol{\Omega}_\tau^t = -\infty\{\boldsymbol{C}(\tau)\} \cdot \boldsymbol{F}^T(t) \tag{3}$$

where p^v is usually an arbitrary pressure, $\boldsymbol{C} = \boldsymbol{F}^T \cdot \boldsymbol{F}$ is the right Cauchy-Green deformation tensor, $\boldsymbol{\Omega}$ is a matrix functional which describes the effect of strain history on stress.

According to the authors of the model, $\boldsymbol{\Omega}$ takes the following form:

$$\boldsymbol{\Omega}_{\tau=0}^t\{\boldsymbol{C}(\tau)\} = \int_0^t \varphi(I_1, I_2) m(t - \tau) \dot{\boldsymbol{E}}(\tau) d\tau \tag{4}$$

where $\dot{\boldsymbol{E}} = \left(\dot{\boldsymbol{F}}^T \cdot \boldsymbol{F} + \boldsymbol{F}^T \cdot \dot{\boldsymbol{F}}\right)$ is the strain rate, while φ and $m(t - \tau)$ are the damping and the relaxation function, respectively. In order to restrain the number of parameters, in the present model, the following expressions:

$$\varphi = A_4 + A_5\left[I_2'(\tau) - 3\right] m(t - \tau) = e^{-(t-\tau)/A_6} \tag{5}$$

have been assumed for the two functions defining the constitutive functional, where $I_2'\left(I_2' = I_2\right)$ is the second invariant of the $\boldsymbol{C}$ tensor and A_4, A_5, and A_6 are a set of material parameters that can be determined by simultaneously fitting two or more $\sigma - \varepsilon$ curves corresponding to different strain rates through a nonlinear multi-fitting procedure [10, 23]. The material constitutive law is obtained by combining Eqs. (1–5). In the monoaxial case, being $\sigma_{yy}^{hyp} = \sigma_{zz}^{hyp} = 0$ and $\sigma_{yy} = \sigma_{zz} = 0$, this takes the following form:

$$\sigma_{xx} = \sigma_{xx}^{hyp} + \lambda^2 \int_0^t \lambda[A_4 + A_5(I_2 - 3)]e^{-(t-\tau)/A_6}\dot{\lambda}d\tau + \frac{1}{2}\lambda^{-1} \int_0^t \lambda^{-2}[A_4 + A_5(I_2 - 3)]e^{-(t-\tau)/A_6}\dot{\lambda}d\tau \quad (6)$$

in this case the stretch term $\lambda(\lambda = 1 + \varepsilon)$ is used as a measure of the material strain. In the Yang model, the material behaviour is thus univocally determined once the set of constitutive parameters $A_1, \ldots, A_6$ is defined. The first three parameters identify the hyperelastic material response and can be calculated by analyzing the quasi-static uniaxial material response according to the procedure devised by the authors. The remaining parameters determine the viscoelastic material response whose strain-rate dependence is incorporated in the constant A_6. The estimation of these parameters can be carried out by analyzing the variation of the uniaxial material response within a given strain-rate interval.

Although the present model is relatively easy in its conception and implementation, it should be noted that such an approach must face two important critical aspects. First, the model is not able to simultaneously interpret tensile and compressive response: this can be a major limit if the material under investigation is characterized by a tensile response different from the compressive one. Secondly, the constitutive parameters estimation is based on nonlinear multi-fitting procedures whose results are strictly correlated to the initial guesses as well as to the adopted minimum search algorithm, as extensively discussed in literature. In other terms, the material behavior may be described by two or more sets of constitutive parameters, thus frustrating every attempt to establish a correlation between the characteristics of this material class and the above-stated parameters.

2.2 Bergstrom–Boyce Nonlinear Hyper-Viscoelastic Constitutive Model

Bergstrom and Boyce [14–16] proposed a representation of the hyper-viscoelastic material behaviour on the basis of microstructure considerations. The basic hypotheses are the same as those of the Yang's model. Even in this case, the material is assumed to be isotropic, homogeneous and incompressible, its response is subdivided into a hyperelastic and a viscoelastic component acting in parallel. The first is meant to interpret the material behavior under equilibrium conditions, the second the material time dependence.

The hyperelastic component is schematized according to the Arruda–Boyce model [24]. In other terms, the quasi-static material response is referred to that of a cubic cell containing eight Langevin chains [24] linking the cell centre with its vertices. If λ_i indicates, as usual, the network stretch along the ith direction, the following correlation [24, 25],

$$\sigma_i^1 = C_R^1 \sqrt{N^1} \frac{\left(\lambda_i^1\right)^2 - \left(\lambda_{chain}^1\right)^2}{\lambda_{chain}^1} L^{-1}\left(\frac{\lambda_{chain}^1}{\sqrt{N^1}}\right) + Bln\left(I_3^1\right) \tag{7}$$

yields the link between the stress along this direction and the corresponding stretch. In Eq. (7), C_R^1, $\sqrt{N^1}$ and B represent the initial elastic modulus, the boundary network stretch and the bulk modulus, respectively. I_3^1 is the third invariant of the deformation tensor while $L^{-1}(x)$ is the inverse of the Langevin function $L(x) = \cot(x) - 1/x$.

According to experimental evidences [14], the viscoelastic component is further subdivided into two contributions, a hyperelastic one described by the Arruda–Boyce model, and a viscous one, both of them acting in series. The latter one is introduced in order to model the strain relaxation in the hyperelastic element, thus capturing the material time dependence. While for the hyperelastic network contribution the correlation between stress and corresponding stretch is immediately established, being this formally identical to Eq. (7),

$$\sigma_i^2 = C_R^2 \sqrt{N^2} \frac{\left(\lambda_i^2\right)^2 - \left(\lambda_{chain}^2\right)^2}{\lambda_{chain}^2} L^{-1}\left(\frac{\lambda_{chain}^2}{\sqrt{N^2}}\right) + Bln\left(I_3^2\right) \tag{8}$$

the formulation of a constitutive time dependent model on the basis of microstructure considerations is pretty difficult. However, if the time dependence is assumed to be caused by reptational motions of the fully or partially inactive polymeric chains, it is possible to model the strain relaxation according to the following model [14],

$$\dot{\gamma} = C_1\left(\lambda_{chain}^2 - 1\right)^{C_2}\left(\frac{\tau_2}{\hat{\tau}_2}\right)^m \tag{9}$$

where $(\dot{\gamma})$ is the effective creep rate, (τ_2) the equivalent shear stress acting on the hyperelastic network. In Eq. (9) C_2 and m are material constants, as well as $C_1/\hat{\tau}_2^m = \hat{C}_1$, while $\lambda_{chain}^2 = I_1^2/3$, being I_1^2 the first invariant of the strain tensor corresponding to the hyperelastic model of the second network. In the present model, the strain-rate dependence of the constitutive law is thus incorporated in this latter parameters set.

In conclusion, once the material parameters $\left(C_R^1, \sqrt{N^1}, C_R^2, \sqrt{N^2}, B, \hat{C}_1, C_2, m\right)$ have been determined, the material behavior is univocally determined by the equation set (7–9). The first five parameters $\left(C_R^1, \sqrt{N^1}, C_R^2, \sqrt{N^2}, B\right)$ can be experimentally assessed, by analyzing the quasi-static uniaxial material response and its behavior during loading/unloading uniaxial cycle at a given strain-rate (for the estimation of these parameters, refer to the appendix in [14]); the evaluation of the remaining parameters, on the contrary, can be only performed through trial and error techniques based on the comparison of the

Table 1 Physical and mechanical properties of APILON 52

Density [kg/m^3]	Glass transition temperature [°C]	Hardness [shore A]	Load at 100% [MPa]	Load at 300% [MPa]	Ultimate tensile strength [MPa]	Strain at break [%]
1,185	30	63	2.2	3.8	27	850

uniaxial material response in a given strain-rate range with that predicted by a numerical model able to simulate the considered experimental test and implementing the constitutive law containing the parameters to be estimated.

3 Experimental Characterization

The implementation of the theoretical models described in the previous section needs the determination of material constants which can be obtained by carrying out a proper experimental campaign. Since any shock-absorber device can undergo both tensile and compressive stress states during the impact transient, it is apparent that both tensile and compressive responses must be experimentally accounted for. Due to the widespread use of polyurethane rubbers in shock-absorbing devices, the investigation was focused onto one of the most popular elastomeric thermoplastic polyurethane (APILON 52, API S.p.A., Italy), which main physical and mechanical properties are collected in Table 1.

The constitutive behavior was determined by performing standard tensile and compressive tests in which the strain-rate was varied between 10^{-3} and 10^2 s^{-1}, thus making it possible to establish both the quasi-static and dynamic material behavior at medium–low strain-rates. Although the considered strain-rates range does not cover all the potential impact conditions, it is representative for a large number of engineering applications. The results of the experimental campaign are reported in the following subsections.

3.1 Tensile Response

The experimental true stress-true strain $(\sigma_t - \varepsilon_t)$ curves obtained by the experimental investigation as regards to the tensile behavior of APILON 52 are summarized in Fig. 2. The $(\sigma_t - \varepsilon_t)$ curves were obtained from a set of standard specimens (ISO 527-2). An universal electro-mechanical testing machine (Instron 4502, ITW Group, US) was used for determining the both quasi-static $(\dot{\varepsilon} = 2.210^{-3} s^{-1})$ and cyclic material behavior.

Engineering strains up to 65% and strain-rates ranging from 10^{-4} and 10^{-2} s^{-1} were enforced to explore the quasi-static cyclic material behavior. By contrast, the dynamic response was explored through universal servo-hydraulic testing machine

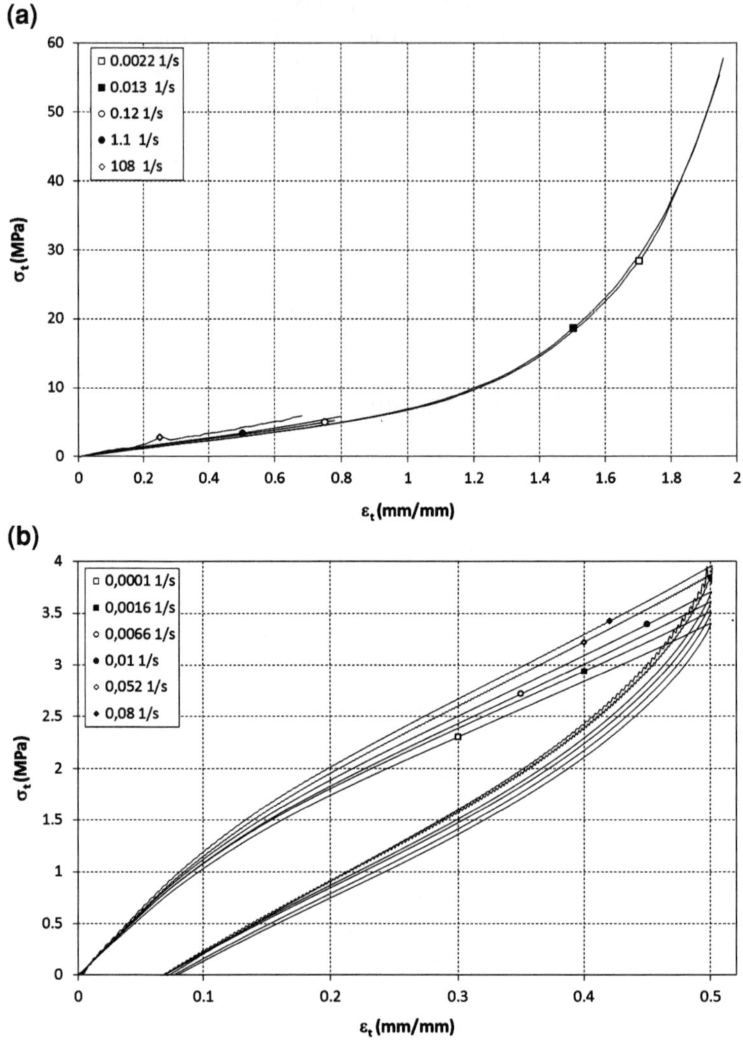

Fig. 2 APILON 52 tensile (**a**) and cyclic (**b**) $\sigma_t - \varepsilon_t$ curves at different strain-rates $\dot{\varepsilon}$, respectively

(DARTEC HA100, Dartec Inc, USA) specifically developed by Politecnico of Torino for investigating the material response at medium strain-rates [26]. As expected, the tested material exhibits a significant elastomeric characteristics: even though large elongations are applied to APILON 52, these are recovered almost completely. The tested material also shows a dependence of the constitutive law on the strain-rate, even at very low strain-rates: the strain-rate effects especially involve the initial elastic modulus, whereas no appreciable shortening of the plateau extension can be detected, at least in the strain-rate range considered in the present work.

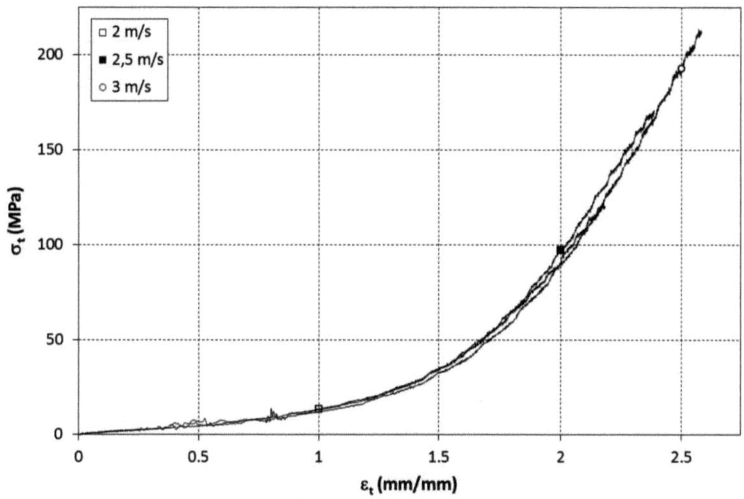

Fig. 3 Tensile impact $\sigma_t - \varepsilon_t$ curve of APILON 52 at strain-rates ranging from 70 and 100 s^{-1}

As will be clarified in the next section, the knowledge of the material tensile impact response is necessary for implementing the theoretical model proposed by Bergstrom and Boyce. Figure 3 shows the average true stress-true strain curves obtained after performing a proper elaboration of the experimental impact force L vs. stretch λ curves (L–λ curve) at three impact velocities: 2, 2.5 e 3 m s^{-1}. In the present investigation the material L–λ curves were determined through instrumented pendulum testing machine (Resil Impactor 6596, CEAST, ITW Group, US) over a set of 15 standard specimens (ISO 527-2). Impact velocities ranging from 2 and 3 ms^{-1} were considered, thus making it possible to infer the material response at strain-rates within 70 and 100 s^{-1}. As shown by Fig. 3, strain-rate effects are not really much meaningful in the considered range. For APILON 52 strain-rate effects become relevant when the variation are of an order of magnitude, especially at very low strain-rate values. These findings suggest that the dependence of the constitutive behavior of the tested material is highly non-linear even at medium–low strain-rates.

3.2 Compressive Response

As regards the characterization of the material compressive behavior, standard uniaxial compressive and cyclic tests were performed. Experimental tests were carried out using the same testing machines employed to characterize the material tensile behavior. A set of 24 cylindrical specimens having height and diameter of 20 mm was used, whereas the strain-rate was varied between 10^{-3} and 10^{1} s^{-1}.

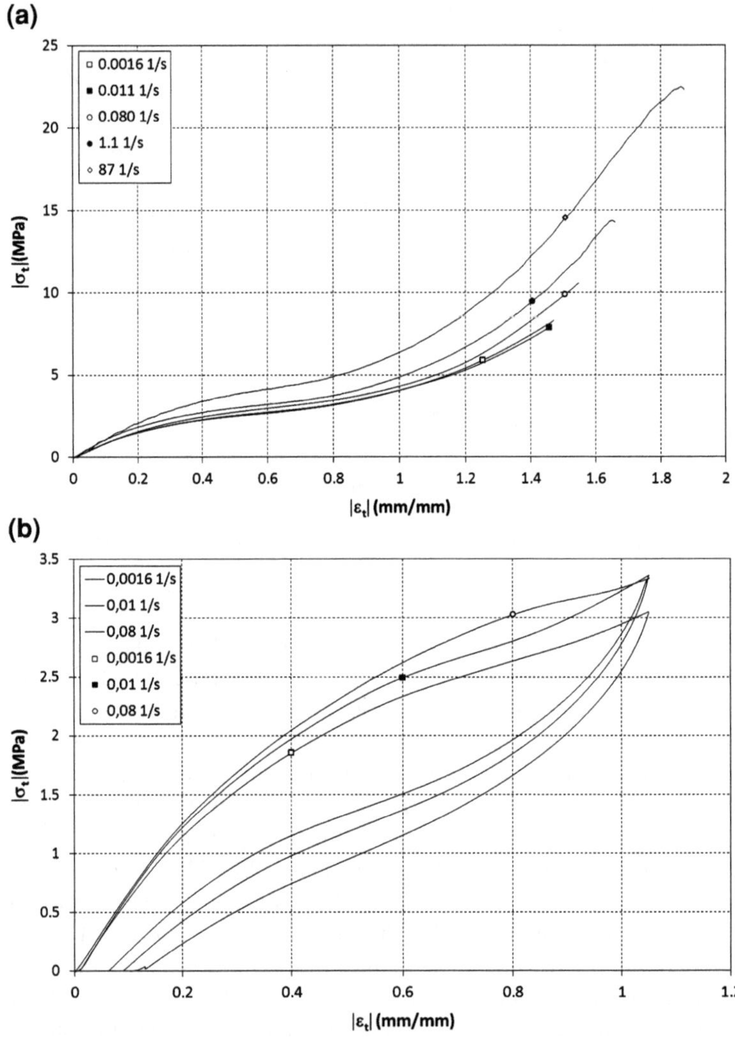

Fig. 4 Strain-rate effects onto the compressive and cyclic $\sigma_t - \varepsilon_t$ curves of APILON 52

The analysis of the experimental trends (Fig. 4) allows to identify same interesting features characterizing the material behavior. The comparison between the results summarized in Figs. 2 and 4 reveals the existence of an asymmetry in terms of structural response. The tested material appears to be more compliant in compression than in tension and the strain-rate effects are much more relevant with respect to what was observed in tensile tests, especially as regards the increase of the initial elastic modulus. In addition, a significant contraction of the plateau can be appreciated. It is apparent that the investigated material is characterized by asymmetric behavior. This experimental evidence is of paramount importance because it must be taken into account in the development or selection of the most proper constitutive model for reproducing the effective material behavior.

4 Constitutive Models Analysis

The qualification of the theoretical models described in the previous section firstly requires the estimation of the corresponding sets of constitutive parameters. While for the determination of the Yang material parameters it is sufficient to fit the uniaxial responses at different strain-rates (tensile and compressive) by Eqs. (1–5) simultaneously, *ad hoc* inverse methods must be built-up for estimating the material constants of the Bergstrom–Boyce model. In the following subsections, the evaluation procedures employed for establishing the material constants in both theoretical models will be presented.

4.1 Yang's Model Qualification

The procedure proposed by Yang et al. [10] was adopted for estimating the constitutive parameters. Hyperelastic constants (A_1, A_2, A_3) were determined fitting the uniaxial tensile and compressive stress–strain curves $\sigma - \varepsilon$ under quasi-static loading conditions by Eq. (1), whereas the viscoelastic ones (A_4, A_5, A_6) by means of nonlinear multi-fitting procedure proposed by Doman et al. [2, 23]. In more detail, the viscoelastic constants were calculated by minimizing the following functional:

$$Err(A_4, A_5, A_6) = N - \sum_{n=1}^{N} \sum_{i=0}^{M-1} \frac{[\hat{\sigma}_n(i\Delta t, A_4, A_5, A_6) - \sigma_n(i\Delta t)]^2}{[\bar{\sigma}_n - \sigma_n(i\Delta t)]^2} \tag{10}$$

in which N and M denote the number of the experimental curves and data points involved into the fitting respectively, while $\hat{\sigma}_n, \sigma_n$ and $\bar{\sigma}_n$ are the estimated stress (refer to Eq. (6)), the experimental stress and the mean experimental stress, respectively. The stress–strain curves provided by the experimental campaign were used to calculate the aforementioned parameters. The time step Δt was chosen according to the criteria defined by Wineman [27] for solving Volterra equation. Table 2 collects the fitting parameters provided by the fitting procedure and inherent the compressive response of APILON 52 for strain-rates ranging from 10^{-3} to 0.8×10^2 s^{-1}. Similar results were obtained from the fitting of the tensile stress–strain curves.

The results obtained from these analyses allow to finely reproduce the stress–strain curves of APILON 52 at strain-rates within the range considered in the present work: relative discrepancies between the predicted curves and the experimental ones were found to be lower than 2% in both tensile and compressive conditions. Although the theoretical model proposed by Yang et al. [10] accounts

Table 2 Hyper-viscoelastic constants reproducing the compressive response of APILON 52 at strain-rates ranging from 10^{-3} and 0.8 10^{2} s^{-1}

A_1 [MPa]	A_2 [MPa]	A_3 [MPa]	A_4 [MPa]	A_5 [MPa]	A_6 [μs]
1.9584	−0.627	0.0434	1.85	−4.2	15.6

for the strain-rate effects excellently for this material, it must be highlighted that the procedure used for estimating the constitutive parameters does not allow to identify the material constants set uniquely once the stress condition (tensile, compressive) has been specified. Depending on the algorithm used for minimizing the objective function and the initial values, different material constants set providing a very good reproduction of the effective material behavior were found both under tensile and compressive loading. It is authors' opinion that the lack of uniqueness inherent in the evaluation procedure developed by Yang et al. represents a very important critical issues. The magnitude of the energy stored by the material was found to be particularly sensitive to the materials constants set $A_4, \ldots, A_6$: small variations in these constants values result in significant variations in the estimation of the energy stored by the material. In addition, the fitting procedure may provide material constants estimation without any physical meaning. From the analogical point of view indeed, the integral formulation of Eq. (4) corresponds to a modified Maxwell model [7, 10] in which the spring and dashpot elements have a stiffness $E = A_4 + A_5(I_2 - 3)$ and viscosity $\eta = A_6[A_4 + A_5(I_2 - 3)]$, respectively. For the material investigated in the present work, it easy to probe that spring stiffness takes always negative values when the compressive constitutive behavior is considered. Thus, any possibility to assume or establish correlations between the model constants set and physical and mechanical material properties is precluded.

4.2 Bergstrom–Boyce Model Qualification

As regards the theoretical model proposed by Bergstrom and Boyce, the constitutive parameters must be desumed through a proper elaboration of the experimental data. However, while the elastic constants $\left(C_R^1, \sqrt{N^1}, C_R^2, \sqrt{N^2}, B\right)$ can be uniquely determined from the experimental results (cf. Sect. 2.2), the time-dependent material constants can be estimated either tentatively or through trial and error techniques based on the comparison of the uniaxial material response in a given strain-rate range with that predicted by a numerical model able to simulate the considered experimental test and implementing the constitutive law containing the parameters to be estimated. Being these parameters particularly crucial in the prediction of the strain energy stored by the material under a specified loading condition, the estimation of such parameters must be accurately carried out. Trial

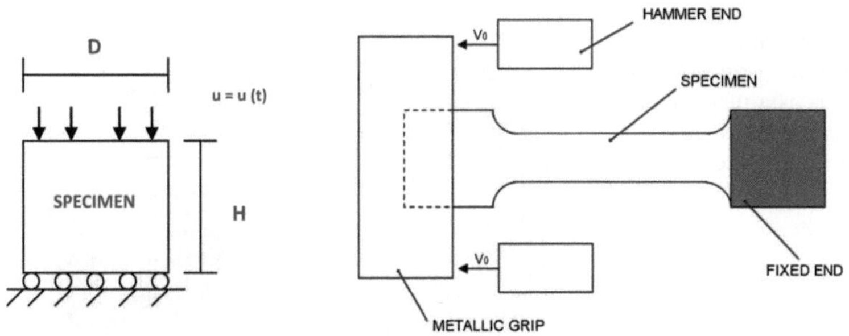

Fig. 5 Conceptual schemes of the computational FE models used for simulating the compressive and tensile impact tests

and error techniques developed by Peroni and Peroni et al. [28, 29] resulted to be particularly accurate and effective and for these reasons were adopted in the present investigation.

As revealed by the experimental campaign, an asymmetric response characterizes the behavior of APILON 52. Accordingly, it is reasonable to expect the existence of two different sets of constitutive parameters: one for the tensile response and one for the compressive response. In order to evaluate the Bergstrom–Boyce parameters as a function of the stress state typology (tension and compression), two different Finite Element (FE) models were built-up, the first one for simulating the uniaxial compressive test at fixed strain-rate and the second one for reproducing the experimental conditions proper of tensile impact tests at specified impact velocity. The conceptual schemes of both FE models with their basic assumptions are illustrated in Fig. 5. With regard to the simulation of the uniaxial compressive test, since the testing machine frame is very stiff, the deformation process solely involves the polymeric material and therefore only the cylindrical specimen was modeled. The supporting plane of the testing machine was introduced into the computational model by constraining the displacements along the loading direction (roller boundary conditions) at the bottom surface of the cylinder. The loading conditions were implemented by enforcing a displacement history $u = u(t)$ along the same direction so as to produce thee prescribed strain-rates in the specimen.

Due to the complexity of the testing apparatus for characterizing the materials tensile impact response, only a part of the specimen and of the instrumented pendulum was implemented into in the FE model.

In more detail, as shown by Fig. 5, only the hammer end and the clamping systems were modeled. In order to conserve the masses values involved into the experimental test, an equivalent mass density was assigned to the modeled hammer end. Specimen was constrained so that to reproduce the effective clamping conditions in the experimental test. By contrast, loading conditions were implemented assigning to the hammer end a prescribed velocity having the same time-law obtained during the experimental tests.

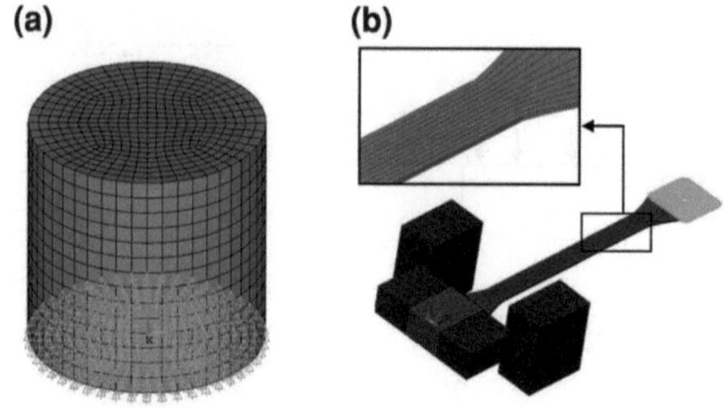

Fig. 6 Meshes used for evaluating the time-dependent Bergstrom–Boyce parameters as a function of the loading configuration. **a** compression, **b** tensile impact

Table 3 Compressive and tensile elastic Bergstrom–Boyce parameters for APILON 52

	C_R^1 [MPa]	$\sqrt{N^1}$ [/]	C_R^2 [MPa]	$\sqrt{N^2}$ [/]	B [MPa]
Compression	0.294	2.04	1.32	2.04	147
Tension	0.233	4.92	1.42	4.92	117

For both FE models proper sensitivity and convergence analyses were carried out in order to establish all the numerical parameters providing solutions model-independent. 3D structural 20-nodes elements were used to discretize specimens and parts of the testing apparatus in both cases. Convergence analyses revealed that mesh-independent results are obtained if the characteristic element size is lower than $H/14$ in the case of uniaxial compressive test modeling and lower the half width of the specimen for the tensile impact test (Fig. 6). The contact between the specimen-fixing grip system and hammer end was implemented by meshing the contacting surfaces with 8-nodes surface-to-surface elements. The contact problem was introduced into the computational model as an associate problem by the Augmented Lagrangian Multipliers (ALM) thus minimizing the effects of contact stiffness onto the model response. Both computational models were developed using the general purpose commercial code ANSYS Rel. 12.0. The Newmark algorithm was used to account for the time-transient.

The first five material constants (Table 3) in both cases (compression and tension) were calculated analyzing the quasi-static and cyclic experimental responses according to the indications and suggestions of the authors (see Appendix in [14]).

The estimation of the remaining material parameters $\left(\hat{C}_1, C_2, m\right)$ was based on the comparison of the experimental and numerical impact force F vs. time t curves in the case of tension and the experimental and numerical evolution of the mean compressive stress in the case of compression. The experimental results referring to the uniaxial compressive test at 88 s^{-1} and the tensile impact test at the impact speed of 3 m s^{-1} were not used for deducing the aforementioned parameters: these results were in fact used as comparison purposes for probing the predictive capabilities of the Bergstrom–Boyce model. In both cases the aforementioned

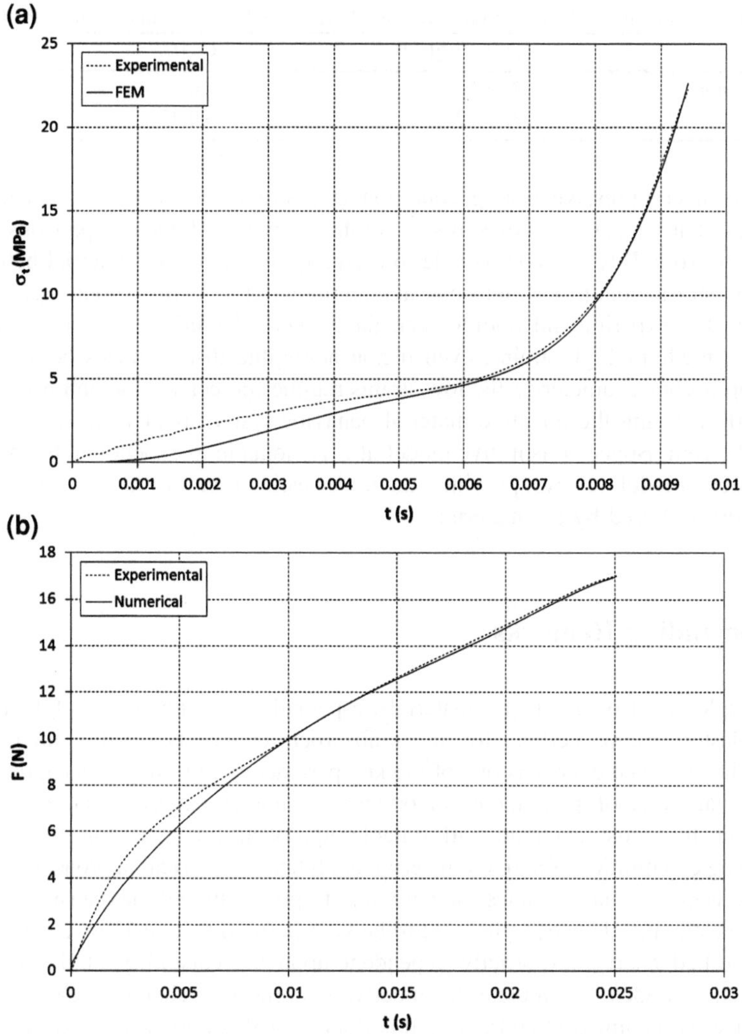

Fig. 7 Comparison between the experimental trends and final numerical predictions obtained by the inverse analysis for compression (*strain-rate*: 88 s^{-1}) (**a**) and tension (*impact speed*: 3 m/s) (**b**)

procedures developed by Peroni and Peroni et al. [28, 29] were employed for deducing the materials parameters $\hat{C}_1, C_2$ and m. A functional akin to that used by Doman et al. [2, 23] (Eq. (10) with $N = 1$) was assumed as the objective function to be minimized.

The comparison between the numerical predictions and the corresponding experimental results is illustrated in Fig. 7 for compression and tension, respectively, while Table 4 collects the numerical values of the corresponding time-dependent constitutive parameters $\left(\hat{C}_1, C_2, m\right)$. The comparisons refer respectively

Table 4 Compressive and tensile time-dependent Bergstrom–Boyce parameters for APILON 52

	$\hat{C}_1$ [$s^{-1}MPa^{-m}$]	C_2 [/]	*m* [/]
Compression	2.5×10^{-5}	−0.80	3.50
Tension	3.1×10^{-3}	−0.97	3.81

to uniaxial compressive testing condition at a strain-rate of 88 s^{-1} and tensile impact test at impact speed of 3 m s^{-1}. As shown by Fig. 7, the comparisons show that Bergstrom–Boyce model is able to correctly reproduce the material behavior in both stress conditions, tension and compression. A very good agreement between the experimental findings and the numerical prediction is found, especially at medium–high strains, even if non negligible discrepancies between the two approaches characterize the low strains regime (percentage deviations greater than 20%). Being the effective material behavior systematically under-estimated by Bergstrom–Boyce constitutive model, it is reasonable to expect that the present theoretical model always provides an under-estimation of the energy storing capabilities offered by the material.

5 Concluding Remarks

The employment of polymeric materials, especially Thermoplastic PolyUrethane with elastomeric properties, for shock-absorber devices is steadily increasing. Since the absorbed energy is one of the key parameters in design of such devices, its estimation is of paramount importance. Although experimental techniques represent the most accurate and reliable approach for assessing the energy absorbing capabilities of a given material, analytical–numerical approaches are an equally valid alternative, since they are able to promptly and inexpensively yield suitable estimates. However, it ought to be kept in mind that accurate estimates of the absorbed energy are strictly dependent upon the capability of the adopted constitutive models to simulate the effective material's mechanical behavior. The present work is aimed at addressing this issue. For this purpose, one of the most used Thermoplastic PolyUrethane has been firstly experimentally characterized and then, among the constitutive models proposed in the literature, those better simulating its dynamic tensile and compressive response have been identified and analyzed.

The hyper-viscoelastic nonlinear approaches proposed by Yang and Bergstrom and Boyce have been identified as the most promising and reliable. Although the two models have been conceived in very different ways, being the first one purely phenomenological and the second one based on microstructure considerations, the predictive capabilities of both of them proved to be adequate. In addition, the Yang's model seems to yield much more accurate results in terms of absorbed energy as compared to the Bergstrom and Boyce's model. The average percentage deviation between the experimental tensile and compressive $\sigma - \varepsilon$ curves and the corresponding ones predicted by the Yang's model is below 2% for the material and the strain rate range (10^{-3}–10^{2} s^{-1}) considered in the present study. Percentage deviations larger than 20%, for both traction and compression, have been found when the experimental curves are compared with those predicted by the Bergstrom and Boyce's model. It should be noted that, although the Yang's model is able to satisfactorily predict the energy absorbed by the material, the method followed to estimate the constitutive law does not permit to establish a biunique correspondence between material behavior and the constitutive parameters. Accordingly, the determination of these parameters is performed through nonlinear fitting procedures whose results proved to be dependent on both the adopted algorithm and the initial guess values. The Yang's approach does not permit to operate on the constitutive parameters in order to optimize for instance the material response. On the contrary, Bergstrom and Boyce's model does not present this limitation: the characteristic constants of the model possess a clear physical meaning and can be uniquely determined through a suitable experimental and numerical campaign.

In conclusion, the limitation of both models shall be remarked. In fact, they have been conceived to separately account for the two possible types of response. However, during an impact transient, it can be expected that the material simultaneously experiences both tensile and compressive loading conditions. Given the critical applications of this class of materials, the development of a constitutive model able to incorporate both responses is surely desirable, especially if the tensile response of the material significantly differs from the compressive one.

References

1. Rao, S., Shim, V.P.W., Quah, S.E.: Dynamic mechanical properties of polyurethane elastomers using a non-metallic Hopkinson bar. J. Appl. Polym. Sci. **66**, 619–631 (1997)
2. Doman, D.A., Cronin, D.S., Salisbury, C.P.: Characterization of polyurethane rubber at high deformation rates. Experimental Mechanics **46**, 367–376 (2006)
3. Yi, J., Boyce, M.C., Lee, G.F., Balizer, E.: Large deformation rate-dependent stress-strain behaviour of polyurea and polyurethanes. Polymer **47**, 319–329 (2006)
4. Sarva, S.S., Deschanel, S., Boyce, M., Chen, W.: Stress-strain behaviour of a polyurea and a polyurethane from low to high strain rates. Polymer **48**, 2208–2213 (2007)
5. Roland, C.M., Twigg, J.N., Lu, Y., Mott, P.H.: High strain rate mechanical behaviour of polyurea. Polymer **48**, 574–578 (2007)
6. Kanyanta, V., Ivankovic, A.: Mechanical characterisation of polyurethane elastomer for biomedical applications. J. Mech. Behavior Bio. Mater. **3**, 51–62 (2010)
7. Lockett, F.J.: Non-linear viscoelastic solids. Academic Press, London, New York (1972)
8. Symonds, P.S.: Survey of methods of analysis for plastic deformation of structures under dynamic loading Report N BU/NSRDC. Brown University (1967)
9. Johnson, G.R., Cook, W.H.: Fracture characteristics of three metals subjected to various strain rates temperatures and pressures. Eng. Frac. Mech. **21**(1), 31–48 (1985)
10. Yang, L.M., Shim, V.P.W., Lim, C.T.: A visco-hyperelastic approach to modelling the constitutive behaviour of rubber. Int. J. Impact Eng. **24**, 545–560 (2000)
11. Shim, V.P.W., Yang, L.M., Lim, C.T., Law, P.H.: A visco-hyperelastic constitutive model to characterize both tensile and compressive behavior of rubber. J. Appl. Polym. Sci. **92**, 523–531 (2004)

12. Hoo Fatt, M.S., Ouyang, X.: Integral-based constitutive equation for rubber at high strain rates. Int. J. Sol. Struct. **44**, 6491–6506 (2007)
13. Li, C., Lua, J.: A hyper-viscoelastic constitutive model for polyurea. Mat. Lett. **63**(11), 877–880 (2009)
14. Bergstrom, J., Boyce, M.: Constitutive modeling of the large strain time-dependent behavior of elastomers. J. Mech. Phys. Solids **46**, 931–954 (1998)
15. Bergstrom, J., Boyce, M.: Large strain time-dependent behaviour of filled elastomers. Mech. Mater. **32**, 627–644 (2000)
16. Bergstrom, J., Boyce, M.: Constitutive modeling of the time-dependent and cyclic loading of elastomers and application to soft biological tissue. Mech. Mater. **33**, 523–530 (2001)
17. Quintavalla, S.J., Johnson, S.H.: Extension of the Bergstrom–Boyce model to high strain rates. Rubber Chem. Tech. **77**(5), 972–981 (2004)
18. Mooney, M.: A theory of large elastic deformation. J. Appl. Phys. **11**, 582–592 (1940)
19. Rivlin, R.S.: Some topics in finite elasticity. In: Proceedings of the First Symposium on Naval Structural Mechanics, pp. 169–198 (1960)
20. Truesdell, C., Noll, W.: The non-linear field theories of mechanics. In: Handbuch der Physik, Fugge, S. (eds.) Springer, Heidelberg (1965)
21. Bernstein, B., Kearsley, A., Zapas, L.J.: A study of stress relaxation with finite strain. Trans. Soc. Rheol. **7**, 391–410 (1963)
22. Bernstein, B., Kearsley, A., Zapas, L.J.: A study of stress relaxation with finite strain. Rubber Chem. Technol. **38**, 76–89 (1965)
23. Doman, D.A.: Modeling of the high rate behaviour of polyurethane rubber. MASc Thesis, University of Waterloo (2004)
24. Arruda, E.M., Boyce, M.C.: A three-dimensional constitutive model for the large stretch behaviour of rubber elastic materials. J. Mech. Phys. Solids **41**(2), 389–412 (1993)
25. Anand, L.: A constitutive model for compressible elastomeric solids. Comput. Mech. **18**, 339–355 (1996)
26. Peroni, M., Avalle, A., Peroni, L.: Caratterizzazione dell'influenza della velocità di deformazione sul comportamento meccanico di materiali polimerici. Atti del XXXVI Convegno Nazionale AIAS (2007)
27. Wineman, A.: Nonlinear viscoelastic solids. A review. Math. Mech. Solids **14**, 300–366 (2009)
28. Peroni, L.: Evaluation of dynamic characteristics by uni-axial tensile impact. In: Proceedings of ICEM12, 12th International Conference on Experimental Mechanics (2004)
29. Peroni, M., Peroni, L., Belingardi, G.: Identification of strain-rate sensitivity of steels with an inverse method. In: Proceedings ICEM13, 13th International Conference on Experimental Mechanics (2007)

A Study on the Displacement Field of Nonlocal Elasticity for Mechanical Analysis of Nano Structures

Emad Jomehzadeh and Ali Reza Saidi

Abstract In this chapter, a study on the displacement field equations of nonlocal elasticity is developed for mechanical analyses of nano structures. Considering the small scale effect, the three dimensional equations of nonlocal elasticity are obtained. At first, three decoupled equations in terms of displacement components and three decoupled equations in terms of rotation components are obtained. These equations are also invariant with respect to the coordinate system. In order to solve a nonlocal elasticity problem based on the presented formulation, one of the three equations in terms of displacement components and corresponding rotation equation should be solved independently. Using some relations, the other two displacement components can be obtained in terms of the mentioned displacement and rotation component. In an illustrative example, these equations associated with simply supported boundary conditions are solved for a nano-plate using the Fourier series technique. In addition, the results are compared with the first order and third order shear deformation theories. It is seen that the natural frequencies of the nonlocal plate theories are not as accurate as the results in classical plate theories.

Keywords Nonlocal elasticity · Displacement field · Nano structures · Small scale

E. Jomehzadeh (✉) · A. R. Saidi
Young Researchers Club, Kerman Branch,
Islamic Azad University, Kerman, Iran
e-mail: jomehzadehe2@asme.org

A. R. Saidi
e-mail: saidi@mail.uk.ac.ir

A. Öchsner et al. (eds.), *Materials with Complex Behaviour II*,
Advanced Structured Materials 16, DOI: 10.1007/978-3-642-22700-4_21,

1 Introduction

Multiple recent experimental results have shown a significant size effect in mechanical properties when the dimensions of the specimen or the probed material volume become small. The classical continuum theories, on the other hand, lack the capability of representing such effects since they do not include any internal length scale. Consequently, these theories are expected to fail when the specimen size becomes comparable with the internal length scale(s) of the material. The common feature of all nonlocal formulations is the elimination of stress field singularities. The gradient elasticity solution shows no singularity in both stress and strain fields at the core of dislocations. Furthermore, the strain energy is finite at sites where local elasticity predicts stress singularities and infinite strain energy density.

The scale effects are accounted by considering the internal size as a material parameter. The most general form of the constitutive relation for nonlocal elasticity involves an integral over the whole body and therefore the governing equations become integro-differential equations (see e.g. [1, 2]). Eringen [3] showed that it is possible to represent the integral constitutive relations of nano-structures in an equivalent differential form. Eringen presented a nonlocal elasticity theory to account the small scale effect by specifying the stress at a reference point is a functional of the strain field at every point in the body. Since then, many studies have been carried out for bending, buckling and vibration analyses of nano-structures based on the gradient elasticity. The nonlocal theory of elasticity has been extensively used to study buckling and vibration analyses of carbon nano-tubes with the help of beam and shell theories [4–6]. Also, the vibration analysis of graphite sheet has been investigated using plate theories. Kitipornchai et al. [7] used the continuum plate model for mechanical analysis of graphene sheets. Pradhan and Phadikar [8] presented the vibration analysis of a simply supported nano-plate based on the first order shear deformation plate theory (FSDT). Aghababaei and Reddy [9] solved bending and vibration of plates based on the nonlocal third order shear deformation plate theory (TSDT) considering the small scale effect.

Apparently, using decoupled forms of governing equations of each theory for solving related problems is much easier than using the original coupled form for both analytical and numerical methods. Thus, decoupling of governing equations is one of the important aspects of solid mechanics problems. Plevako [10] reformulated the elastostatics Navier equations into two uncoupled equations. These uncoupled equations are in terms of two new functions. Levinson [11] presented an exact three dimensional solution for the free vibrations of simply supported rectangular plates by decoupling the equations of elastodynamics. Charalambopoulos et al. [12] represented the Helmholtz decomposition theorem and the method of separation of variables to solve the Navier equations. Poullikkas [13] considered the application of the method of fundamental solutions to isotropic elastostatics problems in three dimensions by the method of fundamental solutions

(MFS). Ozer [14] presented the analytical solutions of axially-symmetric Navier equations in classical elasticity by applying Lie group theory. Recently, Saidi et al. [15] reformulated the Navier equations for solving elastostatic and elastodynamic problems.

Although a few researches studied the decoupling of displacement field of local elasticity, no such work is available in literature in the field of nonlocal elasticity. The primary objective of this chapter is to decouple the displacement field equations of nonlocal elasticity. To this end, three decoupled equations in terms of displacement components and three decoupled equations in terms of rotation components are introduced. Also, some relations for one of the three displacement components are presented. Each of these relations is in terms of other two components of the displacement field and corresponding rotation components. At first, one of the three decoupled equations in terms of displacement components and also its corresponding rotation equation are solved independently. Then, using the presented relations, the other two displacement components can be obtained. Application of this reformulation is presented for a simply supported nano-plate considering the small scale effect. Influence of nonlocal parameter and there dimensional analysis on the vibration characteristics of the nano-plates is studied.

Since the resulted equations are presented in decoupled and invariant form, they can be used for solving the known problems in the field of nonlocal elasticity much easier not only in Cartesian system but also in other coordinate systems. Results for natural frequencies of nano-plates from three dimensional analysis are given for the first time and these can serve as reference values for other numerical analysis.

2 Field Equations of Nonlocal Elasticity

Several modifications of the classical elasticity formulation have been proposed to address the small scale effect. They are of integral non-local or gradient type and, as a common feature, include one or several intrinsic length scales. Their predictions reduce to those of local continuum theories when the specimen size is much larger than the internal length scale. In the integral formulation of non-local elasticity due to Eringen and Edelen [2] the constitutive equation is expressed in terms of a non-local kernel introduced to account for the effect of long-range interatomic forces. The stress at point X is a function of the strain at all points X' in the body, through a weighting kernel α. Specifically, for homogeneous and isotropic elastic solids, the linear theory is expressed by the set of equations as

$$t_{kl,k} + \rho(F_l - \ddot{u}_l) = 0, \tag{1a}$$

$$\sigma_{ij}(X) = \int_V \alpha(|X' - X|, \tau) t_{ij}(X')\, dV \tag{1b}$$

$$\sigma_{kl}(X') = \lambda\varepsilon_{rr}(X')\delta_{kl} + 2\mu\varepsilon_{kl}(X'), \tag{1c}$$

$$\varepsilon_{kl}(X') = \frac{1}{2}\left(\frac{\partial u_k}{\partial x'_l} + \frac{\partial u_l}{\partial x'_k}\right), \tag{1d}$$

where t_{kl}, ρ, F_l and u_l are, respectively, the nonlocal stress tensor, mass density, body force density and the displacement vector at a reference point X in the body at time t. Also, dot above each parameter denotes differentiating with respect to time. $\sigma_{kl}(X')$ is the usual local (classical) stress tensor at X'. λ and μ are the Lame's coefficients which are related to the Young's modulus E and Poisson ratio v as

$$\lambda = \frac{Ev}{(1+v)(1-2v)}; \quad \mu = \frac{E}{2(1+v)} \tag{2}$$

It can be seen that the only difference between the nonlocal equations and the classical elasticity equations is in the constitutive Eq. 1b. The field equations of nonlocal elasticity are obtained by combining Eqs. 1a–1d. Eringen [3] showed that it is possible to represent the integral constitutive relation in an equivalent differential form as

$$(1 - \beta\nabla^2)t_{kl} = \sigma_{kl} \tag{3}$$

where $\beta = (e_0 a)^2$ is nonlocal parameter, a an internal characteristic length and e_0 a constant. Also, ∇^2 is the Laplacian operator.

3 Reformulation of Nonlocal Displacement Field Equations

Based on the Eringen constitutive equation [3], it can be shown that the displacement field equations of motion in three dimensional state are written as

$$\mu\nabla^2\vec{u} + (\lambda + \mu)\nabla(\nabla.\vec{u}) + (1 - \beta\nabla^2)\vec{F} = (1 - \beta\nabla^2)\rho\ddot{\vec{u}} \tag{4}$$

where $\vec{u}$ is the displacement vector and $\vec{F}$ is the body force per unit volume. It is easy to show that the vector Eq. 4 in a Cartesian coordinate system can be written as

$$\begin{aligned}&\eta u_{1,11} + \mu u_{1,22} + (\eta - \mu)u_{2,12} + (\eta - \mu)u_{3,13} + \mu u_{1,33} + (1 - \beta\nabla^2)F_1 \\ &\quad = (1 - \beta\nabla^2)\rho\ddot{u}_1\end{aligned} \tag{5a}$$

$$\begin{aligned}&\eta u_{2,22} + \mu u_{2,11} + (\eta - \mu)u_{1,12} + (\eta - \mu)u_{3,23} + \mu u_{2,33} + (1 - \beta\nabla^2)F_2 \\ &\quad = (1 - \beta\nabla^2)\rho\ddot{u}_2\end{aligned} \tag{5b}$$

$$\mu\nabla^2_{2D} u_3 + (\eta - \mu)(u_{1,1} + u_{2,2})_{,3} + \eta u_{3,33} + (1 - \beta\nabla^2)F_3 = (1 - \beta\nabla^2)\rho\ddot{u}_3 \tag{5c}$$

where ∇^2_{2D} is the two dimensional Laplacian operator, i.e.

$$\nabla_{2D}^2 = \frac{\partial^2}{\partial x_1^2} + \frac{\partial^2}{\partial x_2^2} \tag{6}$$

and η is defined in terms of material properties as

$$\eta = \frac{(1 - v)E}{(1 + v)(1 - 2v)} \tag{7}$$

Differentiating Eqs. 5a and 5b with respect to x_1 and x_2, respectively and adding the results, yields

$$\eta \nabla_{2D}^2 \zeta + \mu \zeta_{,33} + (\eta - \mu) \nabla_{2D}^2 u_{3,3} + (1 - \beta \nabla^2)(F_{1,1} + F_{2,2}) = (1 - \beta \nabla^2) \rho \ddot{\zeta} \tag{8}$$

where the parameter ζ is defined as

$$\zeta = u_{1,1} + u_{2,2} \tag{9}$$

Also, using definition (9), Eq. 5c can be rewritten in the form of

$$\mu \nabla_{2D}^2 u_3 + \eta u_{3,33} + (\eta - \mu) \zeta_{,3} + (1 - \beta \nabla^2) F_3 = (1 - \beta \nabla^2) \rho \ddot{u}_3 \tag{10}$$

Eliminating ζ from Eqs. 8 and 10 yields a partial differential equation in terms of u_3 as

$$\begin{aligned} &\mu \eta \nabla^4 u_3 - \rho(\eta + \mu)(1 - \beta \nabla^2) \nabla^2 \ddot{u}_3 + \rho^2 (1 - \beta \nabla^2)^2 \ddot{\ddot{u}}_3 + \eta \nabla^2 F_3 \\ &- (\eta - \mu)(1 - \beta \nabla^2)(F_{1,1} + F_{2,2} + F_{3,3})_{,3} - \rho(1 - \beta \nabla^2)^2 \ddot{F}_3 = 0 \end{aligned} \tag{11}$$

This equation is a fourth order partial differential equations and is only in term of unknown displacement u_3. Differentiating Eqs. 5a and 5b with respect to x_2 and x_1, respectively and subtracting the results, yields

$$\mu \nabla^2 (u_{1,2} - u_{2,1}) + (1 - \beta \nabla^2)(F_{1,2} - F_{2,1}) = \rho(1 - \beta \nabla^2)(\ddot{u}_{1,2} - \ddot{u}_{1,2}) \tag{12}$$

It can be seen that for finding Eqs. 11 and 12, the third Navier equation (i.e. Eq. 5c) is written in a different form from Eqs. 5a and 5b. Similarly, one can write the first or second Navier equation in different forms. Therefore, two other equations similar to Eq. 11 in terms of u_1 and u_2, and also two equations like as Eq. 12 can be obtained. In other words, one can write these three equations as follows

$$\begin{aligned} &\mu \eta \nabla^4 u_i - \rho(\eta + \mu)(1 - \beta \nabla^2) \nabla^2 \ddot{u}_i + \rho^2 (1 - \beta \nabla^2)^2 \ddot{\ddot{u}}_i + \eta \nabla^2 F_i - (\eta - \mu)(1 - \beta \nabla^2) F_{j,ij} \\ &- (1 - \beta \nabla^2)^2 \rho \ddot{F}_i = 0 \end{aligned} \tag{13}$$

and similarly, Eq. 12 can be written in general form as

$$\mu \nabla^2 \omega_{ij} + (1 - \beta \nabla^2)(F_{i,j} - F_{j,i}) - \rho(1 - \beta \nabla^2) \ddot{\omega}_{ij} = 0 \tag{14}$$

where ω_{ij} is the anti-symmetric rotation tensor (i.e. $\omega_{ij} = (u_{i,j} - u_{j,i})$). Equations 13 and 14 can be rewritten in the invariant form as

$$\mu\eta\nabla^4\vec{u} - \rho(\eta+\mu)(1-\beta\nabla^2)\nabla^2\ddot{\vec{u}} + \rho^2(1-\beta\nabla^2)^2\ddot{\ddot{\vec{u}}} + \eta\nabla^2\vec{F} - (\eta-\mu)(1-\beta\nabla^2)\vec{\nabla}(\vec{\nabla}\cdot\vec{F}) - \rho(1-\beta\nabla^2)^2\ddot{\vec{F}} = 0 \quad (15a)$$

$$\mu\nabla^2\vec{\omega} + (1-\beta\nabla^2)(\nabla\vec{F} - (\nabla\vec{F})^T) - \rho(1-\beta\nabla^2)\ddot{\vec{\omega}} = 0 \quad (15b)$$

where the superscript T denotes the transpose of matrix. Equations 15a and 15b are two vector equations with total order of six. These two vector equations represent six decoupled partial differential equations in terms of displacement and rotation components. Also, these two equations are invariant from the choice of the coordinate system.

In the static case, in the absence of body forces, Eqs. 13 and 14 can be simplified as

$$\nabla^4 u_i = 0 \quad (16a)$$

$$\nabla^2 \omega_{ij} = 0 \quad (16b)$$

Thus, the displacement and rotation components should satisfy biharmonic and harmonic equations, respectively. Looking carefully at Eqs. 13 and 14 it can be found that the total degree of one of the three equations of (13) and the corresponding Eq. 14 is the same as the original displacement field equations. Hence, some relations should be found which describe the other displacement components in terms of the selected displacement component and the corresponding rotation component. For this purpose, differentiating ω_{12} with respect to x_1 and x_2, yields

$$u_{2,12} = u_{1,22} - \omega_{12,2} \quad (17a)$$

$$u_{1,12} = u_{2,11} + \omega_{12,1} \quad (17b)$$

Substituting Eqs. 17a and 17b into 5a and 5b respectively, it can be obtained

$$\eta\nabla^2_{2D}u_1 - (\eta-\mu)\omega_{12,2} + (\eta-\mu)u_{3,13} + \mu u_{1,33} + (1-\beta\nabla^2)F_1 = (1-\beta\nabla^2)\rho\ddot{u}_1 \quad (18a)$$

$$\eta\nabla^2_{2D}u_2 + (\eta-\mu)\omega_{12,1} + (\eta-\mu)u_{3,23} + \mu u_{2,33} + (1-\beta\nabla^2)F_2 = (1-\beta\nabla^2)\rho\ddot{u}_2 \quad (18b)$$

From Eq. 10, it can be found that

$$\zeta_{,3} = \frac{1}{\mu-\eta}(\mu\nabla^2_{2D}u_3 + \eta u_{3,33} + (1-\beta\nabla^2)F_3 - \rho\ddot{u}_3) \quad (19)$$

Also, from definitions of ω_{12} and ζ it can be concluded that

$$\nabla_{2D}^2 u_{1,3} = \zeta_{,13} + \omega_{12,23} \tag{20a}$$

$$\nabla_{2D}^2 u_{2,3} = \zeta_{,23} - \omega_{12,13} \tag{20b}$$

Using Eqs. 19 and 20a, Eq. 18a becomes as

$$\begin{aligned}(\mu u_{1,33} - \rho\ddot{u}_1 + \rho\beta\ddot{u}_{1,33})_{,3} = & \frac{\eta}{\eta-\mu}[\mu\nabla^2 u_3 + \mu(1-\frac{\mu}{\eta})u_{3,33} - \rho(1-\beta\nabla^2)\ddot{u}_3 \\ & + \frac{\rho\beta}{\eta}(\mu\nabla_{2D}^2\ddot{u}_3 + \eta\ddot{u}_{3,33} - \rho(1-\beta\nabla^2)\ddddot{u}_3) \\ & + (1-\beta\nabla^2)(F_3 + \frac{\rho\beta}{\eta}\ddot{F}_3)]_{,1} \\ & - [\mu\omega_{12,2} + (1-\beta\nabla^2)F_1 + \rho\beta\ddot{\omega}_{12,2}]_{,3}\end{aligned} \tag{21}$$

Also, using relations (19) and (20b), Eq. 18b will be simplified to

$$\begin{aligned}(\mu u_{2,33} - \rho\ddot{u}_2 + \rho\beta\ddot{u}_{2,33})_{,3} = & \frac{\eta}{\eta-\mu}[\mu\nabla^2 u_3 + \mu(1-\frac{\mu}{\eta})u_{3,33} - \rho(1-\beta\nabla^2)\ddot{u}_3 \\ & + \frac{\rho\beta}{\eta}(\mu\nabla_{2D}^2\ddot{u}_3 + \eta\ddot{u}_{3,33} - \rho(1-\beta\nabla^2)\ddddot{u}_3) \\ & + (1-\beta\nabla^2)(F_3 + \frac{\rho\beta}{\eta}\ddot{F}_3)]_{,2} \\ & - [-\mu\omega_{12,1} + (1-\beta\nabla^2)F_2 - \rho\beta\ddot{\omega}_{12,1}]_{,3}\end{aligned} \tag{22}$$

Equations 21 and 22 represent displacement components u_1 and u_2 in terms of u_3 and ω_{12}. Thus, for solving a known three dimensional nonlocal elastodynamics problem much simpler, it is sufficient to solve the decoupled Eq. 13, with $i = 3$, and decouple Eq. 14, with $i = 1,\ j = 2$. Substituting the obtained relations for u_3 and ω_{12} into Eqs. 21 and 22 yields the displacement components u_1 and u_2.

Equations 21 and 22, in absence of body forces, are simplified for nonlocal elastostatics problems as

$$u_{1,333} = \left(\frac{\eta}{\eta-\mu}\nabla^2 u_3 + u_{3,33}\right)_{,1} - \omega_{12,23} \tag{23a}$$

$$u_{2,333} = \left(\frac{\eta}{\eta-\mu}\nabla^2 u_3 + u_{3,33}\right)_{,2} + \omega_{12,13} \tag{23b}$$

In a similar way which have used for finding Eqs. 21 and 22, the following equations can be obtained

$$
\begin{aligned}
(\mu u_{2,11} - \rho \ddot{u}_2 + \rho\beta\, \ddot{u}_{2,11})_{,1} = \frac{\eta}{\eta-\mu}&[\mu\nabla^2 u_1 + \mu(1-\frac{\mu}{\eta})u_{1,11} - \rho(1-\beta\nabla^2)\ddot{u}_1 \\
&+ \frac{\rho\beta}{\eta}(\mu(\ddot{u}_{2,22} + \ddot{u}_{2,33}) + \eta\ddot{u}_{2,22} - \rho(1-\beta\nabla^2)\ddddot{u}_2) \\
&+ (1-\beta\nabla^2)(F_1 + \frac{\rho\beta}{\eta}\ddot{F}_1)]_{,2} \\
&- [\mu\omega_{23,3} + (1-\beta\nabla^2)F_2 + \rho\beta\ddot{\omega}_{23,3}]_{,1}
\end{aligned} \tag{24a}
$$

$$
\begin{aligned}
(\mu u_{3,11} - \rho \ddot{u}_3 + \rho\beta\, \ddot{u}_{3,11})_{,1} = \frac{\eta}{\eta-\mu}&[\mu\nabla^2 u_1 + \mu(1-\frac{\mu}{\eta})u_{1,11} - \rho(1-\beta\nabla^2)\ddot{u}_1 \\
&+ \frac{\rho\beta}{\eta}(\mu(\ddot{u}_{2,22} + \ddot{u}_{2,33}) + \eta\ddot{u}_{2,22} - \rho(1-\beta\nabla^2)\ddddot{u}_2) \\
&+ (1-\beta\nabla^2)(F_1 + \frac{\rho\beta}{\eta}\ddot{F}_1)]_{,3} \\
&+ [\mu\omega_{23,2} + (1-\beta\nabla^2)F_3 + \rho\beta\ddot{\omega}_{23,2}]_{,1}
\end{aligned} \tag{24b}
$$

or

$$
\begin{aligned}
(\mu u_{3,22} - \rho \ddot{u}_3 + \rho\beta\, \ddot{u}_{3,22})_{,2} = \frac{\eta}{\eta-\mu}&[\mu\nabla^2 u_2 + \mu(1-\frac{\mu}{\eta})u_{2,22} - \rho(1-\beta\nabla^2)\ddot{u}_2 \\
&+ \frac{\rho\beta}{\eta}(\mu(\ddot{u}_{2,11} + \ddot{u}_{2,33}) + \eta\ddot{u}_{2,22} - \rho(1-\beta\nabla^2)\ddddot{u}_2) \\
&+ (1-\beta\nabla^2)(F_2 + \frac{\rho\beta}{\eta}\ddot{F}_2)]_{,3} \\
&- [\mu\omega_{31,1} + (1-\beta\nabla^2)F_3 + \rho\beta\ddot{\omega}_{31,1}]_{,2}
\end{aligned} \tag{25a}
$$

$$
\begin{aligned}
(\mu u_{1,22} - \rho \ddot{u}_1 + \rho\beta\, \ddot{u}_{1,22})_{,2} = \frac{\eta}{\eta-\mu}&[\mu\nabla^2 u_2 + \mu(1-\frac{\mu}{\eta})u_{2,22} - \rho(1-\beta\nabla^2)\ddot{u}_2 \\
&+ \frac{\rho\beta}{\eta}(\mu(\ddot{u}_{2,11} + \ddot{u}_{2,33}) + \eta\ddot{u}_{2,22} - \rho(1-\beta\nabla^2)\ddddot{u}_2) \\
&+ (1-\beta\nabla^2)(F_2 + \frac{\rho\beta}{\eta}\ddot{F}_2)]_{,1} \\
&+ [\mu\omega_{31,3} + (1-\beta\nabla^2)F_1 + \rho\beta\ddot{\omega}_{31,3}]_{,2}
\end{aligned} \tag{25b}
$$

Equations 24a and 24b represent the displacement components u_2 and u_3 in terms of u_1 and ω_{23}. Therefore, another way for obtaining the solution of a known nonlocal elastodynamics problem is solving Eqs. 13, with $i = 1$ and 14 with $i = 2$ and $j = 3$. Then, Eqs. 24a and 24b are used to find out the remaining displacement components. In a similar way, Eqs. 13 and 14 in terms of u_2 and ω_{31} can be solved and then Eqs. 25a and 25b are used.

4 Application for Free Vibration Analysis of a Nano-Plate

Consider a simply supported rectangular plate with dimensions a and b and thickness h. The transverse deflection and in-plane rotation equations are considered as

$$u_3 = \sum_{m=1}^{\infty}\sum_{n=1}^{\infty} U_{3mn}(x_3) \sin \alpha_m x_1 \sin \gamma_n x_2 \, e^{I\omega^k_{mn} t} \tag{26a}$$

$$\omega_{12} = \sum_{m=1}^{\infty}\sum_{n=1}^{\infty} \Omega_{12mn}(x_3) \cos \alpha_m x_1 \cos \gamma_n x_2 e^{I\omega^k_{mn} t} \tag{26b}$$

where α_m and γ_n denote $m\pi/a$ and $n\pi/b$, ω^k_{mn} is the natural frequency and k denotes the number of mode shape in thickness direction. Substituting Eqs. 26a and 26b into Eqs. 13 for $i = 3$ and 14 for $i = 1, j = 2$ respectively, yields

$$U^{(4)}_{3\,mn} + A\, U''_{3\,mn} + B\, U_{3\,mn} = 0 \tag{27a}$$

$$\Omega''_{12mn} + C\, \Omega_{12mn} = 0 \tag{27b}$$

where the coefficients A, B and C are defined as

$$A = \frac{-2\mu\eta(\alpha_m^2 + \gamma_n^2) + \rho\omega^2(\eta + \mu)(1 + 2\beta(\alpha_m^2 + \gamma_n^2)) - 2\rho^2\omega^4\beta(1 + \beta(\alpha_m^2 + \gamma_n^2))}{\mu\eta - \rho\omega^2\beta(\eta + \mu) + \rho^2\omega^4\beta} \tag{28a}$$

$$B = \frac{\mu\eta(\alpha_m^2 + \gamma_n^2)^2 - \rho\omega^2(\eta + \mu)(\alpha_m^2 + \gamma_n^2)(1 + \beta(\alpha_m^2 + \gamma_n^2)) + \rho^2\omega^4(1 + \beta(\alpha_m^2 + \gamma_n^2))^2}{\mu\eta - \rho\omega^2\beta(\eta + \mu) + \rho^2\omega^4\beta} \tag{28b}$$

$$C = \frac{-\mu(\alpha_m^2 + \gamma_n^2) + \rho\omega^2(1 + \beta(\alpha_m^2 + \gamma_n^2))}{\mu - \rho\omega^2\beta} \tag{28c}$$

The solutions of differential Eqs. 27a and 27b are in the form of

$$U_{3mn}(x_3) = C_1 \sinh J_1 x_3 + C_2 \cosh J_1 x_3 + C_3 \sinh J_2 x_3 + C_4 \cosh J_2 x_3 \tag{29a}$$

$$\Omega_{12mn}(x_3) = C_5 \sin J_1 x_3 + C_6 \cos J_1 x_3 \tag{29b}$$

in which the parameters J_1 and J_2 are defined as

$$J_1 = \sqrt{\frac{(\rho\omega^2\beta - \eta)(\alpha_m^2 + \gamma_n^2) + \rho\omega^2}{\rho\omega^2\beta - \eta}}; \quad J_2 = \sqrt{\frac{(\rho\omega^2\beta - \mu)(\alpha_m^2 + \gamma_n^2) + \rho\omega^2}{\rho\omega^2\beta - \mu}} \tag{30}$$

Substituting Eqs. 29a and 29b into 26a and 26b, and then into (21) and (22) in absence of body forces, yields two ordinary differential equations. Only the homogeneous solutions of the mentioned differential equations are the solution of the displacement components and thus the extra constants assumed to be zero. In such a case, the displacement components u_1 and u_2 can be written as

$$u_1 = \sum_{m=1}^{\infty}\sum_{n=1}^{\infty}[a_1 \sinh J_1x_3 + a_2 \cosh J_1x_3 + a_3 2x_3 \\ + a_4 \cosh J_2x_3] \cos \alpha_m x_1 \sin \gamma_n x_2 e^{I\omega_{mn}^k t} \tag{31a}$$

$$u_2 = \sum_{m=1}^{\infty}\sum_{n=1}^{\infty}[b_1 \sinh J_1x_3 + b_2 \cosh J_1x_3 + b_3 \sinh J_2x_3 \\ + b_4 \cosh J_2x_3] \sin \alpha_m x_1 \cos \gamma_n x_2 e^{I\omega_{mn}^k t} \tag{31b}$$

where a_i and b_i $(i = 1,\ldots,4)$ are the constant coefficients in terms of material properties and coefficients C_i $(i = 1,\ldots,6)$. The six boundary conditions on top and bottom surface of the plate are

$$t_{13} = t_{23} = t_{33} = 0 \quad \text{at } x_3 = \pm h/2 \tag{32}$$

Using strain-displacement and constitutive relations to satisfy the above boundary conditions will result in an eigen-value problem. Setting the determinant of the six order coefficient matrix equal to zero, the natural frequencies of the nano-plate are evaluated.

5 Numerical Results and Discussion

To verify the accuracy of the formulations, a comparison study of the results is performed with the results reported by Levinson [11] for classical rectangular plate $\beta = 0$. The first flexural and breathing (thickness-twist mode) frequencies are presented in Table 1 and it can be seen that the frequencies are accurate.

For numerical results, the following material properties are used throughout the investigation

$$E = 1.2\,\text{TPa},\ \nu = 0.3 \tag{33}$$

The two first non-dimensional frequencies $\Omega = \omega a^2\sqrt{\rho/Eh^2}$ are tabulated in Table 2 for various values of nonlocal parameter, different thickness to length ratios h/a and some aspect ratios a/b.

Based on the results in this tables, it can be concluded that for constant h/a, the frequency parameter decreases for all modes as the nonlocal parameter β increases. The reason is that with increasing the nonlocal parameter, the stiffness of the nano-plate decreases. i.e. the small scale effect makes the nano-plate more flexible

Table 1 Comparison of natural frequencies of classical isotropic rectangular plate ($\beta = 0$)

h/a	Theory	First flexural frequency(ω_{11}^1)	Breathing frequency(ω_{11}^2)
0.05	FSDT [11]	3049	——
	Elasticity [11]	3016	47750
	Present (3D)	3038	47773
0.1	FSDT [11]	5918	——
	Elasticity [11]	5096	47700
	Present (3D)	5928	47717
0.2	FSDT [11]	10820	——
	Elasticity [11]	10880	47460
	Present (3D)	10884	47481
0.4	FSDT [11]	17073	——
	Elasticity [11]	17315	46230
	Present (3D)	17323	46447

Table 2 Two first non-dimensional natural frequencies ($\Omega = \omega a^2 \sqrt{\rho/Eh^2}$) of nonlocal rectangular plate

β	h/a	$b/a = 1$		$b/a = 2$	
		ω_{11}^1	ω_{21}^1	ω_{11}^1	ω_{21}^1
0	0.1	5.7769	13.8051	3.6549	11.8616
	0.2	5.3036	11.6455	3.4513	10.1875
1	0.1	5.1260	10.7794	3.3768	9.5246
	0.2	4.7513	9.3524	3.2040	8.3789
2	0.1	4.6550	9.1451	3.1540	8.1836
	0.2	4.3433	8.0524	3.0035	7.2954
3	0.1	4.2938	8.0832	2.9702	7.2860
	0.2	4.0258	7.1829	2.8366	6.5505
4	0.1	4.0056	7.3029	2.8151	6.6311
	0.2	3.7694	6.5480	2.6949	5.9971

as the nonlocal model may be viewed as atoms linked by elastic springs while the local continuum model assumes the spring constant to take on an infinite value.

The influence of thickness-length ratio on the frequency parameter can also be examined by keeping the nonlocal parameter constant while varying the thickness to length ratio. It can be easily observed that as h/a increases, the frequency parameter decreases. The decrease in the frequency parameter is due to effects of the shear deformation, rotary inertia and use of term a^2/h in the definition of the non-dimensional frequency Ω. These effects are more considerable in the second mode than in the first modes.

A comparison has been carried out with the results of nonlocal classical plate theory (CPT), first order shear deformation theory (FSDT) and third order shear deformation theory (TSDT) reported by Aghababaei and Reddy [9]. The first non-dimensional natural frequency parameter $\Omega = \omega a^2 \sqrt{\rho/Eh^2}$ is listed in

Table 3 Comparison of first non-dimensional frequency ($\omega a^2\pi^4\sqrt{\rho/Eh^2}$) of a square nano-plate with 2-D plate theories ($a = 10$, $E = 30 \times 10^6$, $\nu = 0.3$)

β	h/a	Theory	$\omega a^2\pi^4\sqrt{\rho/Eh^2}$	Error
1	0.05	CPT [9]	0.0220	4.3%
		FSDT [9]	0.0218	3.3%
		TSDT [9]	0.0218	3.3%
		Present(3-D)	0.0211	
	0.1	CPT [9]	0.0880	6.4%
		FSDT [9]	0.0850	2.8%
		TSDT [9]	0.0854	3.3%
		Present(3-D)	0.0827	
2	0.05	CPT [9]	0.0204	6.8%
		FSDT [9]	0.0202	5.8%
		TSDT [9]	0.0202	5.8%
		Present(3-D)	0.0191	
	0.1	CPT [9]	0.0816	8.7%
		FSDT [9]	0.0788	4.9%
		TSDT [9]	0.0791	5.3%
		Present(3-D)	0.0751	
3	0.05	CPT [9]	0.0191	8.5%
		FSDT [9]	0.0189	7.4%
		TSDT [9]	0.0189	7.4%
		Present(3-D)	0.0176	
	0.1	CPT [9]	0.0763	10.3%
		FSDT [9]	0.0737	6.5%
		TSDT [9]	0.0741	7.1%
		Present(3-D)	0.0692	
4	0.05	CPT [9]	0.0180	9.8%
		FSDT [9]	0.0178	8.5%
		TSDT [9]	0.0179	9.1%
		Present(3-D)	0.0164	
	0.1	CPT [9]	0.0720	11.4%
		FSDT [9]	0.0696	7.7%
		TSDT [9]	0.0699	8.2%
		Present(3-D)	0.0646	

Table 3 for some nonlocal parameters. It can be seen that the natural frequencies of the nonlocal plate theories are not as accurate as the results in classical elasticity. Since the plate theories simplify the shear effect in transverse direction, their results are not accurate in nonlocal elasticity. This is because of the small scale effect makes the nano-plate more flexible than local model and it is more convenient to use three dimensional nonlocal elasticity for vibration analysis of nano-plates. Also, the variation of the natural frequencies versus the aspect ratio (h/a) is depicted in Fig. 1 for FSDT and present theories.

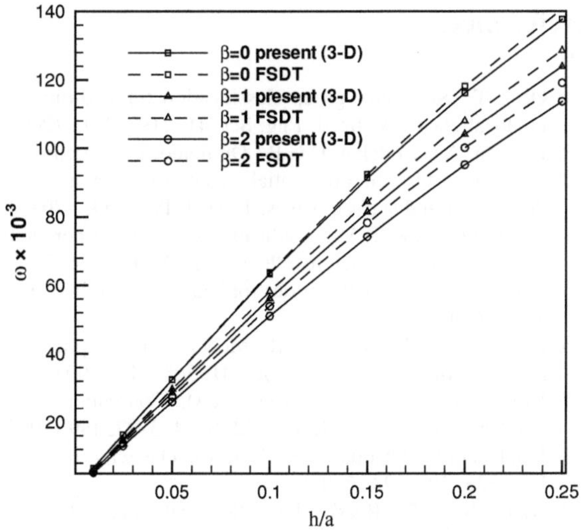

Fig. 1 Variation of natural frequency versus the thickness to length ratio

The shear correction factor for FSDT has been assumed to be 5/6 [8]. As it can be seen, the natural frequencies obtained from first order shear deformation theory are higher than the frequencies of three dimensional nonlocal elasticity. This effect is more significant in higher values of thickness to length ratio. This may be because of the value of 5/6 is no longer appropriate for nonlocal elasticity analysis due to the small scale effect.

6 Conclusion

In this chapter, the displacement field equations of nonlocal elasticity have been reformulated. To this end, a new form of displacement field equations of motion has been developed considering the small scale effect. The main feature of this formulation is converting coupled equations of motion into decoupled equations in terms of displacement and rotation components. These decoupled equations are invariant with respect to choice of coordinate system. One of the decoupled equations of displacement component and corresponding rotation equation can be solved independently and the other unknown problem parameters can be defined in terms of these components. Since the total degree of selected displacement and corresponding rotation equations is the same as original coupled equations, the presented reformulation does not require any complementary equation. Finally, accurate natural frequencies of the nano-plates have been presented for various values of nonlocal parameter.

Acknowledgments The authors wish to acknowledge the financial support from *National Elite Foundation of Iran.*

References

1. Ozer, T.: Symmetry group classification for two-dimensional elastodynamics problems in nonlocal elasticity. Int. J. Eng. Sci. **41**, 2193–2211 (2003)
2. Eringen, A.C., Edelen, D.G.B.: On nonlocal elasticity. Int. J. Eng. Sci. **10**, 233–248 (1972)
3. Eringen, A.C.: On differential equations of nonlocal elasticity and solutions of screw dislocation and surface waves. J. Appl. Phys. **54**, 4703–4710 (1983)
4. Wang, Q., Liew, K.M.: Application of nonlocal continuum mechanics to static analysis of micro- and nano-structures. Phys. Lett. A **363**, 236–242 (2007)
5. Reddy, J.N.: Nonlocal theories for bending, buckling and vibration of beams. Int. J. Eng. Sci. **45**, 288–307 (2007)
6. Peddieson, J., Buchanan, G.R., McNitt, R.P.: Application of nonlocal continuum models to nanotechnology. Int. J. Eng. Sci. **41**, 305–312 (2003)
7. Kitipornchai, S., He, X.Q., Liew, K.M.: Continuum model for the vibration of multilayered graphene sheets. Phys. Rev. B **72**, 075443–075449 (2005)
8. Pradhan, S.C., Phadikar, J.K.: Nonlocal elasticity theory for vibration of nanoplates. J. Sound Vib. **325**, 206–223 (2009)
9. Aghababaei, R., Reddy, J.N.: Nonlocal third-order shear deformation plate theory with application to bending and vibration of plates. J. Sound Vib. **326**, 277–289 (2009)
10. Plevako, V.P.: On the theory of elasticity of inhomogeneous media. J. Appl. Math. Mech. **35**, 806–813 (1971)
11. Levinson, M.: Free vibrations of a simply supported, rectangular plate: an exact elasticity solution. J. Sound Vib. **98**, 289–298 (1985)
12. Charalambopoulos, A., Dassions, G., Fotiadis, D.I., Massalas, C.V.: Note on eigenvector solutions of the Navier equation on cylindrical coordinates. Acta Mech. **134**, 115–119 (1999)
13. Poullikkas A, Karageorghis A, Georgiou G The method of fundamental solutions in three dimensional elastostatics. International conference on parallel processing and applied mathematics, vol. 2328, pp. 747–755. Leczow, Poland (2001)
14. Ozer, T.: The solution of Navier equations of classical elasticity using Lie symmetry groups. Mech. Res. Commun. **30**, 193–201 (2003)
15. Saidi, A.R., Atashipour, S.R., Jomehzadeh, E.: Reformulation of Navier equations for solving three-dimensional elasticity problems with applications to thick plate analysis. Acta Mech. **208**, 227–235 (2009)

A Finite Element Model to Investigate the Stress–Strain Behavior of Single Walled Carbon Nanotube

Ehsan Mohammadpour and Mokhtar Awang

Abstract This chapter describes a finite element (FE) method that is appropriate for the numerical prediction of mechanical behavior of different types of isolated Single walled carbon nanotube (SWCNT). The aim of this research is to develop a FE model based on the modified Morse interatomic potential to evaluate axial Young's modulus of nanotubes. The novelty of the model lies on the use of ANSYS's beam element with non-linear capability, i.e., element type BEAM188 is used to evaluate SWCNT's mechanical properties. In the present modeling work, an individual carbon nanotube (CNT) is simulated as a frame-like structure and the primary bonds between two nearest-neighboring carbon atoms are treated as 3D beam elements. The beam element properties are determined via the concept of energy equivalence between molecular dynamics and structural mechanics using modified Morse potential. The calculated mechanical properties show good agreement with existing works.

Keywords Single walled carbon nanotube (SWCNT) · Finite element model (FEM) · Morse potential functions · Continuum mechanics · Young's modulus

E. Mohammadpour (✉) · M. Awang
Mechanical Engineering Department, Universiti Teknologi PETRONAS,
31750 Seri Iskandar, Perak, MALAYSIA
e-mail: eh.mohammadpour@gmail.com

M. Awang
e-mail: mokhtar_awang@petronas.com.my

A. Öchsner et al. (eds.), *Materials with Complex Behaviour II*,
Advanced Structured Materials 16, DOI: 10.1007/978-3-642-22700-4_22,

1 Introduction

The discovery of carbon nanotubes (CNT) has stimulated considerable experimental and theoretical studies. Various promising applications have been proposed based on their unique geometrical and mechanical properties [1]. They are structurally unique materials that exhibit excellent mechanical, electrical, thermal, and optical properties, e.g., Nanotubes can have diameters ranging from 1 to 100 nm and lengths of up to millimeters. Their densities can be as low as 1.3 g/cm3 and their Young's moduli are superior to all carbon fibers with values greater than 1 TPa [2, 3].

Moreover, the extreme small size makes them suitable to be embedded into any type of light weight and soft materials as reinforcements to form strong and light nanocomposites. Therefore, the potential use of CNTs as reinforcing materials in nanocomposites has originated the need to explore their mechanical properties. To unlock the potential of carbon nanotubes for application in polymer nanocomposites, one must fully understand the elastic and fracture properties of carbon nanotubes as well as the interactions at the nanotube/matrix interface.

However, the true mechanical properties of nanotubes such as their Young's modulus, yield strength, ultimate strength, elastic properties and even fracture behavior are still uncertain to date. This actually induces many arguments in whether the nanotubes are suitable to be used as nano-reinforcements for the nanocomposites or not.

Experiments conducted previously showed that the Young's moduli of nanotubes range from 270 to 950 GPa. Such a large discrepancy was due to the different sizes, lengths and numbers of wall layers used in different tests. However, it is hard to produce identical nanotubes even in the same experiment [4]. At present, since experimental investigation in this area is still a challenging work, numerical studies are another available tool for predicting nanotubes properties. CNTs were simulated extensively using molecular dynamics and continuum mechanics. The atomistic approaches, based on the force field and total potential energy related to the interatomic potentials for CNTs in a macroscopic sense has been used most extensively, include classical molecular dynamics, tight-bonding molecular dynamics and density functional theory [2]. Despite the fact that these approaches can be used for any problem associated with molecular or atomic motions, their huge computational tasks restricted their applications to small number of molecules or atoms. Alternatively, continuum mechanics approaches have also been utilized to estimate physical properties of CNTs with less computational efforts [5]. These approaches mainly involve classical continuum mechanics and continuum shell modeling [4]. Since a CNT can be well described as a continuum solid beam or shell subjected to tension, bending or torsional forces, it is reasonable to model CNT as a frame- or shell-like structure. The mechanical properties of such structure can then be obtained using the FE method for classical continuum mechanics.

Due to the uncertainty of the CNTs characteristics for both of the above modeling techniques, however, the obtained mechanical properties of CNTs are widely scattered. The predicted Young's modulus of CNTs yields a widespread range of about 1.0–5.5 TPa [6]. Experimentally determined Young's modulus of Single walled carbon nanotubes (SWCNT) are also scattered in a relatively large interval of 2.8–3.6 TPa [6, 7]. Furthermore, different configurations of imperfections in CNTs wall such as cracks, vacancies and defects influence the mechanical behavior of carbon nanotubes effectively. These phenomena have been studied using molecular dynamics [8] and FE method [9] to show how the defects and the imperfections affect the CNT structure under different loading conditions.

In this work, the main objectives are to develop a FE model of SWCNTs and to investigate the Young's modulus of SWCNTs based on nanoscale continuum modeling. By employing frame elements to represent carbon-to-carbon bonds, the FE model is developed to predict the elastic modulus of SWCNTs. The effect of CNT chirality, diameter and structure on mechanical properties is also studied in this work.

2 Modeling of Carbon Nanotubes

There are several ways to view a SWCNT. The most widely used is by reference to rolling up graphene sheet to form a hollow cylinder with end caps. The cylinder is composed of hexagonal carbon rings, while the end caps are pentagonal rings. The hexagonal pattern is repeated periodically leading to binding of each carbon atom to three neighboring atoms with covalent bonds. The very strong chemical bond between carbon atoms plays significant role to the impressive mechanical properties of graphene and as a consequence, of all carbon-related nano-structures [7, 10]. The atomic structure of CNTs depends on tube chirality, which is defined by the chiral vector and the chiral angle [7]. In Fig. 1, a section of graphene sheet is presented. One could make a nanotube by cutting of the graphene sheet along the dotted lines and rolling the tube so that the tip of the chiral vector touches its tail. The chiral vector, also known as the roll-up vector, can be described by the following equation:

$$\vec{C}_h = n\vec{a}_1 + m\vec{a}_2 \quad (1)$$

where the integers n and m are the number of steps along the zigzag carbon bonds of the hexagonal lattice and $\vec{a}_1$ and $\vec{a}_2$ are unit vectors. The chiral angle determines the amount of twist in the tube. The limiting values for chiral angle are 0 and 30° which referred to zigzag and armchair CNTs, respectively. Any other form of chiral vector placed between two limiting values is so called Chiral CNT. The roll-up vector of the nanotube also defines the nanotube diameter,

$$D_{\mathrm{CNT}} = \frac{\sqrt{3}L}{\pi}\sqrt{(n^2 + m^2 + nm)} \quad (2)$$

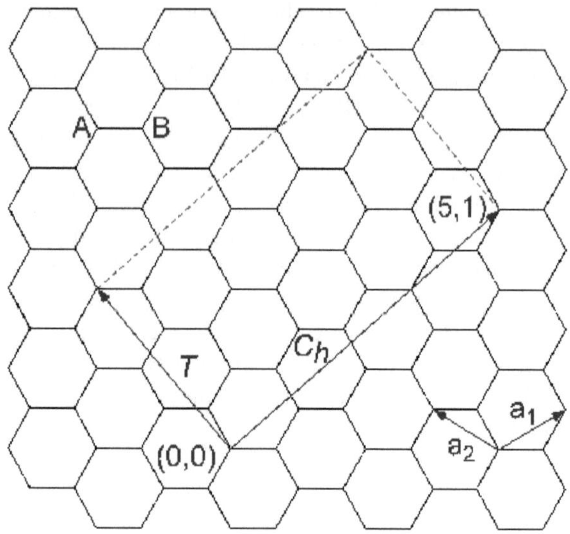

Fig. 1 Schematic of nanotube structure

The physical properties of carbon nanotubes are sensitive to their diameter, length and chirality. In particular, tube chirality is known to have a strong influence on the electronic properties of carbon nanotubes [1]. Graphite is considered to be a semi-metal, but it has been shown that nanotubes can be either metallic or semi-conducting, depending on tube chirality. The influence of chirality on the mechanical properties of carbon nanotubes has also been reported.

The bonds have a characteristic bond length $\alpha_{C\text{–}C}$ and bond angle in the 3D space. The displacement of individual atoms under an external force is constrained by the bonds. Therefore, the total deformation of the CNT is the result of the interactions between the bonds. By considering the bonds as connecting load-carrying elements, and the atoms as joints of the connecting elements, CNTs can be modeled as space-frame structures [4, 11, 12]. In Fig. 2a, a typical CNT in the form of a 3D frame structure is illustrated.

As mentioned before, by treating CNTs as space-frame structures, their mechanical behavior can be analyzed using classical structural mechanics methods. In this work, a 3D FE model is proposed to assess the mechanical properties of SWCNT. The 3D FE model is developed using ANSYS commercial FE code. To model the C–C bonds, 3D beam elements are used, as shown in Fig. 2b.

To calculate the elastic modulus of beam elements, a linkage between molecular and continuum mechanics is used. In its general formula, the potential energy is described as [9].

$$U = \Sigma U_r + \Sigma U_\theta + \Sigma U_\phi + \Sigma U_\omega + \Sigma U_{vdW} + \Sigma U_{el} \tag{3}$$

where, U_r, U_θ, U_φ, U_ω, U_{vdW}, U_{el} are bond stretching, bond angle bending, dihedral angle torsion, inversion terms, van der Walls interaction and electrostatic interaction, respectively. Various functional forms may be used for these potential

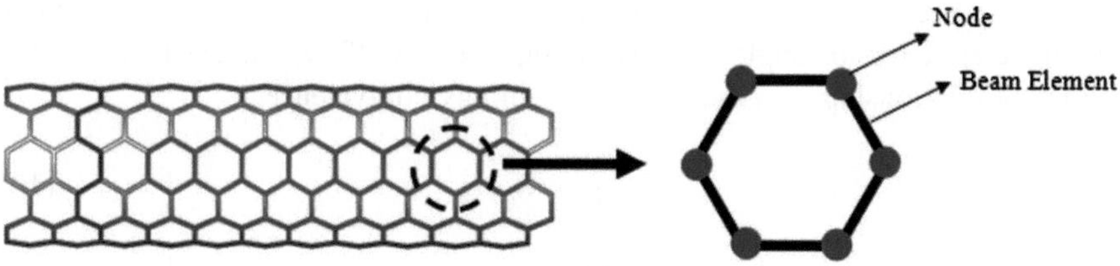

Fig. 2 **a** The hexagonal structure of a typical CNT. **b** FE modeling concept of the hexagonal structure of a CNT

energy terms depending on the particular material and loading conditions. In most cases, the effects of U_{φ}, U_{ω}, U_{vdW}, are neglected under the uniaxial loading and small strain [4, 13, 14]. In addition, in most cases where continuum methods were used to analyze CNTs embedded in an elastic medium, a linear behavior of the reinforcements was assumed such as reported in Fan et al. [14] and Shokrieh et al. [5] works. Therefore, by adopting the simple harmonic form under the assumption of small deformation, and merging U_{φ}, U_{ω} into a single equivalent term, we obtain the energy form of each bonded interaction as

$$U_r = \frac{1}{2}k_r(\Delta r)^2 \tag{4}$$

$$U_\theta = \frac{1}{2}k_\theta(\Delta\theta)^2 \tag{5}$$

$$U_\tau = U_\phi + U_\omega = \frac{1}{2}k_\tau(\Delta\varphi)^2 \tag{6}$$

where k_r, k_θ and k_τ are the associated force constants. Δr, $\Delta\theta$ and $\Delta\varphi$ stand for the bond stretching increment, the bond angle change and the angle of bond twisting, respectively.

In the counterpart, the strain energy forms of a uniform beam due to pure tension, bending and torsion can be represented as

$$U_A = \frac{1}{2}\frac{EA}{L}(\Delta L)^2 \tag{7}$$

$$U_M = \frac{1}{2}\frac{EI}{L}(2\alpha)^2 \tag{8}$$

$$U_T = \frac{1}{2}\frac{GJ}{L}(\Delta\beta)^2 \tag{9}$$

where U_A, U_M and U_T are the axial strain energy, the bending energy and the torsion energy, respectively. L, A, I, J, E,G are defined as the length, the cross section area, the moment of inertia, the polar moment of inertia of the cross section, the Young's modulus and the shear modulus of the beam, respectively.

ΔL, 2α, $\Delta\beta$ are the axial stretching deformation, the total relative rotation angle and the relative torsion angle. Comparing Eqs. (4–6) to (7–9) will easily lead to the following relationships via the concept of energy equivalence,

$$\frac{EA}{L} = k_r \tag{10}$$

$$\frac{EI}{L} = k_\theta \tag{11}$$

$$\frac{GJ}{L} = k_\tau \tag{12}$$

In the FE model, BEAM4 element in ANSYS is selected to simulate the elastic behavior of carbon bonds. This element is a uniaxial element with tension, compression, torsion and bending capabilities. It has six degrees of freedom at each node: translations in the nodal *x, y,* and *z* directions and rotations about the nodal *x, y,* and *z*-axes. The element is defined by two or three nodes as well as its cross cross-sectional area, two moments of inertia, two dimensions and the material properties. The cross sections of the beam elements are assumed to be uniform and circular, and the necessary input data of the BEAM4 element are the Young's modulus E, the Poisson's ratio v and the diameter of the circular cross section d_b. From Eqs. (10) and (11), we have

$$d_b = 4\sqrt{\frac{k_\theta}{k_r}}, \quad E = \frac{Lk_r^2}{4\pi k_\theta} \tag{13}$$

The values of k_r, k_θ, are adopted as those used by other references [4, 5, 11], which are 938 kcal mol^{-1} Å^{-2} (equivalent to 6.53×10^{-7} N/nm) and 126 kcal mol^{-1}/rad^2 (equivalent to 8.79×10^{-10} Nnm/rad^2), respectively. The element length L is set to be equal to the length of a C–C bond, which is 1.42 Å. By using Eq. (13), the values of E and d_b can be computed as 5.5 TPa and 0.147 nm, respectively. It is worth to mention that this assumption leads to accurate predictions only in cases where very small CNT deformations take place. Consequently, this method cannot be used for modeling the mechanical behavior of the composites.

As mentioned above, for larger strains of CNTs the more accurate behavior of carbon bond must be considered. Various functions forms could be used. Regarding the bond stretching, an attempt to represent the experimentally observed bond energy curves of diatomic molecules by simple analytical functions could be utilized. In the following section, the tensile behavior of the isolated carbon CNTs is simulated using a simple analytical Morse function, which can be written as [13],

$$U = U_{stretch} + U_{angle} \tag{14}$$

$$U_{stretch} = D_e\left[\left(1 - e^{-\beta(r-r_0)}\right)^2 - 1\right] \tag{15}$$

$$U_{angle} = \frac{1}{2} k_{\theta} (\theta - \theta_0)^2 \left[1 + k_{sexic} (\theta - \theta_0)^4 \right] \tag{16}$$

where $U_{stretch}$ is the bond energy due to bond stretching and U_{angle} is the bond energy due to bond angle-bending, r is the current bond length and θ is the current angle of the adjacent bonds. The other parameters of the potential are $r_O = 1.421 \times 10^{-10}$ m; $D_e = 6.03105 \times 10^{-19}$ Nm; $\beta = 2.625 \times 10^{10}$ m^{-1}; $\theta_O = 2.094$ rad; $k_\theta = 0.9 \times 10^{-18}$ Nm/rad^2; $k_{sexic} = 0.754$ rad^{-4} [15]. For strains above 10%, as stretching dominates CNT fracture and the effect of angle-bending potential is very small, in the present model only the bond stretching potential is considered [16] to simplify the situation. By differentiating the stretching energy term in Eq. (15), the stretching force of atomic bond, F, is obtained in the molecular force field as

$$F(r) = 2\beta D_e [1 - e^{-\beta(r - r_0)}] e^{-\beta(r - r_0)} \tag{17}$$

The relationship between stress σ and bond strain ε_b for the C–C bonds could be calculated using the element's cross-sectional area equal to 1.69×10^{-20} m for the C–C bond as shown in Fig. 3. The strain of the bond is defined by $\varepsilon_b = (r - r_0)/r_0$. As may be seen, the stress–strain relation is highly non-linear especially at large strains and the inflection point (peak force) occurs at about 19% strain. Therefore, a different element, BEAM188, was chosen to simulate the carbon–carbon atom stretching behavior described by Morse potential. BEAM188 is a two-node beam element in 3D and has six or seven degrees of freedom at each node. So, it is well-suited for linear, large rotation, and/or large strain non-linear applications. Again, we assume that the cross sections of the beam elements are uniform and circular. The initial stiffness is set at 6.5 TPa, according to the initial slope of the C–C bond stress–strain curve (Fig. 3).

3 Numerical Results and Discussion

In this section, we will use the FE to compute the axial Young's modulus of carbon CNTs of various types and sizes. In addition, comparison of these results to those found in the literature will be given. We also discuss the influence of element type and tube size on the mechanical properties obtained.

The CNT is loaded by an incremental force at one end while the other end is being fully constrained, as shown in Fig. 4. To compute the axial Young's modulus from the numerical results, the following equation was used

$$E = \frac{\sigma}{\varepsilon} = \frac{F L_n}{A \Delta L_n} \tag{18}$$

where E is the axial Young's modulus, σ and ε are the axial stress and strain respectively, F is the total force applied on one end of the tube, A is the cross

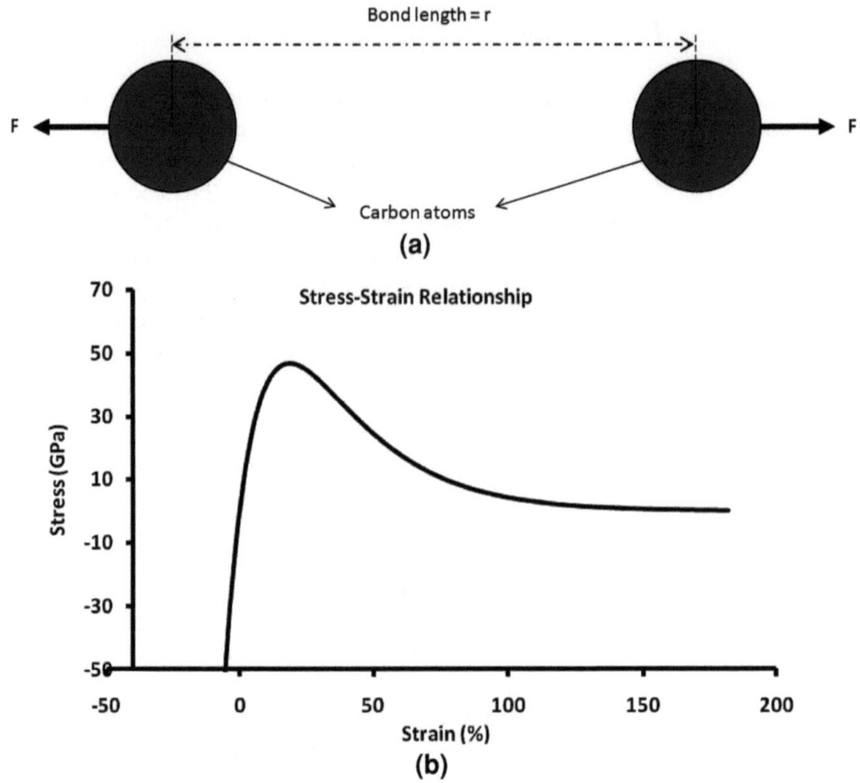

Fig. 3 **a**The C–C bond. **b** stress–strain curve for C–C bond according to the modified Morse potential (Eq. 15)

section area of the CNT, which is defined as $A = \pi\ D_n\ t_n$ (where D_n = CNT diameter, thickness $t_n = 0.34$ nm is the interlayer graphite distance) [9].)

Figure 5 showed the strain of carbon nanotube under axial loading and related stress–strain curve of zigzag (12,0) CNT, respectively. The axial Young's modulus for six zigzag SWCNTs of different sizes and chirality, (6,0), (8,0), (10,0), (12,0), (14,0), (22,0) were simulated using linear elastic theory (Eq. 13) and the results are listed in Table 1 and also depicted in Fig. 6 for comparison. Form the results it could be seen that the axial Young's modulus whose values are about 1.2 TPa for elastic modeling increase slightly with increasing diameter. From Fig. 6, the effect of tube diameter on the Young's modulus is also clearly observed.

Based on modified Morse potential, the stress–strain curve for a six zigzag (6,0), (8,0), (10,0), (12,0), (14,0), (22,0) and six armchair (5,5), (8,8), (10,10), (12,12), (14,14), (16,16) SWCNT were calculated. The value of Young's modulus for each CNT was calculated in the same way and listed in Tables 1 and 2 as well.

Results show that the axial Young's modulus in the nonlinear cases also increase slightly with increasing diameter. For smaller tubes, for example to a

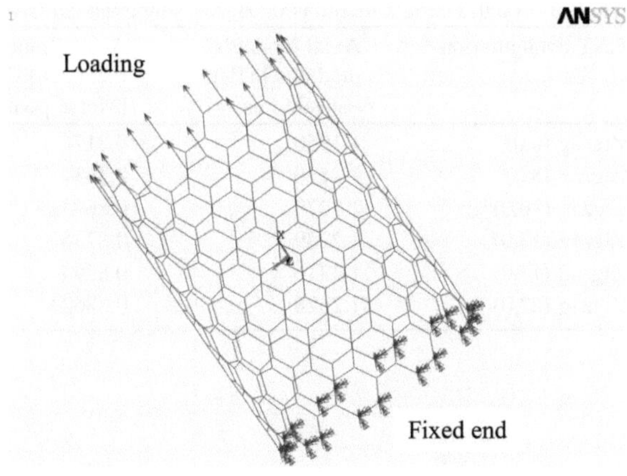

Fig. 4 Isometric view of the FE mesh of the (12,0) SWCNTs along with the applied boundary conditions

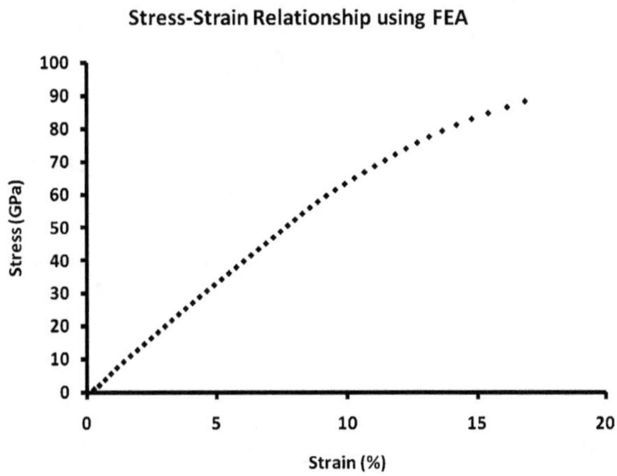

Fig. 5 Uniaxial strain–stress curve for zigzag CNT (12,0) under uniaxial loading

diameter less than 1.0 nm, the Young's modulus exhibits a stronger dependence on the tube diameter.

However, for tube diameters larger than 1.0 nm, this dependence becomes very weak. The general tendency is that the Young's modulus increases with increasing tube diameter.

The lower Young's modulus of zigzag nanotubes at smaller CNT diameter could be attributed to the higher curvature, which results in a more significant distortion of C–C bonds. As the CNT diameter increases, the effect of curvature diminishes gradually [17].

It also could be seen that the Young's modulus calculated from Eq. (15) in all cases is higher than the same values based on Eq. (17). These results show the importance of initial assumption in the FE modeling of carbon nanotubes.

Table 1 Axial Young's modulus of zigzag single-walled CNTs

CNT configuration	Axial Young's modulus [TPa] (Elastic theory)	Axial Young's modulus [TP]) (Morse potential)	CNT Diameter [nm]
Zigzag (6,0)	1.1950	0.8174	0.4697
Zigzag (8,0)	1.2170	0.8657	0.6263
Zigzag (10,0)	1.2270	0.8693	0.7828
Zigzag (12,0)	1.2289	0.8735	0.9394
Zigzag (14,0)	1.2316	0.8794	1.0960
Zigzag (22,0)	1.2328	0.8862	1.7223

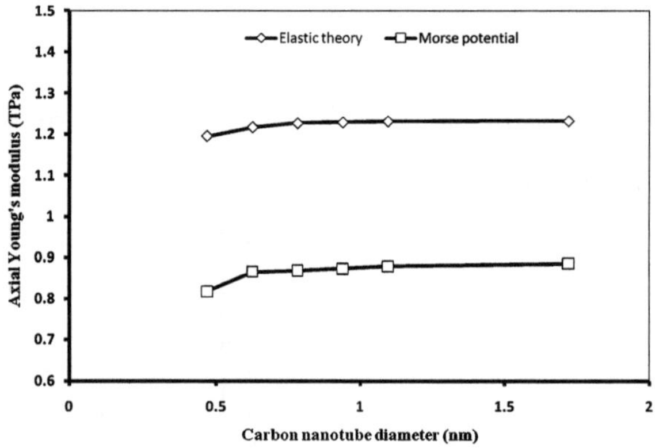

Fig. 6 Axial Young's modulus for zigzag CNTs with various CNT diameters using elastic theory and Morse potential

Table 2 Axial Young's modulus of armchair single-walled CNTs

CNT configuration	Axial Young's modulus [TP]) (Morse potential)	CNT Diameter [nm]
Armchair (5,5)	1.0477	0.553523
Armchair (8,8)	1.0555	0.885637
Armchair (10,10)	1.0573	1.107046
Armchair (12,12)	1.0631	1.328456
Armchair (14,14)	1.0670	1.549865
Armchair (16,16)	1.0675	1.771274

For nonlinear analysis, armchair nanotubes exhibited higher axial modulus in comparison with zigzag nanotubes, as shown in Fig. 7. This fact emphasized the dependence of mechanical properties on the CNT structure which is in agreement with Fan et al. [14], Tserpes et al. [16] and Xiao et al. [18].

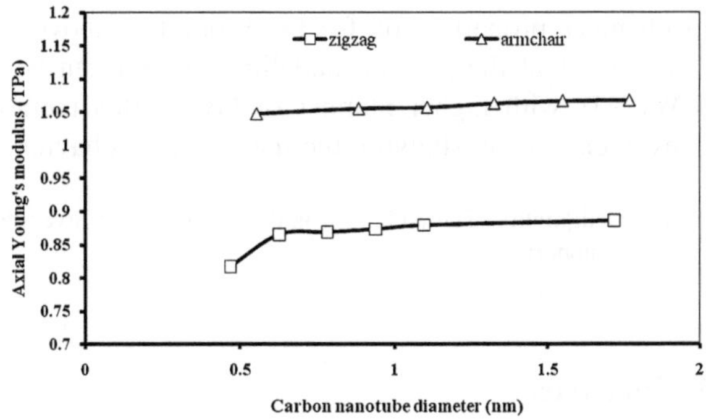

Fig. 7 Axial Young's modulus for zigzag and armchair CNTs with various CNT diameters using Morse potential function (Eq. (17)

Table 3 Comparison between Young's modulus of CNT reported by past researchers

Investigators	Method	CNT type	Young's modulus [TPa]	CNT thickness [nm]
Tserpesa et al. [4]	FE method	SWCNTs	1.029	0.34
Shokrieh et al. [5]	Closed-form solution	SWCNTs	1.033	0.34
Fan et al. [14]	FE method	SWCNTs	1.033	0.34
Georgantzinosa et al. [19]	FE method	SWCNTs	0.936	0.34
Jin et al. [20]	Molecular dynamics	SWCNTs	1.238	0.34
Present work	FE method (Elastic theory)	SWCNTs	1.220	0.34
	(Morse potential)	SWCNTs	0.873	0.34

Our computational results are comparable to those obtained from numerical studies. Comparison between reported data for CNT Young's modulus in literature and obtained results in this study is presented in Table 3.

4 Conclusions

The FE simulation technique for SWCNTs has been developed which can be easily performed by the commercial finite element code ANSYS. The key modeling concept is that molecular bonds are presented as beam elements. We proposed and verified a simplifying method to model non-linear nature of covalent bond between two carbon atoms in the CNT wall. This method can significantly save the modeling and computing effort when FE analysis is performed. Numerical results for axial Young's modulus are presented to illustrate the accuracy of the established FE models. In addition, the relations between these mechanical properties and the CNT size are also investigated to give a better understanding of the variation of mechanical properties of CNTs. From the above results and the outstanding advantage that the present modeling concept can be easily extended to cases of MWCNTs with higher number of layers, this method will be an effective and convenient tool in studying the mechanical behavior of MWCNTs.

Acknowledgments The authors would like to thank Universiti Teknologi PETRONAS for the financial support.

References

1. Li, C., Chou, T.W.: Modeling of elastic buckling of carbon nanotubes by molecular structural mechanics approach. Mech. Mater. **36**, 1047–1055 (2004)
2. Wang, H.: Dispersing carbon nanotubes using surfactants. Curr. Opin. Colloid Interface Sci. **14**, 364–371 (2009)
3. Coleman, J., Khan, U., Blau, W., Gunko, Y.: Small but strong: A review of the mechanical properties of carbon nanotube–polymer composites. Carbon **44**, 1624–1652 (2006)
4. Tserpes, K., Papanikos, P.: Finite element modeling of single-walled carbon nanotubes. Compos. Part B Eng. **36**, 468–477 (2005)
5. Shokrieh, M.M., Rafiee, R.: On the tensile behavior of an embedded carbon nanotube in polymer matrix with non-bonded interphase region. Compos. Struct. **92**, 647–652 (2010)
6. Selmi, A., Friebel, C., Doghri, I., Hassis, H.: Prediction of the elastic properties of single walled carbon nanotube reinforced polymers: a comparative study of several micromechanical models. Compos. Sci. Technol. **67**, 2071–2084 (2007)
7. Thostensona, E.T., Renb, Z., Choua, T.-W.: Advances in the science and technology of carbon nanotubes and their composites a review. Compos. Sci. Technol. **61**, 1899–1912 (2001)
8. Belytschko, T., Xiao, S.P., Schatz, G.C., Ruoff, R.S.: Atomistic simulations of nanotube fracture. Phys. Rev. B **65**, 235430 (2002)
9. Rahmandoust, M., Öchsner, A.: Influence of structural imperfections and doping on the mechanical properties of single-walled carbon nanotubes. J. Nano Res. **6**, 185–196 (2009)
10. Lau, K., Gu, C., Hui, D.: A critical review on nanotube, nanotube/nanoclay related polymer composite materials. Compos. Part B Eng. **37**, 425–436 (2006)
11. Fan, C.W., Huang, J.H., Hwu, C., Liu, Y.Y.: Mechanical properties of single-walled carbon nanotubes—a finite element approach. Adv. Mater. Res. 33–37, 937–942 (2008)
12. Kalamkarov, A., Georgiades, A., Rokkam, S., Veedu, V., Ghaseminejhad, M.: Analytical and numerical techniques to predict carbon nanotubes properties. Int. J. Solids Struct. **43**, 6832–6854 (2006)
13. Shokrieh, M.M., Rafiee, R.: Prediction of Young's modulus of graphene sheets and carbon nanotubes using nanoscale continuum mechanics approach. Mater. Des. **31**, 790–795 (2010)
14. Fan, C.W., Liu, Y.Y., Hwu, C.: Finite element simulation for estimating the mechanical properties of multi-walled carbon nanotubes. Appl. Phys. A **95**, 819–831 (2009)
15. Rossi, M., Meo, M.: On the estimation of mechanical properties of single-walled carbon nanotubes by using a molecular-mechanics based FE approach. Compos. Sci. Technol. **69**, 1394–1398 (2009)
16. Tserpes, K., Papanikos, P., Labeas, G., Pantelakis, S.: Multi-scale modeling of tensile behavior of carbon nanotube-reinforced composites. Theor. Appl. Fract. Mech. **49**, 51–60 (2008)
17. Chou, T.W., Li, C.: A structural mechanics approach for the analysis of carbon Nanotubes. Int. J. Solids Struct. **40**, 2487–2499 (2003)

18. Xiao, J., Gama, B., Gillespiejr, J.: An analytical molecular structural mechanics model for the mechanical properties of carbon nanotubes. Int. J. Solids Struct. **42**, 3075–3092 (2005)
19. Georgantzinos, S.K., Giannopoulos, G.I., Anifantis, N.K.: Investigation of stress–strain behavior of single walled carbon nanotube/rubber composites by a multi-scale finite element method. Theor. Appl. Fract. Mech. **52**, 158–164 (2009)
20. Jin, Y., Yuan, F.G.: Simulation of elastic properties of single-walled carbon nanotubes. Compos. Sci. Technol. **63**, 1507–1515 (2003)

Nonlinear Bending and Post Buckling of Functionally Graded Circular Plates under Asymmetric Thermo-Mechanical Loading

F. Fallah and A. Nosier

Abstract Based on the first-order shear deformation plate theory with von Karman nonlinearity, the nonlinear bending and post buckling of functionally graded circular plates under asymmetric transverse loading and a temperature variation through the plate thickness is investigated. Introducing a stress function and a potential function, the governing equations are uncoupled to form equations describing the interior and edge-zone problems of functionally graded plates. This uncoupling is then used to conveniently present an analytical solution for the nonlinear asymmetric deformation of an FG circular plate. A two parameter perturbation technique, in conjunction with Fourier series method to model the problem asymmetries, is used to obtain the solution for clamped and simply-supported boundary conditions. The material properties are graded through the plate thickness according to a power-law distribution of the volume fraction of the constituents. The results are verified with the existing results in the literature. The effects of nonlinearity, material constant, and boundary conditions on various response quantities in a solid circular plate are studied and discussed. Snap-through buckling is observed in simply-supported FG plates under thermo-mechanical loading. Moreover, it is found that linear theory is inadequate for analyzing FG plates under thermal loading.

F. Fallah (✉)
Aerospace Research Institute, 15th Ave., Mahestan Ave, Shahrake Gharb, Tehran, Iran
e-mail: famida_fallah@hotmail.com

A. Nosier
Sharif University of Technology, Azadi Ave, Tehran, Iran
e-mail: nosier@sharif.edu

A. Öchsner et al. (eds.), *Materials with Complex Behaviour II*,
Advanced Structured Materials 16, DOI: 10.1007/978-3-642-22700-4_23,

1 Introduction

Functionally graded materials (FGMs) were first introduced in 1984 by material scientists in Japan as thermal barrier materials [1] in a high temperature environment. They belong to a new class of materials which are microscopically heterogeneous and their material properties vary continuously. This is achieved by gradually changing the volume fraction of the constituent materials along certain dimension (usually in the thickness direction). Due to the smooth variation of material properties, they offer many advantages over laminated composite materials including improved fatigue resistance, reduction of thermal stresses, residual stresses and interlaminar stresses, and more efficient joining techniques. Thus, FGMs are finding applications in many fields such as aerospace, power generation industries, and energy conversion. Sureh and Mortensen [2] provide an excellent introduction to the fundamentals of FGMs.

Studies on nonlinear behavior of FG plates are, however, rare in comparison with those available on linear analyses of FG rectangular [3–6] and circular [7–9] plates. Using finite element method, Praveen and Reddy [10] investigated the static and dynamic responses of functionally graded plates using first-order shear deformation plate theory (FSDT) and geometric nonlinearity in von Karman sense. Based on the higher-order shear deformation plate theory, Reddy [11] developed Navier's solutions for rectangular plates and finite element models to study the nonlinear dynamic response of FG plates. Based on the classical nonlinear von Karman plate theory, an analytical solution in terms of Fourier series was obtained by Woo and Meguid [12] for the nonlinear bending of functionally graded plates and shallow shells under transverse mechanical loads and a temperature field. Based on the higher-order shear deformation plate theory with the von Karman nonlinearity, Woo et al. [13] provided an analytical solution for the post buckling behavior of functionally graded plates and shallow shells under compressive edge loads and a temperature field. Na and Kim investigated nonlinear bending [14] and thermal buckling and post buckling [15] of FG clamped square plates based on the Green–Lagrange nonlinear strain–displacement relation using a 3-D finite element method. Thermo-mechanical loading is considered in [14]. Yang and Shen [16] investigated the large deflection and post buckling responses of functionally graded rectangular plates with clamped supports on two opposite edges under transverse and in-plane loads using the classical von Karman plate theory. More recently, based on Reddy's higher-order shear deformation plate theory with a von Karman-type of kinematic nonlinearity, Shen presented nonlinear thermal bending [17] and post buckling [18] analysis for a simply-supported FG plate using a two step perturbation technique. Heat conduction is considered in the thickness direction. Based on the same assumptions, Yang and Huang [19] analyzed nonlinear transient response of simply-supported imperfect FG plates in thermal environments. Prakash et al. [20] investigated nonlinear behavior of FG plates exposed to high temperature on the ceramic surface using neutral surface-based first-order shear deformation theory with the von Karman nonlinearity.

They reported snapping phenomenon in FG rectangular and skew plates under thermal loading.

From the literature review, it is evident that analysis of nonlinear bending and post buckling of FG circular plates [21–27] are much less than rectangular plates [10–20], while circular plates made of functionally graded materials are often employed as a part of engineering structures. Gunes and Reddy [21] carried out geometrically nonlinear analysis of FG circular plates with different boundary conditions subjected to mechanical and thermal loads using all the terms in Green–Lagrange strain tensor. Based on the classical nonlinear von Karman plate theory, the axisymmetric bending and post buckling of functionally graded thin circular plates subjected to mechanical and thermal loadings were studied by Ma and Wang [22]. Li et al. [23] studied nonlinear thermo-mechanical post buckling of an imperfect and perfect clamped FG circular plate using shooting method. Several authors have investigated the buckling of functionally graded circular plates [24, 25] in which the existence of bifurcation buckling for both clamped and simply-supported FG plates is assumed and the buckling point is searched using an eigen-value analysis. On the other hand, some other authors [13, 15, 18, 22, 23] have shown that in FG plates, due to edge compression, bifurcation-type buckling happens in clamped FG plates, while bending occurs in simply-supported FG plates.

Prakash and Ganapathi [26] investigated asymmetric free vibration and thermoelastic stability of FG circular plates using finite element procedure. Based on FSDT, Nosier and Fallah presented analytical solution for asymmetric linear [8] and nonlinear [27] bending of FG circular plates with various clamped and simply-supported boundary conditions. In the linear analysis [8] thermo-mechanical loading is considered while only mechanical loading is considered in the nonlinear analysis [27]. Here, asymmetric nonlinear behavior of FG plates subjected to thermo-mechanical loading is investigated. From the review of literature it appears that although axisymmetric buckling and bending of FG circular plates have been taken up by some researchers, very little work is available on asymmetric bending and post buckling of FG circular plates [8, 26, 27]. The authors have so far not come across any paper dealing with asymmetric nonlinear bending of FG circular plates subjected to thermo-mechanical loading.

In the present study, the tenth-order, nonlinear equilibrium equations of FSDT describing the bending-extension problem of FG circular plates subjected to thermo-mechanical loading are studied for analytical solutions. The formulation accounts for moderately large deflection in the von Karman sense. Here, based on the method developed in [27] which uses two potential functions; boundary-layer and stress functions to decouple the governing equilibrium equations, the system of five nonlinear coupled equations are reformulated into three equations; one linear second-order differential equation defining the edge-zone problem of the FG plate in terms of the boundary-layer function Φ, similar to the one obtained in [8], and two nonlinear fourth-order equations in terms of the transverse deflection w, and stress function F. This uncoupling makes it possible to conveniently present an analytical solution for asymmetric nonlinear behavior of clamped and

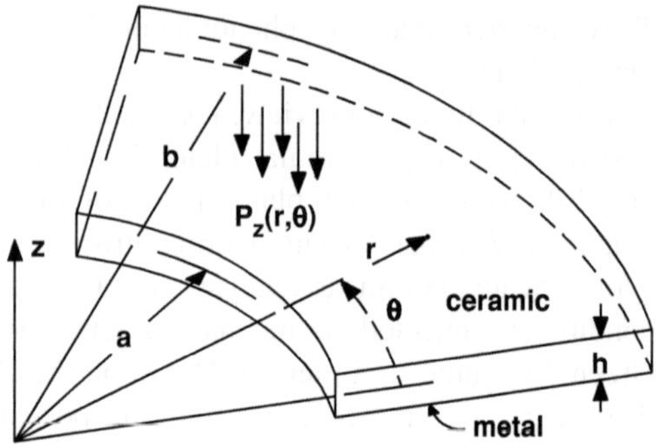

Fig. 1 Geometry of FG circular plate and coordinate system

simply-supported FG circular plates. The effects of nonlinearity, material properties, and boundary conditions on nonlinear behavior of a solid circular plate are studied and discussed in detail.

2 Theoretical Formulation

A functionally graded circular plate of inner and outer radii of, respectively, a and b and thickness h is considered here. The geometry of the plate and the coordinate system are shown in Fig. 1. FGMs are modeled as non-homogenous isotropic or orthotropic materials with different types of material description. Here, they are modeled as a non-homogenous isotropic linear thermoelastic material whose properties P, vary continuously through the plate thickness, as a function of the volume fraction and properties of the constituent materials. Assuming the plate is made from a mixture of ceramic and metal, P can be expressed as [7, 22, 28]:

$$\mathrm{P}(z) = (\mathrm{Pm} - \mathrm{Pc})\left(\frac{h-2z}{2h}\right)^n + \mathrm{Pc} \tag{1}$$

where subscripts c and m refer to ceramic and metal, respectively and n is the power-law index that takes values greater than or equal to zero. To obtain Eq. 1, the linear rule of mixture is used which is the simplest estimate of the effective material properties at a point in a dual-phase metal-ceramic material. A survey on different models to ascertain effective properties of an FGM is included in [3]. In the present study, relation (1) will be used as a model for Young's modulus E, thermal conductivity $\overline{K}$, thermal expansion coefficient α and Poisson's ratio υ of FG plates.

2.1 Thermal Analysis

In thermal analysis it is assumed that the temperature variation is only in the thickness direction and constant surface temperatures at the ceramic and metal rich surfaces are imposed. The one-dimensional steady state heat conduction equation in the z-direction is given by;

$$-\frac{\mathrm{d}}{\mathrm{d}z}\left(\bar{K}(z)\frac{\mathrm{d}T(z)}{\mathrm{d}z}\right)=0 \tag{2}$$

with the boundary condition $T(h/2)=T_c$ and $T(-h/2)=T_m$. Here a stress-free state is assumed to exist at $T_0=25°\text{C}$. The thermal conductivity coefficient $\bar{K}(z)$ is assumed here to obey the power-law relation in (1). Equation 2 is solved analytically for n = 0, 0.5 and integer values of n and its detail is given in [8, 28]. Here, only the final results are presented.

$$T(z)=-C_{1n}\frac{h}{\Delta\bar{K}}A_n(z)+C_{2n} \tag{3a}$$

where $\Delta\bar{K}=\bar{K}_m-\bar{K}_c$ and C_{1n} and C_{2n} are found by imposing the appropriate thermal boundary conditions on the top and bottom surfaces of the plate. The results are as follows:

$$C_{1n}=-\frac{(T_c-T_m)\Delta\bar{K}}{h[A_n(h/2)-A_n(-h/2)]},\quad C_{2n}=\frac{T_mA_n(h/2)-T_cA_n(-h/2)}{[A_n(h/2)-A_n(-h/2)]} \tag{3b}$$

with

$$A_0(z)=\frac{\Delta\bar{K}}{\bar{K}_m}\left(\frac{1}{2}-\frac{z}{h}\right),A_{1/2}(z)=2\sqrt{\frac{1}{2}-\frac{z}{h}}-2\frac{\bar{K}_c}{\Delta\bar{K}}\ln\left(\sqrt{\frac{1}{2}-\frac{z}{h}}+\frac{\bar{K}_c}{\Delta\bar{K}}\right) \tag{4a}$$

Also, for the integer values of n, the quantity A_n appearing in (3a) is given by;

$$A_n(z)=\frac{2}{n\left(\frac{\bar{K}_c}{\Delta\bar{K}}\right)^{\frac{n-1}{n}}}\sum_{k=1}^{\frac{n-1}{2}}\left[\sin\frac{2k\pi}{n}\tan^{-1}\left(\frac{\left(\frac{1}{2}-\frac{z}{h}\right)+\left(\frac{\bar{K}_c}{\Delta\bar{K}}\right)^{\frac{1}{n}}\cos\frac{2k\pi}{n}}{\left(\frac{\bar{K}_c}{\Delta\bar{K}}\right)^{\frac{1}{n}}\sin\frac{2k\pi}{n}}\right)\right]$$

$$+\frac{1}{n\left(\frac{\bar{K}_c}{\Delta\bar{K}}\right)^{\frac{n-1}{n}}}\sum_{k=1}^{\frac{n-1}{2}}\cos\frac{2k\pi}{n}\ln\left[\left(\frac{1}{2}-\frac{z}{h}\right)^2+2\left(\frac{\bar{K}_c}{\Delta\bar{K}}\right)^{\frac{1}{n}}\left(\frac{1}{2}-\frac{z}{h}\right)\cos\frac{2k\pi}{n}+\left(\frac{\bar{K}_c}{\Delta\bar{K}}\right)^{\frac{2}{n}}\right]$$

$$+\frac{\ln\left[\left(\frac{1}{2}-\frac{z}{h}\right)+\left(\frac{\bar{K}_c}{\Delta\bar{K}}\right)^{\frac{1}{n}}\right]}{n\left(\frac{\bar{K}_c}{\Delta\bar{K}}\right)^{\frac{n-1}{n}}}\qquad n=1,3,\ldots \tag{4b}$$

$$A_n(z) = \frac{2}{n\left(\frac{\overline{K}_c}{\Delta\overline{K}}\right)^{\frac{n-1}{n}}} \sum_{k=1}^{\frac{n}{2}} \sin\frac{(2k-1)\pi}{n} \tan^{-1}\left(\frac{\left(\frac{1}{2}-\frac{z}{h}\right)+\left(\frac{\overline{K}_c}{\Delta\overline{K}}\right)^{\frac{1}{n}}\cos\frac{(2k-1)\pi}{n}}{\left(\frac{\overline{K}_c}{\Delta\overline{K}}\right)^{\frac{1}{n}}\sin\frac{(2k-1)\pi}{n}}\right)$$
$$+\frac{1}{n\left(\frac{\overline{K}_c}{\Delta\overline{K}}\right)^{\frac{n-1}{n}}} \sum_{k=1}^{\frac{n}{2}} \cos\frac{(2k-1)\pi}{n}$$
$$\times \ln\left[\left(\frac{1}{2}-\frac{z}{h}\right)^2 + 2\left(\frac{\overline{K}_c}{\Delta\overline{K}}\right)^{\frac{1}{n}}\left(\frac{1}{2}-\frac{z}{h}\right)\cos\frac{(2k-1)\pi}{n} + \left(\frac{\overline{K}_c}{\Delta\overline{K}}\right)^{\frac{2}{n}}\right]$$
$$n = 2, 4, 6, \ldots \tag{4c}$$

2.2 Equilibrium Equations

Within the first-order shear deformation plate theory, the displacement field in polar coordinate is given by:

$$\begin{aligned} &u_1(r,\theta,z) = u(r,\theta) + z\Psi_r(r,\theta), u_2(r,\theta,z) = v(r,\theta) + z\Psi_\theta(r,\theta) \\ &u_3(r,\theta,z) = w(r,\theta) \end{aligned} \tag{5}$$

where u, v, and w denote the displacements of a point on the middle plane of the plate along r, θ , and z coordinates, respectively, and Ψ_r and Ψ_θ represent the small rotations of a transverse normal about the θ- and r- axes, respectively. Upon substitution of Eq. 5 into the von Karman nonlinear strain–displacement relations of elasticity [30] the strain components are obtained as follows:

$$\varepsilon_r = \varepsilon_r^0 + zk_r, \varepsilon_\theta = \varepsilon_\theta^0 + zk_\theta, \varepsilon_z = 0, \gamma_{\theta z} = k_{\theta z}, \gamma_{rz} = k_{rz}, \gamma_{r\theta} = \gamma_{r\theta}^0 + zk_{r\theta} \tag{6}$$

where

$$\begin{aligned} &\varepsilon_r^0 = u_{,r} + \frac{1}{2}(w_{,r})^2, \varepsilon_\theta^0 = \frac{1}{r}(u + v_{,\theta}) + \frac{1}{2}\frac{1}{r^2}(w_{,\theta})^2, \\ &\gamma_{r\theta}^0 = \frac{1}{r}(u_{,\theta} - v) + v_{,r} + \frac{1}{r}w_{,r}w_{,\theta} \end{aligned} \tag{7a}$$

$$k_r = \Psi_{r,r}, k_\theta = \frac{1}{r}(\Psi_r + \Psi_{\theta,\theta}), k_{r\theta} = \frac{1}{r}(\Psi_{r,\theta} - \Psi_\theta) + \Psi_{\theta,r} \tag{7b}$$

$$k_{\theta z} = \Psi_\theta + \frac{1}{r}w_{,\theta}, k_{rz} = \Psi_r + w_{,r} \tag{7c}$$

In Eq. 7 and what follows a comma followed by a coordinate variable indicates partial differentiation with respect to that variable. Based on relations (6) and (7), by using the principle of minimum total potential energy [30] the equilibrium equations are readily found to be:

$$\delta u \; ; \; N_{r,r} + \frac{1}{r}(N_r - N_\theta) + \frac{1}{r}N_{r\theta,\theta} = 0, \; \delta v \; ; \; N_{r\theta,r} + \frac{1}{r}N_{\theta,\theta} + \frac{2}{r}N_{r\theta} = 0 \tag{8a}$$

$$\begin{gathered} \delta \Psi_r \; ; \; M_{r,r} + \frac{1}{r}(M_r - M_\theta) + \frac{1}{r}M_{r\theta,\theta} - Q_r = 0, \\ \delta \Psi_\theta \; ; \; M_{r\theta,r} + \frac{1}{r}M_{\theta,\theta} + \frac{2}{r}M_{r\theta} - Q_\theta = 0 \end{gathered} \tag{8b}$$

$$\delta w \; ; \; rQ_{r,r} + Q_{\theta,\theta} + Q_r = rP_z(r,\theta) - rN_1 \tag{8c}$$

where $P_z(r,\theta)$ is the pressure applied on the top surface of the plate (see Fig. 1) and

$$N_1 = N_r w_{,rr} + N_\theta\left(\frac{1}{r}w_{,r} + \frac{1}{r^2}w_{,\theta\theta}\right) + 2N_{r\theta}\left(\frac{1}{r}w_{,\theta}\right)_{,r} \tag{9}$$

The stress and moment resultants in (8) and (9) are defined as follows:

$$\begin{gathered} (N_r, N_\theta, N_{r\theta}) = \int_{-h/2}^{h/2} (\sigma_r, \sigma_\theta, \sigma_{r\theta})dz, \; (Q_\theta, Q_r) = \int_{-h/2}^{h/2} (\sigma_{\theta z}, \sigma_{rz})dz \\ (M_r, M_\theta, M_{r\theta}) = \int_{-h/2}^{h/2} (\sigma_r, \sigma_\theta, \sigma_{r\theta})z dz \end{gathered} \tag{10}$$

With a normal vector $\vec{e}_n = n_r\vec{e}_r + n_\theta\vec{e}_\theta$ acting on the boundary surface of the plate, the boundary conditions corresponding to Eq. 8 require the specification of (see Fig. 1) either u or $n_r r N_r + n_\theta N_{r\theta}$, v or $n_r r N_{r\theta} + n_\theta N_\theta$, Ψ_r or $n_r r M_r + n_\theta M_{r\theta}$, Ψ_θ or $n_r r M_{r\theta} + n_\theta M_\theta$, and either w or $n_r\, rQ_r + n_\theta\, Q_\theta + Q_{NL}$ where $Q_{NL} = n_r(rN_r w_{,r} + N_{r\theta} w_{,\theta}) + n_\theta(N_{r\theta} w_{,r} + N_\theta w_{,\theta}/r)$.

Using the linear plane-stress thermoelastic constitutive relations of an isotropic material [8, 29], the stress and moment resultants are obtained as:

$$\begin{aligned} N_r &= A_1\varepsilon_r^0 + (A_1 - 2A_2)\varepsilon_\theta^0 + B_1 k_r + (B_1 - 2B_2)k_\theta - N^T \\ N_\theta &= (A_1 - 2A_2)\varepsilon_r^0 + A_1\varepsilon_\theta^0 + (B_1 - 2B_2)k_r + B_1 k_\theta - N^T \\ N_{r\theta} &= A_2\gamma_{r\theta}^0 + B_2 k_{r\theta} \end{aligned} \tag{11a}$$

$$\begin{aligned} M_r &= B_1\varepsilon_r^0 + (B_1 - 2B_2)\varepsilon_\theta^0 + D_1 k_r + (D_1 - 2D_2)k_\theta - M^T \\ M_\theta &= (B_1 - 2B_2)\varepsilon_r^0 + B_1\varepsilon_\theta^0 + (D_1 - 2D_2)k_r + D_1 k_\theta - M^T \\ M_{r\theta} &= B_2\gamma_{r\theta}^0 + D_2 k_{r\theta} \end{aligned} \tag{11b}$$

$$Q_\theta = K^2 A_2 k_{\theta z}, \quad Q_r = K^2 A_2 k_{rz} \tag{11c}$$

where K^2 is a shear correction factor and the stiffness coefficients, and also the thermal stress and moment resultants are defined as:

$$(A_1, B_1, D_1) = \int_{-h/2}^{h/2} \frac{E(z)}{1-\upsilon^2(z)}(1, z, z^2)dz$$
$$(A_2, B_2, D_2) = \int_{-h/2}^{h/2} \frac{E(z)}{2(1+\upsilon(z))}(1, z, z^2)dz \quad (12)$$
$$(N^T, M^T) = \int_{-h/2}^{h/2} \frac{E(z)}{1-\upsilon(z)}\alpha(z)T(z)(1, z)dz$$

where E, α, and υ are the Young modulus, the coefficient of thermal expansion, and the Poisson's ratio which, on the other hand, are assumed here to vary according to the power-law in (1). For constant Poisson's ratio through the thickness of FG plates (an assumption made in the numerical results section of the present study) the integrals in (12) are explicitly evaluated and presented in [8].

2.3 Reformulation of Equilibrium Equations

Here, the five nonlinear coupled equilibrium equations in (8), will be reformulated to yield uncoupled equations. Towards this goal, the force function $F(r, \theta)$ is introduced as follows:

$$N_r = \frac{1}{r}F_{,r} + \frac{1}{r^2}F_{,\theta\theta}, \quad N_\theta = F_{,rr}, \quad N_{r\theta} = -\left(\frac{1}{r}F_{,\theta}\right)_{,r} \quad (13)$$

With the stress resultants in (13), Eq. 8a are identically satisfied. Furthermore, the variable N_1 as introduced in Eq. 9 can be expressed in terms of F and w. Next, it is noted that Eq. 11a may be solved for ε_r^0, ε_θ^0, and $\gamma_{r\theta}^0$ to yield:

$$\varepsilon_r^0 = \frac{A_1}{\bar{A}}(N_r + N_\theta) - \frac{1}{2A_2}N_\theta + \frac{2\bar{C}}{\bar{A}}(k_r + k_\theta) - \frac{B_2}{A_2}k_r + 2\frac{A_2}{\bar{A}}N^T \quad (14a)$$

$$\varepsilon_\theta^0 = -\frac{1}{2A_2}N_r + \frac{A_1}{\bar{A}}(N_r + N_\theta) + \frac{2\bar{C}}{\bar{A}}(k_r + k_\theta) - \frac{B_2}{A_2}k_\theta + 2\frac{A_2}{\bar{A}}N^T \quad (14b)$$

$$\gamma_{r\theta}^0 = \frac{1}{A_2}N_{r\theta} - \frac{B_2}{A_2}k_{r\theta} \quad (14c)$$

where $\bar{A} = 4A_2(A_1 - A_2)$ and $\bar{C} = A_1B_2 - A_2B_1$. Upon substituting Eqs. 7b and 13 into 14 and the ensuing results into the compatibility relation, $\varepsilon_{r,\theta\theta}^0 - r\varepsilon_{r,r}^0 - (r\gamma_{r\theta}^0)_{,r\theta} + (r^2\varepsilon_{\theta,r}^0)_{,r} = N_2$, which exists among the strain components appearing in (7a), the following equation is obtained:

$$\nabla^2\nabla^2 F + \frac{2\bar{C}}{A_1}\nabla^2\left[\frac{1}{r}\Psi_{\theta,\theta} + \frac{1}{r}(r\Psi_r)_{,r}\right] = \frac{\bar{A}}{A_1}\frac{1}{r^2}N_2 \quad (15)$$

where

$$N_2 = \frac{1}{r^2}(w_{,\theta})^2 + (w_{,r\theta})^2 - rw_{,r}w_{,rr} - \frac{2}{r}w_{,\theta}w_{,r\theta} - w_{,rr}w_{,\theta\theta} \tag{16}$$

Also ∇^2 is the two-dimensional Laplace operator in polar coordinates. Substituting (7c) into (11c) and the ensuing results into (8c) yields:

$$\frac{1}{r}\Psi_{\theta,\theta} + \frac{1}{r}(r\Psi_r)_{,r} = -\nabla^2 w + \frac{1}{K^2A_2}(P_z - N_1) \tag{17}$$

Finally, substituting (17) into the compatibility Eq. 15 results in:

$$\nabla^2\nabla^2 F - \frac{2\bar{C}}{A_1}\nabla^2\nabla^2 w = -\frac{2\bar{C}}{A_1K^2A_2}\nabla^2(P_z - N_1) + \frac{\bar{A}}{A_1}\frac{1}{r^2}N_2 \tag{18}$$

Next, upon substitution of Eqs. 7b, 13 and 14 into 11b, 7c into 11c, and the subsequent results into (8b), the following two equations are obtained:

$$\begin{aligned}
\Psi_r \quad ; & -\frac{2\bar{C}}{\bar{A}}(\nabla^2 F)_{,r} + \bar{D}\left[\frac{1}{r}(r\Psi_r)_{,r}\right]_{,r} + \hat{D}\left(\frac{1}{r^2}\Psi_{r,\theta\theta}\right) \\
& - (\bar{D}+\hat{D})(\frac{1}{r^2}\Psi_{\theta,\theta}) + (\bar{D}-\hat{D})(\frac{1}{r}\Psi_{\theta,r\theta}) - K^2A_2(w_{,r} + \Psi_r) = 0 \\
\delta\Psi_\theta \quad ; & -\frac{2\bar{C}}{\bar{A}}\frac{1}{r}\nabla^2(F_{,\theta}) + (\bar{D}-\hat{D})(\frac{1}{r}\Psi_{r,r\theta}) + (\bar{D}+\hat{D})(\frac{1}{r^2}\Psi_{r,\theta}) \\
& + \hat{D}[\frac{1}{r}(r\Psi_\theta)_{,r}]_{,r} + \bar{D}(\frac{1}{r^2}\Psi_{\theta,\theta\theta}) - K^2A_2(\frac{1}{r}w_{,\theta} + \Psi_\theta) = 0
\end{aligned} \tag{19}$$

where $\bar{D} = \left(D_1\bar{A} - 4A_2B_1^2 + 8A_2B_1B_2 - 4A_1B_2^2\right)/\bar{A}$ and $\hat{D} = D_2 - B_2^2/A_2$. Next, by introducing a new function, Φ, which is known as boundary-layer function [8, 27], as follows:

$$\Phi(r,\theta) = \frac{1}{r}[\Psi_{r,\theta} - (r\Psi_\theta)_{,r}] \tag{20}$$

and following the same procedure as in [27], the equations in (19) are reformulated into the following two equations:

$$\nabla^2\Phi - \frac{K^2A_2}{\hat{D}}\Phi = 0 \tag{21}$$

$$-2\bar{C}\nabla^2\nabla^2 F - \bar{D}\bar{A}\nabla^2\nabla^2 w = \bar{A}(P_z - N_1) - \frac{\bar{D}\bar{A}}{K^2A_2}\nabla^2(P_z - N_1) \tag{22}$$

Equations 18 and 22 may, for convenience, be replaced by the following two equations:

$$\nabla^2\nabla^2 w = -\frac{1}{D}(P_z - N_1) + \frac{1}{K^2 A_2}\nabla^2(P_z - N_1) - \frac{2\bar{C}}{A_1 D}\frac{1}{r^2}N_2 \tag{23a}$$

$$\nabla^2\nabla^2 F = -\frac{2\bar{C}}{A_1 D}(P_z - N_1) + \frac{\bar{A}A_1 D - 4\bar{C}^2}{A_1^2 D}\frac{1}{r^2}N_2 \tag{23b}$$

where $D = D_1 - B_1^2/A_1$. Thus, the five nonlinear coupled equilibrium equations in (8) are replaced by a set of three uncoupled equations in (21) and (23). It is to be noted that Eqs. 21 and 23 are known, respectively, as the edge-zone (or boundary-layer) equation and the interior equations of the plate (also see [8, 27]).

Next, it is noted that, by utilizing relation (17), Eq. 19 may be rewritten so as to find Ψ_r and Ψ_θ in terms of Φ, w, and F as follows:

$$\begin{aligned}
\Psi_r &= -w_{,r} - \frac{2\bar{C}}{\bar{A}K^2 A_2}(\nabla^2 F)_{,r} + \frac{\hat{D}}{K^2 A_2}\frac{1}{r}\Phi_{,\theta} - \frac{\bar{D}}{K^2 A_2}\left[\nabla^2 w - \frac{P_z - N_1}{K^2 A_2}\right]_{,r} \\
\Psi_\theta &= -\frac{1}{r}w_{,\theta} - \frac{2\bar{C}}{\bar{A}K^2 A_2}\frac{1}{r}(\nabla^2 F)_{,\theta} - \frac{\hat{D}}{K^2 A_2}\Phi_{,r} - \frac{\bar{D}}{K^2 A_2}\frac{1}{r}\left[\nabla^2 w - \frac{P_z - N_1}{K^2 A_2}\right]_{,\theta}
\end{aligned} \tag{24}$$

Lastly, substituting Eq. 24 into Eq. 7b and the subsequent results, along with (7a) and (13), into Eq. 14 yield:

$$\begin{aligned}
u_{,r} &= -\frac{1}{2}(w_{,r})^2 + \frac{A_1}{\bar{A}}\nabla^2 F - \frac{F_{,rr}}{2A_2} + \frac{B_2}{A_2}\frac{2\bar{C}(\nabla^2 F)_{,rr}}{\bar{A}K^2 A_2} - \frac{B_2}{A_2}\frac{\hat{D}L_2(\Phi)}{K^2 A_2} \\
&\quad + \frac{B_2}{A_2}w_{,rr} + \left(\frac{B_2}{A_2}\frac{\bar{D}}{K^2 A_2}\frac{\partial^2}{2} - \frac{2\bar{C}}{\bar{A}}\right)\left[\nabla^2 w - \frac{P_z - N_1}{K^2 A_2}\right] + \frac{2A_2}{\bar{A}}N^T \\
\frac{1}{r}v_{,\theta} + \frac{u}{r} &= \frac{-1}{2r^2}(w_{,\theta})^2 + \frac{A_1}{\bar{A}}\nabla^2 F - \frac{L_1(F)}{2A_2} + \frac{B_2}{A_2}\frac{2\bar{C}L_1(\nabla^2 F)}{\bar{A}K^2 A_2} + \frac{2A_2}{\bar{A}}N^T \\
&\quad + \frac{B_2}{A_2}\frac{\hat{D}L_2(\Phi)}{K^2 A_2} + \frac{B_2}{A_2}L_1(w) + \left(\frac{B_2}{A_2}\frac{\bar{D}L_1}{K^2 A_2} - \frac{2\bar{C}}{\bar{A}}\right)\left[\nabla^2 w - \frac{P_z - N_1}{K^2 A_2}\right] \\
\frac{1}{r}u_{,\theta} + v_{,r} - \frac{v}{r} &= -\frac{1}{r}w_{,r}w_{,\theta} - \frac{L_2(F)}{A_2} + \frac{B_2}{A_2}\frac{4\bar{C}L_2(\nabla^2 F)}{\bar{A}K^2 A_2} + \frac{2B_2}{A_2}L_2(w) \\
&\quad - \frac{B_2}{A_2}\frac{\hat{D}}{K^2 A_2}\left[L_1(\Phi) - \Phi_{,rr}\right] + \frac{2B_2}{A_2}\frac{\bar{D}}{K^2 A_2}L_2\left[\nabla^2 w - \frac{1}{K^2 A_2}(P_z - N_1)\right]
\end{aligned} \tag{25}$$

where the partial differential operators L_1 and $L_2$are defined in Appendix 1. It is to be noted here that if Poisson's ratio is assumed to be constant through the plate thickness, the identity $\bar{C} = A_1 B_2 - A_2 B_1 = 0$ holds for any assumption arbitrarily made regarding the distribution of Young's modulus (e.g., as in (1)) and the preceding relations become slightly more concise.

2.4 Solution Methodology

The nonlinear partial differential equations in (21), (23a), and (23b) are solved in two steps. First, in Sect. 2.4.1, a two parameter perturbation technique is used to replace the three nonlinear partial differential equations by an infinite set of linear partial differential equations. Second, in Sect. 2.4.2, the Fourier series expansions are used to reduce each set of linear partial differential equations to ordinary ones which are then analytically solved.

2.4.1 Perturbation Technique

To solve the governing equations in (21) and (23) by the perturbation procedure, Φ, w, F, the transverse loading P_z, and the thermal loadings N^T and M^T are expanded in two parameter ascending power series:

$$(\Phi(r,\theta),\, w(r,\theta),\, F(r,\theta)) = \sum_{i=0}^{\infty}\sum_{j=0}^{\infty}(\Phi_{ij}(r,\theta)\,, w_{ij}(r,\theta),\, F_{ij}(r,\theta))\varepsilon_1^i\varepsilon_2^j \tag{26}$$

and

$$\begin{aligned} P_z(r,\theta) &= \sum_{i=0}^{\infty}\sum_{j=0}^{\infty} P_{zij}(r,\theta)\,\varepsilon_1^i\varepsilon_2^j, \quad N^T = \sum_{i=0}^{\infty}\sum_{j=0}^{\infty} N_{ij}^T\,\varepsilon_1^i\varepsilon_2^j, \\ M^T &= \frac{M^T}{N^T}\sum_{i=0}^{\infty}\sum_{j=0}^{\infty} N_{ij}^T\,\varepsilon_1^i\varepsilon_2^j \end{aligned} \tag{27a, b, c}$$

where i and j cannot be zero simultaneously and the perturbation parameters $\varepsilon_1 = w(0,0)/h$ and $\varepsilon_2 = N_r(0,0)b^2/(E_m h^3)$ are the non-dimensional center deflection and non-dimensional radial resultant force at the center of the plate. The definition of perturbation parameters requires that:

$$\begin{aligned} &w_{10}(0,0) = h \quad \text{and} \quad w_{ij}(0,0) = 0, \quad \text{for all other} \quad i,j \\ &N_{r01}(0,0) = \frac{E_m h^3}{b^2} \quad \text{and} \quad N_{rij}(0,0) = 0, \quad \text{for all other } i,j \end{aligned} \tag{28}$$

Substituting (26) and (27) into the governing equilibrium equations in (21) and (23) and collecting the terms having the same order of ε_1 and ε_2 lead to an infinite set of linear equations as follows:

$$\nabla^2\Phi_{ij} - \frac{K^2A_2}{\hat{D}}\Phi_{ij} = 0 \tag{29}$$

$$\nabla^2\nabla^2 w_{ij} = -\frac{1}{D}(P_{zij} - N_{1ij}) + \frac{1}{K^2A_2}\nabla^2(P_{zij} - N_{1ij}) - \frac{2\bar{C}}{A_1 D}\frac{1}{r^2}N_{2ij} \tag{30}$$

$$\nabla^2\nabla^2 F_{ij} = -\frac{2\bar{C}}{A_1 D}(P_{zij} - N_{1ij}) + \frac{\bar{A}A_1 D - 4\bar{C}^2}{A_1^2 D}\frac{1}{r^2}N_{2ij} \tag{31}$$

where N_{1ij} and N_{2ij}, which are determined from the preceding perturbation step, are treated as "pseudo loads" at each perturbation step and they are given as follows:

$$\varepsilon_1 : N_{110} = 0;\ N_{210} = 0\ (i = 1, j = 0);\ \varepsilon_2 : N_{101} = 0;\ N_{201} = 0\ (i = 0, j = 1)$$

$$\varepsilon_1^i \varepsilon_2^j (i, j : 2,0;\ 0,2;\ 1,1;\ 3,0;\ 0,3;\ 2,1;\ 1,2) :$$

$$N_{1ij} = \sum_{q=0}^{j}\sum_{p=0}^{i}\left[\ell_1(F_{pq}, w_{(i-p)(j-q)}) + \ell_1(w_{pq}, F_{(i-p)(j-q)}) + \ell_2(F_{pq}, w_{(i-p)(j-q)})\right]$$

$$N_{2ij} = \sum_{q=0}^{j}\sum_{p=0}^{i}\left[\ell_3(w_{pq}, w_{(i-p)(j-q)}) + \ell_4(w_{pq}, w_{(i-p)(j-q)})\right] \tag{32}$$

where p and q and also $(i-p)$ and $(j-q)$ cannot be simultaneously zero. Also $(i-p)$ and $(j-q)$ cannot be negative and ℓ_1, ℓ_2, ℓ_3, and ℓ_4 are presented in Appendix 1. Finally, in order to determine Ψ_r, Ψ_θ, u, and v, these four displacement field variables are also expressed in two parameter ascending power series as follows:

$$(\Psi_r, \Psi_\theta, u, v)(r,\theta) = \sum_{i=0}^{\infty}\sum_{j=0}^{\infty}(\Psi_{rij}, \Psi_{\theta ij}, u_{ij}, v_{ij})(r,\theta)\,\varepsilon_1^i \varepsilon_2^j \tag{33}$$

Substituting (33), (26), and (27) into (24) and (25) and following a standard perturbation procedure result, respectively, in an infinite set of linear relations as follows:

$$\begin{aligned}\Psi_{rij} &= -w_{ij,r} - \frac{2\bar{C}(\nabla^2 F_{ij})_{,r}}{\bar{A}K^2 A_2} + \frac{1}{r}\frac{\hat{D}\Phi_{ij,\theta}}{K^2 A_2} - \frac{\bar{D}}{K^2 A_2}\left[\nabla^2 w_{ij} - \frac{P_{zij} - N_{1ij}}{K^2 A_2}\right]_{,r}\\ \Psi_{\theta ij} &= -\frac{w_{ij,\theta}}{r} - \frac{1}{r}\frac{2\bar{C}(\nabla^2 F_{ij})_{,\theta}}{\bar{A}K^2 A_2} - \frac{\hat{D}\Phi_{ij,r}}{K^2 A_2} - \frac{\bar{D}}{K^2 A_2}\frac{1}{r}\left[\nabla^2 w_{ij} - \frac{P_{zij} - N_{1ij}}{K^2 A_2}\right]_{,\theta}\end{aligned} \tag{34}$$

and

$$\begin{aligned}u_{ij,r} =& NL_{1ij} + \frac{A_1}{\bar{A}}\nabla^2 F_{ij} - \frac{F_{ij,rr}}{2A_2} + \frac{B_2}{A_2}\frac{2\bar{C}}{\bar{A}K^2 A_2}(\nabla^2 F_{ij})_{,rr} - \frac{B_2}{A_2}\frac{\hat{D}L_2(\Phi_{ij})}{K^2 A_2}\\ &+ \frac{B_2}{A_2}w_{ij,rr} + \left(\frac{B_2}{A_2}\frac{\bar{D}}{K^2 A_2}\frac{\partial^2}{\partial r^2} - \frac{2\bar{C}}{\bar{A}}\right)\left[\nabla^2 w_{ij} - \frac{P_{zij} - N_{1ij}}{K^2 A_2}\right] + \frac{2A_2}{\bar{A}}N_{ij}^T\\ \frac{1}{r}v_{ij,\theta} + \frac{u_{ij}}{r} =& \frac{A_1}{\bar{A}}\nabla^2 F_{ij} - \frac{L_1(F_{ij})}{2A_2} + \frac{B_2}{A_2}\frac{2\bar{C}L_1(\nabla^2 F_{ij})}{\bar{A}K^2 A_2} + \frac{B_2}{A_2}\frac{\hat{D}L_2(\Phi_{ij})}{K^2 A_2}\end{aligned}$$

$$
\begin{aligned}
&\quad + NL_{2ij} + \frac{B_2 L_1(w_{ij})}{A_2} + \left(\frac{B_2}{A_2}\frac{\bar{D}L_1}{K^2A_2} - \frac{2\bar{C}}{\bar{A}}\right)\left[\nabla^2 w_{ij} - \frac{P_{zij} - N_{1ij}}{K^2A_2}\right] + \frac{2A_2}{\bar{A}}N_{ij}^T \\
&\frac{1}{r}u_{ij,\theta} + v_{ij,r} - \frac{v_{ij}}{r} = NL_{3ij} - \frac{L_2(F_{ij})}{A_2} + \frac{B_2}{A_2}\frac{4\bar{C}L_2(\nabla^2 F_{ij})}{\bar{A}K^2A_2} + \frac{2B_2}{A_2}L_2(w_{ij}) \\
&\quad - \frac{B_2}{A_2}\frac{\hat{D}}{K^2A_2}\left[L_1(\Phi_{ij}) - \Phi_{ij,rr}\right] + \frac{2B_2}{A_2}\frac{\bar{D}}{K^2A_2}L_2\left[\nabla^2 w_{ij} - \frac{P_{zij} - N_{1ij}}{K^2A_2}\right]
\end{aligned} \tag{35}
$$

where N_{1ij} is presented in (32) and

$$
\begin{aligned}
&\varepsilon_1;\ NL_{110} = 0;\ NL_{210} = 0;\ NL_{310} = 0,\ \varepsilon_2;\ NL_{101} = 0;\ NL_{201} = 0;\ NL_{301} = 0 \\
&\quad \varepsilon_1^i \varepsilon_2^j\ (i, j : 2,0;\ 0,2;\ 1,1;\ 3,0;\ 0,3;\ 2,1;\ 1,2) : \\
&NL_{1ij} = \frac{-1}{2}\sum_{q=0}^{j}\sum_{p=0}^{i} w_{pq,r} w_{(i-p)(j-q),r};\quad NL_{2ij} = \frac{-1}{2r^2}\sum_{q=0}^{j}\sum_{p=0}^{i} w_{pq,\theta} w_{(i-p)(j-q),\theta} \\
&NL_{3ij} = -\frac{1}{r}\sum_{q=0}^{j}\sum_{p=0}^{i} w_{pq,r} w_{(i-p)(j-q),\theta}
\end{aligned} \tag{36}
$$

Again, it is emphasized that p and q and also $(i-p)$ and $(j-q)$ cannot be simultaneously zero. Furthermore, $(i-p)$ and $(j-q)$ cannot be negative.

2.4.2 General Solutions for a Complete Circular Plate

Here, the general solutions of the perturbed governing equations are obtained for a complete circular plate. For such a plate, the response quantities must be periodic in the θ direction. Thus, to begin with, the boundary-layer function Φ_{ij} within each set of perturbation equations may be represented as:

$$
\Phi_{ij}(r,\theta) = \sum_{m=0}^{\infty} \Phi_{ijm}(r)\cos m\theta + \sum_{m=1}^{\infty} \tilde{\Phi}_{ijm}(r)\sin m\theta \tag{37}
$$

with Φ_{ijm} and $\tilde{\Phi}_{ijm}$ being two unknown functions of r. Substitution of Eq. 37 into 29 yields two modified Bessel equations whose general solutions are given by (e.g., see [31]):

$$
\begin{aligned}
\left\{\begin{matrix}\Phi_{ijm}(r)\\ \tilde{\Phi}_{ijm}(r)\end{matrix}\right\} &= \left\{\begin{matrix}A_{1ijm}\\ \tilde{A}_{1ijm}\end{matrix}\right\} I_m\left(\sqrt{\frac{K^2A_2}{\hat{D}}}\,r\right) \\
&\quad + \left\{\begin{matrix}A_{2ijm}\\ \tilde{A}_{2ijm}\end{matrix}\right\} K_m\left(\sqrt{\frac{K^2A_2}{\hat{D}}}\,r\right) \left\{\begin{matrix}m = 0, 1, 2, \ldots\\ m = 1, 2, \ldots\end{matrix}\right\}
\end{aligned} \tag{38}
$$

where A_{1ijm}, A_{2ijm} ,$\tilde{A}_{1ijm}$, and $\tilde{A}_{2ijm}$ are the integration constants and I_m and K_m are the modified Bessel functions of the first and second kinds, respectively. Next, w_{ij}, F_{ij}, and P_{zij} within each set of perturbation equations may be represented as:

$$w_{ij}(r,\theta)=\sum_{m=0}^{\infty} w_{ijm}(r)\cos m\theta+\sum_{m=1}^{\infty}\tilde{w}_{ijm}(r)\sin m\theta \tag{39a}$$

$$F_{ij}(r,\theta)=\sum_{m=0}^{\infty} F_{ijm}(r)\cos m\theta+\sum_{m=1}^{\infty}\tilde{F}_{ijm}(r)\sin m\theta \tag{39b}$$

$$P_{zij}(r,\theta)=\sum_{m=0}^{\infty} P_{ijm}(r)\cos m\theta+\sum_{m=1}^{\infty}\tilde{P}_{ijm}(r)\sin m\theta \tag{39c}$$

where $(P_{ijm}(r),\,\tilde{P}_{ijm}(r))=\frac{1}{\pi}\int_{-\pi}^{\pi}P_{zij}(r,\theta)(\cos m\theta,\ \sin m\theta)d\theta$ (e.g., see [31]) and w_{ijm}, $\tilde{w}_{ijm}$, F_{ijm} and $\tilde{F}_{ijm}$ are unknown functions of r. Upon substituting (39a) and (39b) into (32) and using trigonometric identities to express the resulting trigonometric functions in terms of $\cos m\theta$ and $\sin m\theta$, the quantities N_{1ij} and N_{2ij} are rearranged as follows:

$$(N_{1ij},N_{2ij})(r,\theta)=\sum_{m=0}^{\infty}(N_{1ijm},N_{2ijm})(r)\cos m\theta+\sum_{m=1}^{\infty}(\tilde{N}_{1ijm},\tilde{N}_{2ijm})(r)\sin m\theta \tag{40}$$

where N_{1ijm}, $\tilde{N}_{1ijm}$, N_{2ijm}, and $\tilde{N}_{2ijm}$ (which are determined from a preceding perturbation step) are known functions of r in each perturbation step. Substitution of Eqs. 39a, 39c, and 40 into Eq. 30 results in what follows:

$$\begin{aligned}&r^{m-1}\frac{d}{dr}\left(r^{1-2m}\frac{d}{dr}\left\{r^{2m-1}\frac{d}{dr}\left[r^{1-2m}\frac{d\{r^m(w_{ijm},\tilde{w}_{ijm})\}}{dr}\right]\right\}\right)=-\frac{2\bar{C}}{A_1D}\frac{(N_{2ijm},\tilde{N}_{2ijm})}{r^2}\\&+\frac{r^{m-1}}{K^2A_2}\frac{d}{dr}\left\{r^{1-2m}\frac{d\left[r^m(P_{ijm}-N_{1ijm},\tilde{P}_{ijm}-\tilde{N}_{1ijm})\right]}{dr}\right\}-\frac{P_{ijm}-N_{1ijm},\tilde{P}_{ijm}-\tilde{N}_{1ijm}}{D}\end{aligned} \tag{41}$$

Direct integration of (41) yields:

$$\begin{aligned}w_{ij0}(r)=&-\frac{q_{1ij0}-\chi_{1ij0}}{D}+\frac{q_{2ij0}-\chi_{2ij0}}{K^2A_2}-\frac{2\bar{C}}{A_1D}\Omega_{1ij0}\\&+D_{1ij0}+D_{2ij0}r^2+D_{3ij0}\ln r+D_{4ij0}r^2\ln r\end{aligned}$$

$$\begin{aligned}\begin{Bmatrix}w_{ij1}\\ \tilde{w}_{ij1}\end{Bmatrix}=&-\frac{1}{D}\begin{Bmatrix}q_{1ij1}-\chi_{1ij1}\\ \tilde{q}_{1ij1}-\tilde{\chi}_{1ij1}\end{Bmatrix}+\frac{1}{K^2A_2}\begin{Bmatrix}q_{2ij1}-\chi_{2ij1}\\ \tilde{q}_{2ij1}-\tilde{\chi}_{2ij1}\end{Bmatrix}-\frac{2\bar{C}}{A_1D}\begin{Bmatrix}\Omega_{1ij1}\\ \tilde{\Omega}_{1ij1}\end{Bmatrix}\\&+\begin{Bmatrix}D_{1ij1}\\ \tilde{D}_{1ij1}\end{Bmatrix}r+\begin{Bmatrix}D_{2ij1}\\ \tilde{D}_{2ij1}\end{Bmatrix}r^3+\begin{Bmatrix}D_{3ij1}\\ \tilde{D}_{3ij1}\end{Bmatrix}r^{-1}+\begin{Bmatrix}D_{4ij1}\\ \tilde{D}_{4ij1}\end{Bmatrix}r\ln r\end{aligned}$$

$$\begin{Bmatrix} w_{ijm} \\ \tilde{w}_{ijm} \end{Bmatrix} = -\frac{1}{D}\begin{Bmatrix} q_{1ijm} - \chi_{1ijm} \\ \tilde{q}_{1ijm} - \tilde{\chi}_{1ijm} \end{Bmatrix} + \frac{1}{K^2 A_2}\begin{Bmatrix} q_{2ijm} - \chi_{2ijm} \\ \tilde{q}_{2ijm} - \tilde{\chi}_{2ijm} \end{Bmatrix} - \frac{2\bar{C}}{A_1 D}\begin{Bmatrix} \Omega_{1ijm} \\ \tilde{\Omega}_{1ijm} \end{Bmatrix}$$
$$+ \begin{Bmatrix} D_{1ijm} \\ \tilde{D}_{1ijm} \end{Bmatrix} r^m + \begin{Bmatrix} D_{2ijm} \\ \tilde{D}_{2ijm} \end{Bmatrix} r^{m+2} + \begin{Bmatrix} D_{3ijm} \\ \tilde{D}_{3ijm} \end{Bmatrix} r^{-m} + \begin{Bmatrix} D_{4ijm} \\ \tilde{D}_{4ijm} \end{Bmatrix} r^{-m+2}$$
$$m = 2,\ 3,\ \ldots \tag{42a}$$

where

$$\begin{Bmatrix} q_{1ijm}, \tilde{q}_{1ijm}, \chi_{1ijm}, \tilde{\chi}_{1ijm}, \\ \Omega_{1ijm}, \tilde{\Omega}_{1ijm} \end{Bmatrix} = \frac{1}{r^m}\int \frac{1}{r^{1-2m}}\int \frac{1}{r^{2m-1}}\int \frac{1}{r^{1-2m}}\int \frac{1}{r^{m-1}} \times \begin{Bmatrix} P_{ijm}, \tilde{P}_{ijm}, N_{1ijm}, \tilde{N}_{1ijm}, \\ N_{2ijm}/r^2, \tilde{N}_{2ijm}/r^2 \end{Bmatrix} drdrdrdr \tag{42b}$$

$$(q_{2ijm}, \tilde{q}_{2ijm}, \chi_{2ijm}, \tilde{\chi}_{2ijm}) = \frac{1}{r^m}\int \frac{1}{r^{1-2m}}\int \frac{1}{r^{m-1}}(P_{ijm},\ \tilde{P}_{ijm}, N_{1ijm}, \tilde{N}_{1ijm})drdr \tag{42c}$$

Next, substituting Eqs. 39b, 39c, and 40 into Eq. 31 and integrating the ensuing result yields:

$$F_{ij0}(r) = -\frac{2\bar{C}}{A_1 D}(q_{1ij0} - \chi_{1ij0}) + \frac{\bar{A}A_1 D - 4\bar{C}^2}{A_1^2 D}\Omega_{1ij0}$$
$$+ B_{1ij0} + B_{2ij0}r^2 + B_{3ij0}\ln r + B_{4ij0}r^2 \ln r$$
$$\begin{Bmatrix} F_{ij1} \\ \tilde{F}_{ij1} \end{Bmatrix} = -\frac{2\bar{C}}{A_1 D}\begin{Bmatrix} q_{1ij1} - \chi_{1ij1} \\ \tilde{q}_{1ij1} - \tilde{\chi}_{1ij1} \end{Bmatrix} + \frac{\bar{A}A_1 D - 4\bar{C}^2}{A_1^2 D}\begin{Bmatrix} \Omega_{1ij1} \\ \tilde{\Omega}_{1ij1} \end{Bmatrix}$$
$$+ \begin{Bmatrix} B_{1ij1} \\ \tilde{B}_{1ij1} \end{Bmatrix} r + \begin{Bmatrix} B_{2ij1} \\ \tilde{B}_{2ij1} \end{Bmatrix} r^3 + \begin{Bmatrix} B_{3ij1} \\ \tilde{B}_{3ij1} \end{Bmatrix} r^{-1} + \begin{Bmatrix} B_{4ij1} \\ \tilde{B}_{4ij1} \end{Bmatrix} r \ln r$$
$$\begin{Bmatrix} F_{ijm} \\ \tilde{F}_{ijm} \end{Bmatrix} = -\frac{2\bar{C}}{A_1 D}\begin{Bmatrix} q_{1ijm} - \chi_{1ijm} \\ \tilde{q}_{1ijm} - \tilde{\chi}_{1ijm} \end{Bmatrix} + \frac{\bar{A}A_1 D - 4\bar{C}^2}{A_1^2 D}\begin{Bmatrix} \Omega_{1ijm} \\ \tilde{\Omega}_{1ijm} \end{Bmatrix}$$
$$+ \begin{Bmatrix} B_{1ijm} \\ \tilde{B}_{1ijm} \end{Bmatrix} r^m + \begin{Bmatrix} B_{2ijm} \\ \tilde{B}_{2ijm} \end{Bmatrix} r^{m+2} + \begin{Bmatrix} B_{3ijm} \\ \tilde{B}_{3ijm} \end{Bmatrix} r^{-m} + \begin{Bmatrix} B_{4ijm} \\ \tilde{B}_{4ijm} \end{Bmatrix} r^{-m+2}$$
$$m = 2,\ 3,\ \ldots \tag{43}$$

where q_{1ijm}, $\tilde{q}_{1ijm}$, χ_{1ijm}, $\tilde{\chi}_{1ijm}$, Ω_{1ijm}, and $\tilde{\Omega}_{1ijm}$ are defined in (42b). The functions Ψ_{rij} and $\Psi_{\theta ij}$ within each set of perturbation relations are expressed as follows:

$$(\Psi_{rij}, \Psi_{\theta ij})(r,\theta) = \sum_{m=0}^{\infty}(\Psi_{rijm}, \Psi_{\theta ijm})(r)\cos m\theta + \sum_{m=1}^{\infty}(\tilde{\Psi}_{rijm}, \tilde{\Psi}_{\theta ijm})(r)\sin m\theta \tag{44}$$

Upon substitution of Eqs. 44, 39, and 40 into 34, the following results are obtained:

$$\begin{aligned}
(\Psi_{rijm},\tilde{\Psi}_{rijm}) &= -\frac{2\bar{C}}{\bar{A}K^2A_2}\frac{d}{dr}L_3(F_{ijm},\tilde{F}_{ijm})+\frac{\hat{D}m(\tilde{\Phi}_{ijm},-\Phi_{ijm})}{rK^2A_2}\\
&\quad-(w'_{ijm},\tilde{w}'_{ijm})-\frac{\bar{D}}{K^2A_2}\frac{d}{dr}L_3(w_{ijm},\tilde{w}_{ijm})\\
&\quad+\frac{\bar{D}(P'_{ijm}-N'_{1ijm},\tilde{P}'_{ijm}-\tilde{N}'_{1ijm})}{(K^2A_2)^2}\\
(\Psi_{\theta ijm},\tilde{\Psi}_{\theta ijm}) &= -\frac{2\bar{C}}{\bar{A}K^2A_2}\frac{m}{r}L_3(\tilde{F}_{ijm},-F_{ijm})-\frac{\hat{D}(\Phi'_{ijm},\tilde{\Phi}'_{ijm})}{K^2A_2}\\
&\quad-\frac{m(\tilde{w}_{ijm},-w_{ijm})}{r}-\frac{m\bar{D}L_3(\tilde{w}_{ijm},-w_{ijm})}{rK^2A_2}\\
&\quad+\frac{m}{r}\frac{\bar{D}[\tilde{P}_{ijm}-\tilde{N}_{1ijm},-(P_{ijm}-N_{1ijm})]}{(K^2A_2)^2}
\end{aligned} \tag{45}$$

where a prime here and in what follows indicates total differentiation with respect to the variable r and L_3 is an ordinary differential operator presented in Appendix 1. Lastly, u_{ij}, v_{ij}, and thermal stress resultant N_{ij}^T within each set of perturbation relations are expressed as follows:

$$(u_{ij},v_{ij})(r,\theta)=\sum_{m=0}^{\infty}(u_{ijm},v_{ijm})(r)\cos m\theta+\sum_{m=1}^{\infty}(\tilde{u}_{ijm},\tilde{v}_{ijm})(r)\sin m\theta \tag{46}$$

$$N_{ij}^T=\sum_{m=0}^{\infty}N_{ijm}^T\cos m\theta+\sum_{m=1}^{\infty}\tilde{N}_{ijm}^T\sin m\theta \tag{47}$$

where $(N_{ijm}^T,\tilde{N}_{ijm}^T)=\frac{1}{\pi}\int_{-\pi}^{\pi}N_{ij}^T(\cos m\theta,\sin m\theta)d\theta$ (e.g., see [31]). Upon substitution of (39a) into (36) and using trigonometric identities to express the resulting trigonometric functions in terms of $\cos m\theta$ and $\sin m\theta$, the quantities NL_{1ij}, NL_{2ij}, and NL_{3ij} are rearranged as follows:

$$\begin{Bmatrix}NL_{1ij}(r,\theta)\\NL_{2ij}(r,\theta)\\NL_{3ij}(r,\theta)\end{Bmatrix}=\sum_{m=0}^{\infty}\begin{Bmatrix}NL_{1ijm}(r)\\NL_{2ijm}(r)\\NL_{3ijm}(r)\end{Bmatrix}\cos m\theta+\sum_{m=1}^{\infty}\begin{Bmatrix}\tilde{N}L_{1ijm}(r)\\\tilde{N}L_{2ijm}(r)\\\tilde{N}L_{3ijm}(r)\end{Bmatrix}\sin m\theta \tag{48}$$

where NL_{1ijm}, $\tilde{N}L_{1ijm}$, NL_{2ijm}, $\tilde{N}L_{2ijm}$, NL_{3ijm} and $\tilde{N}L_{3ijm}$ are known functions of r in each perturbation step (see Appendix 2). Next, substituting (46), (47), (48), (37), (39), and (40) into (35) yields:

$$\begin{Bmatrix} u'_{ijm} \\ \tilde{u}'_{ijm} \end{Bmatrix} = \begin{Bmatrix} NL_{1ijm} \\ \tilde{N}L_{1ijm} \end{Bmatrix} + \left[\frac{A_1}{\bar{A}}L_3 - \frac{d^2}{dr^2}\left(\frac{1}{2A_2} - \frac{B_2}{A_2}\frac{2\bar{C}}{\bar{A}K^2A_2}L_3\right)\right]\begin{Bmatrix} F_{ijm} \\ \tilde{F}_{ijm} \end{Bmatrix}$$
$$-\frac{B_2}{A_2}\frac{\hat{D}m}{K^2A_2}\frac{d}{dr}\left(\frac{1}{r}\begin{Bmatrix} \tilde{\Phi}_{ijm} \\ -\Phi_{ijm} \end{Bmatrix}\right) + \left[\frac{B_2}{A_2}\frac{d^2}{dr^2}\left(1+\frac{\bar{D}}{K^2A_2}L_3\right) - \frac{2\bar{C}}{\bar{A}}L_3\right]\begin{Bmatrix} w_{ijm} \\ \tilde{w}_{ijm} \end{Bmatrix}$$
$$+\left(-\frac{B_2}{A_2}\frac{\bar{D}}{(K^2A_2)^2}\frac{d^2}{dr^2} + \frac{2\bar{C}}{\bar{A}K^2A_2}\right)\begin{Bmatrix} P_{ijm} - N_{1ijm} \\ \tilde{P}_{ijm} - \tilde{N}_{1ijm} \end{Bmatrix} + \frac{2A_2}{\bar{A}}\begin{Bmatrix} N^T_{ijm} \\ \tilde{N}^T_{ijm} \end{Bmatrix} \tag{49a}$$

$$m\begin{Bmatrix} v_{ijm} \\ \tilde{v}_{ijm} \end{Bmatrix} = \begin{Bmatrix} \tilde{u}_{ijm} \\ -u_{ijm} \end{Bmatrix} + r\begin{Bmatrix} -\tilde{N}L_{2ijm} \\ NL_{2ijm} \end{Bmatrix} + \left(\frac{A_1}{\bar{A}}rL_3 - \frac{L_4}{2A_2} + \frac{B_2}{A_2}\frac{2\bar{C}L_4L_3}{\bar{A}K^2A_2}\right)\begin{Bmatrix} -\tilde{F}_{ijm} \\ F_{ijm} \end{Bmatrix}$$
$$+\frac{B_2}{A_2}\frac{\hat{D}mrL_5}{K^2A_2}\begin{Bmatrix} \Phi_{ijm} \\ \tilde{\Phi}_{ijm} \end{Bmatrix} + \left(-\frac{2\bar{C}}{\bar{A}}rL_3 + \frac{B_2L_4}{A_2} + \frac{B_2}{A_2}\frac{\bar{D}L_4L_3}{K^2A_2}\right)\begin{Bmatrix} -\tilde{w}_{ijm} \\ w_{ijm} \end{Bmatrix}$$
$$-\left(\frac{B_2}{A_2}\frac{\bar{D}L_4}{(K^2A_2)^2} - \frac{2\bar{C}r}{\bar{A}K^2A_2}\right)\begin{Bmatrix} -(\tilde{P}_{ijm} - \tilde{N}_{1ijm}) \\ (P_{ijm} - N_{1ijm}) \end{Bmatrix} + \frac{2A_2r}{\bar{A}}\begin{Bmatrix} -\tilde{N}^T_{ijm} \\ N^T_{ijm} \end{Bmatrix} \quad m = 0, 1, 2, \ldots \tag{49b}$$

$$\frac{m}{r}\begin{Bmatrix} -u_{ijm} \\ \tilde{u}_{ijm} \end{Bmatrix} + rL_5\begin{Bmatrix} \tilde{v}_{ijm} \\ v_{ijm} \end{Bmatrix} = \begin{Bmatrix} \tilde{N}L_{3ijm} \\ NL_{3ijm} \end{Bmatrix} + mL_5\left(-\frac{1}{A_2} + \frac{B_2}{A_2}\frac{4\bar{C}L_3}{\bar{A}K^2A_2}\right)\begin{Bmatrix} -F_{ijm} \\ \tilde{F}_{ijm} \end{Bmatrix}$$
$$-\frac{B_2}{A_2}\frac{\hat{D}}{K^2A_2}\left(\frac{1}{r}\frac{d}{dr} - \frac{d^2}{dr^2} - \frac{m^2}{r^2}\right)\begin{Bmatrix} \tilde{\Phi}_{ijm} \\ \Phi_{ijm} \end{Bmatrix} + \frac{2B_2}{A_2}mL_5\left(1 + \frac{\bar{D}L_3}{K^2A_2}\right)\begin{Bmatrix} -w_{ijm} \\ \tilde{w}_{ijm} \end{Bmatrix}$$
$$-\frac{2B_2}{A_2}\frac{\bar{D}mL_5}{(K^2A_2)^2}\begin{Bmatrix} -(P_{ijm} - N_{1ijm}) \\ (\tilde{P}_{ijm} - \tilde{N}_{1ijm}) \end{Bmatrix} \quad \begin{Bmatrix} m = 0, 1, 2, \ldots \\ m = 1, 2, \ldots \end{Bmatrix} \tag{49c}$$

where the ordinary differential operators L_4 and L_5 are displayed in Appendix 1. Integrating (49a) yields:

$$\begin{Bmatrix} u_{ijm} \\ \tilde{u}_{ijm} \end{Bmatrix} = \frac{A_1}{\bar{A}}\int L_3\begin{Bmatrix} F_{ijm} \\ \tilde{F}_{ijm} \end{Bmatrix}dr - \frac{d}{dr}\left(\frac{1}{2A_2} - \frac{B_2}{A_2}\frac{2\bar{C}L_3}{\bar{A}K^2A_2}\right)\begin{Bmatrix} F_{ijm} \\ \tilde{F}_{ijm} \end{Bmatrix} - \frac{B_2\hat{D}}{K^2A_2^2}\frac{m}{r}\begin{Bmatrix} \tilde{\Phi}_{ijm} \\ -\Phi_{ijm} \end{Bmatrix}$$
$$+\int\begin{Bmatrix} NL_{1ijm} \\ \tilde{N}L_{1ijm} \end{Bmatrix}dr + \frac{B_2}{A_2}\frac{d}{dr}\left(1 + \frac{\bar{D}L_3}{K^2A_2}\right)\begin{Bmatrix} w_{ijm} \\ \tilde{w}_{ijm} \end{Bmatrix} - \frac{2\bar{C}}{\bar{A}}\int L_3\begin{Bmatrix} w_{ijm} \\ \tilde{w}_{ijm} \end{Bmatrix}dr$$
$$+\frac{2A_2}{\bar{A}}\int\begin{Bmatrix} N^T_{ijm} \\ \tilde{N}^T_{ijm} \end{Bmatrix}dr - \frac{B_2}{A_2}\frac{\bar{D}}{(K^2A_2)^2}\begin{Bmatrix} P'_{ijm} - N'_{1ijm} \\ \tilde{P}'_{ijm} - \tilde{N}'_{1ijm} \end{Bmatrix}$$
$$+\frac{2\bar{C}}{\bar{A}K^2A_2}\int\begin{Bmatrix} P_{ijm} - N_{1ijm} \\ \tilde{P}_{ijm} - \tilde{N}_{1ijm} \end{Bmatrix}dr + \begin{Bmatrix} C_{ijm} \\ \tilde{C}_{ijm} \end{Bmatrix} \quad \begin{Bmatrix} m = 1, 2, \ldots \\ m = 0, 1, 2, \ldots \end{Bmatrix} \tag{50}$$

The integration constants C_{ijm} and $\tilde{C}_{ijm}$ are determined by substitution of (50) and (49b) into (49c). This completes the solution development for asymmetric

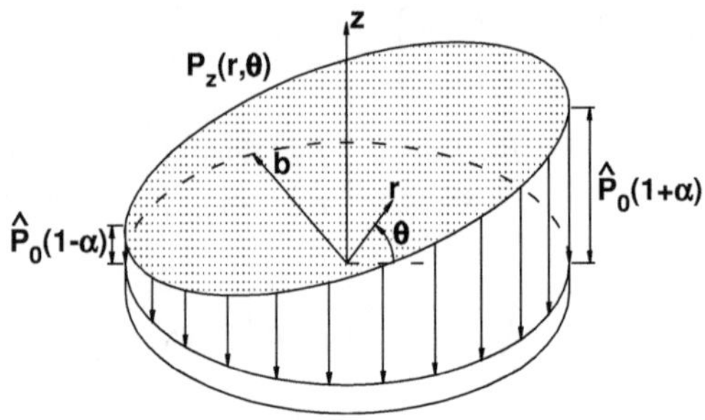

Fig. 2 FG solid circular plate subjected to an asymmetric pressure

nonlinear behavior of functionally graded circular plate under thermo-mechanical loading within the first-order shear deformation theory.

2.5 Solid Circular Plate under Asymmetric Transverse Pressure and Thermal Loading

For numerical purposes, a functionally graded solid circular plate of radius b subjected to the following asymmetric transverse pressure (see Fig. 2) and a temperature variation in the thickness direction (heat transfer in this direction) are considered:

$$P_z(r,\theta) = \hat{P}_0(1 + \alpha\frac{r}{b}\cos\theta) \tag{51a}$$

$$N^T = \text{constant} \tag{51b}$$

where $\hat{P}_0$, α, and N^T are three known parameters. By comparing Eq. 51a with 27a and 39c and also comparing Eq. 51b with (27b) and (47) it can be concluded that:

$$\begin{aligned} &\hat{P}_0 = \sum_{i=0}^{\infty}\sum_{j=0}^{\infty} P_{ij0}(r)\,\varepsilon_1^i\varepsilon_2^j, \hat{P}_0\alpha\frac{r}{b} = \sum_{i=0}^{\infty}\sum_{j=0}^{\infty} P_{ij1}(r)\,\varepsilon_1^i\varepsilon_2^j, \\ &N^T = \sum_{i=0}^{\infty}\sum_{j=0}^{\infty} N_{ij0}^T\,\varepsilon_1^i\varepsilon_2^j \end{aligned} \tag{52a}$$

$$\begin{aligned} P_{ijm}(r) = 0\,,\ m = 2, 3, \ldots, \tilde{P}_{ijm}(r) = 0\ ,\ m = 1, 2, \ldots \text{ for all } i \text{ and } j \\ N_{ijm}^T = \tilde{N}_{ijm}^T = 0\,,\quad m = 1, 2, \ldots \text{ for all } i \text{ and } j \end{aligned} \tag{52b}$$

where the functions P_{ij0}'s and N_{ij0}^T's must be considered here to be unknown constants. Furthermore, from the first two relations in (52a), it can readily be concluded, therefore, that $P_{ij1}(r) = \alpha P_{ij0} r/b$. Since all response quantities must be finite at r = 0, for the loading considered in (51) the response quantities of a solid circular plate within each perturbation step are simplified to what follows (see the appropriate relations of the previous section):

$$(\Phi_{ij}, \Psi_{\theta ij}, v_{ij})(r,\theta) = \sum_{m=1}^{i+j} (\tilde{\Phi}_{ijm}, \tilde{\Psi}_{\theta ijm}, \tilde{v}_{ijm})(r) \sin m\theta$$
$$(w_{ij}, F_{ij})(r,\theta) = \sum_{m=0}^{i+j} (w_{ijm}, F_{ijm})(r) \cos m\theta \tag{53}$$
$$(\Psi_{rij}, u_{ij})(r,\theta) = \sum_{m=0}^{i+j} (\Psi_{rijm}, u_{ijm})(r) \cos m\theta$$

where

$$\begin{aligned}
\tilde{\Phi}_{ijm}(r) &= \tilde{A}_{1ijm} I_m(\mu r)\\
w_{ijm}(r) &= -\frac{q_{1ijm}-\chi_{1ijm}}{D} + \frac{q_{2ijm}-\chi_{2ijm}}{K^2 A_2} - \frac{2\bar{C}\Omega_{1ijm}}{A_1 D} + D_{1ijm} r^m + D_{2ijm} r^{m+2}\\
F_{ijm}(r) &= -\frac{2\bar{C}(q_{1ijm}-\chi_{1ijm})}{A_1 D} + \frac{\bar{A}A_1 D - 4\bar{C}^2}{A_1^2 D}\Omega_{1ijm} + B_{1ijm} r^m + B_{2ijm} r^{m+2}
\end{aligned} \tag{54a}$$

In Eq. 54a $\mu^2 = K^2 A_2/\hat{D}$ and

$$q_{1ij0} = \frac{P_{ij0} r^4}{64},\ q_{1ij1} = \frac{\alpha P_{ij0} r^5}{192 b},\ q_{2ij0} = \frac{P_{ij0} r^2}{4},\ q_{2ij1} = \frac{\alpha P_{ij0} r^3}{8b},$$
$$q_{1ijm} = q_{2ijm} = 0,\ m = 2, 3, \ldots \tag{54b}$$

Also the quantities χ_{1ijm}, χ_{2ijm}, and Ω_{1ijm} appearing in (54a) are in terms of N_{1ijm} and N_{2ijm} and defined in (42b) and (42c). For the first nine perturbation steps, the functions N_{1ijm} and N_{2ijm}, on the other hand, are presented in Appendix 2. Furthermore, substitution of Eqs. 52, 53 and 54a into Eqs. 45, 50, and 49b yield, respectively:

$$\begin{aligned}
\Psi_{rijm}(r) = {} & \frac{q'_{1ijm}-\chi'_{1ijm}}{D} + \hat{C}(q'_{2ijm}-\chi'_{2ijm}) + \frac{2\bar{C}\hat{C}\Omega'_{2ijm}}{A_1} + \frac{m\tilde{A}_{1ijm}}{\mu^2}\frac{I_m(\mu r)}{r}\\
& + \frac{2\bar{C}\Omega'_{1ijm}}{A_1 D} - (m+2)D_{2ijm} r^{m+1} - m\Big[D_{1ijm} + \frac{4\bar{D}(m+1)}{K^2 A_2} D_{2ijm}\\
& + \frac{8\bar{C}(m+1)}{\bar{A}K^2 A_2} B_{2ijm}\Big] r^{m-1}\\
\tilde{\Psi}_{\theta ijm} = {} & -\frac{m}{r}\frac{q_{1ijm}-\chi_{1ijm}}{D} - \hat{C}\frac{m(q_{2ijm}-\chi_{2ijm})}{r} - \frac{2\bar{C}\hat{C}}{A_1}\frac{m}{r}\Omega_{2ijm} - \frac{\tilde{A}_{1ijm}}{\mu^2}\frac{d[I_m(\mu r)]}{dr}\\
& - \frac{2\bar{C}}{A_1 D}\frac{m}{r}\Omega_{1ijm} + m D_{2ijm} r^{m+1} + m\Big[D_{1ijm} + \frac{4\bar{D}(m+1)}{K^2 A_2} D_{2ijm}\\
& + \frac{8\bar{C}(m+1)}{\bar{A}K^2 A_2} B_{2ijm}\Big] r^{m-1}
\end{aligned} \tag{54c}$$

$$
\begin{aligned}
u_{ijm}(r) = & \int NL_{1ijm}dr + (\frac{\bar{C}}{A_1} - B_2)\frac{q'_{1ijm} - \chi'_{1ijm}}{A_2 D} - \frac{B_2\hat{C}}{A_2}(q'_{2ijm} - \chi'_{2ijm}) + \int \Omega_{2ijm}dr \\
& - \frac{B_2}{A_2}\frac{2\bar{C}\hat{C}}{A_1}\Omega'_{2ijm} - \tilde{C}\Omega'_{1ijm} - \frac{B_2}{A_2}\frac{m\tilde{A}_{1ijm}}{\mu^2}\frac{I_m(\mu r)}{r} + \frac{2A_2}{\bar{A}}N^T_{ijm}r + C_{ijm} \\
& + \left\{(\frac{4A_1}{\bar{A}} - \frac{m+2}{2A_2})B_{2ijm} + \left[\frac{B_2(m+2)}{A_2} - \frac{8\bar{C}}{\bar{A}}\right]D_{2ijm}\right\}r^{m+1} \\
& + m\left\{-\frac{B_{1ijm}}{2A_2} + \frac{B_2}{A_2}\left[\frac{8\bar{C}(m+1)}{\bar{A}K^2A_2}B_{2ijm} + D_{1ijm} + \frac{4\bar{D}(m+1)}{K^2A_2}D_{2ijm}\right]\right\}r^{m-1}
\end{aligned}
\quad (54d)
$$

and

$$
\begin{aligned}
\tilde{v}_{ijm} = & -\frac{1}{m}\int NL_{1ijm}dr - (\frac{\bar{C}}{A_1} - B_2)\frac{m}{r}\frac{q_{1ijm} - \chi_{1ijm}}{A_2 D} + \frac{B_2\hat{C}}{A_2}\frac{m}{r}(q_{2ijm} - \chi_{2ijm}) \\
& \times \frac{rNL_{2ijm}}{m} + \frac{\tilde{C}m}{r}\Omega_{1ijm} + \frac{r}{m}\Omega_{2ijm} - \frac{1}{m}\int \Omega_{2ijm}dr + \frac{B_2}{A_2}\frac{2\bar{C}\hat{C}}{A_1}\frac{m}{r}\Omega_{2ijm} \\
& + \frac{B_2}{A_2}\frac{\tilde{A}_{1ijm}}{\mu^2}\frac{d[I_m(\mu r)]}{dr} + \left\{(\frac{4A_1}{\bar{A}} + \frac{m}{2A_2})B_{2ijm} - (\frac{B_2 m}{A_2} + \frac{8\bar{C}}{\bar{A}})D_{2ijm}\right\}r^{m+1} \\
& + m\left\{\frac{B_{1ijm}}{2A_2} - \frac{B_2}{A_2}\left[\frac{8\bar{C}(m+1)}{\bar{A}K^2A_2}B_{2ijm} + D_{1ijm} + \frac{4\bar{D}(m+1)}{K^2A_2}D_{2ijm}\right]\right\}r^{m-1} - \frac{C_{ijm}}{m}
\end{aligned}
\quad (54e)
$$

where $\hat{C} = \frac{1}{K^2A_2}[\frac{4\bar{C}^2}{\bar{A}A_1D} - (1 - \frac{\bar{D}}{D})]$, $\tilde{C} = \frac{1}{A_2A_1D}\left(\frac{\bar{A}A_1D - 4\bar{C}^2}{2A_1} + 2B_2\bar{C}\right)$, and

$$
\Omega_{2ijm} = \frac{1}{r^m}\int \frac{1}{r^{1-2m}}\int \frac{1}{r^{m+1}}N_{1ijm}drdr \quad (54f)
$$

The nonlinear terms NL_{1ijm} and NL_{2ijm} appearing in (54d) and (54e) are presented in Appendix 2 for the first nine perturbation steps. Also, upon substitution of Eqs. 54d and 54e into 49c, it is found that C_{ijm} is zero. It is to be noted that the integration constants in Eqs. 54a, 54c, 54d, and 54e are determined by imposing the appropriate boundary conditions at $r = b$. Here, clamped and simply-supported boundary conditions are considered. In terms of the functions associated with the perturbation power series of the present study, they are defined as:

$$
\begin{aligned}
& C: \Psi_{rij} = \Psi_{\theta ij} = w_{ij} = 0 \text{ and } u_{ij} = v_{ij} = 0 \\
& S: M_{rij} = \Psi_{\theta ij} = w_{ij} = 0 \text{ and } u_{ij} = v_{ij} = 0
\end{aligned}
\quad (55)
$$

By imposing these boundary conditions at $r = b$, the integration constants are found in terms of the unknown constants P_{ij0}'s and N^T_{ij0}'s (appearing in 54b) in each perturbation step. These unknown parameters are determined by solving a set of algebraic equations which, on the other hand, are obtained by utilizing the conditions in (28). Finally, the perturbation parameters $\varepsilon_1$andε_2 are determined by numerically solving the first and the third equations in (52a) which are, in general, two nonlinear polynomial equations in ε_1 and ε_2. It is to be reminded that the solutions of the first-order perturbation steps (associated with ε_1 and ε_2) represent the linear asymmetric solution of the FG plate under the same thermo-mechanical loading assumed in the nonlinear problem.

3 Numerical Results and Discussions

To validate the results of the present study, two validation examples are presented here for linear and nonlinear bending problems of circular plates.

Example 1 The numerical results for the linear asymmetric bending of clamped and simply-supported FG circular plates subjected to mechanical (asymmetric transverse pressure with $\alpha = 1$), and thermal loadings are obtained here and compared with those presented by Nosier and Fallah in [8]. The plate profiles after deformation under mechanical and thermal loadings are shown in Fig. 3. The material properties are the same as those in [8] ($E_m = 70\,\text{GPa}, \upsilon_m = 0.3,\ E_c = 151\,\text{GPa},\ \upsilon_c = 0.3$). Exact agreements are seen to exist between the two results. It is to be mentioned that according to linear theory the responses of the four types of clamped supports identically coincide with each other (see [8] for further discussions).

Example 2 Here, the numerical results for the nonlinear axisymmetric bending of FG circular plates within the classical plate theory ($K^2A_2 \to \infty$) under thermo-mechanical loading (thermal loading $T_c/T_m = 15$ and a uniform transverse pressure ($\alpha = 0$)) are obtained and compared with those presented by Ma and Wang [22]. Figure 4a, b display the load–deflection curves for clamped and simply-supported FG plates for different values of power-law index n, respectively. The material properties are the same as those in [22] ($E_m = 70\,\text{GPa}, \upsilon_m = 0.3,\ E_c = 151\,\text{GPa},\ \upsilon_c = 0.3$) and $D_c = E_c h^3/(12(1-\upsilon^2))$. The trends in the curves of the two results are seen to be similar, although some discrepancies are seen to exist between the two results.

In the remaining of the present work, axisymmetric and asymmetric, linear and nonlinear behavior of FG solid circular plates under a linearly varying transverse mechanical load and a temperature variation in the thickness direction (see Fig. 2 and Eq. 51) are considered. For the purpose of numerical illustrations, Aluminum-Zirconia is considered. The material properties (i.e., Young's modulus, Poisson's ratio, and coefficients of heat conduction and thermal expansion) are assumed to be 70 GPa, 0.3, 204 W/m·K, and 23×10^{-6} /°C for Aluminum and 151GPa, 0.3, 2.09 W/m·K, and 10×10^{-6} /°C for Zirconia, respectively [10, 11]. The Poisson's ratio is assumed to be constant through the plate thickness. Chi and Chung [4, 5] studied the effect of Poisson's ratio on the static response of FG plates. They concluded that the effect of changing Poisson's ratio on the mechanical behavior of

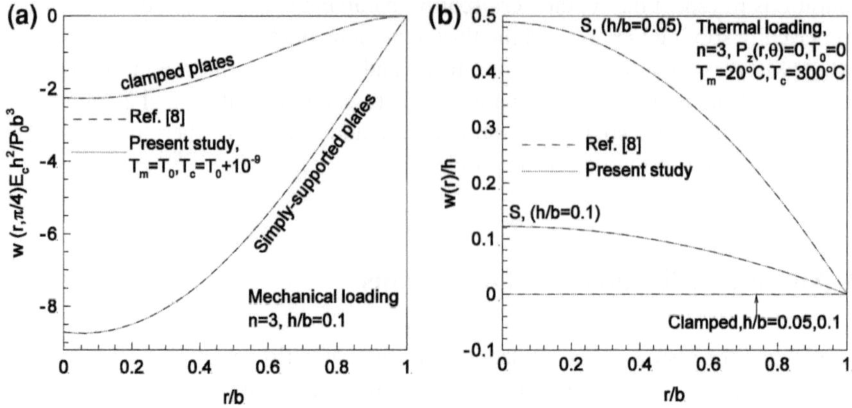

Fig. 3 Comparison of linear transverse deflection of clamped and simply-supported FG circular plate under **a** asymmetric pressure, and **b** thermal loading

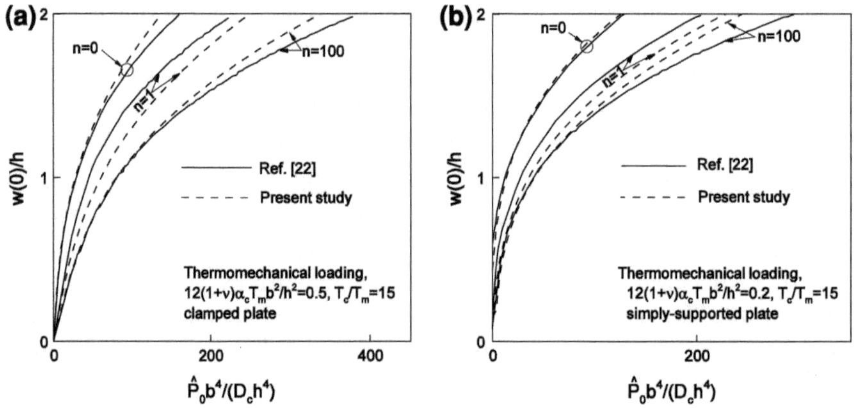

Fig. 4 Comparison of nonlinear results for load–deflection curves in a **a** clamped, and **b** simply-supported FG circular plate under thermo-mechanical loading

the FG plates is very small so that it can be assumed to be constant. In all calculations the shear correction factor is taken to be 5/6. Identification of this factor in FGMs was investigated by Nguyen et al. [6]. Also, unless mentioned otherwise, the value of α appearing in (51) are assumed to be 1 and the results are presented for $\theta = \pi/4$, using third-order perturbation expansions (including nine terms) for the solutions. In thermal loading, it is assumed that $T_m = T_0$ and $T_c = T_0 + 200°C$. For convenience, the following non-dimensional parameters are introduced:

$$\bar{w} = w/h,\ \bar{P} = \hat{P}_0 b^4/(E_m h^4),\ \bar{N}_{ij} = N_{ij} b^2/(E_m h^3) \tag{56}$$

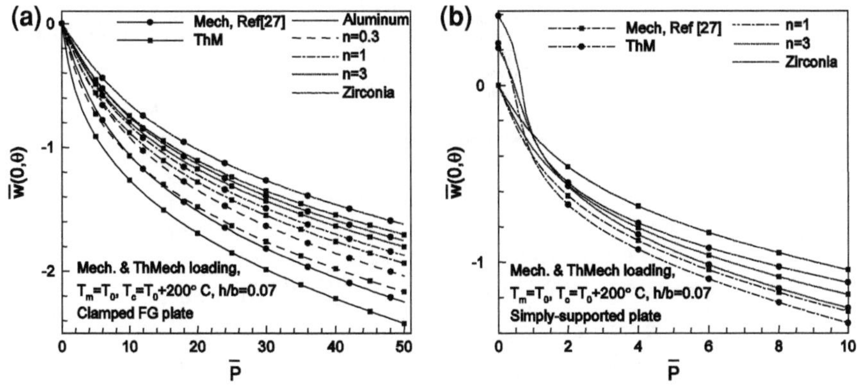

Fig. 5 Center deflection of **a** clamped, and **b** simply-supported FG circular plates versus load

It is to be reminded that the perturbation parameters are defined at the center point of the plate (i.e., $\varepsilon_1 = w(0,0)/h$ and $\varepsilon_2 = N_r(0,0)b^2/(E_m h^3)$) for simplicity and convenience. However, numerical studies, conducted by the present authors [27], indicate that the accuracy of any-order expansion considerably depends on the location chosen for the perturbation parameter. It is, in general, observed that the solution becomes more accurate when the perturbation parameter is specified at the location where its corresponding response quantity is maximum. It is noted that numerical results indicate that the maximum deflection and radial resultant force occur at a point near the center of the plate.

The nonlinear center deflections versus load in clamped and simply-supported FG circular plates for various values of n under thermo-mechanical loading are presented in Fig. 5 and are compared with corresponding results under mechanical loading [27]. In Fig. 5b the results for $n = 0.3$ and Aluminum are not shown since snap-buckling occurs in these two cases and their results are shown in Figs. 6 and 7, respectively. Figure 6a shows the nonlinear center deflection versus load in simply-supported FG circular plates with n = 0.3 under mechanical and various thermo-mechanical loadings. For mechanical loading and also for the temperature change of $\Delta T = 100°C$ or $\Delta T = 150°C$ only one form of equilibrium is possible and the equilibrium is stable (the first and third nonlinear polynomial equations in (52a) converge to only one root for ε_1 and ε_2 under a definite loading). But for $\Delta T = 200°C$, there is a region (surrounding $\bar{P} = 0.7$) for which there exists more than one equilibrium path. In this region, under a definite loading, more than one root (two or three roots) are obtained for the two nonlinear polynomial equations in (52a) and the question of instability arise. In the investigation of stability, the logic to obtain the variation of center deflection versus load is changed. Here, by taking a series of values for ε_2 and under a definite thermal loading, the corresponding values of ε_1 and $\hat{P}_0$ are calculated from the third and first relations in (52a), respectively. The results are displayed in Fig. 6b in which it is seen that the deflection gradually increases with an increase of the load up to point A. At this

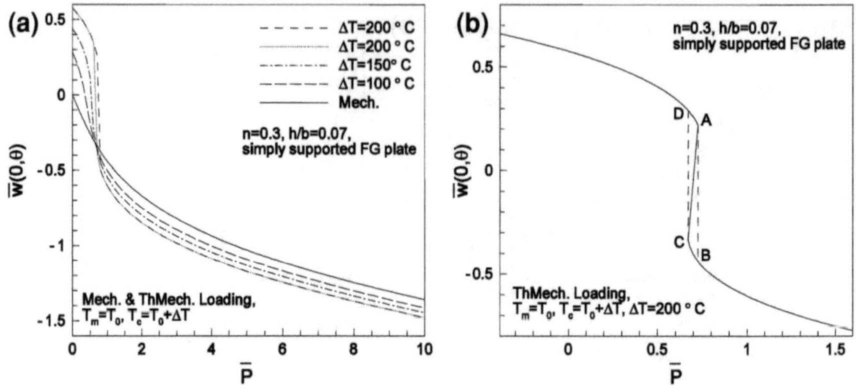

Fig. 6 Center deflection of simply-supported FG circular plates ($n = 0.3$) **a** under various thermo-mechanical loadings, and **b** under $\Delta T = 200\,^{\circ}\mathrm{C}$ versus load

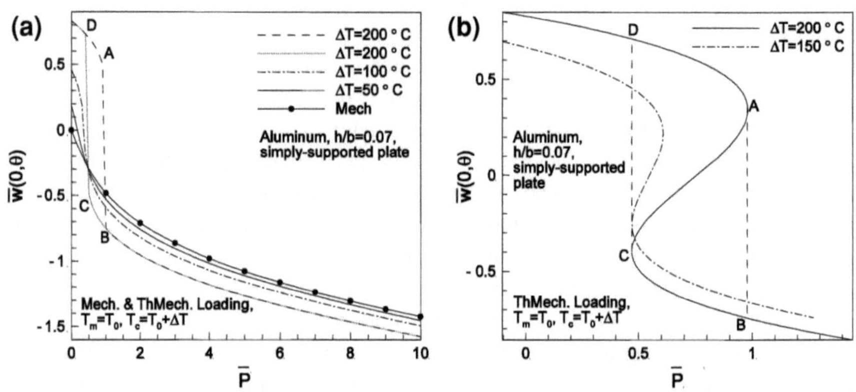

Fig. 7 Center deflection of simply-supported Aluminum circular plates under **a** mechanical and various thermo-mechanical loadings, and **b** thermo-mechanical loading

point, further deflection continues with a decrease of load. This fact indicates that the upward form of the equilibrium of the plate has become unstable and plate buckles downward through the dashed line to point B. At this point, any further increase of load gradually increases the deflection and the new equilibrium form of the plate is stable. On the other hand, by starting from point B, as the load is diminished, the deflection of the plate decreases up to point C. At this point, further decrease of deflection results in a load increase. This fact indicates that the downward form of the plate is unstable and the load is not enough to keep the plate downward and the plate buckles upward through the dashed line to point D. In Fig. 7a the same phenomenon is shown to exist for Aluminum when temperature change of the upper surface of the plate is $\Delta T = 200°\mathrm{C}$. Of course, snap-buckling

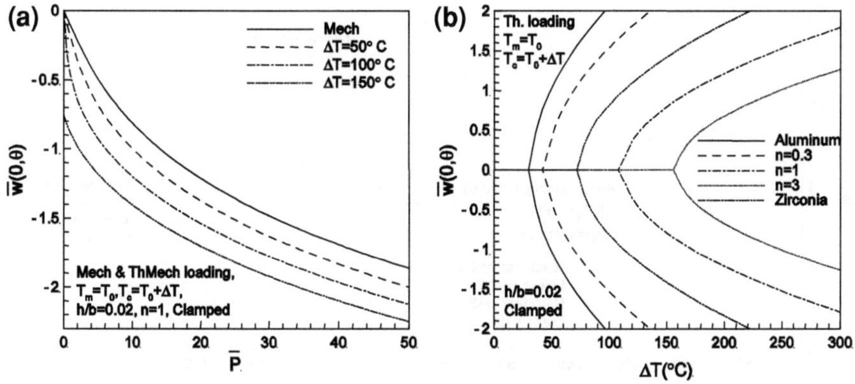

Fig. 8 Center deflection of clamped FG circular plates **a** under mechanical and various thermo-mechanical loadings, and **b** under thermal loading

Fig. 9 Center deflection of simply-supported circular FG plates under thermal loading

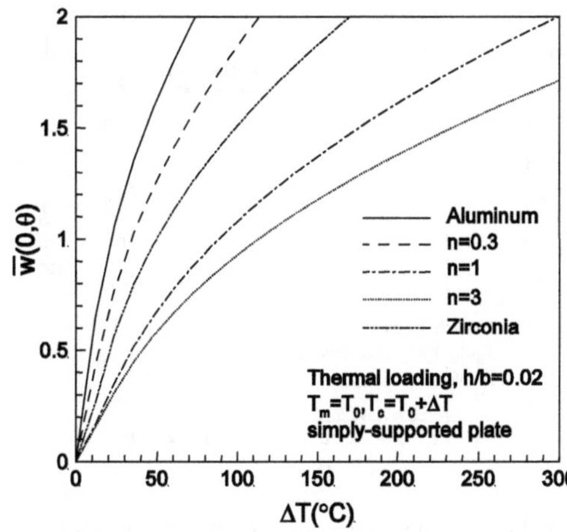

happens in an Aluminum circular plate when $\Delta T = 150°\mathrm{C}$, but for clarity it is shown in Fig. 7b. It is seen that the region in which more than one form of equilibrium is possible becomes larger as ΔT increases. The same phenomenon is seen when axisymmetric behavior of FG circular plates $(\alpha = 0)$ is considered.

In Fig. 8 the behavior of clamped circular FG plates under thermo-mechanical and thermal loading is investigated. It is seen that bending does not occur in clamped FG plates under thermal loading and bifurcation buckling occurs as thermal load is increased. The same phenomenon is seen to be present in clamped FG plates when they are subjected to a uniform temperature rise. The situation is, however, reversed for simply-supported FG plates. The bifurcation buckling is not seen to occur while bending takes place under the thermal loading considered (i.e., heat conduction through plate thickness), as shown in Fig. 9.

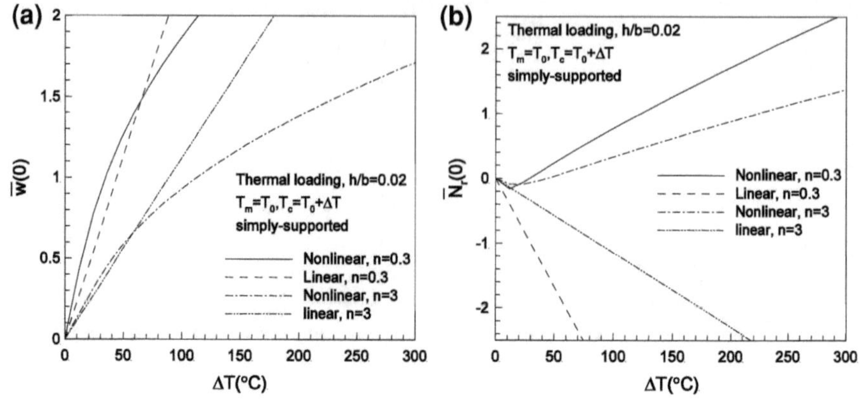

Fig. 10 a Center deflection $\bar{w}$, and b radial stress resultant $\bar{N}_r$ of simply-supported FG circular plate versus thermal load

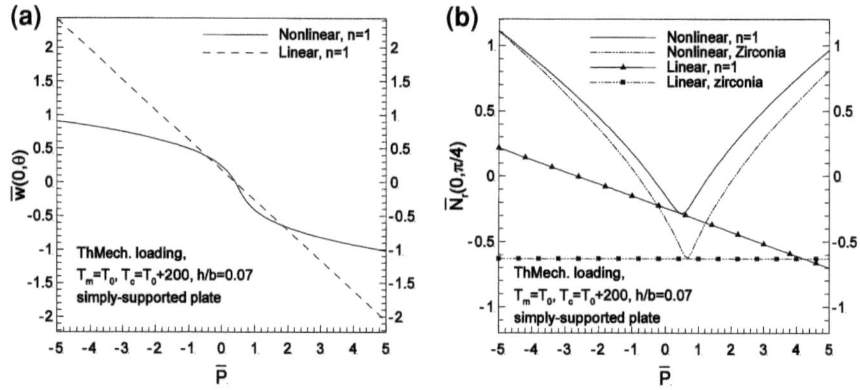

Fig. 11 a Center deflection, and b radial resultant force of simply-supported FG circular plate versus load parameter $\bar{P}$ under thermo-mechanical loading

A comparison of linear (i.e., the first-order perturbation steps) and nonlinear analysis for non-dimensional centre deflection and radial stress resultant N_r of a simply-supported FG plate subjected to thermal loading is displayed in Fig. 10a and b, respectively. Two values of power-law index (namely, $n = 0.3$ and $n = 3$) are considered. In Fig. 10b it is observed that N_r is compressive in the beginning and then becomes tensile as the loading is increased (i.e., as the large deformation effect becomes dominant). This, on the other hand, justifies the observation that the nonlinear deflection is larger than the linear deflection, as the compressive radial resultant force has a softening effect. It must be emphasized that when only thermal load is present, the linear analysis may result in significant errors.

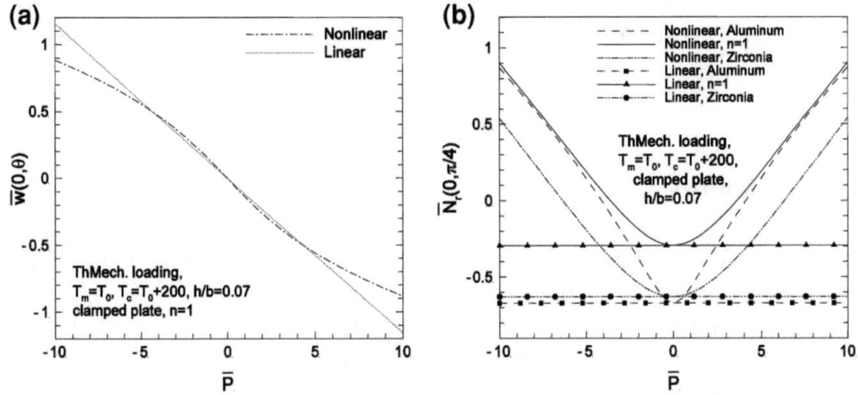

Fig. 12 **a** Center deflection, and **b** radial resultant force of clamped FG circular plate versus load parameter $\bar{P}$ under thermo-mechanical loading

Fig. 13 Nonlinear center deflection of a clamped FG circular plate for different values of α and θ versus load parameter $\bar{P}$ under thermo-mechanical loading

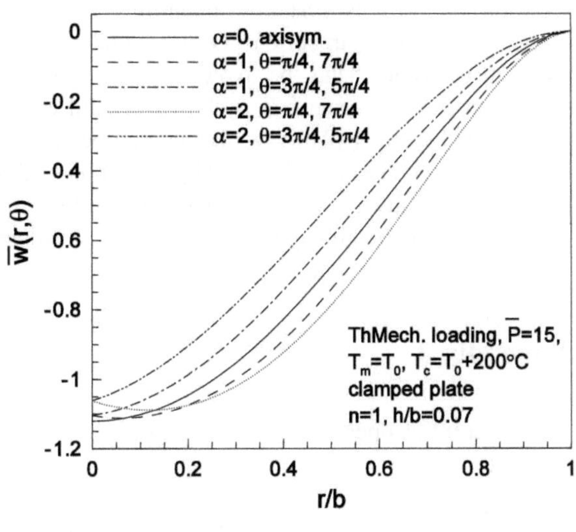

This observation is also reported in the cylindrical bending analysis of simply-supported FG plates [28].

Figures 11 and 12 show a comparison of the linear and nonlinear analyses for the centre deflections and in-plane forces in simply-supported and clamped FG plates under a thermo-mechanical loading, respectively. It is, again, emphasized here that when the thermal loading is dominant, the linear theory becomes inadequate due to the existing compressive in-plane force within the plate.

The results for the deflection of a clamped FG plate (as a function of *r/b*) for different values of α and θ due to load parameter $\bar{P} = 15$ are depicted in Fig. 13. The volume fraction index is taken as n $= 1$. By taking $\alpha = 0$, the results under an axisymmetric loading are obtained and compared with the results of asymmetric loading. The asymmetry effects of transversal load are clearly pronounced in Fig. 13.

4 Conclusions

In the present work, by introducing a stress function and a potential function, the von Karman nonlinear equilibrium equations of FG plates within the first-order shear deformation theory are reformulated into interior and boundary-layer equations. A two parameter perturbation technique, in conjunction with Fourier series method, is employed to develop analytical solutions for simply-supported and clamped FG circular plates under an asymmetric transverse load and a temperature variation through the plate thickness. The results for solid circular plates are validated with the existing results in the literature. Several numerical results are generated to manifest the effects of nonlinearity, material constant, and edge supports on various response quantities. Snap-through buckling behavior is observed in simply-supported FG plates under thermo-mechanical loading which may, on the other hand, be avoided by increasing the amount of ceramic constituent in the FG system. It is shown that bifurcation-type buckling happens in clamped FG circular plates, while bending occurs in simply-supported FG plates. Moreover, it is observed that in FG plates subjected to thermal loadings, since the linear analysis always underestimates the deflection, a nonlinear theory must be used even for deflections that are normally considered small.

Appendix 1

The differential operators appearing in (25), (32), (45), and (49) are defined as follows:

$$L_1 = \frac{1}{r}\frac{\partial}{\partial r} + \frac{1}{r^2}\frac{\partial^2}{\partial \theta^2};\; L_2 = \frac{\partial}{\partial r}\left(\frac{1}{r}\frac{\partial}{\partial \theta}\right);\ell_1(\,,\,) = L_1()\frac{\partial^2()}{\partial r^2},\;\; \ell_2(\,,\,) = -2L_2()L_2();$$

$$\ell_3(\,,\,) = \left(\frac{1}{r}\frac{\partial}{\partial \theta} - \frac{\partial^2}{\partial r \partial \theta}\right)\left(\frac{1}{r}\frac{\partial}{\partial \theta} - \frac{\partial^2}{\partial r \partial \theta}\right),\quad \ell_4(\,,\,) = -\frac{\partial^2}{\partial r^2}\left(r\frac{\partial}{\partial r} + \frac{\partial^2}{\partial \theta^2}\right) \tag{A1}$$

$$L_3 = r^{m-1}\frac{d}{dr}\left\{r^{1-2m}\frac{d}{dr}[r^m()]\right\},\; L_4 = \frac{d}{dr} - \frac{m^2}{r},\; L_5 = \frac{1}{r}\frac{d}{dr} - \frac{1}{r^2} \tag{A2}$$

Appendix 2

The pseudo loads N_{1ijm} and $\tilde{N}_{1ijm}$ in Eq. 40 for the loading considered in (51) for the first nine perturbation steps are found to be as follows:

$$\varepsilon_1: \quad N_{110m} = \tilde{N}_{110m} = 0, \quad \varepsilon_2: \quad N_{101m} = \tilde{N}_{101m} = 0 \tag{B1a}$$

$$\begin{aligned}
\varepsilon_1^2 &: N_{120} = \sum_{m=0}^{2} N_{1gm}(1,0,1,0)\cos(m\theta), \\
\varepsilon_2^2 &: N_{102} = \sum_{m=0}^{2} N_{1gm}(0,1,0,1)\cos(m\theta), \\
\varepsilon_1\varepsilon_2 &: N_{111} = \sum_{m=0}^{2} \left(N_{1gm}(1,0,0,1) + N_{1gm}(0,1,1,0)\right)\cos(m\theta), \\
& N_{1ij3} = 0, \quad for\ i+j=2, \\
\varepsilon_1^3 &: N_{130} = \sum_{m=0}^{3} \left(N_{1gm}(1,0,2,0) + N_{1gm}(2,0,1,0)\right)\cos(m\theta), \\
\varepsilon_2^3 &: N_{103} = \sum_{m=0}^{3} \left(N_{1gm}(0,1,0,2) + N_{1gm}(0,2,0,1)\right)\cos(m\theta), \\
\varepsilon_1\varepsilon_2^2 &: N_{112} = \sum_{m=0}^{3} \begin{pmatrix} N_{1gm}(1,0,0,2) + N_{1gm}(0,2,1,0) \\ +N_{1gm}(0,1,1,1) + N_{1gm}(1,1,0,1) \end{pmatrix}\cos(m\theta) \\
\varepsilon_1^2\varepsilon_2 &: N_{121} = \sum_{m=0}^{3} \begin{pmatrix} N_{1gm}(1,0,1,1) + N_{1gm}(1,1,1,0) \\ +N_{1gm}(0,1,2,0) + N_{1gm}(2,0,0,1) \end{pmatrix}\cos(m\theta) \\
& N_{1ijm}(r) = 0 \quad m = 4,\ 5,\ \ldots,\ \tilde{N}_{1ijm}(r) = 0 \quad m = 1,\ 2,\ \ldots
\end{aligned} \tag{B1b}$$

where

$$\begin{aligned}
N_{1g0}(i,j,k,l) =& \frac{1}{r}F'_{ij0}w''_{kl0} + \frac{1}{r}w'_{kl0}F''_{ij0} + NN_{1ijkl} - NN_{2ijkl}, \\
N_{1g1}(i,j,k,l) =& \frac{1}{r}F'_{ij0}w''_{kl1} + w''_{kl0}\left(\frac{1}{r}F_{ij1}\right)' + \frac{1}{r}w'_{kl0}F''_{ij1} + F''_{ij0}\left(\frac{1}{r}w_{kl1}\right)' \\
& + (Ord-2)\left(NN_{3ijkl} - NN_{4ijkl}\right), \\
N_{1g2}(i,j,k,l) =& NN_{1ijkl} + NN_{2ijkl} \\
& + (Ord-2)\left\{\begin{array}{ll} \frac{1}{r}F'_{ij0}w''_{kl2} + F''_{ij0}\left(\frac{1}{r}w'_{kl2} - \frac{4}{r^2}w_{kl2}\right) & if\ i+j=1 \\ \left(\frac{1}{r}F'_{ij2} - \frac{4}{r^2}F_{ij2}\right)w''_{kl0} + F''_{ij2}\frac{1}{r}w'_{kl0} & if\ i+j=2 \end{array}\right\} \\
N_{1g3}(i,j,k,l) =& NN_{3ijkl} + NN_{4ijkl}
\end{aligned} \tag{B1c}$$

In Eq. B1c and what follows the value of the parameter *Ord* is two for the second-order perturbation steps and it equals three for the third-order perturbation steps and

$$NN_{1ijkl} = \frac{1}{2}\left[(\frac{1}{r}F_{ij1})'w''_{kl1} + F''_{ij1}(\frac{1}{r}w_{kl1})'\right], NN_{2ijkl} = \left[(\frac{1}{r}F_{ij1})'(\frac{1}{r}w_{kl1})'\right]$$

$$NN_{3ijkl} = \frac{1}{2}\begin{bmatrix} \left(\frac{1}{r}F'_{ij(i+j)} - \frac{(i+j)^2}{r^2}F_{ij(i+j)}\right)w''_{kl(k+l)} \\ +F''_{ij(i+j)}\left(\frac{1}{r}w'_{kl(k+l)} - \frac{(k+l)^2}{r^2}w_{kl(k+l)}\right) \end{bmatrix} \tag{B1d}$$

$$NN_{4ijkl} = 2\left[(\frac{1}{r}F_{ij(i+j)})'(\frac{1}{r}w_{kl(k+l)})'\right]$$

Also the pseudo loads N_{2ijm} and $\tilde{N}_{2ijm}$ in Eq. (40) for the loading considered in (51) for the first nine perturbation steps are found to be as follows:

$$\varepsilon_1:\ N_{210m} = \tilde{N}_{210m} = 0,\ \varepsilon_2:\ N_{201m} = \tilde{N}_{201m} = 0 \tag{B2a}$$

$$\begin{aligned}
\varepsilon_1^2 &: N_{220} = \frac{1}{2}\sum_{m=0}^{2} N_{2gm}(1,0,1,0)\cos(m\theta) \\
\varepsilon_2^2 &: N_{202} = \frac{1}{2}\sum_{m=0}^{2} N_{2gm}(0,1,0,1)\cos(m\theta) \\
\varepsilon_1\varepsilon_2 &: N_{211} = \sum_{m=0}^{2} N_{2gm}(1,0,0,1)\cos(m\theta), \quad N_{2ij3} = 0, \quad \mathit{for}\ i+j=2 \\
\varepsilon_1^3 &: N_{230} = \sum_{m=0}^{3} N_{2gm}(1,0,2,0)\cos(m\theta), \\
\varepsilon_2^3 &: N_{203} = \sum_{m=0}^{3} N_{2gm}(0,1,0,2)\cos(m\theta) \\
\varepsilon_1\varepsilon_2^2 &: N_{212} = \sum_{m=0}^{3} \left(N_{2gm}(1,0,0,2) + N_{2gm}(0,1,1,1)\right)\cos(m\theta) \\
\varepsilon_1^2\varepsilon_2 &: N_{221} = \sum_{m=0}^{3} \left(N_{2gm}(1,0,1,1) + N_{2gm}(0,1,2,0)\right)\cos(m\theta) \\
& N_{2ijm}(r) = 0 \quad m = 4, 5, \ldots, \tilde{N}_{2ijm}(r) = 0 \quad m = 1, 2, \ldots
\end{aligned} \tag{B2b}$$

where

$$N_{2g0}(i,j,k,l) = -(rw''_{ij0}w'_{kl0} + rw''_{kl0}w'_{ij0}) + NN_{5ijkl} + NN_{6ijkl}$$

$$N_{2g1}(i,j,k,l) = -\left[rw''_{ij0}(w'_{kl1} - \frac{1}{r}w_{kl1}) + rw''_{ij1}w'_{kl0}\right]$$
$$- \left[rw''_{kl0}(w'_{ij1} - \frac{1}{r}w_{ij1}) + rw''_{kl1}w'_{ij0}\right] + (Ord-2)(NN_{7ijkl} + NN_{8ijkl})$$

$$N_{2g2}(i,j,k,l) = NN_{5ijkl} - NN_{6ijkl}$$
$$- (Ord-2)\left[rw'_{ij0}w''_{kl2} + rw''_{ij0}(w'_{kl2} - \frac{4}{r}w_{kl2})\right]$$

$$N_{2g3}(i,j,k,l) = NN_{7ijkl} - NN_{8ijkl} \tag{B2c}$$

$$\begin{aligned} NN_{5ijkl} &= -\frac{1}{2}\left[rw''_{ij1}(w'_{kl1} - \frac{1}{r}w_{kl1}) + rw''_{kl1}(w'_{ij1} - \frac{1}{r}w_{ij1})\right] \\ NN_{6ijkl} &= (w'_{ij1} - \frac{1}{r}w_{ij1})(w'_{kl1} - \frac{1}{r}w_{kl1}) \\ NN_{7ijkl} &= -\frac{1}{2}\left[rw''_{ij1}(w'_{kl2} - \frac{4}{r}w_{kl2}) + rw''_{kl2}(w'_{ij1} - \frac{1}{r}w_{ij1})\right] \\ NN_{8ijkl} &= 2(w'_{ij1} - \frac{1}{r}w_{ij1})(w'_{kl2} - \frac{1}{r}w_{kl2}) \end{aligned} \tag{B2d}$$

Finally, the nonlinear terms NL_{1ijm}, $\tilde{N}L_{1ijm}$, NL_{2ijm}, $\tilde{N}L_{2ijm}$, NL_{3ijm}, and $\tilde{N}L_{3ijm}$ appearing in (54) (or in (48)) for the nine perturbation steps are found to be as follows:

$$\begin{aligned} \varepsilon_1 &: NL_{110m} = NL_{210m} = NL_{310m} = \tilde{N}L_{110m} = \tilde{N}L_{210m} = \tilde{N}L_{310m} = 0 \\ \varepsilon_2 &: NL_{101m} = NL_{201m} = NL_{301m} = \tilde{N}L_{101m} = \tilde{N}L_{201m} = \tilde{N}L_{301m} = 0 \end{aligned} \tag{B3a}$$

$$\varepsilon_1^2 : NL_{120} = \frac{1}{2}\sum_{m=0}^{2} NL_{1gm}(1,0,1,0)\cos(m\theta)$$

$$\varepsilon_2^2 : NL_{102} = \frac{1}{2}\sum_{m=0}^{2} NL_{1gm}(0,1,0,1)\cos(m\theta)$$

$$\varepsilon_1\varepsilon_2 : NL_{111} = \sum_{m=0}^{2} NL_{1gm}(1,0,0,1)\cos(m\theta), \quad NL_{1ij3} = 0, \quad for\ i+j=2$$

$$\varepsilon_1^3 : NL_{130} = \sum_{m=0}^{3} NL_{1gm}(1,0,2,0)\cos(m\theta),$$

$$\varepsilon_2^3 : NL_{103} = \sum_{m=0}^{3} NL_{1gm}(0,1,0,2)\cos(m\theta)$$

$$\varepsilon_1\varepsilon_2^2 : NL_{112} = \sum_{m=0}^{3}\left(NL_{1gm}(1,0,0,2) + NL_{1gm}(0,1,1,1)\right)\cos(m\theta)$$

$$\varepsilon_1^2\varepsilon_2 : NL_{121} = \sum_{m=0}^{3}\left(NL_{1gm}(1,0,1,1) + NL_{1gm}(0,1,2,0)\right)\cos(m\theta) \tag{B3b}$$

$$NL_{1ijm}(r) = 0, \quad m = 4, 5, \ldots, \tilde{N}L_{1ijm}(r) = 0, \quad m = 1, 2, \ldots$$

$$\varepsilon_1^2 : NL_{220} = \frac{1}{2}\sum_{m=0}^{2} NL_{2gm}(1,0,1,0)\cos(m\theta)$$

$$\varepsilon_2^2 : NL_{202} = \frac{1}{2}\sum_{m=0}^{2} NL_{2gm}(0,1,0,1)\cos(m\theta),$$

$$\varepsilon_1\varepsilon_2 : NL_{211} = \sum_{m=0}^{2} NL_{2gm}(1,0,0,1)\cos(m\theta), \quad NL_{2ij3} = 0, \quad for\ i+j=2$$

$$\varepsilon_1^3 : NL_{230} = \sum_{m=0}^{3} NL_{2gm}(1,0,2,0)\cos(m\theta), \tag{B3c}$$

$$\varepsilon_2^3 : NL_{203} = \sum_{m=0}^{3} NL_{2gm}(0,1,0,2)\cos(m\theta)$$

$$\varepsilon_1\varepsilon_2^2 : NL_{212} = \sum_{m=0}^{3}\left(NL_{2gm}(1,0,0,2) + NL_{2gm}(0,1,1,1)\right)\cos(m\theta)$$

$$\varepsilon_1^2\varepsilon_2 : NL_{221} = \sum_{m=0}^{3}\left(NL_{2gm}(1,0,1,1) + NL_{2gm}(0,1,2,0)\right)\cos(m\theta)$$

$$NL_{2ijm}(r) = 0,\ m = 4, 5, \ldots, \tilde{N}L_{2ijm}(r) = 0 \quad m = 1, 2, \ldots$$

and

$$\varepsilon_1^2 : NL_{320} = \frac{1}{2}\sum_{m=1}^{2} \tilde{N}L_{3gm}(1,0,1,0)\sin(m\theta)$$

$$\varepsilon_2^2 : NL_{302} = \frac{1}{2}\sum_{m=1}^{2} \tilde{N}L_{3gm}(0,1,0,1)\sin(m\theta)$$

$$\varepsilon_1\varepsilon_2 : NL_{311} = \sum_{m=1}^{2} \tilde{N}L_{3gm}(1,0,0,1)\sin(m\theta),\ \tilde{N}L_{3ij3} = 0, \quad for\ i+j=2$$

$$\varepsilon_1^3 : NL_{330} = \sum_{m=1}^{3} \tilde{N}L_{3gm}(1,0,2,0)\sin(m\theta),$$

$$\varepsilon_2^3 : NL_{303} = \sum_{m=1}^{3} \tilde{N}L_{3gm}(0,1,0,2)\sin(m\theta),$$

$$\begin{aligned}\varepsilon_1\varepsilon_2^2 : NL_{312} &= \sum_{m=1}^{3}\left(\tilde{N}L_{3gm}(1,0,0,2)+\tilde{N}L_{3gm}(0,1,1,1)\right)\sin(m\theta)\\ \varepsilon_1^2\varepsilon_2 : NL_{321} &= \sum_{m=1}^{3}\left(\tilde{N}L_{3gm}(1,0,1,1)+\tilde{N}L_{3gm}(0,1,2,0)\right)\sin(m\theta)\\ NL_{3ijm}(r) &= 0 \quad m=0,\,1,\,\ldots,\ \tilde{N}L_{3ijm}(r)=0 \quad m=4,\,5,\,\ldots\end{aligned} \tag{B3d}$$

where

$$\begin{aligned}NL_{1g0}(i,j,k,l) &= -w'_{ij0}w'_{kl0}-\frac{1}{2}w'_{ij1}w'_{kl1}\\ NL_{1g1}(i,j,k,l) &= -(w'_{ij0}w'_{kl1}+w'_{ij1}w'_{kl0})-\frac{Ord-2}{2}w'_{ij1}w'_{kl2}\\ NL_{1g2}(i,j,k,l) &= -(Ord-2)w'_{ij0}w'_{kl2}-\frac{1}{2}w'_{ij1}w'_{kl1},\\ NL_{1g3}(i,j,k,l,) &= -\frac{1}{2}w'_{ij1}w'_{kl2}\end{aligned} \tag{B3e}$$

$$\begin{aligned}NL_{2g0}(i,j,k,l) &= -\frac{1}{2}\frac{1}{r^2}w_{ij1}w_{kl1},\\ NL_{2g1}(i,j,k,l) &= -(Ord-2)\frac{1}{r^2}w_{ij1}w_{kl2},\\ NL_{2g2}(i,j,k,l) &= -NL_{2g0}(i,j,k,l),\quad NL_{2g3}(i,j,k,l)=-NL_{2g1}(i,j,k,l)\end{aligned} \tag{B3f}$$

and

$$\begin{aligned}\tilde{N}L_{3g1}(i,j,k,l) =& \frac{1}{r}(w'_{ij0}w_{kl1}+w'_{kl0}w_{ij1})\\ &+(Ord-2)\frac{1}{r}(w'_{ij1}w_{kl2}-\frac{1}{2}w_{ij1}w'_{kl2})\\ \tilde{N}L_{3g2}(i,j,k,l) =& \frac{1}{2}\frac{1}{r}(w'_{ij1}w_{kl1}+w'_{kl1}w_{ij1})+(Ord-2)\frac{2}{r}w'_{ij0}w_{kl2}\\ \tilde{N}L_{3g3}(i,j,k,l) =& \frac{1}{r}(w'_{ij1}w_{kl2}+\frac{1}{2}w_{ij1}w'_{kl2})\end{aligned} \tag{B3g}$$

References

1. Koizumi, M.: The concept of FGM. Ceram. Trans. Funct. Gradient Mater. **34**, 3–10 (1993)
2. Sureh, S., Mortensen, A.: Fundamentals of Functionally Graded Materials. IOM Communications Limited, London (1998)
3. Vel, S.S., Batra, R.C.: Exact solution for thermoelastic deformations of functionally graded thick rectangular plates. AIAA J. **40**, 1421–1433 (2002)

4. Chi, S.H., Chung, Y.L.: Mechanical behavior of functionally graded material plates under transverse load–Part I: Analysis. Int. J. Solids Struct. **43**, 3657–3674 (2006)
5. Chi, S.H., Chung, Y.L.: Mechanical behavior of functionally graded material plates under transverse load–Part II: Numerical results, Int. J. Solids Struct. **43**, 3675–3691 (2006)
6. Nguyen, T.K., Sab, K., Bonnet, G.: First-order shear deformation plate models for functionally graded materials. Compos. Struct. **83**, 25–36 (2008)
7. Reddy, J.N., Wang, C.M., Kitipornchai, S.: Axisymmetric bending of functionally graded circular and annular plates. Eur. J. Mech. A/Solids **18**, 185–199 (1999)
8. Nosier, A., Fallah, F.: Reformulation of Mindlin-Reissner governing equations of functionally graded circular plates. Acta Mech. **198**, 209–233 (2008)
9. Yun, W., Rongqiao, X., Haojiang, D.: Three-dimensional solution of axisymmetric bending of functionally graded circular plates. Comp. Struct. **92**, 1683–1693 (2010)
10. Praveen, G.N., Reddy, J.N.: Nonlinear transient thermo elastic analysis of functionally graded ceramic-metal plates. Int. J. Solids Struct. **35**, 4457–4476 (1998)
11. Reddy, J.N.: Analysis of functionally graded plates. Int. J. Numer. Meth. Eng. **47**, 663–684 (2000)
12. Woo, J., Meguid, S.A.: Nonlinear analysis of functionally graded plates and shallow shells. Int. J. Solids Struct. **38**, 7409–7421 (2001)
13. Woo, J., Meguid, S.A., Stranart, J.C., Liew, K.M.: Themomechanical post buckling analysis of moderately thick functionally graded plates and shallow shells. Int. J. Mech. Sci. **47**, 1147–1171 (2005)
14. Na, K.S., Kim, J.H.: Nonlinear bending response of functionally graded plates under thermal loads. J. Therm. stresses **29**, 245–261 (2006)
15. Na, K.S., Kim, J.H.: Thermal post buckling investigations of functionally graded plates using 3-D finite element method. Finite Elem. Anal. Des. **42**, 749–756 (2006)
16. Yang, J., Shen, H.S.: Nonlinear analysis of functionally graded plates under transverse and in-plane loads. Int. J. Nonlin. Mech. **38**, 467–482 (2003)
17. Shen, H.S.: Nonlinear thermal bending response of FGM plates due to heat conduction. Compos. Part B. **38**, 201–215 (2007)
18. Shen, H.S.: Thermal post buckling behavior of shear deformable FGM plates with temperature-dependent properties. Int. J. Mech. Sci. **49**, 466–478 (2007)
19. Yang, J., Huang, X.L.: Nonlinear transient response of functionally graded plates with general imperfection in thermal environments. Comput. Methods Appl. Mech. Eng. **196**, 2619–2630 (2007)
20. Prakash, T., Singha, M.K., Ganapathi, M.: Thermal snapping of functionally graded materials plates. Mater Design **30**, 4532–4536 (2009)
21. Gunes, R., Reddy, J.N.: Nonlinear analysis of functionally graded circular plates under different loads and boundary conditions. Int. J. Struct. Stab. Dyn. **8**, 131–159 (2008)
22. Ma, L.S., Wang, T.J.: Nonlinear bending and post buckling of a functionally graded circular plate under mechanical and thermal loadings. Int. J. Solids Struct. **40**, 3311–3330 (2003)
23. Li, S.R., Zhang, J.H., Zhao, Y.G.: Nonlinear thermo-mechanical post buckling of circular FGM plate with geometric imperfection. Thin Wall Struct. **45**, 528–536 (2007)
24. Najafizadeh, M.M., Eslami, M.R.: Higher-order theory for buckling of functionally graded circular plates. AIAA J. **45**, 1153–1160 (2007)
25. Najafizadeh, M.M., Eslami, M.R.: First-order-theory-based thermoelastic stability of functionally graded material circular plates. AIAA J. **40**, 1444–1450 (2002)
26. Prakash, T., Ganapathi, M.: Asymmetric flexural vibration and thermoelastic stability of FGM circular plates using finite element method. Compos. Part B **37**, 642–649 (2006)
27. Nosier, A., Fallah, F.: Nonlinear analysis of functionally graded circular plates under asymmertric transverse loading. Int. J. Nonlin. Mech. **44**, 928–942 (2009)
28. Fallah, F., Nosier, A.: Nonlinear thermo-mechanical cylindrical bending of functionally graded plates. Proc. IMechE, Part C J. Mech. Eng. Sci. **222**, 305–318 (2008)
29. Reddy, J.N.: Theory and analysis of elastic plates. Taylor & Francis, Philadelphia (1999)

30. Fung, Y.C., Tong, P.: Classical and computational solid mechanics. World Scientific, River Edge (2001)
31. Hildebrand, F.B.: Advanced calculus for applications. Prentice-Hall Inc, Englewood Cliffs (1962)

Finite Element Analysis of the Total Forming Force of Thin-Walled Metallic Tube Junctions Employing Elastomers

Cristiano Roberto Martins Foli, Miguel Ângelo Menezes and Lindolfo Araújo Moreira Filho

Abstract Among the forming processes can be emphasized the metallic junction forming using elastomers. This forming process presents a considerable number of variables, as for instance: the definition of the necessary relationship between the progress of the dome formed and the developed pressure in the elastomer, the friction and lubricating conditions during the junction forming, the role of the anisotropy and the material strain-hardening, as well as the influence of the strain-rate in the forming process and the definition of the maximum force to form the junctions. Process modelling has become an effective tool in reducing the lead-time and the cost for designing forming processes for manufacturing automotive and aerospace components. Several research works are being developed seeking to approximate the experimental and mathematical analysis in manufacturing processes. Computer software's for process modelling, and the transfer of this technology to the industry have contributed towards the development of this tool. This chapter aims to calculate the total forming force through numerical simulation process of thin-walled tube junctions using an elastômero, where important parameters are considered such as:

C. R. M. Foli · L. A. M. Filho
ITA-Aeronautic Technological Institute, Praça Mal. Eduardo Gomes, 50-Vila das Acácias-S. J. Campos-SP, São Paulo, CEP 1228-900, Brazil
e-mail: crmfoli@yahoo.com.br

L. A. M. Filho
e-mail: lindolfo@ita.br

M. Â. Menezes (✉)
UNESP-São Paulo State University, Av. Brasil no 56-Centro, Ilha Solteira-SP, CEP 15385000, Brazil
e-mail: miguel@dem.feis.unesp.br

A. Öchsner et al. (eds.), *Materials with Complex Behaviour II*,
Advanced Structured Materials 16, DOI: 10.1007/978-3-642-22700-4_24,

friction, elastomer length and material mechanical properties. Furthermore, in the work the finite element program Deform 3D is employed. Hence, comparisons are made between the obtained numerical results and available experimental results.

1 Introduction

With the development of synthetic elastomers such as polyurethane, urethane and plastiprene, several papers were published using the elastomer metal forming technique, e.g., deep-drawing of metal sheets [11], bending and piercing of tubes [1, 3]. Among the forming processes, stands out the forming of thin-walled tube junctions using elastomer [5, 6, 10, 13, 14].

1.1 Finite Elements Method

Process modeling has become an effective tool in reducing the lead-time and the cost for designing forming processes for manufacturing automotive and aerospace components. Several research programs aimed towards the development of the experimental and mathematical analysis of the mechanics of forming operations, computer software for process modeling, and transfer of this technology to the industry have contributed towards the overall objectives of superior technology for process design in net shape and near net shape forming, [8].

Ribeiro [15] quote as pioneering work on finite elements, the works of [2] and [16]. Zienkiewicz, in his historical paper "The Finite Element method: from Intuition to generality" [18] presents a more detailed description of the finite element method evolution in this initial phase. In the 1970s the finite element method has extended its applications to problems of fluid mechanics and has since been consolidating itself as the more general method of solution of partial differential equations.

The finite element method considers the region to solve the problem formed by small elements interconnected. The area under study is modeled analytically or approximated by a set of predefined discrete elements. Since these elements can be put together in countless different configurations have been modeled complex geometric shapes. It also enables the designer to have good chances in the mode of application of load and boundary conditions.

The finite element method is applicable to a wide range of boundary value problems of engineering. On a boundary value problem, a solution is sought in the body region (domain) as the contours of this region the values of the dependent variables (or their derivatives) are known [4].

1.2 *Finite Elements Method in Forming Process*

The development of reliable analytical procedures and the common applicability to predict the behavior of deformation in forming processes, has encountered some obstacles, such as the nonlinearity of the material, the unstable nature of the processes, the large magnitude of deformation, and involvement of significant effects of friction during forming, which makes the study of forming processes very complex [7].

Between the initial cut and the final shape conformed, the material is subjected to an extremely complex history of deformation, including continuous changes in boundary conditions, large distortions of the membrane, the phenomenon of acceleration due necking (narrowing of the dome), and also decreasing thickness due to the effect of Poisson's ratio and incompressible plastic deformations. Thus, it is necessary to consider geometric and/or material nonlinearity in the process analysis [17].

An elastic–plastic formulation is required due to the fact that the materials also have a portion of non-linear response of stress–strain during processing. What added to the need to simulate parts with geometric and physical nonlinearities, such as those required in the aerospace industry, justify the need to use computer algorithms of large scale [19].

The analysis of forming processes through the finite element method can be classified into two groups, an approach based on rigid-plastic and visco-plastic approach and another based on elastic–plastic. Because of the large deformation in forming processes, the simulation by the finite element method requires special formulations, such as Eulerian, Eulerian Current, Total Lagrangian and Updated Lagrangian [9, 12].

2 Experimental Details

The initial and final geometries of the elastomer (polyurethane of Shore hardness 95 A) and the tested tubes employed in the forming process of thin-walled tube junctions using elastomers are shown in Fig. 1, so as the double stage punch.

Each tube set up in the device with the chamfered part ahead of the die hole was lubricated outside and clean inside with acetone to increase the elastomer-tube friction that improves the forming process. The larger is the elastomer-tube internal wall friction, the larger will be the material flow for the junction formation (dome). Consequently, after the elastomer surface cleaning, it was impregnated with chalk powder (CaCO3), with the purpose of increasing the adherence (friction).

Grease based on colloidal aluminium (Aluminipart) and grease based on sodium (Esso Draw EX-41) were used for the tube external lubrication. The former has presented a better behavior during the deforming process.

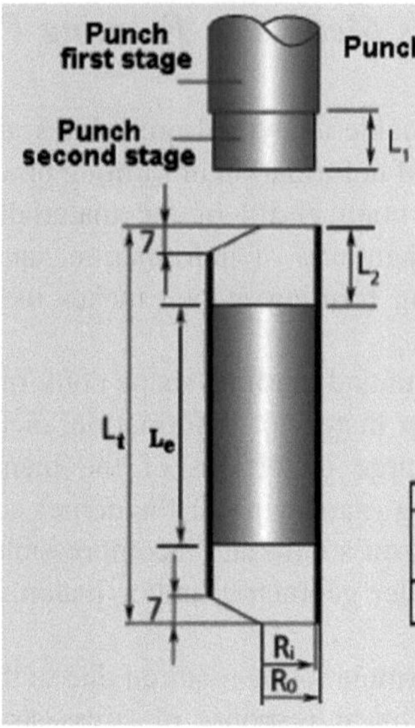

Tube	Elastomer
R_o = 14,3 mm R_i = 13,3 mm	R = 12,5 mm
L_t = 110 mm	L_e = 93 mm

Fig. 1 Description of the tube geometry

The initial elastomer length is determined as a function of the "clearance height" (Z), value determined by the empiric equation obtained by Marreco [10]:

$$Z = L_1 - L_2 = [(15.8 + 0.55X)] \pm 1.8 \text{ mm}, \quad (1)$$

where,

X dome feed (mm);
L_1 second stage length of the punch, Fig. 1, (mm); and
L_2 half of the difference among the tube length (L) and the elastomer length, (mm)

The success of the operation, that is, the forming process without presenting failure by fracture or buckling (Fig. 2a, 2b) depends on application of pressures in adequate proportion in the elastomer as in the tube. The control of the called "clearance height" was possible through the Eq. 1. The clearance height grows with the process development due to the increase of the available space for the elastomer to flow, which is function of the length of the formed dome (X).

In the automated case, a mechanism of four bars was developed, which is presented in Fig. 3 that controls the relative feed of the second stage automatically in relation to the first stage, avoiding fracture or buckling failures (Fig. 2a, 2b) that could happen during the development of the forming process.

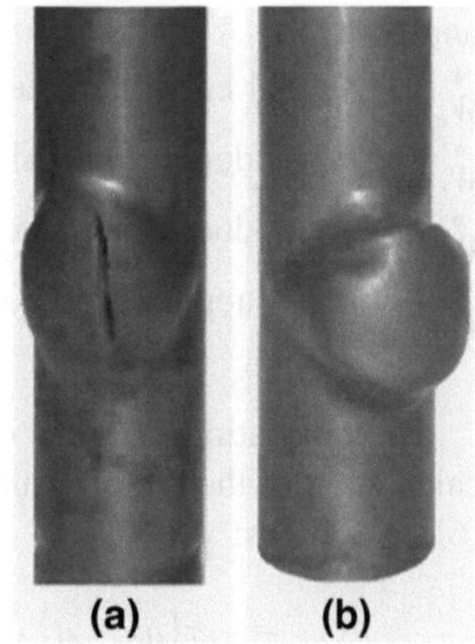

Fig. 2 Types of failure in the forming tubes. **a** Failure by fracture in the dome. **b** Failure by local buckling

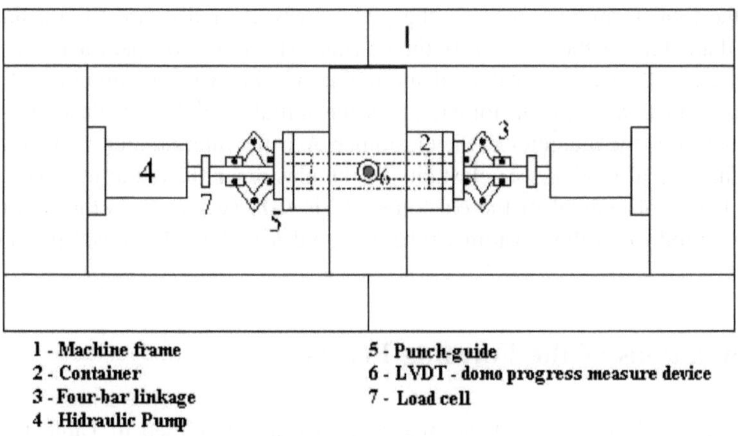

Fig. 3 Scheme of the forming device

3 Theoretical Analytical Model

According to the upper bound theory, the existence of a kinematically admissible velocity field is assumed such that the loads responsible by the formation of this field constitute an upper bound to the loads required for the real solution. The aim of this calculation is to obtain an estimate of the total forming force, and so to define the set of equipments necessary for the test.

The total force required would be determined by the energy balance involved in the process, where the following energy portions take place:

$$\dot{W}_e = \dot{W}_i + \dot{W}_a + \dot{W}_b \tag{2}$$

where,

$\overset{\bullet}{W}_e$ External energy applied to the system;

$\overset{\bullet}{W}_i$ Energy due to internal strain;

$\overset{\bullet}{W}_e$ Energy due to friction losses; and

$\overset{\bullet}{W}_b$ Energy applied to the elastomer

The estimating of these energy portions leads the formula employed for the calculation of the total forming force, which was previously deduced in reference, [6], as follows:

$$F_T = \frac{\frac{\pi}{2\sqrt{3}}.\bar{\sigma}\left\{d_0^2 - d_i^2 + 2.m.(L - Y + X).d_0\right\} + A_0 E_C\left(\frac{d_i^2 - d_r^2}{d_r^2}\right) + \frac{A_i E^*.Y_3}{L'}}{2 - K.E^*} \quad (3)$$

F_T is the total forming force, $\bar{\sigma}$is the yield stress, d_0 and d_i are the outside and inside diameters of the tube, m is the constant shear friction parameter, L is the initial tee length, Y is the total axial shortening of tube, X is the lateral penetration, A_0 is the initial area of elastomer, Ec is the initial modulus of elasticity of the elastomer, d_r is the diameter of the elastomer, A_i is the internal area of the tube, E^* is the theoretical modulus of elasticity of the elastomer, Y_3 is the ram movement due to the displacement of the elastomer, L' is L-Y_1 (where Y_1 is the ram movement required to fill the elastomer/tube gap) and K is the volume compressibility.

4 Simulations of the Forming Process

The mechanical properties of the materials used are presented in Table 1, where C is the material strength coefficient and n is the strain-hardening index. These values were obtained according to ASTM E8-69 standard.

4.1 Finite Element Model

The forming process was simulated using the finite element method using the software Deform 3D Version 5.1 for pre-processing and post processing. The model uses 28,254 tetrahedral type elements to model the tube and 10,809 to model the elastomer. Rigid elements were used to model the matrix and the punches, and only half of the tooling was discretized due to symmetry of the piece. The Newton–Raphson method was used as a method of interaction. Figure 4 presents the finite element model at the beginning in the process.

Due to the "clearance height" previously quoted, it is necessary to stop the simulation process to change the length of the elastomer or punch.

Table 1 Mechanical properties of materials used

Materials	Yield stress $\sigma_e(\mathrm{kgf} / \mathrm{mm}^2)$	Ultimate tensile stress $\sigma_t(\mathrm{kgf} / \mathrm{mm}^2)$	Strain-hardening (n)	Strength coefficient (C) (kgf/mm^2)
Aluminium, annealed	5.00	11.00	0.22	19.50
Copper annealed	6.00	27.00	0.45	41.00

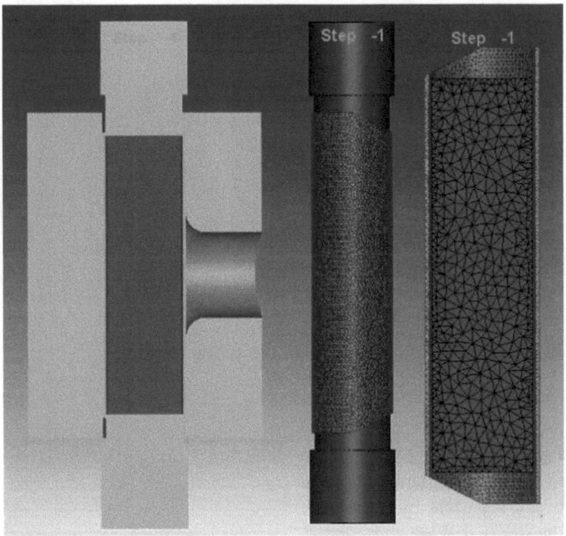

Fig. 4 Finite element model at the beginning in the process

5 Results and Discussion

Figure 5 shows some of the cycles of brass tube forming, so as Fig. 6 shows the cycles of the corresponding numerical simulation for the tube forming.

The numerical simulation of the forming process can also show or follow the success of the operation, that is, the forming process without failure by fracture or buckling (Figs. 7a, 7b) depending on the application of pressures in adequate proportion in the elastomer as well as in the tube. The control of the called "clearance height" was possible through Eq. 1.

Figure 8 presents a "T" junction forming using the finite element method, as an example of using the "T" junction forming process using elastomers.

By analysing Eq. 2, one verifies that it depends on geometric factors, mechanical properties of materials used, operational conditions (dome progress) and on the friction factor (m).

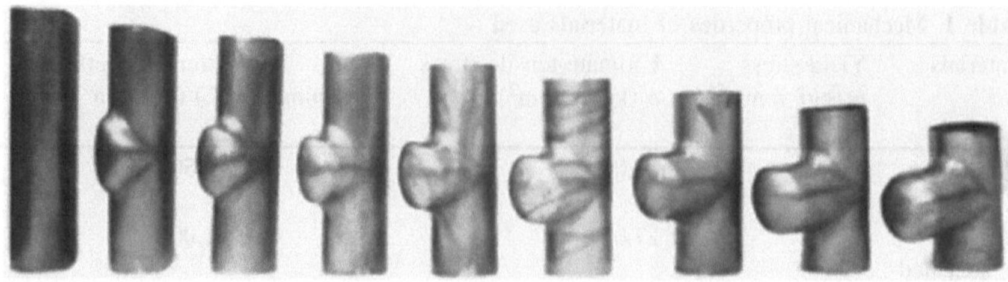

Fig. 5 Cycles of brass tube forming (experimental)

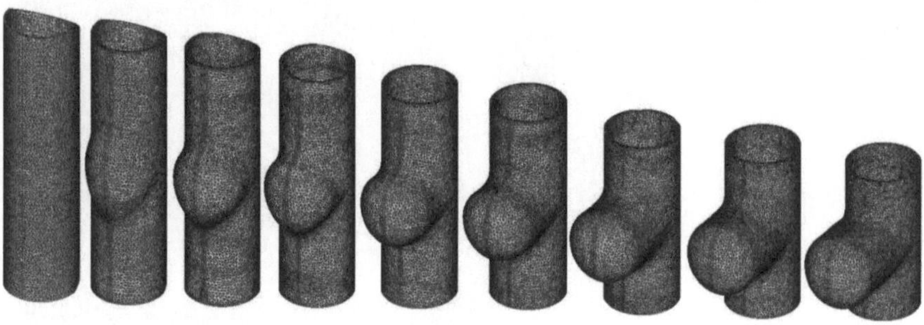

Fig. 6 Cycles of simulation of the tubes forming

Fig. 7 Types of failure in the numerical simulation forming tubes. **a** Failure by fracture in the dome. **b** Failure by local buckling

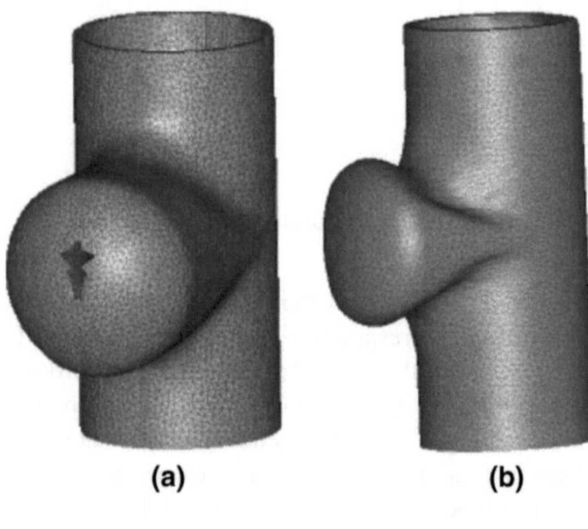

Due to the difficulty for determining previously the friction factor (m) during the process, a series of theoretical curves obtained using Eq. 2, fixing m and varying X up to $X = d_o$ (maximum forming) were built in order to find an average value for the friction by comparing between the theoretical and experimental

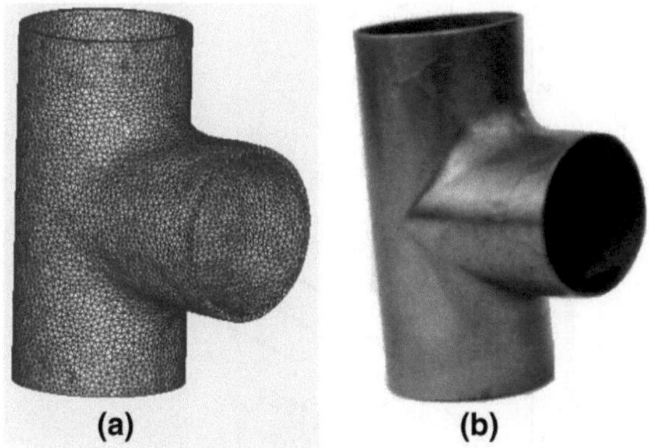

Fig. 8 "T" junction. **a** Finite elements method. **b** Forming using elastomer (experimental)

values. The following points should be taken into account: uniform application of the lubricant along the outside surface of the tube and the inside wall of the die, with the purpose of minimizing the friction variations; and the experimental process should be rigorously followed to avoid great tube variations when rehearsing the forming process of thin-walled metallic tube junctions.

During the simulation of the forming process by the finite element method, the amount of friction between the elastomer and the tube remained constant and equal to 0.08.

Examining Figs. 9 and 10 for the case of aluminium, it is noted that the friction factor values, as well as its variation in the finite elements method simulation is smaller than those found using Eq. 2. This is because of the simplifications adopted in the mathematical model, such as not incorporating the friction factor between the tube and the elastomer.

Examining Fig. 9, for the case of the aluminium, it is noticed an initial friction factor value of $m = 0.05$. After that, it is observed that the friction factor of $m = 0.05$, stays practically constant up to the displacement point of the dome is $X = 10$ mm. It increases afterwards continually from 0.05 up to 0.08; a value corresponding to $X = d_o$ equivalent to the total forming.

For the numerical simulation in Fig. 10, there is a friction factor at the beginning equal $m = 0.04$, where from the point of displacement of the dome of $X = 7$ mm up to 16 mm, values quite similar were observed close a friction factor between $m = 0.04$ and $m = 0.05$, which remain so up to the point $X = 22$ mm. From this point on up to the end of the process there is an increasing in friction factor that can reach a value equal to $m = 0.07$.

This variation can be explained by the fact that, as the process goes on, the lubricant between the external wall of the tube and the die loses efficiency due to its elimination, increasing more the contact metal/metal.

For the copper tube (Fig. 11) and using Eq. 2, it is observed that the friction factor m = 0.07 remains constant up to the displacement of the dome is equal $X = 10$ mm. After a narrow variation in friction factor, it is relevant to note that

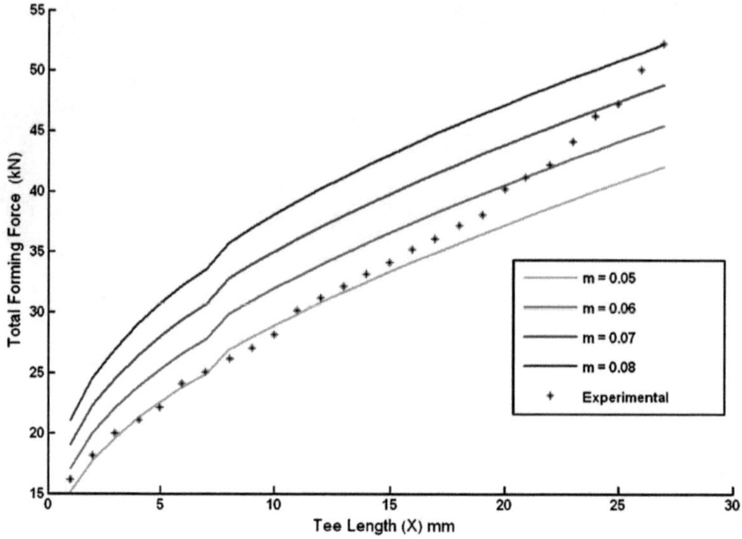

Fig. 9 Experimental and analytical values of the total forming forces using the *Upper* Bond theory–aluminium

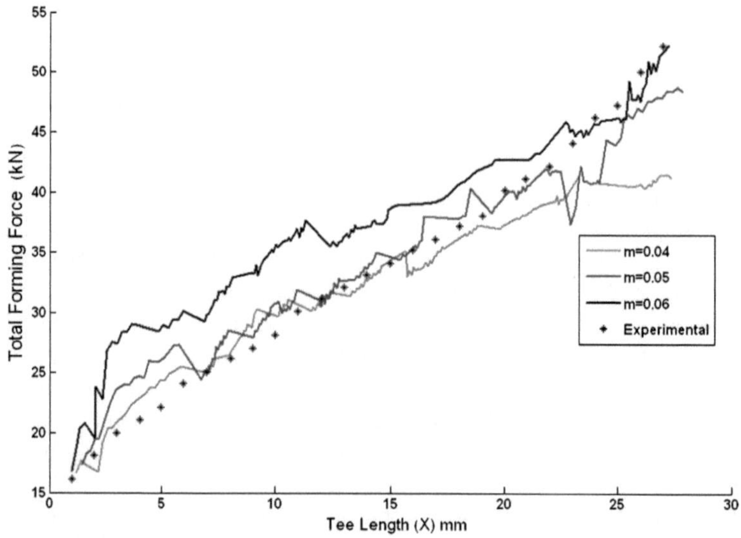

Fig. 10 Experimental and numerical values of total forming forces using the finite element method–aluminium

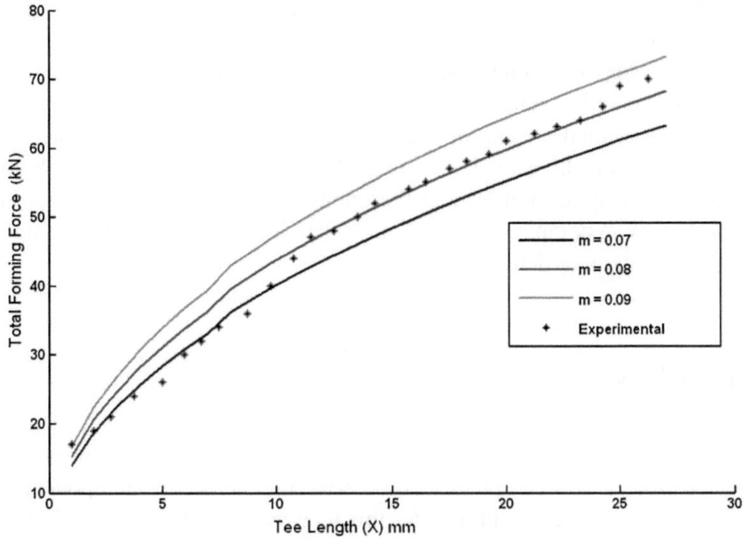

Fig. 11 Experimental and analytical values of total forming forces using the *Upper* Bond theory–copper

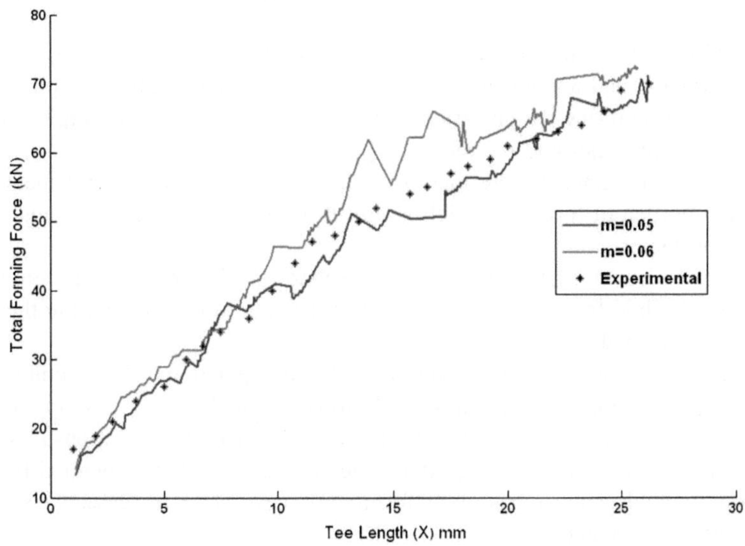

Fig. 12 Experimental and numerical values of total forming forces using the finite element method–copper

between $X > 10$ mm and $X = d_o$ (total conformation), there is no significant variation in the value of friction factor $m = 0.08$.

In the finite elements method simulation in Fig. 12, it is observed that the value of the friction factor has remained virtually constant up to $m = 0.05$. This fact can be related to the smaller loss of the lubricant efficiency and explained by the excellent surface finishing of the copper tube when compared to the aluminium tube. Besides, it can be inferred that the granulation of the copper tube would be more satisfactory in comparison with that of the aluminium tube because it did not present the orange peel defect on its surface.

6 Conclusions

The process of numerical simulation has become an effective tool to reduce time and cost of projects to manufacture automotive and aerospace components. Software to model the process, and transfer this technology to the industry has helped to develops of this tool.

Knowledge of the loads acting on the material during forming, the geometry of flow, and the degree of forming, for instance, constitute a valuable aid to the analysis of possible causes of defects and allows preventing such conveniences. The finite element method simulation can predict failure by fracture or buckling in the dome, helping the designer to relate the pressure between the elastomer and the tube.

The differences found between the values of friction factor using the finite element method and analytical model (upper bound theory) can be explained by better evaluation of the parameters involved during forming, for example, friction between the elastomer and the tube.

The results obtained by the theory of the upper limit constitute a method to assess the burden of forming, which is useful to provide a first approximation of the friction factor.

The finite element method was able to respond satisfactorily, thus becoming an estimation method for calculating the load of forming, useful in defining the tools and devices needed.

The forming process of thin-walled metallic tube junctions, at the actual stage, it is an important area of application of elastomers, which due to its simplicity and low cost of tooling, can be an option of interest in industrial applications involving small-scale production, replacing advantageously the junctions obtained by conventional methods, since it has a cold forming process which increases the strength properties of the material.

Acknowledgments The authors would like to thank CNPq (Brazilian National Research Council), ITA (Technological Institute of Aeronautics) and UNESP (São Paulo State University), for the financial support and for funding the research, as the support during the preparation of this work.

References

1. Al-Qureshi, H.A.: Analytical investigation of ram movement in piercing operation with rubber pads. Int. J. Match. Tool Des. Res. **12**, 229 (1972)
2. Argyris, J.H., Kelsey, S.: Energy Theorems and Structural Analysis. Butterworth Scientific Publications, London (1960)
3. Derweesh, F.L., Mellor, P.B.: Maslennikov's Technique for Forming a Cylindrical Cup, Proc. 10th International MTDR Conference", pp. 499 (1969)
4. Evangelista, S.H.: Diagramas de limite de conformação aplicados à análise por elementos finitos de um processo de estampagem em chapas metálicas, p. 23. Tese de mestrado, Escola de engenharia de São Carlos, São Paulo (2000)
5. Foli, C.R.M., Menezes, M., Moreira Filho, L.A.: Influence of the yielding criterion on total forming force in metallic junctions using elastomers. J.Mater. Process. Technol. **179**, 61–66 (2006)
6. Foli, C.R.M., Menezes, M., Moreira Filho, L.A. Alternative process for unconventional forming of junction in thin-walled metal tubes. In: Proceedings of the 17th International Congress of Mechanical Engineering, 2003, São Paulo. COBEM 2003-Congresso Brasileiro de Engeharia Mecânica, 2003 1:1–10 (2003)
7. Gronostajski, J., Zimniak, Z.: A few methods of analytical calculation of forming limits curves. J. Mater. Process. Technol. **55**, 213–220 (1996)
8. Hattangady, N.V.: Automated modeling and remeshing in metal forming simulation, Thesis of doctor of philosophy (pp. 1), Faculty of Rensselaer Polytechnic Institute. New York (2003)
9. Lee, C.H., Kobayashi, S.: New solutions to rigid-plastic deformation problems using a matrix method. Trans. ASME-J. Eng. Indus. **95**, 865–873 (1973)
10. Marreco, D.B., Al-Qureshi, H.A.: Forming of T-Junction on Metal Tubes by Elastomer Rod Technique, Proceedings 7th North American Metal Working Research Conference, 13–19, pags 107–113 (1979)
11. Maslennikov, N.A.: Deep drawing of sheet metal by friction forces. Eng. Dig. **17**, 336 (1956). September
12. Menezes, M.A.: Strain limit theories, anisotropy in sheet metal forming and simulation of pressing processes. University of Birmingham, PhD. Thesis, p. 361(1995)
13. Moreira Filho, L.A.: Conformação de junções em tubos de parede fina utilizando elastômeros. Instituto Tecnológico de Aeronáutica, CTA. São José dos Campos, Tese de Mestrado (1984)
14. Moreira Filho, L.A.: Modelagem Teórica e Automatização de Processo de Conformação de Junções em Tubos de Parede Fina. Instituto Tecnológico de Aeronáutica, CTA. São José dos Campos, Tese de Doutorado (1998)
15. Ribeiro, F.L.B.: Introdução ao Método de Elementos Finitos. COPPE/UFRJ. http://www.professores.uff.br/rbrunow/arquivos/pdf/Curso_MEF.pdf (2010). Acessado 10 março 2010
16. Turner, M.R., Clough, R., Martin, H., Topp, L.: Stiffness and deflection analysis of complex structures. J. Aero. Sci. **23**(9), 805–823 (1956). September
17. Wang, N.M., Budiansky, B.: Analysis of sheet metal stamping by a finite-element method. J. Appl. Mech. **45**, 73–82 (1978)
18. Zienkiewicz, O.C.: The finite element method: fram intuition to generality. Appl. Mech. Rev. **23**(23), 249–256 (1970). March
19. Zienkiewicz, O.C., Jain, P.C., Oñate, E.: Flow of solids during forming and extrusion: some aspects of numerical solutions. Int. J. Solids Struct. **14**, 15–38 (1978)

On the Stretching and Bending Analyses of Thin Functionally Graded Annular Sector Plates

H. R. Noori, E. Jomehzadeh and A. R. Saidi

Abstract In this chapter, an exact levy solution is presented for the bending analysis of a functionally graded (FG) annular sector plates. The governing equilibrium equations are obtained based on the classical plate theory. Introducing an analytical method for the first time, the three coupled governing equilibrium equations are replaced by an independent equation in terms of transverse deflection. This equation, which is a forth-order partial differential equation, is similar to the governing equilibrium equation of a homogeneous isotropic annular sector plate. Using an equivalent flexural rigidity, the solutions of FG annular sector plates can be easily extracted from the equation of homogeneous annular plates. Finally, the effects of power of functionally graded material (FGM), aspect ratio, inner to outer radius ratio and boundary conditions on the mechanical behavior of a functionally graded annular sector plate are discussed.

Keywords Functionally graded · Annular sector plate · Analytical method

H. R. Noori (✉)
Department of Mechanical Engineering, Islamic Azad University,
Jiroft Branch, Jiroft, Iran
e-mail: hamidr_noori@yahoo.com

E. Jomehzadeh
Young Researchers Club, Department of Mechanical Engineering,
Islamic Azad University, Kerman Branch, Kerman, Iran
e-mail: jomehzadehe2@asme.org

A. R. Saidi
Department of Mechanical Engineering, Shahid Bahonar University of Kerman,
Kerman, Iran
e-mail: saidi@mail.uk.ac.ir

A. Öchsner et al. (eds.), *Materials with Complex Behaviour II*,
Advanced Structured Materials 16, DOI: 10.1007/978-3-642-22700-4_25,

1 Introduction

A new class of materials known as "functionally graded materials" (FGMs) has been introduced in which the material properties vary continuously in one or more directions according to a specific profile. These materials are microscopically heterogeneous and are typically made of isotropic components such as metals and ceramics. FGMs exploit the ideal performance of their composition, e.g. heat and corrosion resistance of ceramics on one side, and mechanical strength and toughness of metals on the other side of a body.

Thin plates are light weight structures with high load-carrying capacity, economy, and technological effectiveness. Because of the distinct advantages, thin plates are extensively used in all fields of engineering. Thus, understanding the behavior of such structures is the main purpose of the researchers in this field. Many models have been developed for static and vibration analysis of homogeneous or non-homogeneous plates with different shapes. Liu and Liew [1] developed the differential quadrature element method (DQEM) for static analysis of the two-dimensional Reissner–Mindlin plate in the polar coordinate system by integrating the domain decomposition method (DDM) with the differential quadrature method (DQM). The axisymmetric bending and stretching analysis of functionally graded solid and annular circular plates was studied by Reddy et al. [2] using the first order shear deformation Mindlin plate theory. Cheng and Batra [3] derived the three-dimensional thermoelastic analysis for functionally graded elliptic plates. The analysis of the functionally graded plates based on the classical plate theory was developed by Chi and Chung [4, 5]. They presented the solution for simply supported S-FG (sigmoid functionally graded) and E-FG (exponential functionally graded) rectangular plates. Abrate [6, 7] showed that no special tools are required to study the analysis of functionally graded rectangular plates because they behave like homogeneous plates. He selected the different reference surface instead of middle surface and showed that based on this surface the static and vibration analysis of FG plates and homogeneous plates related to each other. An approximate closed-form solution is presented for bending of thin isotropic sector plates with clamped edges subjected to uniform and non-uniform loading using the extended Kantorovich method (EKM) by Aghdam et al. [8]. Nie and Zhong [9] investigated the free and forced vibration of functionally graded annular sector plates with simply supported radial edges and arbitrary boundary conditions along the circular edges. They studied an approximate solution along the radial direction using the one-dimensional differential quadrature method (DQM). The meshless local Petrov–Galerkin (MLPG) method was used for analysing two-dimensional static and dynamic deformations of functionally graded materials with material response modeled as either linear elastic or as linear viscoelastic by Gilhooley [10]. Sahraee [11] presented the bending analysis of functionally graded circular sector plates based on the Levinson plate theory. He ignored the middle plane displacement of the FG sector plate and solved the problem similar to the isotropic one. A two-dimensional higher-order deformation theory was presented for the evaluation of displacements and stresses in functionally graded plates subjected to thermal and mechanical loadings by Matsunaga [12]. Zhang and Zhou [13] presented a theoretical analysis to the FG thin rectangular plates based on the physical neutral surface. Jomehzadeh and Saidi [14, 15] studied the vibration analysis of laminated sector and annular sector plates made of transversely isotropic layers. Saidi and Jomehzadeh [16] introduced an analytical method for decoupling the equilibrium equations of Kirchhoff and Mindlin rectangular plates.

Lot of investigations dealing with static and dynamic behavior of isotropic and functionally graded rectangular plates can be seen in literature. However, no such works can be found for analysis of FG annular sector plates. In this study, the static analysis of a functionally graded annular sector plate is presented based on the classical plate theory. Using an analytical method, three coupled stretching and bending equilibrium equations of a FG annular sector plate are decoupled. Solving the decoupled equation, the solution of a FG annular sector plate is obtained. It is found that using an equivalent flexural rigidity, the decoupled equation of a FG annular sector plate becomes like the equilibrium equation of an isotropic plate in polar coordinates. The variations of some physical parameters are shown for different functionally graded materials.

2 Material Properties of the FG Annular Sector Plate

The annular sector plate material is made of a mixture of ceramic and metal. It is assumed that the material properties of the annular sector plate vary as a power law through the thickness as in [17]

$$\eta(z) = \eta_m + \eta_{cm}(1/2 - z/h)^p \tag{1}$$

where $\eta(z)$ is the material property of the annular sector plate such as Young's modulus, h is the total thickness of the plate and p is the power of the FG plate. According to the small range of Poisson's ratio variation, it is assumed to be constant through the thickness of the FG annular sector plate.

3 Governing Equilibrium Equations of the Annular Sector Plate

Consider a FG annular sector plate of inner radius a, outer radius b, uniform medium thickness h and sector angle α, as shown in Fig. 1. The annular sector plate is subjected to transverse loading. It is assumed that the annular sector plate has simply supported radial edges and arbitrary boundary conditions along the circular edges.

The displacement components of the plate in r, θ and z directions are assumed as

Fig. 1 Geometry of the annular sector plate

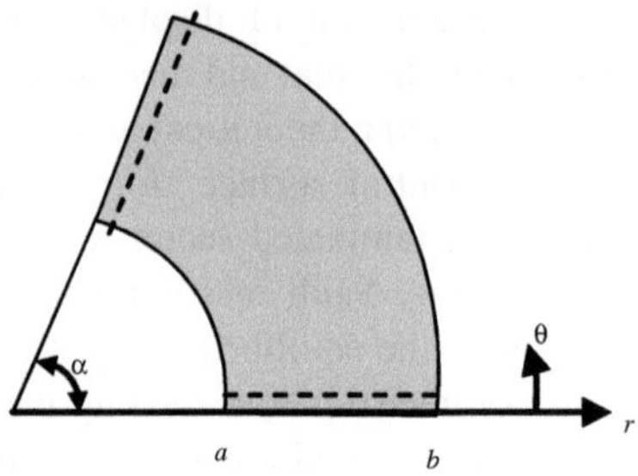

$$u_r(r,\theta,z) = u(r,\theta) - z\frac{\partial w}{\partial r} \tag{2a}$$

$$u_\theta(r,\theta,z) = v(r,\theta) - \frac{z}{r}\frac{\partial w}{\partial \theta} \tag{2b}$$

$$w(r,\theta,z) = w(r,\theta) \tag{2c}$$

where u, v and w are the displacements of the middle surface in r, θ and z directions, respectively. Under the assumption of small deformation and linear strain–displacement relations, the strain components of the FG annular sector plate can be expressed as

$$\begin{aligned} \varepsilon_r &= \frac{\partial u}{\partial r} + z\left(-\frac{\partial^2 w}{\partial^2 r}\right) \\ \varepsilon_\theta &= \frac{1}{r}\frac{\partial v}{\partial \theta} + \frac{u}{r} + z\left(-\frac{1}{r^2}\frac{\partial^2 w}{\partial^2 \theta} - \frac{1}{r}\frac{\partial w}{\partial r}\right) \\ 2\varepsilon_{r\theta} &= \frac{1}{r}\frac{\partial u}{\partial \theta} + \frac{\partial v}{\partial r} - \frac{v}{r} + z\left(-\frac{2}{r}\frac{\partial^2 w}{\partial r \partial \theta} + \frac{2}{r^2}\frac{\partial w}{\partial \theta}\right) \end{aligned} \tag{3}$$

In classical plate theory, it is assumed that the cross section perpendicular to the middle surface of the plate remains normal and unstretched after deformation. Consequently, the transverse shear deformation is neglected and the shear strain components $2\varepsilon_{rz}$ and $2\varepsilon_{\theta z}$ vanish.

Substituting strain components (3) into the principle of minimum potential energy, the equilibrium equations of Kirchhoff plate in polar coordinates are obtained as

$$\frac{\partial N_r}{\partial r} + \frac{1}{r}\frac{\partial N_{r\theta}}{\partial \theta} + \frac{1}{r}(N_r - N_\theta) = 0 \tag{4a}$$

$$\frac{\partial N_{r\theta}}{\partial r} + \frac{2}{r}N_{r\theta} + \frac{1}{r}\frac{\partial N_\theta}{\partial \theta} = 0 \tag{4b}$$

$$\frac{\partial^2 M_r}{\partial r^2}+\frac{2}{r}\frac{\partial M_r}{\partial r}+\frac{1}{r^2}\frac{\partial^2 M_\theta}{\partial \theta^2}-\frac{1}{r}\frac{\partial M_\theta}{\partial r}+\frac{2}{r}\frac{\partial^2 M_{r\theta}}{\partial r\partial \theta}+\frac{2}{r^2}\frac{\partial M_{r\theta}}{\partial \theta}+P=0 \tag{4c}$$

where P is external loading function and the resultant forces N_r, N_θ and $N_{r\theta}$ can be defined by integrating corresponding stresses along the thickness as follows

$$N_r=\int_{-h/2}^{h/2}\sigma_r\,dz,\quad N_\theta=\int_{-h/2}^{h/2}\sigma_\theta\,dz,\quad N_{r\theta}=\int_{-h/2}^{h/2}\sigma_{r\theta}\,dz \tag{5}$$

and the resultant moments M_r, M_θ and $M_{r\theta}$ are

$$M_r=\int_{-h/2}^{h/2}\sigma_r z\,dz,\quad M_\theta=\int_{-h/2}^{h/2}\sigma_\theta z\,dz,\quad M_{r\theta}=\int_{-h/2}^{h/2}\sigma_{r\theta} z\,dz \tag{6}$$

Considering plane stress state for the FG annular sector plate, the stresses are defined as

$$\sigma_x=\frac{E(z)}{1-v^2}(\varepsilon_r+v\varepsilon_\theta) \tag{7a}$$

$$\sigma_\theta=\frac{E(z)}{1-v^2}(\varepsilon_\theta+v\varepsilon_r) \tag{7b}$$

$$\sigma_{r\theta}=\frac{E(z)}{2(1+v)}(2\varepsilon_{r\theta}) \tag{7c}$$

Using Eqs. (3), (7a–c) and the definition of resultant forces and moments in Eqs. (5) and (6), it can be obtained

$$\begin{bmatrix} N_r \\ N_\theta \\ N_{r\theta} \\ M_r \\ M_\theta \\ M_{r\theta}\end{bmatrix}=\begin{bmatrix} A_{11} & A_{12} & 0 & B_{11} & B_{12} & 0 \\ A_{12} & A_{22} & 0 & B_{12} & B_{22} & 0 \\ 0 & 0 & A_{66} & 0 & 0 & B_{66} \\ B_{11} & B_{12} & 0 & D_{11} & D_{12} & 0 \\ B_{12} & B_{22} & 0 & D_{12} & D_{22} & 0 \\ 0 & 0 & B_{66} & 0 & 0 & D_{66}\end{bmatrix}\begin{bmatrix} \partial u/\partial r \\ \partial v/r\partial\theta+u/r \\ \partial u/r\partial\theta+\partial v/\partial r-v/r \\ -\partial^2 w/\partial r^2 \\ -\partial^2 w/r^2\partial\theta^2-\partial w/r\partial r \\ -2\partial^2 w/r\partial r\partial\theta+2\partial w/r^2\partial\theta\end{bmatrix} \tag{8}$$

where the integration coefficients A_{ij}, B_{ij} and D_{ij} $(i,j=1,2,3)$ are defined as

$$A_{11} = \frac{1}{1-\nu^2}\int_{-h/2}^{h/2} E(z)\,dz, \quad A_{12} = \frac{\nu}{1-\nu^2}\int_{-h/2}^{h/2} E(z)\,dz,$$
$$A_{33} = \frac{1}{2(1+\nu)}\int_{-h/2}^{h/2} E(z)\,dz \tag{9a}$$

$$B_{11} = \frac{1}{1-\nu^2}\int_{-h/2}^{h/2} zE(z)\,dz, \quad B_{12} = \frac{\nu}{1-\nu^2}\int_{-h/2}^{h/2} zE(z)\,dz,$$
$$B_{33} = \frac{1}{2(1+\nu)}\int_{-h/2}^{h/2} zE(z)\,dz \tag{9b}$$

$$D_{11} = \frac{1}{1-\nu^2}\int_{-h/2}^{h/2} z^2E(z)\,dz, \quad D_{12} = \frac{\nu}{1-\nu^2}\int_{-h/2}^{h/2} z^2E(z)\,dz,$$
$$D_{33} = \frac{1}{2(1+\nu)}\int_{-h/2}^{h/2} z^2E(z)\,dz \tag{9c}$$

Substituting the resultant forces and moments obtained in Eqs. (8) into Eqs. (4a– c), the governing equilibrium equations are obtained as

$$A_{11}\left(\frac{\partial^2 u}{\partial r^2} + \frac{1}{r}\frac{\partial u}{\partial r} - \frac{u}{r^2} - \frac{1}{r^2}\frac{\partial v}{\partial \theta} + \frac{1}{r}\frac{\partial^2 v}{\partial r\partial \theta}\right) + A_{33}\left(\frac{1}{r^2}\frac{\partial^2 u}{\partial \theta^2} - \frac{1}{r}\frac{\partial^2 v}{\partial r\partial \theta} - \frac{1}{r^2}\frac{\partial v}{\partial \theta}\right)$$
$$B_{11}\left(-\frac{\partial^3 w}{\partial r^3} - \frac{1}{r}\frac{\partial^2 w}{\partial r^2} + \frac{1}{r^2}\frac{\partial w}{\partial r} + \frac{1}{r^3}\frac{\partial^2 w}{\partial \theta^2} - \frac{1}{r^2}\frac{\partial^3 w}{\partial r\partial \theta^2} + \frac{1}{r^3}\frac{\partial^2 w}{\partial \theta^2}\right) = 0 \tag{10a}$$

$$A_{11}\left(\frac{1}{r^2}\frac{\partial^2 v}{\partial \theta^2} + \frac{1}{r}\frac{\partial^2 u}{\partial r\partial \theta} + \frac{1}{r^2}\frac{\partial u}{\partial \theta}\right) + A_{33}\left(\frac{\partial^2 v}{\partial r^2} + \frac{1}{r}\frac{\partial v}{\partial r} - \frac{v}{r^2} + \frac{1}{r^2}\frac{\partial u}{\partial \theta} - \frac{1}{r}\frac{\partial^2 u}{\partial r\partial \theta}\right)$$
$$B_{11}\left(-\frac{1}{r}\frac{\partial^3 w}{\partial r^2\partial \theta} - \frac{1}{r^2}\frac{\partial^2 w}{\partial r\partial \theta} - \frac{1}{r^3}\frac{\partial^3 w}{\partial \theta^3}\right) = 0 \tag{10b}$$

$$B_{11}\left(\frac{\partial^3 u}{\partial r^3} + \frac{2}{r}\frac{\partial^2 u}{\partial r^2} - \frac{1}{r^2}\frac{\partial u}{\partial r} + \frac{u}{r^3} + \frac{1}{r^3}\frac{\partial^2 u}{\partial \theta^2} + \frac{1}{r^2}\frac{\partial^3 u}{\partial r\partial^2 \theta} - \frac{1}{r^2}\frac{\partial^2 v}{\partial r\partial \theta} + \frac{1}{r^3}\frac{\partial v}{\partial \theta} + \frac{1}{r^3}\frac{\partial^3 v}{\partial \theta^3} + \frac{1}{r}\frac{\partial^3 v}{\partial r^2\partial \theta}\right)$$
$$-D_{11}\left(\frac{\partial^4 w}{\partial r^4} + \frac{2}{r}\frac{\partial^3 w}{\partial r^3} - \frac{1}{r^2}\frac{\partial^2 w}{\partial r^2} + \frac{1}{r^3}\frac{\partial w}{\partial r} + \frac{2}{r^2}\frac{\partial^4 w}{\partial r^2\partial \theta^2} - \frac{2}{r^3}\frac{\partial^3 w}{\partial r\partial \theta^2} + \frac{4}{r^4}\frac{\partial^2 w}{\partial \theta^2} + \frac{1}{r^4}\frac{\partial^4 w}{\partial \theta^4})\right) + P = 0 \tag{10c}$$

Equations (10a–c) are three highly coupled equations in terms of in-plane and transverse displacements. For solving such coupled equations, it is reasonable to find a method for decoupling them. Using an analytic method, the three equilibrium equations (10a–c) are decoupled. Equations (10a–c) can be rewritten as follows

$$A_{11}\frac{\partial \phi_1}{\partial r} + A_{33}\frac{1}{r}\frac{\partial \phi_2}{\partial \theta} - B_{11}\frac{\partial}{\partial r}(\nabla^2 w) = 0 \tag{11a}$$

$$A_{11}\frac{1}{r}\frac{\partial \phi_1}{\partial \theta} - A_{33}\frac{\partial \phi_2}{\partial r} - B_{11}\frac{\partial}{r\partial \theta}(\nabla^2 w) = 0 \tag{11b}$$

$$B_{11}\nabla^2 \phi_1 - D_{11}\nabla^2\nabla^2 w + P = 0 \tag{11c}$$

where ∇^2 is two dimensional Laplace operator in polar coordinate ($\nabla^2 = \partial^2/\partial r^2 + \partial/r\partial r + \partial^2/r^2\partial\theta^2$) and the variables ϕ_1 and ϕ_2 are defined as

$$\phi_1 = \frac{\partial u}{\partial r} + \frac{u}{r} + \frac{1}{r}\frac{\partial v}{\partial \theta} \tag{12a}$$

$$\phi_2 = \frac{1}{r}\frac{\partial u}{\partial \theta} - \frac{\partial v}{\partial r} - \frac{v}{r} \tag{12b}$$

Differentiating of Eq. (11b) with respect to θ and dividing by r and differentiating Eq. (11a) with respect to r and then adding the two results, yields

$$A_{11}\nabla^2\phi_1 = B_{11}\nabla^4 w \tag{13}$$

Upon substitution of Eq. (13) into the last governing equilibrium equation (11c), yields

$$\left(D_{11} - \frac{B_{11}^2}{A_{11}}\right)\nabla^4 w = P \tag{14}$$

Equation (14) is an independent forth-order partial differential equation in terms of transverse displacement w. This equation is very similar to the equation of isotropic homogenous Kirchhoff plate. Introducing an equivalent flexural rigidity, the decoupled governing equilibrium equation of the FG annular sector plate (Eq. (14)) becomes as the governing equation of an isotropic homogeneous annular sector plate. Equation (14) can be rewritten as

$$\hat{D}\nabla^4 w = P \tag{15}$$

where $\hat{D}$ is the equivalent flexural rigidity of the FG annular sector plate which is equal to $D_{11} - B_{11}^2/A_{11}$. The coefficients A_{11}, B_{11} and D_{11} can be defined in terms of the material properties of the FG annular sector plate from Eqs. (1) and (9a–c) as follows

$$A = \frac{h}{1-\nu^2}\left(E_m + \frac{E_{cm}}{n+1}\right) \tag{16a}$$

$$B = -\frac{E_{cm}h^2}{1-\nu^2}\left(\frac{n}{2(n+1)(n+2)}\right) \tag{16b}$$

$$D = \frac{E_m h^3}{12(1-\nu^2)} + \frac{E_{cm}h^3(n^2+n+2)}{4(1-\nu^2)(n+1)(n+2)(n+3)} \tag{16c}$$

Also, differentiating of equation (11) with respect to θ and dividing by r and differentiating equation (11b) with respect to r and then subtracting the two results from each other, yields

$$\nabla^2\phi_2 = 0 \tag{17}$$

Due to Eqs. (13), (17) and the definition of variables ϕ_1 and ϕ_2 in Eqs. (12a, b), it can be concluded that these equations are satisfied by letting the in-plane displacements as follows

$$u = \frac{B_{11}}{A_{11}}\frac{\partial w}{\partial r} \tag{18a}$$

$$v = \frac{B_{11}}{A_{11}}\frac{\partial w}{r\partial\theta} \tag{18b}$$

It is easy to show that the relations (18a, b) satisfy Eqs. (13) and (17) and all boundary conditions along the edges of the plate. Substituting these middle plane displacements into Eqs. (2a–c), the proposed displacement field of the FG Kirchhoff annular sector plate becomes as

$$u_r(r,\theta,z) = (B_{11}/A_{11} - z)\frac{\partial w}{\partial r} \tag{19a}$$

$$u_\theta(r,\theta,z) = (B_{11}/A_{11} - z)\frac{1}{r}\frac{\partial w}{\partial\theta} \tag{19b}$$

$$w(r,\theta,z) = w(r,\theta) \tag{19c}$$

It can be found from Eqs. (18a, b) that the in-plane displacements of the FG annular sector plate do not vanish at the middle surface ($z = 0$). Unlike the isotropic homogeneous annular sector plate, the FG plate is not symmetric with respect to the middle plane. Because of the variable properties through the thickness, the neutral surface of the FG annular sector plate is not located at the middle plane. It can be seen from Eqs. (19a–c) that the in-plane displacements of the plate at $z = B_{11}/A_{11}$ vanish. The surface located at $z = B_{11}/A_{11}$ is the real neutral surface of the FG annular sector plate. In fact, the neutral surface of the FG plates depends on the variation of material properties in the thickness direction and the location of the neutral surface is independent of the geometric parameters.

Based on the obtained displacement field for the FG annular sector plate (Eqs. (19a–c)), the resultant moments can be obtained in the form of

$$M_r = \hat{D}\left(\frac{\partial w}{\partial r} + v\left(\frac{\partial w}{r\partial r} + \frac{1}{r^2}\frac{\partial^2 w}{\partial \theta^2}\right)\right) \tag{20a}$$

$$M_\theta = \hat{D}\left(\frac{\partial w}{r\partial r} + \frac{1}{r^2}\frac{\partial^2 w}{\partial \theta^2} + v\frac{\partial w}{\partial r}\right) \tag{20b}$$

$$M_{r\theta} = \hat{D}(1 - v)\left(\frac{1}{r}\frac{\partial^2 w}{\partial r \partial \theta} + \frac{1}{r^2}\frac{\partial w}{\partial \theta}\right) \tag{20c}$$

where the parameter $\hat{D}$ is the equivalent flexural rigidity as mentioned before.

4 Solution

For static analysis of the FG annular sector plate which is simply supported at two radial edges, the transverse displacement and uniformly distributed load can be represented as

$$w(r,\theta) = \sum_{m=1,3,..}^{\infty} w_m(r)\sin(\beta_m \theta) \tag{21a}$$

$$P = \sum_{m=1,3,..}^{\infty} \frac{4p_0}{m\pi}\sin(\beta_m \theta) \tag{21b}$$

where β_m denotes $m\pi/\alpha$. Substituting the proposed series solutions (21a, b) into Eq. (15) and solving the resulted ordinary differential equation, yields

$$w_m(r) = \left(C_1 r^{\beta_m} + C_2 r^{-\beta_m} + C_3 r^{-\beta_m+2} + C_4 r^{\beta_m+2} + \frac{4p_0 r^4}{\pi n \hat{D}(\beta_m^4 - 20\beta_m^2 + 64)}\right)\sin(\beta_m \theta) \tag{22}$$

Imposing arbitrary boundary conditions at inner ($r = a$) and outer ($r = b$) circular edges, the four unknown coefficients (C_1, C_2, C_3, C_4)can be determined.

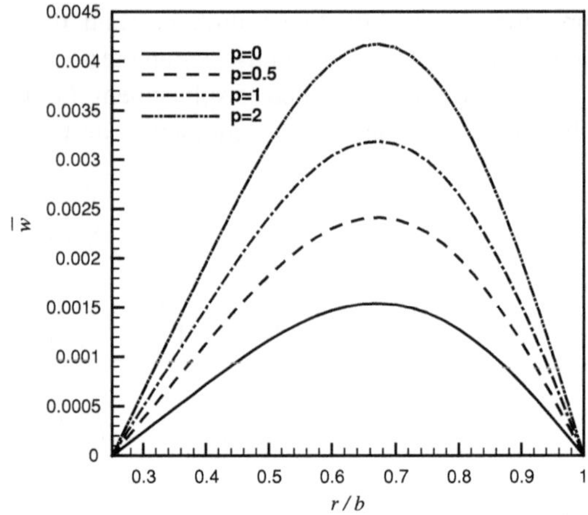

Fig. 2 Non-dimensional deflection along radial direction (S–S)

5 Numerical Results and Discussion

For simplicity and generality, the following non-dimensional terms are introduced

$$\bar{w} = \frac{w(r, \alpha/2)}{h}, \quad \bar{\sigma}_r = \frac{\sigma_r\left(\frac{a+b}{2}, \alpha/2\right)h^2}{p_0 b^2}, \quad \bar{\sigma}_\theta = \frac{\sigma_\theta\left(\frac{a+b}{2}, \alpha/2\right)h^2}{p_0 b^2} \tag{23}$$

In numerical calculation, the FG annular plate is assumed to have simply supported radial edges. The inner radius, outer radius and thickness of the annular plate are considered to be $a = 2.5$, $b = 10$ and $h = 0.2$, respectively. The Poisson's ratio of the plate is assumed to be constant through the thickness and equal to 0.3. The functionally graded material used is composed of aluminum with $E_m = 70$ GPa and Silicon Carbide with $E_c = 420$ GPa. The plate is subjected to uniformly distributed load with intensity of $p_0 = 1 \times 10^4\ \text{N/m}^2$.

The boundary conditions are identified according to the inner and outer radius of the annular sector plates (e.g. **F–C** denotes free inner and clamped outer edges).

In Fig. 2, the non-dimensional deflection is shown along the radial direction for the FG annular sector plate with two circular edges simply supported. The variation is depicted for various power of FGM. As the power of FGM decreases, the FG annular plate becomes stiffer and this leads to lower deflection. The non-dimensional radial and circumferential stresses of the FG annular sector plate with two circular edges simply supported are shown in Figs. 3 and 4 for different power of FGM. It can be found that unlike the homogeneous plates $(p = 0)$, the stresses $\bar{\sigma}_r$ and $\bar{\sigma}_\theta$ do not vanish at the mid-plane of the FG annular sector plates. The normal stress in r direction is equal to zero at the neutral surface which is located at $z = B/A$ as discussed. Also, the stress components of the FG annular sector plates along the thickness direction are not linearly proportional to z direction.

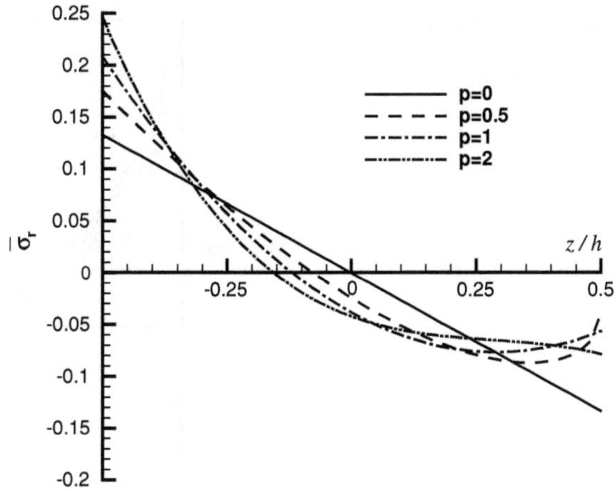

Fig. 3 Non-dimensional radial stress along thickness direction (S–S)

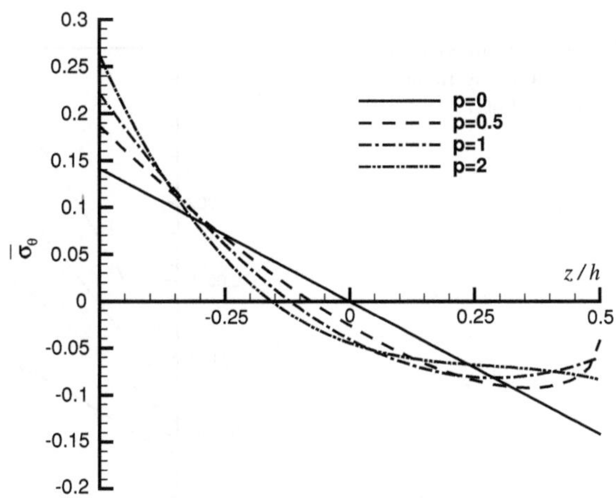

Fig. 4 Non-dimensional circumferential stress along thickness direction (S–S)

The variation of non-dimensional radial stress ($\bar{\sigma}_r$) is depicted in Fig. 5 through the thickness direction for different powers of functionally graded material (p). The plate is assumed to have clamped edges in two circular edges. It can be seen that for p equal to 3.5, the non-dimensional stress vanishes at a distance far from the middle plane. It can be easy to show that the parameter B/A has a maximum value at $n = 3.5$.

The non-dimensional transverse displacement is shown for $a/b = 0.25$ and $a/b = 0.5$ in Figs. (6) and (7), respectively. It can be seen that when the inner edge of the annular plate is close to the center of the annular sector plate, the maximum deflection does not occur at the free edge. However, approaching the inner edge to outer one, the maximum non-dimensional deflection occurs at the free inner edge.

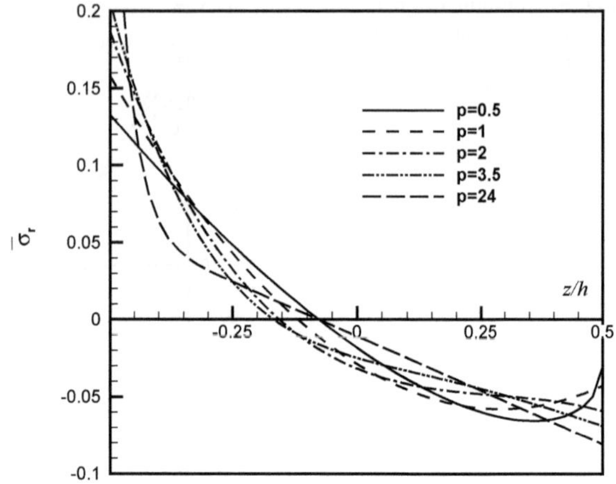

Fig. 5 Non-dimensional radial stress along thickness direction (C–C)

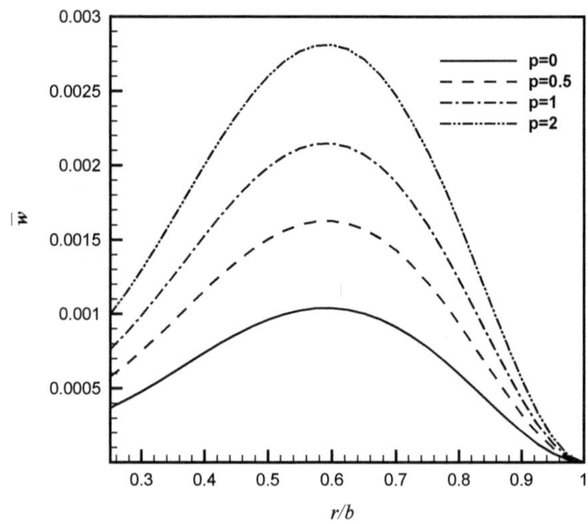

Fig. 6 Non-dimensional deflection along radial direction (F–C)

This is acceptable because with decreasing the inner radius, the length of the free inner edge decreases and the annular sector plate becomes stiffer.

The maximum non-dimensional transverse deflection is tabulated for different sector angles and some aspect ratios and inner to outer radius ratios in Table 1. In this case, the plate has free inner and clamped outer edges and power of FGM is assumed to be 0.5.

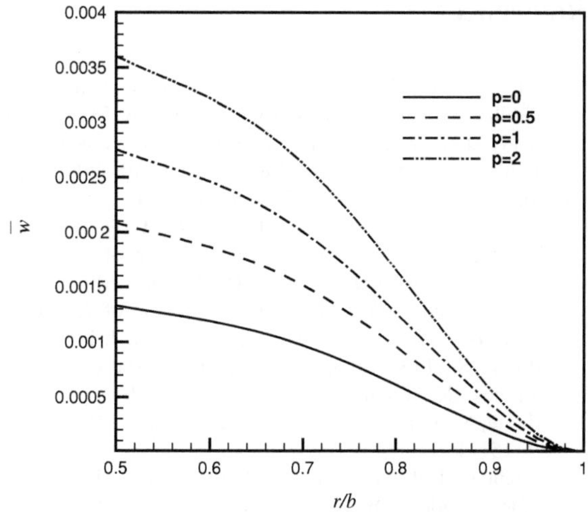

Fig. 7 Non-dimensional deflection along radial direction ($\mathrm{F}-\mathrm{C}, a/b = 0.5$)

Table 1 Maximum Non-dimensional deflection for C–F plate, $p = 0.5$

α	h/b	$a/b = 0.25$	$a/b = 0.5$
30°	0.01	0.0341	0.0321
	0.02	0.0021	0.0020
45°	0.01	0.1590	0.1158
	0.02	0.0099	0.0072
60°	0.01	0.4567	0.2184
	0.02	0.0285	0.0136
120°	0.01	2.1001	0.3734
	0.02	0.1313	0.0233

6 Conclusion

In this chapter, an exact analytical solution has been presented for static analysis of functionally graded annular sector plate. The annular sector plate is assumed to have simply supported radial edges and arbitrary boundary conditions along the circular edges. Three coupled governing equilibrium equations of FG annular sector plate have been converted to a decoupled equation in terms of transverse displacement. Introducing a flexural rigidity, the decoupled equation of FG plate has become similar to the governing equilibrium equation of isotropic homogeneous plate. Finally, the variation of some parameters has been shown in figures and in a table.

The exact solution for functionally graded annular sector plate has been obtained for the first time and the present results can be regarded as a database in the field of FG annular sector plates.

References

1. Liu, F.L., Liew, K.M.: Differential quadrature element method for static analysis of Reissner–Mindlin polar plates. Int. J. Solids Struct. **36**, 5101–5123 (2000)
2. Reddy, J.N., Wang, C.M., Kitipornehai, S.: Axisymmetric bending of functionally graded circular and annular plates. Eur. J. Mech. A/Solids **18**, 185–199 (1999)
3. Cheng, Z.Q., Batra, R.C.: Three-dimensional thermoelastic deformation of a functionally graded elliptic plate. Composites Part B **31**, 97–106 (2000)
4. Chi, S., Chung, Y.: Mechanical behavior of functionally graded material plates under transverse load Part1: Analysis. Int. J. Solids Struct. **43**, 3657–3674 (2006)
5. Chi, S., Chung, Y.: Mechanical behavior of functionally graded material plates under transverse load Part2: Numerical results. Int. J. Solids Struct. **43**, 3675–3691 (2006)
6. Abrate, S.: Free vibration, buckling and static deflections of functionally graded plates. Compos. Sci. Technol. **66**, 2383–2394 (2006)
7. Abrate, S.: Functionally graded plates behave like homogeneous plates. Composites Part B **39**, 151–158 (2008)
8. Aghdam, M.M., Mohammadi, M., Erfanian, V.: Bending analysis of thin annular sector plates using extended Kantorovich method. Thin-Walled Struct **45**, 983–990 (2007)
9. Nie, G.J., Zhong, Z.: Vibration analysis of functionally graded annular sectorial plates with simply supported radial edges. Compos Struct **84**, 167–176 (2008)
10. Gilhooley, D.F., Xiao, J.R., Batra, R.C., McCarthy, M.A., Gillespie, J.W.: Two-dimensional stress analysis of functionally graded solids using the MLPG method with radial basis functions. Comput. Mater. Sci. **41**, 467–481 (2008)
11. Sahraee, S.: Bending analysis of functionally graded sectorial plates using Levinson plate theory. Compos. Struct. **88**, 548–557 (2009)
12. Matsunaga, H.: Stress analysis of functionally graded plates subjected to thermal and mechanical loadings. Compos. Struct. **87**, 344–357 (2009)
13. Zhang, D.G., Zhou, Y.H.: A theoretical analysis of FGM thin plates based on physical neutral surface. Comput. Mater. Sci. **44**, 716–720 (2008)
14. Jomehzadeh, E., Saidi, A.R.: Analytical solution for free vibration of transversely isotropic sector plates using a boundary layer function. Thin-Walled Struct. **47**, 82–87 (2009)
15. Jomehzadeh, E., Saidi, A.R.: Accurate natural frequencies of transversely isotropic moderately thick annular sector plates. J. Mech. Eng. Sci. **223**, 307–317 (2009)
16. Saidi, A.R., Jomehzadeh, E.: On analytical approach for the bending/stretching of linearly elastic functionally graded rectangular plates with two opposite edges simply supported. J. Mech. Eng. Sci. **223**, 2009–2016 (2009)
17. Bao, G., Wang, L.: Multiple cracking in functionally graded ceramic/metal coatings. Int. J. Solids Struct. **32**, 2853–2871 (1995)

Multi-Walled Carbon Nanotubes Effect on Mechanical Properties of High Performance Fiber/Epoxy Nanocomposite

Mehdi Taghavi Deilamani, Omid Saligheh and Rouhollah Arasteh

Abstract In this study, the effect of multi-walled carbon nanotubes (MWNTs) and functionalized multi wall carbon nanotubes having covalent attachments of carboxylic groups (MWNTs-COOH) on mechanical properties of carbon fiber-reinforced epoxy were investigated. Multi-walled carbon nanotubes (MWNTs) were used as a nanofiller to enhance the mechanical properties of carbon fiber reinforced epoxy composite. This was performed through a high speed mechanical stirring as well as high intensity ultrasonic technique to obtain a homogeneous mixture of epoxy resin and carbon nanotubes. It was found that the CNTs give enhanced strength and elastic modulus.

Keywords Three phase nanocomposite · Carbon nanotubes · Carbon fibers · Epoxy matrix

M. Taghavi Deilamani (✉)
Department of Textile Engineering, Islamic Azad University, Science and Research Branch, Tehran, Iran
e-mail: mehdi.taghavi.d@gmail.com

O. Saligheh
Department of Textile Engineering, Islamic Azad University, South Tehran Branch, Young Researchers Club, Tehran, Iran
e-mail: omid.saligheh@gmail.com

R. Arasteh
Department of Textile Engineering, Islamic Azad University, South Tehran Branch, Tehran, Iran
e-mail: ro.arasteh@gmail.com

A. Öchsner et al. (eds.), *Materials with Complex Behaviour II*,
Advanced Structured Materials 16, DOI: 10.1007/978-3-642-22700-4_26,

1 Introduction

The enhanced mechanical properties of carbon fibre-reinforced composites make them attractive for structural applications in the aerospace, civil, marine, military, etc. Carbon fiber composites have attracted great attention due to the unique properties introduced by nanofillers, which typically refer to carbon blacks, silicas, clays, or carbon nanotubes (CNTs).

Compared to other nanofillers, the unique structures of CNTs potentially provide superior mechanical, electrical, and thermal properties which was discovered by Iijima, a Japanese scientist, in 1991 [1, 2]. They are known to possess the mechanical properties that, when used in conjunction with conventional structural materials such as carbon fiber composites, have the potential to enhance or contribute to increases in mechanical performance without the penalties of excessive weight gain. Carbon nanotube, a novel crystalline carbon form is basically a cylindrical tubule containing sp^2 hybridized carbon–carbon bond along its lengths without any defect as far as structure is concerned because of its nano size [3]. There are two main types of carbon nanotubes that can have high structural perfection. Single walled nanotubes (SWNT) consist of a single graphite sheet seamlessly wrapped into a cylindrical tube, and multi walled nanotubes (MWNT) comprise an array of such nanotubes that are concentrically nested-like rings of a tree trunk [4, 5].

Adding CNTs to the matrix is expected to improve the stiffness, toughness, electrical conductivity and the interlaminar shear strength of composites. The effective utilization of carbon nanotubes in composite applications strongly depend on the ability to homogeneously disperse them throughout the matrix without destroying their integrity. Furthermore, good interfacial bonding is required in order to achieve load transfer across the CNT–matrix interface and a necessary condition is needed for improving the mechanical properties of the composite [6]. Load transfer from matrix to CNTs plays a key role in the mechanical properties of composites. If the adhesion between the matrix and the CNTs is not strong enough to sustain high loads, the benefit of the high tensile strength of CNTs are lost [7].

In this study, MWNTs and MWNTs-COOH were mixed with the epoxy resin to fabricate carbon fiber/CNTs/epoxy nanocomposites. The tensile strengths and the Young's modulus of the three kinds of composites were measured. The purpose of this paper is to show the effect of carbon nanotubes on the mechanical properties of conventional carbon fiber reinforced epoxy matrix as well as to improve compatibilty between epoxy and nanofiller by introducing functional groups onto the surface of carbon nanotubes.

2 Experimental

2.1 Materials

The fabric used in this work was 2D plain weave carbon fibers. MWNTs used in this study were provided by the Research Institute of Petroleum Industry (RIPI) and were synthesized by a chemical vapor deposition (CVD) with a purity >95%. MWNTs were prepared with diameter ranging from 10 to 50 nm, lengths varying from 1 to 3 μm. For preparation of MWNTs-COOH, MWNTs were dispersed into a flask containing the mixture of sulfuric acid and nitric acid with 3:1 volumetric ratio and sonicated for 3 h. After that the MWNTs were washed with deionized water until the pH of the solution reached approximately 7 and then the mixture was filtered and dried at 120°C. According to previous investigations, this procedure is designed to achieve functionalized multi wall carbon nanotubes having covalent attachments of carboxylic groups (MWNTs-COOH) with specific concentration. The epoxy matrix utilized in this study is based on two parts: part A: Aralditely-556 based on bisphenol A (DGEBA) and part B: a hardner HY-951 (aliphatic amine) in wt. ratio 100/12 (Ciba-Geigy company, Switzerland).

2.2 Experimental Procedure

In this study, for fabrication of the nanocomposite, at first the amount of 0.5 wt% nanofillers with epoxy resin, Aralditely-556 were sonicated, and then the hardner HY-951, was added to the mixture. This blend was mixed using a high intensity ultrasonic processor and manual stirrer simultaneously in three 30-min phases and for better dispersion between these phases a high speed mechanical stirrer was used. Ultrasonic process is an efficient method of dispersing carbon nanotubes into epoxy resin when CNTs weight fractions are 0.5 wt% or lower than this amount. Above this concentration, a CNTs agglomerate. The carbon fiber/CNTs/epoxy nanocomposite laminate were prepared using hand lay-up process, and the laminates were cured under vacuum at 80°C for 4 h and then post-cured at room temperature overnight before the mechanical testing. Carbon fiber/epoxy composite laminates without carbon nanotubes were prepared for comparision as well. The specimens were cut to $25 \times 250 \times 2.5\ mm^3$, and mechanical properties of samples were investigated according to ASTM D3039, at room temperature with a constant cross-head rate of 2 mm/min by an Instron mechanical testing machine. Five specimens were tested for each sample and the average value was obtained from the data of these measured specimens.

Fourier-transform infrared spectroscopy (FT-IR) was used to analyze the changes in the surface chemical bonding and structure in the frequency range of 4000–400 cm^{-1}. Spectrometer used was a Bruker Equinox 55 FTIR.

Morphologies of the nanocomposites were studied using a SEM (SEM XL 30; Philips Instruments). SEM micrographs were taken from cryogenically fractured surfaces of specimens. The specimens were fractured manually after immersion in liquid nitrogen and were then coated with a thin layer of gold prior to SEM investigation.

3 Results and Discussion

Figure 1a and b shows the FT-IR spectra of pristine and functionalized MWNTs. The spectra of the acid-treated MWNTs showed peaks at around 1,701 and 1,187 cm^{-1} apparently corresponding to the stretching mode of carboxylic C=O and C–O, respectively. The peak at around 1,570 cm^{-1} corresponds to the C=C graphene sheet bonds of the MWNTs [8].

Intrinsic van der waals attraction among tubes, in combination with their high surface area and high aspect ratio often leads to significant agglomeration, and thus prevents efficient transfer of their superior properties to the matrix. To unlock the potential of carbon nanotubes for application in polymer nanocomposites, the dispersion property becomes very important when carbon nanotubes are added into matrix. Carbon nanotubes tend to remain agglomerated and a homogenous dispersion is not easily obtained, therefore, the dispersion of the nanotubes in the resin is a key parameter. Additional processing difficulties for carbon nanotubes reinforced epoxy composites come significantly in viscosity when the nanotubes are directly added to the epoxy.

Figure 2a and b are SEM images of the MWNT-COOH/Epoxy. It can be seen that the nanotubes were dispersed with good quality. The white spots indicate MWNT's ends that were pulled out of the epoxy matrix.

The mechanical properties of carbon fiber-reinforced epoxy matrix composites and carbon fiber/CNTs/epoxy nanocomposites such as Young's modulus, tensile strength, and strain at break were investigated and listed in Table 1.

The modification of the mechanical properties of composite laminates made of long carbon fibers and epoxy matrix modified by the addition of MWNTs depends on many parameters. Adding some MWNTs can enhance the properties but adding more MWNTs may not guarantee more improvement. This is due to the increase in viscosity of the epoxy on the addition of the MWNTs and the augmentation of the amount of air bubbles during the mixing process.

The MWNTs–matrix interfacial adhesion is an important factor determining the mechanical properties of composites. Interfacial engineering was extensively applied to improve the mechanical and thermo-mechanical properties of fiber reinforced epoxy composites (including MWNT/epoxy, SWNT/epoxy and carbon nanofiber (CNF)/epoxy systems) [9–17]. These reported results indicated that the formation of covalent bonding between functionalized MWNTs and epoxy matrix would lead to a more effective stress transfer and form denser crosslink structure which might limit the mobility of matrix backbone, so mechanical properties of

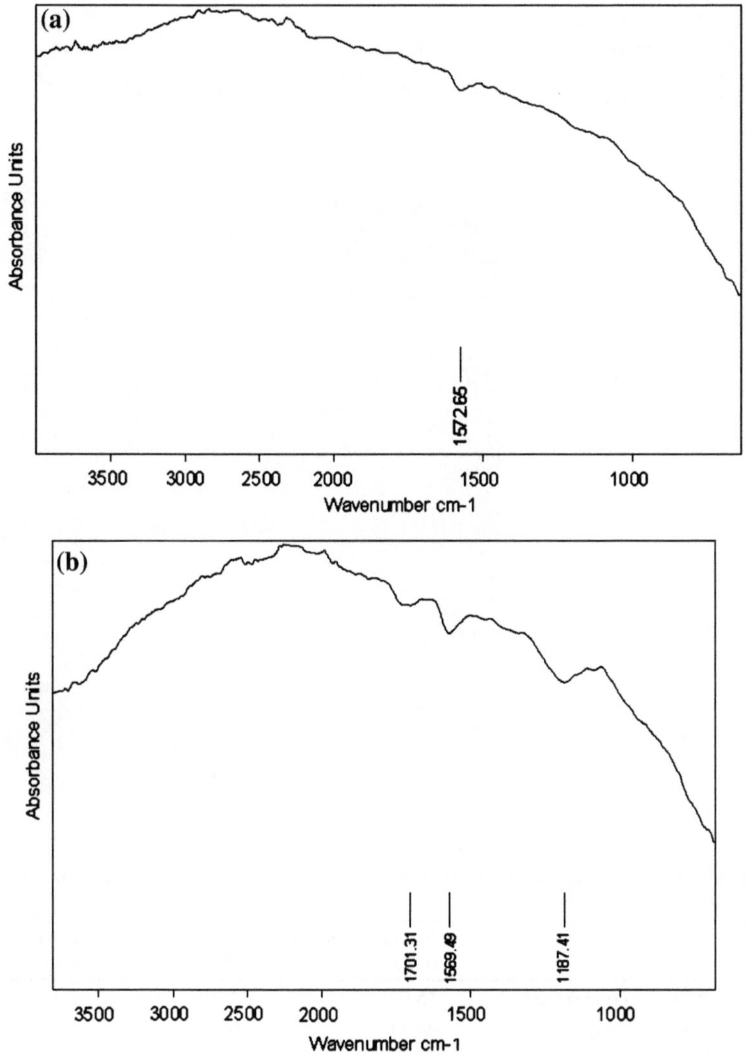

Fig. 1 FT-IR spectra of (**a**) pristine MWNTs and (**b**) MWNTs-COOH

carbon fiber/epoxy composites could be improved. Improvements in Young modulus and tensile strength have been achieved with 0.5 wt% nanotubes. These improvements in mechanical properties are attributed to both the improved dispersion of the nanotubes and grafting of epoxy resin to MWNTs by an esterification reaction. The formation of covalent bonds between the MWNTs and epoxy resin facilitates load transfer between the MWNT and epoxy matrix and contributes to the improvement in the mechanical properties of the composites.

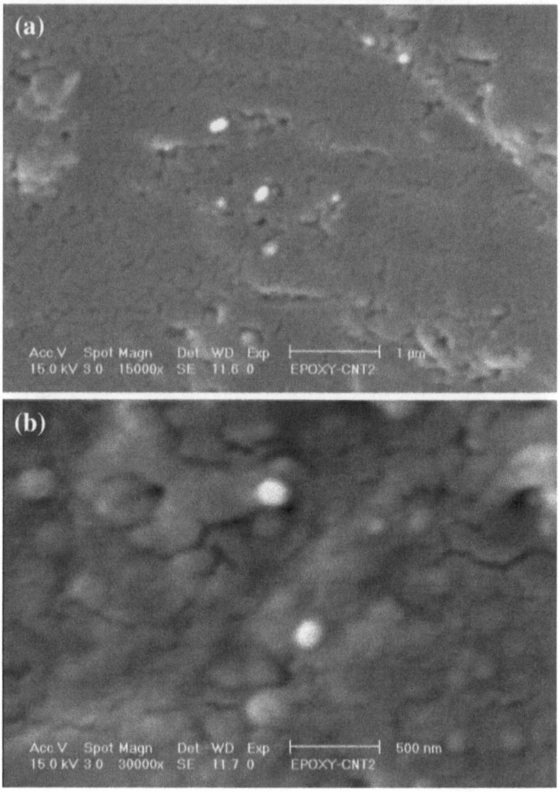

Fig. 2 SEM image of the MWNT-COOH/Epoxy nanocomposite, magnification is 15000× (**a**), 30000× (**b**)

Table 1 Mechanical properties of carbon fiber/epoxy composites and carbon fiber/MWNTs/epoxy nanocomposites

Samples	Young' s modulus (GPa)	Tensile strength (MPa)	Strain at break (%)
Carbon fiber/Epoxy composite	45.04	340.18	1.44
Carbon fiber/MWNTs/Epoxy nanocomposite	46.17	365.84	1.53
Carbon fiber/MWNTs-COOH/Epoxy nanocomposite	49.58	372.21	1.45

4 Conclusions

In order to obtain uniform dispersion of MWNTs in the epoxy matrix and combine the design of MWNT–epoxy interfacial interaction with the demand for the improvement of mechanical properties of carbon fiber/MWNT/epoxy composites, we carried out sidewall carboxylic acid functionalization of MWNTs. In the functionalization process, acid treatment of MWNTs should be executed to effectively generate carboxyl groups on their surface. FTIR analyses indicate that H_2SO_4/HNO_3 treatment would vary the feature of the MWNT surface and efficiently introduce carboxyl groups onto it.

Sonication was an efficient method of infusing carbon nanotubes into epoxy resin when CNT weight fractions are lower than 0.5 wt%. Above the 0.5 wt%, CNTs agglomerated. The morphology characterization of the fracture surfaces of MWNT-COOH/epoxy nanocomposite shows that their fracture patterns are brittle fracture and ductile fracture, respectively. In addition, MWNTs are homogeneously dispersed in epoxy matrix and they possess an intimate contact with matrix. The mechanical properties of MWNTs and MWNTs-COOH reinforced carbon fibers/epoxy composites were investigated at the same nanotube loading of 0.5 wt%. A hybrid laminate is presented that demonstrates enhancement of mechanical properties. The Young's modulus and tensile strength properties of carbon fiber/MWNT-COOH/epoxy composites are better than those of carbon fiber/MWNT/epoxy composites. Mechanical properties improvement indicating that there is efficient load transfer between the matrix and the functionalized carbon nanotubes.

References

1. Gogotsi Y.: Nanotubes, Nanofibers. Taylor & Francis, Drexel University, Philadelphia, pp. 1–37 (2006)
2. Ijima, S., Ichihashi, T.: Single-shell carbon nanotubes of 1-nm diameter. Nature **363**, 603–605 (1993)
3. AH Cohen, S.R., Kenig, S., Wagner, H.D.: Interfacial fracture energy measurements for multi walled carbon nanotubes pulled from a polymer matrix. Compos. Sci. Technol. **64**, 2283–2289 (2004)
4. Baughman, R.H., Zakhidov, A.A., Heer, W.A.D.: Carbon nanotubes-the route toward applications. Science **297**, 787–792 (2002)
5. Rupesh, K., Suryasarathi, B.: Carbon nanotube based composites—a review. J. Miner. Mater. Charact. Eng. **4**(1), 31–46 (2005)
6. Andrews, R., Jacques, D., Minot, M., Rantel, T.: Fabrication of carbon multiwall nanotube/ polymer composites by shear mixing. Macromol. Mater. Eng. **287**(6), 395–403 (2002)
7. Schadler, L.S., Giannaris, S.C., Ajayan, P.M.: Load transfer in carbon nanotube epoxy composites. Appl. Phys. Lett. **73**(28), 3842 (1998)
8. Arasteh, R., Masoumi, M., Rashidi, A.M., Moradi, L., Samimi, V., Mostafavi, S.T.: Adsorption of 2-nitrophenol by multi-wall carbon nanotubes from aqueous solutions. Appl. Surf. Sci. **256**, 4447–4455 (2010)
9. Gojny, F.H., Schulte, K.: Functionalisation effect on the thermomechanical behaviour of multi-wall carbon nanotube/epoxycomposites. Compos. Sci. Technol. **64**(15), 2303–2308 (2004)
10. Chen, X., Wang, J., Lin, M., Zhong, W., Feng, T., Chen, X., et al.: Mechanical and thermal properties of epoxy nanocomposites reinforced with amino-functionalized multi-walled carbon nanotubes. Mater. Sci. Eng. A **492**(1–2), 236–242 (2008)
11. Zhou, W., Wang, B., Zheng, Y., Zhu, Y., Wang, J., Qi, N.: Effect of surface decoration of CNTs on the interfacial interaction and microstructure of epoxy/MWNT nanocomposites. Chem. Phys. Chem. **9**(7), 1046–1052 (2008)

12. Yaping, Z., Aibo, Z., Qinghua, C., Jiaoxia, Z., Rongchang, N.: Functionalized effect on carbon nanotube/epoxy nanocomposites. Mater. Sci. Eng. A **435–436**, 145–149 (2006)
13. Geng, Y., Liu, M.Y., Li, J., Shi, X.M., Kim, J.K.: Effects of surfactant treatment on mechanical and electrical properties of CNT/epoxy nanocomposites. Composite Part A **39**(12), 1876–1883 (2008)
14. Gojny, F.H., Wichmann, M.H.G., Fiedler, B., Schulte, K.: Influence of different carbon nanotubes on the mechanical properties of epoxy matrix composites—a comparative study. Compos. Sci. Technol. **65**(15–16), 2300–2313 (2005)
15. Sun, L., Warren, G.L., O'Reilly, J.Y., Everett, W.N., Lee, S.M., Davis, D., et al.: Mechanical properties of surface-functionalized SWCNT/epoxy composites. Carbon **46**(2), 320–328 (2008)
16. Prolongo, S.G., Buron, M., Gude, M.R., Chaos-Moran, R., Campo, M., Urena, A.: Effects of dispersion techniques of carbon nanofibers on the thermo-physical properties of epoxy nanocomposites. Compos. Sci. Technol. **68**(13), 2722–2730 (2008)
17. Yang, K., Gu, M., GuoY, Pan.X., Mu, G.: Effects of carbon nanotube functionalization on the mechanical and thermal properties of epoxy composites. Carbon **47**(7), 1723–1737 (2009)

Effects of Single-Walled Carbon Nanotubes on Fiber Diameter Distribution of Poly (Butylene Terephthalate) Electrospun Composite Nanofibers

Mehdi Forouharshad, Omid Saligheh, Rouhollah Arasteh and Reza Eslami Farsani

Abstract In this Study, Poly (butylene terephthalate) (PBT)/Single-walled carbon nanotubes composite nanofibers mats were prepared by electrospinning, being directly deposited in the form of a random fibers web. The effect of different loads of single-walled nanotubes on the morphology of the electrospun PBT/SWCNTs composite nanofibers was investigated with scanning electron microscopy (SEM). SEM observations indicated that the presence of SWCNTs resulted in finer nanofibers for lower loading; however, a broader distribution, especially for the higher diameter ranges was found for nanofibers with higher amounts of carbon nanotubes.

Keywords Composite nanofibers · Poly butylenes terephthalate · Single wall carbon nanotube · Electrospinning · Morphology

M. Forouharshad (✉) · O. Saligheh · R. Arasteh · R. Eslami Farsani
Young Researchers club, Islamic Azad University, South Tehran Branch, Tehran, Iran
e-mail: Forouharmahdi@gmail.com

O. Saligheh
e-mail: Omid.saligheh@gmail.com

R. Arasteh
e-mail: R.arasteh@gmail.com

R. Eslami Farsani
Faculty of Mechanical Engineering, K.N.Toosi University of Technology, Tehran, Iran
e-mail: reslami@yahoo.com

A. Öchsner et al. (eds.), *Materials with Complex Behaviour II*,
Advanced Structured Materials 16, DOI: 10.1007/978-3-642-22700-4_27,

1 Introduction

Carbon nanotubes (CNTs) with a high aspect ratio and low density have been shown to possess excellent mechanical, thermal and electronic properties [1]. These characteristics make them an ideal candidate as a filter to develop potentially revolutionary composites with light weight and enhanced mechanical [2], electrical [3] or thermal properties [4]. Polymer fibers reinforced with CNTs are of particular interest [5–7]. Several parameters can affect the properties of the CNTs composites, such as the filler dispersion, orientation and interfacial bonding [8].

Recently, electrospinning has been used to produce ultrafine CNT composite fibers. Electrospinning is a novel and efficient tool for producing nanotube-polymer composites. The electrospinning set-up consists of a bipolar high voltage source, a syringe injector coupled to a needle and a conducting collector to obtain randomly orientated or aligned nanofibers. The electrospinning process involves stretching a polymer solution under a strong electric field to form dry or semi-dry fibers with diameters on the nanometre scale [9]. From solution to dry fiber, the fiber stretching process takes just tens of milliseconds [10]. With such a fast fiber-stretching speed and high aspect ratio of the resultant nanofiber, an alignment of CNTs along the axis of the nanofiber could be achieved when a polymer solution containing well-dispersed carbon nanotubes is electrospun [11, 12]. Dror et al. reported that one advantage of electrospinning was that carbon nanotubes or their bundles orient parallel to the main nanofiber axis during fiber formation due to combination of dielectrophoretic forces caused by dielectric or conductivity mismatch between CNTs and the polymer solution and high shear forces induced by the electrospinning [13]. Carbon nanotubes generally tend to exist as bundles or even networks of aggregates because of strong non-bonded interaction. Therefore for exploiting their full potential as fillers, particular attention need to be paid to achieving a good dispersion of nanotubes in the spinning solution, and achieving a high degree of axial orientation of CNTs to maximize mechanical properties of nanofibers. Experimentally, this is achieved by either mechanical methods (using either high shear mixers) or more commonly by sonication [14]. Although nanofibers have wide range of applications in various fields, their mechanical properties are comparatively low, but they may be improved by the incorporation of reinforcing nanomaterials.

PBT, a linear polyester of aromatic nature, is a thermoplastic of excellent mechanical properties. Like PET, electrospun fibers of PBT can be used in blood vessel tissue engineering applications as scaffolds for endothelial cells [15]. Lately, electrospun composite nanofiber of carbon nanotubes with different polymers such as polycaprolactone [16], nylon [17], polyvinyl alchohol (PVA) [18], polycarbonate (PC) [19], polyacrylonitrile (PAN) [20, 21], etc. have been reported, which show enhanced mechanical properties of nanofibers compared to their pure polymer nanofibers. In this study, we used single-wall carbon nanotubes (SWCNTs)/PBT composite nanofibers as model material to examine the effect of loading of carbon nanotubes on the diameter of composite fibers.

2 Experimental

2.1 Materials

The PBT (PBT 1800, MFI = 20 g/10 min) used in this study was supplied by Shenzhan Youhua Engineering Plastic Co., China, in the form of pellets. Trifluoroacetic acid (TFA) and dichloromethane (DCM) (Merck, Germany) were obtained and used without further purification. A volume ratio of 50/50 TFA/DCM was used as a solvent for dissolving PBT to prepare the polymer solution for electrospinning. The single walled carbon nanotubes (SWCNTs) (diameter < 2 nm, length < 20 μm, purity > 95%) were obtained from Shenzhen Nanotechnologies Co., Ltd. (NTP) and were used as-received without purification and surface treatment.

2.2 Samples Preparation and Electrospinning

For making nanocomposite solutions, quantitative SWCNTs (0.5, 1, 2, 4 wt%) were added to TFA/DCM, 50/50 (v/v); and ultrasonicated with a probe sonicator for 10 min at an amplitude of 50% using a 24 kHz ultrasonic processor (Model UP200H, Hielscher Ultrasonics GmbH, Stuttgart, Germany) at room temperature to produce uniformly dispersed SWCNT suspensions. Then, the powdered polymer was added to the solution while the mechanical mixer was working. After adding PBT, the mixture was stirred via the heater stirrer (1,700 rpm) for 30 min to obtain a uniform composite solution for electrospinning. A concentration of 14 wt% pure PBT solutions was found to be optimum for PBT nanofiber fabrication [22]. Thus, the concentration of nanocomposite PBT/CNT solutions was 14 wt% with a variation in the weight of the nanotubes of 0.5, 1, 2 and 4 wt%. The electrospinning process consisted of (1) a polymer solution delivery system, (2) a power supply to generate an electrical field, and (3) a fiber collection device. Electrospinning dopes (pure PBT and SWCNTs/PBT solutions) were placed in a syringe with a capillary tip of 0.7 mm diameter, with pressure applied by a syringe pump (KDS 100, KD Scientific Inc., USA); the syringe pump speed was adjusted to a flow rate 0.5 mL/h. The electric field was provided by a high voltage power supply (model RR30-1.25P, Gamma High Voltage Research, Inc., USA). The applied voltage was 20 kV, the tip to collector distance was set to be 15 cm and the electrospining was carried out according to [23]. When the voltage rises to a critical value (20 kV), the charged jet is ejected; as the jet travels in air, most of solvent is evaporated and the fibers hit the surface on an aluminum foil which was placed on the metal collector and solidified rapidly because of the low boiling point of DCM and TFA.

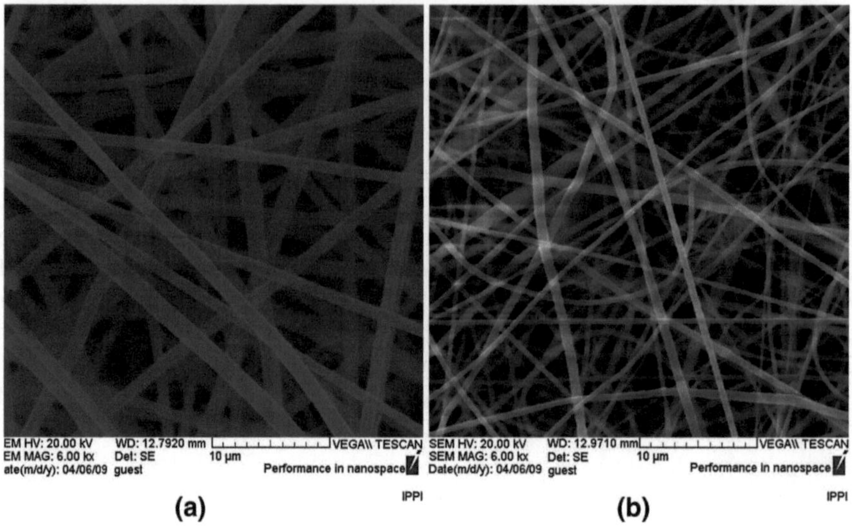

Fig. 1 SEM images of electrospun SWCNTs/PBT composite nanofibers

2.3 Scanning Electron Microscopy (SEM)

The morphology of the electrospun composite nanofibers was observed by scanning electron microscopy (VEGA TESCAN, Czech Republic). Samples were fixed to SEM holders and coated with a thin layer of gold prior to SEM investigation.

3 Results and Discussion

3.1 Morphology

The SEM images of electrospun SWCNT/PBT composite nanofibers are shown in Fig. 1. They demonstrate that the PBT/CNT composite nanofibers were successfully spun by electrospinning at 20 kV. The mats are composed of numerous, randomly oriented, continuous fibers which have a distribution in diameter without abnormal morphology, such as beads.

Figure 2 shows the fiber diameter distribution of the electrospun composite nanofibers; the diameter of the pure PBT electrospun fiber from the 14 wt% concentration solution ranged from 300 to 800 nm with average diameter 549 nm [22], compared to PBT/CNT composite nanofibers for which the average fiber diameters were 473, 422 and 462 nm for 0.5, 1 and 2% SWCNTs, respectively. The presence of SWCNTs resulted in finer nanofibers and the fiber distribution became narrow as well. The reduction in fiber diameter suggests that the presence

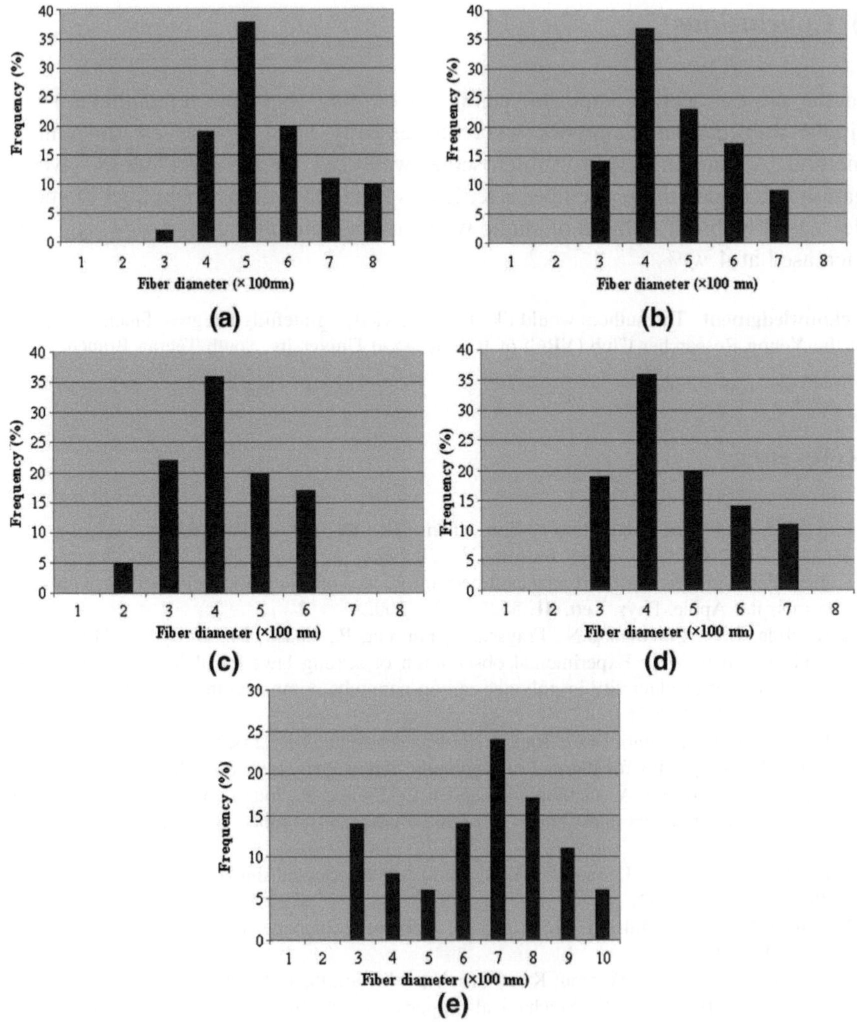

Fig. 2 Fiber diameter distribution of electrospun SWCNTs/PBT composite nanofibers. **a** PBT neat, **b** PBT-SWCNT 0.5%, **c** PBT-SWCNT 1%, **d** PBT-SWCNT 2%, **e** PBT-SWCNT 4%

of carbon nanotubes in the polymer solution affected the fiber stretching process due to the increased electrical conductivity; the consequent increase in surface charge in the jet may have led to smaller average nanofiber diameters according to references [23]. However, a broader distribution, especially in the higher diameter ranges, was found for PBT/CNT with 4% of SWCNTs. The diameter of the electrospun composite nanofibers with 4% SWCNTs ranged from 300 to 1000 nm with average diameter 651 nm. The increase in diameter can be explained by the increase in solution viscosity with addition of more CNTs. This kind of solution viscosity to the diameter effect has been extensively studied [24, 25].

4 Conclusions

In the present work composite nanofibers of poly (butylene terephthalate) and single wall carbon nanotubes were successfully spun by the electrospinning method. A scanning electron microscope was used for the morphology investigation and to determine the fiber size. It was found that the average fiber diameter decreased with the addition of single wall carbon nanotubes until 2 wt%, and then increased at 4 wt%.

Acknowledgment The authors would like to acknowledge gratefully the great financial support of the Young Researcher Club (YRC) of Islamic Azad University, South Tehran Branch.

References

1. Ajayan, P.M.: Nanotubes from carbon. Chem. Rev. **99**, 1787–1799 (1999)
2. Cadek, M., Coleman, J.N., Barron, V., Hedicke, K., Blau, W.J.: Morphological and mechanical properties of carbon nanotube-reinforced semicrystalline and amorphous polymer composite. Apple. Phys. Lett. **81**, 5123–5125 (2002)
3. Kilbride, B.E., Coleman, J.N., Fraysse, J., Fournet, P., Cadek, M., Drury, A., Hutzler, S., Roth, S., Blau, W.J.: Experimental observation of scaling laws for alternating current and direct current conductivity in polymer-carbon nanotube composite thin film. J. Appl. Phys. **92**, 4024–4030 (2002)
4. Biercuk, M.J., Liaguno, M.C., Radosavljevic, M., Hyun, J.K., Fischer, J.E., Johnson, A.T.: Nanotube composites for thermal management. Appl. Phys. Lett. **80**, 2767–2769 (2002)
5. Vigolo, B., Penicaud, A., Coulon, C., Sauder, C., Pailler, R., Journet, C., Bernier, P., Poulin, P.: Macroscopic fibers and ribbons of oriented carbon nanotubes. Science **290**, 1331–1334 (2000)
6. Poulin, P., Vigolo, B., Launois, P.: Films and fibers of oriented single wall nanotubes. Carbon **40**, 1741–1749 (2002)
7. Munoz, E., et al.: Multifunctional carbon nanotube composite fibers. Adv. Eng. Mater. **6**, 801–804 (2004)
8. Park, C., Ounaied, Z., Watson, K.A., Crooks, R.E., Smith, J., Lowther, S.E., Connell, J.W., Siochi, E.J., Harrison, J.S.: Mechanical properties of electrospun PVA/MWNTs composite nanofibers. Chem. Phys. Lett. **364**, 303 (2002)
9. Reneker, D.H., Chun, I.: Nanometre diameter fibres of polymer, produced by electrospinning. Nanotechnology **7**, 216–223 (1996)
10. Shin, Y.M., Hohman, M.M., Brenner, M.P., Rutledge, G.C.: Electrospinning: a whipping fluid jet generates submicron polymer fibers. Appl. Phys. Lett. **78**, 1149–1151 (2001)
11. Gao, J., Yu, A., Itkis, M.E., Bekyarova, E., Zhao, B., Niyogi, S., Haddon, R.C.: Large-scale fabrication of aligned single-walled carbon nanotube array and hierarchical single-walled carbon nanotube assembly. J. Am. Chem. Soc. **126**, 16698–16699 (2004)
12. Salalha, W., Dror, Y., Khalfin, R.L., Cohen, Y., Yarin, A.L., Zussman, E.: Single-walled carbon nanotubes embedded in oriented polymeric nanofibers by electrospinning. Langmuir **20**, 9852–9855 (2004)
13. Dror, Y., Salalha, W., Khalfin, R.L., Cohen, Y., Yarin, A.L., Zussman, E.: Carbon nanotubes embedded in oriented polymer nanofibers by electrospinning. Longmuir **17**, 7012–7020 (2003)

14. Sundaray, B., Subramanian, V., Natarajan, T.S., Krishnamurthy, K.: Electrical conductivity of a single electrospun fiber of poly(methyl methacrylate) and multiwalled carbon nanotube nanocomposite. Appl. Phys. Lett. **14**, 143114–1 (2006)
15. Ma, Z., Kotaki, M., Yong, T., He, W., Ramakrishna, S.: Surface engineering of electrospun polyethylene terephthalate (PET) nanofibers towards development of a new material for blood vessel engineering. Biomaterials **26**, 2527–2536 (2005)
16. Saeed, K., Park, S.Y., Lee, H.J., Baek, J.B., Huh, W.S.: Preparation of electrospun nanofibers of carbon nanotube/polycaprolactone nanocomposite. Polymer **23**, 8019–8025 (2006)
17. Saeed, K., Park, S.Y., Haider, S., Baek, J.B.: In situ polymerization of multi-walled carbon nanotube/nylon-6 nanocomposites and their electrospun nanofibers. Nanoscale. Res. Lertt. **4**, 39–46 (2009)
18. Zhou, W.P., Wu, Y.L., Wei, F., Luo, G.H., Qian, W.Z.: Elastic deformation of multiwalled carbon nanotubes in electrospun MWCNTs-PEO and MWCNTs-PVA nanofibers. Polymer **26**, 12689–12695 (2005)
19. Kim, G.M., Michler, G.H., Poetschke, P.: Deformation processes of ultrahigh porous multiwalled carbon nanotubes/polycarbonate composite fibers prepared by electrospinning. Polymer **18**, 7346–7351 (2005)
20. Ge, J.J., Hou, H.Q., Li, Q., Graham, M.J., Greiner, A., Reneker, D.H., Harris, F.W., Cheng, S.Z.D.J.: Assembly of well-aligned multiwalled carbon nanotubes in confined polyacrylonitrile environments: electrospun composite nanofiber sheets. Am. Chem. Soc. **48**, 15754–15761 (2004)
21. Hou, H.Q., Ge, J.J., Zeng, J., Li, Q., Reneker, D.H., Greiner, A., Cheng, S.Z.D.: Electrospun polyacrylonitrile nanofibers containing a high concentration of well-aligned multiwall carbon nanotubes. Chem. Mater. **17**, 967–973 (2005)
22. Forouharshad M., Saligheh O., Arasteh R., Farsani, E.R.: Manufacture and characterization of poly (butylenes terephthalate) nanofibers by electrospinning. J. Macromol. Sci. Part B: Phys. **49**, 833–842 (2011)
23. Ra, E.J., An, K.H., Kim, K.K., Jeong, S.Y., Lee, Y.H.: Anisotropic electrical conductivity of MWCNT/PAN nanofiber paper. Chem. Phys. Lett. **413**, 188–193 (2005)
24. Kim, J.S., Reneker, D.H.: Polybenzimidazole nanofiber produced by electrospinning. Polym. Eng. Sci. **39**, 849–854 (1999)
25. Doshi, J., Reneker, D.H.: Electrospinning process and applications of electrospun fibers. J. Electrostat. **35**, 151–160 (1995)

Structure, Surface and Hardness Properties of $Y_xAg_yBa_2Cu_3O_{7-\delta}$ Composites Superconductor

Nurulhawa Ali Hasim and Mohd Rafie Johan

Abstract Samples of the nominal composition, $Y_xAg_yBa_2Cu_3O_{7-\delta}$ where $x = (0.10, 0.30, 0.50, 0.70, 0.90)$ and $y = (0.12, 0.36, 0.60, 0.84, 1.09)$ were prepared using the conventional solid state reaction. XRD patterns show that silver was incorporated into the Y123 orthorhombic structure up to 1.09 molar ratios and released the excess Ag_2O. Silver was present within the grains of Y123 composite as confirmed by field emission scanning electron microscopy (FESEM) images and energy dispersive spectroscopy (EDS) analysis. Silver atoms were agglomerated among the Y123 intercrystalline grain growth as shown by FESEM images and filling the voids within the grain boundaries. As a result, the microhardness properties were improved.

Keywords Solid state reaction · Y123 · Silver · Microhardness · Grain boundaries

1 Introduction

The discovery of the high-T_c superconducting materials Yttrium Barium Copper Oxide (Y123) in 1986 set into motion an extraordinary worldwide outburst of superconductivity research [1, 2]. However, it shows relatively poor mechanical

N. A. Hasim (✉) · M. R. Johan
Advanced Materials Research Laboratory, Department of Mechanical Engineering, Faculty of Engineering, University of Malaya, Lembah Pantai, 50603 Kuala Lumpur, Malaysia
e-mail: nurulhawa@siswa.um.edu.my

M. R. Johan
e-mail: mrafiej@um.edu.my

A. Öchsner et al. (eds.), *Materials with Complex Behaviour II*,
Advanced Structured Materials 16, DOI: 10.1007/978-3-642-22700-4_28,

properties such as brittleness and others superconducting properties. With respect to that addition of other doping material such as silver (Ag) may provide a large number of beneficial effects to the Y123 composite.

Ag can be incorporated into the Y123 matrix by different methods: mixing with metallic Ag [3–5] or Ag_2O [6, 7] and electrochemical ways [8–10]. The former produced a random non-uniform distribution of Ag in the composite while the later produced a non-random distribution of Ag on the grain surface of the composite.

Ag was combined with Y123 oxide in many solid states sintering process, which shows signs of superior superconducting and mechanical properties [10–19]. It has been used either as an isolated phase filling voids within the solid-state sintered Y123 bulk [10–14], as the substrate material in the Y123/Ag composite wire or tape [16–18]. The hardness properties of Y123 bulk at room temperature were mostly obtained by conventional Vickers measurements in the range of 5–8 GPa [19, 20].

In this chapter, we synthesize Ag doped Y123 superconductor by using conventional solid state reaction technique. We show that the addition of Ag into the Y123 matrix has improved the microstructure and hardness properties of the composites.

2 Experimental Method

The current investigation involved a series of sample preparation by conventional solid-state reaction. High-purity powders of Y_2O_3, $BaCO_3$, CuO and Ag_2O were mixed in the stoichiometric proportions. The mixtures were then calcined at 930°C for 24 h and subsequently pressed into pellet form under 5 tonne of pressure. It follows a sintering process at 950°C under oxygen flow for 20 h. The samples were then cooled and oxygenation to 500°C with a cooling rate of 1°C/min. It continues until reaching 300°C with a higher cooling rate of 10°C/min. The structural properties of the samples were tested by Philips X'Pert MPD PW3040 XRD with CuKα radiation at 1.5406 Å. The surface morphology and elemental analysis were carried out using Zeiss AURIGA and EDAX TSL, respectively. The microhardness measurements were performed by Mitutoyo MVK-H2.

3 Results and Discussion

XRD patterns shown in Fig. 1 indicate that the addition of Ag_2O does not change the superconducting structures nor form an undesirable second phase. The peak appearing at 2θ value of 44.2° be indexed as (200) reflection of Ag and appears as a separate phase at the grain boundaries of Y123. The (111) plane of Ag_2O which has the highest intensity, overlapped with the (013) plane of Y123 orthorhombic structure at $2\theta = 32.9°$. The (111) plane of Ag_2O is clearly seen in the diffraction

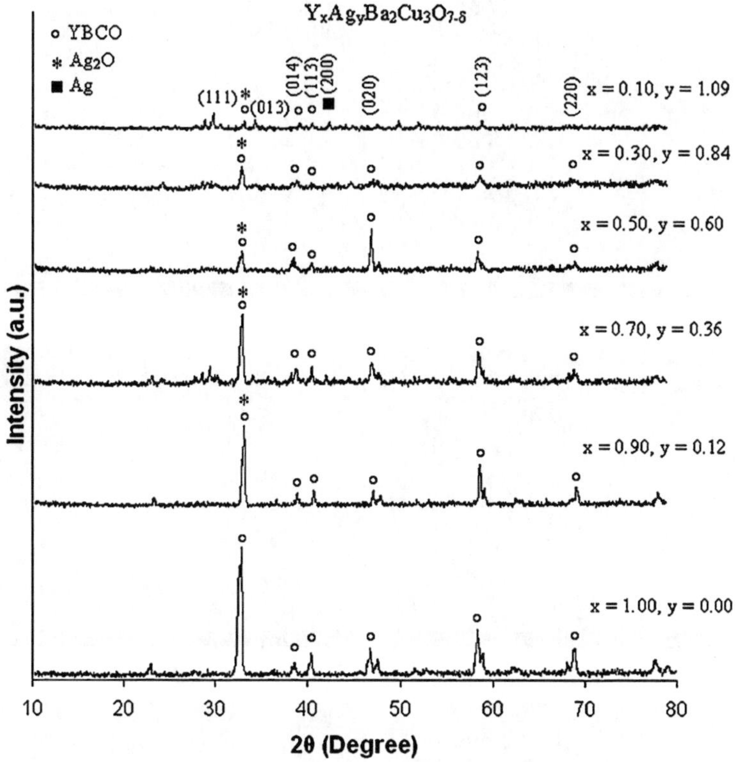

Fig. 1 XRD patterns of $Y_xAg_yBa_2Cu_3O_{7-\delta}$

pattern indicating that excess of Ag_2O remains in a separate phase and has not been decomposed to Ag and O.

Figure 2 shows the FESEM micrograph of pure Y123 and several composites materials. The granular and porosity of the sample are clearly seen. The pure Y123 exhibits a non-uniform structure and relatively with large number of pores. The surface of the Ag_2O diffused sample is much denser than that of the pure material. These results indicate that the surface morphology of the Y123 sample is improved by Ag diffusion doping. The grain growth enhanced with Ag concentration up to $y = 0.12$. Beyond this concentration, the grain growth is inhibited, as illustrated in Fig. 2d. The prominent change of grain growth can be seen at $y = 0.36$ as shown in Fig. 2c. Ag atoms were agglomerated among Y123 composites and formed a bulk sample up to 0.47 μm size. The EDAX result shows that Ag existed in the bulk sample of Y123 as shown in Fig. 3.

In Fig. 4, the microhardness of samples increases with the increase Ag_2O content with the exception of $y = 1.09$. This indicates that effect of Ag in the Y123 superconducting phase on strengthening the composites. It could be inferred that this could be due to the reduction of pores in the samples as shown in Fig. 2b–f.

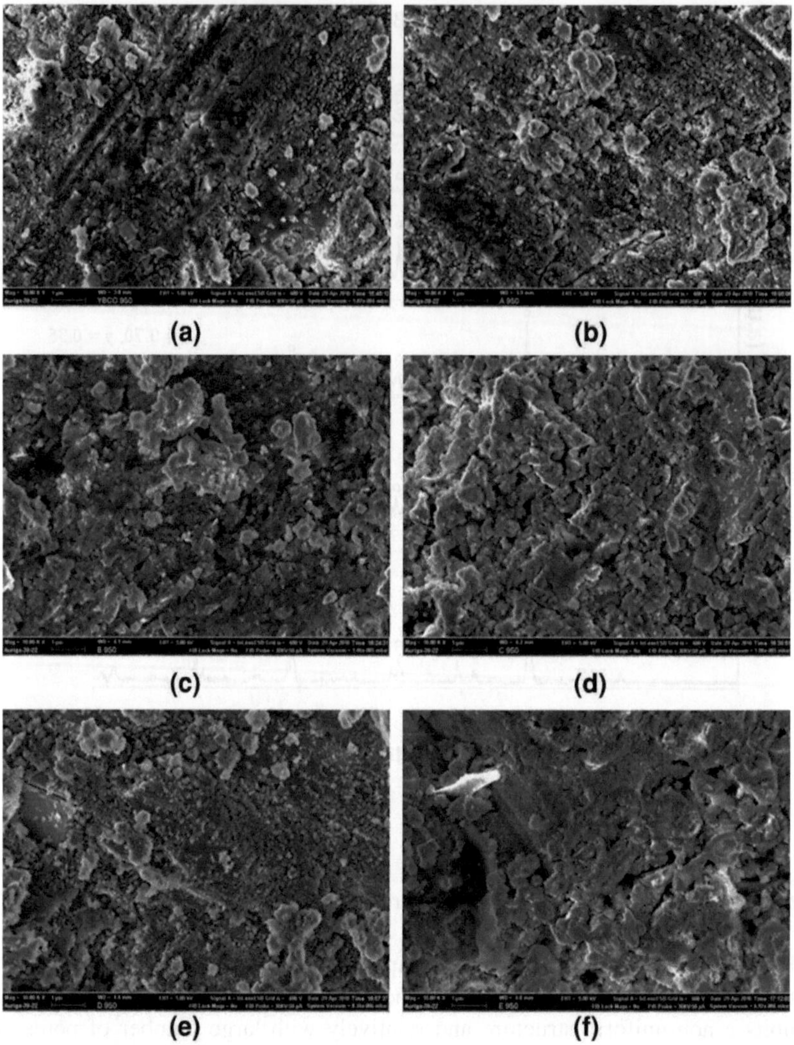

Fig. 2 FESEM micrographs of **a** Y123, **b** $Y_{0.90}Ag_{0.12}Ba_2Cu_3O_{7-\delta}$, **c** $Y_{0.70}Ag_{0.36}Ba_2Cu_3O_{7-\delta}$, **d** $Y_{0.50}Ag_{0.60}Ba_2Cu_3O_{7-\delta}$, **e** $Y_{0.30}Ag_{0.84}Ba_2Cu_3O_{7-\delta}$ and **f** $Y_{0.10}Ag_{1.09}Ba_2Cu_3O_{7-\delta}$

Furthermore, the existence of Ag produces a compressive stress field in the Y123 matrix. However, the standard deviation was 42.14 HV with Ag addition does not show any clear tendency and the strength depend on the amount of Ag and yttrium in the composites.

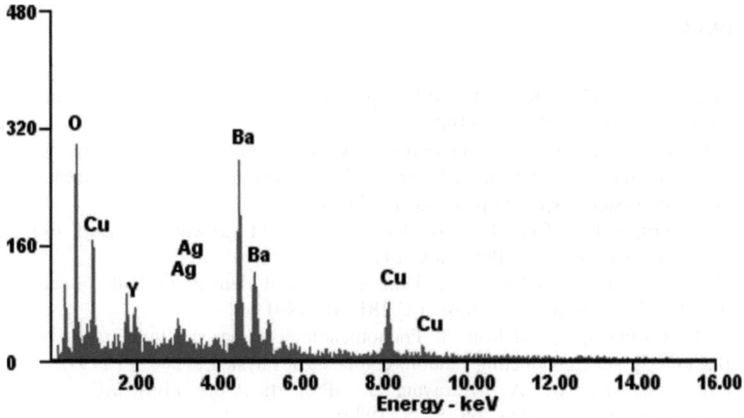

Fig. 3 EDAX spectrum of $Y_{0.5}Ag_{0.6}Ba_2Cu_3O_{7-\delta}$

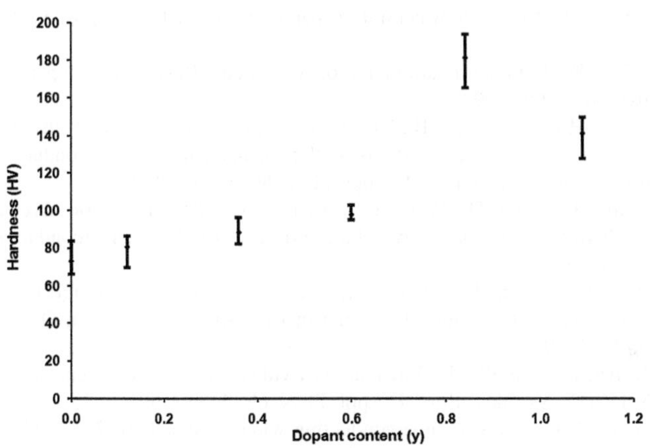

Fig. 4 Microhardness of composites as a function of dopant content

4 Conclusion

In summary, we have demonstrated that the addition of Ag_2O to Y123 matrix does not only increases the microhardness, but also affects their microstructures. It was found that Y123 sites are substituted by Ag ions at 1.09 molar ratios, as well as that Ag_2O also fills the pores in the grain boundaries. In addition, Ag has promoted good grain growth among the multiphase composition.

The authors are grateful to the University of Malaya for financial support under PPP [PS068/2008C] and UMRG [RG038/09AET] grants.

References

1. Bednorz, J.G., Müeller, K.A.: Possible high Tc superconductivity in the Ba–La–Cu–O system. Z. Phys. B **64**, 189–193 (1986)
2. Burns, G.: High Temperature Superconductivity. Academic Press, Inc, Boston (1992)
3. Petrisor, T., Giugiu, A., Ciontea, L.: Magnetic field dependence of critical current in YBCO–Ag composite system. Appl. Supercond. **1**, 1219 (1993)
4. Wu, N., Zern, H.H., Chen, C.: Low-temperature melt growth of $YBa_2Cu_3O_{7-x}$/silver composite in partial vacuum. Physica C **241**, 198 (1995)
5. Lee, D.F., Chaud, X., Salama, K.: Transport current density in bulk oriented-grained $YBa_2Cu_30_x$/ silver composites. Physica C **181**, 81 (1991)
6. Ropers, B., Carmona, F., Flandrois, S.: Phenomenological approach to the resistive transition of YBaCuO–Ag superconducting random composites. Physica C **204**, 71 (1992)
7. Ganapathi, L., Kumar, A., Narayan, J.: Properties of $YBa_2Cu_3O_{7-\delta}$ composite superconductors. J. Appl. Phys. **66**, 5935 (1989)
8. Rodrigues Jr., P., Ghivelder, L., Pureur, P., Reich, S.: Resistive transition and fluctuating conductivity in YBaCuO–Ag composites. Physica C **211**, 13 (1993)
9. Knoll, P., Reich, S.: J. Morphological and magnetic effects of compaction pressure on $YBa_2Cu_3O_{7-x}$-silver nonrandom composites. Appl. Phys. **71**, 1035 (1992)
10. Reich, S., Felner, I.: Nonrandom ceramic superconductor-metal composites. J. Appl. Phys. **67**, 388 (1990)
11. Wei, W.C., Lee, W.H.: Grain delamination of Ag-doped $YBa_2Cu_3O_{7-x}$ superconductor. Jpn. J. Appl. Phys. **31**, 1305 (1992)
12. Singh, J.P., Leu, H.J., Poeppel, R.B., Van Voorhees, E., Goudey, G.T., Winsley, K., Shi, D.: Effect of silver and silver oxide additions on the mechanical and superconducting properties of $YBa_2Cu_3O_{7-\delta}$ superconductors. J. Appl. Phys. **66**, 3154 (1989)
13. Singh, J.P., Joo, J., Singh, D., Warzynski, T., Poeppel, R.B.: Effect of silver additions on resistance to thermal shock and delayed failure of $YBa_2Cu_3O_{7-\delta}$ superconductors. J. Mater. Res. **8**, 1226 (1993)
14. Kuppermen, D.S., Singh, J.P., Faber Jr., J., Hitterman, R.L.: Application of neutron diffraction to the characterization of residual thermal strains in $YBa_2Cu_3O_{7-\delta}$/Ag. J. Appl. Phys. **66**, 3396 (1989)
15. Dwir, B., Kellet, B., Mieville, L., Pavuna, D.: Evidence for improvement of critical current by Ag in YBaCuO–Ag thick films. J. Appl. Phys. **69**, 4433 (1991)
16. Shi, D., Goretta, K.C.: Swaged superconducting wires. Mater. Lett. **7**, 428 (1989)
17. Shi, D., Xu, M., Chen, J.G., Umezawa, A., Lanan, S.G., Miller, D., Goretta, K.C.: Mater. Superconducting coil made of filamentary composites. Lett. **9**, 1 (1989)
18. Kullberg, M.L., Lanagan, M.T., Wu, W., Poeppel, R.B.: A sol-gel method for preparing oriented $YBa_2Cu_3O_{7-\delta}$ films on silver substrates. Supercond. Sci. Technol. **4**, 337 (1991)
19. Yoshino, Y., Iwabuchi, A., Noto, K., Sakai, N., Murakami, M.: Vickers hardness properties of YBCO bulk superconductor at cryogenic temperatures. Physica C **796**, 357–360 (2001)
20. Lubenets, S.V., Natsik, V.D., Fomenko, L.S., Kaufmann, H.J., Bobrov, V.S., Isotov, A.N.: Influence of oxygen content and structural defects on low-temperature mechanical properties of high-temperature superconducting single crystals and ceramics. Low Temp. Phys. **678**, 23 (1997)

Transport Phenomena in an Evaporated Black Hole

R. Leticia Corral-Bustamante and Aarón Raúl Rodríguez-Corral

Abstract In this chapter, the study of the behavior of a gigantic mass is presented, the mass collapses becoming a mini hole in the continuous space–time, like the ones that were formed in the first instants of the Universe, when not yet elapsed a billionth of second in the Big Bang, due to an enormous concentration of energy in a tiny region of the Space. To detect the presence of mini-hole of infinite surface gravity, were studied the transport phenomena during the evaporation stage of the mass until its disappearing by means of theoretical measurements of energy, in matter of General Relativity, originating from a proposed metric. The Hawking's radiation emissions from the hole whose loss is measured by its entropy, are the ones responsible for the loss of hole energy. The radiation emitted by the black hole at the end of its life could be similar to gamma radiation. The results obtained show evidence that the final state of the hole would be able to be treated as a pure quantum state, that is, in the period of its evaporation, the information of its interior escaped with the Hawking's radiation, nevertheless, the entropy that its emissions show in indefinite form, allows us to include in the calculation some of the possible evolutions of the last scenes of this physical phenomenon, with some of the possible angular positions (energy states) of the hole. The lost information measured by means of entropy is calculated with the relationship of the natural logarithm of the wave length of the mass to consider the situation in which the hole not even has been formed.

R. Leticia Corral-Bustamante (✉)
Technological Institute of Cuauhtémoc City, Tecnológico Ave. S/N, 31500 Cuauhtémoc City, Chihuahua, Mexico
e-mail: leticia.corral@cimav.edu.mx

A. R. Rodríguez-Corral
Universidad Autónoma de Chihuahua, Av. Escorza No. 900, 31000 Chihuahua, Mexico
e-mail: aaronrrc@hotmail.com

A. Öchsner et al. (eds.), *Materials with Complex Behaviour II*,
Advanced Structured Materials 16, DOI: 10.1007/978-3-642-22700-4_29,

Keywords Parameters black hole · Superficial gravity · Entropy · Energy · Evaporation time · Hawking radiation · Event horizon area · Naked singularity · Cosmic censorship

1 Introduction

Einstein's theory of the General Relativity came to revolutionize and to give sense and light to the human mind upon permitting us to examine this unprecedented bequest in magnitude: the matter behavior in the continuous space–time. This chapter makes use of the mathematics of Einstein's theory proposing a metric that permits to obtain the information of the mass in measured transportation phenomena terms through different types of energy: kinetic, potential and internal. The true anomaly associated with the orbit; the entropy (quantized energy) of the hole of enormous superficial gravity, measured by means of the semi-parameter of the conic section of a gigantic mass; the enthalpy measured through the angular coordinate for a relativistic particle; as well as the arbitrary starting anomaly of the mass were calculated. The behavior of the mass studied is seemed to matter that existed early in the Universe [1, 2].

The Hawking radiation process reduces the mass of the black hole and is therefore also known as black hole evaporation [3–5].

In this work, determinations were made of the temperature of the mass of huge surface gravity, which collapsed to a mini-black hole that probably exploded (as predicted by evaporation results in some energy states). Measurement were done with and without charge and angular momentum.

Because Hawking radiation [6] allows black holes to loose mass, this black hole looses more matter than it gains through other means and is expected to dissipate, shrink, and ultimately vanish. Forming a black hole is the most efficient way to compress mass into a region, and this entropy [7, 8] is also a bound on the information content of any sphere in space time. The radiation emitted by the event horizon, seems to resemble the gamma rays [9].

In prior work [10], the entropy of the mass was calculated before collapsing, as well as the entropy of the mass collapsed to a mini black hole, with the purpose of considering the lost information during the Hawking's radiation, like a precedent to confirm that the temperature and the entropy of a black hole is due to purely quantum effects [6–8, 11]. The results obtained permit to show an increase of the total entropy and can confirm that the radiation of the last phases of evaporation of the hole is indefinite [12].

Such black hole might possibly be formed in an early stage of the evolution of the universe, just after the Big Bang, when densities were extremely high. Therefore, this hypothetical baby black hole is called primordially black hole [1, 2, 13, 14].

1.1 Nomenclature

a	Angular momentum or spin of a black hole, $a = 0.9965$, 0.99616, 0.995 ($cm^2\ s^{-1}$). Semi-major axis of an elliptical orbit. [For the planet Mercury, $a = 5.78 \times 10^{12}$ [cm])
A	Event horizon's area (cm^2)
c	Speed of light [$cm\ s^{-1}$]
_Cn, $n = 1, 2$	Constants to be determined
G	Universal Gravitation Constant, $G = 6.67 \times 10^{-8}$ (dyne $cm^2\ g^{-2}$)
G_{ij}, $i, j = 1, 2, 3, 4$	Riemann-Christoffel tensor or contracted tensor of Einstein
$\hbar$	Reduced Planck's constant, $\hbar = 6.6260689633 \times 10^{-27}$ (erg s)
k	Boltzman constant, $k = 1.380650424 \times 10^{-16}$ (erg K^{-1})
l_p	Planck length, $l_p \approx 1.61625281 \times 10^{-33}$ (cm)
M	Mass of a black hole (g)
$M_{\odot}$	Sun mass (g); 1solar mass = 1 $M_{\odot}$ = 1.99 × 1033 (g)
P	Power of the emissions of event horizon (W)
Q	Electric charge of a black hole (cm $dyne^{1/2}$)
r_+	Radius of external horizon (cm)
r_-	Radius of internal horizon (cm)
s	Arch element (cm)
S	Entropy (erg K^{-1}). Semi parameter of the conic section of a mass or spatial coordinate (cm)
t	Temporal coordinate (s), evaporation time (s)
T	Temperature of a black hole (K)
x^i, $i = 1,2,3,4$	Spatial and temporal coordinates in a four-dimensional space

1.2 Greek letters

κ	Surface gravity of a black hole ($cm\ s^{-2}$)
ϕ	Energies: kinetic, potential, internal, quantized (erg). True anomaly associated with the orbit (spatial coordinate) (rad, degrees)
ϕ_0	Arbitrary starting anomaly (rad, degrees)
	Enthalpy (erg $g^{-1}\ K^{-1}$). Angular coordinate for a relativistic particle (spatial coordinate) (rad, degrees)

2 Modelling

This chapter the mathematical model described by the metric is proposed

$$\begin{aligned} ds^2 = & -e^{-\cos(u(S))}(dS)^2 - S^2(d\theta)^2 \\ & - S^2\sin^2(\theta)(d\phi)^2 + e^{-\cos(v(S))}(dt)^2 \end{aligned} \tag{1}$$

where $u(S)$ and $v(S)$ are functions to be determined.

Modeling the behavior of a matter in space–time continuum suggests a spherical symmetry with 3 spatial coordinates and 1 temporal, in this case: $(x^1, x^2, x^3, x^4) = (S, \theta, \phi, t)$, in a four-dimensional space. For the proposed metric, the differential equations that contain the information of the behavior of matter correspond to the non-zero components of the Einstein tensor

$$G_{11} = \frac{e^{-\cos(u)}}{S^2}$$

$$G_{22} = -\frac{1}{4e^{-\cos(u)}}\left(\left(\begin{matrix} 2\cos(v)\left(\frac{dv}{dS}\right)^2 + 2\sin(v)\left(\frac{d^2v}{dS^2}\right) + \left(\frac{dv}{dS}\right)^2 - \\ \left(\frac{dv}{dS}\right)^2\cos^2(v) - \sin(u)\left(\frac{du}{dS}\right)\sin(v)\left(\frac{dv}{dS}\right) \end{matrix}\right)S^2\right)$$

$$G_{33} = \frac{1}{4e^{-\cos(u)}}\left(\sin^2(\theta)S^2\left(\begin{matrix} -2\cos(v)\left(\frac{dv}{dS}\right)^2 - 2\sin(v)\left(\frac{d^2v}{dS^2}\right) - \left(\frac{dv}{dS}\right)^2 + \\ \left(\frac{dv}{dS}\right)^2\cos^2(v) + \sin(u)\left(\frac{du}{dS}\right)\sin(v)\left(\frac{dv}{dS}\right) \end{matrix}\right)\right)$$

$$G_{44} = \frac{e^{-\cos(v)}}{S^2} \tag{2}$$

The differential equations representing the geodesics are found to be

$$\frac{d^2t}{ds^2} - \frac{\_C1\frac{dS}{ds}\frac{dt}{ds}}{(\_C1 - c^2S)S} = 0 \tag{3}$$

$$\frac{d^2\phi}{ds^2} + \frac{2\frac{dS}{ds}\frac{d\phi}{ds}}{S} + \frac{2\cos\theta\frac{d\theta}{ds}\frac{d\phi}{ds}}{\sin\theta} = 0 \tag{4}$$

$$\frac{d^2\theta}{ds^2} + \frac{2\frac{dS}{ds}\frac{d\theta}{ds}}{S} + \sin\theta\cos\theta\left(\frac{d\phi}{ds}\right)^2 = 0 \tag{5}$$

$$\frac{d^2S}{ds^2}+\frac{1}{2}\frac{\_C1\left(\frac{dS}{ds}\right)^2}{(\_C1-\_C2S)S}+\frac{(\_C1-\_C2S)\left(\frac{d\theta}{ds}\right)^2}{\_C2}$$
$$-\frac{(\_C1-\_C2S)\sin^2\theta\left(\frac{d\phi}{ds}\right)^2}{\_C2}-\frac{1}{2}\frac{(\_C1-\_C2S)\_C1\left(\frac{dt}{ds}\right)^2}{\_C2S^3}=0 \quad (6)$$

where $S = 1/u$.

According to Hawking [5, 6], a black hole behaves like black body radiation in quantum mechanics. The temperature, T, of the mass, M, of a black hole, without electric charge neither angular momentum or spin, can be calculated with the relationship [15]

$$T=\frac{\hbar c^3}{16\pi^2 GMk}=\frac{10^{23}}{M} \quad (7)$$

The time that would take a back hole in being evaporated completely can be calculated; this time results to be of the order of

$$t = G^2M^3/\hbar c^4 \quad (8)$$

or, approximately, $t = 8.40716 \times 10^{-17}\ M^{3\ s}$ (s), if the mass M is measured in kilograms (kg). Black hole entropy is

$$S=\frac{\pi c^3 k}{2\hbar G}A \quad (9)$$

where the event horizon area, A, is given by three parameters: mass, M, electric charge, Q, and angular momentum, a [15].

$$A=4\pi G\left(2GM^2-Q^2+2M\sqrt{G^2M^2-GQ^2-c^2a^2}\right)/c^4 \quad (10)$$

The electric charge of the hole is calculated clearing Q from Eq. 10.

Temperature, T, can be related without ambiguity with the superficial gravity, κ, of a black hole with mass, charge and angular momentum, by means of the relationship

$$T = \hbar\kappa/4\pi^2 ck \quad (11)$$

where κ is given by Kerr–Newmann relationship

$$\kappa = 4\pi\sqrt{G^2M^2-GQ^2-a^2c^2}/A \quad (12)$$

The power of the emissions of the event horizon of the hole is given by

$$P = 3.56345 \times 10^{32}(1/M)^2 \quad (13)$$

for a mass in kilograms.

To verify the natural law of cosmic censorship, the radii of the external horizon (sign +) and internal (sign −) were computed through the relationship

$$r_{\pm} = \sqrt{\pi(Ac^2 - 4\pi a^2)}\Big/2\pi c \tag{14}$$

to the values of angular momentum or spin of 0.9965, 0.99616 and 0.995. Event horizon area was calculated with the expression

$$A = \frac{4\ell_p^2 S}{k}; \quad \ell_p = \sqrt{\frac{G\hbar}{c^3}} \tag{15}$$

3 Results and Discussions

Modeling results by means of the metric proposal, Eq. 1, for the study of the behavior of the collapsed mass, are shown through Tables 1, 2 and 3.

In Table 1, results of mass, temperature, initial power emissions, and the black hole evaporation time to diverse angular positions are shown by means of: (i) The spatial coordinate called true anomaly associated with the orbit, ϕ, to which at the same time, is associated with different energy types (kinetic, potential, internal); and (ii) The arbitrary starting anomaly, ϕ_0.

For a hole mass of 1.481411892×10^6 g, the temperature increases to $6.750317082 \times 10^{19}$ K, which is indicative of a high degree of entropy in the event horizon of the hole (see Table 1). Then the Hawking effect takes importance. This evidence that although the hole seems naked, is actually wrapped by his event horizon which shining by the radiation emitted it as in [10].

Thus, the black hole radiates, at the cost of losing its own energy or, what is equivalent, its mass. When temperature increases, the mass of black hole decreases, and the radiation is more intense. Consistently, the mass diminishes more each time and faster, until the black hole evaporates completely in a true explosion. The black hole emits more than it absorbs, and thereby looses mass.

For mass of 0.I and 0, infinite temperatures are observed because the hole lost its mass. The quantity Float (∞) for temperature is used to indicate a floating-point value that is too large to be otherwise represented.

The temperatures of the collapsed mass in angular positions of 0 to $\pi/2$ in Table 1, as well as in the huge mass before collapse ($5.027399326 \times 10^{-8}$ K calculated with Eq. 7) are greater than absolute zero, which supports the third law of thermodynamics.

Also, for the position $\phi = \pi$ (energetic state in Table 1), the temperature of –Float(∞) is an infinitely small quantity above the absolute zero.

This unexpected value that apparently indicates absence of quantum effect in the hole, seems to indicate that the position measurement or energy state was

Table 1 Mass, temperature, emissions power and black hole evaporation time without charge and angular momentum or spin determination

ϕ_0 (rad)	ϕ (rad)	M (g)	T (K) Eq. 7	P (W) Eq. 13	t (s) Eq. 8
0	0	1.48×10^6	6.750×10^{19}	1.624×10^{20}	573.323
−0.04	0.39×10^{-135}	0.I	−Float(∞)I	−Float(∞)	−0.I
-1.11×10^{243}	6.63×10^{-14}	0	1/0 = +∞	+∞	0.
-1.1×10^{270}	$\pi/2$	0	1/0 = +∞	+∞	0.
-1.7×10^{270}	π	−0.	−Float(∞)	Float(∞)	−0.
-6.85×10^{270}	2π	Undefined	Undefined	Undefined	Undefined

Table 2 Temperature and superficial gravity of the black hole with mass, charge and angular momentum calculation

M (Solar masses)	T (K) Eq. 11	k (cm s^{-2}) Eq. 12
7.45×10^{-28}	$3.35 \times 10^{-7} - 2.02 \times 10^{38}$ I, $\mid T \mid = 2.02 \times 10^{38}$	$5.19 \times 10^{16} - 3.12 \times 10^{61}$ I, $\mid \kappa \mid = 3.12 \times 10^{61}$
0.I	−Float(∞)I	−Float(∞)I
0	1/0 = + ∞	Float(∞)I
0	1/0 = + ∞	Float(∞)I
−0.	−Float(∞)	−Float(∞)I
Float(undefined) + Float(undefined)I	*Float(undefined) + Float(undefined)I*	*Float(undefined) + Float(undefined)I*

Table 3 Calculation of black hole temperature with charge, angular momentum and mass for other angular positions or energetic states

ϕ_0 (rad)	ϕ (rad)	M [g]	T (K), Eq. 11
1.712E + 270	$-\pi$	−0.	−Float(∞)
1.070e + 269	$\pi/4$	0.	Division by zero
9.631e + 269	$3\pi/4$	Float(undefined) + Float(undefined)I	Float(undefined) + Float(undefined)I
3.852e + 270	$3\pi/2$	0.	Division by zero

carried out in a part where the two horizons disappear, leaving apparently the naked singularity. In this case, a prior study [10] shows that, for the mass of this hole complies the relation $G^2M^2 \ll GQ_I^2 + a^2c^2$ according to the first and second thermodynamics laws. Besides, in this case, the superficial gravity is infinite (−Float(∞)), so this physical process does not produce a naked singularity according to the third law of thermodynamics, which confirms the natural law of cosmic censorship. Another evidence of the existence of the quantum effect in the hole, is the presence of the emissions of the event horizon that are shown by means of the graphic of entropy of Fig. 1 in which $y = S$, where we can observe that for determined positions, the quantum effect is important, while in a rank of

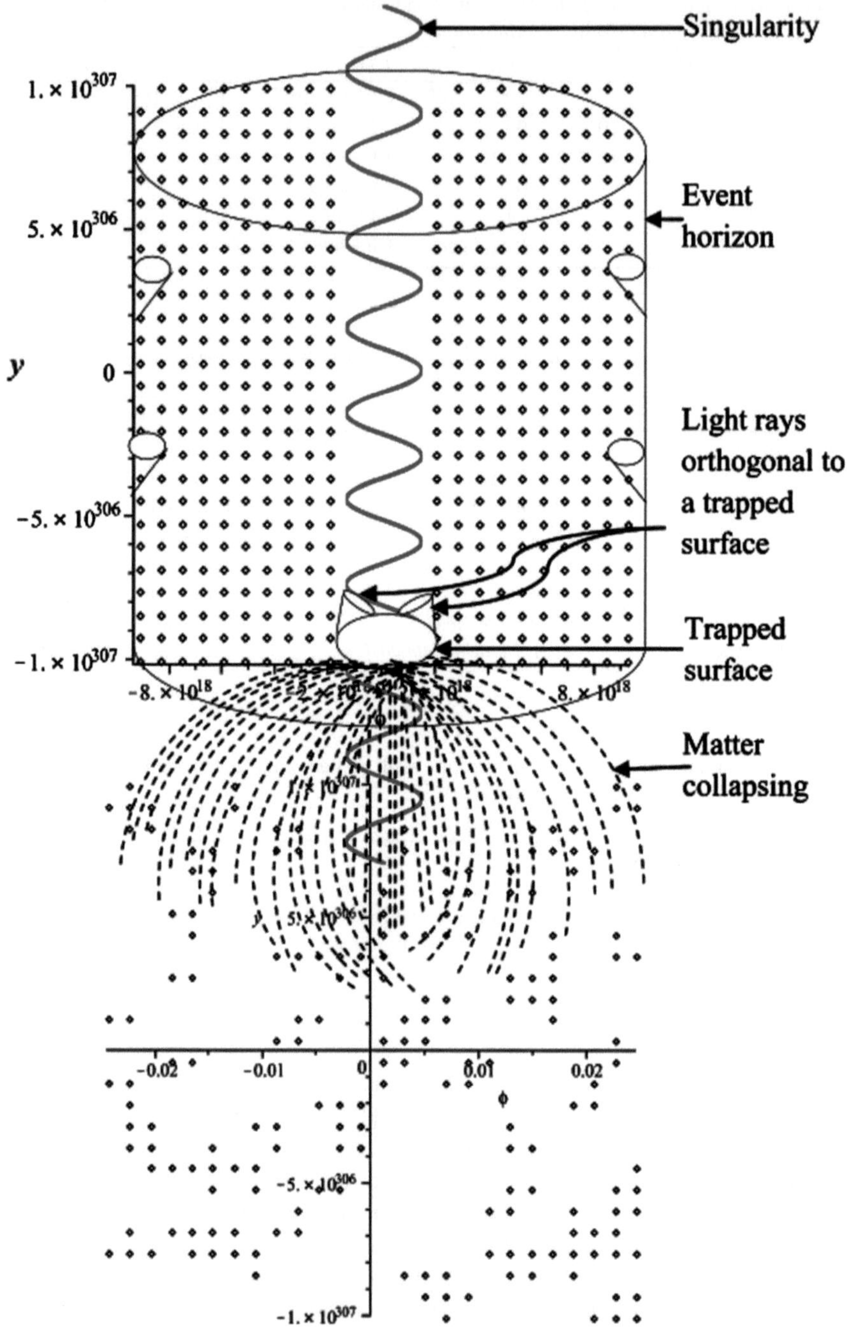

Fig. 1 Entropy ($y = S$) of black hole corresponding to its event horizon of cylindrical symmetry, which illustrates a trapped surface that implies the existence of a singularity and whose area is not reduced along the light rays that are initially orthogonal to it. The solution of Einstein's equations that describe the black hole, allow a glimpse into a collapsing dust cloud, inside which there is a singularity that is not visible from the outside, since it is surrounded by the event horizon

approximately $-2 \times 10^{18} \leq \phi \leq 2 \times 10^{18}$ radians the quantum effect seems not to exist. In fact, the entropy is quantized in that range [10].

Thermodynamically (third law), the temperature of absolute zero is unattainable, therefore, the temperature of π energy state in Table 1 is above this reference value. The entropy (or disorder) of the hole would be null in the absolute zero. For the time being, the unique thing that can be argued is that the mass presents strange properties when is chilled to very low temperatures.

In π energy state, the angular momentum is bigger than the mass of the hole (0.99616 > −0.) [10], according with the predictions of Petters and Werner [16] for naked singularity.

Figure 1 shows a powder cloud in the collapse of the black hole similar to that of Oppenhieimer and Snyder [17] that illustrates a trap surface. This model, unlike Oppenhieimer and Snyder's model, possesses a trapped surface that corresponds to a surface whose area doesn't decrease along the rays of light that are initially perpendicular to it.

The solution of the equations of Einstein that describe the black hole, allows us to glimpse a cloud of powder collapsed with a singularity hidden by the event horizon.

You can infer, as made by Roger Penrose in 1965 [18] that the existence of a trapped surface implies the existence of a singularity on the base of reasonable suppositions of causation supposing a spherical symmetry.

Figure 2 has three graphs in 2D for: (a) internal event horizon radius, (b) external event horizon radius and (c) both radii, all in centimeters for three values of the angular momentum calculated with Eq. 14. From these graphs, we can conclude that there is a naked singularity, due to the distance between both radii that is of the order of 6.644×10^{-11} cm, which can be verified in Fig. 2c. Until this moment, this singularity alters the well-known theoretical physics. Also, the area of event horizon shown in Fig. 3 calculated with Eq. 15, allows to confirm the nakedness of the hole, because is of the order of 10^{-51} cm^2.

Finally, when the hole has mass Float(undefined + Float(undefined)I, it has an undefined temperature. In computing, floating point describes a system for representing numbers that would be too large or too small to be represented as integers. The quantity undefined in Table 1 for the position 2π comes from the math expression: Float(undefined) + Float(undefined)I, where undefined $= \infty/\infty$.

Table 1 also shows that when the black hole has a mass of 1.4814119×10^6 g, equivalent to an energy of $1.3332707028 \times 10^{23}$ J, this could be liberated with a initial power of $1.62374944711 \times 10^{20}$ W. When the mass is null and indefinite, the power of the emissions of the horizon is infinite or indeterminate and indefinite, respectively, as was to be expected.

For different angular positions (states of energy), the time of evaporation of the black hole is zero, almost zero, an imaginary number or is undefined, as can be appreciated in Table 1. For a black hole with a mass of a hundred million tons (1×10^{11} kg), evaporation time results in about 2.66×10^9 years, comparable to the universe age (10^{10} years). It is possible that black holes with this characteristics

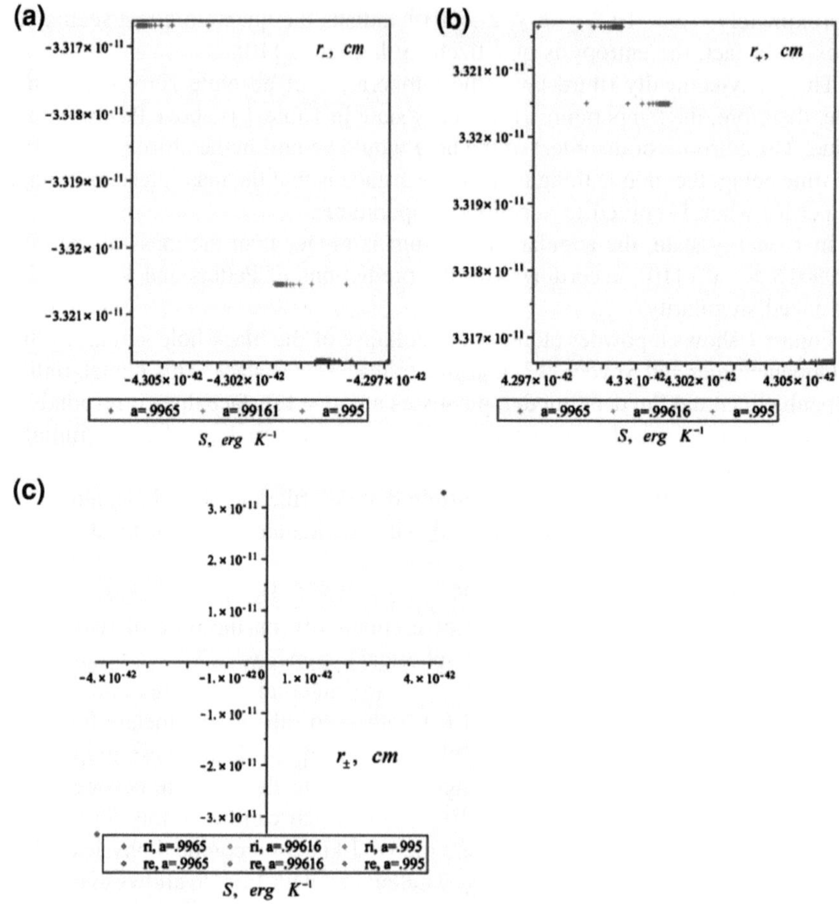

Fig. 2 Radius (**a**) of the inner horizon, (**b**) the outer horizon (**c**) of both horizons, as a function of the entropy of the hole, S, for three values of angular momentum

had form with relatively small masses in the first instants of the Universe. A fundamental black hole that was born in those remote epochs would be found currently exactly in the final phase of evaporation. Due to its highest temperature, such black hole would be observed probably as an intense source of gamma rays, produced by its explosion. If this turned out to be certain, we can see for the hole studied here, with a mass approximately of 67,503 times less than 1×10^{11} kg, that currently it has evaporated (evaporation times of 0., −0. and 573.32 s), but, the results of undefined and imaginary time in Table 1 allow us to show black hole radiations for a process with undefined evaporation.

Collapsed mass of 1×10^{11} kg at $1.481411892 \times 10^{6}$ g may have evaporated in 9.999961×10^{9} years with emissions indefinitely that shine through the entropy of event horizon (see Fig. 1).

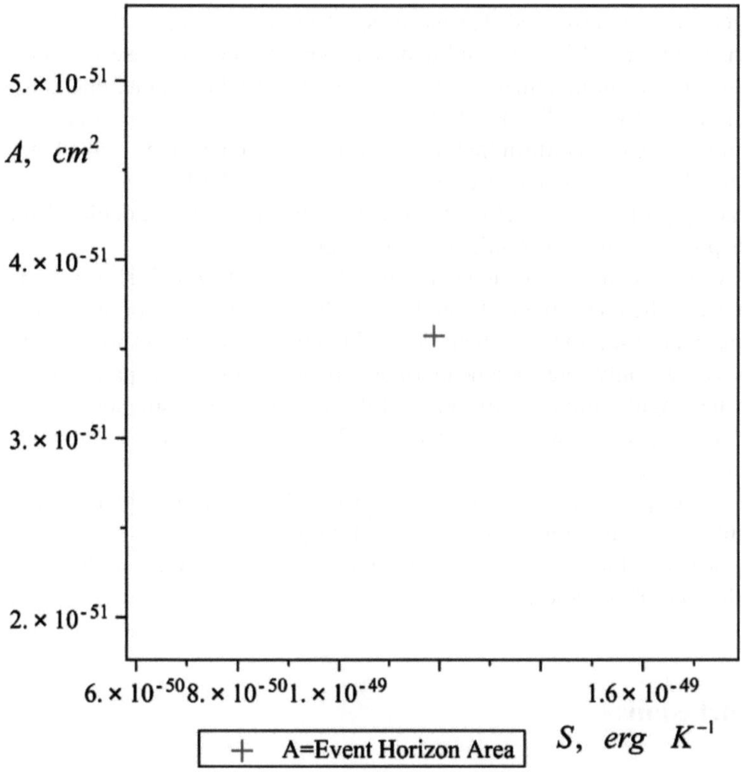

Fig. 3 Magnitude of the event horizon area of black hole with respect to its entropy

The fact that a black hole have a different temperature from the absolute zero (Table 1) implies a precise relation between the entropy and the area of the horizon. The temperature and the entropy of a black hole are due to purely quantum effects [7, 8].

Finally, we indicate that the evaporation and eventual disappearance of a black hole does not contradict the second law of thermodynamics: though the area—entropy—of the hole diminishes, the radiation produced possesses a very high entropy, therefore, the total entropy increases, which confirms the existing radiation in the last phases of the evaporation of the hole [7, 8].

Temperature and superficial gravity of the black hole were calculated with charge, angular momentum and mass for three values of angular momentum, *a*, to know: 0.9965, 0.99616 and 0.995. The results for $a = 0.99616$ are the ones shown in Table 2 for the same cases of Table 1, only that in Table 2 the conversion from grams to solar masses were made for the hole mass. Here, the fact is confirmed that small primordial black holes would emit more than they absorb, and thereby loose mass.

In general, it is observed that the mass of the black hole in the course of the time has a superficial infinite, and undefined gravity. For the case in which κ is a real number plus an imaginary number, taking its absolute value, that is to say, if $\kappa = 5.194164713 \times 10^{16} - 3.124791380 \times 10^{61}I$, then, to $|\kappa| = 3.124791380 \times 10^{61}$ cm s^{-2}, we can confirm that the law zero of the mechanics of the black holes is complied where κ is likewise infinite anywhere in the horizon of a black hole independently of the time. Hole mass and temperature were calculated for other angular positions and the results are shown in Table 3.

It is verified that the hole temperatures obtained with Eq. 7, shown in Table 1, are similar to the ones obtained with Eq. 11, shown in Table 3 for different angular positions, which signifies that their determination is independent from the fact that hole possesses only one parameter (mass) or possesses three parameters (mass, charge and angular momentum). From Tables 1 and 3, the same temperatures can be seen when $\phi = \pi$ and $\phi = -\pi$; $\phi = \pi/2$, $\phi = 6.628 \times 10^{-14}$, $\phi = 3\,\pi/2$ and $\phi = \pi/4$; $\phi = 2\pi$ and $\phi = 3\pi/4$.

The model presented here, deserves a detailed study of the quantum geometry of the black hole horizon using loop quantum gravity.

This leads to the computation of quantum gravity corrections to the entropy and radiation of the hole [11].

4 Conclusions

The behavior of the mass (1.98892×10^{30} kg) collapsed to a black mini-hole studied here, corresponds to an evaporated hole, exhibiting Hawking's radiation indefinitely.

From the calculation of some of the possible evolutions of the last scenes, during the time of the process of evaporation of the micro black hole, with some of its possible angular positions (states of energy), the postulated metric allows to predict the entropy transportation phenomenon, in very approximate form to the scientific predictions carried out by Hawking, Penrose, Bekenstein, Werner and Petters and Oppenheimer and Snyder, as well as the phenomena of transportation by means of enthalpy and other energy types, according to the system of coordinates used.

The results of the mass, the temperature, the initial emissions power, the superficial gravity and the black hole evaporation time for diverse angular positions or energy states, show evidence of a fundamental black hole evaporated, that emits more matter than the one it absorbs loosing its mass. Given time results in complex and undefined numbers permit to argue that, although the hole must have evaporated since 10^9 years ago, its emissions could continue in indeterminate form for indefinite time. The entropy betrays the presence of the hole, while the event horizon hides the singularity exposing the naked hole, confirming thus, the natural law of cosmic censorship.

Of the analyzed cases, the cases when $\phi = \pi$ and $\phi = -\pi$, are against the natural law of cosmic censorship. The angular momentum is greater than the mass, so it is said that the hole is naked, but the surface gravity keeps it wrapped, validating this way the third law of the thermodynamics.

The relativist metric proposed here for the calculations of energies, enthalpy and entropy predicts similar results for the cases of one parameter (mass) and three parameters (mass, charge, angular momentum) of the mini-hole in the continuous space–time. Except for the only evidence of mentioned nakedness, in all the cases the three laws of thermodynamics are confirmed, as well as law zero of the black holes mechanics. It is verified that the temperature and the entropy of the hypothetical mini-hole are due to purely quantum effects and the results obtained show a fundamental black hole similar to the ones that arose in the first instants after the Big Bang.

Black hole's temperature increases as it radiates away mass. The rate of temperature increase is infinite, with the most likely endpoint being the dissolution of the black hole in a violent burst of gamma rays. A complete description of this dissolution requires a model of quantum gravity, however, as it occurs when the black hole approaches Planck mass ($2.17644(11) \times 10^{-8}$ kg) and Planck radius ($1.616252(81) \times 10^{-35}$ m). The results presented here are estimates made during the evaporation time, when the event horizon of the black hole emits Hawking radiation.

In this study, a strange value for the hole temperature was obtained when mass in state of collapse had $-0.$ kg for $\phi = \pi$ and $\phi = -\pi$, which is indicative of null entropy and violation of the third law. In the π energy state, the hole has an infinite surface gravity, which confirms the cosmic censorship law, which has not been proven scientifically.

References

1. Carr, B.J., Hawking, S.W.: Links between primordial black holes and the early universe. Mon. Not. Roy. Astron. Soc. **168**, 399 (1974)
2. Barrow, J.D., Carr, B.J.: Formation and evaporation of primordial black holes in scalar-tensor gravity theories. Phys. Rev. D. **54**, 3920 (1996)
3. Page, D.: First detailed studies of the evaporation mechanism. Phys. Rev. D. **13**, 198 (1976)
4. Teitelboim, C., Zanelli, J.: Black Holes and the Structure of the Universe. World Scientific Publishing Co. Pte. Ltd., Singapore (2000)
5. Hawking, S.W.: The evaporation of primordial black holes. In: Contribution to the Proceedings of the 3rd RESCUE International Symposium, pp. 185–197 (1998)
6. Brout, R., Massar, S., Parentani, R., Spindel, Ph.: Hawking radiation without trans-Planckian frequencies. Phys. Rev. D. **52**, 4559 (1995)
7. Bekenstein, J.D.: Sci. Am. **289**(2), 61 (2003)
8. Bekenstein, J.D., Schiffer, M.: Int. J. Mod. Phys. **1**, 355 (1990)
9. Dermer, Ch.D., Menon, G.: High Energy Radiation from Black Holes: Gamma Rays, Cosmic Rays, and Neutrinos. Princeton University Press, Princeton (2009)

10. Corral-Bustamante, R.L.: Review talk given at the 6th International Conference on Diffusion in Solids and Liquids DSL-2010: Cosmic Censorship!: Thermal Transport in a "Naked" Black Hole, Paris, France, 5–7 July 2010
11. Barrau, A., Grain, J., Alexeyev, S.O.: Gauss–Bonnet black holes at the LHC: beyond the dimensionality of space. Phys. Lett. B **584**, 14 (2004)
12. Barrau, A., Féron, C., Grain, J.: Astrophysical production of microscopic black holes in a low-Planck-scale world. Astrophys. J. **630**, 1015 (2005)
13. Barrau, A.: Experimental searches for primordial black holes thanks to the emitted antimatter. Astron. Astrophys. **388**, 676 (2002)
14. Barrau, A.: Antideuterons as a probe of primordial black holes. Astron. Astrophys. **398**, 403 (2003)
15. Hacyan, S.: Los hoyos negros y la curvatura del espacio-tiempo. Biblioteca digital-ILCE. http://bibliotecadigital.ilce.edu.mx/sites/ciencia/volumen1/ciencia2/50/html/sec_8.html (1998–2009). Accessed 10 July 2009
16. Werner, M.C., Petters, A.O.: Magnification relations for Kerr lensing and testing cosmic censorship. Phys. Rev. D **76**, 064024 (2007)
17. Oppenheimer, J.R., Snyder, H.: On continued gravitational contraction. Phys. Rev. **56**, 455–459 (1939)
18. Penrose, R.: Gravitational collapse and space-time singularities. Phys. Rev. Lett. **14**, 57–59 (1965)

Calendering Process Numerical Simulation with an Elastoviscoplastic Polymer Blend Model

Baya Madani and Ahmed Ouibrahim

Abstract The actual earth environment constitutes the major world preoccupation. This involved situation is linked to several parameters among which the fossil energy consumption. Thereby, the recycling of material waste takes on great importance. For example, brand new polymers are then substituted by polymer blends using appropriate wasted materials. With this idea, we propose here a rheological model having elastoviscoplastic properties. This model is elaborated as a result of a blend between a material having a viscoplastic behaviour modelled by a generalized power law type and a viscoelastic material the behaviour of which is provided by Oldroyd B type. As an application, we use then this model in the case of the calendering processing of (elastoviscoplastic) sheets of a finite thickness through two counter rotating cylinders (a two roll mills). Numerical analysis had to be used to solve the whole associated equations. The conducted numerical simulations provided then the pressure distribution of the flow field, the separating force and the power required for the calendering processing at different values of the relaxation and retardation times for the viscoelastic model and at different values of the consistency and the power law index for the viscoplastic model. The effect of concentration of the two blend components is considered as well.

Keywords Calendering · Elastoviscoplastic model · Polymer blend · Rheology · Two roll mills · Viscoelastic

B. Madani · A. Ouibrahim (✉)
Laboratoire d'Energétique Mécanique et Matériaux-Université de Tizi-Ouzou, Tizi-Ouzou, Algeria
e-mail: ouibraah@live.fr

B. Madani
Departement Maintenance, Faculté des Sciences de l'Ingénieur, Université de Boumerdès, Boumerdès, Algeria
e-mail: hnmb3@yahoo.fr

A. Öchsner et al. (eds.), *Materials with Complex Behaviour II*, Advanced Structured Materials 16, DOI: 10.1007/978-3-642-22700-4_30,

Nomenclature

D_{ij}	Components of the deformation rate tensor
H_E	Exit sheet thickness (m)
H_F	Entry sheet thickness (m)
X_E	Exit sheet distance (m)
X_F	Entry sheet distance (m)
H_0	Half of nip opening (m)
K	Material consistency (Pa s^{1-n})
n	Fluidity index
p	Pressure (Pa)
R	Roll radius (m)
U	Tangential velocity of the roll (m/s)
$V(x,y)$	Velocity of the blend melt in the two rolls mill (m/s)
u, v	Velocity components in x and y direction (m/s)
w	Width of the sheet (m)
τ_{ij}	Components of the deviatoric stress tensor (N/m^2)
λ	Relaxation time (s)
λ_ρ	Retardation time (s)
μ	Newtonian viscosity (Pa s)
μ_α	Apparent viscosity (Pa s)
$\dot{\gamma}$	Shear rate (s^{-1})
ρ	Volumic mass of the blend melt (kg/m^3)
α	Concentration of the viscoelastic part in the blend
Ω	Rotation of the rolls (rad/s)

1 Introduction

Global warming and the environment protection are actually the main problems to be urgently handled. They are related to many factors including the consumption of energy fossil. As an issue of possible solutions participating to these two problems is the recycling of material wastes leading for example to the substitution of brand new polymers by polymer blends to be used in many industrial processes such as calendering.

On the other hand, some of these polymer blends, because involving several materials having different mechanical properties, may therefore need brand new rheological models to take into account their whole properties.

Calendering is a continuous process used in many industries such as the production of sheets or films of uniform thickness of paper, rubber, steel, etc.

The process was extensively studied by many researchers. Starting with Ardichvili [5], the work was extended to Newtonian and Bingham plastics by

Gaskel [8]. His analysis was based on the assumption that the diameter of the rolls was large enough compared to the gap between the two rotating rolls.

Most models proposed in the literature are based on Gaskel's model. It is a one dimensional, a rather restrictive model, and to use the model, the location X_E, where the rolls come in contact with the polymer, must be known; which is tantamount to the knowledge, a priori, of the exiting sheet thickness $2H_E$. Mc. Kelvey [11] pointed out that X_E must be considered to be an experimentally determined parameter of the model.

Following Gaskel's work, a great deal of effort was invested by numerous workers in the field to improve his model. Most of this effort, however, concentrated on solving basically the Gaskell model with the original realistic constitutive equations and attempts to account for nonisothermal effects.

Mc. Kelvey [11] and Brasinsky et al. [6] extended the model to power law fluids; while Alston and Astill [4] investigated fluids whose shear rate dependent viscosity can be represented by a hyperbolic tangent function. Flow of viscoelastic fluids in the roll geometry was considered by Paslay [14] who obtained an approximate solution essentially for an Oldroyd's fluid model with three rheological constants. He analysed the interrelation of the parameters of the constitutive equation with the flow kinematic but he neglected the normal stress in the equation of motion. Tokita and White [15] related experimental observations on milling of elastomers to rheological parameters of a second order Rivlin-Ericksen asymptotic expansion fluid and pointed out the significance of the Deborah and Weissenberg numbers in milling and calendering. However, velocity and pressure were not obtained by them. Chong [7] analysed a power law model fluid, a three constants Oldroyd fluid and a modified second order Rivlin-Ericksen equation; the velocity profile for the Oldroyd fluid cannot be obtained analytically. Therefore, he [7] obtained an approximation pressure distribution by assuming Newtonian flow kinematics, and he analysed the flow pattern with the Rivlin-Ericksen flow in terms of dimensionless groups only. He also measured the separating force at various calendering conditions of cellulose acetate.

Calendering defects with polyvinyl chloride (PVC) were also studied by Agassant and Avenas [1] using Newtonian and power law models. All these analyses are based on lubrication approximation theory of Reynolds. Lifting this assumption leads to a two-dimensional analysis, as was done by Mitsoulis and Safou [12, 13] and Agassant and Espy [2]. These works have shown interesting results with intricate patterns dominated by large vortices in the melt bank before the rolls, found both experimentally and computationally.

However, most of the fluid used in calendering only exhibit either pseudo plastic (inelastic) and viscoplastic or viscoelastic; but never all together elastoviscoplastic. Moreover, besides the increase of the equations complexity to be solved, it seems there is no full numerical solution available for a viscoelastic calendering problem, while materials such as polymeric fluids and foods processed in calendars, mostly exhibit viscoelastic behaviour.

In the present work, we gather all together these properties by proposing an elastoviscoplastic rheological model. We then use this model by undertaking here a 2D analysis of calendering processing of a blend melt displaying the above mentioned mechanical properties.

2 Rheological Model

2.1 Constitutive Equations

As a result of a blend of recycled materials having different mechanical properties, an elastoviscoplastic rheological model has been developed by considering together a viscoplastic model of a generalised power law type and a viscoelastic model of Oldroyd-B type, at a given concentration α. This rheological model has the following constitutive equations, Madani and Ouibrahim [9]; Madani [10]:

$$\tau_{ij} + \lambda \frac{D\tau_{ij}}{Dt} = \left[2\alpha\mu + (1-\alpha)\mu_a + \lambda(1-\alpha)\frac{D\mu_a}{Dt}\right] D_{ij} + [2\alpha\mu\lambda_r + \lambda(1-\alpha)\mu_a]\frac{DD_{ij}}{Dt} \tag{1}$$

where $\frac{D}{Dt}$ is the convective or Oldroyd derivative, D_{ij} the shear rate tensor components, α the viscoelastic concentration in the blend, λ and λ_r the relaxation and the retardation time, respectively, μ the Newtonian viscosity of the viscoelastic part, while the apparent viscosity μ_a is given by:

$$\mu_a = K\left(\frac{I_2}{2}\right)^{n-1} \tag{2}$$

in which K and n are the consistency and the flow index, respectively, I_2 is the second invariant of the shear rate tensor.

3 Application–Calendering Processing

3.1 Governing Equations

We now test the above rheological model in the case of a polymer blend submitted to a calendering processing.

A schematic of the associated geometry, two counter rotating rolls of radius R, is illustrated in Fig. 1.

The geometric parameters and variables involved in the model are included in Fig. 1 and they do not need any further description.

The variables X_E, and X_F, however, need to be defined: X_E, represents the distance at which the sheet separates from the rolls and it is known as the separation point or the exit sheet distance, and X_F represents the entry sheet distance.

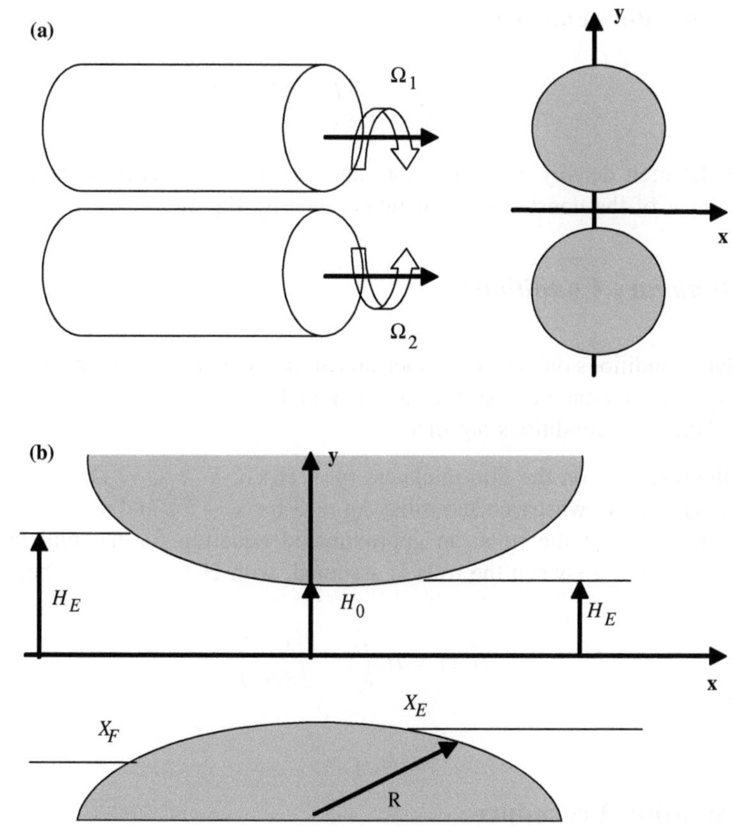

Fig. 1 Schematic representation of calendering processing **(a)**, with associated geometric parameters **(b)**

The motion of the blend melt is assumed to be 2D, incompressible, isotherm and stationary.

So that we have the following set of governing equations as described in below.

3.1.1 Continuity Equation

$$\frac{\partial V_i}{\partial x_i} = 0 \tag{3}$$

with the velocity $\vec{V} = \{u(x,y), v(x,y)\}$

3.1.2 Momentum Equations

$$\rho \frac{dV_i}{dt} = -\frac{\partial p}{\partial x_i} + \frac{\partial \tau_{ij}}{\partial x_j} \tag{4}$$

ρ is the melt density, p the pressure and τ_{ij} the components of the deviatoric stress tensor of the constitutive equations given by Eq. 1.

3.2 Boundary Conditions

Boundary conditions on the roll surface and at the inlet and outlet are specified. No slip conditions on each roll surface are assumed.

The boundary conditions are then:

- On the rolls and on the film thickness (y = H(x)): V = U = ΩR
- Upstream and downstream the rolls: Δp = 0 for $x \leq X_F$ and $x \geq X_E$
- On both sides of the rolls: an approximated equation for the film thickness *H*(*x*) prevailing between the rolls is assumed, with H(x) < < R, to be:

$$H(x) = H_o\left(1 + \frac{x^2}{2RH_o}\right) \tag{5}$$

4 Resolution Procedure

The above involved equations were numerically solved. The transport equations for mass and momentum were solved with the help of the Finite Volume software Fluent 6.3 completed by appropriate user defined functions (UDF) to take into account the formulation of the constitutive equations and then the associated contributions in the momentum equations.

5 Results and Discussion

5.1 Newtonian Material

In order to check the validity of the proposed model, we compare first our numerical solution with the results obtained by Agassant and Hinault [3] in the case of a Newtonian model with the same dimensions and data they used.

To simulate the Newtonian behaviour with the rheological model here presented, we set in the above constitutive Eq. 1, $\alpha = 0$ (no viscoelastic fluid) and n = 1 (value of the flow index corresponding to a Newtonian fluid).

As Agassant and Hinault [3], we have here used the following data: $H_o = 0.2$ mm, $U = 0.15$ m/s (tangential velocity of the rolls), $\mu_a = K = \mu = 10^3$ Pa.s, and $\rho = 10^3$ kg/m^3

Pressure distributions along the center line are shown in Fig. 2a for different values of the attachment of the entry sheet relative thickness H_F/H_o, a quite significant factor on the pressure. The results of Fig. 2a are all in very good agreement with those obtained by Agassant and Hinault [3] represented in Fig. 2b. The pressure increases from the entry to reach a maximum located close to the minimum gap after which it decreases drastically, obviously at the beginning of the diverging region. The magnitude of the pressure, as well as the maximum reached, increases with the increase of the relative sheet thickness.

5.2 Power Law Material

For the power law analysis, we use the same procedure as indicated above for the Newtonian material with the only difference that the flow index $n \neq 1$. The pressure distributions corresponding to different values of n are shown in Fig. 3a.

It is important to notice the significant decrease of the pressure by decreasing the flow index n of the power law. The pressure is as low as the material is shear thinning. These results are also in very good agreement with those of Agassant and Hinault [3] shown in Fig. 3b.

5.3 Elastoviscoplastic Material

5.3.1 Elastic Contribution in the Blend

We are here mainly interested in the viscoelastic response of the blend via the influence of the viscoelastic concentration α. Simulations were carried out for several values of the concentration α. As previously, the same geometrical and kinematic parameters were used while the rheological ones for the blend are as follows: $\mu = 10^3$ Pa.s, $\lambda = 0.01$ s, $\lambda/\lambda_r = 0.4$, $\rho = 10^3$ kg/m^3, $K = 10^3$ Pa.s^{1-n} with $n = 0.7$.

Figure 4 represents the pressure distribution for various values of the concentration α. It can be seen in Fig. 4 that increasing the viscoelastic contribution from $\alpha = 0$, to $\alpha = 1$ leads to increase the magnitude of the pressure as well as its maximum.

The value of $\alpha = 0$ corresponds to a power law material pressure distribution. It is then interesting to notice, by comparison to Fig. 3a, that we obtain exactly the same value of the maximum pressure for the corresponding flow index $n = 0.7$. The case of $\alpha = 1$ represents a purely viscoelastic material, no longer a blend.

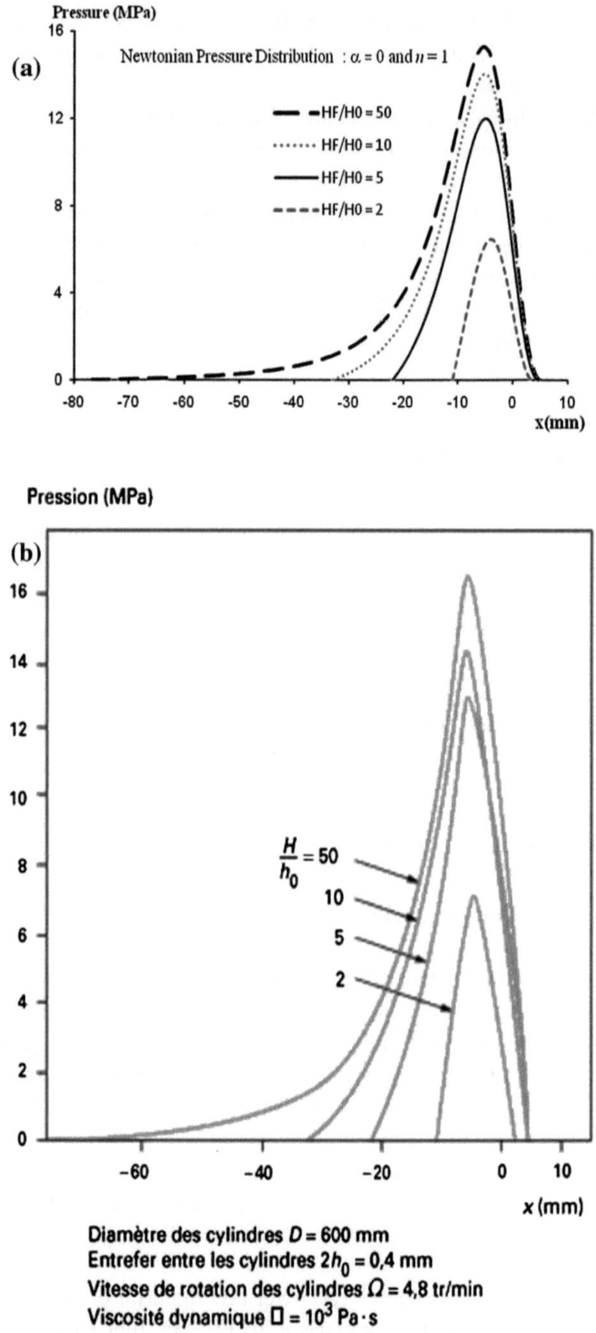

Fig. 2 **a** Pressure distribution between the inlet and the outlet for different values of the entry relative sheet thickness H_F/H_O, for a Newtonian material ($\alpha = 0$ and $n = 1$). **b** Newtonian pressure distribution between the inlet and the outlet following Agassant and Hinault [3]

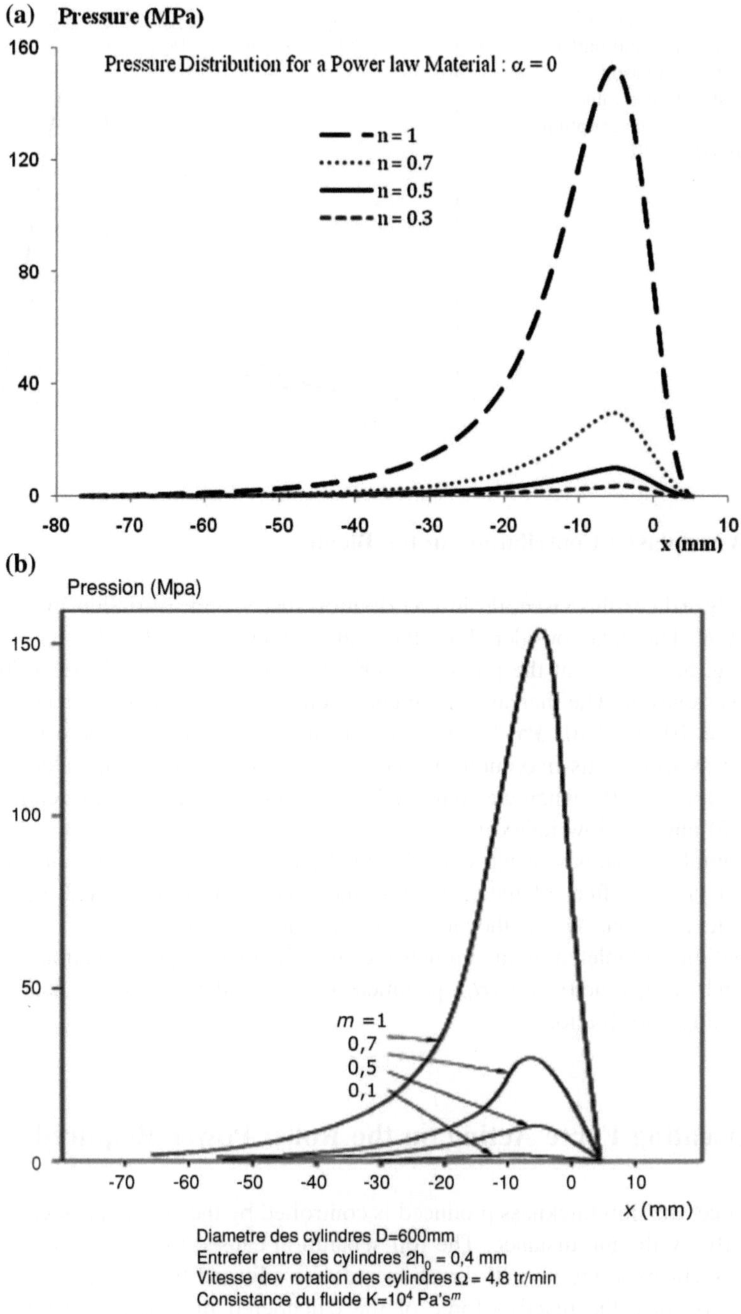

Fig. 3 **a** Pressure distribution between the inlet and the outlet for a Power law material ($\alpha = 0$).at different values of the flow index n **b** Pressure distribution between the inlet and the outlet of a Power law material following [3]

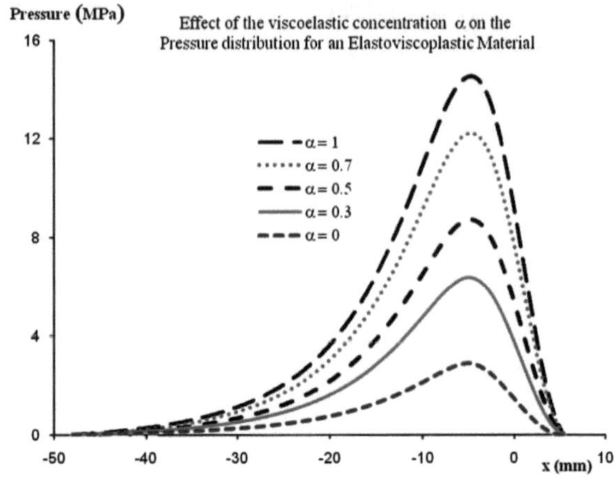

Fig. 4 Pressure distribution between the inlet and outlet for a elastoviscoplastic material at different values of the viscoelastic concentration α with $H_F/H_O = 20$

5.3.2 Viscoplastic Contribution in the Blend

We analyze here the viscoplastic contribution just via the variation of the consistency K. The data considered are those given above in Sect. 5.3.1.

In Fig. 5, we present the pressure distribution obtained for different values of the consistency K. The increase of the consistency of the viscoplastic contribution from $K = 10^3$ to 2.10^3 Pa.s$^{1\text{-}n}$ ($n = 0.7$) leads to a significant increase of the pressure as well as its maximum. Of course, we observe the same values of the pressure for $K = 10^3$ when we compare Figs. 4 and 5, at the same concentration α ($\alpha = 0.3$) and the flow index n ($n = 0.7$).

A complete analysis, when considering the blend melt, with the elastic contribution via the effect of the relaxation and retardation times λ and λ_r, the viscoplastic contribution via the flow index n and the consistency K, these two contributions completed by the influence of the kinematic (speed rotation Ω of the rolls) and the geometric (H_F/H_o) parameters, is considered in more details in a forthcoming contribution.

6 Separating Force Acting on the Rolls: Power Required

In practice, the film thickness produced is controlled by the geometry of calenders, especially by the nip distance. The roll separation caused by the stress acting on the roll surfaces in the normal direction must be adjusted by the loading force in order to balance the reactive force of the calendered fluid. The roll separating force can be calculated by integrating the product of the total normal stress, which is represented here by σ_{yy} following Fig. 1, to the roll surface area in contact

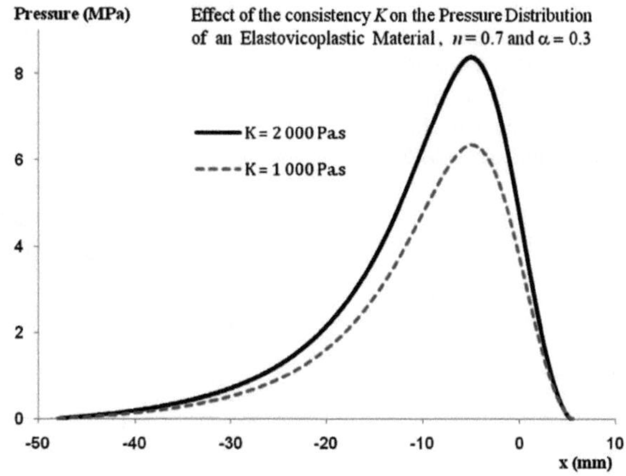

Fig. 5 Pressure distribution between the inlet and outlet for a elastoviscoplastic material at different values of the consistency K with $n = 0.7$ and $a = 0.3$ while $H_F/H_0 = 20$

with the melt. Then, the normal force F acting on the roll surfaces can be determined from the following relation:

$$F = -\int_{X_F}^{X_E} \sigma_{yy} dS \tag{6}$$

where

$$\sigma_{yy} = -p + \tau_{yy} \tag{7}$$

while

$$dS = wdx \tag{8}$$

is the elementary surface of the calendered sheet of width w. We consider here the force F per unit length (i.e. $w = 1$).

Thus, it follows that the power input into each roll can be calculated by integrating the product of shear stress σ_{xy} at the roll surface with the area of contact with the blend melt and the tangential velocity of the roll, that is to say:

$$P = \int_{X_F}^{X_E} \sigma_{xy} V dS \tag{9}$$

Figure 6 clearly indicates that the roller separating force, calculated from Eqs. 6 and 7, is affected by the concentration α of the viscoelastic part in the blend melt. More precisely, it can be noticed that this separating force: (a) increases with the concentration α and (b) this increase varies linearly with α. Moreover, the magnitude of this increase appears to be more pronounced with the increase of the entry relative thickness H_F/H_O; it increases drastically with the increase of H_F/H_O but slightly for high enough values of H_F/H_O.

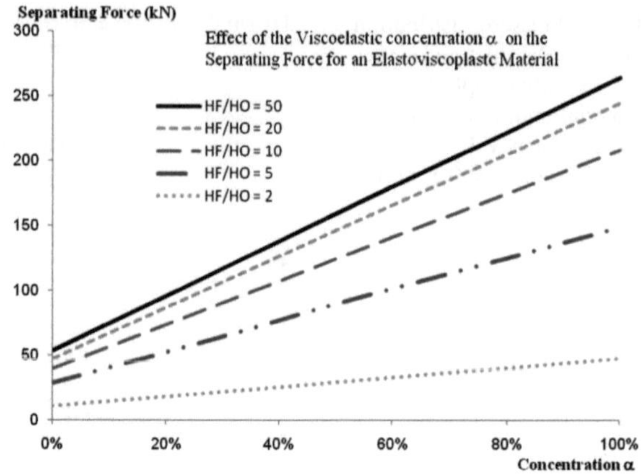

Fig. 6 Separating force versus viscoelastic concentration α for an elastoviscoplastic material at different values of the entry relative thickness H_F/H_O

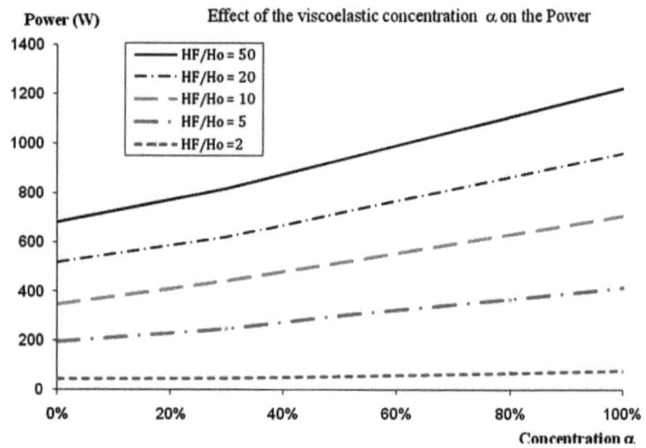

Fig. 7 Power versus viscoelastic concentration α for different values of H_F/H_O

In the case of the power required for the motion of each roll, Fig. 7, provided by the calculations with Eq. 9, also indicates that the power depends on the concentration α, and that it increases linearly with α with a magnitude and a slope increasing with the increase of the relative sheet thickness H_F/H_O.

7 Conclusion

A new rheological model for elastoviscoplastic polymer is here proposed.

It is elaborated, as a result of a blend of different materials governed for a part by a viscoplastic model of a generalized power law type and for the other part, at a given concentration α, by a viscoelastic model of Oldroyd B type.

As an application, we use this model in the case of calendering processing.

Numerical solutions have been obtained for this model for different values of its characteristic parameters using, for comparison purpose geometric and kinematic parameters found in literature. It is important to note that a finite volume software, Fluent 6.3, not especially devoted for a non Newtonian elastoviscoplastic model, has been adapted for our purpose via user defined functions (UDF).

This technique has been tested first, from this proposed model, for a Newtonian ($n = 1$ and $\alpha = 0$ in the model), then a power law model ($\alpha = 0$ in the model) to show that the obtained results are in agreement with those existing in the literature for Newtonian and power law models. This allows us to validate the numerical procedure and the obtained results.

In the case of the entire elastoviscoplastic model, the numerical simulation indicates that the concentration α of the viscoelastic contribution has a significant effect on the distribution of the pressure as well as on the separating force on the rolls.

As expected, the increase in the concentration α of the viscoelastic part of the blend leads to the increase of the magnitude of the pressure along the line between the rolls as well as the maximum reached by the pressure. Such an increase with α is also observed for the roll separating force and the power required which are seen to increase linearly with the concentration.

References

1. Agassant, J.F., Avenas, P.: Calendering of PVC: Prediction stress and torque. J. Macromol. Sci. Phys. **14**, 345–365 (1977)
2. Agassant, J.F., Espy, M.: Theoretical and experimental study of the molten polymer flow in the calender bank. Polym. Eng. Sci. **25**, 113–121 (1985)
3. Agassant, J.F., Hinault, R.: Calandrage, Techniques de l'Ingénieur,Référence AM3663. http://www.techniques-ingenieur.fr/basedocumentaire/materiaux-th11/plastiques-et-composites-ti100/calandrageam3663/3 (2001)
4. Alston, W.W., Astill, K.N.: Calendering of non newtonian fluid. J. Appl. Polym. Sci. **17**, 3157–3174 (1973)
5. Ardichvili, G.: An attempt at a rational determination of theambering of calender rolls, Kautschuk. **14**, 23–28 (1938)
6. Brasinsky, I., Cosway, H.F., Vall, C.F., Jones, R.C., Story, J.A.: Theoretical study of liquid-film spread heights in the calendering of newtonian and power law fluids. J. Appl. Polym. Sci. **14**, 2771–2784 (1968)
7. Chong, J.S.: Calendering thermoplastic materials. J. Appl. Polym. Sci. **12**, 191–212 (1968)
8. Gaskel, R.E.: The calendering of plastic materials. J. Appl. Mech. **17**, 334–336 (1950)
9. Madani, B., Ouibrahim, A.: Modélisation Rhéologique d'un Mélange Elastoviscoplastique via un Mélangeur Physique à Chaud. Conférence Internationale sur le Génie des Procédés, CIGP'07, Beja Octobre (2007)
10. Madani, B.: Étude Numérique d'un Mélange Polymérique dans un Mélangeur Physique à Chaud. Thèse de doctorat en cours, Université de Boumerdès, Algérie (2011)
11. Mc. Kelvey, J.M.: Polymer Processing. Wiley, New York (1962)
12. Mitsoulis, E., Safou, S.: Calendering pseudoplastic and viscoplastic fluids with slip at the roll surface. J. Appl. Mech. **73**, 291–299 (2006)
13. Mitsoulis, E., Safou, S.: Numerical of calendering viscoplastic fluids. J. New. Fluid. Mech. **157**, 77–88 (2008)
14. Paslay, P.R.: Calendering of viscoelastic fluids. Phd Dissertation, MIT (1955)
15. Tokita, N., White, J.L.: Milling behaviour of gum elastomers: experiment and theory. J. Appl. Polym. Sci. **10**(7), 1011–1026 (1966)

Condensation Issues in Ventilated Façades

Matthias Haase and Tore Wigenstad

Abstract The design of ventilated glass–glass facades in large buildings seems a promising technology to enhance energy efficient building design. In order to describe the optical and thermal heat transfer mechanism in glass–glass configurations with vertical airflow regimes radiation, conduction and convection models have been developed and validated by Arasteh et al. (ASHRAE Trans 95:2, 1989), Manz (Energy Buildings 35(3):305–311, 2003). In application to existing buildings, condensation on various surfaces in the construction is always a risk. A prediction model that takes weather conditions for a specific site into consideration is needed. In this work measured data of condensation on external window panes was used and a mathematical model for predicting condensation depending on air temperature, humidity and airflow regimes was developed by Thyholt (SINTEF Byggforsk, 2006). The validation shows very good agreement and gives confidence in using the model for further analysis of condensation times over the year in this type of façade. The model was implemented in a programme for heat transfer calculations and used to evaluate condensation and energy issues for various façade material configurations.

Keywords Double façade · Condensation risk · Renovation · Energy · Thermal comfort

M. Haase (✉)
Department of Architectural Design, History and Technology,
Norwegian University of Science and Technology, 7491 Trondheim, Norway
e-mail: Matthias.haase@sintef.no

M. Haase · T. Wigenstad
Department for Energy and Architecture, SINTEF Building and Infrastructure,
7465 Trondheim, Norway
e-mail: Tore.wigenstad@sintef.no

A. Öchsner et al. (eds.), *Materials with Complex Behaviour II*,
Advanced Structured Materials 16, DOI: 10.1007/978-3-642-22700-4_31,

1 Introduction

Energy usage for room heating, cooling and ventilation still accounts for more than one third of the total primary energy demand in the industrialized countries, and it is in this way a major polluter of the environment with CO_2 and greenhouse gas emissions. To successfully achieve the targets set out in the Kyoto protocols, it is necessary to identify innovative energy technologies and solutions for the medium and long term. They should facilitate the implementation and integration of low carbon technologies, such as renewable power generation devices within the built environment [1].

One focus has to be put on the energy efficient refurbishment of the existing building stock. Here, appropriate solutions have to be identified and possible technologies have to be developed that are integrated into the building. One possibility might be an advanced façade system. A lot of developments in façade design has focused on ventilated double façade systems for new buildings [2, 3]. However, there exists very little work on exploring the possibility of energy efficient refurbishment by applying a ventilated double façade system to an existing building.

1.1 What is a Double Façade System?

A double façade system consists of an exterior facade in glass added outside the original facade. The space is usually expected to be 50 cm deep, and usually holds a sunshade, and any other technical installations such as ventilation ducts, cooling installations or daylight systems.

The internal façade of new buildings can thus be made easier, or in the case of rehabilitation, they can be maintained or repaired by simple means. Ventilated double facades are often categorized by the type of ventilation principle they are designed for. The ventilation solution is important, because this greatly affects the energy consumption for heating, ventilation and cooling, comfort, air quality, sound insulation and fire safety [4].

1.2 Type of Double Façades

It is usual to divide double facades into five different categories as illustrated in Fig. 1:

- Exhaust air facade,
- Supply air facade,
- Static air buffer,
- External air curtain,
- Internal air curtain.

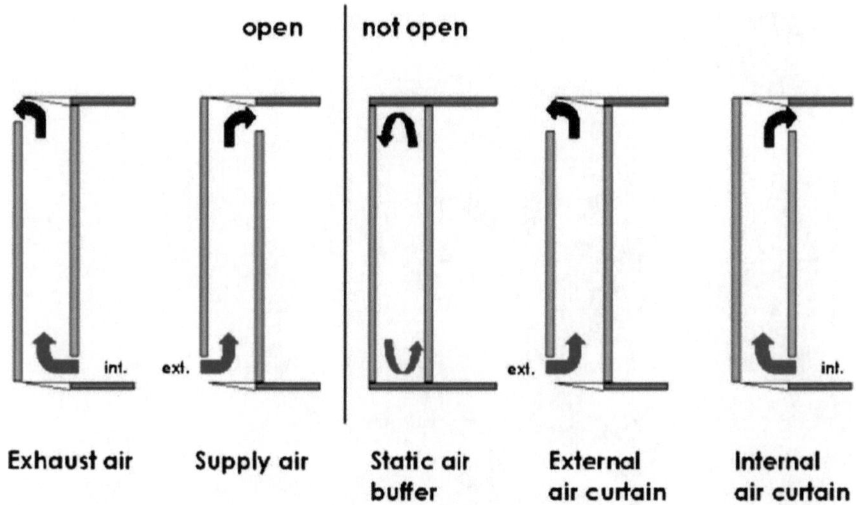

Fig. 1 Simple ventilation principle of double facades [5]

In addition there are various combinations of these.

Closed Buffer

For this type of double façade solution, there is no air exchange between the residence zone and the gap in the double façade. Exterior facades may have permanent or controllable openings. The double facade works here as a thermal buffer, and will reduce heating needs in relation to a single facade. It will also provide better sound insulation against the exterior. Ventilation of the residence zone and ventilation within the double facade are, in this case, two independent systems.

Open Buffer

In this concept, the exterior facade windows can be opened for ventilation or utilization of the space for preheating the ventilation air. If the free height of the double facade is large, i.e more than third and fouth floors, the air temperature in the gap can become uncomfortably high. The free height of the double façade can also be limited by requirements for sound insulation between floors and fire requirements. Therefore, most projects realized with this type of facade close the air gap between floors. In this project it was important to find out how high temperatures get in the gap and to estimate the potential of problems with thermal comfort in the upper floors.

Fig. 2 Situation plan for building complex (a textile covers the facade that is considered in the project; *Source* Gulesider.no)

1.3 Description of the Building

The authors were engaged by the Convention Centre people house (AOF), to conduct a detailed study of energy and ventilation concepts by adding a double façade to an existing building complex in Krambuveita, Trondheim.

The existing building is a convention and meeting center with cinema halls, restaurants and hotels. The façade is in the west, east and south as shown in Fig. 2. Western and east facing facades are facing other surrounding buildings while the south facing facade is facing a small square. The existing facade has a thermal resistance of U-value = 0.6 W/m^2K. Windows consists of two layers glazing with a U-value = 2.6 W/m^2K.

1.4 Weather Data

Outside temperature and solar radiation on vertical surfaces are the most important parameters to describe the effect of a double façade. Figure 3 shows the outside temperature and solar radiation on a vertical south facing facade situated in Oslo every hour throughout the year.

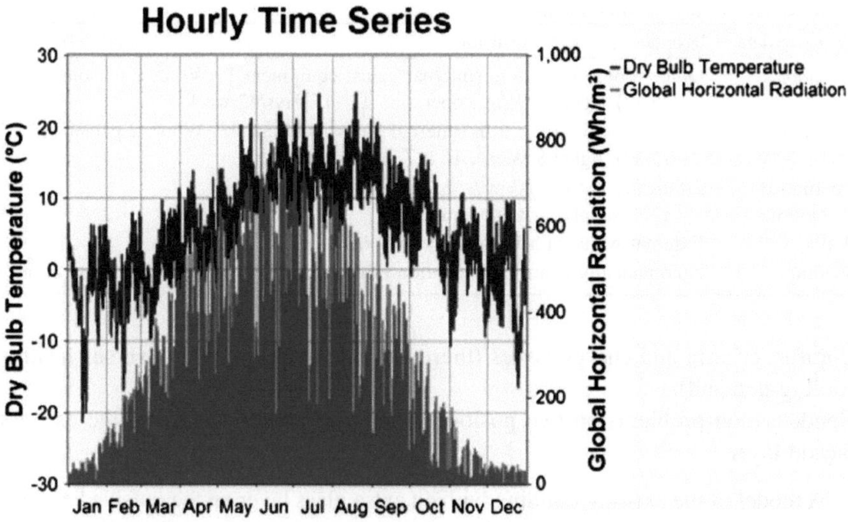

Fig. 3 Outdoor temperature and vertical solar radiation for each hour throughout the year. (*Source* Meteonorm)

1.5 Objectives

AOF in Trondheim wants to upgrade parts of the existing facade at the Congress people house. The aim of the study was to find the energy consequences of the use of a double facade system (dfs) on the south facing facade.

2 Method

With the help of dynamic computer simulations of energy and indoor environment for a case building in Norway, the impact of an additional ventilated glass facade on the energy demand and indoor environment was analyzed. A focus was put on a comparison of energy demand and thermal comfort levels of various cases. Main parameters to study were:

Different construction standards (air tightness, thermal bridges, and facade design) and their energy demand implications

Simulation robustness in dependence of different assumptions (thermal bridges in and air tightness of the existing building)

Airflow control strategies and their energy demand implications

Table 1 Description of simulation model

Climatic data	Trondheim (meteonorm file)
3rd floor	Office room: 5.3 × 6 m (internal gains: equipment 11 W/m^2, 2 persons (2 × 75 W), lights 8 W/m^2, operation 12 h/5 days/52 weeks)
5th floor	hotel room: 5.3 m × 6 m (internal gains: equipment 1 W/m^2, 2 persons (2 × 75 W), lights 8 W/m^2, 16 h/7 days/52 weeks)
Ventilation system	2 fans per room (120 m^3/s, balanced ventilation), 17°C supply air temperature
Walls	External walls with U-value = 0.6 $W/(m^2\ K)$
Shading:	Automatically controlled venation blinds in cavity (no shading in base case)

Comfort criteria and energy issues (thermal versus visual comfort versus heating cooling demand).
Condensation problems on two positions; inner and outer surface of the exterior façade layer

A model of the existing building with an extra glass layer on the outside has been developed. Dynamic thermal building simulation has been coupled with airflow network in order to simulate the airflow through the ventilated double-skin façade.

2.1 Thermal Model

Three different models were developed using dynamic building thermal performance tools TRNSYS and airflow network model TRNFLOW [6–8]:

Existing facade (base case) model
Ventilated double façade system with insulated glass (dfs (1))
Ventilated double façade system with single laminated glass (dfs (2))

Two different rooms were taken to compare the results; an office room in the 3rd and a hotel room in the fifth floor. The model description is detailed in Table 1.

2.2 Airflow Model

Airflow modeling was coupled to the thermal model (see Fig. 4). Here, the double facade system consists of 16 different zones that were linked using the specifications in Table 2.

In order to evaluate the winter and summer performance the following parameters were examined:

Temperature (inside window surface),
Energy (power),
Thermal comfort (with values as described in Table 3; according to ISO 7730 [9]).

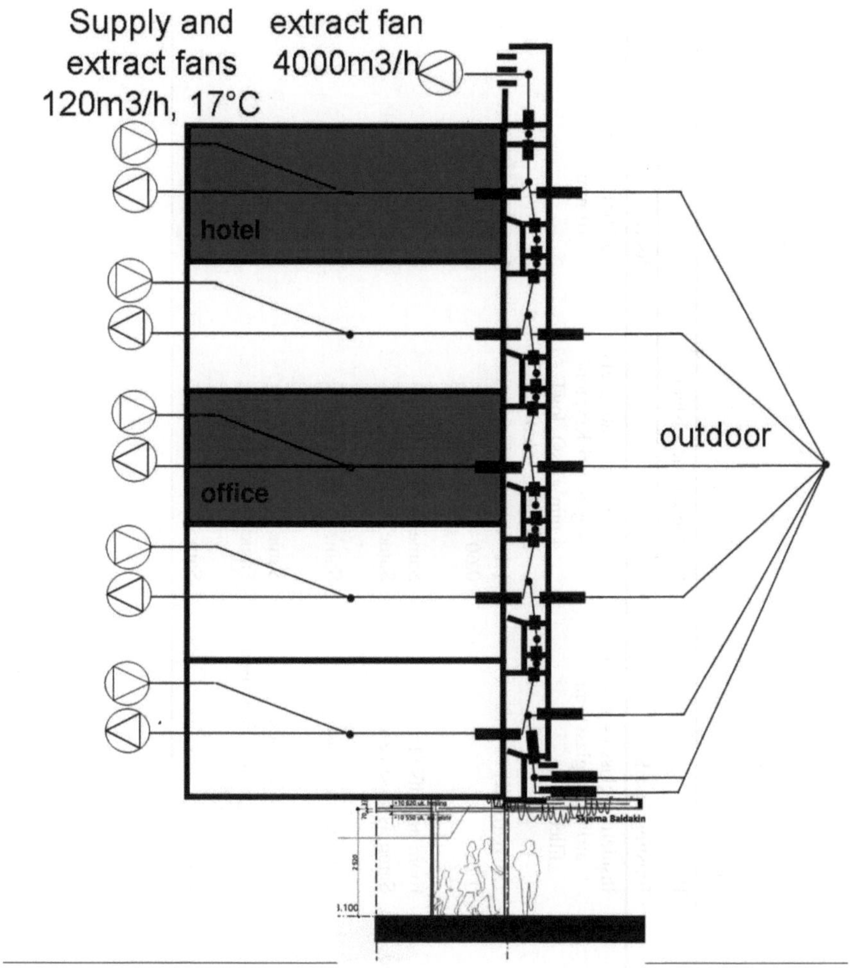

Fig. 4 Model for air flow between the different types of rooms

The simulations are done with TRNSYS connected with TRNSFLOW [10]. Figure 4 shows a schematic representation of the air currents.

3 Results

Results can be seen in Figs. 5, 6, 7 and 8. They are divided into temperature, energy, and thermal comfort in the following sections.

Table 2 Description of airflow and leakage

Specifications		Base case	Double façade system	
		Existing fasade	dfs (1)	dfs (2)
Window properties	Glass layers	Insulating glass (4 glass/16 air/4 mm glass), air filled	Additional insulating glass (4 glass/16 Krypton/4 mm glass), lowE, Krypton filled	Additional single glass 10 mm
	U-value	2.6 W/(m^2K)	1.1 W/(m^2K)	5.46 W/(m^2K)
	g-value	0.76	0.60	0.77
Air leakage, m = Cm x $(\Delta p)^n$				
Leakage between	room and outside	room and dfs (1)	Same as dfs (1)	
Cm	0.0128 kg/s @ 1 Pa (based on 0.6 h^{-1})	Same as base case	Same as dfs (1)	
n	0.65	Same as base case	Same as dfs (1)	
Other leakage, m = Cm x $(\Delta p)^n$				
Leakage between	–	dfs (1) and outside	Same as dfs (1)	
Cm	–	0.0021 kg/s at 1 Pa (based on 0.1 h^{-1})	Same as dfs (1)	
n	–	0.65	Same as dfs (1)	

Table 3 Criteria for thermal comfort. Values for NS ISO 7730

Parameter	value
Clothes level [CLO]	1
Metabolic rate (metabolic rate) [MET]	1
Activity [W/m^2]	0
Air speed [m/s]	0.1

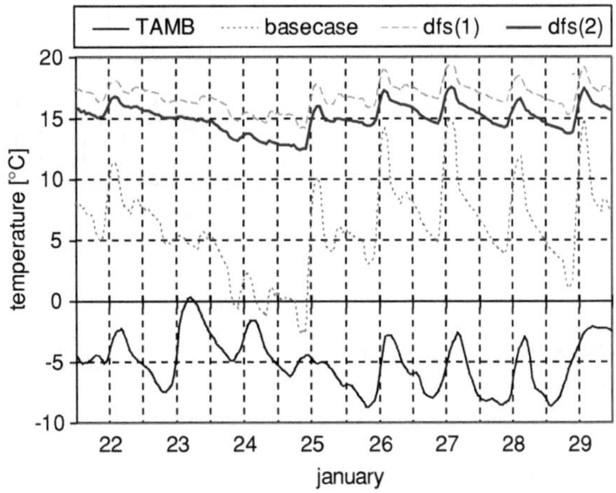

Fig. 5 Temperature distribution in hotel room (fifth floor) winter

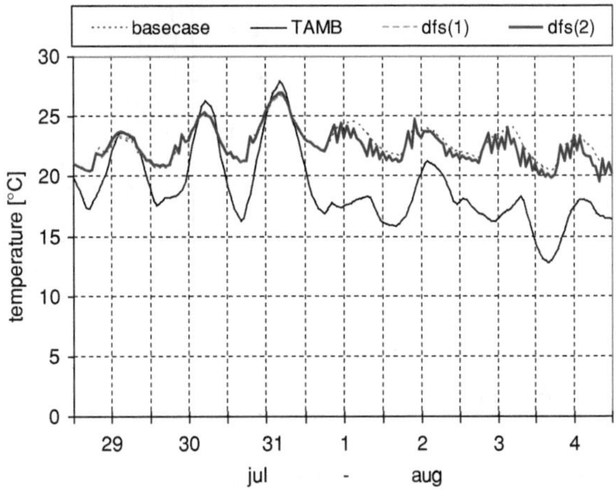

Fig. 6 Temperature distribution in hotel room (fifth floor) summer

3.1 Temperatures

Figure 5 presents the temperatures of the window at the inside of the hotel room for a typical winter week It can be seen that window temperatures are higher for the dfs than the base case in the winter. Figure 6 presents the temperatures of the window at the inside of the hotel room for a typical summer week. In the summer, temperatures are lower than in the base case.

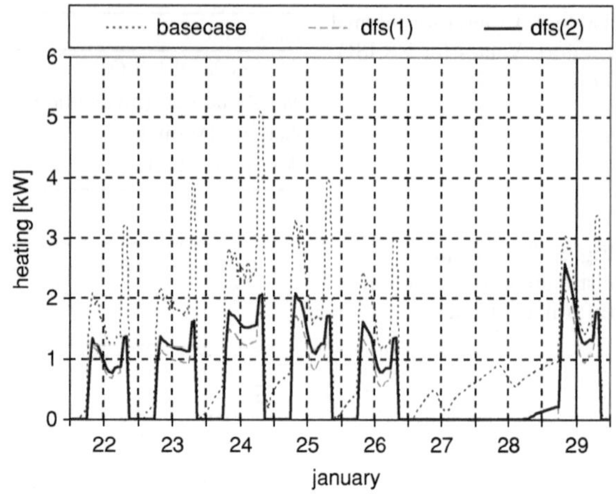

Fig. 7 Power distribution for office room

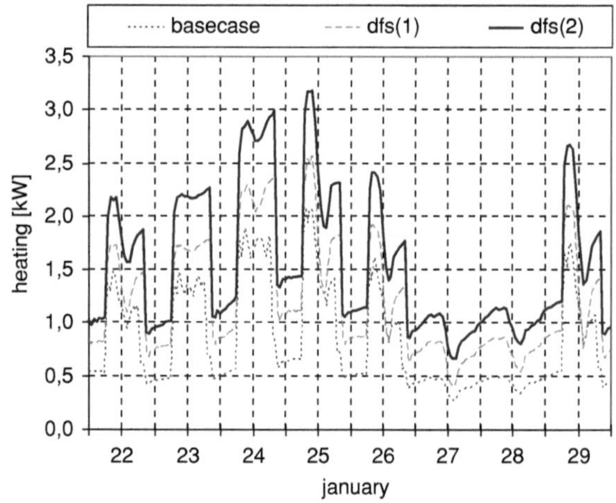

Fig. 8 Power distribution for hotel room

3.2 Energy

Figures 7 and 8 present the power needed for the office and the hotel room for a typical winter week. It can be seen that power distribution in the office (left) is reduced for both dfs types (1 and 2) with slightly more reduction for dfs (1). The power distribution in the hotel room (right) shows an increase for both dfs types (1 and 2) with higher increase for dfs (2). Figures 5 and 6 show the net energy demand results for both rooms (left) and a reduction in energy demand in percentage compared to base case for all rooms (right). It can be seen that energy demand in the different rooms vary. Energy demand for heating in the office room (third floor) is reduced by 59% while energy demand in the hotel room (fifth floor) is increased by 89% (Figs. 9 and 10).

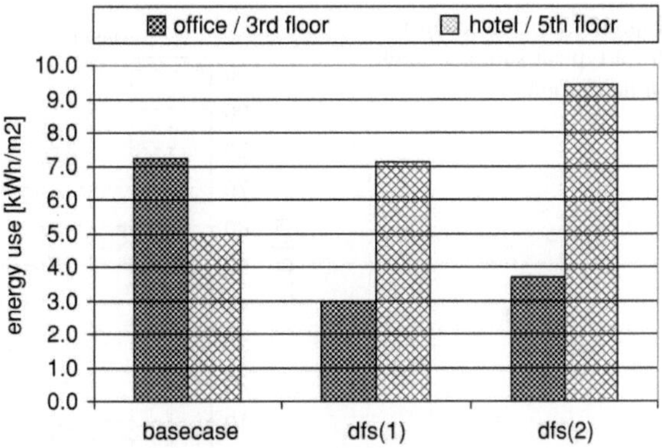

Fig. 9 Summed net energy demand for heating in office and hotel room

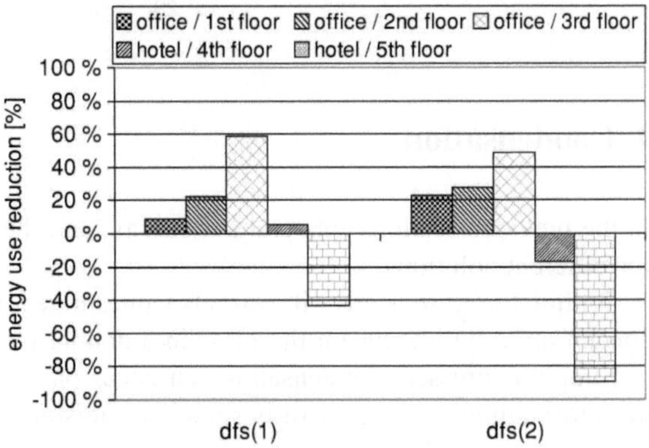

Fig. 10 Percentage energy demand for heating compared to base case in all rooms

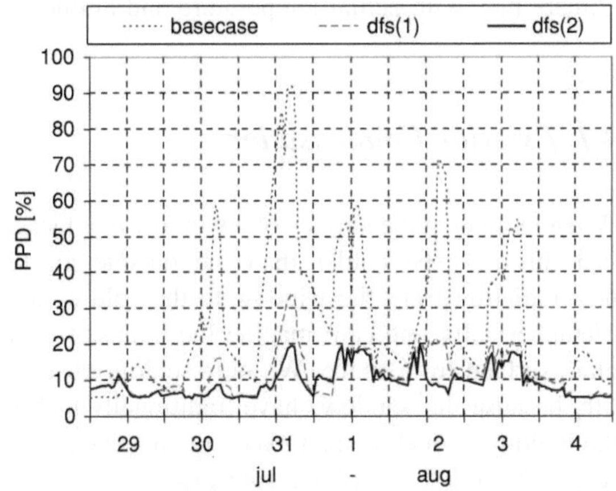

Fig. 11 Thermal comfort for a typical summer week in office room

3.3 Thermal Comfort

Figure 11 presents thermal comfort of the office and Fig. 12 of the hotel room for a typical summer week.

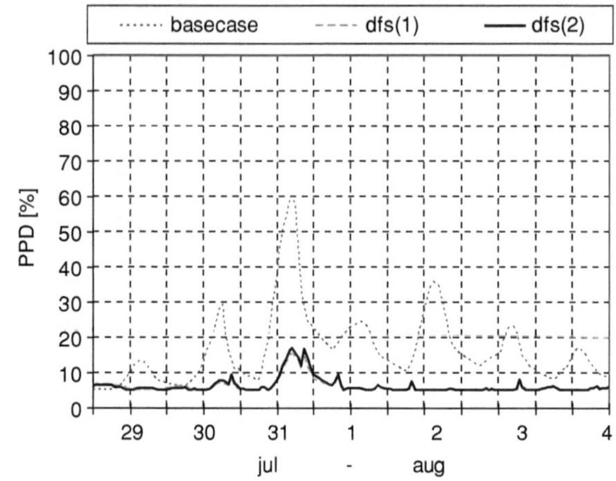

Fig. 12 Thermal comfort for a typical summer week in hotel room

4 Condensation

In the previous section, temperature, energy, and thermal comfort were evaluated for different solutions.

Within the case study, it was also questioned whether internal or external condensation will occur on the glass in a double façade.

Simply expressed, condensation can occur on a surface when it is colder than the surrounding air, and there is sufficient moisture available in the air.

To determine the probability of condensation, it is therefore important to use climate files with actual temperature and humidity values.

4.1 Exterior Condensation

When looking at the surface of a window in a facade one can ask how it is possible that this may be colder than the outside air temperature? Physical cause is heat radiation from the window to the cold atmosphere as illustrated in Fig. 13. Metaphorically one can imagine the opposite course; cold exposure towards the glass surface from space. We only seldom observe this phenomenon in the daily life, because the windows have traditionally had a large heat transfer from inside the building, which helps to keep the outside surface temperature of the glass at a relatively high level. (After cold exposure)

The conditions of condensation on the outside of windows depends, therefore, on a combination of a variety of climatic conditions, including outdoor air temperature, the environment (landscapes, buildings, sky) radiation temperature, relative humidity and wind speed. However, it is not sufficient for a cold, clear and

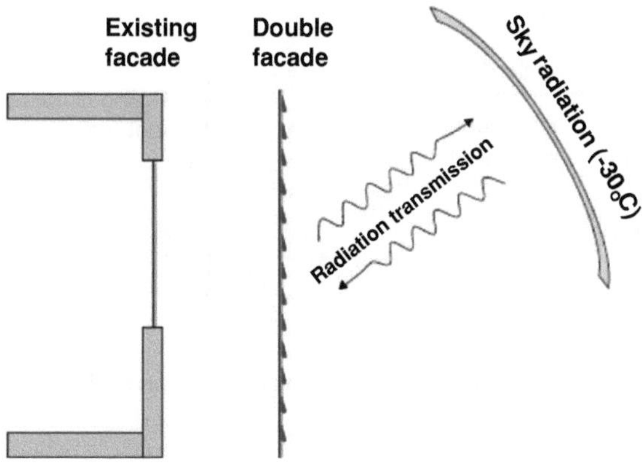

Fig. 13 Exterior condensation

calm night for the condensation to occur. Outdoor air relative humidity must also be so high that the dew point temperature is higher than the outer surface temperature of the window pane. Surface temperature is influenced in turn by the heat insulation ability of the window glazing and the indoor temperature level.

4.1.1 Practical Experiments

Measurements conducted in Borås in Sweden show some conditions (individually) that must be met in order for condensation to occur:

The difference between the outdoor air temperature and the ambient radiation temperature (sky, landscapes, buildings, etc.) was between 4 and 7°C
The relative humidity was a minimum of 95% (average winter is about 85%)
External condensation performed at an outside air temperature of 0°C in the spring and at approximately 10°C in the autumn
External condensation appeared especially at low wind speeds (0–3 m/s), only a few instances condensation performed at wind speeds between 3 and 4 m/s.

4.1.2 Previous Calculations

In connection with the report "Exterior condensation on window panes", a number of computer calculations were carried out in order to investigate the effect of a number of parameters contributions for condensation to occur.

Computer calculations showed otherwise good accordance with measurements referenced.

In the simulations, there was no climate data of Trondheim available. Varnes airport, some 40 km from Trondheim is therefore used as the nearest station.

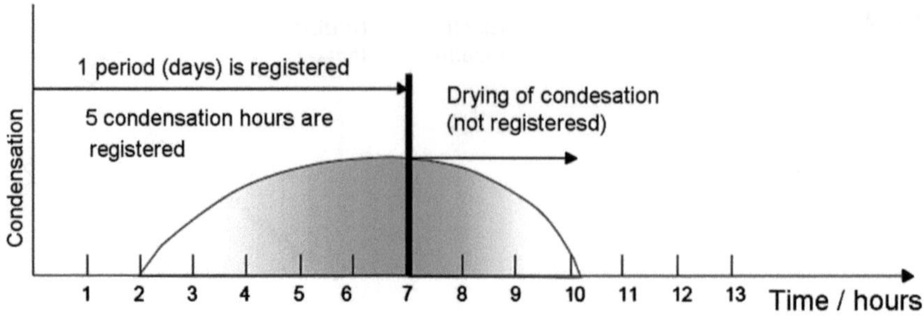

Fig. 14 The assessment of the condensation-based simulations

Moreover, the year 1985 was selected as most representative of typical weather (temperature, wind, humidity).

Figure 14 shows a typical result for an arbitrary day. This shows that the conditions for condensation is present in 2, 3, 4, 5 and 6 h. Each of these hours are recorded together with the record that the condensation occurred in this day. In the subsequent period when the conditions for condensation is no longer present, there will still be condensation on the glass. The 'recovery' period is not recorded.

Simulation

Table 4 summarizes the data that can be extracted from the simulations:

It is important to note that condensation ring would normally occur over consecutive periods, i.e. condensation extends over several hours. Table 5 can shed light on how this can be understood:

Shielding

In the simulations, it is assumed that the vertical glass surface "sees" 40% of the sky. The remaining 60% represented in this case, shielding and radiation from the surrounding "hot" building/environment.

The value of 40% is relatively high, since 50% is the maximum value that a vertical-scale window can have.

The façade will often be exposed to varying shielding. This means that different parts of the façade will be exposed to different proportion of cold air exposure. Typically, the lower floors are very sheltered, and in practice do not get condensation, while the upper floors are considerably more exposed to cold exposure.

Based on simulations for Tromsø, an increase from 40 to 50% (window with free horizon) sky radiation results in an increase of about 150% of the values in Tables 1 and 2. Conversely, a reduction to 30% sky radiation results in a reduction

Table 4 Number of hours (hours/events) with condensation

Varnes (1985), the number of hours of condensation rates over hours of day																									
Glazing type	1	2	3	4	5	6	7	8	9	10	11	12	13	14	15	16	17	18	19	20	21	22	23	24	SUM
U = 0.8	35	24	19	15	12	14	21	5	0	0	0	0	2	0	0	0	0	2	9	8	6	8	10	13	203
U = 1.1	18	10	8	6	5	5	5	2	0	0	0	0	0	0	0	0	0	0	2	1	1	2	3	6	74
U = 1.4	10	7	4	3	4	4	4	1	0	0	0	0	0	0	0	0	0	0	0	0	1	1	2	4	45

Accumulated over years

Table 5 Number of hours/events with condensation

Varnes (1985)								
Range of day	01–08		09–20		21–24		Sum	
Window type	Total number of hours	Number of times	Total Number of hours	Number of times	Total Number of hours	Number of times	Total Number of hours	Number of times
Us = 0.8 W/m^2K	145	49	21	11	37	14	203	64
Us = 1.1 W/m^2K	59	20	3	2	12	6	74	27
Us = 1.4 W/m^2K	37	12	0	0	8	4	45	16

Accumulated over years

to about 50%. The effect of sky radiation is thus very large, and in practice is crucial for condensation to actually occur, even if climatic and temperature conditions are present.

Effect of Internal Temperature

In the simulations above, it is assumed a constant temperature in the room behind the double façade at 20°C.

In the double façade, the "indoor temperature" varies and could be down to 5°C at design outdoor temperature in winter.

Based on simulations for Tromsø, a reduction from 20 to 17°C results in an increase of the values in Tables 4 and 5, by approximately 50%. Temperature levels thus have a great effect on condensation risk.

4.1.3 Calculations Using TRNSYS

It was decided to proceed on the issue of current temperature conditions within the double façade, and there was done/performed calculations using the program TRNSYS.

The starting point is climate data in the same winter week as used in the Sects. 3.1 and 3.2

Three different cases are calculated:

The existing façade, U-value new window: 1.1 W/(m^2 K)
This metric is chosen as a reference, and to test the model.
New double façade, U-value 2-layer glass: 1.1 W/(m^2 K)
This corresponds to previously simulated double façade, type: DFS (3). See Sect. 3.1
New dual front, U value of one-layer glass: 5.7 W/(m^2 K),
This is equivalent to previously simulated double façade, type: DFS (4). See Sect. 3.1

Surface temperature of the glass exterior is always higher than the condensation temperature (dew point temperature) for outdoor air. The result is that it will not cause condensation on the outside window through the simulation period as shown in Fig. 15.

Surface temperature of the glass exterior is lower than the condensation temperature (dew point temperature) for outdoor air in much of the period. The result is that the climatic conditions are present that condensation will occur significantly during night time as shown in Fig. 16.

Figure 17 shows large heat losses from the building and through the double façade, leading to a relatively high surface temperature of the exterior glass layer. In practice, the temperature never falls below the dew point temperature for outside air and the conditions for condensation do not appear.

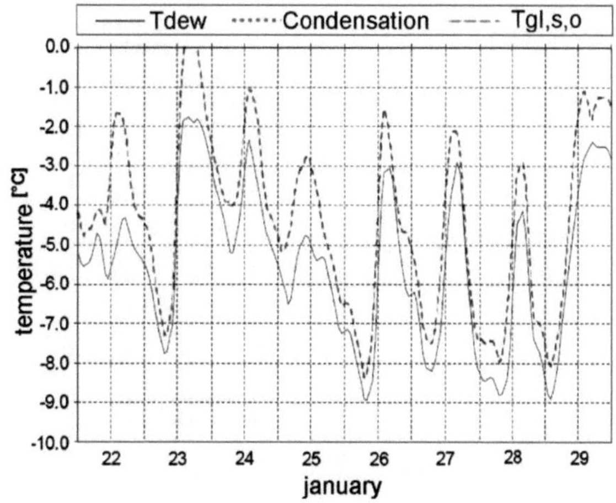

Fig. 15 The existing façade with a new glass ($U = 1.1$ W/(m^2 K)). Typical winter week

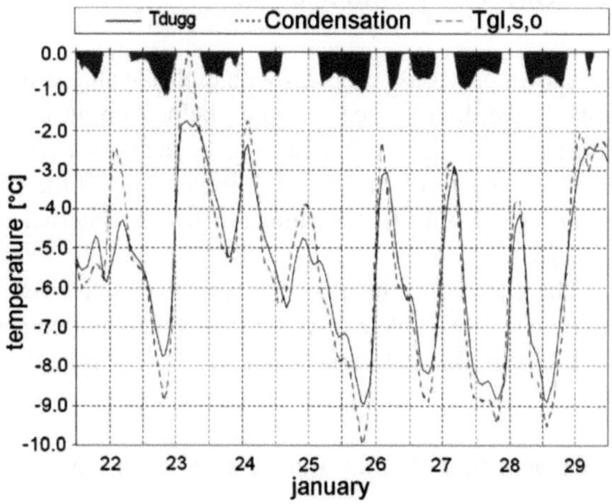

Fig. 16 New double façade, U-value 2-layer glass: 1.1 W/(m^2 K), DFS (3). Typical winter week

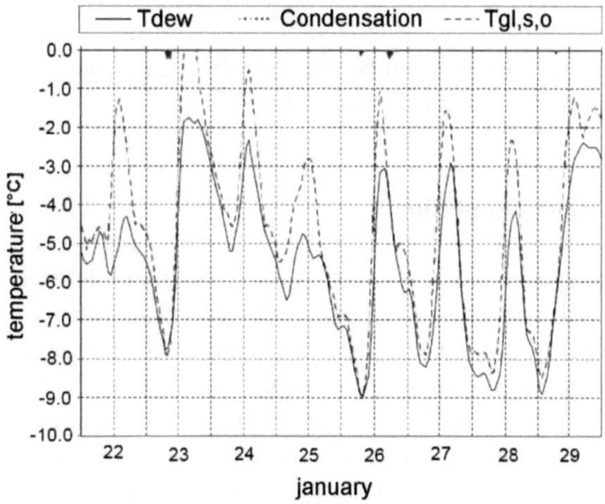

Fig. 17 New double facade, U-value one-layer glass: 5.7 W/(m^2 K), DFS (4). Typical winter week

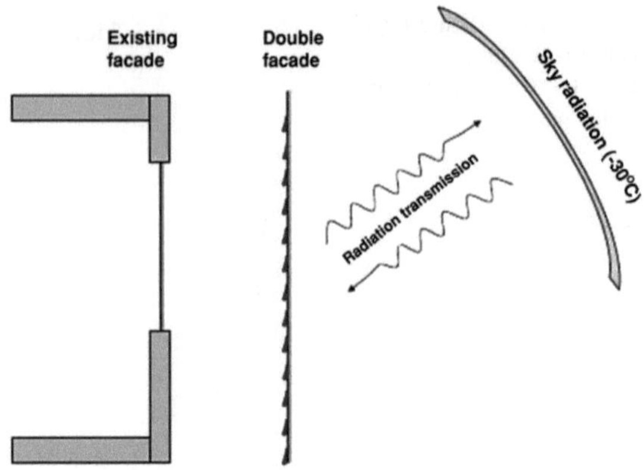

Fig. 18 Internal condensation

4.2 Internal Condensation

As shown previously, a glass with high U-value gives a relatively high temperature outside the double façade, which helps to reduce the external condensation. The same glass quality will lead to more heat losses which results in cavity temperatures that are similar to outside temperatures. If there is critical moisture conditions inside the double facade, this can lead to internal condensation as illustrated in Fig. 18.

The calculation results of a typical winter week are shown in Fig. 19 (simulation No. 3 in Table 6). As we see the calculation shows large periods (sum equal to 130 h/week) with large amounts (= very likely) of interior condensation.

Measures that Reduce Indoor Condensation

Simulation results must be considered to describe unacceptable conditions, and therefore we have looked at various measures to improve conditions.

In Table 6 it can be seen that the use of double glazing (Simulation No. 1 and 2) improve conditions somewhat.

Night Ventilation

Measures for one-layer glass show small effects, except for night venting the cavity between 21 pm and –05 am (simulation No. 7). The measure increases the energy use, because hot (and humid) air is replaced with cold (but dry) air.

Condensation risk as shown in Fig. 20 is a conservative estimate. The reason for this is that the calculation program does not take into account the effect of air movements on the glass surface. Halving the probability of the risk of condensation in relation to the figures may be a reasonable assumption.

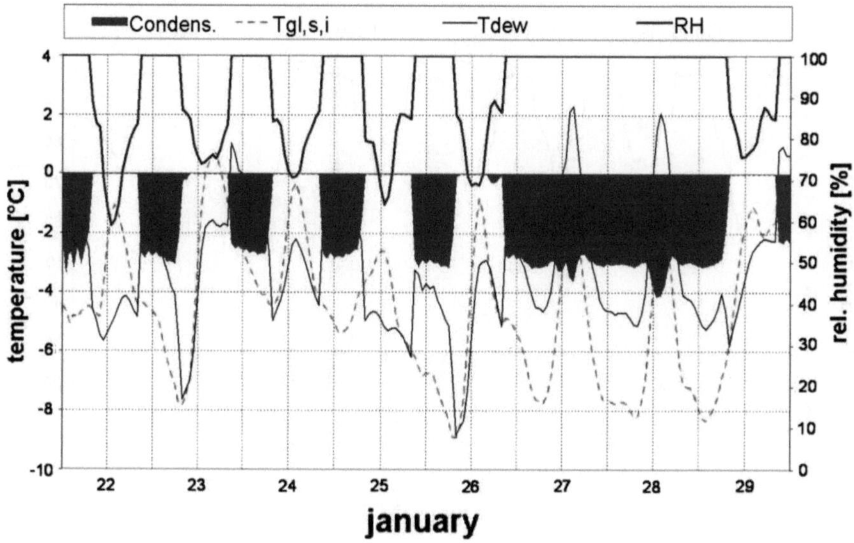

Fig. 19 Internal condensation. New double facade, U-value of 1 layer glass: 5.7 W/m^2 K, DFS (4). Typical winter week

Table 6 Number of hours/events with internal condensation. Typical winter week

Simulation no.	Code	Type glazing	U-value	measures	Condensation hours/week	Figure
1	iso	Double	1.1	Air tightened on first and second floor	108	
2	iso (2)	Double	1.1	Not air tightened	108	
3	esg	Single	5.5	Air tightened on first and second floor	130	5.7
4	esg2	Single	5.5	operation times ventilation office = hotel	130	
5	esg3	Single	5.5	operating times ventilation and heating office = hotel	130	
6	esg4	Single	5.5	DF open (23-5)	118	
7	esg5	Single	5.5	DF open (21-5)	108	5.8
8	esg6	Single	5.5	DF open (6-20)	134	
9	esg7	Single	5.5	DF with cracks in first and second flor, DF closed	130	

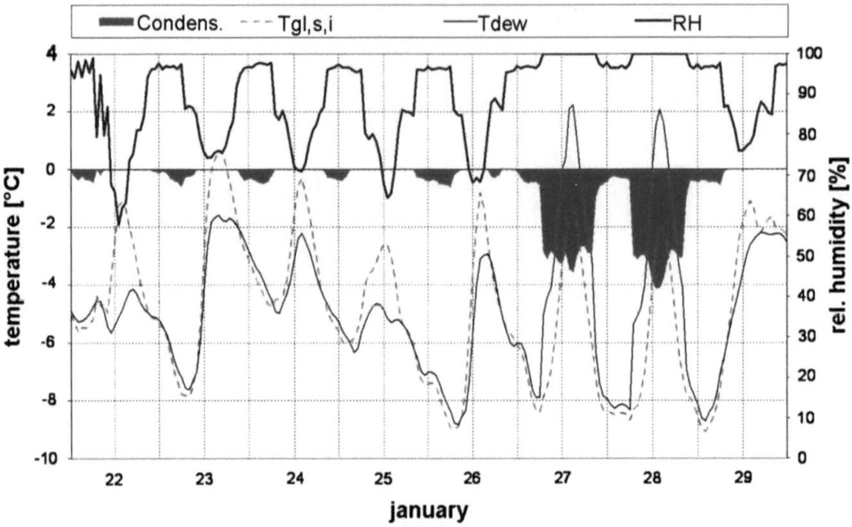

Fig. 20 Internal condensation. New double facade, U-value one-layer glass: 5.7 W/m^2 K, DFS (4). Typical winter week. Night Ventilated

Fig. 21 Overhang. Sketch of the overhang with depth 2 m installed at the cornice of the sixth floor

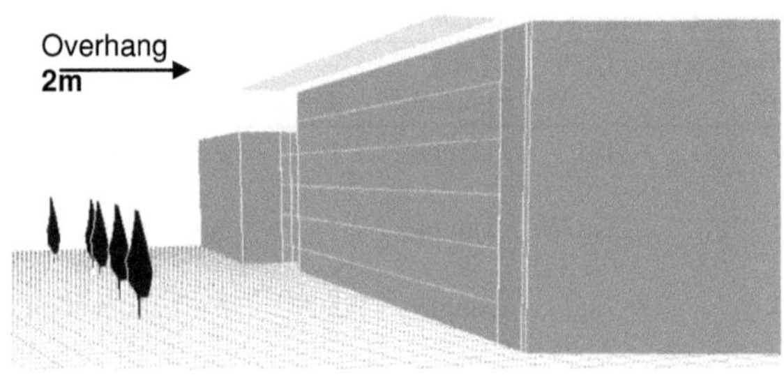

However, probability and extent of internal condensation is significant and the increase of energy consumption so large that the use of one-layer glass as an outer part of a double façade cannot be recommended.

4.3 Shading as a Measure Against the Outside Condensing

Among the measures that are possible to apply is exterior shading to reduce sky radiation. This measure is combined with the use of double glazing.

The purpose of the overhang is to shield the external radiation and thereby reduce the external condensation we have previously identified for this glass quality (Fig. 21).

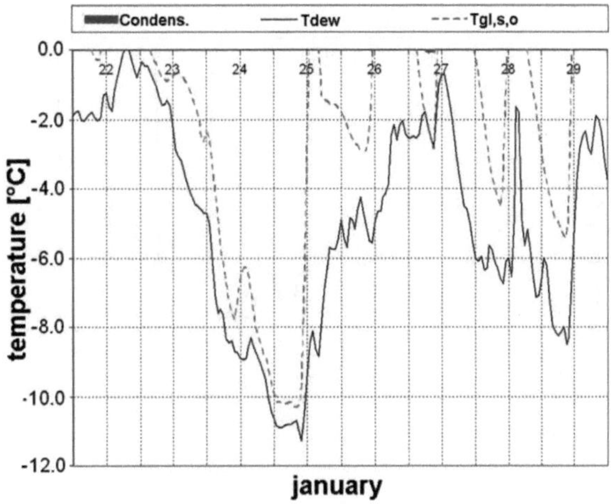

Fig. 22 Exterior condensation. New double facade, U-value 2-layer Glass: 1.1 W/m^2 K. Typical winter week. Shielded

With an overhang in this shape, external condensation risk reduces practically to zero as shown in Fig. 22.

Figure 23 shows that internal condensation will occur (106 h/week), but in small quantity. Given the conservative method in the model, the condensation may in praxis be near zero.

5 Summary and Conclusion

This report deals with thermal indoor climate, energy and condensation challenges for a planned double façade of the Convention Centre (AOF) in Trondheim.

For the calculations, the simulation program TRNSYS, in combination with TRNSFLOW, wais used [8]. The software modules were developed within a PhD study by Haase [7]. The use of a double façade is not very widespread in Norway. As far as SINTEF knows, there exist only 10–12 buildings to date where it is used. Of these projects, only a couple of the double façades are used in connection with the renovation of the façade. Applying a double façade system this way, renovation work on existing façades can be reduced, in addition to the building's energy, and power performance being improved. The use of dfs must be done properly, and problems are far more complex—the energy aspect, fire risk, escape paths, noise issues, operation and security of the system must also be included in the design of an optimal solution. The reduction of condensation risk is an issue that must be considered.

Within these areas there is still lack of knowledge, calculation tools and of course the experience and feedback from completed projects.

In the case of condensation, we have also relied on earlier work by SINTEF report, "Exterior condensation on window panes" [11].

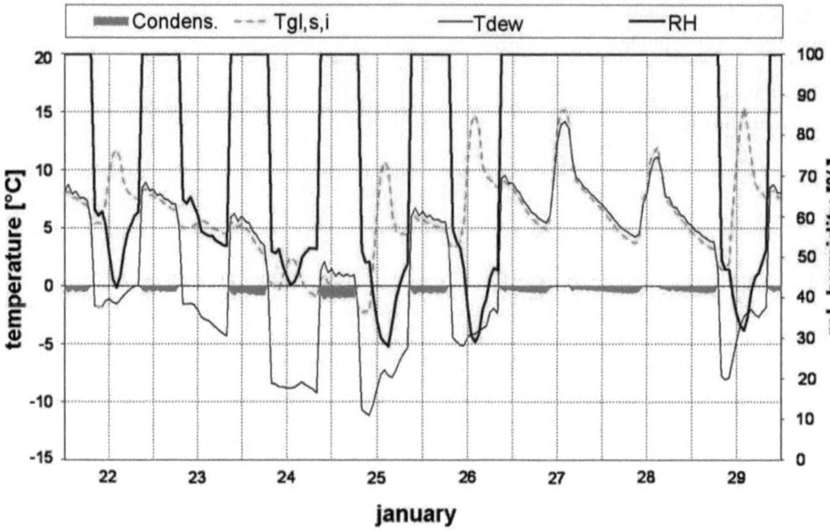

Fig. 23 Internal condensation. New double facade, U-value 2-layer Glass: 1.1 W/M^2 K. Typically winter week. Shielded

The results of the calculations and the conclusions will necessarily be specific to the actual case of AOF Convention Centre in Trondheim. But the problems discussed can be applied to other commercial buildings in Norway and countries with similar climates.

5.1 Summary

Energy

The results show that energy efficient refurbishment of an existing façade with double façade system is possible. Temperatures on the inside of the windows as well as thermal comfort are improved with both types of dfs (1 and 2). Energy savings seem to depend on the vertical airflow within the dfs and range between 59 and −89% for the different rooms.

The construction of a double façade (df) in combination with high air leakages in the old façade results in airflows between rooms and the df cavity. The amount of airflow increases because of the df, which leads to an increase in energy demand for heating. Especially, the hotel room in the fifth floor needs between 43 and 89% more heating (with an additional insulated glass layer (dfs 1) and single glass layer (dfs 2) respectively).

The solution (1) with insulated glass seems to perform better with respect to glass temperatures, thermal comfort, and energy savings than solution (2).

More work is needed in order to optimize the construction of a ventilated dfs in respect to operational energy savings. One possibility could be to reduce air leakages in the old façade construction.

Condensation of humid cold air on the outside and inside of the dfs layer could lead to unwanted effects and was therefore evaluated.

Exterior Condensation

The calculations show, however, that the likelihood of condensation outside is great, both when we count the number of hours and condensed amount, when using double glazing.

With one-layer glass the calculations show that the exterior condensation is nearly absent.

Internal Condensation

By using 2-layer glass, the calculations show that the probability of condensation inside is large in terms of hours, but not in quantity.

Using one-layer glass calculations show that internal condensation is very likely in terms of both number of hours and quantity. Ventilation of the cavity will improve the situation significantly but the consequences are huge energy consumption.

Shelter From the Sky Radiation

If the building had mounted external shading/overhang (depth 2 m), this would have a major impact on the outside condensation of double layered glass. Simulations show that the condensation is reduced to near zero. The measure is recommended.

5.2 Conclusion

Based on the simulations we have conducted, we recommend that it be used double layer glass U-value equal to 1.1 W/(m^2 K). Moreover, we recommend that low emission coating facing outwards on the outer glass should be used.

This solution is combined with horizontal shading/overhang (depth 2 m) located at the cornice of the top floor.

Elements of applying a double façade system that are not considered in this report:

Can result in reduced light levels in the room due to several layers of glass and overhangs

Can provide greater sound transmission between adjacent rooms (also vertically)
Can provide greater danger of fire and smoke spreading.

Acknowledgments This chapter has been written within the ongoing SINTEF project "Low Energy Commercial Buildings". The authors gratefully acknowledge the financial support of the Research Council of Norway.

References

1. Annex 44.: State of the art review of methods and tools for designing integrated building concepts. In: Andresen, I., Kleiven, T., Knudstrup, M.-A., Heiselberg, P. (eds.) Integrating Responsive Building Elements in Buildings, International Energy Agency I. E. A., Energy Conservation in Buildings and Community Systems Programme, vol. 2006. Aalborg University, http://www.ecbcs.org/docs/Annex_44_SotAr_IBC_Vol_2B.pdf (2008)
2. Arasteh, D., Reilly, M.S., Rubin, M.: A Versatile procedure for calculating heat transfer through windows. ASHRAE Trans. **95**, 2 (1989)
3. Dorer, V., Weber, A.: Air, contaminant and heat transport models: Integration and application. Energy Buildings **30**(1), 97–104 (1999)
4. Haase, M., Amato, A.: Performance evaluation of three different façade models for sustainable office buildings. J. Green Building **1**(2), 361–373 (2006)
5. Haase, M., Marques da Silva, F., Amato, A.: Simulation of ventilated facades in hot and humid climates. Energy Buildings **41**(4), 361–373 (2009)
6. Haase, M., Wong, F., Amato, A: Double-skin facades for Hong Kong. Surv. Built environ. **18**(2), 17–32 (2007)
7. Manz, H.: Numerical simulation of heat transfer by natural convection in cavities of facade elements. Energy Buildings **35**(3), 305–311 (2003)
8. NS-ISO7730.: Ergonomics of the thermal environment—analytical determination and interpretation of thermal comfort using calculation of the PMV and PPD indices and local thermal comfort criteria, vol. 7730, Standard Norge, Lysaker (2005)
9. Oesterle, E., Lieb, R.-D., Lutz, G., Heusler, B.: Double-skin facades: Integrated planning: Building physics, construction, aerophysics, air-conditioning, economic viability. Prestel, Munich (2001)
10. Poirazis, H.: Double-skin Facades for Office Buildings, Division of Energy and Building Design, Department of Construction and Architecture, Lund Institute of Technology. Lund University, Lund, p. 196 (2004)
11. Saelens, D.: Energy performance assessment of single storey multiple-skin facades, PhD, Department of Civil Engineering. Katholieke Universiteit Leuven, Leuven (2002)
12. Thyholt M.: Utvendig kondens på vindusruter. In: Sintef, (ed.) SINTEF Byggforsk, (2006)
13. TRNSYS.: Trnsys16 user manual, vol. 2005, Solar Energy Laboratory, University of Wisconsin-Madison, Madison (2004)

Structural Design and Experimental Investigation of a Carbon Fibre Wheel for Low Consumption Vehicle

Massimiliana Carello and Alessandro Scattina

Abstract Today, in the design of a new vehicle, one of the most important challenge is the weight reduction. This item is even more important in the design of prototype vehicles aimed to low consumption competitions, where it is necessary to minimize the weight of all components. For this reason the carbon fibre composite materials appear the best solution in terms of low density and mechanical properties. In this work the attention is focused on the wheel of the IDRA prototype, which participated in the Shell Eco Marathon competition. The wheel rim is a lenticular single part made of carbon fibre materials, without the use of structural adhesive. The different design steps, from the concept phase to structural analysis made by means of finite element code, are discussed. At the end the innovative manufacturing production process is presented.

1 Introduction

The weight reduction is, together with aerodynamics, frictions reduction and propulsion efficiency, a key factor in the design of new vehicles [1] and in particular it is even more important for design of prototype vehicles aimed to low consumption competition, such as the Shell Eco Marathon [2]. The target is very

M. Carello (✉) · A. Scattina
Mechanics Department, Politecnico di Torino, Corso Duca degli Abruzzi 24,
10129 Turin, Italy
e-mail: massimiliana.carello@polito.it

A. Scattina
e-mail: alessandro.scattina@polito.it

A. Öchsner et al. (eds.), *Materials with Complex Behaviour II*,
Advanced Structured Materials 16, DOI: 10.1007/978-3-642-22700-4_32,

sophisticated, the vehicle's weight is about 30–35 kg. For this reason it is necessary to search the minimum weight not only on the major parts, such as the chassis and the body, but also to minimize the weight of different components with smaller size. In particular in this work the attention was focused on the wheels. Moreover, in this type of prototype, the wheels have also an important aerodynamic function. For this reason a spoked wheel does not appear as a right solution, while a lenticular geometry should be an optimum solution to obtain also the best aerodynamic efficiency.

A good technological process to obtain a lenticular geometry is the use of carbon fibre composite materials, which have excellent performance in terms of low density and mechanical properties if compared to traditional mechanical construction materials like aluminium and steel [3]. Usually in the automotive field the carbon fibre is used for high level applications, where there is need for high performance and low weights while the higher costs are fewer important [4–9].

A typical manufacturing technology used to make carbon fibre parts for the automotive applications, provides the use of prepregs [10, 11]. They are fabric composite material already impregnated with resin. The prepregs (unidirectional or multi directional carbon fibre) are shaped by hand on moulds that are covered by the vacuum bags and then polymerized in autoclave, with pressure and temperature defined a priori. With this process it is possible to obtain product with high resistance, stiffness and good surface finish.

For these reasons, during the construction of IDRA prototype, which participated in the Shell Eco Marathon 2008, 2009 and 2010 in the prototype category with fuel cell propulsion, the H_2politO Team adopted lenticular carbon fibre wheel rims. These wheel rims are specifically designed to use the particular Michelin's tyres with very low rolling coefficient. These tyres are specific for the Shell Eco Marathon competition, and, on the market, there are not wheel rims that can adopt these tyres. Moreover the wheel rims must have also different requirements, such as the low weight, the compatibility with very low friction ceramic ball bearings, small hubs with particular design and, at last but not least, the lenticular shape.

Up to now other Italian and European teams for the same competition adopt aluminium or nylon wheels specially designed, or different types of carbon fibre solutions [12, 13]. However, in this last case, the wheel rim is made of two lenticular parts and by a central groove joined together with structural adhesive.

In this work the structural design and the production of the lenticular carbon fibre wheel rim for the IDRA prototype will be examined in detail. The main innovation of this wheel rim described in the following is the monolithic architecture, moreover the wheel does not need the use of tube being a tubeless solution. In particular this work is focused on the wheel rim for the IDRA09 and IDRA10 prototype, which was born as an evolution of the wheel rim for the IDRA08 prototype.

2 Lenticular Wheel Rim

2.1 Geometry

The design of the wheel rim started from two important points: the dimensions and the shape of the tyre groove and the wheel hub which have specific dimensions and geometry. With a benchmarking analysis, different solutions were examined. In particular the first idea was to make a nylon wheel rim in a single part. This solution led to an estimate weight of about 5 kg. In order to reduce thickness and consequently weight, a second solution completely made of aluminium was considered. A lenticular aluminium solution should lead to a total weight of about 2.2 kg, while a spoked solution should lead to a total weight of 1.5 kg. However, this last solution is not very good from the aerodynamic point of view. A third solution considered the use of a groove and a hub made of aluminium and two lenticular carbon fibre discs. The lenticular discs are joined with the groove with structural adhesive. This wheel should have a weight of about 1.2 kg, but it should not ensure the tightness in pressure if the bonding shows some irregularities. From this overview it was decided to design a monocoque solution completely made of carbon fibre. The use of structural adhesive is avoided with consequently improvements in term of total weight and better tightness, strength and stiffness. This solution should ensure a total weight lower than 1 kg. In this way was made the wheel rim for the IDRA08 prototype.

The wheel rim for the IDRA09 prototype (Fig. 1) was born as evolution of the wheel rim of the previous prototype. The use during competition of the firt solution put in evidence some weakness, in particular a vibration of the lenticular wall along the wheel hub axis. It is due to the too large section in the central zone and to the presence of plate part not well connected to the pending wall. These vibrations caused loss of aerodynamic efficiency and could lead to structural problems.

Starting from these considerations the geometry of the new wheel rim was redesigned. Figure 2 shows the two sides of the IDRA09 wheel rim while Fig. 3 shows also a section view of the wheel rim. Comparing the new design with the previous one (Fig. 3 on the left), the IDRA09's wheel rim has a lower frontal section (about 25% less). This choice has been done to further reduce the aerodynamic vehicle resistance of vehicle and to solve the vibration problems. On the front view, the design of the wheel rim wall is characterized, on one side, by the hole for the tyre valve, while, on both sides, the central zone has a particular shape necessary to introduce and to fasten the wheel hub. This last zone was redesigned to use an improved wheel hub, and to facilitate the assembly. The section view (Fig. 3) allows to better understand the profile shape, which is substantially defined by the tyre's constructor with specific dimensions. Also in order to use the same parts of the previous die, and consequently to save money, the design of the groove profile has not been changed from that of the previous solution.

The new solution is also a tubeless wheel. This improvement was made using an Easytubeless system which consists in a specific valve and liquid applied along

Fig. 1 Idra09 prototype and its wheel rim

Fig. 2 *Left* and *right* side of the IDRA09 wheel rim design

the groove profile to make the tightness. In fact the first wheel rim had some problems of sealing due to leaks both through the fibres and through the valve.

2.2 Materials

Starting from the calculation of equivalent properties for the lamina [14–16] and considering the wheel rim as a circular plates where a symmetrical bending load is applied [17] a series of composite materials and a specific ply lay-up were chosen to make the wheel rim. The attention was pointed to prepregs materials, in order to simplify the production process. Many considerations were done during this phase,

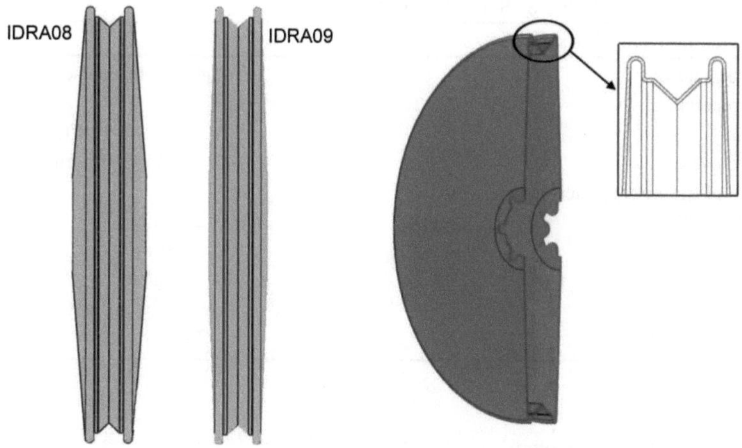

Fig. 3 On the *left*: comparison between the first and the last version of the wheel rim. On the *right*: diametrical section view of the wheel rim with a profile detailed view

Table 1 Evonil Degussa Rohacel polymethacrylimide foam

Density	0.0521	(g/cm^3)
Tensile strength	1.6	(MPa)
Modulus of elasticity	0.075	(GPa)
Compressive strength	0.8	(MPa)

but mechanical performance and weight containment were the key points which led to the choice. In particular a unidirectional carbon fibre material (STS-150-DT-36 F) and a bidirectional fabric were selected (Epoxy resin system DT150). In order to improve the performance of the component, the use of a middle layer of polymethacrylimide foam (Evonil Degussa Rohacell) positioned between the prepregs layer was also taken into consideration, in this way a sandwich structure is obtained. The main properties of these materials are summarized in Tables 1, 2 and 3.

Always considering previous experience a first ply lay-up was defined. The target was to obtain a symmetric ply lay-up in order to balance the stress due to pressure and temperature variation, which may occur during cure cycle in autoclave. For the ply lay-up the wheel rim was divided in three different zones: the lateral walls of the wheel rim (part A in the Fig. 4), the zone around the wheel hub (part B in the Fig. 4), and the groove profile (part C in the Fig. 4). Examining in more details the different ply lay-up, for the lateral walls, from the external side to the interior one, the different layers have been illustrated in Table 4.

For the zone around the wheel hub, the layers were the same as shown in Table 4, but between the fourth and fifth layers, two layers of bidirectional fibres, with 90° orientation and a thickness of 0.2 mm were added to improve stiffness

Table 2 Unidirectional fibre STS-150-DT-36 F

Features of the resin matrix	
Chemical nature	Toughered thermosetting epoxy
Cure temperature	120 ÷ 150°C
Gel time	5 ÷ 9 min @ 120°C
Tg (cure cycle)	140 ÷ 145°C (90 min @ 120°C)
Features of the impregnated tape	
Standard width	600 ± 3 mm
Standard length	150 ± 10 m
FAW	150 ± 5 g/cm^3
Resin content	36 ± 3% by weight
Laminate thickness	0.152 mm

Table 3 Bidirectional fibre epoxy resin system DT150

Chemical nature	Toughered thermosetting epoxy	
Cure temperature	120 ÷ 150°C	
Gel time	5 ÷ 9 min @ 120°C	
Tg (cure cycle)	140 ÷ 145°C (90 min @ 120°C)	
Plate mechanical properties		
Test	Strength (MPa)	Modulus (GPa)
Traction	831	57.4
Flexion	964	50.3
Compression	682	49.2

and strength of this part. At the end, the groove had a structure completely different. It was made of four layers of bidirectional fibres with a 0° orientation and a thickness of 0.2 mm.

Due to the symmetric design of the wheel rim and its circular shape, it is not very important to define the main direction of fibres for the side walls and for the zone around the wheel hub. For what concerns the groove profile the main direction was defined along the wheel rotation axis, which is the direction perpendicular to the side walls.

3 Structural Analysis

3.1 FEM Model

The FEM model of the wheel rim was obtained starting from a CAD geometry using the Altair Hypermesh© pre-processor. The mesh of the wheel rim model is shown in Fig. 4. The mesh of the wheel rim was made using shell elements and thanks to the geometry of the wheel rim it was possible to obtain a mesh

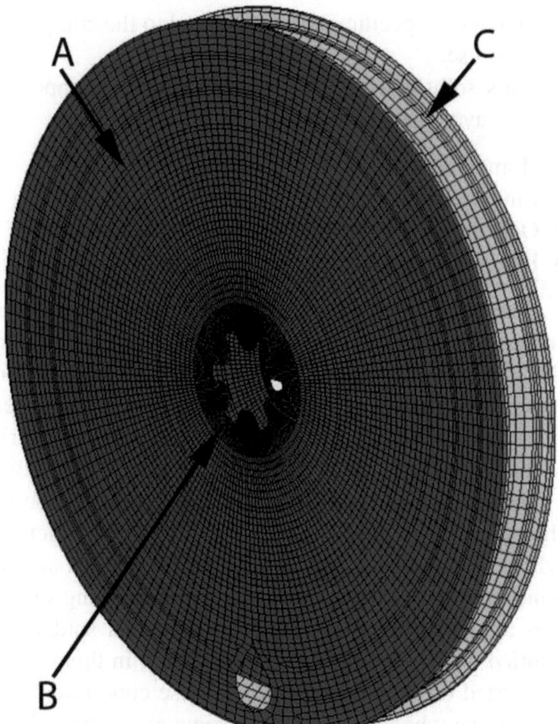

Fig. 4 The FEM model of the wheel rim:lateral walls (A); zone around wheel hub (B); groove profile (C)

Table 4 Ply lay-up for the side walls of the wheel rim

N°	Layer quantity	Thickness (mm)	Type	Orientation (°)
1	1	0.2	Bidirectional fibres	0
2	1	0.15	Unidirectional fibres	45
3	1	1	Structural foam reinforcement (Rohacell)	–
4	1	0.15	Unidirectional fibres	−45
5	1	0.2	Bidirectional fibres	90

completely made of four node elements. Only few three node elements (less than 0.5% of the total model) were used near the valve's hole and in the zone around the wheel hub. The medium size of element was about 4 mm for the side walls and the groove and about 2 mm for the wheel hub zone due to its more defined geometry and because in this zone the loads were applied. The complete model was made of about 20,000 elements. The geometry of the wheel rim suggested to model only a quarter of the wheel but the little dimensions of the model made possible to consider the entire wheel rim, the calculation time was not substantially influenced. The solutor software adopted for this analysis is the Radioss© Linear

10.0, it has a specific module oriented to the composite materials quite reliable and easy to use.

This solutor uses the classic theory of composite laminate, in particular the single layer was considered:

- Homogeneous;
- Linear elastic;
- Orthotropic at macroscopic level;
- Fibres and matrix are perfect integrated.

The main parameters required by the solutor software for the carbon fibre materials card were: the elastic modulus in longitudinal and transversal direction, the Poisson's ratios and the allowable stresses or strains in the longitudinal and transversal directions. These last parameters are used to make the failure calculations. It is possible to select different failure criteria on the element properties.

For what concerns the boundary conditions, two different load cases were examined. The first one was the calculation of side stiffness of the wheel rim and the second one was a static load condition which simulates the real use of the wheel. For the first case, the load (200 N) was applied on the walls on a circular zone (with a diameter of about 150 mm) along the wheel rotating axis. The load was applied using a rigid one dimensional spider element. The constraints were applied on the opposite side of the wall, in this way the wheel rim was uniformly supported on its lateral surface. These conditions allow to measure the side stiffness of the wheel rim and reply some experimental tests done. These experimental tests will be illustrated in the following, they are aimed to validate the finite element model of the wheel rim. In the second load case, vertical loads were applied in the rotational centre of the two side walls of the wheel rim. The nodes of the wheel hub zone were connected to a single central node using rigid one dimensional spider elements. For each side a load of 900 N was applied. This value was obtained considering the worst load condition and applying a safety coefficient to consider short disconnectedness of the ground. To calculate the loads the single rear wheel of the prototype was considered, applying on this wheel the 60% of the moving vehicle mass (1,000 N).

The wheel was considered supported to the ground in its work position. For this reason some nodes on the perimeter of the wheel rim, in the support area were constrained. Moreover, to consider the worst condition, the vertical loads, and consequently the nodes constrained were applied in radial direction in correspondence of the hole for the valve.

3.2 FEM Results

For what concerns the lateral stiffness, it was calculated as the ratio between the load applied and the displacement obtained in the load direction and in the application point of load. For this wheel a value of about 90 N/mm was obtained.

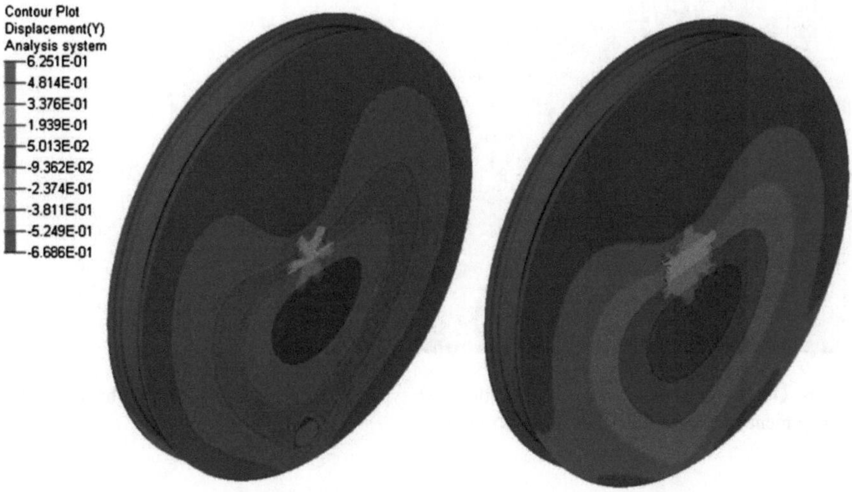

Fig. 5 Absolute displacement on both side of the wheel rim along Y axis

This value of stiffness refers only to the monocoque shell of carbon fibre and it does not consider the presence of the wheel hub which drastically increases the stiffness of the complete wheel.

For the second load case first of all were examined the displacements and in particular the attention was focused on component along the wheel rotating axis (Y axis, Fig. 5). The maximum value obtained is about 0.6 mm, which is far from 1 mm, which was considered the maximum accepted value for this application. Moreover the displacements are well distributed along the lateral wall, while in the wheel rim for IDRA08 the displacements were concentrated in correspondence of the slope change in the lateral wall. This is a first important result because in this zone, during the practice on the track, the wheel rim showed important dangerous vibrations. With this design, this phenomenon is avoided. This result was obtained redesigning the profile of the side wall of the wheel rim and introducing a layer of reinforcement foam. As it is possible to see in the Fig. 5, the presence of the valve hole created an important discontinuity on the wheel's walls, for this reason the displacements were not perfectly symmetrical. However the magnitude of the difference is very little. To solve this problem a further development of the wheel rim should erase the discontinuity created by the valve hole, using a specific little rubber pipe. In this way the number of composite layers could be further reduced.

For what concerns the structural behaviour, for each composite layer the failure index was examined. It is a parameter calculated by the solver on the basis of the material parameters introduced and on the failure criteria selected. It is based on energy criteria [18] and if the value obtained is higher than one it is possible to have delamination or break of lamina. For these analyses the Tshai–Hill criteria was adopted [18–20]. The different parts of the wheel rim, which have different ply

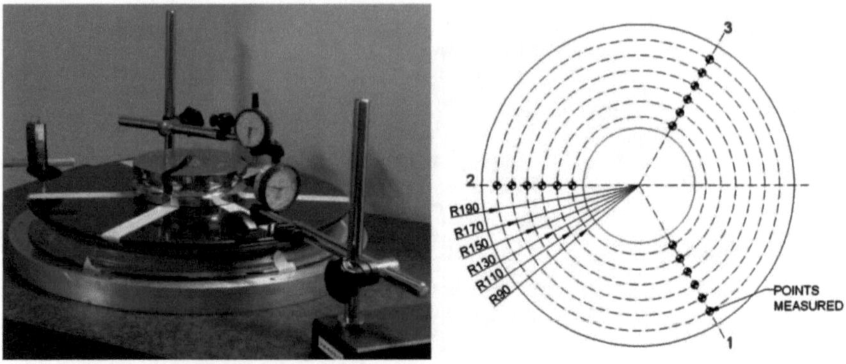

Fig. 6 On the *left*: the test layout for the first type of experimental test. On the *right*: the arrangement of measured points in a top view

lay-up, were separately examined. The values obtained from simulation were always very low and far from the critical value of one. Only few elements in correspondence of the constraints and near the wheel hub, where the loads were applied, needed more attention. However these elements were not very significant because their behaviour was high influenced by the boundary conditions which were worst than reality. The contribution of the tyre pressure and of the wheel hub could be considered in more detail.

4 Experimental Tests

On the lenticular wheel rim the experimental tests were done in order to validate the FEM model. In particular three different types of static and quasi static compression tests were done on the lateral wall of the wheel rim.

The test layout for the first type of test is shown in Fig. 6. The wheel rim is supported by one of its semi shell die (Fig. 6 on the left). On the other side, in the central zone, different defined masses were applied. In particular starting from 5 kg, the masses was increased to 20 kg with steps of 5 kg. The vertical displacements of the lateral walls were measured by means of comparators. In particular three different positions, with the same angular distance were considered. The measures were replicated at six different radius with steps of 20 mm from the wheel rim external edge. The arrangement of the measured points is shown, in Fig. 6 on the right. Both sides of the wheel rim were measured. Three replications for all measurements were done both increasing and then decreasing the weight. The average results of the vertical displacement measurements are shown in Fig. 7. It is possible to note that the displacements obtained on one side (figure on the left—side opposite valve) are higher than those on the other side (figure on the right—side with valve). This is due to a partial overlapping of

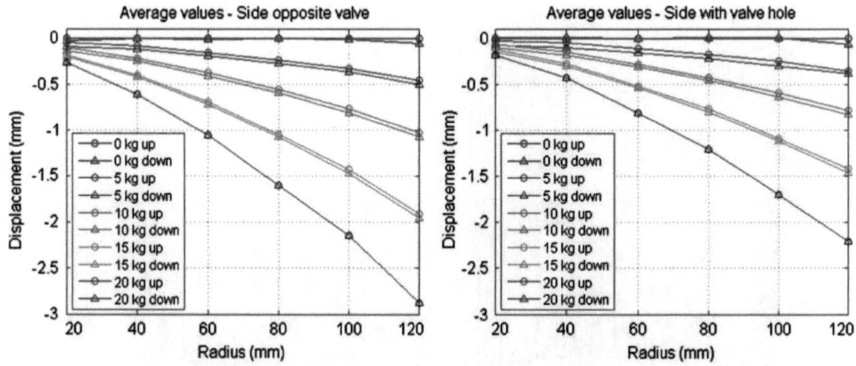

Fig. 7 *Vertical* displacements of the side walls for the first type of static compression test

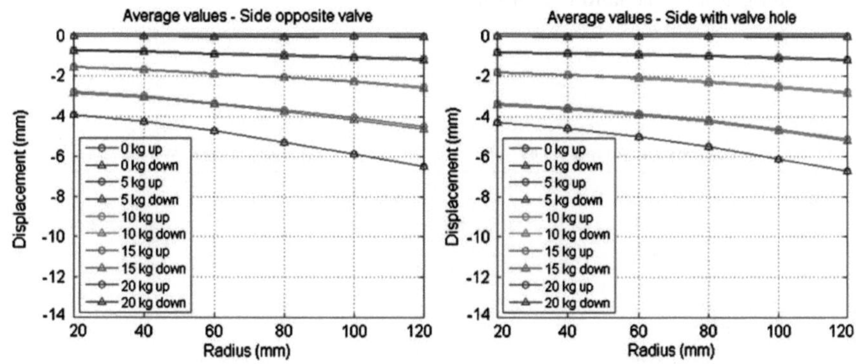

Fig. 8 *Vertical* displacements of the side walls for the second type of static compression test

prepregs layers on one side wall when the die is closed during the production process. However this partial asymmetry does not cause unbalance phenomena during wheel rotation.

The second type of test is very similar to the first one, but in this case the wheel rim is not supported by the semi shell die but it is supported on a rigid plane. In this way the contribution of both sides of walls is taken into consideration. The same points and the same replications were made also for this test. The results in terms of vertical displacements are shown in Fig. 8. The values obtained are much higher than those of first test, because there is the displacement also of the opposite side.

The third type of test was performed to better investigate the behaviour of the wheel in conditions that reproduce the real work situation. In particular the wheel hub and the tyre, pressurized at 6.5 bar, was assembled on the wheel rim. For this purpose a specific test bench was prepared (Fig. 9). The wheel is fixed on its shaft and a lateral force is applied by means of a pneumatic actuator. In this way the

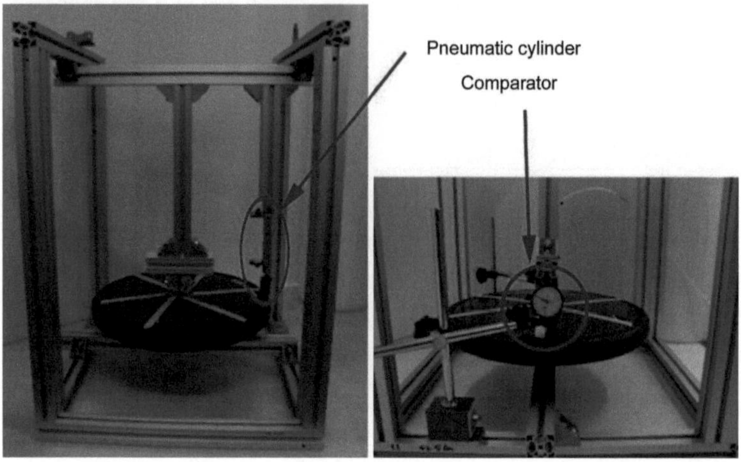

Fig. 9 Test bench for the third type of test

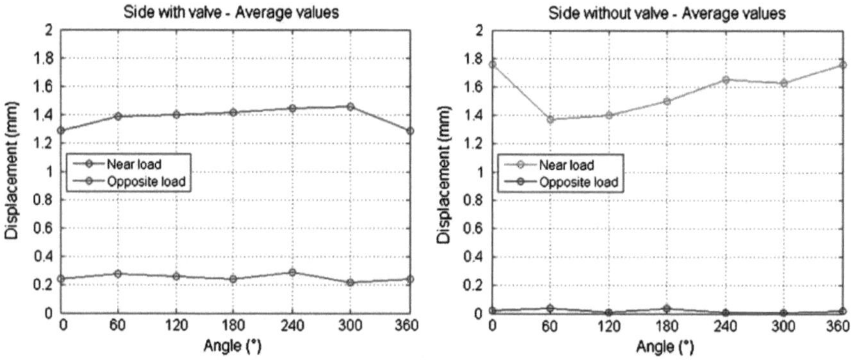

Fig. 10 *Vertical* displacements of the side walls for the third type of static compression test

work condition of the vehicle along a curve can be simulated. A force of 200 N (which correspond approximately to 0.5 g of acceleration) is applied on the tyre. The vertical displacements were measure always by means of two comparators: the first in correspondence of the load application and the second in a opposite diametrical position. The measurements were repeated every 60°. The tests results, in terms of displacements as a function of the angle are shown in Fig. 10. It is possible to note that the displacements are very low and the assembly wheel appears to be very stiff.

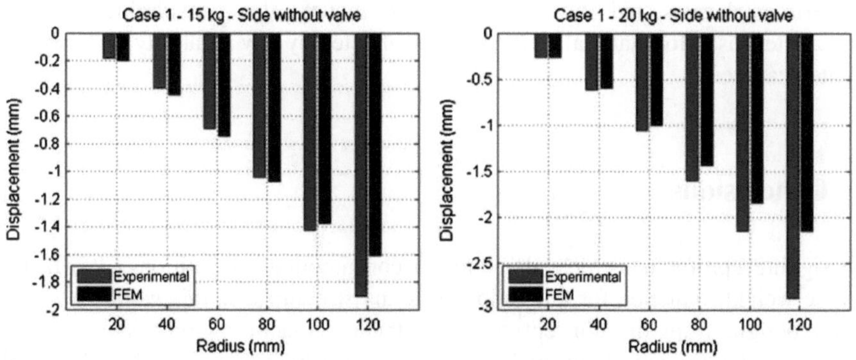

Fig. 11 Comparison between experimental and numerical results for *vertical* displacements of the side walls obtained with the first type of static compression test

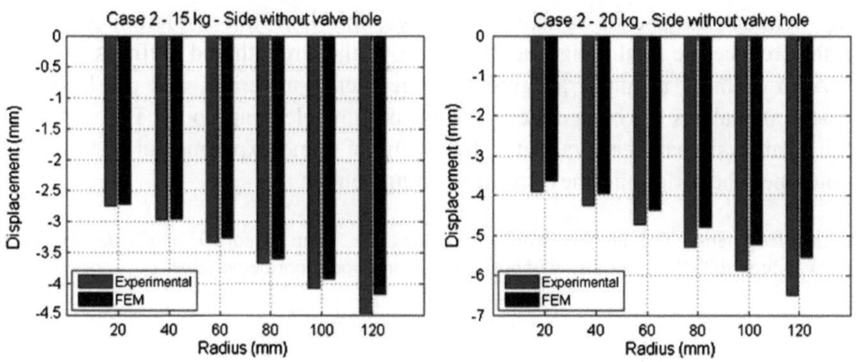

Fig. 12 Comparison between experimental and numerical results for *vertical* displacements of the side walls obtained with the second type of static compression test

5 FEM Model Validation with Experimental Data

The first two types of compression tests were simulated also by means of finite element model techniques, using the same model described in the paragraph 3, in order to validate this model for further improvements and more accurate analysis. In the model the constraints were applied like in the experimental tests, and the application of load was done by means of rigid one dimensional elements. The vertical displacements were evaluated on nodes positioned in the same points evaluated with the experimental tests. The results obtained with the finite element model are compared with the experimental ones in Fig. 11 and 12, considering two different masses (15 and 20 kg). Examining the results, excluding the points which are influenced by boundary conditions (in particular those at high radius) it was possible to conclude that there is a good correlation between the results of the experimental tests and those of the FEM model, also considering that the parameters used for material cards could be affected by low reliability and they are quite scattered .

6 Conclusions

Designing a prototype vehicle aimed to low consumption competitions a key factor is the weight reduction for all components and structures. In this work the structural design, from the concept phase to the manufacturing process, has been illustrated for the wheel rim of IDRA prototype, which participated in the Shell Eco Marathon competitions. The wheel rim for this type of vehicle has also an important aerodynamic function, for this reason a lenticular geometry has been chosen. To minimize the weight and to obtain a lenticular geometry, carbon fibre materials have been used. The wheel rim has been made in a single monocoque part, without the use of structural adhesive, this innovative solution allows to further reduce the final weight and to improve the strength and stiffness.

After defining a first arrangement for geometry, materials, and ply lay-up, a structural analysis has been done by means of finite element model. The structural design analysis has been validated by means of some experimental tests on the prototype wheel rim and they are in good agreement.

Acknowledgments The authors would to thanks the Team H_2politO—*molecole da corsa* (www.polito.it/h2polito) because without it, this work would not be possible.

References

1. Scattina, A.: Structure lightweight: a key priority for the automotive design. Ph.D. thesis, Politecnico di Torino (2009)
2. http://www.shell.com/eco-marathon. Accessed 10 January 2010
3. Halpin, J.: Primer on Composite Materials: Analysis. Technomic Publishing, USA (1984)
4. Frost & Sullivan: Future Automotive Materials Assessment (5, 10, 15 Year Outlook) (2006)
5. Edwards, K.L.: Strategic substitution of new materials for old: applications in automotive product development. Mater. Des. **25**, 529–533 (2004)
6. Frost & Sullivan: Advances in Automotive Plastics (2007)
7. Wall, E., et al.: Progress report for automotive lightweightinh materials I & II. Energy efficiency and renewable energy, Washington (2007)
8. Krajewski, P.F., Morales, A.T.: Materials for automotive and ground vehicles. ASM Int. J. Mater. Eng. Perform. **17**, 287 (2008)
9. Frost & Sullivan: Lightweight Materials for Automotive Applications (2005)
10. Trucker, N., Lindsey, K.: An Introduction to Automotive Composites. Rapra technology limited, UK (2002)
11. Harper, C.A.: Handbook of Plastics, Elastomers & Composite. McGraw-Hill, New York (2002)
12. Santin, J.J., Honder, C.H., Bernard. J., et al.: The world's most famous fuel efficiency vehicle (2007)

13. Galmarini, G., Gobbi, M., Mastinu, G., et al.: Construction of a CFRP wheel. In: ETDCM 8th Seminar on Experimental Techniques and Design in Composite Materials, Castiadas (2007)
14. Cristescu, N., Craciun, E.M., Soos, E.: Mechanics of elastic composites. Chapman and Hall, USA (2004)
15. Clarke, J.K.: Structural Design of Polymer Composites: EUROCOMP Design Code and Handbook. E & FN Spoon, UK (1996)
16. Garbo, S.P.: Composite Materials: Testing and Design, vol. 9. ASTM, USA (1990)
17. Timoshenko, S.: Theory of Plates and Shells. McGraw-Hill, New York (1969)
18. Altair Engineering: Radioss Theory Manual 10.0 Version. Troy, USA (2009)
19. Jones, R.M.: Mechanics of Composite Materials. Taylor & Francis, New York (1999)
20. Decolon, C.: Analysis of Composite Structures. Taylor & Francis, New York (2000)

Non Linear Shear Effects on the Cyclic Behaviour of RC Hollow Piers

P. Delgado, A. Monteiro, A. Arêde, N. Vila Pouca, A. Costa and R. Delgado

Abstract Reinforced concrete (RC) hollow section piers have been the subject of several studies in the recent past, from which one important drawn conclusion is the significant influence of shear effects on these piers behaviour, particularly under cyclic loading. In that framework, tests were carried out at Laboratory of Earthquake and Structural Engineering (LESE) from Faculty of Engineering of University of Porto (FEUP) on several reduced scale (1:4) RC hollow section bridge piers under lateral cyclic loading with constant axial force reported by Delgado (Bull Earthquake Eng 7:377−389, 2009). This work aims at presenting the numerical simulations performed for some of the referred piers in order to better understand phenomena associated with its cyclic behaviour, as observed in the experimental campaign. The numerical strategy was based on refined 3D Finite

P. Delgado (✉)
Escola Superior de Tecnologia e Gestão, Instituto Politécnico de Viana do Castelo, Av. Atlântico, 4900-348, Viana do Castelo, Portugal
e-mail: pdelgado@estg.ipvc.pt

A. Monteiro · A. Arêde · N. Vila Pouca · R. Delgado
Departamento de Engenharia Civil, Faculdade de Engenharia, Universidade do Porto, Rua Dr. Roberto Frias, s/n, 4200-465, Porto, Portugal
e-mail: abessa@fe.up.pt

A. Arêde
e-mail: aarede@fe.up.pt

N. Vila Pouca
e-mail: nelsonvp@fe.up.pt

R. Delgado
e-mail: rdelgado@fe.up.pt

A. Costa
Departamento de Engenharia Civil, Universidade de Aveiro, Campus Universitário de Santiago, 3810-193, Aveiro, Portugal
e-mail: agc@ua.pt

A. Öchsner et al. (eds.), *Materials with Complex Behaviour II*,
Advanced Structured Materials 16, DOI: 10.1007/978-3-642-22700-4_33,

Element Mesh (FEM) discretization using a two-scalar variable damage model for the concrete constitutive law and a suitable cyclic behaviour law for steel bars represented by uniaxial elements. As is well known, the shear effects are complex phenomena involving the global behaviour of the structural elements and where the non linear effects have a crucial role. In this chapter a detailed modelling was used allowing for realistic simulations of the non linear behaviour, which was found particularly suitable when significant shear effects are involved. The bond between the bars and the embedding concrete, by incorporating a bond stress—slip behaviour law in the numerical calculations is considered, an effect particularly important when thin plain steel bars are considered, as is the case of the tested piers. The adopted formulation is similar to the well known Eligehausen proposals although with slightly modified cyclic behaviour parameters.

Keywords Numerical modelling · Non-linear cyclic behaviour · RC hollow piers · Damage model · Shear effects · Bond stress-slip behaviour

1 Experimental Campaign

The test setup, shown in Fig. 1, makes use of a 500 kN actuator to apply lateral loads and a 700 kN actuator to apply axial loads. The specimen and reaction frame are bolted to the strong floor with high strength prestressed rods. A constant axial load was applied during the tests, herein described, while the lateral loads were cycled, under displacement controlled conditions. A special sliding device consisting of two steel plates, shown also in Fig. 1, was used to minimize the friction created by the axial loads. The lower plate was bonded to the specimen top, whereas the upper was hinged to the vertical actuator, allowing top-end displacements and rotations on the specimens to take place when lateral loading was imposed during the test. The upper plate was also connected to a load cell to measure the residual frictional force between the two plates. During the tests, the hydraulic system of the vertical actuator was designed to keep constant the oil pressure, in order to maintain constant the axial force.

The horizontal actuator control is done using a PXI controller system from National Instruments (NI) and specifically home developed control routines based on the LabVIEW software platform (also from NI). The data acquisition is also based on another PXI system equipped with acquisition and signal conditioning cards and allows direct reading of data from strain gauges, load cells, Linear Voltage Displacement Transducers (LVDTs) and other types of amplified analogical or digital sensors.

The specimens presented in this chapter correspond to the third group of piers tested within this framework, being the results of the first and second group already shown in previous reports, [1–3]. This set of specimens was based on square piers tested at the Laboratory of Pavia University, Italy, [4, 5]. The model

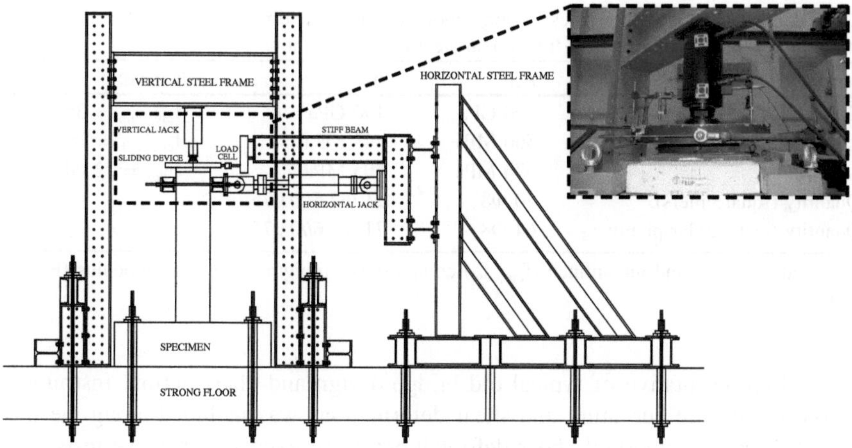

Fig. 1 Schematic layout of the test setup at LESE laboratory and sliding device used to apply the axial load

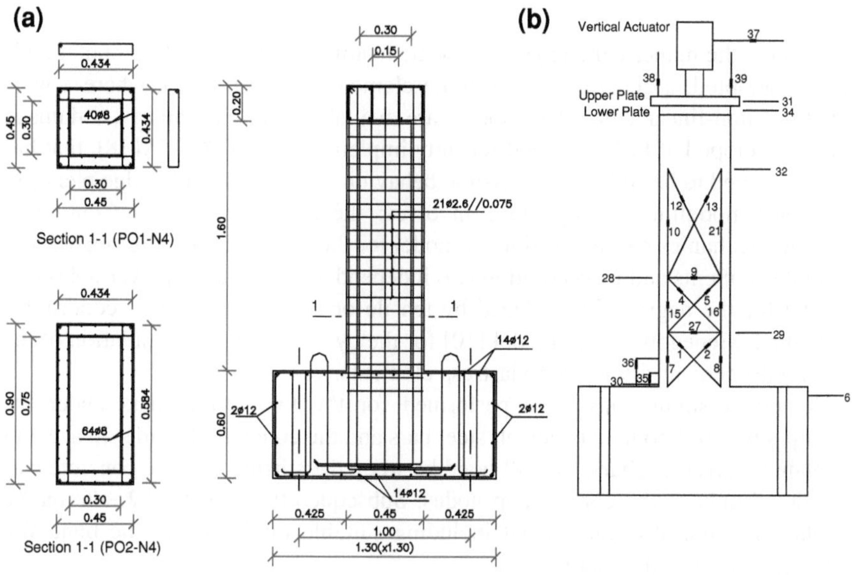

Fig. 2 Hollow RC piers: (**a**) model schemes (dimensions in meters) and (**b**) lateral LVDT layout

schemes shown in Fig. 2a correspond to ¼ scale representations of hollow section bridge piers, herein referred to as PO1-N4 for square section and PO2-N4 for rectangular section, with the material characteristics presented in Table 1. The transverse reinforcement details consisted of a single stirrup in each pier wall,

Table 1 Material characteristics: Young modulus (E_s), yielding strength (f_s) and ultimate strength (f_s) for longitudinal and transversal steel reinforcement

Steel (A_{sl}; A_{st})	A_{sl}	A_{st}	Concrete
E_{sl}; E_{sw}	200 GPa	190 GPa	$f_{cm,\ cub} = 35.7$ MPa
f_{sy}; f_{swy}	560 MPa	443 MPa	$f_{cm,\ cyl} = 28.5$ MPa
f_{su}; f_{swu}	670 MPa	505 MPa	$E_{cm} = 22$ GPa
Quantity (square piers):	40 Φ8	21 Φ2.6//0.075	
Quantity (rectangular piers):	64 Φ8	21 Φ2.6//0.075	

Confined ($f_{cm,\ cub}$) and unconfined ($f_{cm,\ cyl}$) compressive strength and elastic modulus (E_c) for concrete

deemed representative of typical old bridge design and construction. Instrumentation to measure curvature and shear deformations was included along the pier height, because important shear deformations were expected in these tests. The LVDT configuration used in both specimens is shown in Fig. 2b.

2 Numerical Simulations

Concerning the numerical analysis for cyclic loading, the CAST3M computer code [6] was adopted, a general purpose finite element based program, where a wide variety of non-linear behaviour models are available and, particularly, a damage model developed at FEUP [7] and recently implemented in CAST3M [8], that has already proved to be suitable for seismic behaviour analysis of RC bridge piers [9]. This later modelling strategy thus involves: the above mentioned Continuum Damage Mechanics based constitutive model for the concrete zone discretized into 3D finite elements and incorporating two independent scalar damage variables that account for the degradation induced by tensile or compressive stress conditions; the Giuffré-Menegotto-Pinto model [10] for the cyclic behaviour simulation of the steel reinforcement discretized via truss elements.

Numerical simulations were performed for the two tested piers under the assumption of full bond between all steel bars and the concrete. In order to account for plain transverse rebars, the full bond between the stirrups and the concrete was considered only at the corner loop nodes. Subsequently, in Sect. 2.2, alternative simulations were also carried out including suitable modelling of the bond-slip between stirrups and concrete.

The modelling strategy was applied to the experimentally tested specimens and the analysis of obtained results was focused first on the response comparison in terms of force–displacement plots concerning both experimental and numerical findings. The flexural and shear capacity curves are also depicted as calculated by Delgado [11] where, the flexural capacity cross-section was obtained by performing a moment–curvature analysis of the section, while for shear strength a methodology suggested by Priestley et al. [12] was adopted, which can be conveyed by Eq. (1):

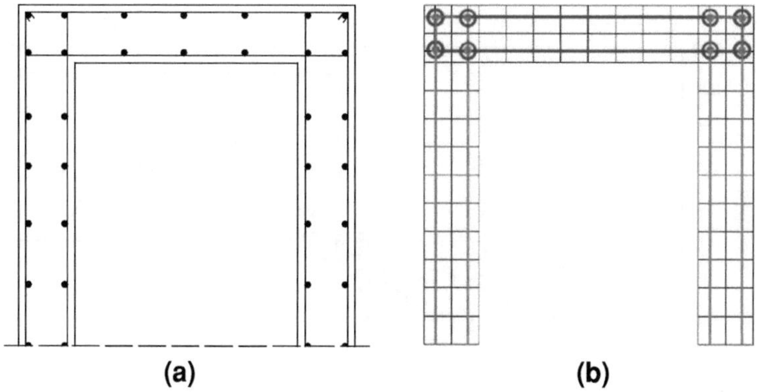

Fig. 3 Rectangular hollow pier cross-sections. (**a**) "as designed" and (**b**) finite element discretization

$$V_d = V_c + V_s + V_p \quad (1)$$

where V_c, V_s and V_p are the shear force components accounting, respectively, for the nominal strength of concrete, the transverse reinforcement shear resisting mechanism and the axial compression force.

Regarding those plots, four drift levels are also included. Those levels, D1, D2, D3 and D4 represent 0.21, 0.93, 1.43 and 2.14% respectively, which were the reference drift levels used in all the numerical applications for comparisons sake throughout the entire campaign. In addition, some other representative parameters were analysed, namely deformed configurations, maps of principal stress values and directions, strain and stress distributions in steel reinforcement and maps of concrete damage variables.

2.1 Piers Considering Full Bond Conditions

The square and rectangular cross-section piers, respectively PO1-N4 and PO2-N4, have the same transverse reinforcement ratio and single loop type (shown in Fig. 2), the same dimension in the loading direction but a larger cross-section width.

These piers transverse reinforcement was made with plain bars, thus leading to a weaker bond mechanism which is likely to have an influence on the strength and, particularly, on the deformation capacity of the specimen. In order to account for the low bond interaction between the transverse reinforcement (stirrups) and the surrounding concrete, the stirrup mesh was connected to the concrete mesh only at the corner loop nodes (Fig. 3).

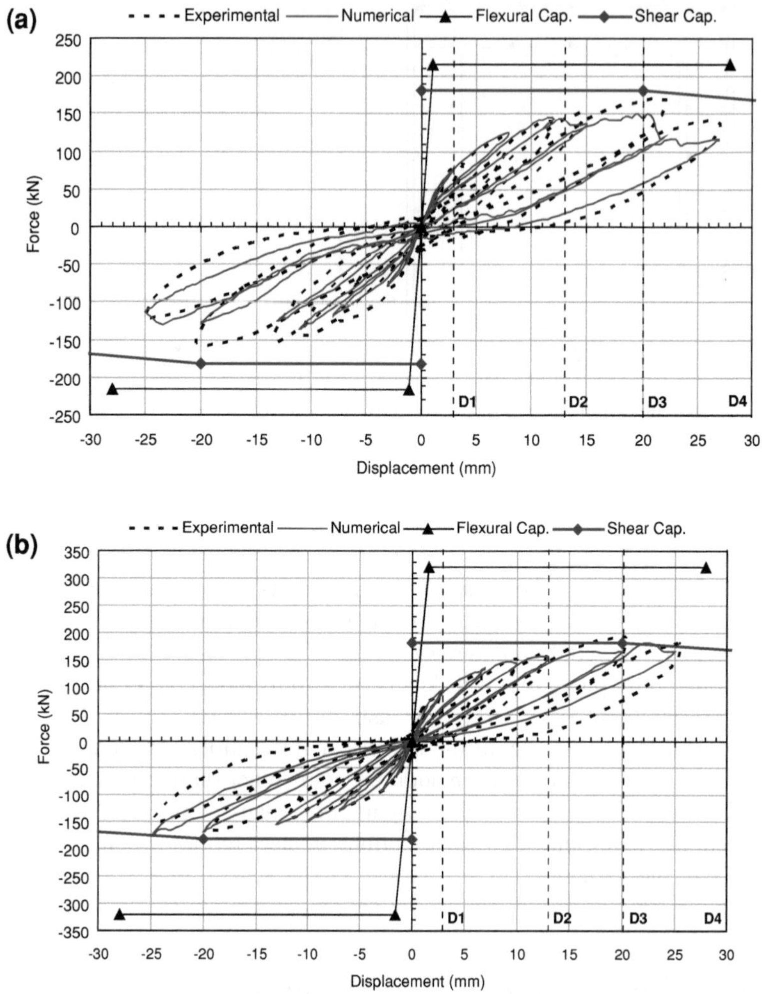

Fig. 4 Force-displacement for (**a**) PO1-N4 and (**b**) PO2-N4 piers. Experimental versus numerical comparison

Therefore, the flexural capacity of PO2-N4 pier was increased (due to larger steel contents and compression web area) while the shear capacity of both remained the same. As observed in the experimental tests, shear effects showed a clear influence on the piers numerical response, wherein the expected shear force levels were satisfactorily reached (Fig. 4).

It should be remarked, however, that a sudden force decay is also exhibited by the numerical response of PO1-N4 pier for the 21 mm cycle; in the numerical model this effect was found to be due to local concrete crushing near the footing caused by the stress accumulation in that zone induced by the footing, without an

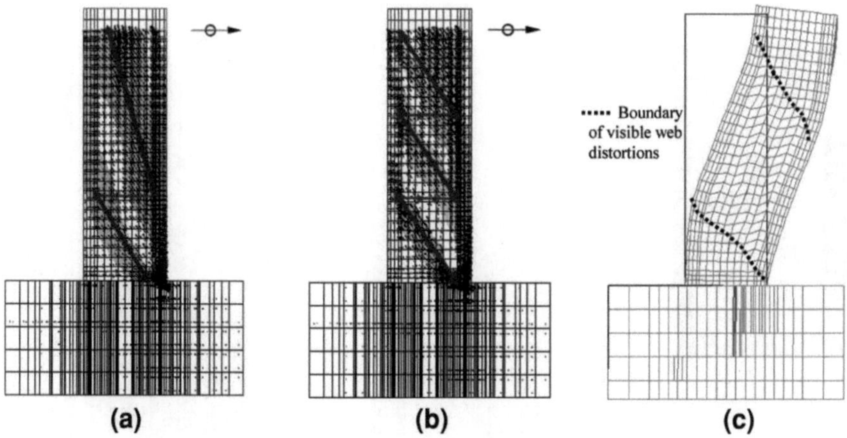

Fig. 5 PO1-N4 pier. Compressive principal stress directions (*solid diagonal lines*) for: (**a**) 12 mm displacement, (**b**) 20 mm displacement and (**c**) corresponding deformed shape

appropriated concrete confinement, as certainly occurred in the experimental test. However, although this effect has prevented a better response simulation for that displacement level, it was found irrelevant for the development of the observed shear effects.

In fact, shear effects are well captured by the numerical model as evidenced in Fig. 5a, b where compressive principal stress directions are illustrated for two drift levels (solid diagonal lines), showing the formation of the well-known strut-and-tie mechanism typical of a shear dominant stress state; as exhibited in Fig. 5b, this effect becomes more apparent for larger drifts. In accordance with the latter plot, Fig. 5c shows the corresponding deflected shape evidencing significant distortions in the PO1-N4 pier web where approximate borderlines of the distorted zone are sketched by the thick dotted lines on the deformed shape.

The numerically simulated flexural effects showed little influence on this pier response. Indeed, the plastic hinge mechanism was not mobilized as evidenced by the longitudinal reinforcement stress maps which indicate that yielding was not reached.

The above observations point again to shear-driven collapse of both piers, which is readily supported by experimental evidence. In fact, generalized yielding of transverse reinforcement web bars was achieved, which agrees with strong diagonal cracking observed in the experimental test (illustrated in Fig. 6a for pier PO2-N4, also available in [11]), eventually causing some concrete spalling near the longitudinal steel bars. Accordingly, these experimental results are supported by numerical simulation, where the shear stress map (τxz) clearly suggests the referred web distortions (Fig. 6b) and the large concrete principal strains was also able to detect the cracking pattern in the pier web (Fig. 6c).

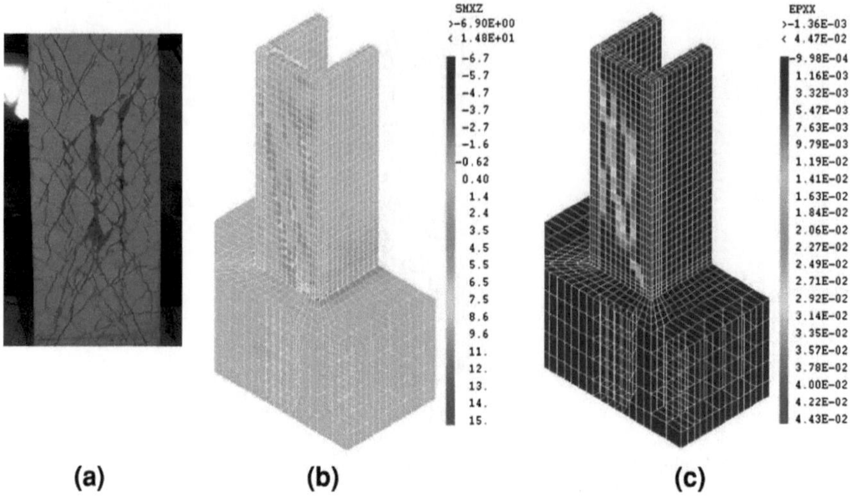

Fig. 6 PO2-N4 pier for 20 mm displacement: (**a**) web damage, (**b**) Numerical shear stress map (MPa) and (**c**) Numerical concrete principal strains map

Fig. 7 Refinement scheme of cross-section finite element discretization

2.2 Piers Including Stirrup Bond-Slip Behaviour Modelling.

From previously presented results, although the global behaviour was satisfactorily simulated, none of the cases reached the peak force recorded in the experimental tests. Therefore, some considerations were made regarding possibilities of refining their numerical models in order to improve results.

A careful analysis of both pier results showed that the generalized yielding of transverse reinforcement was reached for small drift levels, typically within the

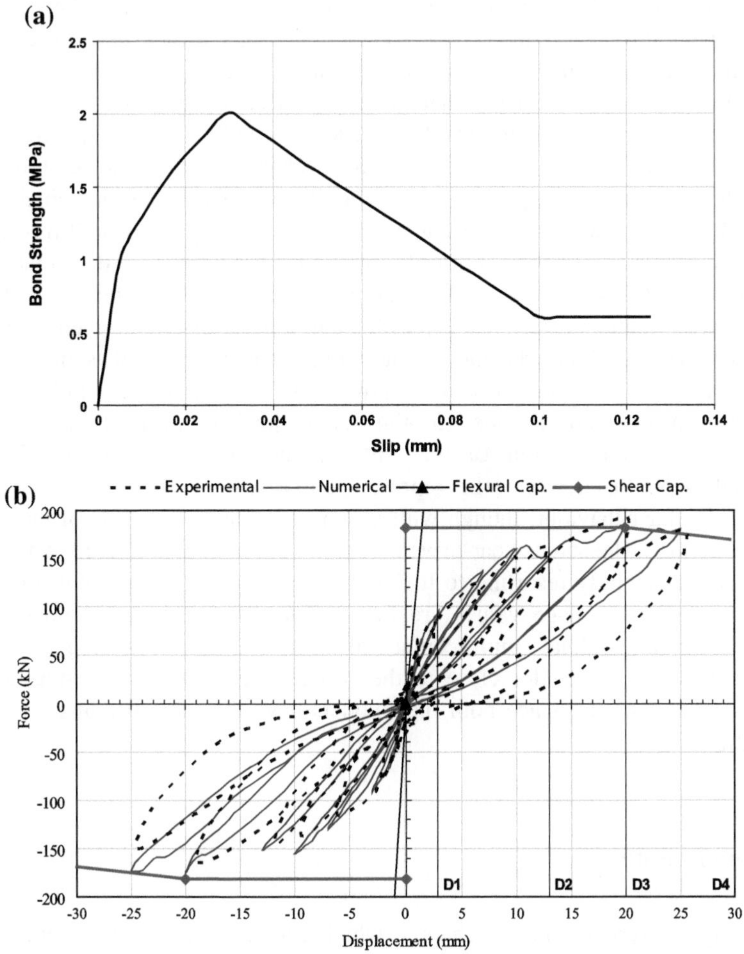

Fig. 8 Model improvement for pier PO2 simulation: (**a**) Adopted bond stress-slip law:$\tau(s)$ and (**b**) Force–displacement response with model 2 (refined model with bond-slip law)

range of 10−15 mm. Since the numerical peak force was expected to be achieved for a displacement level around 20 mm (by comparison with the experimental results), when both piers were already strongly governed by their shear resisting mechanisms, it is reasonable to assume that the early spread of stirrup yielding might have restricted the global pier behaviour, namely by not allowing it to reach larger peak forces.

Moreover, since stirrups are made of plain steel bars, once bond strength is reached and destroyed for a given loading direction, they are no longer effective in the other direction because this type of bond strength does not recover during load reversals.

Additionally, it should be reminded that each stirrup bar was modelled with just one element throughout the whole web length. This means that, once stirrup yielding occurs, the entire web length (attached to the transverse bar end nodes) might have a reduction on the shear resisting mechanisms. Thus, a new model discretization was carried out, which consisted in dividing each unique transverse bar element into several elements, so that a better spread of the reinforcement strength can be achieved, particularly in the yielding phase, but including a bond stress-slip behaviour law to the bar elements simulating the web stirrups (bars delimited by circles, in Fig. 7) in order to control each pier web deformability.

Therefore, the improved model was developed based on the original PO2 numerical simulation model, namely with the new transverse reinforcement refinement scheme (Fig. 7) and the added bond-slip law (herein denoted as model 2).

The adopted bond-slip behaviour law, $\tau(s)$, was defined according to the parameters proposed by Tassios [13], thus taking the form illustrated in Fig. 8a for plain steel bars embedded in concrete with the characteristics included in Table 1.

Results from this new model is represented in Fig. 8b. As evidenced, the model yielded similar behaviour, though with an improved approximation of the experimental results. The same shear-driven response previously described can also be observed, particularly resorting to the compressive damage maps (not illustrated here) where low damage values are found. The distribution of strain ductility demands in longitudinal reinforcement also showed small yielding excursion. Nonetheless, for the cycles close to the collapse drift level, the results are, as expected, very similar since both pier models have the same bending and shear strengths.

3 Conclusions

The analyses presented in this chapter, which comprised both square and rectangular cross-sections, have highlighted the adequacy of the modelling strategy adopted to simulate the most important features of the cyclic behaviour of reinforced concrete hollow piers. Particularly, shear effects were herein thoroughly addressed with the Continuum Damage Model that was found to be suitable for numerically capturing concrete behaviour under strong shear forces.

Concerning the square piers, shear effects were also seen to be of key importance and governed the global numerical response. However, larger bending influence was detected (in comparison with the rectangular section cases) since curvatures were more concentrated near the pier base and significant yielding spread was found along the longitudinal reinforcement.

As described in the chapter, an improvement was proposed for the modelling strategy regarding the anchorage of the transverse reinforcement in order to account for the interface slippage between concrete and stirrups. The refinement included in this model showed better ability to represent a shear dominated behaviour, where insufficient shear strength mechanisms were observed due to premature generalized yielding of stirrups. Unlike the initial modelling strategy, the proposed model with the referred improved anchorage scheme was found to provide an improved approximation of the experimental test, namely the peak force level obtained.

In particular, by coupling a more detailed scheme of transverse rebar anchorage to the surrounding concrete (which adds on increased web stiffness) with a bond-slip behaviour law for the steel–concrete interface, more satisfactory results are obtained comparatively to experimental ones. Thus, considering the obtained numerical results, the adopted modelling strategy can be considered good enough for the purposes of the cyclic simulations herein presented.

Acknowledgements This chapter refers research work made with financial contribution of "FCT - Fundação para a Ciência e Tecnologia", Portugal, that is gratefully acknowledged.

References

1. Delgado, R., Delgado, P., Pouca, N.V., Arêde, A., Rocha, P., Costa, A.: Shear effects on hollow section piers under seismic actions: experimental and numerical analysis. Bull Earthq. Eng. **7**: 377–389 (2009) (DOI: 10.1007/s10518-008-9098-x)
2. Delgado, P., Rocha, P., Rodrigues, V., Santos, M., Arêde, A., Vila Pouca, N., Costa, A., Delgado, R.: Experimental cyclic tests and retrofit of RC hollow piers. In: Proceedings of the 13th European conference on earthquake engineering (13ECEE), Geneva, Switzerland, Paper N. 1205, 3–8 September 2006
3. Delgado P., Rocha P., Pedrosa J., Arêde A., Vila Pouca N., Santos M., Costa A., Delgado R.: Retrofitting of bridge hollow piers with CFRP. In: COMPDYN 2007-ECCOMAS thematic conference on computational methods in structural dynamics and earthquake engineering. Rethymno, Crete, Greece, Paper N. 1492, 13–16 June 2007
4. Calvi, G.M., Pavese, A., Rasulo, A., Bolognini, D.: Experimental and numerical studies on the seismic response of R.C. hollow bridge piers. Bull. Earthq. Eng. **3**(3), 267–297 (2005)
5. Pavese, A., Bolognini, D., Peloso, S.: FRP seismic retrofit of RC square hollow section bridge piers. J. Earthq. Eng. 1 SPEC. ISS., 225–250 (2004)
6. Pasquet, P. (ed.): CEA Manuel d'utilisation de Cast3 m. Commissariat à l'Énergie Atomique. (2003)
7. Faria, R., Oliver, J., Cervera, M.: A strain based plastic viscous damage model for massive concrete structures. Int. J. Solids Struct. **35**(14), 1533–1558 (1998)
8. Costa, C., Pegon, P., Arêde, A., Castro, J.: Implementation of the damage model in tension and compression with plasticity in cast3 m. Report EUR, ISPC, CEC, JRC, Ispra (VA), Itália, (publication stage) (2005)
9. Faria, R., Vila P.N., Delgado, R.: Simulation of the cyclic behaviour of R/C cectangular hollow section bridge piers via a detailed numerical model. J. Earthq. Eng. **8**(5), 725–748 (2004)
10. Giuffrè, A., Pinto, P.: Il comportamento del cemento armato per sollecitazione ciclice di forte intensitá. Giornale del Genio Civile (1970)
11. Delgado, P. Avaliação da segurança sísmica de pontes". Ph.D., FEUP, 2009 (in Portuguese), (http://ncrep.fe.up.pt/web/artigos/PDelgado_PhD_Thesis.pdf)
12. Priestley, M.J.N., Seible, F., Calvi, G.M.: Seismic design and retrofit of bridges. Wiley, New York (1996)
13. Tassios, T.: Properties of bond between concrete and steel under load cycles idealizing seismic actions. Boletin d'Information du CEB. **131**, 67–122 (1979)

Heat Transfer in a Packed-Bed Cylindrical Reactor of Elliptic Cross Section: Mathematical Modeling and Simulation

Antonio Gilson Barbosa de Lima, Laércio Gomes de Oliveira and Wagner Celso Paiva Barbosa de Lima

Abstract This chapter provides information about heat transfer in porous media such as definition and modeling, with particular emphasis to packed bed reactor. The packed-bed reactors with heated or cooled walls are frequently used to carry out solid–gas reactions in many engineering applications. The design of a fixed bed reactor requires an extensive knowledge of heat transfer characteristics within the equipment. Here, application to chemical industry has been given to predict the steady-state three-dimensional heat transfer inside a packed-bed elliptic cylindrical reactor including chemical reaction term. The energy conservation equation, written in the elliptical cylindrical coordinate system, was discretized by using finite-volume method. Effects of the fluid-wall heat transfer coefficient, inlet reactant fluid concentration, inlet fluid temperature, and pre-exponential factor of the Arrhenius equation on the temperature profiles within the reactor are presented. To validate the proposed model, numerical results of the heat transfer obtained for a circular cylindrical reactor are compared with experimental data and a good agreement was obtained.

A. G. B. de Lima (✉) · W. C. P. B. de Lima
Department of Mechanical Engineering, Center of Science and Technology,
Federal University of Campina Grande (UFCG), PO Box 10069,
Campina Grande-PB, 58429-900, Brazil
e-mail: gilson@dem.ufcg.edu.br

W. C. P. B. de Lima
e-mail: gentil_wagner@hotmail.com.br

L. G. de Oliveira
Department of Chemical Engineering, Center of Science and Technology,
Federal University of Campina Grande (UFCG), PO Box 10057,
Campina Grande-PB, 58429-900, Brazil
e-mail: laercio@deq.ufcg.edu.br

A. Öchsner et al. (eds.), *Materials with Complex Behaviour II*, 549
Advanced Structured Materials 16, DOI: 10.1007/978-3-642-22700-4_34,

Notation

C	Concentration (mol/m^3)
c_p	Specific heat of the fluid (J/kg/K)
d_p	Particle diameter (m)
Ea	Activation energy (J/mol)
h_w	Wall-bed heat transfer coefficient (W/m^2/K)
H	Height of the bed (m)
$\Delta\hat{H}$	Heat of reaction (J/mol)
J	Jacobian of the transformation (1/m^2)
k	Effective thermal conductivity of the bed (W/m/K)
L	Focal length of the ellipse (m)
L_1	Minor axis of the ellipse (m)
L_2	Major axis of the ellipse (m)
$\dot{q}$	Heat generation (W/m^3)
R	Radius of the bed, m
$\bar{R}$	Universal constant of gas (J/mol/K)
t	Time (s)
T	Temperature (°C)
To	Temperature in the entrance of the reactor (°C)
T_w	Wall temperature of the reactor (°C)
Tabs	Absolute temperature (K)
u_z	Interstitial velocity of fluid (m/s)
z	Axial position (m)
α and β	Coefficients
ρ	Density of the fluid (kg/m^3)
ε	Porosity of the bed
ξ, η	Radial and angular elliptical coordinates

1 Heat Transport and Forced Flow in Reactive Porous System

1.1 Fundamental of Porous Media

In fluid–solid systems, either both phases are moving or the solid is stationary. The stationary solid with the fluid moving through it is considered as a porous media and the solid phase is envisioned as being made of particles. These particles can be consolidated (for example, rock, and foams) or nonconsolidated as in packed beds of pellets (for example, granular material) [26, 48]. According to Nield and Bejan [37], Bejan [5] and Nield [36], porous media is a material consisting of a solid matrix with an interconnected void. In the usual situation, the solid matrix is either rigid or it undergoes small deformation. The interconnectedness of the void (pores) allows the flow of one or more fluids through the material. In a natural porous media the distribution of pores with respect to shape and size is irregular.

Transport phenomena in porous media represent an important research area related to heat and mass transfer and fluid fields. The practical interest in convective heat transport in porous media is expanding quickly, due to a wide range of applications in several areas of science and technology for several geometries and different processes. This interest occurs due to the porous media to be present in grains, fruits, vegetables, soils and plants, as well as in different branches of agricultural, food, civil, mechanical, petroleum and chemical engineering such as: combustion, filtration, distillation, absorption and adsorption in filled columns, drying and catalytic reactions in fixed and fluidized beds as well as in petroleum reservoir, storage of geothermal energy, recoverable systems, pollutant dispersion in aquifers, irrigation systems, etc. to name just a few applications of this area. Many of these systems are complex and involve two-phase flow, phase change, heterogeneity, property variations, localized effects among others characteristics [8, 9, 13, 16, 20, 23, 27, 45, 46, 51, 52, 56, 57, 61, 62].

1.2 Heat Transfer in Packed Bed Reactor

The packed bed tubular reactors of heated or cooled walls are often used in the industry to carry out homogeneous or heterogeneous catalytic reactions (gas–solid) which can be exothermic or endothermic. In solid–gas transport-reactions systems, reactions can occur within the fluid phase, within the solid phase, or at solid–fluid interface. Reactions between molecules of different constituents can occur during their collision and in some cases a third and still different molecule may be needed for a reaction to occur. There are many chemical reactions which proceed only very slowly, on not at all, except in the presence of a catalyst. In despite of its importance, chemically reactive flow both viscous (non-porous) fluids and fluid-saturated porous media have received for less attention than those of non-reactive systems [45]. The design of a packed bed reactor requires an extensive knowledge of heat transfer inside the equipment. However, for a good design of such equipment theoretical and experimental studies and an adoption of efficient models based on experiments is required. Models describing transport phenomena in packed bed reactors are generally presented in a form of partial differential equations containing temporal and spatial derivatives. The complexity of real processes taking place in a packed bed reactor not only leads to uncertainties and difficulties with their mathematical formulation, but also cause considerable problem with the numerical and analytical treatment of resulting equation.

Numerical and analytical solutions of the heat diffusion/convection equation in fixed bed reactor, for several cases have been reported by many researchers in early works [1–3, 6, 7, 11, 18, 21, 24, 34, 40, 42, 50, 54, 58, 60]. However, all the studies are limited to the use of rectangular or circular cylindrical geometries.

Hence, rigorous solutions to the heat transfer in other geometrical shapes are essential, for the predictions of both the performance and control of the reactor.

For ducts and fixed beds with a more complex geometry, or exhibiting boundary conditions that involve time and space dependence, or for temperature dependent fluid properties, etc., analytical solutions, and even approximate solutions are more difficult to be obtained and, therefore, are not often found in the literature.

In recent years, special attention has been given to ducts of elliptic cross-section which have increased heat transfer rates compared to circular pipe [30, 31, 43, 49]. The problem of forced convection heat transfer from duct with elliptic cross-section is important because of its numerous direct applications in heat exchangers, energy conservation and many others [4]. More recently, Oliveira et al. [41] reports a 3D theoretical study of heat transfer in a fixed bed tubular reactor of elliptic cross section, but no reference to chemical reaction was given.

2 Modeling Conduction/Convection Heat Transfer in Homogeneous Porous Media

2.1 Heat Transfer Model

Heat transfer through fully-saturated matrices, as with heat conduction/convection through any heterogeneous media, depends on the structure of the matrix and the thermal properties of each phase (fluid and solid). One of the most difficult aspects of the analysis is structural modeling. Since the thermal properties of the solid phase is generally different than that of the fluid, the manner in which the solid is interconnected influences the conduction/convection heat transfer significantly.

In packed bed reactor, the evaluation of the heat transfer rates and temperature distribution is essential in the control and performance of the reactor [12]. The heat transfer in porous media can be modeled of two ways: local thermal equilibrium (LTE), and local thermal non-equilibrium (LTNE). In the local thermal equilibrium approach, both solid and fluid phase temperatures are represented by the unique value, then we have the absence of any interfacial heat transfer or heat transfer between phases occurs quickly (temperature reach equilibrium instantaneously). In this case, $T_s = T_f = T$, where T_s and T_f are the temperature of the solid and fluid phases, respectively [8, 10, 14–16, 26, 29, 37, 38, 47, 57, 61] and the multiphase system can be treated as a single continuum (single media representing all the phase). This assumption is an idealization and can simplify the theoretical treatment and be used when the dynamic and thermal behaviors are relatively slow, thus providing ample time for efficient heat transfer between the individual phases [57]. However for many industrial applications such approach cannot be considered, mainly in high-speed flows transient situations and when there is significant heat source in any one of the two phases [16, 27]. Rees and Pop [47] report the conditions to validate the LTE approach for some physical problems.

The heat transfer phenomenon in porous media by assuming LTE approach is given by the macroscopic conservation equation of energy as follows [25, 48]:

$$\frac{\partial}{\partial t}(\lambda\Phi) + \nabla \cdot (\lambda\vec{v}\Phi) = \nabla \cdot (\Gamma^{\Phi}\nabla\Phi) + S^{\Phi} \tag{1}$$

In the Eq. 1, we have $\lambda = \rho c_p \varepsilon$; $\Phi = T$ and $\Gamma^{\Phi} = k$, where ρ, c_p, T and k, represent density, specific heat, temperature and effective thermal conductivity, respectively, while ε, it is the porosity of bed, and t is the time. In this same equation, $\vec{v}$ is the velocity vector and S^{Φ} represents the source term given by:

$$S^{\Phi} = \zeta T \frac{DP}{Dt} + \mu\varphi + \dot{q} \tag{2}$$

where DP/Dt is the substantive derivative of the pressure, $\mu\phi$ is the viscous dissipation term (local production of thermal energy by internal friction), $\dot{q}$ is the heat internal generation, and ζ is the coefficient of thermal expansion. For an ideal gas $\zeta = 1/T$. The porosity of a porous media is defined as the fraction of the total volume of the porous media that is occupied by the void space.

2.2 Chemical Reaction Model

The special interest in chemical reactions as a heat source is in the temperature dependence of the rate of reaction [26].

Besides, if the consumption of one mole of reagent causes the heat energy to increase by an amount $-\Delta\hat{H}$ due to the reaction, then the increase in energy per unit volume of the fluid mixture is

$$\dot{q} = \Delta\hat{H}\frac{dC}{dt} \tag{3}$$

where $\Delta\hat{H}$ is the heat of reaction and dC/dt represents the reaction rate. For exothermic reactions, $\Delta\hat{H}$ is negative and for endothermic reaction, $\Delta\hat{H}$ is positive. This heat source is the major source of the non-uniformity of the temperature distribution across the reactor.

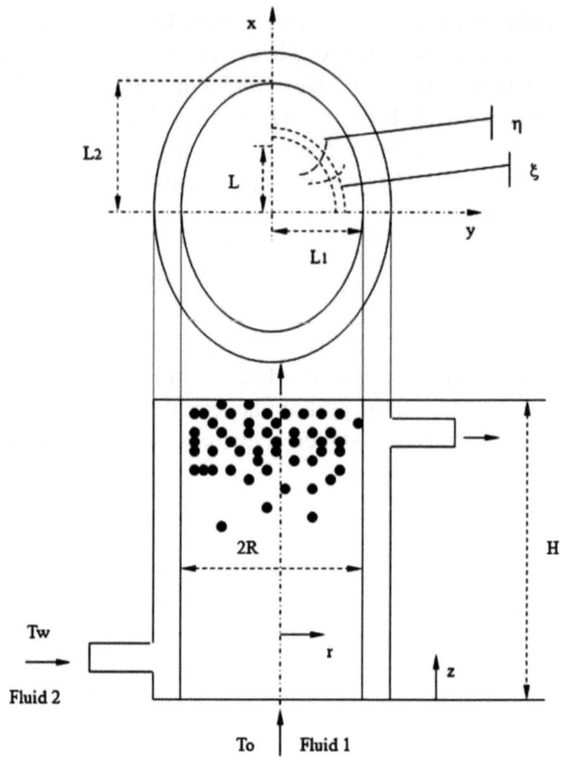

Fig. 1 Scheme of the packed-bed elliptic cylindrical reactor

3 Heat Transfer in Packed Bed Elliptical Cylindrical Reactor

3.1 Governing Equations

Now, consider a fluid mixture flowing through the interstices of a porous media with heat being removed from the system by a percoolant flowing on the outside of the equipment, according to Fig. 1. This figure presents a scheme of the fixed bed reactor with an elliptic cylindrical shape. Fluid 2 maintains the wall temperature of the reactor with a constant value. In this figure, ξ and η are the elliptic cylindrical coordinates. Information of the elliptic cylindrical coordinates system and its relationships with the Cartesian coordinate system can be found in literature [22, 33, 53].

To model heat transfer inside the packed-bed cylindrical reactor of elliptic cross section including chemical reaction, the following assumptions were used:

(a) Steady-state operation;
(b) The packed bed is isotropic;
(c) The thermo-physical properties of the fluid are constant during the process;

(d) The process occurs with internal energy generation due to chemical reaction in the fluid;
(e) Temperature and velocity distributions are constant and uniform at the entrance of the bed;
(f) Symmetry of the temperature profile at the center of the reactor (homogeneous flow);
(g) The fluid velocity and porosity are uniform and constant in the bed;
(h) Convective boundary condition at the surface of the equipment;
(i) Fluid velocity is lower than sonic velocity, so the viscous dissipation and work done by pressure changes are negligible;
(j) No distinction is made between the solid and fluid phases and local thermal equilibrium between the phases occurs;
(k) Cooling fluid temperature is constant along the equipment.

Based on those hypotheses and geometry presented in Fig. 1, the energy conservation equation, written in elliptic cylindrical coordinates, for a three-dimensional case, is given as follows:

$$\frac{\partial}{\partial z}\left(\frac{\rho c_p \varepsilon u_z T}{J}\right) = \frac{\partial}{\partial \xi}\left(k\sqrt{\frac{\xi^2-1}{1-\eta^2}}\frac{\partial T}{\partial \xi}\right) + \frac{\partial}{\partial \eta}\left(k\sqrt{\frac{1-\eta^2}{\xi^2-1}}\frac{\partial T}{\partial \eta}\right) + \frac{\partial}{\partial z}\left(\frac{k}{J}\frac{\partial T}{\partial z}\right) + \dot{q} \tag{4}$$

where u_z is the interstitial velocity of the fluid and J is the Jacobian of the transformation obtained by:

$$J = \frac{1}{L^2}\frac{\sqrt{(\xi^2-1)(1-\eta^2)}}{\xi^2-\eta^2} \tag{5}$$

with $L = \sqrt{L_2^2 - L_1^2}$ being the focal length, and L_2 and L_1 the major and minor axis of the ellipse (see Fig. 1).

To model effects of a chemical reaction inside the reactor, we suppose that the rate equation for the first order reaction is given as follows [37]

$$\frac{dC}{dT} = -\hat{k}C \tag{6}$$

where C is the reactant concentration in the fluid mixture.

The rate coefficient $\hat{k}$ is a function of the absolute temperature T given by the Arrhenius equation as follows

$$\hat{k} = A_0 \exp\left(-\frac{E_a}{\bar{R}T_{abs}}\right) \tag{7}$$

where E_a is the activation energy of the reaction (energy per mole), $\bar{R}$ is the universal constant of gas, and A_0 is a constant called the pre-exponential factor.

We assume further that the solid material of the porous media is inert, that the reaction produces a product whose mass can be neglected, and that there is no change in volume.

3.2 Symmetry and Boundary Conditions

Due to the symmetry condition at the reactor center (see Fig. 1), the following symmetry and boundary conditions are used:

(a) Symmetry conditions:

$$\frac{\partial T}{\partial \xi}(\xi = 1, \eta, z, t) = 0 \tag{8}$$

$$\frac{\partial T}{\partial \eta}(\xi, \eta = 0, z, t) = 0 \tag{9}$$

$$\frac{\partial T}{\partial \eta}(\xi, \eta = 1, z, t) = 0 \tag{10}$$

(b) Boundary condition at the wall of the reactor:

$$-\frac{k}{L}\sqrt{\frac{(\xi^2 - 1)}{(\xi^2 - \eta^2)}}\frac{\partial T}{\partial \xi}\bigg|_{\xi=\xi_n} = h_w[T(\xi = \xi_n, \eta, t) - T_w] \tag{11}$$

where $\xi_n = L_2/L$ at the wall and h_w is the fluid-wall convective heat transfer coefficient.

$$T(\xi, \eta, z = 0, t) = To \tag{12}$$

$$\frac{\partial T}{\partial z}(\xi, \eta, z = H, t) = 0 \tag{13}$$

The domain of the variables ξ, η and z are: $1 \leq \xi \leq L_2/L$; $0 \leq \eta \leq 1$ and $0 \leq z \leq H$. The value $\xi = 1$ correspond to the line $0 \leq x \leq L$ in $y = 0$; $\xi = L_2/L$ corresponds to the surface of the reactor; $\eta = 0$ correspond to the line $0 \leq y \leq L_1$ in $x = 0$, and $\eta = 1$ correspond to the line $L \leq x \leq L_2$ at $y = 0$ (see Fig. 1).

3.3 Numerical Procedure

Several numerical methods (finite-difference, finite-element, finite-volume, etc.) are used to solve the problem of the heat transfer and fluid flow (CFD problems). In this work, the finite-volume method was used, considering a fully implicit scheme, practice B (nodal points in the center of the control-volume), and the WUDS interpolation scheme for the convective terms. In the method to be developed, integrating the governing differential equation over a volume and in the time derives a system of linear algebraic equations [32, 44, 59]. The resulting discretized linear equation is given by:

$$A_P T_P = A_E T_E + A_W T_W + A_N T_N + A_S T_S + A_F T_F + A_T T_T + B^* \tag{14}$$

where:

$$A_E = \sqrt{\frac{1-\eta_e^2}{\xi_e^2-1}}\frac{k\beta_e \Delta\xi\Delta z}{\delta\eta_e}$$

$$A_W = \sqrt{\frac{1-\eta_w^2}{\xi_w^2-1}}\frac{k\beta_w \Delta\xi\Delta z}{\delta\eta_w}$$

$$A_S = \sqrt{\frac{\xi_s^2-1}{1-\eta_s^2}}\frac{k\beta_s \Delta\eta\Delta z}{\delta\eta_s}$$

$$A_N = \begin{cases} 0, & \text{for boundary control–volume} \\ \frac{\beta_n k_n}{\delta\xi_n}\Delta\eta\Delta z\sqrt{\frac{\xi_n^2-1}{1-\eta_P^2}}, & \text{for internal control–volume} \end{cases}$$

$$A_F = \frac{L^2(\xi_f^2-\eta_f^2)}{\sqrt{(\xi_f^2-1)(1-\eta_f^2)}}\left[(0.5-\alpha_f)\rho\varepsilon c_p u_z + \frac{\beta_f k}{\delta z_f}\right]\Delta\xi\Delta\eta$$

$$A_T = \frac{L^2(\xi_t^2-\eta_t^2)}{\sqrt{(\xi_t^2-1)(1-\eta_t^2)}}\left[(0.5-\alpha_t)\rho\varepsilon c_p u_z \frac{\beta_t k}{\delta z_t}\right]\Delta\xi\Delta\eta$$

$$A_P = \sum A_K + \overline{SM}$$

$$\overline{SM} = \begin{cases} \dfrac{L\Delta\eta\Delta z\sqrt{\frac{\left(\xi_n^2-\eta_P^2\right)}{\left(1-\eta_P^2\right)}}}{\left[\frac{1}{h_w}+\frac{L\delta\xi_s}{k\sqrt{\frac{\left(\xi_n^2-1\right)}{\left(\xi_n^2-\eta_P^2\right)}}}\right]}, & \text{for boundary control - volume} \\ 0, & \text{for internal control - volume} \end{cases}$$

$$\mathrm{B}^* = \frac{\mathrm{L}\sqrt{\frac{\xi_n^2-\eta_P^2}{1-\eta_P^2}}\Delta\eta\Delta z \mathrm{T}_w}{\frac{1}{\mathrm{h}_w} + \frac{\mathrm{L}\delta\xi_s}{k\sqrt{\frac{(\xi_n^2-1)}{(\xi_n^2-\eta_P^2)}}}} + \dot{\mathrm{q}}\frac{\mathrm{L}^2(\xi_P^2-\eta_P^2)}{\sqrt{(\xi_n^2-1)(1-\eta_P^2)}}\Delta\xi\Delta\eta\Delta z$$

The α and β coefficients in the equations are given by:

$$\alpha = \frac{\hat{r}^2}{(10+2\hat{r}^2)} \tag{15}$$

$$\beta = \frac{(1+0.005\hat{r}^2)}{1+0.05\hat{r}^2)} \tag{16}$$

where $\hat{r}$ is the ratio between the convective and diffusive flux in the coordinate direction. Figure 2 illustrates a typical control-volume and the neighbor grid points used in this work. The A_K coefficients, with $K \neq P$, represent the contributions of the diffusive and convective transports of T due to the neighbor points in the direction to the point P. The linear equations system was solved interactively using the Gauss–Seidel algorithm. After grid refinement a structured grid of $20 \times 20 \times 20$ nodal points was chosen.

The convergence criterion adopted, to be satisfied at all the points of the computational domain, it is given by:

$$\left|T^{n+1} - T^n\right| \leq 10^{-8} \tag{17}$$

where n represents the nth iteration. The values of the parameters used in the model are given in Table 1.

The proposed model has been applied to determine the temperature distribution inside a packed-bed reactor. The effects of parameters: feed temperature, reactant concentration, pre-exponential factor of the Arrhenius equation and heat transfer coefficient on the temperature along the bed are presented. The fluid properties used were estimated by the mean temperature between the initial value T_0 and the wall temperature of the equipment, T_W. A computational program codified in the Mathematica® ambient was elaborated to solve the set equations and to obtain numerical results.

3.4 Model Validation

To validate the numerical methodology used in the present work, a comparison between numerical ($L_2/L_1 = 1.001$) and experimental data of the temperature both obtained by the authors for a packed-bed circular cylindrical reactor was made (no chemical reaction). The comparison with analytical data for this situation has been presented in previous work [41].

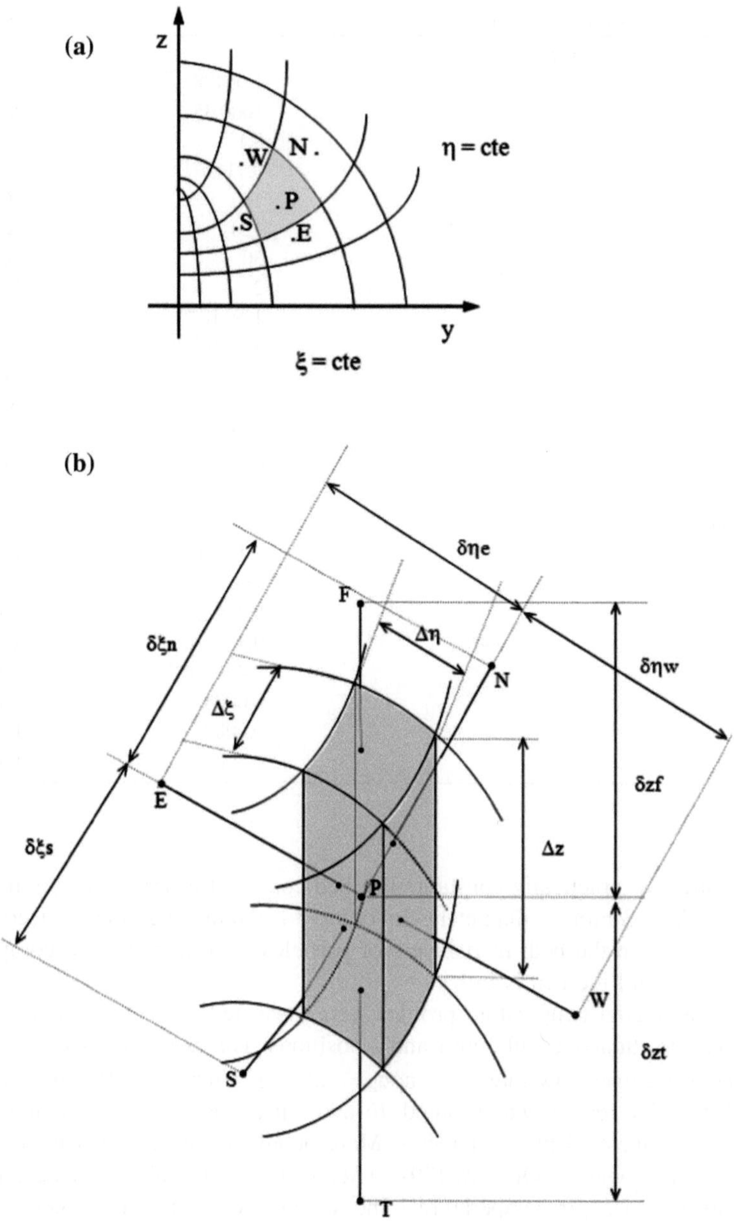

Fig. 2 **a** Grid points and **b** control-volume

The experimental reactor made in brass (copper–zinc alloy) consisted of a column 57 mm in diameter with a packed-bed height of 100 mm. The wall was surrounded by water cooling flowing at ambient temperature (22.5°C). The cooling

Table 1 Parameters used in the simulation

Physical properties and bed characteristics		
ρ	1.09488	kg/m^3
c_p	1000.35	J/kg/K
k	0.1	W/m/K
u_z	0.11417	m/s
T_o	100	°C
	120	°C
T_w	30	°C
h_w	2	$W/m^2/K$
	1×10^{30}	$W/m^2/K$
d_p	4×10^{-3}	m
ε	0.4	m
H	0.2	m
L_1	0.05	m
L_2	0.10	m
$\bar{R}$	8.3143	J/mol/K
Kinetic parameters		
$\Delta\hat{H}$	−104.55	MJ/kmol
E_a	158.92	MJ/kmol
A_0	9.4×10^{19}	1/s
	9.4×10^{23}	1/s
	9.4×10^{25}	1/s
C	0.0008	$kmol/m^3$
	0.0080	$kmol/m^3$
	0.8000	$kmol/m^3$

temperature was practically constant. The bed consisted of zinc oxide cylindrical particles ($d_p = 3$ mm in diameter). In order, to obtain plug flow and constant porosity (0.40) in the bed, inert layers of particles were inserted and compacted successively until the bed is full.

Radial and axial temperature profiles were measured with Fe-constantan thermocouples positioned at different r and z positions. The experiments were carried out at constant inlet flow rate ($G = \varepsilon\rho u_z = 0.63\,kg/s/m^2$) and inlet air temperature (70°C). The reactor was isolated, to avoid heat losses to surroundings. The flow rate was metered by a rotameter. More details about the experimental procedure can be found in Oliveira [39]. After metering, the air was lead into the heater to increase air temperature. The comparison between numerical and experimental data of the cross section mean temperature of the equipment along the bed is shown in Fig. 3. In this figure $\frac{R_e = Gd_p}{\mu}$ represents the Reynolds number.

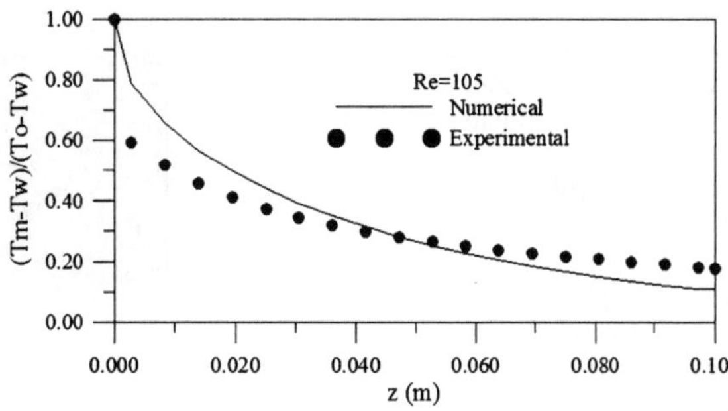

Fig. 3 Comparison between numerical and experimental dimensionless cross section mean temperature along of the fixed bed circular reactor for $\dot{q} = 0$, $L_2/L_1 = 1.001$, $T_0 = 70°C$ and h_w infinity

3.5 Temperature Distribution in the Reactor

The study about the geometry of packed bed reactors is a fundamental requirement for the knowledge of the fluid flow and heat transfer through porous media. Here, the packed bed is treated as quasi-phase as if it were wholly homogeneous.

The knowledge of the temperature distribution within the reactor is very important for verifying the stability of the equipment, to increase thermal efficiency and to minimize operational cost. In this sense, the developed model was used to investigate the effects of some operational conditions such as inlet fluid temperature, reactant fluid concentration, Arrhenius pre-exponential factors and convective heat transfer coefficient on the steady-state heat transfer within the packed bed. Dimensionless temperature profiles $[(T - T_w)/(T_o - T_w)]$ inside the packed bed, at positions z = 0.00556 m, 0.03889 m, 0.10556 m and 0.19444 m are shown in Figs. 4, 5, 6, 7, 8 and 9, for aspect ratios $L_2/L_1 = 2.0$ and 5.0 and several operational conditions.

In any of the figures, it is possible to see that the temperature gradients exist at different scales. The different values of dimensionless temperature in the bed particles and gas flowing through the interstices of the bed indicate that the heat flux occurs from center to wall (angular and radial directions) and from the entrance to outline of the reactor, as expected [17–19, 28, 55]. It can be seen, in accordance with the isothermal lines, that the highest temperature gradient occurs near the wall. There is a strong heating also near the focal point. This is true for other cross sections presented in early work [39, 41].

The rate of heat transfer in porous media with a flowing fluid is determined by a combination of several mechanisms. When heat transfer takes place during fluid flow through a packed bed, dispersion and diffusional mixing occurs. The heat axial dispersion may occur, for example, by molecular conduction, irregularities in the fluid flow through packed-bed due to irregularities in the packing of the bed, alternatively converging and diverging void passages, and fluid flow rate. When the pores are small, convection can be neglected. Then, the contribution of the heat transfer through the solid phase is much larger; however, this situation is not used in practice. The axial diffusion in packed bed industrial equipment is of minor importance.

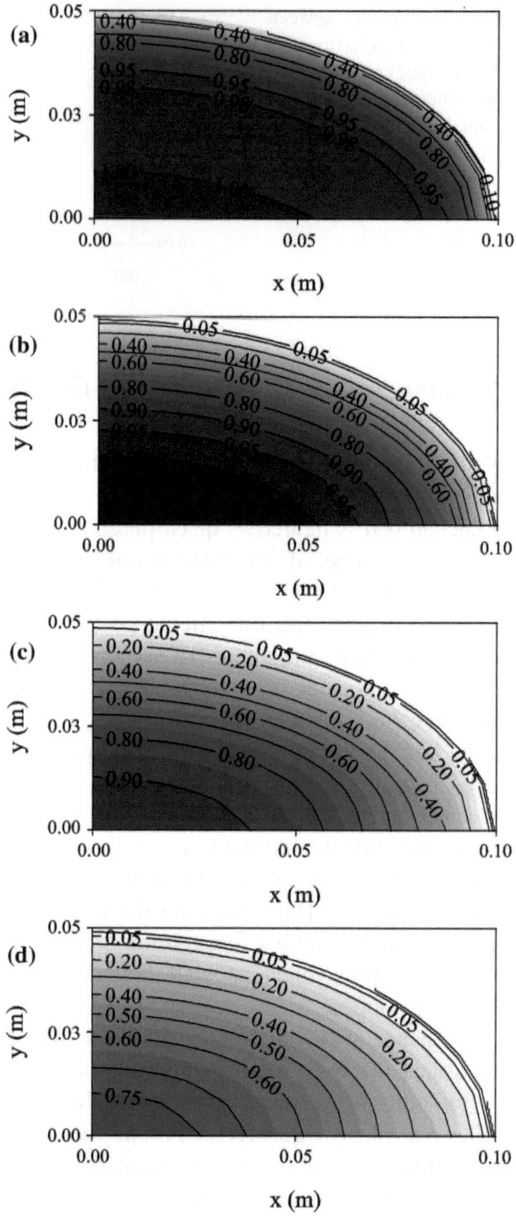

Fig. 4 Dimensionless temperature profile inside the bed for an aspect ratio $L_2/L_1 = 2.0$, h_w infinity, $m = 1$, $A_0 = 9.4 \times 10^{19}\ s^{-1}$, $C = 0.8\ kmol/m^3$ and $T_0 = 100°C$. **a** z = 0.00556 m, **b** z = 0.03889 m, **c** z = 0.10556 m, **d** z = 0.19444 m

Fig. 5 Dimensionless temperature profile inside the bed for an aspect ratio $L_2/L_1 = 2.0$, $h_w = 2.0$ W/m^2 K, $m = 1$, $A_0 = 9.4 \times 10^{19}$ s^{-1}, $C = 0.8\,\text{kmol/m}^3$ and $T_0 = 100$°C. **a** z = 0.00556 m, **b** z = 0.03889 m, **c** z = 0.10556 m, **d** z = 0.19444 m

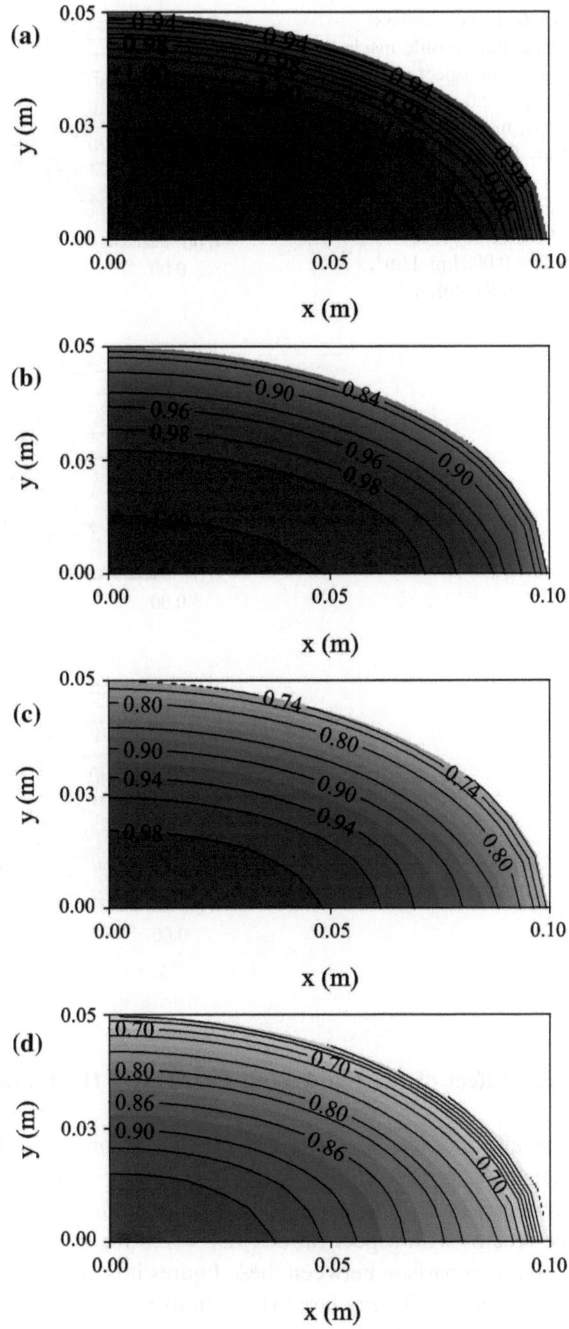

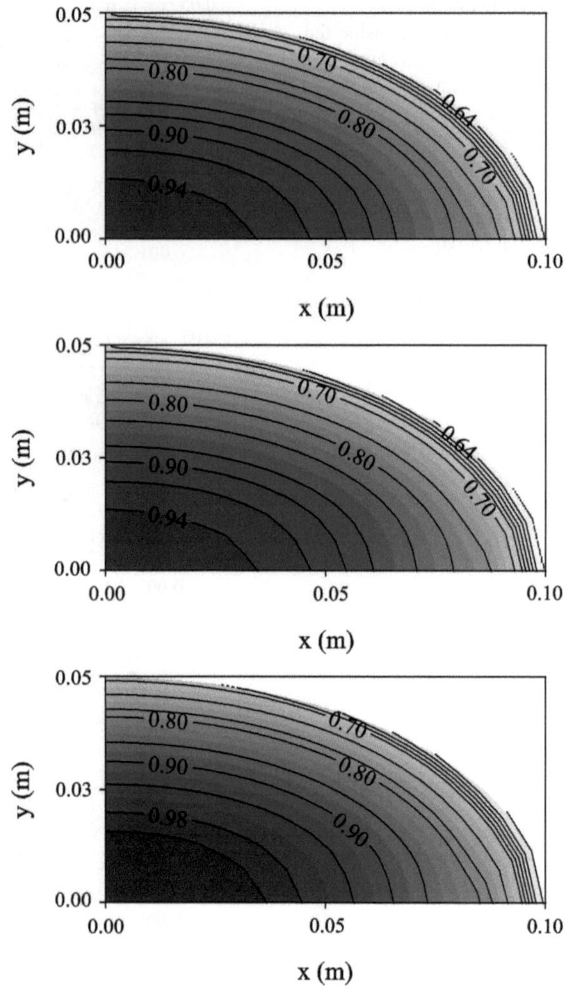

Fig. 6 Dimensionless temperature profile inside the bed for an aspect ratio $L_2/L_1 = 2.0$, $h_w = 2.0$ W/m^2 K, $m = 1$, $A_0 = 9.4 \times 10^{19}$ s^{-1}, $z = 0.19444$ m and $T_0 = 120$°C. **a** $C = 0.0008\,\text{kmol/m}^3$, **b** $C = 0.008\,\text{kmol/m}^3$, **c** $C = 0.8\,\text{kmol/m}^3$

3.5.1 Effect of the Fluid-Wall Convective Heat Transfer Coefficient

The effect of the heat transfer coefficient on the diffusive and convective heat transfer within the elliptic cylindrical reactor at the same operational conditions is presented in Figs. 4 and 5. These figures illustrate temperature distribution inside the reactor with aspect ratio $L_2/L_1 = 2.0$ for $h_w = 2.0$ W/m^2 K and h_w infinity.

For comparison between these figures it is verified high thermal gradients to the case where h_w is very large (h_w infinity), so, in this case, convective heat transfer mechanisms is dominant when compared to heat transfer by conduction. Then, heat transfer depends strongly of the fluid velocity in agreement with experimental observations.

Fig. 7 Dimensionless temperature profile inside the bed for an aspect ratio $L_2/L_1 = 2.0$, h_w infinity, $m = 1$, $C = 0.0008$ kmol/m^3, $z = 0.10556$ m and $T_0 = 100°C$. **a** $A_0 = 9.4 \times 10^{19}\ s^{-1}$, **b** $A_0 = 9.4 \times 10^{23}\ s^{-1}$, **c** $A_0 = 9.4 \times 10^{25}\ s^{-1}$

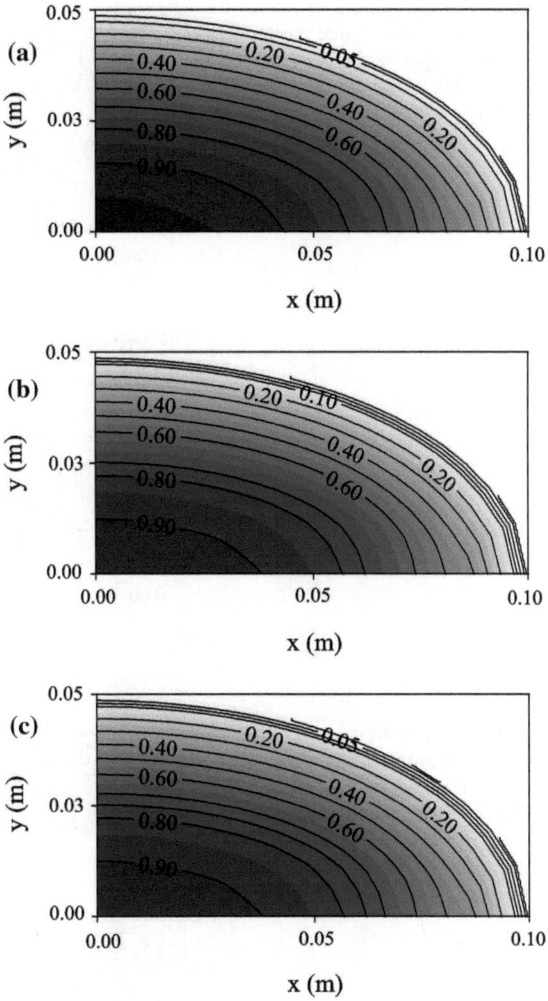

3.5.2 Effect of the Reactant Fluid Concentration

Figure 6 shows the effect of the reactant concentration on the heat transfer within the fixed-bed reactor. By comparing Fig. 6a–b it is verified which an increase of 10 times in the reactant concentration (from 0.0008 to 0.008 kmol/m^3) temperature inside the bed was not modified significantly. However, when reactant concentration changes from 0.008 to 0.8 kmol/m^3, a big modification is found. Then, for highest reactant concentration, we have the highest number of molecules within the bed, consequently we have more collisions between the molecules and consequently, an increase of the local temperature is verified due to high chemical reaction rate (volumetric energy conversion by chemical reaction).

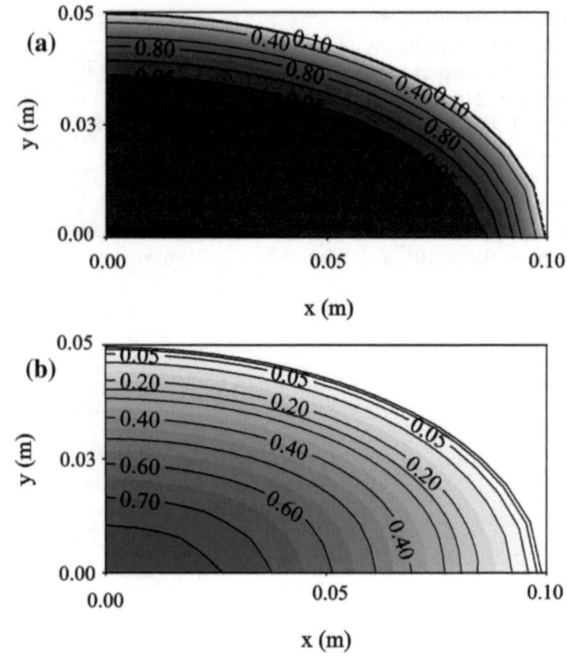

Fig. 8 Dimensionless temperature profile inside the bed for an aspect ratio $L_2/L_1 = 2.0$, h_w infinity and $\dot{q} = 0$. **a** z = 0.00556 m, **b** z = 0.19444 m

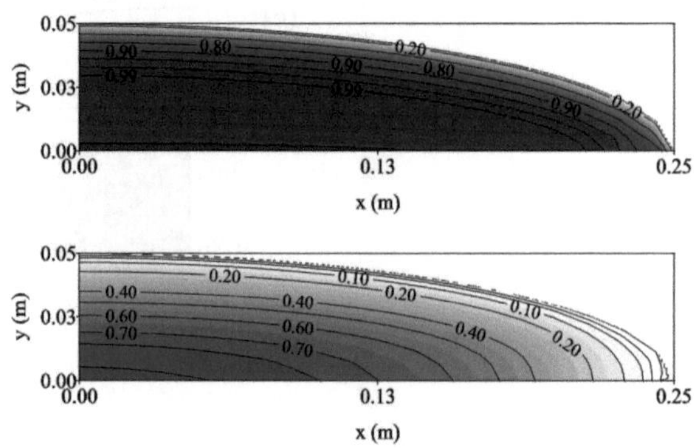

Fig. 9 Dimensionless temperature profile inside the bed for an aspect ratio $L_2/L_1 = 5.0$, h_w infinity and $\dot{q} = 0$. **a** z = 0.00556 m, **b** z = 0.19444 m

3.5.3 Effect of the Pre-exponential Factor of the Arrhenius Equation

Figure 7 shows the effect of the pre-exponential factor on the Arrhenius equation on the heat transfer within the fixed-bed reactor. By comparing Fig. 7a–c it is verified which an increase of this term (from 9.4×10^{18} to 9.4×10^{25} s^{-1}) temperature inside the bed was not modified significantly.

3.5.4 Effect of the Inlet Mixture Temperature

The effect of the inlet mixture temperature on the temperature distribution inside the reactor is shown in Figs. 5d and 6c for the same operational conditions (fixed values of the heat transfer coefficient, reaction order and reactant concentration). It can be seen that the value of the inlet mixture temperature has an insignificant effect on the bed temperature (dimensionless) inside the reactor. However, it is important to cite that inlet temperature affects reaction rate.

3.5.5 Effect of the Heat Generation Term by Chemical Reaction

Figures 8 and 9 present temperature distribution inside the reactor for aspect ratios $L_2/L_1 = 2.0$ and 5.0 and h_w infinity, respectively, for the case without to consider chemical reaction effect. By comparing Figs. 4 and 8, higher thermal gradients are verified in comparison to the case where heat generation is considered as expected.

A comparison of Figs. 8 and 9 indicates that as the aspect ratio L_2/L_1 at cross section increases from 2.0 to 5.0, the temperature profile acquires higher gradients in radial and angular directions. This is true for other cross sections presented in previous work [39, 41].

In this research, the shape, size and directions of the interstices and the detailed locations of the particles inside the bed are neglected and the bulk flow of the fluid in the bed in the axial direction (z-axis) is assumed to be plug flow (gas velocity is assumed constant) and there is axial, radial and angular dispersion of heat. In reality, the velocity distribution across the bed depends on the variations in the flow resistance over different parts of the bed, and changes from zero at the wall to maximum nearly one particle diameter from the wall and falls to a minimum at the center of the bed. The porosity is higher near the wall and lower towards the centre of the bed and the particles are not monosize. Also, the shape of the particles may vary and the surface roughness of the particles could increase the flow resistance [34]. The local changes in porosity can lead to large variations in predicted velocity and temperature profiles and the non-uniform heat loss along the packed bed.

According to Nemec and Levec [35], in commercial packed bed reactors, porosity does not vary a lot and a uniform flow distribution within the bed is encountered. These reactors are made up of roughly uniform particles in terms of both shape and size, where the possible porosity span encountered is relatively narrow ($0.35 < \varepsilon < 0.55$), and the wall effect is negligible. Since in reality two phases are involved in the system, the thermal properties are termed as effective which is dependent on the nature of the individual phases comprising the packed bed.

4 Concluding Remarks

In this chapter, conduction/convection heat transfer in porous media (packed bed reactor) has been explored. Interest in this type of problems is motivated by its importance in many practical situations, such as in hydrology, food processing, petroleum, mechanical, civil and chemical engineering, where heat and mass convections is involved. Here, our attention is focused on steady-state and local thermal equilibrium conditions.

A consistent three-dimensional mathematical model for the conduction/convection heat transfer inside a packed bed reactor of elliptic cross section is proposed. A general numerical formalism for the governing equation, which is based on the finite-volume method, has been applied to the solution of the heat transfer equation in porous media. The analyses of the obtained results using this model following predictions are made by including agreement with experimental data. The heat flow occurs from center to wall (angular and radial directions) and from the entrance to outlet, as expected. The heat transfer rate increases with the increase in the axial position for fixed values of the aspect ratio. The region near the focal point of the ellipse presents a higher heat flow rate. For all axial positions of the bed, temperature increases as the radial position increases but it decreases as the angular position increases. The model may be used for heat transfer problems, such as diffusion and convection in fixed bed equipment with geometry that changes from circular cylinder to elliptic cylinder and also having rectangular channel.

In a general way, we can say that fluid-wall convective heat transfer coefficient affect strongly heat transfer inside the reactor. However, reactant concentration, inlet fluid temperature and pre-exponential factor of the Arrhenius equation have minor importance on the heat flux inside the equipment. Obviously this comment is dependent of the values of these parameters.

Finally, we would like to cite the fact that models, such as this outlined in this work, can be used with great confidence to elucidate unknown features of several complex systems in porous media.

Acknowledgments The authors would like to express their thanks to Brazilian agencies CNPq, FINEP and CAPES for supporting this work, and are also grateful to the authors of the references in this chapter that helped in the improvement of quality. The authors acknowledge the opportunity given by the Editors to present our work in this chapter.

References

1. Akpan, E., Sun, Y., Kumar, P., Ibrahim, H., Aboudheir, A., Idem, R.: Kinetics, experimental and reactor modeling studies of the carbon dioxide reforming of methane (CDRM) over a new Ni/C_eO_2–ZrO_2 catalyst in a packed bed reactor. Chem. Eng. Sci. **62**, 4012–4024 (2007)
2. Alopaeus, V., Laavi, H., Aittamaa, J.A.: Dynamic model for plug flow reactor state profiles. Comput. Chem. Eng. **32**, 1494–1506 (2008)

3. Andrigo, P., Bagatin, R., Pagani, G.: Fixed bed reactors. Catal. Today **52**, 197–221 (1999)
4. Badr, H.M.: Forced convection from a straight elliptical tube. Heat Mass Transf. **34**, 229–236 (1998)
5. Bejan, A.: Convection Heat Transfer. Wiley, New York (1995)
6. Berendsen, W.R., Lapin, A., Reuss, M.: Non-isothermal lipase-catalysed kinetic resolution in a packed bed reactor: modeling, simulation and miniplant studies. Chem. Eng. Sci. **62**, 2375–2385 (2007)
7. Borkink, J.G.H., Westerterp, K.R.: The significance of axial heat dispersion for the description of heat transport in wall-cooled packed beds. Chem. Eng. Technol. **15**, 371–384 (1992)
8. Bradean, R., Heggs, P.J., Inghan, D.B., Pop, I.: Convective heat flow from suddenly heated surfaces embedded in porous media. In: Ingham, D.B., Pop, I. (eds.) Transport Phenomena in Porous Media. Elsevier, Oxford (1998)
9. Chang, W.J., Weng, C.I.: Heat and mass transfer in porous materials. In: Ingham, D.B., Pop, I. (eds.) Transport Phenomena in Porous Media II. Elsevier, Oxford (2002)
10. Cheng, P., Hsu, C.: Heat conduction. In: Ingham, D.B., Pop, I. (eds.) Transport Phenomena in Porous Media. Elsevier, Oxford (1998)
11. Cong, T.N., He, Y., Chen, H., Ding, Y., Wen, D.: Heat transfer of gas–solid two-phase mixtures flowing through a packed bed under constant wall heat flux conditions. Chem. Eng. J. **130**, 1–10 (2007)
12. Cooussirat, M., Guardo, A., Mateos, B., Egusquiza, E.: Performance of stress–transport models in the prediction of particles-to-fluid heat transfer in packed beds. Chem. Eng. Sci. **62**, 6407–6897 (2007)
13. Cotta, R.M., Luz Neto, H., Alves, L.S.B., Quaresma, J.N.N.: Integral transforms for natural convection in cavities filled with porous media. In: Ingham, D.B., Pop, I. (eds.) Transport Phenomena in Porous Media. Elsevier, Oxford (2005)
14. de Lemos, M.J.S.: The double-decomposition concept for turbulent transport in porous media. In: Ingham, D.B., Pop, I. (eds.) Transport Phenomena in Porous Media III. Elsevier, Oxford (2005)
15. de Lemos, M.J.S.: Mathematical modeling and applications of turbulent heat and transfer in porous media. In: Vafai, K. (ed.) Handbook of Porous Media, 2nd edn. CRC Press/Taylor & Francis, Boca Raton (2005)
16. de Lemos, M.J.S.: Turbulence in Porous Media: Modeling and Applications. Elsevier, Amsterdan (2006)
17. De Wash, A.P.E., Froment, G.F.: Heat transfer in packed beds. Chem. Eng. Sci. **27**, 567–576 (1972)
18. Dixon, A.G.: The length effect on packed bed effective heat transfer parameters. Chem. Eng. J. **31**, 163–173 (1985)
19. Dixon, A.G.: An improved equation for the overall heat transfer coefficient in packed beds. Chem. Eng. Process. **35**, 323–331 (1996)
20. Freire, J.T.: Heat transfer in porous media. Doctor Thesis, Federal University of Rio de Janeiro, Rio de Janeiro (1979) (In Portuguese)
21. Guardo, A., Coussirat, M., Recasens, F., Larrayoz, M.A., Escaler, X.: CFD study on particle-to-fluid heat transfer in fixed bed reactor: convective heat transfer at low and high pressure. Chem. Eng. Sci. **61**, 4341–4353 (2006)
22. Happel, J., Brenner, H.: Low Reynolds Number Hydrodynamics with Special Applications to Particulate Media. Kluwer Academic Publishers, London (1991)
23. Howle, L.E.: Convection in ordered and disordered porous layers. In: Ingham, D.B., Pop, I. (eds.) Transport Phenomena in Porous Media II. Elsevier, Oxford (2002)
24. Iordanidis, A.A., van Sint, A.M., Kronberg, A.E., Kuipers, J.A.M.: A numerical method for the solution of the wave model and convection dominated diffusion type models for catalytic packed bed reactors. Comput. Chem. Eng. **28**, 2337–2349 (2004)
25. Kakaç, S., Shah, R.K., Aung, W.: Handbook of Single-Phase Convective Heat Transfer. Wiley, New York (1987)

26. Kaviany, M.: Principles of Convective Heat Transfer, 2nd edn. Springer, New York (2001)
27. Kuznetsov, A.V.: Thermal nonequilibrium forced convection in porous media. In: Ingham, D.B., Pop, I. (eds.) Transport Phenomena in Porous Media. Elsevier, Oxford (1998)
28. Li, C.H., Finlayson, B.A.: Heat transfer in packed beds: a reevaluation. Chem. Eng. Sci. **32**, 1055–1066 (1977)
29. Liu, S., Masliyah, H.: Dispersion in porous media. In: Vafai, K. (ed.) Handbook of Porous Media, 2nd edn. CRC Press/Taylor & Francis, Boca Raton (2005)
30. Maia, C.R.M., Aparecido, J.B., Milanez, L.F.: Heat transfer in laminar forced convection inside elliptical tube with boundary conditions of first kind. In: Proceedings of the 3th European Thermal Sciences Conference. Heidelberg, vol. 1, p. 347 (2000)
31. Maia, C.R.M., Aparecido, J.B., Milanez, L.F.: Heat transfer in laminar forced convection inside elliptical cross section ducts with boundary conditions of second kind. In: Proceedings of the 22nd Iberian Latin-American Congress on Computational Methods in Engineering, Campinas (2001) (in CD-ROM)
32. Maliska, C.R.: Computational Heat Transfer and Fluid Mechanics. LCT, Rio de Janeiro (2004). (In Portuguese)
33. Morse, P.M., Feshbach, H.: Methods of Theoretical Physics. Mc Graw-Hill Book Company, Inc., New York (1953)
34. Natarajan, S., Zhang, C., Briens, C.: Numerical simulation and experimental verification of gas flow through packed beds. Powder Technol. **152**, 31–40 (2005)
35. Nemec, D., Levec, J.: Flow through packed bed reactors: I. Single-phase flow. Chem. Eng. J. **60**, 6947–6957 (2005)
36. Nield, D.A.: Modeling fluid flow in saturated porous media and at interfaces. In: Ingham, D.B., Pop, I. (eds.) Transport Phenomena in Porous Media II. Elsevier, Oxford (2002)
37. Nield, D.A., Bejan, A.: Convection in Porous Media. Springer, New York (1992)
38. Nield, D.A., Kuznetsov, A.V.: Heat transfer in bidisperse porous media. In: Ingham, D.B., Pop, I. (eds.) Transport Phenomena in Porous Media III. Elsevier, Oxford (2005)
39. Oliveira, L.G.: Heat transfer in a fixed bed elliptic cylindrical reactor: geometric and thermofluidynamic aspects. Doctor Thesis, Federal University of Campina Grande, Campina Grande (2004) (In Portuguese)
40. Oliveira, L.G., Teruel, B.J., Lima, A.G.B.: Heat transfer in a cylindrical reactor: new analytical solution of the three-parameter model. Informaci6n Tecnol6gica **14**(1), 145–151 (2003) (In Spanish)
41. Oliveira, L.G., Swarnakar, R., de Lima, A.G.B.: Steady state model for heat transfer in a packed bed reactor of elliptic cylindrical shape. Int. J. Chem. React. Eng. **6**(A71), 1–20 (2008)
42. Özdural, A.R., Alkan, A., Webb, C.: Derivation of a simple equation for the determination of kinetics coefficients in packed-bed reactors. Chem. Eng. J. **118**, 17–22 (2006)
43. Papadopoulos, P.K., Hatzikonstantinou, P.M.: Numerical study of laminar fluid flow in a curved elliptic duct with internal fins. Int. J. Heat Fluid Flow. **29**, 540–544 (2008)
44. Patankar, S.V.: Numerical Heat Transfer and Fluid Flow. Hemisphere Publish Cooporation, New York (1980)
45. Pop, I., Merkin, J.H., Ingham, D.B.: Chemical driven convection in porous media. In: Ingham, D.B., Pop, I. (eds.) Transport Phenomena in Porous Media II. Elsevier, Oxford (2002)
46. Pop, I., Ingham, D.B., Merkin, J.H.: Transient convection heat transfer in a porous medium: external flows. In: Ingham, D.B., Pop, I. (eds.) Transport Phenomena in Porous Media. Elsevier, Oxford (1998)
47. Rees, D.A.S., Pop, I.: Local thermal non-equilibrium in porous medium convective. In: Ingham, D.B., Pop, I. (eds.) Transport Phenomena in Porous Media III. Elsevier, Oxford (2005)
48. Rohsenow, W.M., Hartnett, J.P., Cho, Y.I.: Handbook of Heat Transfer, 3rd edn. MacGraw Hill, Boston (1998)

49. Sakalis, V.D., Hatzikonstantinou, P.M.: Thermally developing flow in elliptic ducts with axially variable wall temperature. Int. J. Heat Mass Transf. **45**, 25–35 (2002)
50. Shahamiri, S.A., Wierzba, I.: Modeling catalytic oxidation of lean mixtures of methane–air in a packed-bed reactor. Chem. Eng. J. **149**, 102–109 (2009)
51. Škerget, L., Jecl, R.: Boundary element method for transport phenomena in porous media. In: Ingham, D.B., Pop, I. (eds.) Transport phenomena in porous media II. Elsevier, Oxford (2002)
52. Storesletteen, L.: Effects of anisotropy on convective flow through porous media. In: Ingham, D.B., Pop, I. (eds.) Transport Phenomena in Porous Media. Elsevier, Oxford (1998)
53. Stratton, J.A., Morse, P.M., Chu, L.J., Hutner, R.A.: Elliptic Cylinder and Spheroidal Wave Functions. Wiley, New York (1941)
54. Theuerkauf, J., Witt, P., Schwesig, D.: Analysis of particle porosity distribution in fixed beds using the discrete element method. Powder Technol. **165**, 92–99 (2006)
55. Thoméo, J.C., Freire, J.T.: Heat transfer in fixed bed: a model non-linearity approach. Chem. Eng. Sci. **55**, 2329–2338 (2000)
56. Vadasz, P.: Free convection in rotating porous media. In: Ingham, D.B., Pop, I. (eds.) Transport Phenomena in Porous Media. Elsevier, Oxford (1998)
57. Vafai, K., Amiri, A.: Non-Darcian effects in confined forced convection flows. In: Ingham, D.B., Pop, I. (eds.) Transport Phenomena in Porous Media. Elsevier, Oxford (1998)
58. Valipour, M.S., Saboohi, Y.: Modeling of multiple noncatalytic gas–solid reactions in a moving bed of porous pellets based on finite volume method. Heat Mass Transf. **43**, 881–894 (2007)
59. Versteeg, H.K., Malalasekera, W.: An Introduction to Computational Fluid Dynamic: The Finite Volume Method. Prentice Hall, London (1995)
60. Wakao, N., Kaguei, S.: Heat and Mass Transfer in Packed Beds. Gordon and Breach Science Publishers, New York (1982)
61. Wang, C.Y.: Modeling multiphase flow and transport in porous media. In: Ingham, D.B., Pop, I. (eds.) Transport Phenomena in Porous Media. Elsevier, Oxford (1998)
62. Wang, B.X., Du, J.H., Peng, X.F.: Internal natural, forced and mixed convection in fluid-saturated porous media. In: Ingham, D.B., Pop, I. (eds.) Transport Phenomena in Porous Media. Elsevier, Oxford (1998)

49. Sokolis, V. E., Hubbard, ..., P. M.: Thermally developing flow in ... with variable wall temperature. Int. J. Heat Mass Transf. ...
50. ... catalytic ... Chem. Eng. J. 149, ...

Study of Thermal and Rheological Behavior of Polystyrene/TiO_2, Polystyrene/SiO_2/TiO_2 and Polystyrene/SiO_2 Nanocomposites

A. Mohebbi, M. Dehghani and A. Mehrabani-Zeinabad

Abstract In this study, polystyrene/TiO_2, polystyrene/SiO_2/TiO_2 and polystyrene/SiO_2 nanocomposites were initially prepared by solution method. Their characteristics were investigated by Thermo Gravimetric Analysis (TGA), and parallel plate rheometer. Then, their rheological behaviors were studied through analysis either by shearing alone, or in combination with temperature. Results of experiments show that the type of nano-filler and the method of composite preparation affect the thermal and rheological properties of the composites. Results demonstrate that the produced composites had various phase morphologies with nano-scales and distinct behaviors of a percolation network structure under certain conditions. Also the linear viscoelastic region of the nanocomposites is much narrower than that for Polystyrene (PS) matrix. Moreover, nanocomposites present the nature of temperature independence of storage modulus versus loss modulus. Furthermore, PS/TiO_2 had a strain-scaling stress response to the startup of steady shear that indicate formation of more connections sites between the nano particles and polystyrene matrix.

Keywords Rheological behaviors · SiO_2/TiO_2 nano particles · Polystyrene/TiO_2 · Polystyrene/SiO_2/TiO_2 · Polystyrene/SiO_2 nanocomposites

A. Mohebbi (✉)
Department of Polymer Engineering, Darab Branch, Islamic Azad University, Darab, Iran
e-mail: mohebbi@iaudarab.ac.ir

M. Dehghani · A. Mehrabani-Zeinabad
Department of Chemical Engineering, Isfahan University of Technology, Isfahan 8415683111, Iran
e-mail: Maziariut@gmail.com

A. Mehrabani-Zeinabad
e-mail: Arjomand@cc.iut.ac.ir

A. Öchsner et al. (eds.), *Materials with Complex Behaviour II*, Advanced Structured Materials 16, DOI: 10.1007/978-3-642-22700-4_35,

1 Introduction

The field of nanocomposites exists in many diverse topics including composite reinforcement, barrier properties, flame resistance, electro-optical properties, cosmetic applications, and bactericidal properties [1]. These properties are usually increased by increasing the amount of added fillers. Addition of fillers in polymers also alters their flow behaviors and various processing abilities. Studies on rheological properties of particle filled polymers matrix have been attracted researchers attention [2–5].

Dynamic rheological tests are believed to be a preferential method for dealing with the structure/morphology of materials, as the structure of exposing materials to testing processes is not destroyed under small-strain amplitude [6]. The so-called second plateau, a phenomenon in dynamic viscoelastic functions, for example, the storage modulus G' and loss modulus G'', which exhibit a special response corresponding to linear viscoelasticity, is thought to give important information concerning the viscoelastic behavior of multicomponent/multiphase polymer systems at very low frequencies (terminal region) [7]. Moreover, the appearance of a second plateau in the terminal region is considered to be induced by formation of infra-structures of particles, such as, agglomerated structures, skeletons, or network structures [8].

Compared with the rheological behavior of the pure polymer matrix, polymer composites generally exhibit pronounced elastic properties and longer relaxation times [9–11].

Rheological behaviors of a filled polymer composite is initially governed by properties of both polymer matrix and incorporated fillers, and then is influenced by the volume fraction and agglomerations of fillers particles and the interfacial adhesion between the filler and matrix [12–14]. On the other hand, dynamic rheological measurement has been an effective method to characterize the morphological and structure of the multicomponent and/or multiphase polymer composites [15–17].

Rheological properties of a polymer matrix induced by fillers is thought to be affected by two factors. The first one is the mechanical coupling between fillers and polymer matrix [13, 14]. The other one is the interfacial adhesion or adsorption between the phases. But, in addition to these factors, the state of fillers, solid or liquid, has also a significant influence on the rheological behaviors of composite.

In recent years, researches have been performed on dynamic rheological behaviors of polymers composites mixtures containing either traditional fillers or nanoscope fillers, such as high density polyethylene (HDPE) filled with conductive particle, PS filled with Sn–Pb alloy, low density polyethylene (LDPE)/montmorillonite (MMT-Clay) nanocomposites.

An important effect of addition of nano particles in polymer matrix is the variation of the T_g property. There are reports on increasing and decreasing of T_g

depending to the interaction between matrix and added particles. Silica has been demonstrated to be an ideal inorganic additive for composites [18–20].

In this present work, an investigation of rheological and thermal properties of Polystyrene in presence of nano-materials such as nano-silica and nano-titan has been conducted. Preparations of nanocomposites were preformed by solution method and by comparison of the effect of each nanoparticle alone and together.

2 Experimental

2.1 Materials

In this study, the main materials for the preparation of nanocomposites were polystyrene, nano SiO_2, and nano TiO_2 with following grades and manufacturer.

- Polystyrene: grade GPPS-1540 from IR-TABRIZ Petrochemical Co.
- Nano SiO_2: grade F-110 from Fadak Chemical Co. (Non-Surface treated)
- Nano TiO_2: grade P-25 from Evonik Germany Co. (Surface treated with HNO_3)

2.2 Nanocomposite Preparation

Silica (SiO_2) and titanium dioxide (TiO_2) are inorganic materials and have no tendency for agglomeration. The surface energy of nano-scale size of these particles is increased which leads their aggregation. Therefore, physical methods cannot be implemented for preparation of highly dispersed nanocomposite based on nano particles of silica and titan. Thus, the nanocomposites of silica and titan were synthesized based on the solution method using toluene as a solvent. Initially, 16 g of silica was dispersed in the solvent by a magnetic mixer. The mixing process was performed at temperature of 70°C for a period of 30 min. Then, an ultrasonic mixer was used for having a higher dispersion of nano particles in the solvent. This process was performed at a frequency of 50 MHz for 15 min. The dispersed product was carefully mixed with a prepared solution of 30 wt% Polystyrene in toluene. The final compound was prepared by removal of the solvent in a vacuum oven. This master-batch contained 17 wt% of nano silica particles. By using a twin screw extruder, the nanocomposite samples with 1 wt% and 3 wt% of the nano particles were prepared from the produced master-batch.

The samples of PS-TiO_2 and PS-TiO_2-SiO_2 nanocomposites were produced similarly. The amounts of each particle in the mixed nanocomposite samples, i.e. PS-ST1 and PS-St3, were the same. The labels and specifications of all prepared samples are shown in Table 1.

Table 1 Specifications of prepared samples

Sample-label	Nano filler	Content (wt%)
PS-S1	SiO_2	1
PS-ST1	$SiO_2 + TiO_2$	1
PS-T1	TiO_2	1
PS-S3	SiO_2	3
PS-ST3	$SiO_2 + TiO_2$	3
PS-T3	TiO_2	3

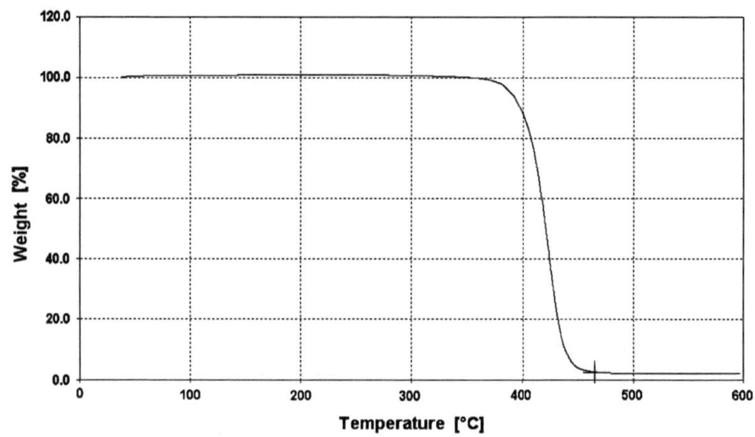

Fig. 1 TGA curve of sample PS-S3

2.3 Thermogravimetric Measurements Analysis (TGA)

TG analysis was conducted through a DuPont 2010 system in the range of ambient temperature to 600°C under a steady flow of nitrogen.

2.4 Rheological Analysis

Prior to the rheological measurement, all the samples were tested with a capillary rheometer at 100°C. Melt flow behavior of the samples was characterized at an angular frequency ranging from 10^{-2} to 10^{2} s^{-1}.

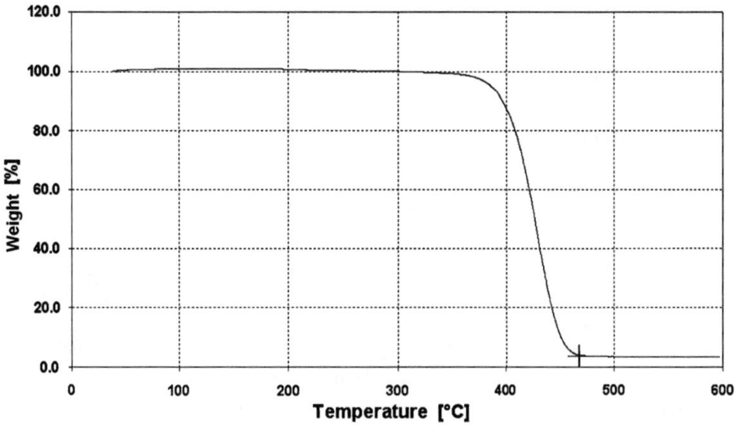

Fig. 2 TGA curve of sample PS-ST3

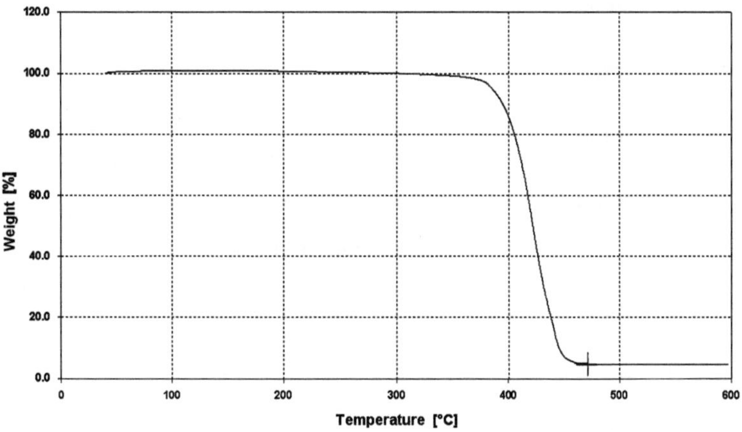

Fig. 3 TGA curve of sample PS-T3

3 Results and Discussion

Thermal stability is an important property in processing and application of materials. Various factors such as the structure of the repeating unit, and crystallinity influence thermal stability of polymers. Figures 1, 2 and 3 illustrate TGA curves of samples having 3 wt% of nano particles under application of thermal energy at a rate of 10°C/min by using a steady flow of nitrogen.

Thermogravimetric analysis of PS-nanocomposites shows that the presence of nano particles has a clear influence on the thermal stability of PS. Thermal stability

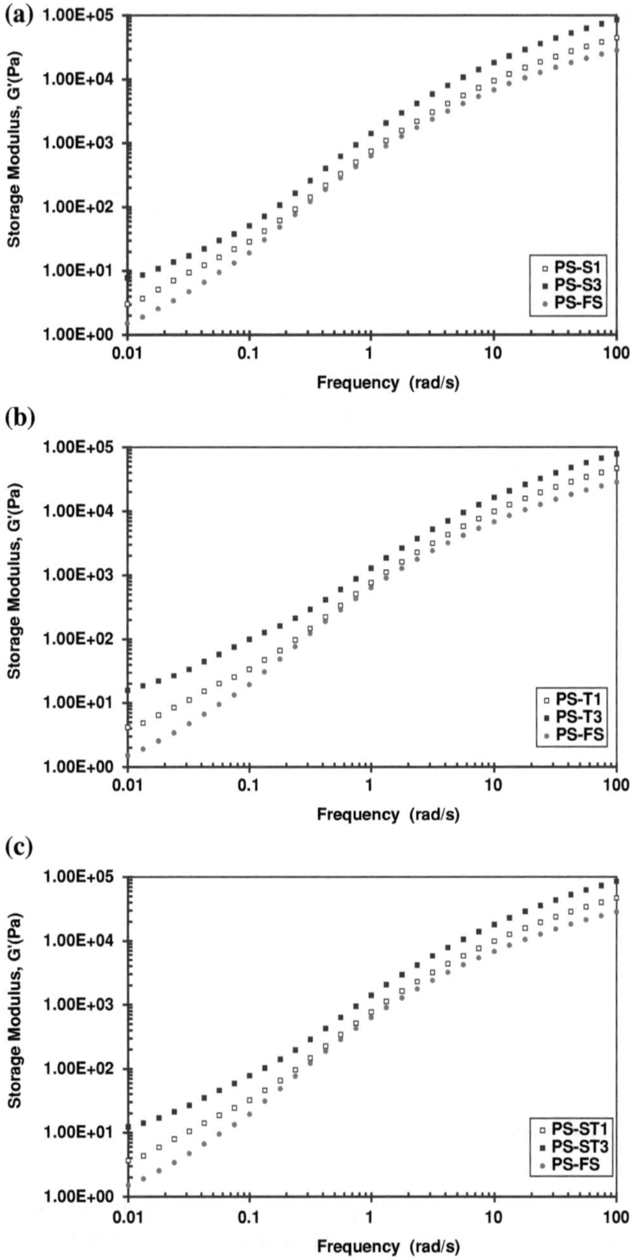

Fig. 4 Comparison of storage modulus of samples: (**a**) PS-SiO_2 nanocomposites, (**b**) PS-TiO_2 nanocomposites, and (**c**) PS-SiO_2-TiO_2 nanocomposites

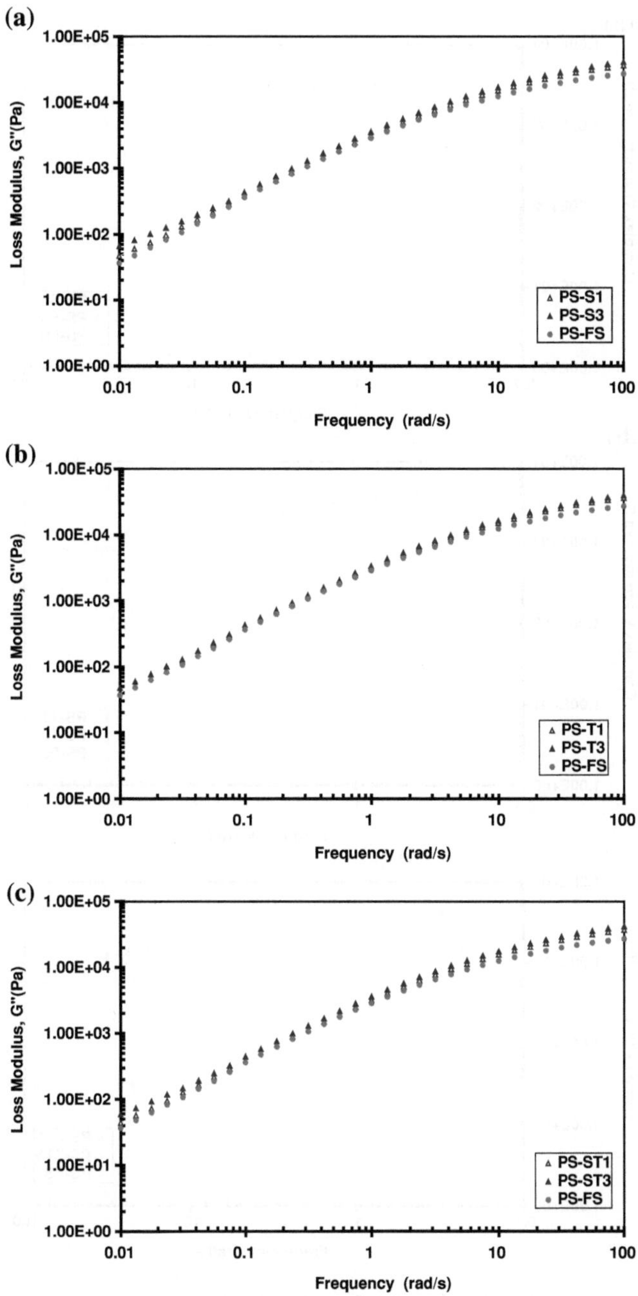

Fig. 5 Comparison of loss modulus of samples: (**a**) PS-SiO_2 nanocomposites, (**b**) PS-TiO_2 nanocomposites, and (**c**) PS-SiO_2-TiO_2 nanocomposites

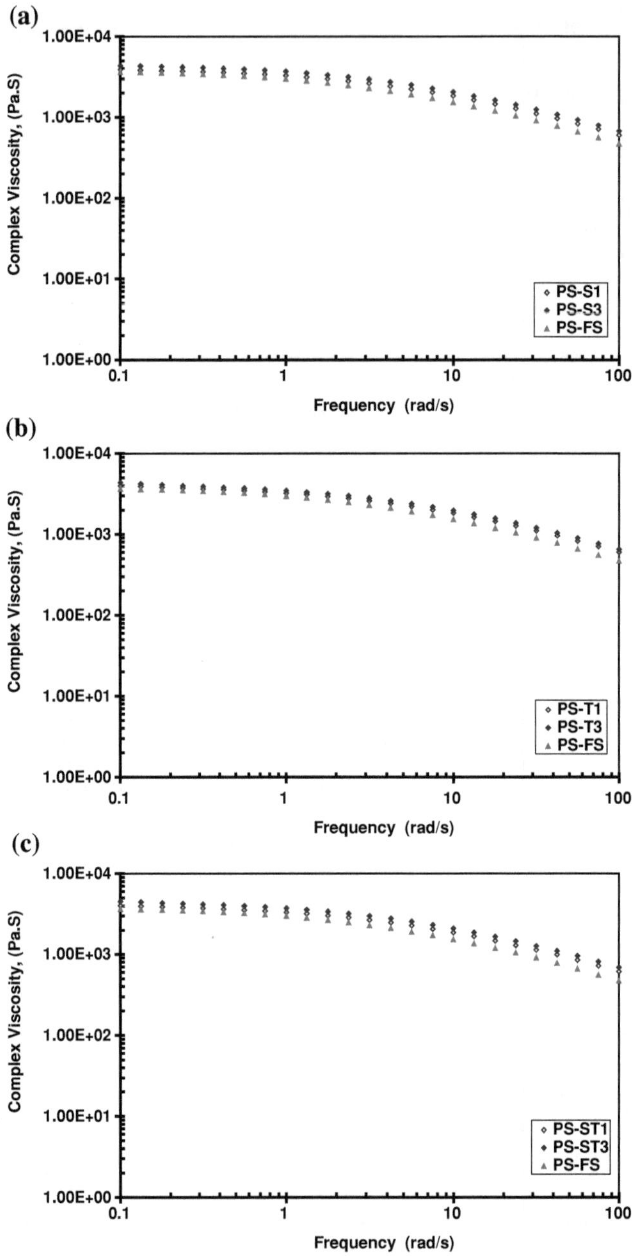

Fig. 6 : Comparison of complex viscosity of samples: (**a**) PS-SiO_2 nanocomposites, (**b**) PS-TiO_2 nanocomposites, and (**c**) PS-SiO_2-TiO_2 nanocomposites

of polystyrene in presence of either nano TiO_2, nano SiO_2, particles or both of them was increased around 50°C. Results demonstrate that the presence of one the nano particles or both of them in polystyrene matrix had the same effect on thermal stability.

The investigations of rheological behavior of samples at 100°C were carried out using capillary rheometer. Effect of nano particle loading on storage modulus, G', of nanocomposites are shown in Fig. 4. The magnitude of G', increased monotonically by increasing the frequency. Also, the storage modulus was increased by increasing of nano particles content. The enhancement for the samples with nano SiO_2 was lower than that for the samples with nano TiO_2. The storage moduli of samples with a mixture of two nano particles were about the average of the storage modulus of samples with a single kind of particles. The same trend was also observed for G'', Fig. 5.

Variations of complex viscosity of samples with of frequency are illustrated in Fig. 6. It exhibits a shear-thinning phenomenon. The complex viscosity of PS-nanocomposite melts were decreased sharply by increasing of shearing rate. This is due to the existence of entangled points in molecular chains (between the surface of nanoparticles and polymer matrix). By increasing the shearing rate, a slide between molecular chains occurred, and the number of tangled points decreased. Therefore, low flowing obstacle and low apparent viscosity are experienced [21].

4 Conclusion

PS-SiO_2 and PS-TiO_2 nanocomposites melt are shear-thinning fluids. The apparent viscosity of these melts was decreased by an increasing shearing rate (Fig. 6). These were as a result of high hydrolysis rate. The thermal stability and rheological properties of nanocomposites were obviously influenced in presence of nano particle.

Also, the linear viscoelastic region of the nanocomposites was much narrower than the one for PS matrix. Moreover, PS/TiO_2 had a strain-scaling stress response to the startup of steady shear. This issue indicates formation of more connecting sites between the nano particles and polystyrene matrix. Furthermore, the storage modulus property was augmented by increasing the nano particles content. The increasing of this property for nano SiO_2 particles was lower than that for TiO_2 particles. This property for nanocomposites containing a mixture of the two nano particles was about the corresponding average of the property for nanocomposites containing the single particles. This subject shows a similar structure upon presence of TiO_2 and SiO_2 in PS matrix.

Therefore, the type of nano particles and their amounts of in a PS-nanocomposite is an effective method for controlling PS properties.

Acknowledgments Financial support of Islamic Azad University Darab Branch has greatly acknowledged.

References

1. Paul, D.R., Robeson, L.M.: Polymer **49**, 3187 (2008)
2. Aoki, Y.: Macromolecules **20**, 2208 (1987)
3. Li, L., Masuda, T.: Polym. Eng. Sci. **30**, 841 (1990)
4. Bousmina, M., Muller, R.: J. Rheol. **37**, 663 (1993)
5. Memon, N.A., Muller, R.: J. Polym. Sci. Part B Polym. Phys. **36**, 2623 (1998)
6. Utracki, L.A.: Polymer Alloys and Blends, pp. 131–174. Carl Hanser, New York (1989)
7. Takahashi, M., Li, L., Masuda, T.: J. Rheol. **33**, 709 (1989)
8. Onogi, S., Matsumoto, T.: Polym Eng Rev **1**, 45 (1981)
9. Graebling, D., Muller, R.: Colloids Surf. **55**, 89 (1991)
10. Graebling, D., Benkira, A., Gallot, Y., Muller, R.: Macromolecules **30**, 301 (1994)
11. Graebling, D., Muller, R., Palierne, J.F.: Macromolecules **26**, 320 (1993)
12. Chazeau, L., Paillet, M., Cavaille, J.Y.: J. Polym. Sci. Phys. **37**, 2151 (1999)
13. Liang, J.Z., Li, R.K.Y., Tjong, S.C.: Polym. Int. **48**, 1068 (1999)
14. Alberola, N.D., Fernagut, F., Mele, P.: J. Appl. Polym. Sci. **63**, 1029 (1997)
15. Zheng, Q., Du, M., Yang, B.B., Wu, G.: Polymer **42**, 5743 (2001)
16. Peng, M., Zheng, Q.: Chin. J. Polym. Sci. **18**, 565 (2000)
17. Han, C.D.: J. Appl. Polym. Sci. **32**, 3809 (1986)
18. Mascia, L., Kioul, A.: Polymer **36**, 3649 (1995)
19. Maier, G.: Prog. Polym. Sci. **26**, 3 (2001)
20. Cornelius, C.J., Marand, E.: Polymer **43**, 2385 (2002)
21. Yang, K.K., Wang, X.L., Wang, Y.Z., Huang, H.X.: Mater. Chem. Phys. **87**, 218 (2004)

Characterizing the Anisotropic Hardening Behavior of Aluminum Bonding Wires

Holm Altenbach, Christian Dresbach and Matthias Petzold

Abstract In power electronic devices the electrical connections of different components are mainly realized by heavy aluminum wire bonding. When a device heats up or cools down during use, there is a relative displacement between the first and the second contact because of differences in thermal expansion coefficients of the components and the housing of the device. This cyclic thermo mechanical loading can lead to fatigue failure of the bonding wire. Especially when placed near mechanical working components (e.g. automotive engine) additional vibrations can increase damage evolution and heating can accelerate ageing effects of the bonding wire. In the last few years there have been multiple publications presenting experimental and numerical results for high and low cycle fatigue of wire bonded devices. But all of these publications did not consider the mechanical properties of the wire in an adequate way. We present results of a micro-compression test that allows the determination of the hardening behavior parallel and perpendicular to the wire axis at moderate and large strains of small

H. Altenbach (✉)
Martin-Luther-University Halle-Wittenberg, Kurt-Mothes-Str. 1, 06120 Halle, Germany
e-mail: holm.altenbach@ovgu.de

H. Altenbach
Otto-von-Guericke-University Magdeburg, Universitaetsplatz 2, 39106 Magdeburg, Germany

C. Dresbach · M. Petzold
Fraunhofer Institute for Mechanics of Materials IWM, Walter-Huelse-Str. 1, 06120 Halle, Germany
e-mail: christian.dresbach@dlr.de

M. Petzold
e-mail: matthias.petzold@iwmh.fraunhofer.de

C. Dresbach
German Aerospace Center, Linder Hoehe, 51147 Cologne, Germany

A. Öchsner et al. (eds.), *Materials with Complex Behaviour II*, Advanced Structured Materials 16, DOI: 10.1007/978-3-642-22700-4_36,

wire cylinders. The hardening behavior in compression parallel to the wire axis correlates very well to the hardening behavior determined by tensile tests at moderate strains. The hardening behavior perpendicular to the wire axis shows an anisotropic behavior of the aluminum wires depending on the drawing texture which was also analyzed by electron backscatter diffraction methods. The results for different wire materials show a dependence of the yield stress on the grain size. With the determined hardening parameters it is possible to consider the hardening of the material during the bonding process.

keywords Aluminum wire bonding · Electron backscatter diffraction · Micro compression test · Hall-Petch relation

1 Introduction

The electrical connection of different components in power and automotive electronic devices are mainly realized by heavy aluminum wire bonding. Typical dimensions of heavy aluminmum bonding wires are between 125 and 500 μm in diameter. An example of a contacted microelectronic device is shown in Fig. 1a.

In the wedge/wedge wire bonding process, a very pure aluminum wire is carried through a wedge bond tool. The wire is pressed with the wedge on the pad surface by a bond force F and the connection is formed under the presence of additional ultrasonic power (US), see Fig. 2. After that, the loop is formed by moving the wedge to the second pad. The second contact is formed by pressing the wire on the pad surface and applying additional ultrasonic power. When the wire is cut, the next wire bond connection can be realized in the same way [1].

In some cases the connected bond pads are on different substrates which are mounted in a housing. When the device heats up ($+125$°C) or cools down (-20°C) during use, there is a relative displacement between the first and second contact because of differences in thermal expansion coefficients of the components and the housing of the device. The relative displacement leads to a bending of the bonding wire and the yield stress can be exceeded locally, thus a local plastic deformation results. The geometrically most critical regions are the heel and the loop top, where also additional pre-deformation has occured during the bond process. The cyclic thermo mechanical loading can lead to fatigue failure of the bonding wire. Especially when placed near mechanical working components (e.g. automotive engine) additional vibrations can increase damage evolution and heating can accelerate aging effects of the bonding wire. Fig. 1b shows a fatigue crack in an aluminum bonding wire, which has not failed at that moment, because the electrical connection is still remaining. But additional movement will increase crack growth or damage evolution until the component fails. In the last few years there have been multiple publications presenting experimental and numerical results for high and low cycle fatigue of wire bonded devices [2–4]. But all of these publications did not consider the mechanical properties of the wire in an adequate way.

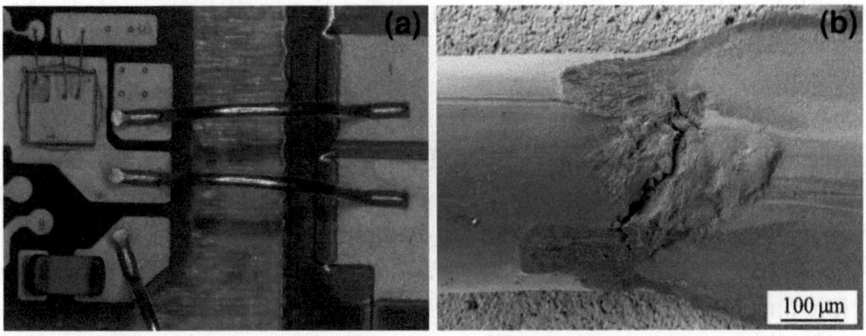

Fig. 1 Heavy aluminum wire bond contact of a microelectronic component **a**; detail of a fatigue crack in the heel of an aluminum wedge contact **b**

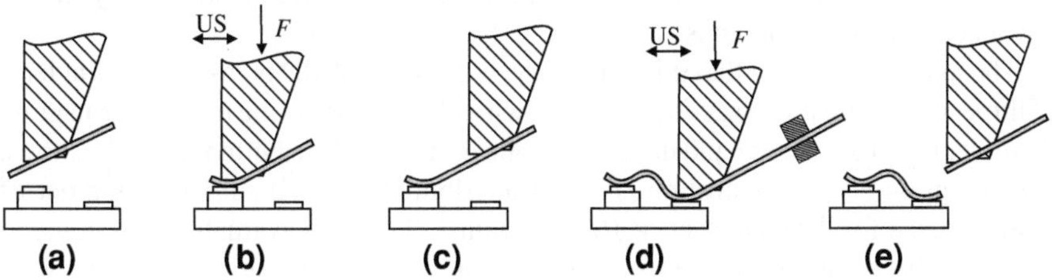

Fig. 2 Principle of wedge/wedge bonding for heavy aluminum bonding wires after [1]; positioning **a**; first contact **b**; loop forming **c**; second contact **d**; cut of wire **e**

Table 1 Overview of aluminum bonding wires used in this investigation

Code	Supplier	Diameter [μm]
A	A	125, 150, 200, 300, 400, 450, 500
B	A	150, 300, 450
C	A	150, 300, 450
D	B	125, 250, 350, 500

We present results of a micro-compression test that allows the determination of the hardening behavior parallel and perpendicular to the wire axis at moderate and large strains of small wire cylinders. The results are compared to microstructure and mechanical properties of typical heavy aluminum bonding wires. An overview of the used bonding wires is shown in Table 1.

2 Microstructure and Macroscopic Mechanical Properties

For verification and interpretation of the results described in the following sections, the macroscopic mechanical properties and the microstructure of typical heavy aluminum bonding wires will be discussed in this section. The microstructure was determined by the electron backscatter diffraction (EBSD) method. Multiple cross-sections of each wire were produced by metallographic preparation techniques in combination with ion polishing. A nearly artifact-free surface over the whole cross-section had to be assured.

The samples were analyzed by a TSL-EBSD system with a DigiViewIII detector in a scanning electron microscope Zeiss SUPRA VP55. The specimens were tilted 20° to the incident electron beam and the diffraction patterns were analyzed at 15 kV acceleration voltage and 12 mm working distance. From the diffraction patterns of each scan point the corresponding orientations were calculated. Neighboring scan points with only small differences in the orientation (5°) were merged to one grain. The area of each grain was used for the calculation of the mean grain size. In the following figures, the microstructure is displayed in an inverse pole figure color code. The grain size distribution was analyzed by advanced statistical methods.

The analyses showed that it is necessary to perform EBSD measurements at least of three to ten wire cross-sections for each wire, depending on the ratio of mean grain diameter to wire diameter and the width of the grain size distribution. It can be shown that the grain size is logarithmic normal distributed, when a sufficient number of grains is used for the analysis ($>10,000$ grains). The used evaluation procedure to calculate the mean grain size has also an influence on the statistical distribution. A more detailed discussion of this aspect can be found in [5].

In this study we quantified the microstructure by an equivalent mean grain diameter D_{A_w} calculated from the expected value of the grain area A_w :

$$D_{A_\mathrm{w}} = 2\sqrt{\frac{A_\mathrm{w}}{\pi}} \qquad A_\mathrm{w} = \sum_{i=1}^{n} A_i \frac{A_i}{A_\mathrm{tot}} \qquad A_\mathrm{tot} = \sum_{i=1}^{n} A_i \tag{1}$$

A_i is the area of each individual grain and A_tot is the total analyzed area, respectively the sum of A_i.

The investigations showed that all heavy aluminum bonding wires can be divided into two classes of Al wires: coarse-grained and fine-grained wires. The coarse-grained aluminum wires consist of very large grains compared to the wire diameter, see Fig. 3. The majority of grains is $\langle 100\rangle$ orientated.

The fine-grained aluminum wires consist of very small grains compared to the wire diameter. The grains are mainly $\langle 111\rangle$ orientated, even though the grains in the center of the wire are mainly $\langle 100\rangle$ orientated, see Fig. 4.

The macroscopic mechanical properties of the aluminum bonding wires were determined by tensile tests. Ten to twenty specimens were tested for each wire material and wire diameter. From the recorded force/diplacement data the nominal stress/strain behavior was directly calculated from the inititial area A_0 and the initial specimen length L_0 by:

$$\sigma_\mathrm{n} = \frac{F}{A_0} \qquad \text{and} \qquad \varepsilon_\mathrm{n} = \frac{\Delta L}{L_0} \tag{2}$$

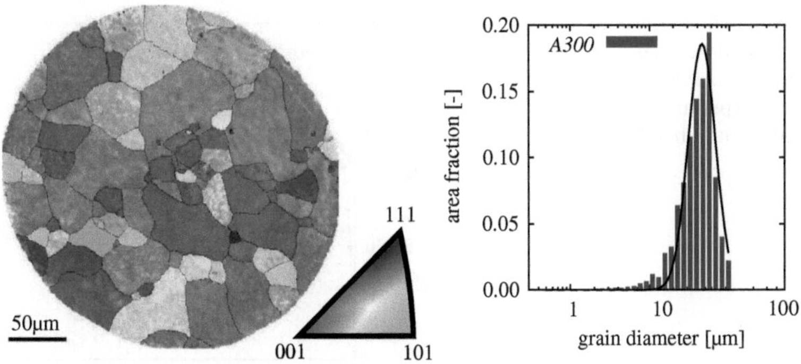

Fig. 3 Microstructure of a coarse-grained heavy Al bonding wire (*A* 300) and corresponding grain size distribution of seven cross sections

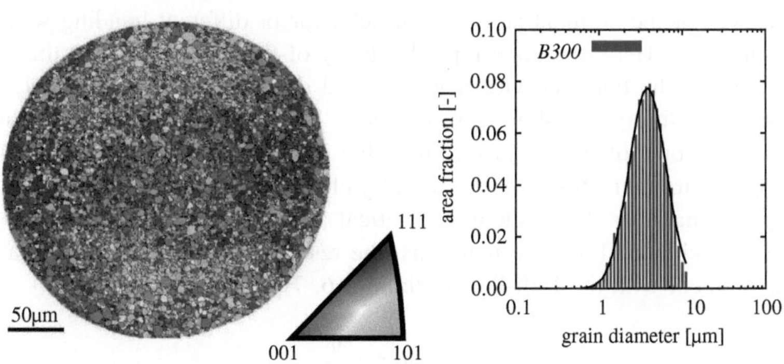

Fig. 4 Microstructure of a fine-grained heavy Al bonding wire (*B* 300) and corresponding grain size distribution of three cross sections

Possible experimental artifacts were corrected by determining the point of maximum slope in the first part of the stress/strain curve and shifting the strain data by the value of the intersection from a tangent at the point of maximum slope with the x-axis. The initial specimen length was also corrected by the shifting strain [5]. Considering volume constancy and actual specimen length, the flow curves (true stress σ_t to plastic logarithmic strain ε_p) were calculated by:

$$\sigma_t = \sigma_n(1 + \varepsilon_n) \qquad \text{and} \qquad \varepsilon_p = \ln(1 + \varepsilon_n) - \frac{\sigma_t}{E} \tag{3}$$

E is the Young's modulus of the material. Exemplary nominal stress/strain curves and flow curves of tensile test experiments of a 450 μm aluminum bonding wire are shown in Fig. 5.

Reduced mean curves of the nominal stress strain curves and the flow curves were calculated and will be used in the following. The initial yield stresses

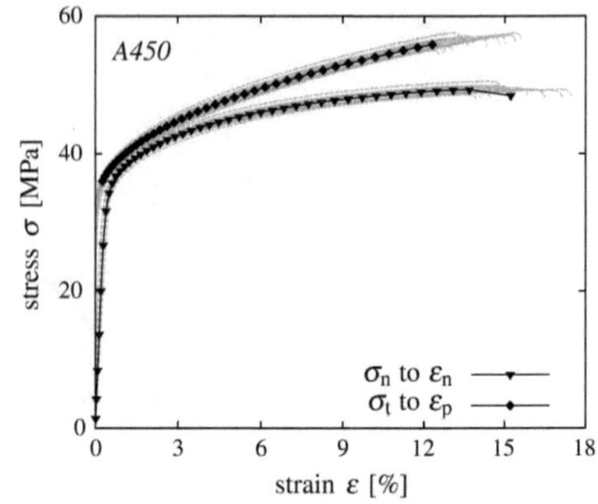

Fig. 5 Nominal stress/strain curves, flow curves and reduced mean curves of tensile test experiments of a 450 μmaluminum bonding wire

K_0 were determined in terms of nominal stresses at 0.2% plastic strain. Exemplary mean curves of the nominal stress/strain behavior of different bonding wires are shown in Fig. 6. Here, the good reproducibility of the experiment and the differences between the bonding wire materials and diameters become evident. Most aluminum bonding wires show a pronounced hardening and no distict point of yield, except some of these wires with higher initial yield stresses.

Compared to microstructure, the initial yield stresses increase with decreasing mean grain diameter. Even though a critical value of mean grain size to wire diameter is achieved for these materials, the results evaluated here are in a very good agreement with the Hall-Petch relation [6, 7]:

$$\sigma = \sigma_0 + \frac{k_0}{\sqrt{D}} \tag{4}$$

The value k_0 is material constant, σ_0 is the initial yield stress of the single crystal and D is the grain diameter. The Hall-Petch plot (σ to $1/\sqrt{D}$) is discussed in Sect. 4 in more detail.

3 Local Hardening in an Axial Micro-Compression Test

The local hardening behavior of heavy aluminum bonding wires can be investigated using a special test setup called axial micro-compression test. Therefore, it is necessary to produce small cylindrical samples with a length to diameter ratio of $L/D \approx 2$ out of the region of interest. When the geometry is nearly an ideal cylinder, the specimen can be placed upright on a flat steel support and loaded in the wire direction by a steel flat punch using an material testing machine, like it is shown in Fig. 7.

Using special metallographic preparation techniques, it is possible to produce samples with high geometric quality for axial micro-comperssion tests from bonding wires between 250 μm and 500 μm in diameter. The specimen should be

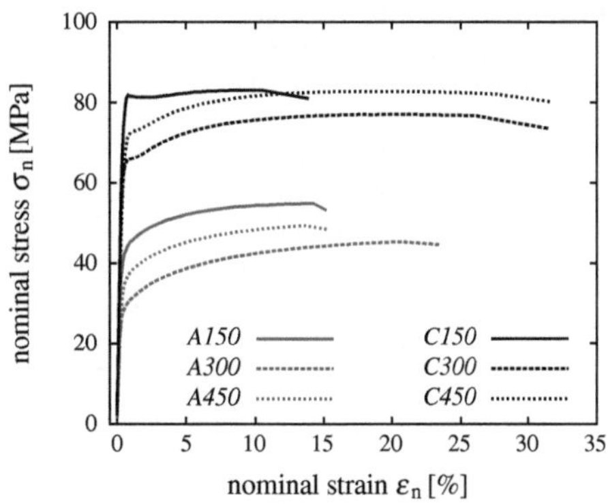

Fig. 6 Comparison of nominal stress/strain behavior of the bonding wires *A* and *C* with three different wire diameters (150, 300, 450 μm)

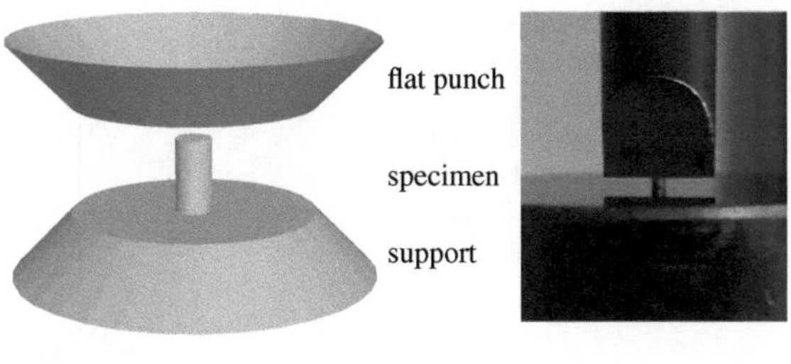

Fig. 7 Principle of axial micro-compression test for determining local hardening behavior of small cylindrical samples from bonding wires with diameters between 250 and 500 μm

handled with care to avoid damage or hardening due to sample preparation and handling. With a plane-parallel arrangement of support and punch, the cylindrical sample can be loaded in a displacement controlled experiment while force and displacement are recorded. The nominal stress/strain curves, the flow curves and the reduced mean curves can be calculated as described for the tensile test experiments. In Fig. 8 the results of a 450 μm aluminum bonding wire are shown exemplarily. Here, the strong influence of considering actual specimen length and volume constancy at higher strains becomes evident.

In compression experiments, it is necessary to regard strain constraints due to friction between support and specimen. For this purpose the Siebel's correction [8, 9] can be applied:

$$\sigma_s \approx \sigma_t \left(1 + \frac{2\mu R_0}{3h_0} e^{3/2\varepsilon_p}\right)^{-1} \tag{5}$$

Here, R_0 is the initial cylinder radius, h_0 is the initial cylinder height and μ is the friction coefficient between the sample and the support, respectively the sample and the flat punch. Because the effects of friction depend on the ratio of specimen radius and specimen height R_0/h_0, the unknown friction coefficient can be identified from experiments with cylinders of different heights. For this reason, specimens with four different heights were prepared from one wire material and tested in axial micro-compression experiments.

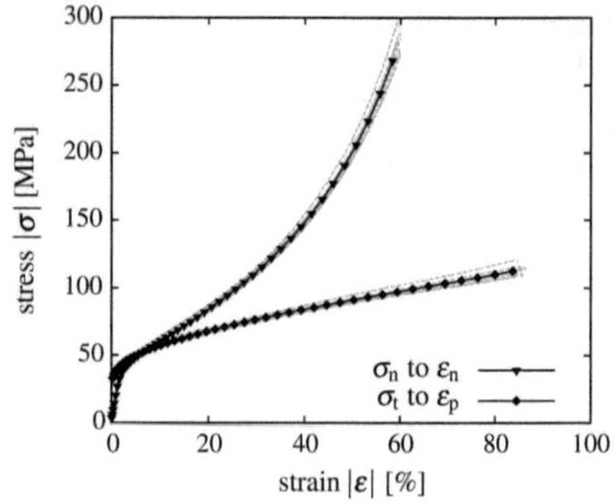

Fig. 8 Nominal stress/strain curves, flow curves and the reduced mean curves of axial micro-compression test experiments of a 450 μm aluminum bonding wire with a cylinder length of 600 μm

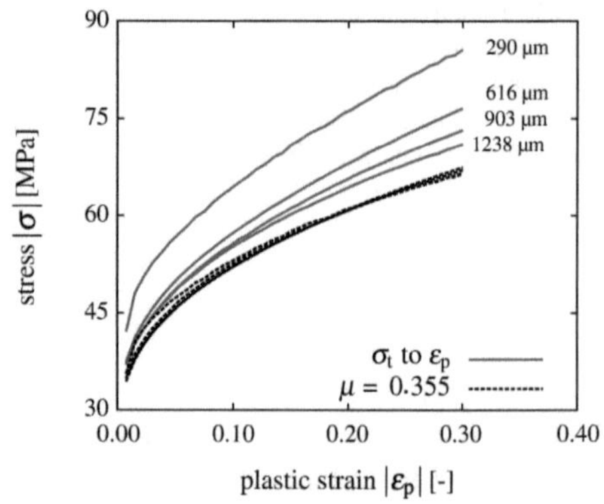

Fig. 9 Identification of friction coefficient using Siebel's correction on compression experiments on a 450 μm aluminum bonding wire with different cylinder heights h_0 : 290, 616, 903, 1238 μm

The friction coefficient was identified by an optimization routine, which minimizes the sum of the error square of the variation coefficient between all siebel corrected stress/strain curves. Considering all four specimen heights of 290, 616, 903 and 1238 μm for a 450 μm wire the friction coefficient is identified to be $\mu \approx 0.355$, the corresponding flow curves and the Siebel-corrected stress/strain data are shown in Fig. 9. Because the height of the smallest cylinder is smaller than the wire diameter and, therefore, the assumptions made for this evaluation are questionable [8], additional identification of the friction coefficient was performed with the results from the three biggest cylinder heights. Here, the friction coefficient is identified to be $\mu \approx 0.185$. So, the friction coefficient is assumed to be between 0.18 and 0.36 and will be discussed in the following.

Ten different bonding wires with diameters between 250 μm and 500 μm from four different wire materials were analyzed using the described axial micro-compression test. Five to ten samples were tested for each bonding wire. The flow curves were corrected by a friction coefficient of $\mu = 0.2$ and $\mu = 0.3$ and the results compared to each other and tensile test experiments. In Fig. 10 the mean flow curves of three different aluminum bonding wires are shown exemplarily.

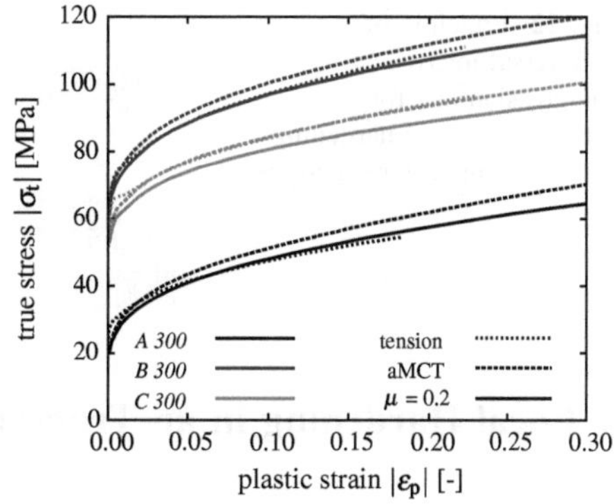

Fig. 10 Mean flow curves of axial micro-compression tests (aMCT) and friction corrected flow curves ($\mu = 0.2$) compared to flow curves in tensile tests (tension) of three different 300 μm bonding wires *A* 300, *B* 300 and *C* 300

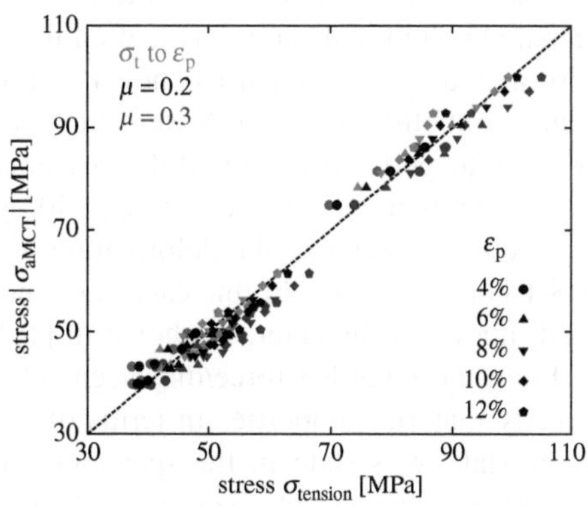

Fig. 11 Correlation of yield stresses at discrete plastic strains for all tested aluminum bonding wires

Here, it becomes evident that considering a friction coefficient lowers the determined yield stresses, but the differences are rather small compared to the hardening and the differences between the wire materials.

Furthermore, using a Siebel's correction with $\mu = 0.2$ results in a good correlation of the yield stresses with tensile test results, whereas the stresses without considering friction are slightly higher and the stresses assuming a friction coefficient of $\mu = 0.3$ are slightly lower, see Fig. 11. Consequently, the aluminium bonding wires behave isotropic in tension and compression when loaded parallel to the wire axis and a friction coefficient of $\mu = 0.2$ is an appropriate assumption for this experiment.

The axial micro-compression test is the first methodology that allows a quantitative characterization of local mechanical properties also for samples taken from processed bonding wires and real field returns. Furthermore, it is the first approach which allows a characterization of quantitative meaningful mechanical properties for bonding wires up to plastic strains of $\varepsilon_p = 0.6$; which is necessary when considering the bond deformation for loop simulations and lifetime estimations.

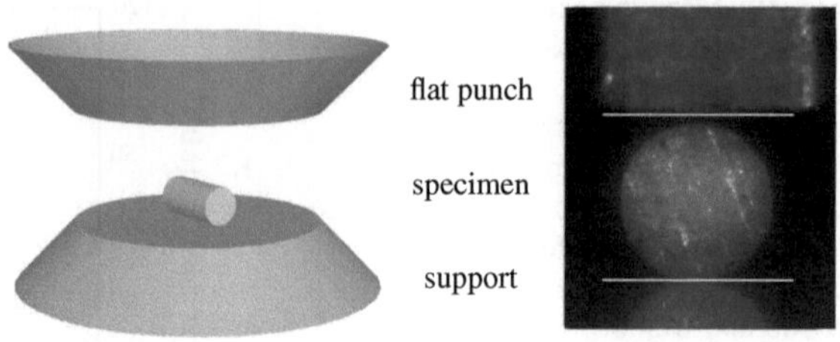

Fig. 12 Principle of transversal micro-compression test for determining local hardening curves of small wire sections

4 Local Hardening in an Transversal Micro-Compression Test

The axial micro-compression test, as described in the former section, is limited to testing wires with a diameter equal or above 250 μm. To overcome this drawback, a transversal micro-compression test was developed. Here, the metallographically prepared cylindrical sample is laid on the support and loaded perpendicular to the wire axis by a flat punch as shown in Fig. 12. Because the requirements to the plane-parallelity are much lower than for the axial micro-compression test, it is possible to perform meaningful transversal micro-compression tests on bonding wires with diameters between 18 and 500 μm. So this methodology can further be applied to characterize the deformation behavior also of smaller copper, gold and AlSi1 bonding wires. In this case, a diamond flat punch and diamond support are used in a micro indenter, as shown in [10].

From the recorded force/displacement behavior, it is not possible to calculate directly material properties in terms of stress/strain relationships, because of the multiaxial stress state in the specimens. It is necessary to identify the material properties by using optimization strategies with finite element simulations.

The force/displacement data in Fig. 13 show a very good reproducibility of the experiment. For parameter identification, mean curves with reduced data points were calculated from the loading part of the force/displacement data. When testing small samples, it is very important to correct the data by the compliance of the experimental setup. Therefore, the system stiffness is determined from every unloading curve by:

$$S_{\text{system}} = \frac{\mathrm{d}F}{\mathrm{d}u} \tag{6}$$

Assuming an equivalent model of a serial connection, the machine stiffness can be identified if the stiffness of the specimen S_{specimen} is known:

$$S_{\text{machine}} = \left(\frac{1}{S_{\text{system}}} - \frac{1}{S_{\text{specimen}}} \right)^{-1} \tag{7}$$

When unloading the specimen after a nominal compression of more than 50%, an uniaxial stress state can be assumed for the elastic unloading, thus the specimen stiffness can be calculated easily, if the elastic properties of the material are known:

Fig. 13 Force/displacement curves and reduced mean curves of the wires *A* 450 and *C* 450 in a transversal micro-compression test

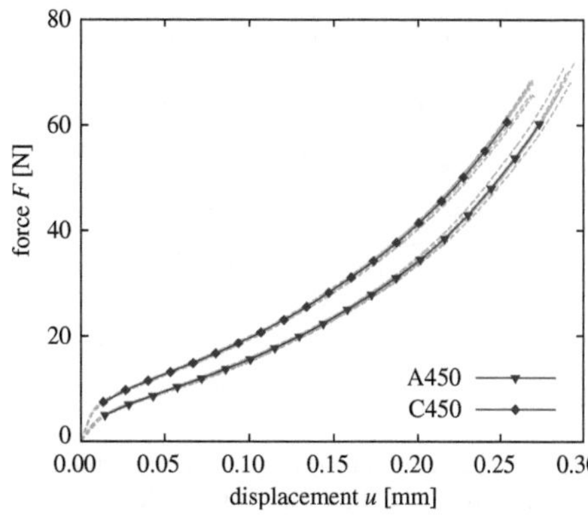

$$E = \frac{\sigma}{\varepsilon} = \frac{\mathrm{d}F/A_c}{\mathrm{d}u/D_{min}} = \frac{\mathrm{d}F}{\mathrm{d}u}\frac{D_{min}}{A_c} \quad \Rightarrow \quad S_{specimen} = \frac{EA_c}{D - u_{max}} \tag{8}$$

With a specimen height of $D_{min} = D - u_{max}$ when starting the unloading, the only unknown value is the contact area A_c at that point. Because of the dominating plastic deformation in a transversal micro-compression test when loading up to 60% strain, there is nearly no difference between the real contact area when starting the unloading and the resulting area after deformation. Consequently, A_c can be determined by light optical inspection after the experiment. But when testing a very small cylindrical specimen ($18\,\mu m \leq D \leq 32\,\mu m$), a lightoptical inspection after the test is not always possible, due to adherence of the specimen at the flat punch.

To overcome this problem, finite element simulations with typical bonding wire material properties and geometries [5] were performed using Abaqus 6.9 (Dassault Systèmes Simulia Corp., Providence, RI, USA). The used finite element model is shown in Fig. 14a. In the simulations, a wire cylinder was deformed by the translation of an analytical rigid body perpendicular to the wire axis and the contact area was calculated. The results indicate that material properties have nearly no influence on contact area, whereas friction coefficient and length to diameter ratio are very important. Considering the former verified friction coefficient of $\mu = 0.2$, a normalized contact area function A_{nc} was calculated and implemented into a Scilab-scribt as shown in Fig. 14b, so that the contact area could be determined by $A_c = A_{nc}LD$.

For parameter identification, the plastic properties have to be defined in a parameterized material model, whereas the elastic properties have to be constant. An isotropic hardening law for finite strain was used. The flow curve was defined in a modified Chaboche model for isotropic hardening [11]. An additional term $C{\varepsilon_p}^2$ was added for reflecting the flow curve also at large strains in an adequate way [5]:

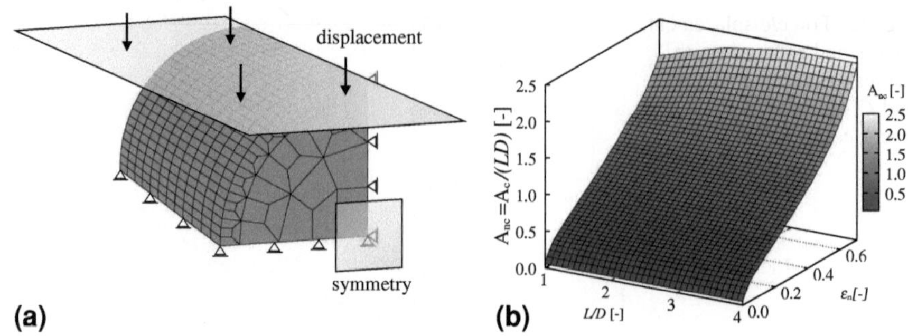

Fig. 14 Finite element model for determining the area function and identifying the plastic material properties in a transversal micro-compression test (**a**); normalized contact area function for $\mu = 0.2$ determined via finite element simulations (**b**)

$$\sigma_y = K_0 + K_\infty\left(1 - e^{-b\varepsilon_p}\right) + H\varepsilon_p + C{\varepsilon_p}^2 \tag{9}$$

The parameters of the hardening law were varied by an adaptive response surface methodology (ARSM) using the optimization software optiSLang [12] until the minimum in the error square sum between the experimental and simulated force/displacement data was achieved. The reduced data points of the displacement defined the load steps, whereas the resulting reaction forces were used for the calculation of the error square sum:

$$\min \sum_{i=1}^{n} \left(F_{\text{exp}}(u_i) - F_{\text{sim}}(u_i)\right)^2 \tag{10}$$

$F_{\text{exp}}(u_i)$ is the force at a defined displacement increment u_i of the reduced experimental results and $F_{\text{sim}}(u_i)$ is the corresponding force calculated by finite element simulation. The good correlation between experiment and final simulation with identified material properties is exemplarily shown in Fig. 15.

The identified flow curves and corresponding tensile test results of three different bonding wires are shown in Fig. 16. The results clearly indicate different material properties determined in transversal micro-compression tests and tensile tests. These differences can be explained by microstructure, as will be discussed in the following.

The results show that the initial yield stresses determined in a transversal micro-compression test and a tensile test are nearly the same, if the bonding wire is a coarse-grained material (relatively to the wire diameter). In contrast, the hardening is a bit higher for that kind of material, when testing perpendicular to the wire axis. The initial yield stresses of the fine-grained material identified in the transversal micro-compression test are significantly lower than those determined in tensile test or axial micro-compression test. Here, the hardening is comparable for all testing directions.

Fig. 15 Comparison of experimental and simulated force/displacement data for the identified material parameters of a 300 µm bonding wire

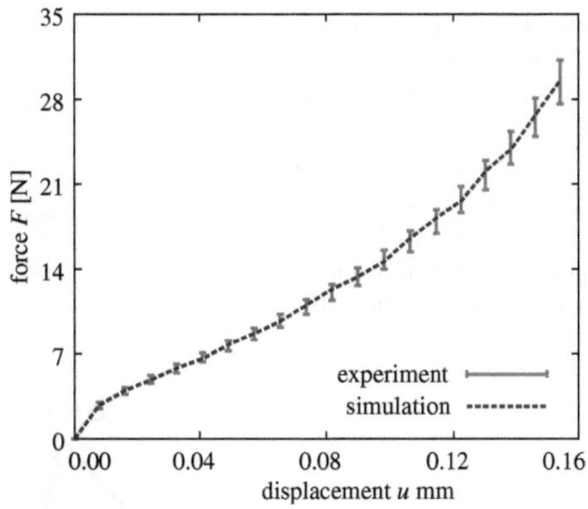

Fig. 16 Flow curves identified from transversal micro-compression tests (tMCT) compared to corresponding flow curves from tensile test experiments (tension) of three typical 300 µm bonding wires

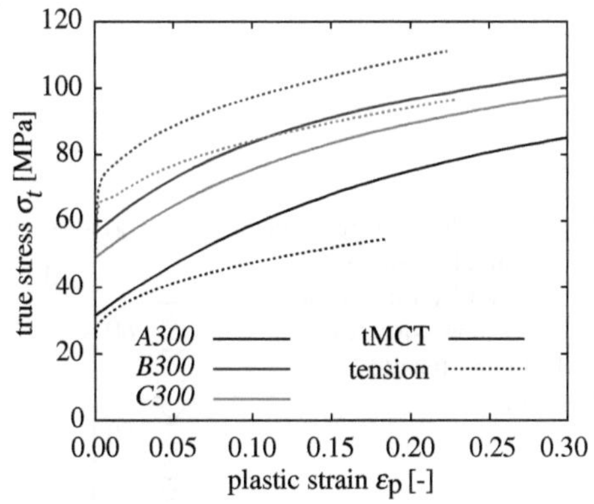

Microstructural investigations of lengthwise cross-sections from typical wire materials show that the grains of the coarse grained materials are nearly globular, whereas the fine-grained materials show a distinct drawing texture with grains much longer than wide, see Fig. 17. These elongated grains result in a bigger mean grain diameter in a longitudinal cross-section compared to a perpendicular cross-section. Thus, the initial yield stress must be lower when testing perpendicular to the wire axis.

This effect is shown in the Hall-Petch plot in Fig. 18. The values investigated from transversal micro-compression experiments and EBSD measurements of lengthwise cross-sections are in the range expected from the tensile test results and EBSD measurements on perpendicular cross-sections. Consequently, the mean grain size is the dominating factor for the initial yield stress of the material. The small differences in hardening of the coarse-grained wire materials can be explained by the decreasing free surface to grain boundary ratio during the transversal micro-compression test, which results in a small amount of additional hardening.

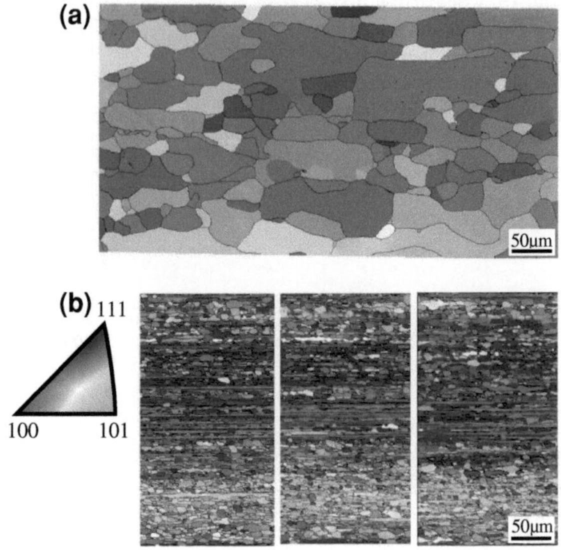

Fig. 17 Microstructure (inverse pole figure color code in wire direction) of longitudinal cross sections from a coarse grained wire *A* 300 (**a**) and a fine grained wire *B* 300 (three individual scans of neighboring regions) (**b**)

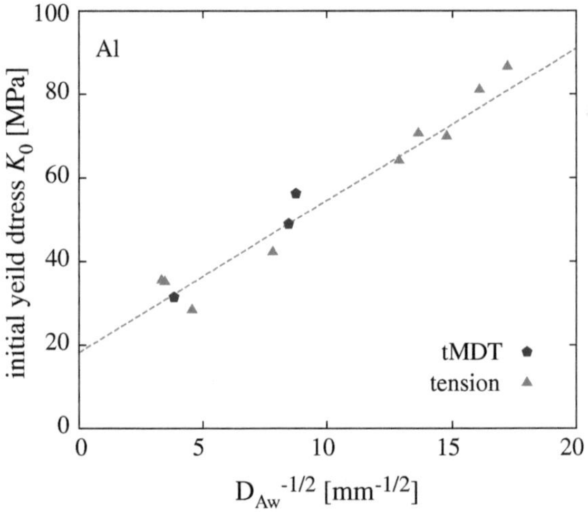

Fig. 18 Hall-Petch relation of the initial yield stress determined by transversal micro-compression tests (tMCT) and tensile tests (tension)

The shown results clearly demonstrate that the transversal micro-compression test is the first methodology to quantify the material properties of small wire sections. It can also be applied to characterize sections of real bond loops. Furthermore, the results show that the plastic material properties of the aluminum bonding wires depend on the loading direction and can be explained by the drawing texture.

5 Summary

For life-time estimations and failure diagnostics of wire bonded power and automotive electronic devices, it is necessary to know and to consider the mechanical properties of the bonding wires in an adquate way. The most critical regions for low cycle fatigue failure of heavy aluminum wire bonded devices are the heel and the loop top, where the wire is predeformed.

To determine the mechanical properties of the bonding wires parallel and perpendicular to the wire axis at moderate and large strains, a micro-compression test for small wire cylinders was developed. The results clearly indicate that the mechanical behavior is isotropic when loading the wires in tension or compression along the wire axis. On the other hand, due to the drawing texture, the wires show an aniosotropic plastic behavior depending on the loading situation. The determined initial flow stresses correlate very well with the mean grain size in terms of the well known Hall-Petch relation, wich was verified by tensile tests and electron backscatter diffraction analyses.

Using the developed methodologies, it is now possible to determine the static hardening parameters of bonding wires depending on the loading direction in a meaningful way. The micro-compression test can also be applied to characterize the material properties of wire sections from real components and field returns. The micro-compression test is the basis for further developments of new cyclic tension/compression tests, which can be applied for characterizing the cyclic hardening behavior, so that more reliable lifetime estimations can be performed.

References

1. Scheel, W.: Baugruppentechnologie der Elektronik. Verlag Technik GmbH, Berlin (1997)
2. Mehrotra, V., et al.: Wirebond reliability in IGBT-power modules: application of high resolution strain and temperature mapping. International Symposium on Power Semiconductor Devices and ICs ISPSD, Toronto, pp. 113–116 (1999)
3. Ramminger, S., et al.: Reliability model for Al wire bonds subjected to heel crack failures. J. Microrel. **40**, 1521–1525 (2000)
4. Wilde, J.: Lebensdauerprognose von Drahtbond-Verbindungen für die Mechatronik mittels FEM. Elektronische Baugruppen, Aufbau und Verbindungstechnik, DVS/GMM-Fachtagung, Fellbach, 40 (2002)
5. Dresbach, C.: Ermittlung lokaler mechanischer Kennwerte mikroelektronischer Drahtkontaktierungen. Ph.D.-thesis Martin-Luther-University Halle-Wittenberg (2010)
6. Hall, E.O.: The deformation and ageing of mild steel: III discussion of results. Proc. Phys. Soc. **64**, 747–753 (1951)
7. Petch, N.J.: The cleavage strength of polycrystals. J. Iron Steel Inst. 25–28 (1953)
8. Hill, R.: The Mathematical Theory of Plasticity. Oxford University Press Inc., New York (1950)
9. Charalambides, M.N. et al.: The analysis of the frictional effect on stress—strain data from uniaxial compression of cheese. J. Mater. Sci. **365**, 2313–2321 (2001)

10. Dresbach, C., et al.: Local hardening behavior of free air balls and heat affected zones of thermosonic wire bond interconnections. European Microelectronics and Packaging Conference EMPC, Rimini (2009)
11. Lemaitre, J., Chaboche, J.L: Mechanics of solid materials. Cambridge University Press, Cambridge (2002)
12. Dynardo: optiSLang the optimization structural language. Dynamic Software and Engineering GmbH (2008)

Part III
New Technologies

Numerical Computation of Melting Efficiency of Aluminum Alloy 5083 During CO_2 Laser Welding Process

Joseph Ifeanyichukwu Achebo and Oviemuno Oghoore

Abstract This chapter is aimed at determining the melting efficiency of aluminum alloy 5083 during CO_2 laser welding process. Theoretical models were used for the melting efficiency determination as proposed by other investigators which also included an examination of the fluid flow pattern of the alloy. The results obtained indicate that the acceptable melting efficiency calculated was 38%. This value compares well with and falls within the range of other values reported in other literature. The theory of metal melting as it relates to laser welding depends on the thermal state of the material under investigation. Applying high laser power under a controlled environment would achieve deeper penetration with fewer heat affected zones; therefore a deep understanding of the chemo-physical properties of a metal is required to determine its melting efficiency and these properties have been adequately treated in this study.

Keywords Melting efficiency · Coupling · Surface tension · Melting temperature · CO_2 laser welding process

1 Introduction

Okon et al. [1] wrote that laser welding is at the frontier of welding technology and the use of keyhole welding has been increasingly adopted by various sectors of the manufacturing industry. They stated that the possibility of welding materials of

J. I. Achebo (✉) · O. Oghoore
Department of Production Engineering, University of Benin, Benin City, Edo, Nigeria
e-mail: josephachebo@yahoo.co.uk

O. Oghoore
e-mail: oviemuno2002@yahoo.co.uk

A. Öchsner et al. (eds.), *Materials with Complex Behaviour II*, Advanced Structured Materials 16, DOI: 10.1007/978-3-642-22700-4_37,

varying thickness quickly, efficiently and with resultant small heat affected zones continues to attract more attention in manufacturing and industrial complexes.

Weston [2] posited that in laser welding, sufficient power density is necessary to produce a high heat input that should cause some of the volatile alloying elements such as magnesium to vaporize and become ionized to form a plasma. The plasma contains electron and ions that would further increase the temperature that should lead to the further melting of the metals to be joined because the plasma is itself responsible for transferring the energy from the beam to the weld.

In the melting of aluminium alloys using the laser welding process, a lower energy of between 10^6 and 10^8 W/cm^2 is desirable since a higher energy above the threshold value would possibly lead to unnecessary increase in vaporization of alloy elements. It must be remembered that aluminium alloys have a higher heat absorption rate (lower heat losses), than pure aluminium [3].

Olander [4] stated that absorbed power density together with the laser pulse duration and the reflectivity of the surface, jointly determine the absorbed energy density. He went further to state that energy absorption depends on thermal diffusivity, the vapor pressure and optical absorbance. These properties govern the depth of penetration of the thermal front, the maximum surface temperature, and the density of the vapor just above the surface.

In fusion welding, metals are joined together by the melting and subsequent solidification of the adjacent areas of the two separate parts [5]. These separate parts can only be melted, when temperatures much higher than the ambient temperature are achieved and applied. This critical temperature, capable of agitating the molecules and transforming them from the solid phase to the liquid phase, is known as the melting temperature. Mahfoud et al. [6] were of the opinion that melt quality assessment is based on thermal analysis of the material investigated. David and DebRoy [7] also observed that the achieved melting rate is an important factor in determining the composition change of weld metal [8]. This situation happens during the melting process of aluminium alloys when highly volatile alloying elements such as magnesium would normally vaporize. The chemical composition of the solidified weldment is therefore altered in the process, and this alteration would affect its mechanical properties which includes the corrosion resistance and hot crack susceptibility of the weldment, as well as its load bearing capacity [9, 10].

Researchers have investigated the thermal behavior of aluminium alloys which assists to further explain the melting process of these alloys during laser welding [11, 12, 13]. Weston [2] noted the changes in the melting pattern of metal under increased temperature. He observed that at temperatures above half the melting of a metal, the flow is very plastic and of great significance when determining future strains that have resulted from thermal changes during the welding. At above 80% of the melting point, the metal looses its plasticity and viscous behavior occurs.

This chapter aims at determining the melting efficiency of aluminum alloy 5083 using available data obtained from experimental results, and applying these results to models proposed by Fuerschbach [14] and Weston [2]. Their resultant melting efficiencies are compared with experimentally determined standard ranges of values.

2 Classification of the Laser Welding Process by Mode

There are two major modes of laser welding, the conduction mode and keyhole mode.

2.1 Conduction Mode

The heat flow present in this mode is governed by thermal conduction from a surface heat source that causes the melting of the portions of the base metal to be joined. Trautmann [15] wrote extensively on the conduction mode of the laser welding process. The laser beam once focused is designed to generate a power density on the base metal of the order of magnitude of 10^3 W mm^{-2}. A power density of this order of magnitude is used to join materials to form a joint, no significant vaporization occurs as there is not a high enough heat input to form a significant keyhole.

Trautmann [15] observed that heat flow in Laser welding mode is measured by the determination of the Peclet number, Pe. The Peclet number is a measure of the distribution of the radiation power absorbed through the melt pool present in the conduction and convectional flow process. For a typical case of aluminium welding, Pe scales with 54.8u when Pe is 5, meaning that heat transportation is by conduction as well as convection. Note here that, u denotes the welding speed. This further reveals that when Pe is less than 5, it is in the conduction mode welding process but above this value signifies that the welding process is convective, that is, the convection mode is dominant. In the conduction mode, flow is restricted near the fusion boundary, which is a cooler region away from the weldpool. In this mode, the Marangoni effect which is the spatial temperature gradient of the surface tension, creates a driving force that pulls the melts away from the hotter region towards the cooler region. During laser welding, this surface tension force dominates the flow of the melt. In conduction mode welding, a hemispherical weld bead, a large heat affected zone (HAZ), and a low aspect ratio are formed.

Zhao et al. [16] measured the value of Peclet number as expressed as

$$Pe = \frac{u\rho c_p L}{k} \tag{1}$$

where u is the average velocity of the weldpool which is in the order of 0.1 m s^{-1}, less than the maximum velocity. ρ is the density, c_p is the specific heat at constant pressure, L is the characteristic length, and k is the thermal conductivity of the melt.

2.2 Keyhole Mode

Al-Kazzaz et al. [17] wrote on the keyhole mode of the laser welding process. Using this method, a keyhole is formed when the laser power absorbed by the metal is greater than the material specific intensity threshold value [8, 18]. The heating power is obtained by focusing the laser beam into a very small spot that generates a very high power [18, 19, 20]. Okon et al. [1] and Weston [2] said that with a high laser power, maximum penetration and rapid welding can be achieved. The maximum power density depends on the maximum machine power output and the minimum focused spot size. The exact spot focusing of the beam is further dependent on both the wavelength of the light and the beam quality. The wavelength of the laser light, limits the focused spot size through diffraction effects. Duley [21] said that the minimum power density that can form a keyhole should be about 10^6 W cm^{-2} for the CO_2 laser welding of aluminum alloys. Zhao and Debroy [8] were of the opinion that the power density of the laser beam is Gaussian in nature. The distribution of absorbed laser power density on the weld pool surface as proposed by Mouroulis and Macdonald [22] is as in Eq. 2

$$J(r) = \frac{2\eta_{abs}P}{\pi r_b^2} e^{\left(\frac{-2r^2}{r_b^2}\right)} \tag{2}$$

where η_{abs} is the absorption coefficient, P is the laser power, r_b is the beam radius, and r is the radial distance from the beam axis. Howard et al. [23] said that absorptivity is an important factor in laser welding processes, as greater absorptivity will have a strong influence on the process efficiency and weld cross sectional area attainable at a given travel speed.

The beam radius is defined at different locations along the beam axis z as

$$r_b = r_o \left[1 + \left(\frac{\lambda z M^2}{\pi r_o^2}\right)\right]^{\frac{1}{2}} \tag{3}$$

where r_o is the beam radius at the focal point, λ is the beam wavelength, z is the beam defocusing, i.e, the distance from the focal point to the top surface of the weldment, and M^2 is the dimensionless beam quality figure of merit expressed as

$$M^2 = \frac{\pi r_o \beta}{\lambda} \tag{4}$$

where β is the half angle of beam divergence. The value of M^2 varies with increasing laser power due to the intrinsic distortion of the laser beam [8].

The concentration of the laser power on a localized spot on the workpiece causes the alloying elements, like Mg which is very volatile and highly susceptible to heat, to evaporate and creating a cavity under the beam. This ablation according to Weston [2] causes a recoil pressure on the liquid surface, further depressing it. When the cavity is sufficiently deep, reflections of laser light from the inclined sides of the cavity can impinge upon the other side of the cavity rather than being lost to the environment. The repeated contact of the beam with the material in the cavity leads to increase in energy adsorption and an increase in the rate of material loss. Thus, the cavity deepens to form a deep narrow keyhole.

As the keyhole hole is formed, Al-Kazzaz et al. [17] wrote that its stability depends on the force balance between the keyhole wall and the molten metal around it. This means that the stability of the keyhole depends on the force balance between the vapor pressure and surface tension pressure. The vapor pressure tends to open the keyhole whereas the surface tension pressure tends to close it at the threshold intensity of the material ([8, 24]).

Al-Kazzaz et al. [17], noted that the threshold intensity, I_s, is determined by materials vaporization temperature, T_v, thermal conductivity, k and absorptivity at normal incidence, η_n [25]. This is expressed as

$$I_s \,\alpha\, \frac{T_v k}{\eta_n} \tag{5}$$

Leong et al. [26] proposed a model for the irradiance necessary for melting and forming a keyhole as expressed in Eq. 6

$$I_m = \frac{k(T_{melt} - T_o)}{\eta_s d J_{\max}} \tag{6}$$

where k is the thermal conductivity of the metal; $(T_{melt} - T_o)$ is the difference in melting and ambient temperature. η_s is the absorptivity of the surface, d is the diameter of the beam at the surface and J_{max} is a function of the ratio of the thermal diffusivity to the product of the reverse speed and diameter of the incident beam [27]. Leong et al. [26] used a value of 5% for the absorptivity and traverse speed of 12.7 ms^{-1}, and obtained irradiance values of 2.7 and 1.3 MW cm^{-2} for the 200 and 400 μm beam diameter, respectively.

However, the increase in power intensity could also increase weld penetration. The depth at which temperature reaches melting point, Z_m was measured by Okon et al. [1]

$$Z_m = \frac{1.2KT_m}{H}\left(\frac{T_b}{T_m} - 1\right) \tag{7}$$

where $H = I(1 - R_s)$, I is the beam intensity, R_s is the surface reflectance, Von Allmen [28] obtained a reflectance value of 94% for 1.06 μm light (Nd:YAG laser output) and 98% for 10.6 μm light (CO_2 laser output), K is the thermal conductivity, T_b is the boiling temperature and T_m is the melting temperature.

Okon et al. [1] wrote that the temperature at a depth z below the surface of a semi infinite solid at a time t after heat flow starts is expressed as

$$T_{(z,t)} = \frac{2H}{k}(\alpha t)^{\frac{1}{2}} \operatorname{ierfc}\left(\frac{Z}{2(\alpha t)^{\frac{1}{2}}}\right) \tag{8}$$

where α is thermal diffusivity and t is the time.

Grong [29] looked at the study by Rosenthal in 1948 of a thermal field theory around a moving heat source and considered a two dimensional form for thin sheet welding of a given point in the workpiece at a given time. The temperature rise in a metal welding scenario was given as

$$T - T_0 = \frac{P/d}{2\pi\lambda}\exp\left(\frac{-ux}{2a}\right)k_0\left(\frac{ur}{2a}\right) \tag{9}$$

where P = absorbed power; d = sheet thickness; λ = thermal conductivity; u = welding speed x = distance travelled along weld; a = thermal diffusivity = $(\lambda/\rho C_p)$; k_o = modified Bessel function of the second kind and zero order; r = radial distance from weld; ρ = density; C_p = specific heat.

3 Melting Efficiency Computation

Pure aluminium, from experimental procedures, has been found to melt at 660°C and its alloys (including their oxides) at about 1,926°C. Some alloys melt at temperatures below 1,926°C. The melting of aluminium alloys using the CO_2 laser welding process requires a high laser power density which produces a temperature high enough to melt the surfaces of the aluminium metals to the joined.

Zhao et al. [16] defined melting efficiency as the ratio of the heat required to melt the base metal to the heat absorbed by the workpiece. Fuerschbach [14] suggested an equation for melting efficiency, η_m in the CO_2 laser welding of 1,018 steel, 304 stainless steel and Tin correlated with a modified Rykalin number, Ry, as expressed in Eq. 10

$$\eta_m = 0.48 - 0.29\exp\left(\frac{-\text{Ry}}{6.8}\right) - 0.17\exp\left(\frac{-\text{Ry}}{59}\right) \tag{10}$$

where $Ry = \frac{q_{in}U}{\alpha^2 \Delta H_m}$ q_{in} is the net power absorbed by the workpiece, U is the welding speed, ΔH_m is the latent heat and α is the thermal diffusivity. Weston [2] proposed an equation for determining the melting efficiency of aluminum alloys which is expressed in Eq. 11

$$\eta_m = \frac{V\left[\rho c_p(T_m - T_o) + \Delta H_m\right]}{\eta_c P} \tag{11}$$

where, η_c is the coupling. The coupling is a dimensionless term used to describe the efficiency of energy transfer from a laser beam to a weld. Coupling can be defined as:

$$\text{Coupling} = \frac{\text{energy absorbed by weld}}{\text{energy output by laser source}} \tag{12}$$

in this study, the value for coupling is taken to be equivalent to the absorptivity energy value. P is the laser power; C_p, the specific heat, T_m is the melting temperature; T_o the ambient temperature, ρ is the liquid density and V is the volume of metal droplet.

Al-Kazzaz et al. [17] defined coupling efficiency, η_c, as the portion of the laser power available to the workpiece which is equal to the ratio of the absorbed power, P_{abs} by the keyhole wall to the total laser power, P. The absorbed power as written by Zhao et al. [16] and developed by Bramson [30] using the series expansion method is expressed as

$$\eta_{abs}(T) = 0.365\left(\frac{r}{\lambda}\right)^{\frac{1}{2}} - 0.0667\left(\frac{r}{\lambda}\right) + 0.006\left(\frac{r}{\lambda}\right)^{\frac{3}{2}} \tag{13}$$

where r is the resistivity (Ω cm) at temperature, T; λ is the wavelength in cm for CO_2 laser welding, $\lambda = 10.6$ μm.

$$\eta_c = \frac{\eta_{abs}(T)}{P} \tag{14}$$

From Eq. 13, it can be seen that the absorbed power is dependent on temperature. This indicates that coupling efficiency depends on the thermal conductivity of the base metal material. The thermal efficiency, η_{th}, is defined as the portion of the absorbed laser power that produced the weld seam and is equal to the ratio of the heat of melting, $H_{melting}$ divided by the absorbed laser power, η_{abs} [17].

$$\eta_{th} == \frac{H_{melting}}{\eta_{abs}} \tag{15}$$

The heat of melting, $H_{melting}$ can be calculated from the equation proposed by Kaplan [31] and Lampa et al. [32] for heat flow in keyhole mode as expressed in Eq. 16

$$H_{melting} = \frac{1}{r}(T - T_o)kPe\left(\cos\varphi + \frac{k_1 Pe}{k_0 Pe}\right) \tag{16}$$

where q is the heat flow (W m^{-2}), k_1 is the modified Bessel function of second kind and first order, and r, φ are the polar coordinates. At the boundary between the PMZ and HAZ, r is equal to $\pm$ half the average width of fusion zone, and $\varphi \pm \pi/2$.

The laser process efficiency is calculated as

$$\eta_p = \eta_c \eta_{th} \tag{17}$$

The laser process efficiency, η_p, can be defined as the ratio of the melting power to the total laser power

$$\eta_p = \frac{H_{\text{melting}}}{P} \tag{18}$$

Punkari et al. [24] and Swift-Hook and Gick [33] defined melting ratio (MR) as the fraction of total incident laser power that is used to melt the weld metal.

$$MR = \frac{VA_w \rho \left(C_p \left(T_{mp} - T_0\right) + \Delta H_f\right)}{P} \tag{19}$$

where A_w is the melted area (i.e., FZ area), ρ is the density of the molten metal, C_p is the specific heat, ΔH_f is the latent heat of fusion, V is the welding speed, T_{mp} is the melting point temperature, T_o is the room temperature and P is the incident laser power.

Kwon and Weckman [34] wrote that the melting ratio, MR is used to determine the effects of welding process conditions on the overall energy coupling efficiency. Swift-Hook and Gick [33] were of the opinion that the melting ratio is the fraction of the total incident power which is used to heat and melt the weld metal. They proposed the following equation.

$$MR = \frac{\left[\mu C_p \left(T_{mp} - T_0\right) + \Delta H_f\right]}{L_w \left(\frac{P_l}{v_w}\right)} L_w A_w \tag{20}$$

where L_w is the length of weld metal, A_w is the weld metal area, ρ is density, C_p is the specific heat, ΔH_f is the latent heat of fusion, V_{ws} is welding speed and T_{mp} is the solidus temperature. T_o is the room temperature and P_L is the power from the power supply. Kwon and Weckman [34] used the thermo physical properties of A5182, as obtained from ASM Handbook [35] for their calculations, given $\rho = 2.65$ mg m^{-3}, $C_p = 0.904$ kJ kg^{-1} K^{-1}; $\Delta H_f = 397$ kJ kg^{-1} and $T_{mp} = 850$ K.

Increasing laser power leads to increased heat input per unit length. A high melting ratio variation is obtained when the welding speed is similarly increased to match the increasing heat, because the quicker the weld process the lower the time spent or lost. The heat is, therefore, more localized within the heat affected zone as high conductivity is curbed and a reasonably sized weld pool is formed and maintained. The weld pool has accumulated to a reasonable size due to prolonged localized heating, and quickly enough too. Thus, heat is immediately dissipated around the weld environment through the reflection of the multiple incident laser beam rays out of the surface of the weld, and this heat is not absorbed by the base metal. It means that the threshold value for the laser power that would ensure a good melting ratio has been attained.

However, in the cases where further increase in laser power continues unnecessary, it could reduce the melting ratio. It means that the threshold value for the laser power that was available in the melting ratio has been exceeded [34]. Swift-Hook and Gick [33] have actually proposed a threshold value for the melting ratio as 0.48 for 2D keyhole mode laser beam welds.

Kwon and Weckman [34] obtained a maximum value of 0.45 at 2.8 kW with improved cathode cleaning and good weld quality using the double sided arc welding (DSAW) process. Fuerschbach and Knorovsky [36], in studying the melting efficiency of plasma arc welding (PAW) and gas tungsten arc welding (GTAW) of 304 stainless steel and Ni 200 specimens reported their melting efficiency of 0.46 at the highest welding speed. Punkari [37] and Deutsch [38] as reported by Kwon and Weckman [34] found that the maximum melting ratio of single beam Nd:YAG laser welded AA5182 alloy sheet was approximated to be 0.3, while that for dual beam laser welds was 0.2, and for variable polarity plasma arc welding (VPPAW) welds full penetration was obtained at a maximum threshold value of 0.2.

Quintino et al. [39] derived the equations as proposed by Ready [40] for determining melting efficiency. They considered a case of full efficiency when the power, *P*, available is used to melt a unit of material volume, *V*, the weld bead is cylindrical in shape and no losses were observed. This expression is shown in Eq. 21

$$P = VE \tag{21}$$

where *P* is the laser power in watts, *V* is the volume of metal melted per unit of time (m^3/s) and *E* is the energy required for melting.

Paleocrassas and Tu [41] studied the critical fluence per weld length, E_c which they said indicates the amount of energy per area for a unit length of weld that is required to form a weld in an aluminium alloy. They assume that at a specific laser power and travel speed, a burnt mark starts to form when the fluence per weld length reaches a threshold. This threshold value is determined by

$$E_c = \frac{P}{\pi r_i^2} \cdot \frac{l_i}{v_c} \cdot \frac{1}{l_i} = \frac{P}{\pi r_i^2 v_i} \tag{22}$$

i = 1, 2, …., n where E_c is in J/cm^3. *P* is the incident power in watts, r_i is the beam radius at the onset of the burn mark in centimeters, l_i is the total length of the burning mark, and v_i is the processing speed.

$$\text{Total welding time } = \frac{total\ weld\ length}{welding\ speed} = \frac{l_w}{v} \tag{23}$$

Paleocrassas and Tu [41] observed that the smaller width has a higher aspect ratio, deeper penetration and higher welding efficiency, E_w

$$E_w = \rho . A_{\text{weld}} \frac{\zeta}{\pi r^2_{\text{profile}}} \tag{24}$$

where

$$\zeta = C_p \Delta T + \Delta H_f \tag{25}$$

where A_{weld} is the cross sectional area of the weld, ρ is the density, r_{profile} is half the weld's cross sectional width at the top of the weld. C_p is the specific heat, ΔT is the temperature rise from room temperature to melting temperature and ΔH_f is the latent heat of fusion.

Equation 21 is further expanded to form Eq. 26,

$$P = V . \rho \left[C_p(\Delta T) + H \right] \tag{26}$$

where ρ is the density in kg/m^3, C_p is the specific heat in J/kg. K, ΔT is the difference between the melting temperature and the ambient temperature in Kelvin and H is the latent heat of melting in J/kg.

The volume of the metal melted per unit of time, V is given in Eq. 27

$$V = v_t bh \tag{27}$$

where b and h is the weld bead width and depth or penetration respectively.

Substituting the value of V and thermal conductivity, $k = \rho C_p \alpha$ (W/mK) in Eq. 26. P becomes

$$P = \frac{v_t b}{\alpha} h . k \left(T_m - T_o + \frac{H}{C_p} \right) \tag{28}$$

Assuming that $d = T_m - T_o + \frac{H}{C_p}$

Equation 28 becomes

$$P = \left(\frac{v_t b}{\alpha} \right) (hkd) \tag{29}$$

Rearranging Eq. 29

$$\frac{v_t b}{\alpha} = \frac{P}{hkd} \tag{30}$$

Quintino et al. [39] used Eq. 30 to compute the melting efficiency. They classified Eq. 30 as welding speed parameter ($v_t b/\alpha$) which forms the vertical axis of the graph and laser power parameter (P/hkd) which constituted the horizontal axis of the graph. The slope of this graph gave a value of 0.66 which resulted to a melting efficiency of 66%. This was regarded as good when compared to conventional lasers. What energy losses that may have occurred, is attributed to conduction losses.

4 Determination of Volume of the Melted Metal, V

4.1 Filler Wire in Laser Welding

Filler wire is important to laser welding because it is used to increase the volume of molten metal used to achieve deep penetration. Al-Kazzaz et al. [17] stated the advantages of using filler metals to include improving weld properties, increasing the gap between the welding parts and welding thick sections using a multi pass technique. Haferkamp et al. [18] observed that during laser welding, underfill, and notching effect and porosity of the welded joint can be reduced using filler metal. The filler wire is positioned at the intersection of laser beam and top surface of the work piece at an angle of 60°

$$Wire\,feed\,rate = \frac{welding\ speed\ x\ gap\ area}{filler\ wire\ area} \tag{31}$$

To determine the volume of melted material, the equations proposed by Padmanobham et al. [42], were used to determine the volume of the melted metal as expressed in Eqs. 32–36.

Volume of wire melted,

$$V_f = w_f \times \pi r w^2 \tag{32}$$

where, w_f is the wire feed rate and r_w is the wire radius.

Laser beam pulse frequency,

$$F = \frac{1}{T} \tag{33}$$

where T is the cycle time to make one droplet

$$F = \frac{V_f}{V_s} \tag{34}$$

V_s is the volume of a single droplet.

The radius of the spreading droplet, r_s was also determined using the equation suggested by De Gennes [43] written in Eq. 35

$$r_s = V_s^{\frac{3}{10}} \left(\frac{\gamma T}{\mu} \right)^{\frac{1}{10}} \tag{35}$$

where γ is the surface tension and T is the time taken to make one drop of aluminum metal. μ is the viscosity

$$V = \frac{4}{3} \pi r_d^3 \tag{36}$$

where r_d is the radius of metal droplet which, in this study is equivalent to the radius of spreading droplet, $r_{s.}$

The value of V, from Eq. 36 is input into Eq. 27 for the determination of melting efficiency, η_m of aluminum alloy.

The viscosity, μ as proposed by Smithells and Brandes [44] can be determined computationally using Eq. 37

$$\mu = 0.1492\exp\left(\frac{1984.5}{T_m}\right)\ \mathrm{mNs/m^2} \tag{37}$$

T_{m} is the melting temperature in degree kelvin

5 Surface Tension Determination

5.1 Forces Acting on the Keyhole in Laser Welding of Aluminium Alloys

Weston [2] outlined and discussed as follows the forces acting on the keyhole of aluminium alloy

1. Surface tension and surface tension gradients;
2. Hydrostatic and hydrodynamic forces due to liquid metal (fluid flow);
3. Gaseous pressure in the keyhole; and
4. Ablation pressure from evaporation.

The first two forces act to close the keyhole, the last two act to open it [45]. Arata [46] has claimed that the fourth force, the ablation pressure of evaporation, balances the surface tension and is responsible for the keyhole stability. As the temperature of the surface of the weld pool is not always constant, it would lead to surface tension gradients which cause liquid flow. This process is known as the marangoni effect. However, the liquid surface is often in contact with rapidly flowing ionized gas and plasma (extremely high temperature is present in the plasma) at pressures above ambient and this subjects the liquid metal to a steep temperature gradient, taking into account the different temperatures that exist at different locations in the weld pool. These temperatures range from the boiling point at the irradiated surface [47] to the liquidus at the solid/liquid boundary. Since these temperatures are not constant across its width, the change in surface tension with temperature as well as the steep curvature of the surface are responsible for flow. Heiple and Burgardt [48] were of the opinion that Marangoni forces can influence fluids to flow in various patterns in the weld pool.

In conduction and keyhole modes, Weston [2] observed that buoyancy forces exist. They arise from the volume increase of molten metal and corresponding density decrease as the material expands on heating. Howard et al. [23] said that buoyancy forces are due to density changes of the molten metal and emphasized that in weld pools of uniform composition, liquids at higher temperatures become less dense and thus tend to flow upward while cooler and therefore denser liquids flow downward. The density differences due to the different alloying compositions overcome buoyancy due to heating, and thus have a stronger influence on fluid motion. At this point, the gravity effect seems to be dominant as it eventually leads to achieving deep weld penetration. The more plastic molten metal are usually found nearer the cooler part of the weld pool, that is, the boundary of the melt/solid interface. Whereas, the liquid metal exist at the center of the weld pool where convection mode is dominant. Berkmanns et al. [49] have found that in diode laser welding of steel Marangoni forces are the dominant force on the weld pool, and combined with the effect of buoyancy, affect the direction of fluid flow in the weld pool. Fujii et al. [50] studied the effect of gravity on fluid motion of weld pools and found that surface tension and buoyancy had the greatest effects of fluid motion in electron beam welds. Goumiri et al. [51] observed that the surface tension of pure aluminium reduces with the increase in temperature and also stated that it is difficult to determine the true surface tension of molten aluminium because of the formation and effect of thin oxide layers on its surface.

Aerodynamic effects do arise mainly due to the high rate of evaporation of alloying elements from the keyhole. The ablation pressure tends to produce a rapid flow of liquid out of the keyhole [2].

Howard et al. [23] wrote that Lorentz forces (Electromagnetic forces) are present in arc welding processes where the current density gradient in the arc and weld pool interact with the magnetic field generated by the current to generate forces acting away from the surface of the weld pool. Since there is no arc in laser welding processes there is no Lorentz force.

Paul and DebRoy [52] observed that the primary driving force for fluid flow in the laser melted pools is the surface tension gradient.

The surface tension, γ, of aluminium alloy 5083 was calculated using the model proposed by (Sahoo and DebRoy [53]) for non ferrous metals. This equation is expressed in Eq. 38

$$\gamma = 1.39 - 3.9 \times 10^{-4}(T_b - 1356)\text{Nm}^{-1} \tag{38}$$

where T_b is the boiling temperature in Kelvin.

6 Results and Discussion

The melting efficiency of aluminium alloy 5083 was investigated in this study by applying the melting efficiency equations given by Fuerschbach [14] and Weston [2] and comparing the obtained results with acceptable standard ranges of values.

Table 1 Parameters used for calculation

Absorptivity $\eta = 12\%$ (as received aluminum alloy)
Welding speed, $v = 6$ m min^{-1}
Thermal diffusivity, $a = 0.68$ cm^2 s^{-1}
Latent heat, $\Delta H_m = 0.80$ J mm^{-3}
Net power absorbed by the workpiece, $q_{\text{in}} = 35$ J mm^{-1}
Wire feed rate, $w_f = 30$ cm min^{-1}
Wire diameter, $D_f = 2$ mm
Cycle time for making one drop of metal, $T = 0.06$ s
Liquid aluminum density, $\rho = 2{,}385$ kg m^{-3}
Viscosity, $\mu = 0.0013$ kg m^{-1} s^{-1}
Melting temperature, $T_m = 2{,}482$°C
Ambient temperature, $T_o = 25$°C
Specific heat at constant pressure, $Cp = 1{,}295$ J kg^{-1}°C^{-1}

The welding parameters used for calculating the results discussed, are shown in Table 1.

A CO_2 laser heat source of 35 J mm^{-1} was used to conduct an experimental procedure, and the obtained results were used to calculate the melting efficiency of aluminium alloy 5083 by applying the Fuerschbach Equation, giving a melting efficiency of 6%. The 6% low melting efficiency can be attributed to the claim made by Zhao et al. [16] that aluminum has a much higher thermal diffusivity than other common metals; therefore, under the same welding conditions, the dimensionless, number Ry is expected to be lower for aluminum than for other alloys, resulting in a lower melting efficiency.

However, the low melting efficiency using Fuerscbach equation was not quite acceptable; therefore, the Weston equation for determining melting efficiency was considered. The melting efficiency obtained using Weston's equation was 38%. This result is more acceptable and is close to the threshold value of 48%. Swift-Hook and Gick [33] claimed that no melting efficiency should exceed this threshold value, whatever the welding situation. However, the melting efficiency calculated by Weston in 1999 fell within the range of 24 and 46% with only two values above the theoretical limit. From this analysis, the Weston equation has proved to be more dependable than the Fuerschbach equation.

The melting behavior of the wire (bare electrode) was also investigated. From the study, the volume of filler wire melted per second was 15 mm^3 s^{-1}. The volume of single droplets is 0.942 mm^3 and the radius of the spreading droplet was 0.356 mm at an average speed of 0.1 m/s in the weldpool, whereas, the volume of a metal droplet was 0.189 mm^3. These values compare very well with the ones obtained by Sun and Wu [54], although in the case of Sun and Wu, the MIG welding process was the process actually applied and considered. The surface tension of the weldpool surface was calculated to be 0.84 Nm^{-1}. This value of surface tension accurately compared with 0.84 Nm^{-1} being the value obtained by Leong et al. [27]. Tang et al. [55] were of the opinion that surface tension along an interface arises as the result of attractive forces between molecules in a fluid and they further said that in a droplet surface, the net force is radially inward, and the combined effect of the radial components of forces across the entire spherical surface is to make the surface contract, thereby increasing the pressure on the concave side of the surface. At equilibrium, the opposing pressure gradient and cohesive forces balance each other out to form spherical drops. Surface tension acts to balance the radially inward intermolecular attractive force with the radially outward pressure gradient across the surface. However, the increase of temperature by a corresponding increase in heat input reduces the effect of surface tension, which initiates flow. Low surface tension causes high amount of spatter when welding aluminium [27].

7 Conclusion

The melting behavior of aluminium alloys under the conduction and convection modes, in general was investigated. The effect of using filler metal to increase molten metal volume used to achieve deep penetration during backfills or notching effect was emphasized. The scenario where volume increments as they relate to the effects of surface tension forces was also looked at.

The melting efficiency of aluminum alloy 5083 was investigated in this study. The equation proposed by Fuerschbach [14] was applied and the value obtained for the melting efficiency was very low, although Fuerschbach applied his equation to determine the melting efficiencies of steel and tin and not aluminum. However, this value was unacceptable when compared to the one obtained by employing the equation provided by Weston [2]. From Weston's equation, the melting efficiency calculated was closer to the threshold value provided by Swift-Hook and Gick [33] and falls within the range provided by Weston [2].

References

1. Okon, P., Dearden, G., Watkins, K., et al.: Laser welding of aluminium alloy 5083, In: 2nd ICALEO, Scottsdale, 14–17 October 2002
2. Weston, J.: Laser welding of aluminum alloys. Ph.D. thesis, Department of Material Science and Metallic Engineering, University of Cambridge (1999)
3. TALAT Lecture 4300: Beam welding processes of aluminium. EAA, www.aa.net/eaa/education/TALAT/Lectures/4300.pdf (1994)
4. Olander, D.R.: Laser-pulse vaporization of refractory materials. Pure Appl. Chem. **62**(1), 123–138 (1990)
5. Yeh, R.-H., Liaw, S.-P., Yu, H.-B.: Thermal analysis of welding on aluminum plates. J. Marine Sci. Tech. **2**(4), 213–220 (2003)
6. Mahfoud, M., Mucciardi, F., Gruzleski, J.E.: Detection of solidification reactions in aluminum alloys using heat pipe technology. AFS Trans. **105**(25), 473–480 (1997)
7. David, S.A., DebRoy, T.: Current issues and problems in welding science. Science **257**, 497–502 (1992)

8. Zhao, H., Debroy, T.: Weld metal composition change during conduction mode laser welding of aluminium alloy 5182. Met. Mater. Trans. B **32B**, 163–172 (2001)
9. Cieslak, M.J., Fuerschbach, P.W.: On the weldability, composition, hardness of pulsed, continuous Nd:YAG laser welds in aluminum alloys 6061, 5456 and 5086. Metall. Trans. **19B**(4), 319–329 (1988)
10. Hatch, J.E.: Aluminum: Properties and Physical Metallurgy. The American Society for Metals, Metals Park (1984)
11. Allen, C.M.: Laser welding of aluminium alloys—principles and applications, TWI Report No. 795/2004. The Welding Institute (2004)
12. Buschenhenke, F., Seefeld, T., Schulz, A., et al.: Hot cracking during welding of aluminium alloys using spray formed filler wire with high silicon content. In: Proceedings of 4th International WLT Conference, Lasers in Manufacturing, LIM2007. Munich, pp. 25–29 (2007)
13. Verhaeghe, G., Allen, C., Hilton, P.: Achieving low-porosity laser welds in 12.7 mm thickness aerospace aluminum using a Yb-Fiber laser. In: Proceedings of 4th International WLT-Conference, Lasers in Manufacturing Limited 2007, pp. 17–24, Munich (2007)
14. Fuerschbach, P.W.: Measurement and prediction of energy transfer efficiency in laser beam welding. Weld. J. **75**(1), 24s (1996)
15. Trautmann, A.: Bifocal Hybrid Laser Welding: A Technology for Welding of Aluminium and Zinc-Coated Steels. Herbert Utz Verlag GmbH, München (2009)
16. Zhao, H., White, D.R., DebRoy, T.: Current issues and problems in laser welding of automotive aluminum alloys. Int. Mater. Rev. **44**(6), 238–266 (1999)
17. Al-Kazzaz, H., Medraj, M., Cao, X., et al.: Nd: YAG laser welding of aerospace grade ZE4IA magnesium alloy: modeling and experimental investigations. Mater. Chem. Phys. **109**, 61–76 (2008)
18. Haferkamp, H., Von Alvensleben, F., Burmester, J., et al.: The characteristics of laser beam welded magnesium alloys. In: Proceedings of the Laser Material Processing Conference ICALEO'97, pp. G/140–G/149 (1997)
19. Duley, W.W.: Laser Welding. Wiley, New York (1998)
20. Tsirkas, S.A., Papanikos, P., Kormanidis, Th.: Numerical simulation of the laser welding process in butt-joint specimens. J. Mater. Process. Technol. **134**, 59–69 (2003)
21. Duley, W.W.: Laser Processing and Analysis of Materials, p. 71. Plenum Press, New York (1983)
22. Mouroulis, P., Macdonald, J.: Gaussian Optics. In: Geometrical Optics and Optical Design. Oxford University Press, New York (1997)
23. Howard, K., Lawson, S., Zhou, Y.: Welding aluminum sheet using a high power diode laser. Weld. J. **5**, 101s–110s (2006)
24. Punkari, A., Weckman, D.C., Kerr, H.W.: Effects of magnesium content on dual beam Nd:YAG laser welding of Al-Mg alloys. J. Sci. Tech. Weld. Join **8**(4), 269–281 (2003)
25. Dausinger, F., Rapp, J., Beck, M., et al.: Welding of aluminum: a challenging opportunity for laser technology. J. Laser Appl. **8**(6), 285–290 (1996)
26. Leong, K.H., Sabo, K.R., Albright, C.E.: Laser beam welding of 5182 aluminum alloy sheet. J. Laser Appl. **11**(3), 109–118 (1999)
27. Leong, K.H., Sabo, K.R., Sanders, P.G., et al.: Laser welding of aluminium alloys. Proc. SPIE **2993**, 37 (1997)
28. Von Allmen, M.: Laser Beam Interactions with Materials. Springer, Bern, Table A.1 (1987)
29. Grong, Θ.: Metallurgical modeling of welding. Materials Modeling Series. In: Bhadeshia, H.K.D.H. (ed.) The Institute of Materials (1994)
30. Bramson, M.A.: Infrared Radiation: A Handbook for Applications. Plenum Press, New York (1968)
31. Keplan, A.: A model of deep penetration laser welding based on calculation of the keyhole profile. J. Phys. D. **27**, 1805–1814 (1994)
32. Lampa, C., Kaplan, A.F.H., Powell, J., et al.: An analytical thermodynamic model of laser welding. J. Phys. D Appl. Phys. **30**(9), 1293–1299 (1997)

33. Swift-Hook, D.T., Gick, A.E.F.: Penetration welding with lasers. Weld. J. **52**, 492s–499s (1973)
34. Kwon, Y., Weckman, D.C.: Double sided arc welding of AA5182 aluminium alloy sheet. Sci. Tech. weld. Join **13**(6), 485–495 (2008)
35. ASM Handbook: Properties and Selection: Nonferrous Alloys and Special Purpose Materials, vol. 2(95), 10th edn. ASM International, Materials Park, (1990)
36. Fuerschbach, P.W., Knorovsky, G.A.: A study of melting efficiency in plasma arc and gas tungsten arc welding. Weld. J. **70**(11), 287s–297s (1991)
37. Punkari, A.: Variable polarity plasma arc welding and dual beam Nd: YAG laser welding of aluminum alloys. Master's dissertation, University of Waterloo, Waterloo (2002)
38. Deutsch, M.: Effects of Nd: YAG laser welding and VPPAW welding process variables on weld metal, geometry and defects of 1.6 mm thick 5182 aluminum. Master's dissertation. University of Waterloo, Waterloo (2002)
39. Quintino, I., Costa, A., Miranda, R., et al.: Welding with high power fiber lasers-a preliminary study. Mater. Design **28**(4), 1231–1237 (2007)
40. Ready, J.F.: LIA Handbook of Laser Material Processing. Published by the Laser Institute of America (2001)
41. Paleocrassas, A.G., Tu, J.F.: Low speed laser welding of aluminum alloy 7075–T6 using a 300 W, single mode, ytterbium fiber laser. Weld. J. **86**, 181s–182s (2007)
42. Padmanobham, G., Schaper, M., Pandey, S., et al.: Tensile and fracture behaviour of pulsed gas metal arc welded Al-Cu-Li. Weld. J. **86**(6), 149 (2007)
43. De Genes, P.G.: Wetting: statics and dynamics. Rev. Mod. Phys. Part 1 **53**(3), 827–863 (1985)
44. Smithells, C.J., Brandes, E.A. (eds.) Metals Reference Book, 5th edn, p. 944. Butterworths, Boston (1976)
45. Kroos, J., Gratzke, U., Simon, G.: Towards a self consistent model of the keyhole in laser beam welding. J. Phys. D. **26**, 474–480 (1993)
46. Arata, Y.: Plasma, Electron and Laser Beam Technology. The American Society for Metals, Metals Park (1986)
47. Ducharme, R., Williams, K., Kapadia, P., et al.: The laser welding of thin metal sheets: an integrated keyhole and weldpool model with supporting experiments. J. Phys. D Appl. Phys. **27**(8), 1619–1627 (1994)
48. Heiple, C.R., Burgardt, P.: Fluid flow phenomena during welding. ASM Handbook. Welding, Brazing and Soldering, vol. 6, pp. 19–24. ASM Internetional, Materials Park (1993)
49. Beckmanns, J., Danzer, W., Hartl, J.: Influence of process gases in welding with diode laser. In: Yao, Y.L. (ed.) Proceedings of ICALEO2002; Laser Institute of America, Orlando, pp. 1413–1422 (2002)
50. Fujii, H., Sumi, Y., Yamamoto, T., et al.: Effect of gravity and surface tension on convection in molten pool. Trans. JWRI Special Issue HTC **2000**(30), 523–528 (2001)
51. Goumiri, L., Joud, J.C., Hicter, J.M.: Surface tension of binary Al alloys. Surf. Sci. **83**, 471 (1979)
52. Paul, A.J., Debroy, T.: Free surface flow and heat transfer in conduction mode laser welding. Metall. Trans. B **19B**, 851–858 (1988)
53. Sahoo, P., DebRoy, T.: Interfacial tension between low pressure argon plasma and molten copper and iron. Metall. Trans. B **18B**, 597–601 (1987)
54. Sun, J., Wu, C.: The effect of welding heat input on the weldpool behavior in MIG welding science in china. Series E. **45**(3), 291–299 (2002)
55. Tang, H., Wrobel, L.C., Fan, Z.: Hydrodynamic analysis of binary immiscible metallurgical flow in a novel mixing process: rheomixing. Appl. Phys. A **81A**, 549–559 (2005)

The Credibility Measure of Probabilistic Approaches in Life-Cycle Assessment of Complex Systems: A Discussion

Elsa Garavaglia

Abstract The life-cycle assessment of complex systems over time suffers of uncertainty and its probabilistic modeling is required. Two kinds of uncertainties reside in a probabilistic model: epistemic and aleatory. Both of these uncertainties can be reduced but cannot be eliminated. In this chapter the effect of the epistemic uncertainty on the service life prediction of a steel truss subjected to deterioration will be investigated. For this purpose, an opportune credibility indicator will be introduced. This indicator will be able to compare two models and decide which one is the most reliable. The results prove that the credibility index is able to recognize, in qualitative and quantitative terms, the most reliable modeling. Using the case study proposed it is shown how the influence of the epistemic uncertainty can have a relevant influence on the lifetime prediction and, consequently, on the possible maintenance strategies and their related costs.

Keywords Life-cycle assessment · Structural reliability · Predictable models · Markovian processes · Credibility of modeling

1 Introduction

The assessment of the life-cycle performance of deteriorating structures can be formulated as a reliability problem where a loss of performance greater than prescribed threshold values is considered as a failure. Therefore, when a failure is

E. Garavaglia (✉)
Department of Structural Engineering,
Politecnico di Milano, Milan, Italy
e-mail: elsa.garavaglia@polimi.it

A. Öchsner et al. (eds.), *Materials with Complex Behaviour II*,
Advanced Structured Materials 16, DOI: 10.1007/978-3-642-22700-4_38,

reached, the system passes from the current state into another characterized by a lower level of performance. On the other hand, the structural performance can also be improved by maintenance and/or rehabilitation interventions. In this case the system may move from the current state into another characterized by a higher level of performance. In both cases the failure process may be defined as a *transition process* through different service states due to environmental attacks and/or maintenance actions. Since the problem is affected by several sources of uncertainty, the assessment of the life-cycle performance must be based on a suitable damage modeling and on a probabilistic analysis able to model the main features of the time-variant deterioration process.

Under this point of view the deterioration process can be assumed to be a transition process and modeled as a semi-Markov process. This modeling, as all probabilistic modeling, suffers of epistemic and aleatory uncertainties. In this chapter the influence of the epistemic uncertainty on service life prediction is investigated with the application of a *relative* criterion of validation introduced by Grandori et al. [1] and based on an estimator called *credibility*. The aim of this criterion is the construction of an *index* able to compare two models and decide which one is the most reliable. In this chapter Gamma, Weibull and Exponential distributions are put in competition in the modeling of the failure times of a steel truss subjected to damage process and their credibility evaluated assuming as *true* failure times process the failure time process reproduced by Monte Carlo simulation.

The credibility is built comparing the probability of failure obtained by the Gamma, Weibull and Exponential processes with the probability of failure representing the *truth*. The comparison shows the importance to have information concerning which between two models is the most reliable to assess a given process and, consequently, the important role of credibility indicator. Moreover, it is also shown as the wrong model used can have a relevant influence on the lifetime prediction and, consequently, on the maintenance strategies and their related cost [2].

2 Uncertainty in Prediction

Two kinds of uncertainties reside in a probabilistic model: epistemic and aleatory. Epistemic uncertainty concerns the formal model. Even if the model is correct, there is an uncertainty related to the parameters involved into the model. Epistemic uncertainty can be reduced if the physical aspect of the phenomenon dealt with is well known. The phenomenon is, seldom, known a priori; often the law governing the phenomenon must be deduced by experimental tests and monitoring, but, usually, they are reduced in number and extension, therefore the level of uncertainty could remain high.

Aleatory uncertainty is difficult to reduce because the estimation of the model's parameters is made starting from the available data, but usually the available samples of data are few among all performable samples.

In structural performance prediction, the presence epistemic uncertainty cannot be excluded. In fact, in structural performance analysis, the available samples are the monitoring results. The monitoring of a given significant damage parameters is usually made over the lifetime of the structure at discrete time instants, only at the end of a long monitoring it is possible to describe the variability of each quantity over time. Moreover, the monitoring concerns a single structure, therefore, the development of an evolution law, generally valid for each structure of the same type, becomes a difficult matter.

In this context, it is important to develop a procedure having general validation and being able to predict the possible time dependent behavior of significant damage parameters, starting from few monitoring data. It is clear that the prediction must be made in probabilistic terms, but the choice of the "true" model is a debated question. Moreover, what is called "statistical validation" is usually not reachable. To be fully statistically substantiated, a proposed model must be developed on the basis of a sufficient number of directly observed successes and failures so to establish its performance at different levels of agreement [3]. In addition, it should be noted that a probabilistic model can not be validated in absolute sense [1, 4]. A probabilistic model consists in a continuous way towards more robust degree of knowledge [5]. The Bayesian approach indicates how, starting from a basic guess (a basic prior probability) the confidence grows towards a posterior more credible probability when more information becomes available and can be incorporated in the probabilistic model.

Usually, in statistical analysis the choice among different models can be supported by classic statistical tests if other evaluations are done. Typically, comparison between models and methods is done with relative likelihood functions or least squares method. They help to explore the reliability of one model (or hypothesis). The physics of the hazard deserves particular care: a "model" is not a pure statistical consequence of a data set. Each distribution contains some physical interpretations. Therefore, the modeling over time of the behavior of a structure affected by deterioration will require the physical knowledge about the mechanism of damage like crack propagation, carbonation, corrosion, diffusive attacks [6, 7].

In conclusion, a (probabilistic) model can be supported by a wide data interpretation in such a way as to get a certain degree of confidence. Even if it contains always a subjective choice, a careful analysis can make reasonable the choice. Obviously the most unsatisfactory residual field of uncertainty remains in problems concerning small samples. However, when the physic of the phenomenon is not clearly defined, the modeling can suffer of epistemic uncertainty; in order to reduce this uncertainty an opportune index of credibility must be assumed.

Concerning uncertainty in prediction, in 1998 Grandori et al. [1, 4] developed an index Δ, called *credibility*, that has the merit to focus the possible error in the estimation of a given parameter and, in this way, to decide, between two candidate models, which one is the best. This index has been developed in seismic engineering to investigate the peak ground acceleration at a given site and corresponding to given return period, but it can be extended also to other engineering problems [8].

Based on the approach proposed by Grandori et al. [1, 4] the credibility for the semi-Markov approach, already used by the author in prediction of reliability and planning maintenance of structures subjected to deterioration [9, 10] is, here, investigated.

3 Improvement in Credibility

To investigate credibility of two models put in competition in the evaluation of a quantity P starting from a single data set means to shift the attention from the data fitting to the error in estimating the quantity P of interest. This, if well approached, leads to catch which between the two models is more reliable.

In this section, the Grandori's method is extended to the structural lifetime prediction. Supposing that in a given structure the time evolution of a given parameter is completely defined, it is possible to suppose that also the "truth" is completely defined and the quantity P assumes a precise value P° in each instant t^{*}. In other words, if the variation law F° of a given quantity P is given in both its form and parameters, the value assumed by P in each instant t^{*} is known and in the follow will be marked with an upper index as all the values or function considered as the *statistical truth*. On the contrary, lower index will indicate all estimate values or functions with estimated parameters.

Therefore, if the distribution F° is known, it will be assume like the statistical truth, and following a Monte Carlo simulation, it will be possible to draw from it many samples of size v on which the credibility analysis can be developed.

The proposed procedure follows the following steps: each sample v will be modeled with a chosen r – model, the parameters of the cumulative distribution F_r will be evaluated through the likelihood method and a value P_r of the parameter investigated will be estimated (one for each sample). All the P_r values, obtained by the size v samples drawn by the truth F°, together, will form the random variableP_r, which distribution is the sampling distribution of the parameter considered (in sequence: waiting time into a performance state, failure time, transition probability through two consequent states of performance, conditioned probability of occurrence of an event, and so on).
The index:

$$\Delta_r^{\circ} = \Pr\{ P^{\circ} - h < P_r \leq P^{\circ} + h \} \tag{1}$$

is the probability that the estimated value P_r with the r – model falls in a given interval around the "true" value P° (h defines a conventional interval around P°). The index Δ_r° takes into account both epistemic and aleatory uncertainty. It measures the credibility of r – model with respect to F° when P is the estimable quantity (credibility based not on data fitting, but looking at the results).

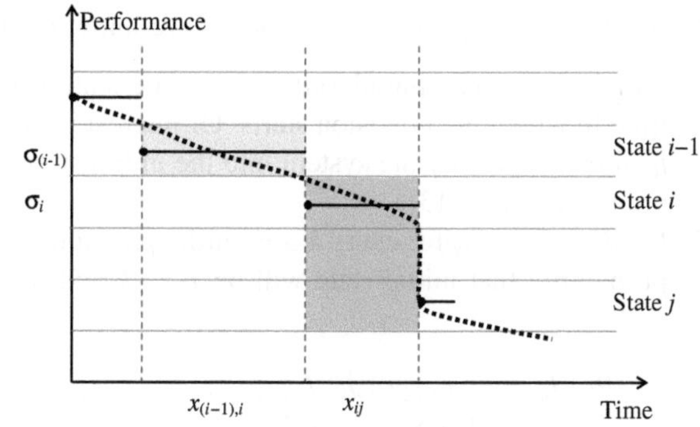

Fig. 1 Schematic transitions process between different states of performance. Transition can happen between contiguous states (*light gray area*) or between not contiguous states (*dark gray area*)

Briefly: Δ_r° is the probability that, on the basis of a random size v sample drawn from F°, the model F_r leads to estimate the random variable with an error $\varepsilon_r^\circ \leq h$ (as absolute value)

$$\Delta_r^\circ = \Pr\left\{ \left| \varepsilon_r^\circ \right| \leq h \right\}. \quad (2)$$

An analogous procedure is followed for the s – model, with F_s cumulative distribution. The difference:

$$\Delta_{rs}^\circ = \Delta_r^\circ - \Delta_s^\circ \quad (3)$$

is the *index of the relative credibility* between the two models in competition. Assuming that r – model is more reliable than s – model for estimation of P, the value of (1) will be $\Delta_{rs}^\circ > 0$ and vice versa. Then the sign of Δ is of crucial importance: it decides which model is the most reliable.

4 The Deterioration Process Modeling

Aggressive environmental attacks can induce damage into the members of a structural system. Each time a significant damage level is reached, the system reliability decreases and the system reaches a performance level and it remains in this level until a new dangerous damage will happen (Fig. 1).

Following this approach the damage evolution due to environmental aggressive attacks can be modeled as a transition of the structure through different states of performance, where each state's transition is due to the reaching of a specific significant damage level (Fig. 1). Therefore, the evolution of the system performance can be considered as a *transition process* described by a random variable representing the waiting time t_i spent by the system into a state i before the transition into another state j. The transition can happen between contiguous states or not contiguous states.

A transition process can be modeled as Semi-Markov process (s-MP) [11].

A s-MP is completely determined if we know its initial laws [12]:

1. initial conditions: initial *state* J_1, i.e. the state occupied by the system at the time in which the prevision starts. Of interest is also the knowledge of the time t_0 already spent by the system into the initial state before the time in which the prevision starts [13].
2. Pr $(J_0 = k) = a_k$ for every $k \in E$, initial probability distribution **a** describing the probability that initial state will be $J_0 = k$ and its semi-Markov kernel
3.

$$\Pr\left(J_n = k,\ \tau_n \leq t \mid J_0, J_1, \tau_1, \ldots, J_{n-1}, \tau_{n-1}\right) = p_{J_{n-1},k} F_{J_{n-1},k}(t)\ \forall\ t \in (0, +\infty),\quad k \in E. \tag{4}$$

$p_{J_{n-1},k}$ are the transition probabilities of the Markov chain $(J_n)_{n\geq 0}$ and $F_{J_{n-1},k}(x)$ are the distribution function associated to waiting times in state J_{n-1} before going in state $J_n = k$. In other words, Eq. 4 describes the probability that transition into the next state J_n, happens by time x, if the present state is J_{n-1}.

Some consequences of the definition now introduced are:

i. $(J_n)_{n\geq 0}$ is a E-valued Markov chain with transition matrix **P** and initial distribution **a**.

ii. for every $n > 1$, $\tau_1, \ldots, \tau_n$ are conditional independent, given $(J_n)_{n\geq 0}$, and

$$\Pr\left(J_n = j \mid J_0, J_1, \ldots, J_{n-1} = i\right) = p_{ij}, \quad \text{for all } n > 0, \tag{5}$$

$$\Pr\left(\tau_n \leq t \mid J_0, \ldots, J_{n-1} = i,\ J_n = j\right) = F_{ij}(t), \quad \text{for all } n \geq 0 \text{ and } t \geq 0, \tag{6}$$

$$\Pr\left(\tau_1 \leq t_1, \ldots, \tau_n \leq t_n \mid J_n, n \geq 0\right) = \prod_{i=1}^{n} F_{J_{i-1},J_i}(t_i), \quad \text{for all } n \geq 0 \text{ and } t_i \geq 0. \tag{7}$$

Equations 5–7 describe respectively the transition probabilities between states, the distribution functions of waiting times between two transitions and the marginal distribution function of the waiting times t_i describing the probability of transition into each state J_i by time t_i.

In a transition process the role played by the transition rate function or hazard rate function is crucial. For all i and j in E, the immediate transition rate function $\lambda_{ij}(x)$ is given by:

$$\begin{aligned} \lambda_{ij}(t) &= \lim_{\Delta t \to 0} \frac{1}{\Delta t} \Pr\left(J_n = j, t < \tau_n \leq t + \Delta t \mid J_{n-1} = i, \tau_n > t\right) \\ &= \begin{cases} \frac{q_{ij}}{[1 - H_i(t)]} & \text{if } p_{ij} > 0 \text{ and } [1 - H_i(t)] < 1 \\ 0 & \text{otherwise} \end{cases} \end{aligned} \tag{8}$$

where $q_{ij} = p_{ij} f_{ij}(t)$ and $H_i = \sum_{j=1}^{s} p_{ij} F_{ij}(t)\ \forall t \in \Re_+$.

The estimation of transition rate is not a simple matter. Ouhbi and Limnios [14] consider a history of s-MP and maximize its likelihood function to obtain the maximum likelihood estimator of hazard rate function.

Let (J_0, J_1, t_1,..., t_{w-1}, $J_{w)}$ be a realization of the s-MP on the time window [0,T]; w represents the number of states visited in [0, T] and for the lost event J_w we have the censored data $t_w > [T - (t_1 + \cdots + t_{w-1})]$.

Then, the conditional likelihood is

$$L(T) = \left[\prod_{i=0}^{w-1} p_{J_i,J_{i+1}} f_{J_i,J_{i+1}}(t_i) \; 1_{w>0} + 1_{w=0} \right] \cdot \sum_{k=1}^{s} p_{j_w,k} \left[1 - F_{j_\tau,k}(t_w) \right] \tag{9}$$

where $1_{w>0}$ is equal one if $w > 0$ and zero otherwise, and vice versa for $1_{w=0}$.

4.1 Crossing State Prediction

In the damage transition process, when the system passes from the performance state i into the performance state j it already has spent some time, t_0, into i before the crossing into j. The time t_0 plays an important role in the prediction model and the probability of transition becomes a conditional probability strictly connected to t_0.

The probability of transition into the next state j, if the present state is i and t_0 is the time already passed by the last transition occurred, can be defined as follow:

$$\Pr(state\; j, \tau_{ij} \le t_0 + \Delta t |\; state\; i, \tau_{ij} > t_0), \quad i,j = 1, s \tag{10}$$

where i is the state of the present performance; j is the state of next lower performance; τ_{ij} is the waiting time spent by the system into the state i before to have the transition into j, under the condition that no transition has already happened (defined through the condition: $state\;\; i,\; \tau_{ij} > t_0$); Δt is the discrete time in which the prediction would be obtained.

If the distribution functions $F_{ij}(t)$ are defined and $\tau_{ij} = t_0$, Eq. 10 can assume the following form:

$$\mathrm{P}^{(ij)}_{\Delta t|\, t_0} = \frac{\left[F_{ij}(t_0 + \Delta t) - F_{ij}(t_0) \right] p_{ij}}{\sum_{k=1}^{s} \left[1 - F_{ik}(t_0) \right] p_{ik}}. \tag{11}$$

5 The Deterioration of a Steel Truss Modeled as a Transition Process

To investigate the credibility of a probabilistic approach, the procedure introduced will be applied on a statically indeterminate steel truss (Fig. 2) immerged into an

Fig. 2 Statically indeterminate truss system with members area A = 2500 mm^2 and total volume V = 0.0723 m^3

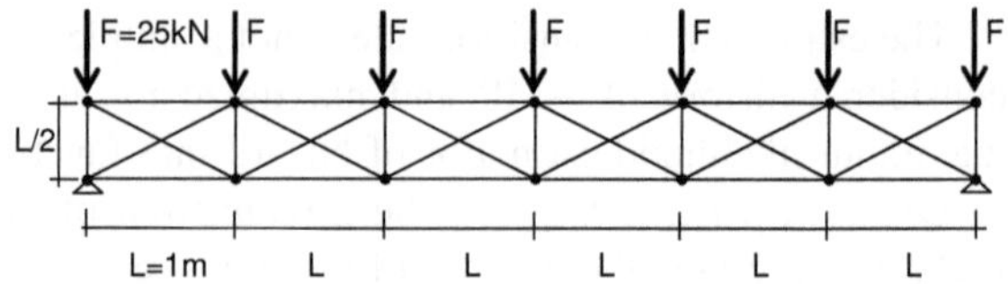

aggressive environment and subjected to a damage process inducing loss of performance over time.

To describe the deterioration process of the system here proposed, the random variable chosen is the stress value σ recorded/evaluated at every monitoring and considered as the ratio between the internal forces and the deteriorated area of the elements' cross sections. In order to model the time evolution of such variable within s-MP, the following assumptions are introduced [9]:

- the structure is undamaged at the initial time $t_0 = 0$.
- the damaged structure is considered to be in a state $i > 0$ when $\sigma_i \leq \sigma \leq \sigma_{(i+1)}$, where σ_i and $\sigma_{(i+1)}$ are the lower and upper thresholds, respectively, which characterize the state i.
- the structure evolves from a state $i > 0$ into another state $j > i$, characterized by a lower level of performance σ_i $\sigma_i < \sigma_j$, during a time interval τ_{ij}. Of course, the condition $j < i$, with $\sigma_i > \sigma_j$, is also possible if some maintenance will be operated.

Under the hypothesis of s-MP, the time evolution of the structural behavior is then represented as transitions between different states of performance.

For each transition, the waiting time τ_{ij} must be modeled by choosing an appropriate probability density function (p.d.f). This choice is not a simple choice and it can be "not unique". It should be made on the basis of physical knowledge of phenomenon and on the characteristic of the distributions in their tails where, usually, no much data are recorded.

The physic suggest that the deterioration follows an increasing law over time, therefore the distributions that can satisfy this tendency are distributions with increasing hazard rate functions $\lambda(t)$ (Eq. 8). Distributions obeying to this law are, for example, the Weibull distributions where, if $t \to \infty$, $\lambda(t)$ tends to an infinite value, and Gamma distributions where, if $t \to \infty$, $\lambda(t)$ tends to an asymptotic value.

Gamma and Weibull distributions will be put in competition with the "true" distribution, obtained by application of a Monte Carlo simulation, to investigate which distribution is the most reliable in the modeling of the failure time of the system in Fig. 2.

Exponential distributions present a constant hazard rate function and seem to be not able to catch the real behavior of a damage process due to environmental aggressive attacks. However, the Poisson behavior described by the Exponential distribution is often used to model stochastic damage processes, therefore also the exponential distribution will be put in competition with the statistical truth obtained by the Monte Carlo simulation.

5.1 Deterioration Modeling of a Steel Truss: A Monte Carlo Simulation

The credibility method introduced in the previous sections requires the definition of a *statistical truth* (true process) with which different models can be compared with the aim to catch the most reliable one. The statistical truth can be built through a Monte Carlo simulation where the structural performance evolution is assumed depending on an adequate time depending damage model. In the following, a general approach to deterioration modeling of structural members is presented and an adequate damage index introduced [10].

By denoting K a generic material property, the deterioration over time t of it can be measured by the following law:

$$\mathrm{K}(t) = \mathrm{K}_0[1 - \delta(t)] \tag{12}$$

where "0" denotes the initial undamaged state, and the time-variant damage index $\delta = \delta(t) \in [0;1]$ is the measure of deterioration over time. The index δ is assumed to follow the damage model proposed by Biondini et al. [10]:

$$\delta(\tau) = \begin{cases} \omega^{1-\rho}\tau^{\rho} & , \tau \leq \omega \\ 1 - (1-\omega)^{1-\rho}(1-\tau)^{\rho} & , \omega < \tau < 1 \\ 1 & , \tau \geq 1 \end{cases} \tag{13}$$

where $\tau = t/T_C$, T_C is the normalized time instant of reaching the failure threshold $\delta = 1$, ω and ρ are shape parameters of the damage curve. The damage parameters ρ and ω must be chosen according to the actual evolution of the damage process. Damage rates may be associated with the aggressiveness of the environment, as well as with the level of acting stress through the following linear relationship [10]:

$$\rho = \rho_a + (\rho_b - \rho_a)\,\xi \quad \omega = \omega_a + (\omega_b - \omega_a)\,\xi \tag{14}$$

where the subscript "a" refers to damage associated with environmental aggression, and the subscript "b" refers to damage associated with loading effects and ξ refers to ratio between the level of acting stress and the limit state value. The law (13) is able to model damage mechanisms induced both by environmental deterioration, like carbonation of concrete, corrosion of steel or material fatigue, and by changing in mechanical aspects, like, for example, the changing of live loads during the system service life. Generally, these mechanisms are present and interacting, and a proper calibration of the damage parameters is required based on experimental observations and/or laboratory accelerated test data.

As said before, the calibration of each probabilistic approach requires many experimental data usually difficult to collect. In this study, the data samples for parameter estimation are obtained by means of a Monte Carlo simulation of the life-cycle structural performance based on the damage modeling previously introduced. In this simulation process, the damage parameters ρ_a, ρ_b, ω_a, ω_b, and T_C, are modeled as random variables with prescribed probability distribution: lognormal distribution for ρ parameters, normal for ω parameters, and Gamma for the failure time T_C. Each distribution here used is chosen on the basis of physical knowledge of the phenomena investigated [15, 16].

The Monte Carlo procedure has been applied to investigate the performance of the statically indeterminate truss system in Fig. 2 with members' area $A = 2500\ \text{mm}^2$ and total volume $V = 0.0723\ \text{mm}^3$. The allowable material strength is $\bar{\sigma} = 140\text{MPa}$. Buckling failures are assumed to be avoided. The structure is subjected to a set of forces $F = 25$ kN. The initial value of member cross-sectional area A and of the material strength $\bar{\sigma}$, as well as the force F, are assumed as deterministic.

The deterioration process follows the damage model (13), it induces in each system's member a reduction of both the cross-sectional area A and material strength $\bar{\sigma}$ of each structural member. Without any loss of generality, in this study it is assumed that such properties undergo the same damage process:

$$A(t) = A_0[1 - \delta(t)] \quad \bar{\sigma}(t) = \bar{\sigma}_0[1 - \delta(t)] \tag{15}$$

where "0" denotes the initial undamaged state, and the damage rates may be associated with the aggressiveness of the environment, as well as with the level of acting stress σ with respect to the material strength $\bar{\sigma}$, or $\xi = \sigma/\bar{\sigma}$.

The results obtained by Monte Carlo simulation will be assumed in the following credibility analysis as *statistical truth*, as the *true* deterioration *process*.

6 Credibility in Life–Cycle Prediction

The prediction of the failure time of structures subjected to environmental attacks, usually, is based on monitoring data of several parameters able to describe the damage evolution over time. However, the physical behavior of a structural system affected by deterioration is hard to model also if the monitoring is very long and detailed. Therefore, the prediction can be approached only in probabilistic terms with all the uncertainty properly modeled.

As shown in the previous sections, the deterioration process of a structural system in an aggressive environment can be modeled as a transition process through different states of performance and where the probability of failure is described as the probability of transition from a given state i into another state j characterized by a lower degree of performance. The random variable assumed to describe it was the waiting time τ_{ij} spent by the system into the state i before transition into j. In this study, the state i is the *safe state* and j the *unsafe state*. The transition process will present just two states and τ_{ij} will be equal to t_0, the age of the system at the instant of analysis. In this case Eq. 11 becomes:

$$\mathrm{P}_{\Delta t|t_0} = \frac{[\, F(t_0 + \Delta t) - F(t_0) \,]}{1 - F(t_0)} \tag{16}$$

where $F(t)$ is the cumulative distribution chosen to model the waiting time in state i (safe state) before transition and t_0 is the time already spent by the system into i. Eq. 16 describes the probability to have a failure in the next Δt if no failure happens until the current instant $t = t_0$.

Of course, the probability $\mathrm{P}_{\Delta t|\,t_0}$ depends on the model, $F(t_0)$, chosen to model the failure time of the system. The *credibility index* Δ (2.1), here, will have the aim to measure the error made in the evaluation of the probability (16) in the comparison of different models $F(t_0)$ with the *true process* F° .

The proposed procedure uses Monte Carlo simulation to generate a large sample of failure times. The large population obtained by 5000 simulations can be considered in statistical way the *true* failure time process F°. On the basis of F° the probability $\mathrm{P}^\circ_{\Delta t|\,t_0}$ can be evaluated and assumed as the statistical *true process* in credibility analysis.

The large population will be modeled with a theoretical distribution too. Clearly, the choice of distribution cannot be based only on the data fitting but require physical knowledge of the phenomenon investigated and of the distributions in the tails, where usually less monitoring data are present. As already discussed, Gamma and Weibull distribution seem to be suitable choices to model the deterioration process of structural systems and elements. The true failure process is modeled with a Gamma distribution and with a Weibull distribution and, for both the distributions used, the probability $\mathrm{P}_{\Delta t|\,t_0}$ is evaluated. To complete the analysis, the failure time process is supposed following a Poisson process, therefore, it was modeled with an exponential distribution.

For the three distributions: Gamma, Weibull and Exponential, the credibility index Δ° is built following Eq. 2. The range of error h in (2) is defined as $h = 0.2{\cdot}\mathrm{P}^\circ$ where P° is the probability (16) evaluated for the *true* process F°, therefore Eq. 2 assumes the form:

$$\Delta_r^\circ = \Pr\left\{ \left| \varepsilon_r^\circ \right| \leq (0.2 \cdot \mathrm{P}^\circ) \right\}, \tag{17}$$

where as absolute error $|\varepsilon|$ is assumed the difference (in absolute value)

$$\left|\varepsilon_r^\circ\right| = \mathrm{P}^\circ - \mathrm{P}_r \tag{18}$$

In (18), P_r is the probability (16) evaluated for the models Gamma, Weibull or Exponential. The probabilities (16) are evaluated for different t_0 and $\Delta t = 1$ year.

Figure 3 shows the comparison between the probability of failure, $\mathrm{P}_{\Delta t|\,t_0}$, evaluated for the models (Gamma, Weibull and Exponential), with respect to $\mathrm{P}^\circ_{\Delta t|\,t_0}$, evaluated for the *true* process. The radius of each circle is proportional to

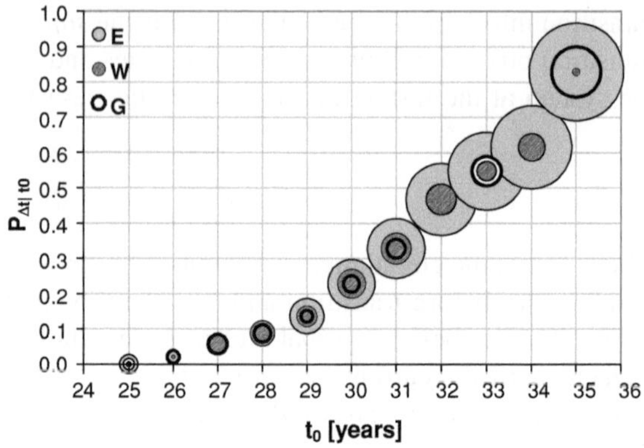

Fig. 3 Comparison between $P_{\Delta t|\,t_0}$ of Exponential (E), Weibull (W) and Gamma (G) models with respect to $P^{\circ}_{\Delta t|\,t_0}$ of the *true* process through the absolute error $|\varepsilon|$. Error interval around $P^{\circ}_{\Delta t|\,t_0}: h = 0.2$. The radius of each circle is proportional to the value of $|\varepsilon|$. Each prediction is made for $\Delta t = 1$ year

the value of the absolute error $|\varepsilon|$. Fig. 3 shows that the error $|\varepsilon|$ connected with the Exponential model increases with t_0, whereas the errors connected with Gamma and Weibull are very similar for each t_0 considered. Only for the last $t_0 = 35$ years the Weibull seems to be more reliable than Gamma.

Table 1 shows the values of Δ°_r from Eq. 18, which is the average result obtained by considering all the t_0 investigated.

Table 1 highlights that the Gamma model seems to be the most reliable model in the modeling of the failure process of deteriorated structural systems. Conversely, the Exponential model appears to be the worst model between the three models in competition.

The results shown in Fig. 3 and Table 1 suggest that the Gamma model is the most reliable model for the failure process. Instead, the interpretation of the failure process as a Poisson process leads to predictions affected by rather large error.

After this first step of analysis, the question now is: if the prediction is based on a limited number of monitoring which model is the most reliable? And, which error is made if the model chosen is wrong?

Supposing that the *true* process F° is known and that it is a Gamma process; with a random procedure, a lot of samples of v size will be generated from F° and modeled with the *right* model, Gamma, and with other *wrong* models. The credibility index Δ°_r, obtained by Eq. 1 or 2 with $h = 0.05$, and the relative credibility index Δ°_{rs} obtained by Eq. 3 will highlight which model will be the most reliable in the failure time prediction.

Assuming a small error's interval, $h = 0.05$, around the true value P°, Fig. 4 shows the credibility of the three different models investigated to predict, for different t_0, the system's failure in the next $\Delta t = 1$ year. This figure shows, it is evident that the samples drawn by the *true* process Gamma and models with Gamma and Weibull lead to very similar values of credibility. On the contrary, when the samples are modeled with the Exponential, the reliability in prediction significatively decreases. These results highlight the importance of the knowledge of the physics of the phenomenon studied and on the theoretical frame of

Table 1 Credibility index Δ_r° evaluated for the Exponential, Weibull and Gamma models in comparison with the *true* process F° ($h = 0.2$). For each model the Δ_r° represent the average value obtained considering all the t_0 investigated

r-model	Δ_r°
Exponential	0.00
Weibull	0.36
Gamma	0.54

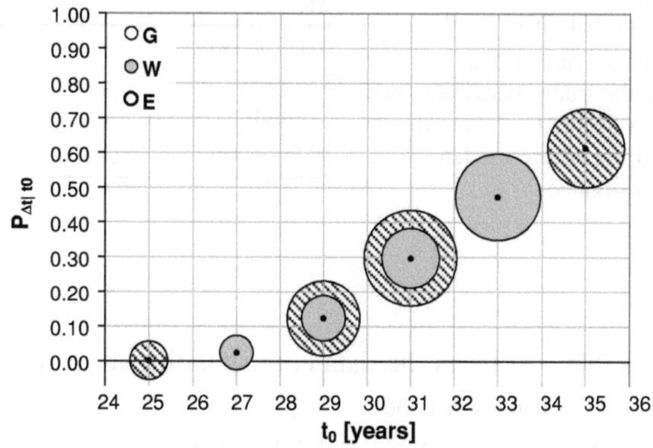

Fig. 4 Comparison between $P_{\Delta t|\,t_0}$ related to Gamma (G), Weibull (W) and Exponential (E) models with the *true* $P^\circ_{\Delta t|\,t_0}$ through the index Δ_{rs}° for $\Delta t = 1$ year, $v = 100$ and $h = 0.05$. The radius of each circle is proportional to the value of the credibility index Δ_r°

distribution's hazard rate: Gamma and Weibull present a hazard rate function increasing with time; therefore, the prediction can be very similar for both the distribution until extreme values of t_0 where, instead, the probability could significantly diverge.

In Table 2 the values of the relative credibility index Δ_{rs}° obtained by Eq. 3 are presented. Each time that $\Delta_{rs}^\circ > 0$ the r-model is the most reliable in the prediction of probability $P_{\Delta t|\,t_0}$.

7 Influence of the Model on Maintenance Strategies

As it is well known, the modeling of a process influences the planning of maintenance strategies. Each time the probability of failure reaches a dangerous level, two maintenance actions are taken in consideration: replacing of the whole members or repairing of the members heavy damaged (selective maintenance), however crucial is the prediction of the instant in which one of these actions must be made. The analysis presented in the previous section has highlighted that the prediction of the failure in the next year becomes important between 25 and 30 years of system lifetime. To plan maintenance in advance the failure, the probability $P_{\Delta t|\,t_0}$ is assumed as a dangerous threshold when its magnitude is higher or equal to $n \times 10^{-4}$ and t_0 is the instant in which the failure could happen.

Table 2 Relative credibility index Δ°_{rs}, r – model: Gamma, s – model: Weibull, Exponential, evaluated for $\Delta t = 1$ year and different t_0 ($h = 0.05$)

r – model	s – model	Δ°_{rs}					
		$t_0 = 25$	$t_0 = 27$	$t_0 = 29$	$t_0 = 31$	$t_0 = 33$	$t_0 = 35$
Gamma	Weibull	0.05	−0.01	0.12	0.19	−0.02	0.21
Gamma	Exponential	0.05	0.03	0.19	0.31	0.23	0.21

Table 3 Instant t_0 in which the probability $P_{\Delta t|\, t_0}$ becomes equal or higher than $n \times 10^{-4}$ for the true process obtained by a Monte Carlo simulation, and for all the models investigated

Model	$P_{\Delta t|\, t_0}$	t_0(years)
True process	2.22E−04	25
Gamma	2.63E−04	24
Weibull	1.82E−04	19
Exponential	3.11E−02	1

Table 3 collects the data corresponding to of the dangerous threshold $P_{\Delta t|\, t_0} = n \times 10^{-4}$ for each model.

When $P_{\Delta t|\, t_0}$ reaches the dangerous threshold, the maintenance can be applied to improve the performance of the system.

Maintenance can be applied:

- on the whole system. In this case the system comes back into its initial state of performance, but, usually, without reaching the characteristics of the *new* system, so the new failure will happen after a time t_0 different by the t_0 of the previous failure.
- only on some members (selective maintenance). In this case the system improves its performance but it does not reach the initial state of performance, therefore, so the new failure will happen after a time t_0 different by the t_0 of the previous failure.

The scenario here investigated involves the improvement of performance of the whole system. In Fig. 5 the maintenance scenarios of the *true* process and of the Gamma and Weibull models are plotted.

For the Exponential model, the choice of the maintenance instant cannot be decided on the basis of the constant $P_{\Delta t|\, t_0}$. The choice must be based on expert judgments (here $t_0 = 20$ years will be assumed).

As it is well known, each maintenance scenario must be associated to maintenance cost and the convenience of the choice must take into account this important aspect. For example, for a prescribed maintenance scenario, the total cost of maintenance C_m can be evaluated by summing the costs C_q of the individual interventions [2, 17],

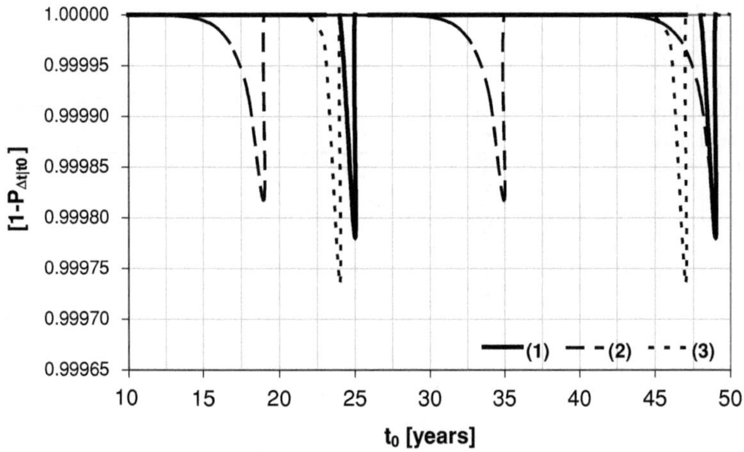

Fig. 5 Comparison of $[1-P_{\Delta t|\,t_0}]$ for different maintenance scenarios: (1) *true* process (2) Weibull model (3) Gamma model. For the Exponential model, maintenance will be made every $t_0 = 20$ years on the basis of expert judgment, however the probability $P_{\Delta t|\,t_0}$ will remain always constant, even after the maintenance

$$C_m = \sum_{q=1}^{n} \frac{C_q}{(1+v)^{t_q}} = \sum_{q=1}^{n} C_{0q} \tag{19}$$

where the cost C_q of the qth rehabilitation has been referred to the initial time of construction by taking a proper discount rate of money v into account [17]. Here, the cost C_q of the individual intervention is assumed as follows:

$$C_q = C_f + \sum_{k=1}^{m} \delta_k \cdot V_k \cdot c_{qk} \tag{20}$$

where $C_f = \alpha C_0$ is a fixed cost computed as a percentage α of the initial cost C_0, δ_k is the damage index of element k of the structural system, V_k is the volume of element k, and c_{qk} is the volume unit cost for restoring the element k. Therefore, the total cost C will be considered as the sum of the initial cost C_0 and the maintenance cost C_m, defined in (19).

Based on this cost model, the maintenance scenarios previously introduced are now compared with each other under the economic point of view.

Figure 6 shows the total cost C, computed as the sum of initial cost C_0 and maintenance cost C_m, of the scenario (2): Weibull model, scenario (3): Gamma model and scenario (4): Exponential model, normalized to the cost C of scenario (1): *true* process. All normalized costs are evaluated for different discount rates v, and $\alpha = 0$ of the fixed cost of maintenance $C_f = \alpha C_0$ (no fixed cost admitted). The primacy in economic terms of scenario (3) over scenario (2) and (4) is evident for

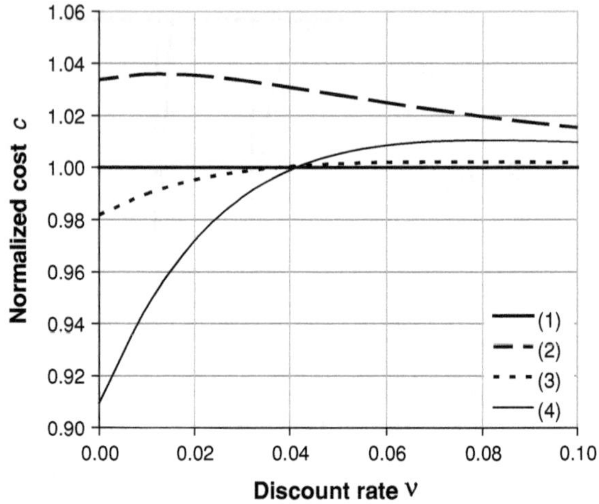

Fig. 6 Total cost C, computed as the sum of initial cost C_0 and maintenance cost C_m, of the scenarios (2) Weibull (3) Gamma, and (4) Exponential normalized to the cost C of the scenario (1) *true* process, versus the discount rate v, for $\alpha = 0$ of the fixed cost of maintenance $C_f = \alpha\ C_0$ (no fixed cost admitted)

all the discount rates v. For discount rate $v < 0.04$ scenario (3) seems to be more convenient even then the *true* process (scenario (1)).

The influence of the fixed cost α on the total costs of maintenance is shown by Fig. 7 where the comparison between scenario (3), normalized with scenario (1), is made for different values of α. The differences recorded are not so heavy, however if α increases the total cost increases for each value of discount rate considered.

8 Conclusions

The prediction of the failure time of a structural system affected by deterioration due to environmental aggression cannot be approached in deterministic way. Each probabilistic approach suffers of uncertainty. Two kinds of uncertainties reside in a probabilistic model: epistemic and aleatory. Both these uncertainties can be reduced but cannot be eliminated. Epistemic uncertainty concerns the modeling of a process. For example, the wrong modeling of a failure process can suggest inadequate strategies of maintenance, therefore, it is crucial to reduce the epistemic uncertainty in the modeling of any kind of phenomena.

In this chapter, the influence of the epistemic uncertainty on service life prediction has been investigated with the application of a *relative* criterion of validation introduced by Grandori et al. [1] and based on an estimator called *credibility*.

To investigate the reliability of the probabilistic modeling, the main features of a procedure for life-cycle reliability assessment and maintenance planning of deteriorating structural systems have been presented. This procedure is based on an effective modeling of structural damage and takes advantage from the

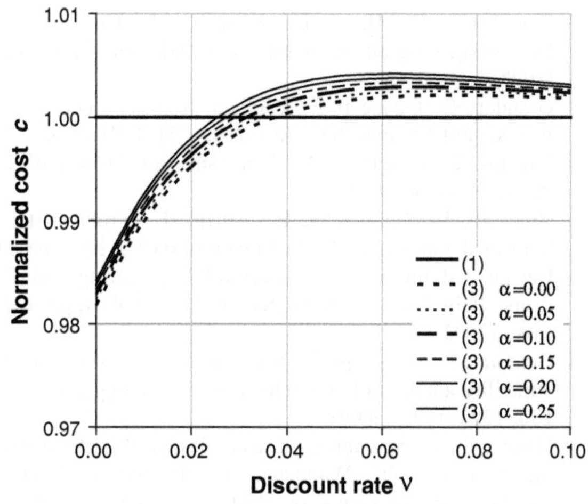

Fig. 7 Total cost C, computed as the sum of initial cost C_0 and maintenance cost C_m, of the scenario (3) Gamma normalized to the cost C of the scenario (1) true process, versus the discount rate v, for different values α of the fixed cost of maintenance $C_f = \alpha\ C_0$

generality of the Markov renewal process assumed to model the deterioration process when it can be dealt as a transition process between performance states. To support the s-MP with many experimental data, a Monte Carlo simulation is adopted. The use of Monte Carlo simulation is finalized to collect many samples representing a significant population for a probabilistic approach.

In the credibility analysis, the results developed with the Monte Carlo simulation and concerning the failure time process has been assumed as the *true* process. Different models were put in competition with each other to identify which of them was the most reliable in the modeling of the failure process investigated.

The small size of the samples experimentally collected can compromise a probabilistic analysis of their time variant behavior. Following the procedure introduced, a credibility analysis using small samples is made to investigate the magnitude of the epistemic uncertainty in presence of small populations.

In the chapter the comparison of different scenarios of maintenance is approached. In each scenario, the maintenance is always made on the whole system. The results prove that the failure process modeled following the knowledge on the physic of the phenomenon investigated is reliable and the error made is limited. Moreover, to make maintenance before reaching failure threshold can be convenient until the fixed costs of maintenance do not become important.

References

1. Grandori, G., Guagenti, E., Tagliani, A.: A proposal for comparing the reliabilities of alternative seismic hazard models. J. Seismol. **2**, 27–35 (1998)
2. Frangopol, D.M., Lin, K.Y., Estes, A.C.: Life-Cycle Cost Design of Deteriorating Structures. J. Struct.Eng. ASCE **123**, 1390–1401 (1997)

3. Vere-Jones, D., Harte, D., Kosuch, M.: Operational requirements for an earthquake forecasting programme in New Zealand. Bull N. Z. Natl. Soc. Earthq. Eng. **31**, 194–205 (1998)
4. Grandori, G., Guagenti, E., Tagliani, A.: Magnitude distribution versus local seismic hazard, BSSA. Bull Seismol. Soc. Am. **2003**(93), 1091–1098 (2003)
5. Lakatos, I., Musgrave, A.: Criticism and Growth of Knowledge. Cambridge University Press, Cambridge (1974)
6. Guagenti, E., Garavaglia, E., Petrini, L.: Probabilistic Hazard Models: Is it Possible a Statistical Validation? In: F. Bontempi (ed) System–based Vision for Strategic and Creative Design 2nd International Structural Engineering and Construction Conference (ISEC02), Rome, Italy, U.E., 23–26 September 2003, vol. II, pp. 1211–1216. A.A. Balkema Publishers, Lisse (2003)
7. Biondini, F., Bontempi, F., Frangopol, D.M., Malerba, P.G.: Cellular automata approach to durability analysis of concrete structures in aggressive environments. J. Struct. Eng. ASCE **130**, 1724–1737 (2004)
8. Garavaglia, E., Sgambi, L.: The credibility of lifetime assessment based on few experimental measurements. In: Watanabe, E., Frangopol, D.M., Utsunomiya, T. (eds.) Second International Conference on Bridge Maintenance, Safety and Management (IABMAS'04), Kyoto, Japan, October 18–22 2004, vol. CD-ROM. A.A. Balkema Publishers, Leiden (2004)
9. Biondini, F., Garavaglia, E.: Probabilistic service life prediction and maintenance planning of deteriorating structures. In: Augusti, G., Schuëller, G.I., Ciampoli M. (eds.) International Conference on Structural Safety and Reliability (ICOSSAR'05), Rome, Italy, 19–22 June 2005, vol. CD-ROM, pp. 495–501. Millpress, Rotterdam (2005)
10. Biondini, F., Frangopol, D.M., Garavaglia, E.: Life-cycle reliability analysis and selective maintenance of deteriorating structures. In: Biondini, F., Frangopol, D.M. (eds.) First International Symposium on Life-Cycle Civil Engineering (IALCCE'08), Varenna, Lake Como, Italy, 10–14 June 2008, vol. I, pp. 483–488. CRC Press, Taylor & Francis Group, London (2008)
11. Garavaglia, E., Pavani, R.: About earthquake forecasting by Semi–Markov renewal processes. Methodol. Comput. Appl Probab. Spring Ed., US. doi: 10.1007/s11009-009-9137-3 (2009)
12. Limnios, N., Oprisan, G.: Semi-Markov Processes and Reliability. Birkhauser, Boston-Basel-Berlin (2001)
13. Howard, R.A.: Dynamic Probabilistic Systems. Wiley, New York (1971)
14. Ouhbi, B., Limnios, N.: Non-parametric estimation for semi-Markov processes based on it's hazard rate. Stat. Inf. Stoc. Proc. **2**, 151–173 (1999)
15. Ciampoli, M.: A probabilistic methodology to assess the reliability of deteriorating structural elements. Comput. Meth. Appl. Mechan. Eng. **168**, 207–220 (1999)
16. Gupta, A., Lawsirirat, C.: Strategically optimum maintenance of monitoring enabled multi-component systems using continuous-time jump deterioration models. In: Kachitvichyanukul, V., Purintrapiban, U., Utayopas, P. (eds.) Proceeding of International Conference on Simulation and Modeling, SimMod2005, Nakornpathom, Asian Institute of Technology, 17–19 January 2005, Bankgok, Thailand, CD-ROM (2005)
17. Kong, J.S., Frangopol, D.M.: Life-Cycle reliability-based maintenance cost optimization of deteriorating structures with emphasis on bridges. J. Struct. Eng. ASCE **129**, 818–824 (2003)

Simulation of Blunt Defect Behaviour in a Thin Walled Cylinder for the Development of Structural Health Monitoring Techniques for Pipeline Repairs

M. A. Murad and F. P. Brennan

Abstract The objective of this chapter is to analyse numerically the effects of artificial notch defect size on a pressurised straight pipe (thin walled cylinder) subjected to limited testing pressures within an elastic working region. The chapter reports part of a wider study into the use of composite wrap repair techniques to extend the fatigue life of pipelines containing flaws and cracks. The methodology employed has been to verify the numerical stress analysis using strain gauge measurements made before during and after application of the composite repair. The measurement system will be further developed as a permanently deployed structural health monitoring system to ensure structural integrity during service. Finite element analysis using ABAQUS was employed to simulate stress and strain behaviour of the blunt defects. A comparative study was carried out in terms of stress concentration factor, relative notch length, Poisson's ratio and loading. The chapter reports the parametric analysis of the flaws and of the physical notched specimen.

Keywords Notch defect · Pressurised straight pipe · Structural health monitoring techniques · Stress concentration factors · Finite element analysis

M. A. Murad (✉) · F. P. Brennan
Department of Offshore, Process and Energy Engineering School of Engineering, Cranfield University, Building 52, Bedfordshire MK43 0AL UK
e-mail: m.abdmurad@cranfield.ac.uk

F. P. Brennan
e-mail: f.brennan@cranfield.ac.uk

A. Öchsner et al. (eds.), *Materials with Complex Behaviour II*, Advanced Structured Materials 16, DOI: 10.1007/978-3-642-22700-4_39,

1 Introduction

In oil and gas applications, pipes containing flaws and cracks detected during inspection are sized and assessed using fracture mechanics techniques [1]. In order to assess whether or not a flaw is critical, the structural integrity engineer needs to understand the ability of the structure to resist further loading and the critical amount of damage that the structure can sustain before remedial action is required [2]. If repair is required, then composite materials may be used to contain flaws or mitigate against other mechanical damage in pipelines [3].

The flaw type considered here is a blunt flaw (notch) due to local wall thinning caused by erosion or corrosion. The presence of a notch in a structural system is not only a common occurrence in engineering but any geometric discontinuity can initiate a surface crack due to the stress concentration effects. The methodology for strength and fatigue life assessment of notched components is well known [1].

It is important to understand this notch effect in terms of stress concentration factors (SCFs). SCFs are normally obtained analytically from elastic theory, computationally from Finite Element Analyses (FEA), and experimentally using methods such as photoelasticity or strain gauges [4]. Thus, by conducting a three-dimensional (3D) FEA, the elastic stress concentration factor for a pressurised pipe with a circumferentially arc-shaped notch on the external surface was determined in this study as a precursor to further FEA analysis of the stress–strain distribution in the composite based pipeline repair. Knowing the external loading, the stress concentration factor can be obtained in the 3D pipeline repair as a function of defect geometry and Poisson's ratio [5]. The effect of stress gradient due to different sizes of notch is also presented.

This geometry presents an interesting biaxial situation changing in significance with notch length and depth and future experimental work will be limited to only a small number of configurations. The FEA model once validated will allow the study of the biaxial SCFs so that an optimum repair scenario can be investigated.

Nomenclature	L = Notch length	t = Nominal thickness
L/t = Relative notch length	d = Defect depth	P = Internal pressure
$\sigma^h_{\max}$ = Maximum hoop stress	σ^h_{nom} = Nominal hoop stress	K^h_t = Stress concentration factor based on hoop stress
$\sigma^a_{\max}$ = Maximum axial stress	σ^a_{nom} = Nominal axial stress	K^a_t = Stress concentration factor based on axial stress

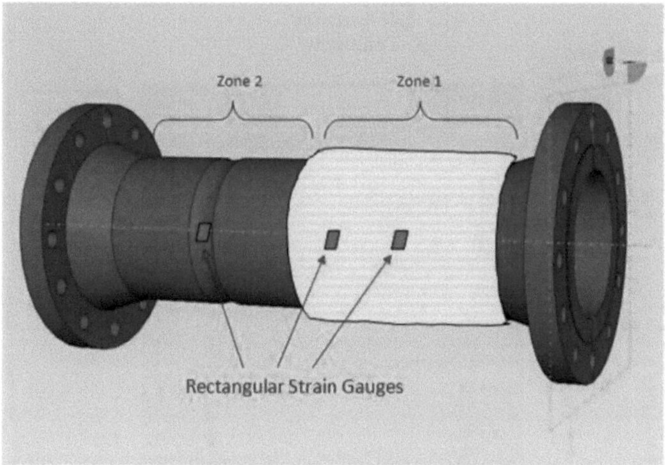

Fig. 1 The experimental test set up using strain gauges technique within a composite pipeline repair

2 Geometry and Finite Element Analyses

An experimental test rig configuration fabricated from API 5L Grade B carbon steel pressurized pipe containing artificial arc-shaped notches (machined defects) within two zones is illustrated in Fig. 1. The physical experiment and test parameters are described in Ref. [6].

This chapter only presents the preliminary finite element work on the unrepaired notches. The analysis assumes a perfectly straight arrangement without the effect of bending moments induced by misalignment. Loading is solely by internal pressure within the elastic working region. Symmetry was applied as shown in Fig. 2. Point 1 (at the root notch) and point 2 (away from the discontinuity), as shown in Fig. 3, indicate the local and remote reference stress positions respectively and partitioning and seeding were used to gain additional control over the mesh generation process [7]. Five FE models of various arc-shaped notches are shown in Fig. 4 and Table 1. The notch length is denoted *L*, nominal thickness, *t*, and defect depth, *d*, respectively. The nominal thickness is 8.2 mm and defect depth, *d* is 3.5 mm and remains constant throughout the whole process.

Element types including eight-node hexahedrons led to more reliable FEA solutions. There are many reasons why the eight-node hexahedral elements produce more accurate results than other elements (e.g. four-node tetrahedrons). These elements capture the singularities of the model at much less cost since they consume much less computer time and memory compared to the processing of higher order of Polynomial (p) or making element size smaller (h) or both. Hence, in the present work, the linear eight node hexahedral elements ($p = 1$) with reduced integration (C3D8R) were used. For example, for the N40 model, 4452

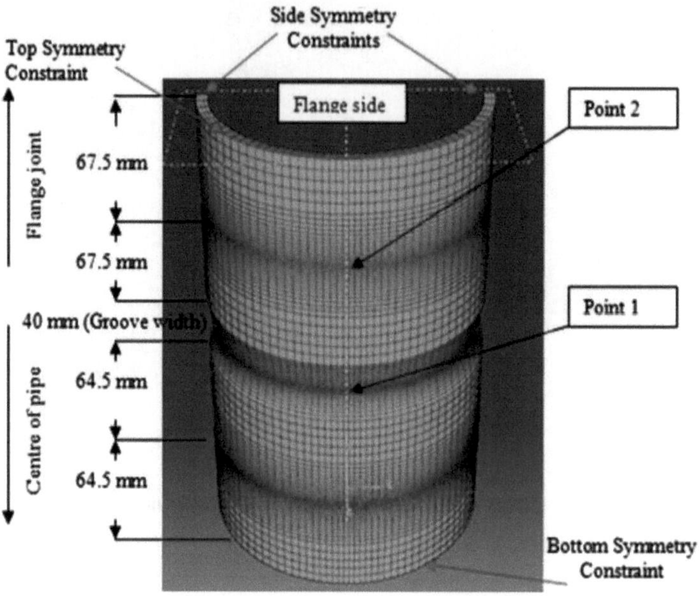

Fig. 2 A finite element model; the location of points used in stress concentration factors calculation

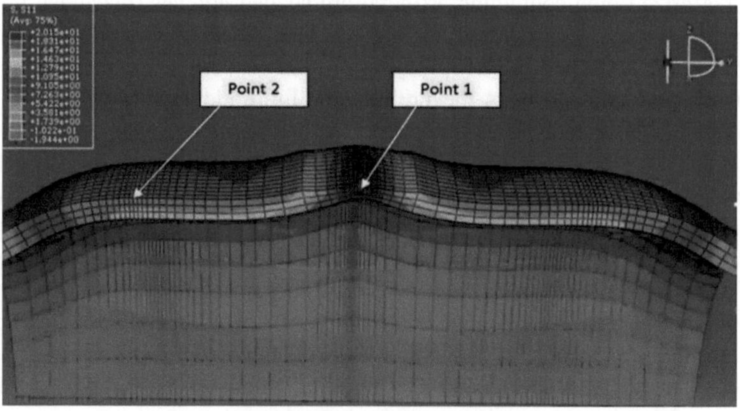

Fig. 3 Cross sectional view of the model that illustrates the smallest elements

elements up to 14504 were used and resulting the smallest element size (h) = 0.996 mm. Although tetrahedral elements are linear they were not used in this FEA model. They can have more discretization errors since they have a constant strain [8].

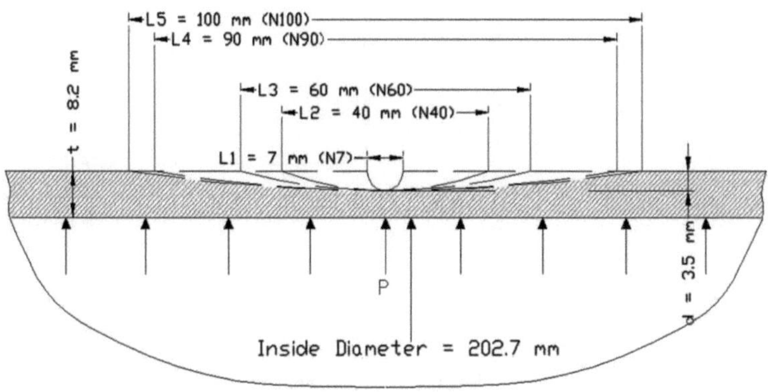

Fig. 4 The detailed dimensions of different arc-shaped notch sizes

Table 1 Convergence tests on various sizes of notch; test pressure at 50 bar

No of elements	$\sigma^h_{\max}$	σ^h_{nom}	K^h_t	$\sigma^a_{\max}$	σ^a_{nom}	K^a_t
Relative notch, L/t = 0.85 (N7); Arc shaped (semi circle)notch						
4576	90.86	62.61	1.45	55.45	15.44	3.59
5120	90.94	61.36	1.48	55.65	15.37	3.62
11872	90.86	61.23	1.48	55.47	15.51	3.57
14000	90.82	61.25	1.48	55.33	15.63	3.54
Relative notch, L/t = 4.88 (N40); Arc shaped notch						
4454	98.04	66.86	1.61	48.69	17.80	2.73
6800	98.94	60.99	1.62	49.94	17.69	2.82
11664	100.73	61.01	1.65	56.21	17.95	3.13
14504	100.74	60.90	1.66	56.18	17.73	3.17
Relative notch, L/t = 7.32 (N60); Arc shaped notch						
3840	104.98	61.76	1.70	50.05	19.46	2.57
7200	105.04	61.82	1.70	48.91	19.32	2.53
11520	105.07	61.59	1.71	48.54	19.86	2.44
14190	105.08	61.75	1.70	48.31	19.77	2.44
Relative notch, L/t = 10.98 (N90); Arc shaped notch						
3840	107.88	61.49	1.75	42.89	22.53	1.90
7200	107.31	61.47	1.75	41.34	22.36	1.85
11520	107.17	61.21	1.75	40.91	22.33	1.83
14448	107.14	61.15	1.75	40.78	22.35	1.83
Relative notch, L/t = 12.20 (N100); Arc shaped notch						
3936	108.17	61.17	1.77	40.79	23.65	1.72
7200	107.29	61.08	1.76	39.22	23.43	1.67
11520	107.10	60.81	1.76	38.80	23.40	1.66
14448	107.10	60.75	1.76	38.68	23.42	1.65

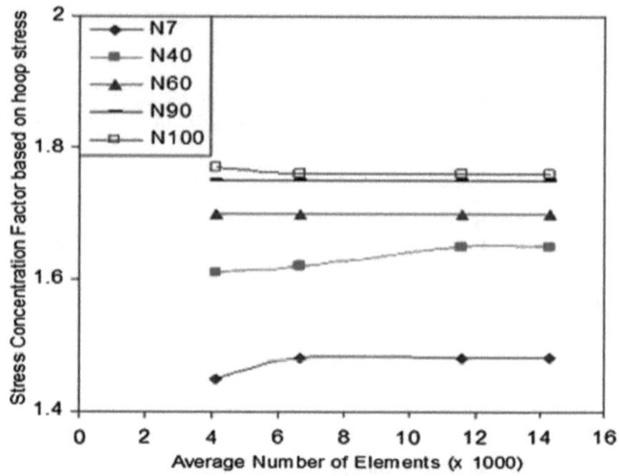

Fig. 5 Convergence test for SCF: notched pipe under hoop stress

3 Results

3.1 Internal Pressure and Stress Concentration

For a pipe under internal pressure, the hoop stress, σ_h, and the axial stress, σ_a are the two major stress components that are always considered. In this FE model, the maximum hoop and axial stresses were taken at Point 1 and nominal hoop and axial stresses were taken at Point 2 which gives the stress concentration factor, K_t by

$$K_t^h = \frac{\sigma_{\max}^h}{\sigma_{nom}^h} \quad \text{and} \quad K_t^a = \frac{\sigma_{\max}^a}{\sigma_{nom}^a}$$

The problem of SCF for circumferentially notched pipe has been examined by several authors [9], yet the study of arc-shaped notches in a pipe has not been carried out so far. The outer radius of the external notch is difficult to obtain because of the nature of its geometry and this makes comparison against other types of notches (e.g. those given in Ref. [4]) impractical. Prior to making experimental measurements, confidence can be attributed to the FEA by considering the results of the parametric variations and using engineering experience and judgment.

3.2 Convergence Study

Convergence tests were carried out on all notch geometries as represented in Table 1 and Figs. 5 and 6. These show all model stress concentration factor results converging, based on both hoop and axial stresses as the number of element

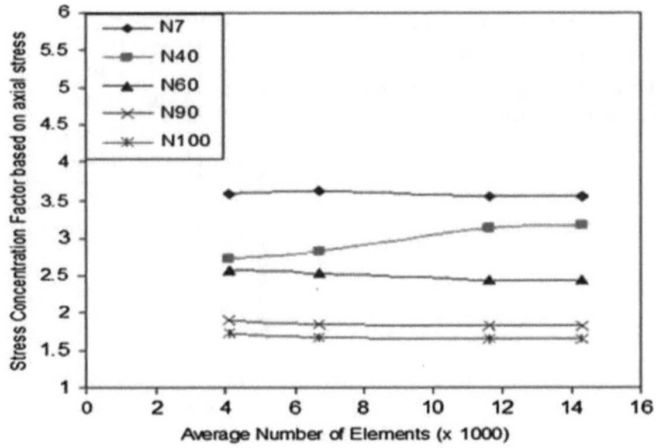

Fig. 6 Convergence test for SCF: notched pipe under axial stress

Table 2 Poisson's ratio influence on hoop stress concentration of various notches; pressure at 50 bar

v	N7	N40	N60	N90	N100
K_t^h					
0.26	1.42	1.62	1.68	1.74	1.75
0.30	1.48	1.65	1.70	1.75	1.76
0.34	1.55	1.70	1.73	1.76	1.77
0.38	1.63	1.74	1.75	1.78	1.78
K_t^a					
0.26	3.65	3.29	2.51	1.83	1.64
0.30	3.54	3.17	2.44	1.83	1.65
0.34	3.46	3.07	2.39	1.82	1.66
0.38	3.39	2.99	2.35	1.81	1.66

increases and the size of element becomes smaller. Notch N100 has the highest K_t^h (i.e. 1.76) and the lowest K_t^a (i.e. 1.65).

Notch N100 shows a similar trend, as the Poisson's ratio increases, as shown in Table 2 and Figs. 7 and 8. Due to the increased biaxial effect, all notches show a converging increment in SCF values based on hoop stress. It can also be observed that for notch N100, the SCF based on both stresses remains constant as the Poisson's ratio increases. So, it can be concluded, as expected, that the Poisson's ratio effect has a lesser influence on stress concentration factors if the arc-shaped notch is made longer.

Table 3 and Fig. 9 show that as the relative notch increases, K_t^h increases and K_t^a decreases. Figure 10 confirms that the increasing pressure in the elastic working region has no influence in the increment of SCF values for any type of FE model.

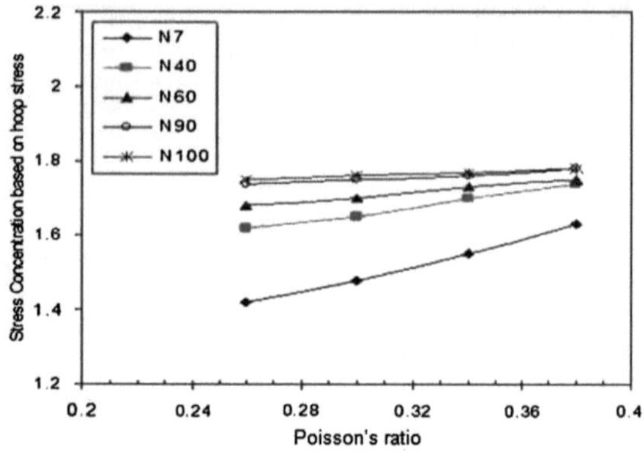

Fig. 7 Variations of the stress concentration based on hoop stress against increasing Poisson's ratio

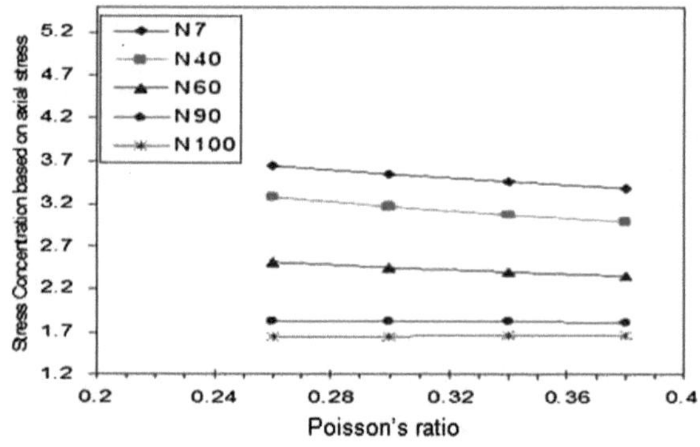

Fig. 8 Variations of the stress concentration based on axial stress against increasing Poisson's ratio

Table 3 Loading influence on stress concentrations of two types of notches

P	K_t^h		K_t^a	
	N7	N40	N7	N40
10	1.48	1.66	3.54	3.17
20	1.48	1.66	3.54	3.17
30	1.48	1.66	3.54	3.17
40	1.48	1.66	3.54	3.17
50	1.48	1.66	3.54	3.17

Fig. 9 Effects of notch length to hoop and axial stresses

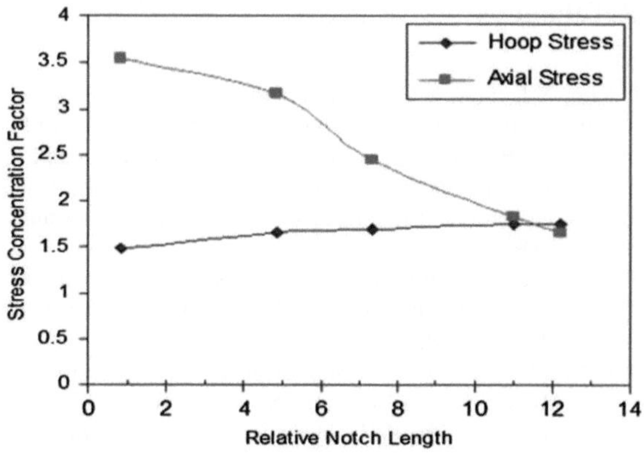

Fig. 10 The influence of increasing pressure in the stress concentration study

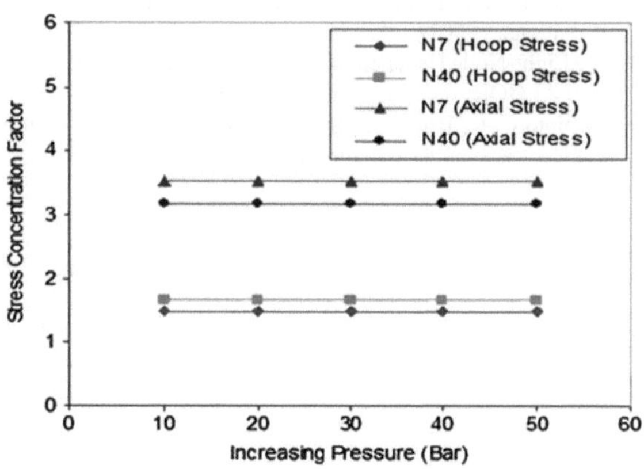

4 Conclusions

A biaxial notch situation has been studied in an internally pressurised pipe as a precursor to a repair method study. Without published results to compare with, results have been examined to consider whether or not the FEA model is acting sensibly. As predicted the relationship between hoop and longitudinal stress concentration factors varies with notch geometry and Poisson's ratio effectively altering the degree of biaxiality. Future work will be to validate the models against strain gauge results and then to apply a composite wrap repair technique.

References

1. Kim, Y.J., Son, B.G.: Finite element based stress concentration factors for pipes with local wall thinning. Int. J. Press. Vessel. Pip. **81**, 897–906 (2004)
2. Brennan, F.P., Leeuw, B.De.: The use of Inspection and monitoring reliability information in criticality and defect assessment of ship and offshore structures. In: Proceedings of the ASME 27th International Conference on Offshore Mechanics and Artic Engineering, Portugal, 15–20 June 2008
3. Alexander, C.: Strain based design methods for composite repair systems. In: Proceedings of IPC2008, 7th International Pipeline Conference, Calgary, Alberta, Canada, September 29–October 3 2008
4. Pilkey, W.D.: Peterson's Stress Concentration Factors, 2nd edn. Wiley, (1997)
5. Abd Murad, M.: An Integrated Structural Health Monitoring Approach to Composite Based Pipeline Repair, the Ph.D. thesis to be submitted, June 2011, Cranfield University, UK
6. Abd Murad, M.: The Development of Structural Health Monitoring (SHM) procedures for the structural integrity and maintenance repair of offshore ageing pipelines In: 4th European–American Workshop on Reliability of NDE, Berlin, Germany, 24–26 June 2009
7. Abaqus Inc and DS: Abaqus CAE User Manual, Version 6.7 ed, Abaqus Inc, DS, United State of America (2007)
8. The Eight-Node Hexahedral "Brick" Element in Finite Element Analysis ALGOR Development Staff ALGOR, Inc., Pittsburgh, PA, USA
9. Carpinteri, A., Brighenti, R., Vantadori, S.: Circumferentially notched pipe with an external surface crack under complex loading. Int. J. Mechan. Sci. **45**, 1929–1947 (2003)

High-Low Sequence Loading Effect on the Crack Growth Rate using UniGrow Model

N. Nik Abdullah, Mohammad Hadi Hafezi and S. Abdullah

Abstract Fatigue crack growth (FCG) behaviour under variable (VA) amplitude loading is a subject that has been studied since 1970, with Elber's contribution to fracture mechanics. Simple VA load sequences are most commonly used to study the load interaction effect phenomenon. Changes in the loading amplitude, so-called high-low sequence loading experiments, can introduce profound effects on fatigue crack growth. The purpose of this research is to test the applicability of a unified two-parameter FCG model—called UniGrow model—for the above mentioned loading conditions. Finally, the authors have made efforts to identify some uncertainty parameters to evaluate the prediction accuracy of the model.

Keywords Fatigue crack growth · Load sequence · UniGrow model

1 Introduction

Fatigue crack growth models under variable amplitude (VA) loading with consideration of the preceding cyclic loading histories (memory effects) caused by various load sequences have had a high significance for investigation among researchers during the past decade. At the same time, FCG models under VA loading are involved in load interaction effects. As a result, control on the retardation and/or acceleration of fatigue crack growth rate (FCGr) has been a

N. Nik Abdullah (✉) · M. H. Hafezi · S. Abdullah
Department of Mechanical and Material Engineering, Universiti Kebangsaan Malaysia, 43600, Bangi, Selangor, Malaysia
e-mail: enikkei@eng.ukm.my; enikkeister@gmail.com

M. H. Hafezi
e-mail: hafezi.hadi@gmail.com

A. Öchsner et al. (eds.), *Materials with Complex Behaviour II*, 647
Advanced Structured Materials 16, DOI: 10.1007/978-3-642-22700-4_40,

Table 1 Monotonic and Fatigue Material Properties

Elastic modulus (MPa)	E	121000
Poisson's ratio	v	0.3
Fatigue strength coefficient (MPa)	σ'_f	1276
Fatigue strength exponent	b	−0.139
Fatigue ductility coefficient	ε'_f	0.174
Fatigue ductility exponent	c	−0.415
Cyclic strength coefficient(MPa)	K'	2275
Cyclic strength hardening exponent	n'	0.334
Notch tip radius or elementary material block size (m)	ρ^*	1.1×10^{-05} 1.45×10^{-04}

challenging subject. In other words, understanding the transient effect is important. A new achievement in this area is represented by Glinka and Noroozi's so-called UniGrow model [1]. This chapter is intended to show the sensitivity of UniGrow model in different load sequences mathematically, as an introduction for applicability of the UniGrow model to reliable life prediction of components subjected to random loads. A simple deterministic loading function as an input for the UniGrow model was defined, with the aim of investigating the response of the model to probable analysis outcomes. As a reasonable assumption, we have taken constant frequency, with variations only in amplitudes. High-low sequence effects have been investigated. Study on the behaviour of fatigue crack growth rate can be an introduction to extend the scope of the work to application of reliability techniques. These are involved in random variables which are applied to calculate local stress/strain at the crack tip. It is clear that Noroozi's work has shown load interaction effects with experimental verification on metallic materials. However, this research is made in relation to the use of UniGrow for load sequences studies. The rationale is that a few models can explain the load sequence effect. So this is the first time the authors have tried to do this using the UniGrow model.

Mikheevskiy [2] presented a stepwise procedure for fatigue crack growth analysis based on the UniGrow fatigue crack model. Significant questions have been raised which have prompted the author to carry out this research into what will happen if the UniGrow model is used for investigating the successive block of cycles. We selected AISI 304 stainless steel, which belongs to a class of 300-type austenitic stainless steels, widely used in engineering applications. Monotonic material properties and Coffin–Manson fatigue material properties have to be taken from available sources [3]. These properties are listed in Table 1. Constant amplitude fatigue crack growth data obtained at several mean stresses needs to be selected as a base for subsequent fatigue crack growth and fatigue life analysis. One important parameter in the UniGrow model is elementary material block or notch tip radius. The magnitude of the elementary material block size parameter ρ^* needs to be obtained according to three methods, which are suggested by Mikheevskiy [2]. The elementary material block size ρ^* can be determined for each particular point of the experimental fatigue crack growth rate curve. The iteration procedure is complete when approximately the same value of ρ^* parameter is obtained for all experimental data points. Determination of ρ^* parameter that is initially unknown is an important factor in the accuracy of the UniGrow model. But because the main focus in this chapter is to investigate load sequence effects, and we would like to consider parameter ρ^* as a random variable, it will be logical if we just assume the parameter ρ^* in the range for steel materials. The present work covers the addition of further loading condition information to the UniGrow model proposed by Noroozi et al. [4], in such a way that FCG relations are applied for a complex load spectra. Traditionally, the crack length vs. number of cycles to failure are plotted. As mentioned briefly above, the problem involved some approximate calculations, and most importantly, a complete understanding of all parameters used in the equations. Methodologies of calculation, mathematical as well as physical, and mechanical interpretation are both important. This is a primary objective, and will help in proving the UniGrow model.

2 Fatigue Crack Growth Simulation

The analytical concept is selected based on the framework which has been represented by authors in another work [5]. In fact, this framework has obviously decreased the degree of difficulty for analyses of fatigue crack growth model under variable amplitude loading conditions. All stages of this study have been carried out based on the following stepwise procedures: Material selection and specimen geometry, Loading conditions (load spectra), Data collection, Using UniGrow model, and Results and discussion respectively.

2.1 Material and Specimen Geometry

The material used in this study is stainless steel 304. A lot of work has been done in the past on crack growth behaviour in stainless steel 304, which is used for high temperature components. We are interested in analyzing components under as realistical as possible loading conditions, including mixed mode. That is to say, fatigue and creep modes for fracture in a high temperature environment. The reasons for choosing this material can be listed as follows:

(1) Welding: Excellent weld ability by all standard fusion methods, both with and without filler metals
(2) Heat resistance: Good oxidation resistance in intermittent service to 870°C and in continuous service to 925°C.
(3) Corrosion resistance: Excellent in a wide range of atmospheric environments and many corrosive media.

Table 1 shows the material property values which will be needed for the UniGrow model calculations. The selected geometry—used successfully in

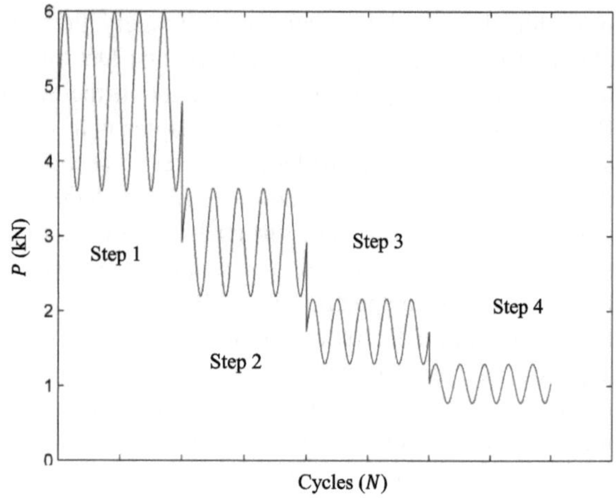

Fig. 1 Schematic high-low sequence loading identical *R*-ratio of 0.6

Table 2 Four succesive block of cycles, identical *R*-ratio of 0.6

f (Hz)	*R* (ratio)	$\Delta P_1/2$(kN)	$\Delta P_2/2$(kN)	$\Delta P_3/2$(kN)	$\Delta P_4/2$(kN)
0.05	0.6	1.2	0.72	0.432	0.2592

South Korea fatigue crack growth tests [6]—is a compact tension specimen [7] with a width of 50 mm and a thickness of 12.5 mm.

2.2 Loading Conditions (Load Spectra)

It has been shown by a number of researchers that the application of a high-low sequence loading consisting of two loading steps with higher loading amplitude in the first step, and lower loading amplitude in the second step, can cause noticeable effects in fatigue crack growth rate. In our case, the four steps have the same *R*-ratio of 0.6. The load amplitude is 1.2 kN in the first step, 0.72 kN in the second step, 0.432 in the third step and 0.2596 in the last step. Figure 1 shows an illustration of high-low sequence loading histories, with identical *R*-ratio of 0.6. Table 2 lists the details of load generation which were conducted to study the high-low sequence effect.

The sequence load effect perturbs steady state fatigue crack growth conditions and affects the growth rates by retarding or accelerating growth. Understanding these transient effects is important for the reliable life prediction of a component subjected to random loads. As an example, in crack propagation experiments conducted on 6061-T6 al alloy by Kumar [8] in the low–high sequence, acceleration is observed in the crack growth rate. For the high–low load sequence, retardation was observed in the crack growth rate. To give another example, as Kalnaus et al. [9] has mentioned: "in a two-step high-low loading experiment, the first loading step has significant influence on the subsequent crack growth in the second loading step, if the minimum load or the *R*-ratio is the same in the two loading step".

2.3 Stress Intensity Factor

The strain-based approach is still used in many situations. In fact the application of fracture mechanics to analysis of fatigue crack growth comes from the strain-based approach. Fracture mechanics is a most promising approach to assess sequence load effects. In particular the stress intensity factor (SIF) used in the regime of linear elastic fracture mechanics has been applied. The stress intensity factor for a compact tension specimen [7] is given by:

$$K = \frac{PF(\alpha)}{(BW^{0.5})} \tag{1}$$

Where

$$\alpha = \frac{a}{w},\ F(\alpha) = \frac{(2+\alpha)(0.886 + 4.64\alpha - 13.32\alpha^2 + 14.72\alpha^3 - 5.6\alpha^4)}{(1-\alpha)^1.5} \tag{2}$$

and α, *W*, *B* are the crack length, specimen width and thickness, respectively, and *P* is the applied load. Using the formulation $\Delta K_{appl} = K_{\max, appl(1-R)}$ in order to calculate the applied stress intensity is the best way. Note that for analysis in high stress ratios—for stress ratios more than 0.5—the residual stress effect can be ignored. On the other hand, the residual stress intensity will be zero at the crack tip. This means the total value of stress intensity is equal to the applied value. This point has been clarified in the following discussion regarding the UniGrow model and the principles used in this model.

3 Discussion on UniGrow Model

As could be found in the original work by Noroozi et al. [4], the two-parameter fatigue crack model is based on the following assumptions. In relation to using Coffin–Manson in the UniGrow model, the question is what will influence the accuracy of the model. As we know, Coffin–Manson does not show the mean-stress effect. In response to this question, let us examine some phenomenological reasons. First, macroscopic phenomenological equations such as Coffin–Manson strain life approach and/or Smith–Watson–Topper (SWT) equation, elaborated in [10] are ways to interlink fundamental concepts to give engineering solutions. Second, the accuracy of the UniGrow model does not depend on the Coffin–Manson strain life approach. As Noroozi mentioned, it (Coffin–Manson) has been used only to derive the form of the UniGrow model. The main feature of the UniGrow model is the correct estimation of residual stresses produced by all previous loading cycles and corresponding residual SIF, because this becomes one of the most important and complicated parts of the UniGrow model. Moreover, two models can be used to show the mean-stress effect. The first one is the closure model and the second is the two-parameter model. Recent experimental and analytical evaluation of crack closure has confirmed its negligible contribution to crack growth, and has demonstrated that changes in the stress ahead of the crack tip are more important than closure behind the crack tip [11]. However, for crack growth under constant-amplitude loading, the weakness is in those methods that changed from local to remote methods, and not the *crack-closure concept itself.* On a side note, the load-shedding test method in ASTM E647 causes remote closure, as may happen under variable-amplitude loading. Yamada found that remote crack-mouth-opening-displacement (CMOD) and/or Backface Strain (BFS) gauge records are satisfactory at low stress ratios (R), but at high R the remote methods do not measure the correct opening loads [12]. They have gone back to a local method, similar to what Elber originally used. In the UniGrow model, SWT parameter is used to accumulate fatigue crack growth in each cycle, and the strain–stress material behaviour was modelled by the Ramberg–Osgood expression (for each particular material used in analysis). Therefore, the following phenomenological assumptions are reasonable:

- Ramberg–Osgood rule for stress/strain behaviour
- Neuber plasticity rule for calculation of stress/strain field in the crack trip
- Coffin–Manson strain life approach to derive a form for UniGrow expression
- Smith–Watson–Topper fatigue model to determine the fatigue damage accumulation at the crack tip

Using strain-life test data which is obtained with various mean stress damage parameters (SWT), the following expression can be derived [13]. Equation (3) has been used to handle mean-stress effects. It is shown to correlate mean-stress data better for a wider range of materials and is therefore regarded as more promising for general use. Monotonic material properties such as the modules of elasticity, E and the Poisson ratio, v and cyclic stress–strain material properties such as the cyclic hardening coefficient, K' as well as cyclic strain hardening exponent, n', and Coffin–Manson fatigue material properties have to be provided for simulations using UniGrow model.

$$\sigma^a_{\max}\frac{\Delta\varepsilon^a}{2}=\frac{\left(\sigma'_f\right)}{E}\left(2N_f\right)^{2b}+\sigma'_f\varepsilon'_f\left(2N_f\right)^{b+c} \quad (3)$$

Where

$$\sigma_{\max}=\sigma_m+\sigma_a \quad (4)$$

In this chapter, to collect the constant amplitude FCG data, the authors tried to use the literature review. In order to do that, the range of the applied SIF reported by [6] for the SS 304 has been used. It was also necessary to collect sufficient data

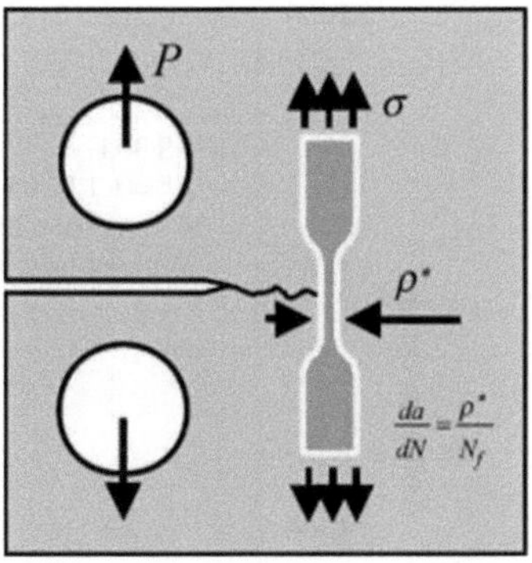

Fig. 2 Relationship between fatigue and fracture Source: Ref. [14]

to explain the fatigue behaviour of material around threshold region—elastic fracture mechanics. Besides, in Appendix I, all formulations used in the UniGrow model are sequenced.

In using the UniGrow model, it was assumed that the material at the crack tip was of discrete composition, and the material block size was assumed to be an inhomogeneous block, of average dimension, but which still retained the behavior characteristics of the bulk material. Thus, The crack tip also was modeled as a notch with finite radius of ρ^*. This assumption has been indicated in Fig. 2 whereby the fatigue crack growth rate and fracture mechanics are related [14]. The determination of elementary block size is most important to the effectiveness in the UniGrow model, because the accuracy of the model strongly depends on that. The authors take the value of parameters based on Noroozi's work. This will not be fully accurate, but because the goal is to investigate the sensitivity of this model to sequence loading, it will suffice for our purpose.

There are some complexities in calculating elementary material block size. The size of the elementary material blocks ρ^* has an effect on the calculated crack tip residual stresses σ_r and on the resulting residual stress intensity factor. Consequently, the residual stress intensity factor, when factored into the applied value of stress intensity factor (SIF), influences predicted fatigue crack growth rate. Besides, the reversed plastic deformation around the crack tip, induced at relatively high stress ratio $R > 0.5$ and relatively small stress intensity ranges (near threshold FCG), are usually not sufficient to produce compressive residual stresses. Therefore, the residual stress intensity factor is close to zero ($K_r = 0$) and the total SIFs are the same as the applied ones (See Sect. 3, Appendix I).

4 Results and Discussion

Regarding the crack front passing through a component from a cross section of micro crack, and the modeling of material in the UniGrow feature based on parameter ρ^* as a bulk property, this model has good potential for inititation

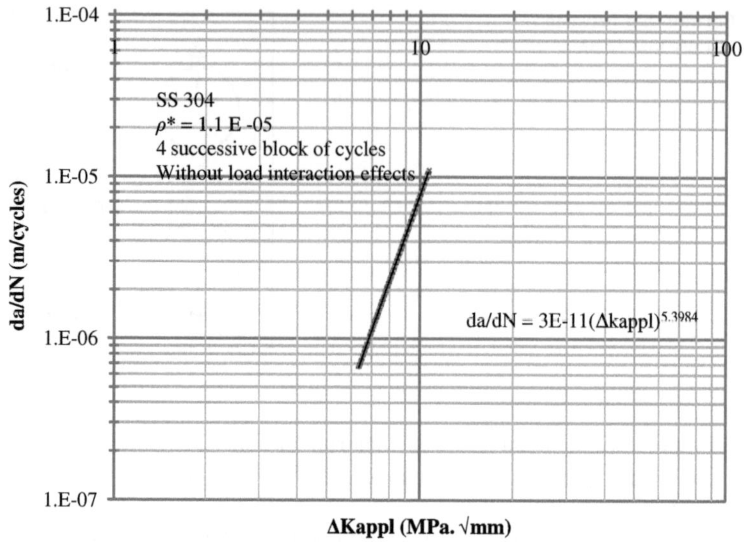

Fig. 3 Fatigue crack growth data calculated using the UniGrow model for the stainless steel as a function of the applied stress intensity factor. Block of cycles are considered separately without load interaction effects, the parameter ρ^* is assumed 11 micrometers

analysis, and it is well known that "in the crack initiation period, fatigue is a material surface phenomenon". But, attention should be given to something else as well. First, in some structural applications, the fatigue threshold is very important, and the dependency of determination of ρ^* on this data can be a source of uncertainty in the prediction process. The smallest change in the value of parameter ρ^* will make a tangible change in the fatigue life prediction. Figure 3 shown the results when the parameter ρ^* are taken 13 times lower. Secondly, cyclic strain fields in the plastic zone are greatly influenced by crack closure, so the model is missing a major contributor to cyclic damage in a basic sense. This work is approached with consideration of different mean stresses. However, it was based on using the identical R-ratios in constant-amplitude of each block of cycles. Table 3 shows different numbers of cycles in the same stress–strain field (equality in the applied SIF range) in the transition lengths. Physically this phenomeonon is called retardation. Another conclusion is that in the fourth step we can observe many more cycles to failure in the same crack length by comparing to other step loadings. This means that when the amplitude of loads is decreasing in relation to prior high amplitude loads, a retardation can be observed in fatigue propagation. Fatigue crack growth data for stainless steel as a function of the applied stress intensity factor using the UniGrow model are shown in Fig. 4. But the load interaction effects are not considered. The range of rates is marked for four blocks of cycles one-by-one. Using the regression technique the distribution of data is captured in the Paris law form. By comparing Figs. 3 and 4 it was found that,

Table 3 Material behaviour in the pre-cracked length and transition lengths before and after immediately after changing the intensity of loads. Number of cycles to failure of the material element N_f and the local stress–strain values are summarized

a (mm)	P(kN)	N_f	K_{max}(MPa. $\sqrt{mm}$)	Δk(MPa. $\sqrt{mm}$)	$\Delta\varepsilon$	σ_{max} (MPa)
22.7	6	9344	17.844	7.137	3.19×10^{-3}	547.6
29 B	6	1050.5	26.75	10.7	4.78×10^{-3}	670.6
29 A	3.6	271962.3	10.7	4.282	2.87×10^{-3}	519.3
34.7B	3.6	612882.2	26.48	10.6	4.74×10^{-3}	667.3
34.7A	2.196	630359.2	6.42386	2.569	2.84×10^{-3}	516.7
38.9B	2.196	887117.9	26.55	10.62	4.75×10^{-3}	668
38.9A	1.296	904353.9	15.93	6.37	2.85×10^{-3}	517.37

A and B are denoted as "After" or "Before" status

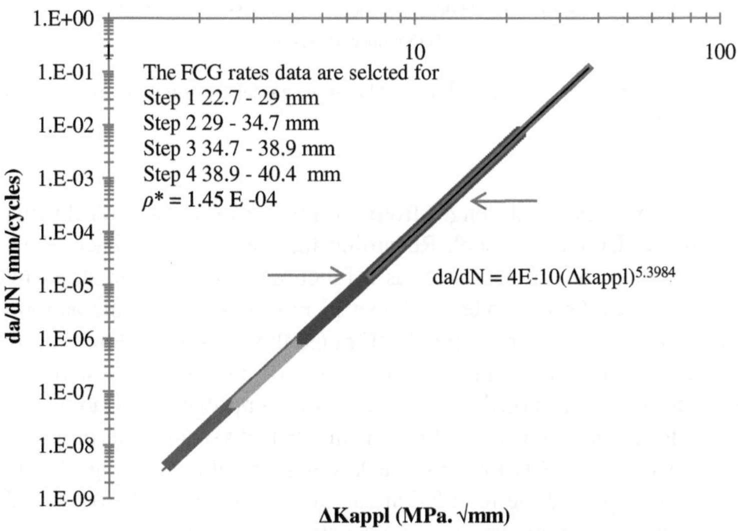

Fig. 4 Fatigue crack growth data calculated using the UniGrow model for the stainless steel as a function of the applied stress intensity factor. Block of cycles are considered separately without load interaction effects, the parameter ρ^* is assumed 145 micrometers

surprisingly, the FCG exponent of data was constant $m = 5.3984$, and the FCG parameter, C is changed. It can be understood that in the same load intensity range the rates are constant, this means the model dose not make sense when only the current number of cycles to failure are assumed to calculate the rates. This is number of cycles that are needed for the crack to propagate to a certain incremental length. This number of cycles should be added to the next corresponding number of cycles to failure in the new position of the crack, and then the rates can be calculated.

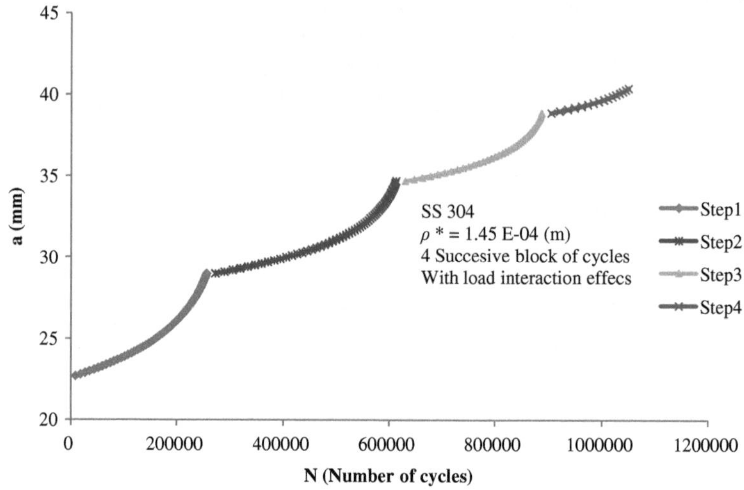

Fig. 5 Crack growth length calculated using UniGrow model from the notch root for a given loading condition

Figure 5 shows the crack length from the notch root for a given loading condition using the UniGrow model. Regarding the similarity approach, the fatigue crack growth data for stainless steel as a function of the applied stress intensity factor using the UniGrow model is shown in Fig. 6. The load interaction effects can be felt where in the same applied SIF range the rates are different in different load steps. Although the load intensity increases in each load step, the rates decreases. It should be noted that in the lower amplitude of loading the rates approaches to a constant value, which means that physically, the propagation of the crack is slow. The FCG rates vs. crack length are plotted in Fig. 7. The crack length in each steps and the retardation can also be understood from this figure.

In a simulation program we defined a constant interval for crack propagation as 0.1 mm. It has been observed that we have sufficient data from a numerical point of view. But why cannot enough data be observed in all three regions of fracture. The answer is because the mechanism of failure depends on the number of cycle to failure. In fact, this is also the nature of the failure mechanism in the UniGrow model. A deeper look can refer to the choice of analytical concept [5] that is used to derive the final expression of the fatigue crack growth model. However, although the investigators tried to study all aspects, in some areas they were also forced to simplify. This does not mean that it is wrong, but also it is proof that we need a new way to cover all the reasonable possibilities. Therefore, the problem will involve the study of the threshold effect and some other complicated aspects, if we intend to complete the process of fatigue failure. If we have enough R-ratio data all the way from very low (near threshold ΔK), to very high ΔK, we can make a complete plot of $\Delta K - K_{\max}$. We should pick out the ΔK data at each R-ratio and plot ΔK vs. $K_{\max}$. It is predicted that we can see the L-shape diagram, just as

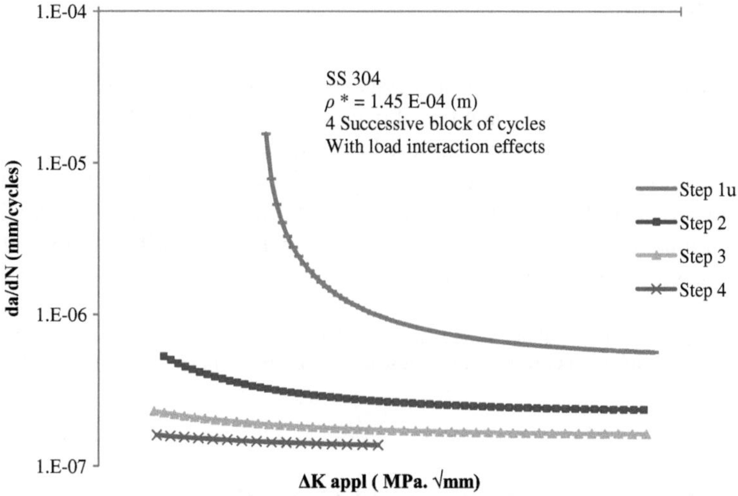

Fig. 6 Fatigue crack growth data calculated using the UniGrow model for the stainless steel as a function of the applied stress intensity factor. 4 successive block of cycles from higher (1u) to the lower amplitude (4u) are applied

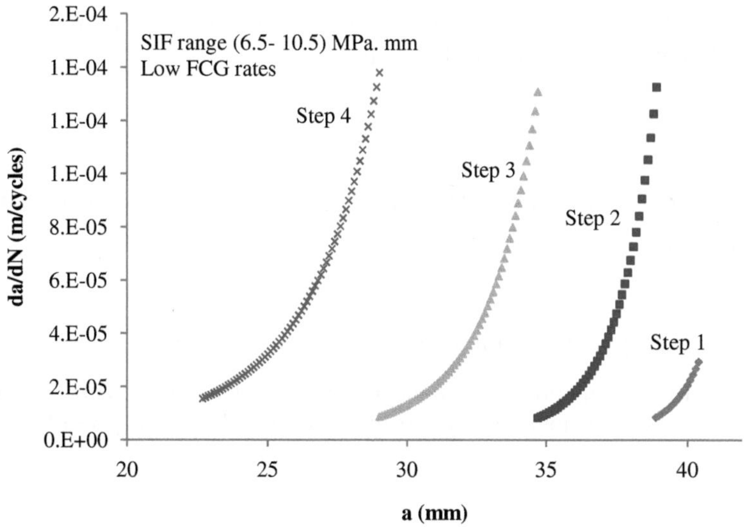

Fig. 7 The crack length for low fatigue crack growth rates and high stress ratio (0.6) with an immediate changing of mean stress in switching the block of cycles in the certain crack length

Sadananda et al. investigated [11]. This is a link between unified approach [11] and UniGrow model [4]. Thus, the threshold relationships with $K_{\max}$ that are affected with residual stress effect will be clear.

5 Conclusion

The present work tested the 4 successive block of cycles using the UniGrow model. The crack length vs. number of cycles to failure and FCG rates vs. applied SIF range are plotted. For the high-low load sequence, significant retardation was observed in the crack growth rate. In the UniGrow model the determination of parameter ρ^* can be investigated as a most uncertain value in calculation. Attention must be given to the size of this parameter, as it will directly influence the accuracy of the UniGrow model.

Acknowledgments The authors would like to acknowledge Universiti Kebangsaan Malaysia (UKM) and the Malaysian Government for research sponsorship under project UKM-GUP-NBT-08-26-100.

A.1 6 Appendix I—UniGrow FCG relations

A.1.1 Bi-axial Neuber's Rule

It is shown that Neuber's rule is actually a particular case of equivalent strain energy density (ESED), namely when the dissipation of the plastic strain energy at the notch root is neglected in ESED method.

1. Plane stress

The normal and shear stress components in the Z-direction are zero or negligible.

2. Plane strain

Plane strain state is obtained if strains in the z direction are precluded.

Plane stress assumptions were made for all calculations. The relationship between elastic behaviour and local (actual) stress and strain is given by:

$$\sigma^e \varepsilon^e = \sigma^a \varepsilon^a$$

A.1.2 Elastic–Plastic Material Behaviour at the Crack Tip

We studied the Elastic–Plastic material behaviour at the crack tip. There are many important classes of materials that are too ductile to permit description of their behaviour by LEFM: the crack tip plastic zone is simply too large in comparison to the cracked body. For this case we must apply EPFM. Therefore, the maximum stress and strain range at the crack tip can be determined in a closed form given by [2] :

$$\sigma^a_{\max} = \left[\frac{(K')^{\frac{1}{n'}}(\psi_{y,1})^2}{2\pi E\rho^*}\right]^{\frac{n'}{n'+1}} \left(K^2_{\max,\,tot}\right)^{\frac{n'}{n'+1}}$$

$$\Delta\varepsilon^a = \frac{1}{E}\left(\frac{\Delta K_{tot} \times \psi_{y,1}}{\sqrt{2\pi\rho^*}}\right)$$

A.1.3 Calculation of Stress Intensity Factor for High Stress Ratios (R > 0.5):

For some materials, there are some small differences between $R = 0.5$ and higher/ lower R. However, special consideration was made in order to omit the residual stress intensity factor. Thus, the stress intensity factor is given by [4] :

$$K_{\min,\,tot} = K_{\min,\,net} = K_{\min,\,appl}$$
$$K_{\max,\,tot} = K_{\max,\,net} = K_{\max,\,tot}$$
$$\Delta K_{tot} = K_{\max,\,tot} - K_{\min,\,tot} = \Delta K_{net} = \Delta K_{appl}$$

A.1.4 Number of Cycles to (N_f) Needed to Fail the Elementary Material Block at the Crack

Smith, Watson and Topper (SWT) equation [10] was used to calculate the number of cycles to failure. The plastic term in the Manson–Coffin together with the SWT was omitted [13].

$$\sigma^a_{\max}\frac{\Delta\varepsilon}{2} = \frac{\left(\sigma'_f\right)^2}{E}\left(2N_f\right)^{2b}$$

References

1. NooroziSani, A.: Development of a two-parameter model (K(max), delta-K) for fatigue crack growth analysis, p. 192. University of Waterloo, Canada (2007)
2. Mikheevskiy, S.: Elastic-plastic fatigue crack growth analysis under variable amplitude loading spectra, p. 167. University of Waterloo, Canada (2009)

3. Ralph, I., Stephens, A.F.: Metal Fatigue in engineering, 2nd edn, p. 497. Willey–Interscience, New York (2001)
4. Noroozi, A.H., Glinka, G., Lambert, S.: A two parameter driving force for fatigue crack growth analysis. Int J Fatigue **27**(10–12), 1277–1296 (2005)
5. Abdullah N.N., Hafezi, M.H., Abdullah, S.: Analytical concepts for recent development in fatigue crack growth under variable amplitude loading, part I: qualitative interpretation. Key Eng. Mat. **462–463**, 59–64 (2011)
6. Baik, Y.M., Kim, K.S.: The combined effect of frequency and load level on fatigue crack growth in stainless steel 304. Int. J. Fatigue **23**(5), 417–425 (2001)
7. ASTME647-08e1, Standard test method for measurement of fatigue crack growth rates (2004)
8. Kumar, R.: An invcstigation of the influence of high-low and low-high amplitude block loadings on fatigue crack growth. Eng. Fract. Mech. **38**(2–3), 225–230 (1991)
9. Kalnaus, S., et al.: An experimental investigation of fatigue crack growth of stainless steel 304L. Int. J. Fatigue **31**(5), 840–849 (2009)
10. Dowling, N.E.: Mean stress effects in strain-life fatigue. Fatigue Fracture Eng Mater Struct **32**(12), 1004–1019 (2009)
11. Sadananda, K., et al.: Analysis of overload effects and related phenomena. Int. J. Fatigue **21**(Suppl. 1), S233–S246 (1999)
12. Newman, J., Yamada, Y.: Crack-closure behavior of 2324–T39 aluminum alloy near-threshold conditions for high load ratio and constant Kmax tests. Int. J. Fatigue **31**, 1780–1787 (2009)
13. Noroozi, A.H., Glinka, G., Lambert, S.: Prediction of fatigue crack growth under constant amplitude loading and a single overload based on elasto-plastic crack tip stresses and strains. Eng. Fract. Mech. **75**(2), 188–206 (2008)
14. Stoychev, S., Kujawski, D.: Analysis of crack propagation using Δk and Kmax. Int. J. Fatigue **27**(10–12), 1425–1431 (2005)

Molecular Dynamics Simulation on Nano-Machining of Single Crystal Copper with a Void

Seyed Vahid Hosseini, Mehrdad Vahdati and Ali Shokuhfar

Abstract Nowadays, the ultra-precision machining with single diamond tools can remove materials at nanometer scale, which has been used to produce surface with high quality finishing. As far as the conventional finite-element method becomes impossible for numerical analysis, as an alternative, molecular dynamics (MD) method is significantly implemented in the field of nano-machining process to investigate cutting mechanism. Although it is well known that even the purest real material contains a large number of defects within its crystal structure, in conventional MD simulation of nano-cutting process the workpiece is assumed as a perfect single crystal. So, there is a need to check the effect of defect inclusion in the workpiece on nano-machining process.

In this chapter molecular dynamics simulations of the nano-metric cutting on single-crystal copper were performed with the embedded atom method (EAM) potential. To investigate the effect of the void on the workpiece machining characteristic, a comparison was done between perfect single crystal and a single crystal with a certain void. The numerical results reveal that the void inclusion can decrease the tool forces and affect the chip formation mechanism. Also, in defected workpiece, the value of pressurized atoms is decreased in front of the tool tip. In addition, the plastic zone becomes larger in a workpiece with a void defect compared to a pure workpiece, which can affect the surface integrity. Finally, results show that the internal surface of a void that is positioned under the machined surface is crumpled due to high compressive hydrostatic pressure.

S. V. Hosseini (✉) · M. Vahdati · A. Shokuhfar
Mechanical Engineering, K. N. Toosi University of Technology, Tehran, Iran
e-mail: mscmechanic@hotmail.com

M. Vahdati
e-mail: vahdati@kntu.ac.ir

A. Shokuhfar
e-mail: shokuhfar@kntu.ac.ir

A. Öchsner et al. (eds.), *Materials with Complex Behaviour II*, 661
Advanced Structured Materials 16, DOI: 10.1007/978-3-642-22700-4_41,

Keywords Nano-metric cutting · Molecular dynamic simulation · Material defect · Nano void

1 Introduction

With the increasing demand on scaling down surface roughness of mechanical and electronic components, ultra precision machining (UPM) has attracted significant interest to date. This requires a deeper understanding of the mechanical behavior of materials on the nanometer scale. However, the deformation mechanism in the nano-cutting has not been completely understood. The experimental investigation on the nano scale is constrained by high costs, long time and large uncertainty of measurements. With the increase of computer power, molecular dynamics (MD) simulation has become an important complementary tool that may appropriate help to interpret nano-cutting experimental data.

Maekawa and Itoh [1] have developed a MD simulation model to investigate friction and tool wear in nano-scale machining. They revealed that as the bonding energy at the interface increases, both chip thickness and contact length between the chip and tool become larger, leading to increases in average cutting forces and temperature. Fang et al. [2] carried out the scratching experiments and MD simulations to study the mechanical characteristics of an Atomic Force Microscope (AFM) based lithography process. They found that rougher surfaces are produced at larger loads and surface roughness can be improved by adjusting the scratching feed to a small value. The simulated results showed that the wear depth of gold decreased as the scratching velocity was increased and the temperature was decreased. Komanduri et al. [3, 4] used a MD simulation model to investigate the effect of tool geometry in nano-metric cutting of single crystal aluminum. They reported that an increase in the magnitudes of forces, the ratio of the thrust to the cutting force, specific energy and sub-surface deformation were observed with increase in the negative rake. Unlike in the conventional cutting, the edge radius plays an important role in nano-metric cutting.

As the cutting proceeds, in real condition in nano-scale, chip formation mechanism is triggered by initial material defects such as precipitates, dislocations, grain boundaries, voids etc. [5]. Consequently, plastic deformation in cutting process leads to a multiplication of defects, a considerable number of which will remain in the sub-surface layer and will thus influence the surface integrity [3]. Most MD research papers, reported in the literature, only address perfect materials such as pure single crystal metals with no defects. Although these MD simulations offer many advantages, it is important to bear in mind that the defects in base material of workpiece have a significant effect on approximation of true cutting characteristics. So, to understand the effect of material defects, there is a need to compare pure single crystal workpiece with a workpiece containing defects in nano-machining process.

Some studies were done on the effects of material defect using uni-axial loading and nano-indentation. Chang investigated the mechanical properties of nano-scale copper with vacancies under static and cyclic loading using MD simulation [6]. The Young's modulus decreases with increasing vacancy fraction, and it also decreases with increasing temperature. From the fatigue test, fatigue limit of the nano-scale copper increases with decreasing vacancy fraction and increasing temperature, when the material is under a less-than-critical value of applied stress. Kim et al. and Jang et al. studied the interaction of lattice dislocations with a grain boundary during nano-indentation simulation [7, 8]. The results showed that the dislocation transmitted across the grain boundary during nano-indentation and left a step in the boundary plane. Burgers vector analysis suggested that a partial dislocation in first grain merged into the grain boundary and it was dissociated into another partial dislocation in second grain and a grain boundary dislocation, introducing a step in the grain boundary. Tan and Jeng simulated the nano-indentation on copper with a void [9]. They observed that the void disappears when the indentation depth is sufficiently large. Also, stress concentration is observed at the internal surface of the void in all simulation cases. In addition, results indicate that the presence of a void has a significant influence on the nano-hardness extracted from the nano-indentation tests.

Although it is well known that even the purest real material contains a large number of defects within its crystal structure, it is impractical to simulate all of these defects at the atomic level. Hence, this study analyzes the interesting and computationally more straightforward problem of how the nano-cutting of single crystal copper is influenced by the presence of a single void.

2 MD Simulation Model

Figure 1 shows a schematic of the MD simulation model. The dimension of the workpiece is 20(a), 30(a) and 30(a) along x, y and z directions respectively that (a) is lattice constant.

The nanomachining is done along the [010] direction of the (001) surface of the workpiece. The material of the workpiece is a single crystal copper that has a Face Centered Cubic (FCC) lattice with 3.62 Å lattice constant (a). The outer three layers of atoms on the bottom side and left side of the substrate are fixed (magenta color). The periodic boundary condition is applied along the x direction and the top surface (machined surface) is free. Also, three layers of atoms on the bottom side and left side of the workpiece, next to the fixed zone, are considered as a thermostatic boundary which temperature was maintained at 300(°K) with a standard nose thermostatic algorithm (green color). The diamond tool has an edge with 5(a) radius. The rake and flank angles of the tool are 15° and 10° respectively. Also, the tool is assumed to be rigid (blue color). All atoms in the tool move together during nanocutting. Their relative distances are fixed.

Details of the computational model are summarized in Table 1. Two cases of workpiece have been simulated in this study. In the first case (Case I), the workpiece is a pure single crystal that is assumed in most researches. In the second

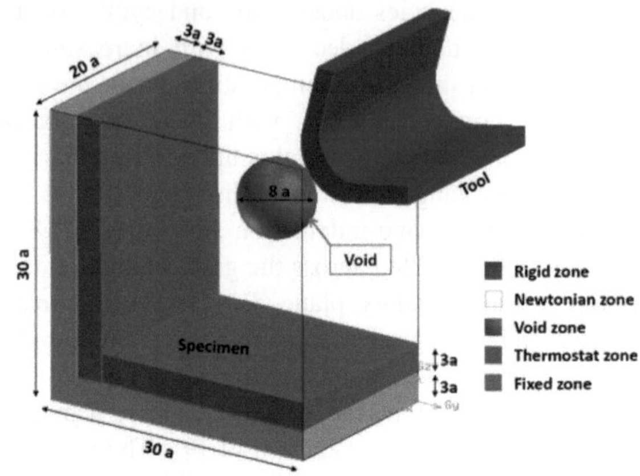

Fig. 1 Schematic of the MD model with a void

Table 1 Computational parameters used in the MD simulation of nanocutting process

Type of Workpiece	Case I: Pure Single Crystal	Case II: Single Crystal with a Void
Potential type	Embedded atom method (EAM) for Cu–Cu; Cut off distance = 5 (Å); Morse for Cu–C; Cut off distance = 6.5 (Å)	
Workpiece material	Single crystal of copper; FCC lattice; lattice constant (a) = 3.62 (Å)	
Tool properties	Rake angle: 15°; Clearance angle: 10°; Tool edge radius = 5 (a)	
Cutting parameters	Cutting depth = 10 (Å); Cutting speed = 100 (m/s); Directions: [010] on (001) surface	
Simulation parameter	Time step = 1 (fs); Boundary temperature = 300 (°K)	
Defect properties	No defect	Spherical void Void diameter = 8 (a) Void position = [10 (a), −10 (a), 22 (a)]
Number of atoms	4553 Rigid tool atom 58320 Newtonian atoms 14420 Thermostatic atoms 16100 Fixed atoms	4553 Rigid tool atom 57263 Newtonian atoms 14420 Thermostatic atoms 16100 Fixed atoms

case (Case II), the workpiece is a single crystal with a specified void. The void has a spherical shape with 8(a) diameter that is placed under the machined surface and there is no atom in void zone (yellow color in Fig. 1).The model consists of 144807 atoms for Case I and 143750 atoms for Case II.

In the MD simulation, the motion of atoms is governed by Newton's second law, which is integrated by the Velocity-Verlet algorithm [10]. The force exerting on each atom is calculated by the summation of interaction from the neighbor atoms based on an inter-atomic potential. In this chapter, pair-wise Morse potential is used for interaction of tool atoms with workpiece atoms (C–Cu bonding). Also, having more accuracy in atoms interaction of the metallic workpiece, Embedded Atom Method (EAM) potential is used for the Cu–Cu interaction. The EAM potential is based on molecular dynamics principles developed initially by Foiles et al. [11]. EAM allows the calculation of the thermodynamic forces and stress tensors for the atoms in the lattice based on the notion of embedded energy.

The total energy of an atomic system (U_{metal}) is calculated by summing the individual embedding energy (F) of each atom i in the atomic aggregate with short-range repulsive potential (V_{ij}) energy, as shown in Eq. (1).

$$U_{metal} = \frac{1}{2}\sum_{i=1}^{N}\sum_{j\neq i}^{N} V_{ij}(r_{ij}) + \sum_{i=1}^{N} F(\rho_i) \quad (1)$$

The embedding energy is the energy to place an atom i in a host electron density (ρ_i) at the site of that atom. The electron density (ρ_i) at any point is well described by a sum of the individual atomic densities. EAM potential has been very successful in modeling the elastic properties, defect formation energies and fracture mechanisms of various bulk metals [11]. The EAM has also been widely applied to surface properties as well, successfully describing surface energies, surface reconstructions and adsorption on metal surfaces.

3 Results and Discussions

To validate MD simulation, a comparison is made between results on pure workpiece and other MD simulations and also experimental results of Ultra Precision Machining (UPM) in the terms of specific cutting energy versus nominal depth of cut normalized by the tool edge radius that shows a good agreement [12].

3.1 Tool Forces During the Nano-cutting Process

Figures 2 and 3 show the behavior of tool forces during the ploughing process in pure and defected workpieces.

Although the cutting force is not changed significantly, the trust force is decreased dramatically. As shown in Fig. 3, the trust force reduction is intensified after 2 nm tool travel. At this moment, the tool is positioned on the top of the void. So, the inner wall of the void is deformed by tool pressure. Table 2 shows the comparison of pure and defected workpiece on averaged tool forces. Results show that the cutting force and trust force are declined about 7 and 27% respectively in workpiece with a specific void. Also, the resultant force is decreased about 17%.

3.2 Distribution of Hydrostatic Stress

Figure 4 compares the hydrostatic pressure in pure workpiece and workpiece with a void at 3.5 (nm) ploughing distance. The area that is highlighted with dashed line indicates the deformed space of the spherical void during the nano-cutting process.

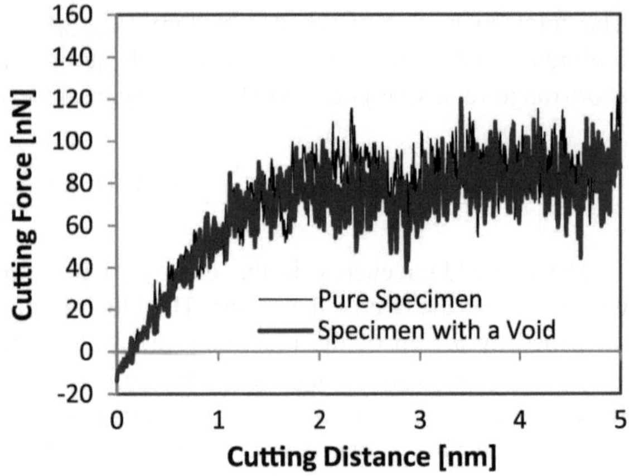

Fig. 2 Behavior of cutting force during the machining process in pure and defected workpiece

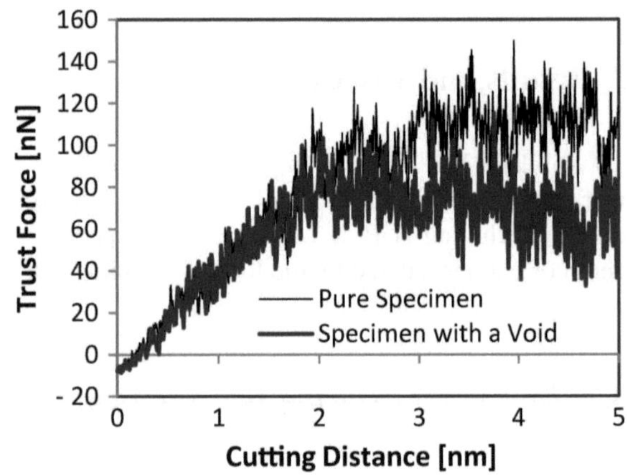

Fig. 3 Behavior of trust force during the machining process in pure and defected workpiece

Table 2 Comparison of pure workpiece with a known defected workpiece on averaged tool forces

	Width-direct Force (Fx) [nN]	Cutting Force (Fy) [nN]	Trust Force (Fz) [nN]	Resultant Force [nN]
Pure workpiece	1.0	71.3	79.7	108.6
Workpiece with a void	−1.0	66.4	58.5	90.1
Difference (%)	–	7%	27%	17%

During machining, material ahead of the tool is subjected to great compressive hydrostatic stress that leads to plastic flow when the flow criterion is fulfilled. During the penetration of tool edge, material separated on a specific location on the rounded tool edge known as the stagnation point. Under such a condition, material above the stagnation point was separated from the workpiece as chip. It can be

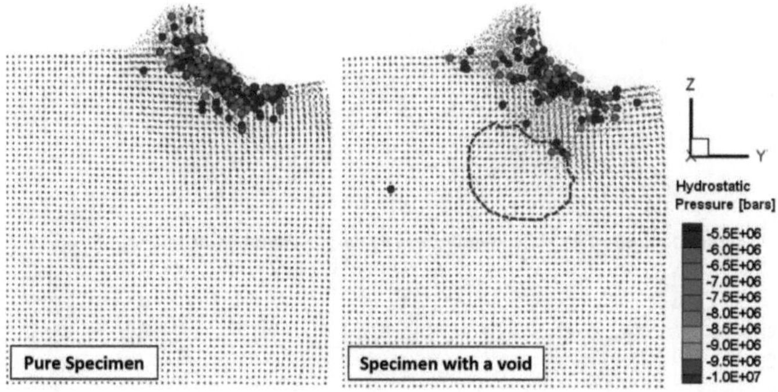

Fig. 4 Comparison of hydrostatic stress in pure and defected workpiece [cutting distance = 3.5 (nm)]

seen that in case of a pure workpiece, the hydrostatic pressure is distributed uniformly only in front of the tool tip. But in the case of a workpiece with a void, the number of pressurized atoms decreases which leads to tool force reduction (Table 2). Also, the high pressurized atoms are distributed in a bigger space that is confined between the tool and the void.

3.3 Atomic Displacement in the Nano-machining Process

Figure 5 indicates the atomic deformation in pure and defected workpiece during the nano-cutting. To better understand the deformation mechanism, the values lower than 2 (Å) are eliminated. In the case of machining with defect-less material, the large deformation zone is confined only near the machined surface. But in defected specimen, this zone is started from tool tip and continued to void surface. For example with 10 (Å) depth of cut at 5 (nm) cutting distance, the large deformation zone depth is about 22 (Å) in pure workpiece compared with 30 (Å) in defected workpiece.

3.4 Mechanism of Void Crumpling

To check the deformation mechanism of the workpiece with a void, centrosymmetry parameter [13] is used, as shown in Fig. 6. To highlight the surface atoms, centrosymmetry parameter is plotted in range of upper 12. It can be seen that during the nano-machining process, the upper internal surface of void is folded due to high hydrostatic pressure. In this case, the spherical void is shrunk about 41% of its volume at 5 (nm) cutting distance.

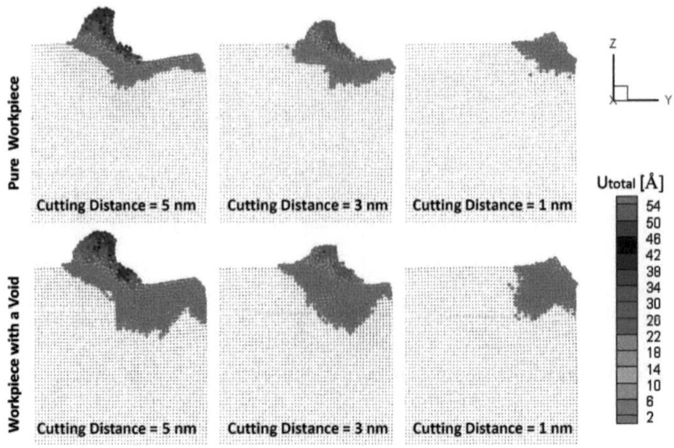

Fig. 5 Atomic displacement in pure and defected workpiece during the ploughing process

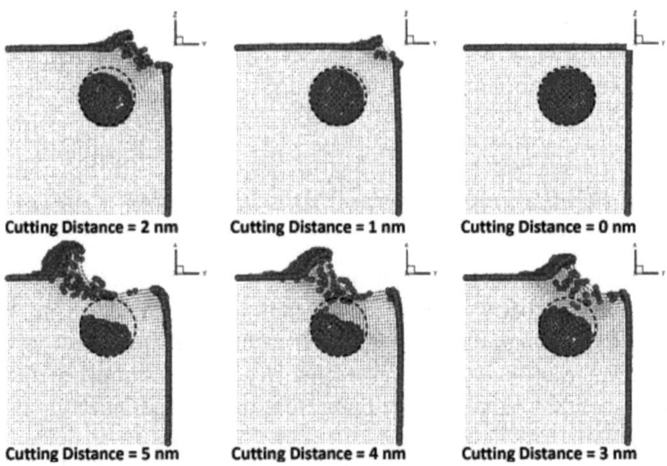

Fig. 6 Mechanism of void crumpling in nanomachining process

4 Conclusion

The three-dimensional MD simulations for nano-cutting process were performed to evaluate void defect influenced on the nature of chip formation, void deformation and tool forces. From the current study, the specific conclusions were deduced that is summarized as follows:

- Void inclusion in workpiece can influence on tool forces. The cutting and trust force in a workpiece with a specific void are reduced, compared with pure material, about 7 and 27% respectively. Also, resultant force decreased about 17%.

- Results show that in a pure workpiece, the hydrostatic pressure area is distributed uniformly in front of the tool tip. However in a defected workpiece, the area of hydrostatic pressure is expanded while the number of high pressurized atoms is decreased. This can affect on tool forces.
- The plastic zone in a defected workpiece is larger than that of a pure workpiece. That can affect surface integrity. With a specific void, the depth of plastic zone is increased from 22 (Å) to 30 (Å) compared with a pure workpiece using 10 (Å) depth of cut.
- Because of high compressive hydrostatic pressure during the nano-cutting, the void is crumpled. In this case, the shrinkage of void is about 41% of its volume.

References

1. Maekawa, K., Itoh, A.: Friction and tool wear in nano-scale machining-a molecular dynamics approach. Wear **188** (1995)
2. Fang, F.Z., Wu, H., Liu, Y.C.: Modelling and experimental investigation on nanometric cutting of monocrystalline silicon. Int. J. Mach. Tools. Manuf. **45**, 1681–1686 (2005)
3. Komanduri, R., Raff, L.M.: A review on the molecular dynamics simulation of machining at the atomic scale. Proc. Inst. Mech. Eng. **215**, 1639–1672 (2001). Part B
4. Komanduri, R., Chandrasekaran, N., Raff, L.M.: Effect of tool geometry in nanometric cutting: an MD simulation approach. Wear **219** (1998)
5. Sandy, T.S.: Effect of crystallographic orientation on material behavior in ultraprecision diamond turning of single crystals. Ph.D. thesis, Hong Kong Polytechnic University, October (2000)
6. Chang, W.J.: Molecular dynamics study of mechanical properties of nanoscale copper with vacancies under static and cyclic loading. J. Microelectron. Eng. **65**, 239–246 (2003)
7. Kim, K.J., Yoon, J.H., Cho, M.H., Jang, H.: Molecular dynamics simulation of dislocation behavior during nanoindentation on a bicrystal with a $\Sigma = 5$ (210) grain boundary. Int. J. Mater. Lett. **60**, 3367–3372 (2006)
8. Jang, H., Farkas, D.: Interaction of lattice dislocations with a grain boundary during nanoindentation simulation. Int. J. Mater. Lett. **61**, 868–871 (2007)
9. Tan, Ch.M., Jeng, Y.R.: Computer simulations of nanoindentation on Cu (111) with a void. Int. J. Sol. Struct. **46**, 1884–1889 (2009)
10. Haile, J.M.: Molecular Dynamics Simulation: Elementary Methods. Wiley Pub, (1992)
11. Foiles, S.M., Daw, M.S., Baskes, M.I.: Embedded-atom-method functions for the fcc metals Cu, Ag, Au, Ni, Pd, Pt, and their alloys. Phys. Rev. B **33**, 12 (1986)
12. Hosseini, S.V., Vahdati, M.: Effect of tool nose radius on nano-machining process by molecular dynamics simulation. In: 6th International Conference on Diffusion in Solids and Liquids-DSL2010, Paris-France (2010)
13. Kelchner, C.L., Plimpton, S.J., Hamilton, J.C.: Dislocation nucleation and defect structure during surface indentation. Phys. Rev. Part B **58**(17), 11085–11088 (1998)

Dynamic Modeling and Control Study of the NAO Biped Robot with Improved Trajectory Planning

E. Hashemi and M. Ghaffari Jadidi

Abstract Motion study of bipedal robots necessitates correct solutions of the forward and inverse kinematics with optimized and fast closed form computations which justifies an accurate kinematic model. On the other hand, dynamic modeling and stability analysis are essential for control study of humanoid robots to reach robust walk. This chapter is focused on dynamic modeling of the Nao humanoid robot, made by Aldebaran Co., in the RoboCup standard platform league. Moreover, trajectory approximation with a cubic Spline and kinematic analysis are described in brief here in this chapter. Main constraints such as inertial forces and joint angles for the given position and nominal conditions are simulated, mathematically described, and verified through experimental results from the real robot sensory data. The above mentioned modifications on the solution together with the dedication of other physical properties in dynamic modeling results in more precise acceleration and torque values as it is concluded in this work.

Keywords Bipedal locomotion · Dynamic simulation · Inverse kinematics · Motion planning · Trajectory approximation

E. Hashemi (✉)
Faculty of Industrial and Mechanical Engineering,
Islamic Azad University, Qazvin Branch, Qazvin, Iran
e-mail: e.hashemi@qiau.ac.ir

M. Ghaffari Jadidi
Faculty of Electrical, Computer and IT Engineering,
Islamic Azad University, Qazvin Branch, Qazvin, Iran
e-mail: m.ghaffari@qiau.ac.ir

A. Öchsner et al. (eds.), *Materials with Complex Behaviour II*,
Advanced Structured Materials 16, DOI: 10.1007/978-3-642-22700-4_42,

1 Introduction

Bipedal locomotion has been a topic of great attention in a various researches performed on legged robots and is probably the most suitable method for robots to execute assigned maneuvers in a real environment with various obstacle conditions and geometry [1–3]. Extensive studies have been conducted on various control methods of bipedal walking and stability, and now biped robots are capable of walking with a certain amount of stability [4–9]. Combined Forward and Inverse Kinematics models are utilized to specify the reliable method to control motion and preserve stability as discussed in [10–12]. The humanoid bipedal locomotion needs reasonable solutions of the inverse kinematics and localization problems with optimized computations as illustrated in [13]. Since the end effector configurations and its exact locations are related to the above mentioned joint parameters with nonlinear characteristics, inverse kinematics problems are usually complicated. It is essential to do a transformation between the Cartesian end effector orientation and location vector and the corresponding vector of joint angular position which leads to determination of actuating signals and driver torque values in the dynamic model. Trajectory approximation, kinematic, and dynamic modeling are described in the following sections of this material.

2 NAO Walking Trajectory Generation

Kinematic and dynamic study of linkage systems are key and fundamental issues for trajectory control, motion planning and locomotion modeling which are discussed in [4, 5]. In order to attain a stable and reliable walk in different ground conditions such as regions containing obstacles, rough terrain and slopes, it is required to maintain the stability of the biped robot and adapt to such environmental conditions with a suitable foot motion and a smooth hip shift which justifies comprehensive study of trajectory generation and transformation to desirable joint angles and develop some specific controllers such as torque control methods.

Zero Moment Point (ZMP) stability criteria is introduced for the stability assessment of the bipedal robots during walking [14] and consideration of this criterion for trajectory generation has an essential role to reach a dependable trajectory for both foot and hip joints as studied in [15–17]. The presented method in [15] uses a stability margin to achieve a certain degree of stability instead of marginal stability. Walking cycles falls into two main stages including the single-support phase (SSP) and the double support phase (DSP). In the SSP, one foot is stationary on the ground and the other foot swings from the rear to the front, but both feet are in contact with the ground in DSP. DSP begins with the heel of the forward foot touching the ground, and ends with the toe of the rear foot taking off the ground [18]. The walking pattern can therefore be denoted uniquely by both foot and the hip trajectories as presented in [19, 20]. If foot and hip trajectories are identified, all joint trajectories of the robot will be determined by the kinematic constraints and geometrical relations.

Trajectories considered for dynamic modeling are derived with the assumptions of 10 cm/s walking velocity and a 5 cm step length. Implementing velocity and kinematic constraints specified in Eq. A.1, Appendix brings about Eqs. 1 and 2 which comprehensively offered a cubic Spline for foot movement in *sagittal* plane in x and z directions.

$$x_f(t) = \begin{cases} -L_{step}, & t_1 < t < t_2 \\ -1.2848t^3 + 1.4202t^2 - 0.1938t - 0.0435, & t_2 < t < t_3 \\ -1.5144t^3 + 1.5855t^2 - 0.2334t - 0.0404, & t_3 < t < t_4 \\ L_{step}, & t_4 < t < T_{cycle} \end{cases} \quad (1)$$

$$z_f(t) = \begin{cases} FootHeight, & \\ -19.724t^3 + 10.2116t^2 - 1.2359t + 0.0412, & t_1 < t < t_2 \\ & t_2 < t < t_3 \\ 8.1169t^3 - 9.8338t^2 + 3.575t - 0.3437, & t_3 < t < t_4 \\ FootHeight, & t_4 < t < T_{cycle} \end{cases} \quad (2)$$

Torso trajectory with execution of the cubic Spline could be written as bellow in which coefficients are calculated by the method presented in [18–20] and mentioned in Eq. A.2, Appendix.

$$x_t(t) = \begin{cases} 1.4687t^3 - 0.1762t^2 + 0.0554t - 0.0228 & t_1 < t < t_2 \\ -0.3672t^3 + 0.3745t^2 + 0.0003t - 0.0209 & t_2 < t < t_4 \end{cases} \quad (3)$$

$$y_t(t) = \begin{cases} 1.3782t^3 - 0.0292t^2 - 0.239t, & t_1 < t < t_3 \\ -1.2079t^3 + 1.8328t^2 - 0.6859t + 0.0358, & t_3 < t < t_4 \\ -1.3768t^3 + 2.0862t^2 - 0.8126t + 0.0569, & t_4 < t < t_1 + T_{step} \\ 1.2066t^3 - 3.6491t^2 + 3.4353t - 0.9928, & t_1 + T_{step} < t < T_{cycle} \end{cases} \quad (4)$$

$$z_t(t) = \begin{cases} -0.8413t^3 + 0.2436t^2 + 0.0317t + 0.2 & t_1 < t < t_3 \\ 0.7766t^3 - 0.9214t^2 + 0.3113t + 0.1776 & t_3 < t < t_4 \end{cases} \quad (5)$$

Equations 3–5 are tailor made for a Nao robot's step length and velocity scenario and modified for the robot's torso within the above mentioned periods.

3 Kinematic and Dynamic Modeling

This section deals with the model derivation for Nao robot of the MRL team. The purpose of the model is to give the necessary insight of the system to design, test and simulate controllers for the above mentioned biped robot. The models developed are a kinematic model to determine joint angles, link positions, velocities, and acceleration for different torso and foot trajectories and a dynamic model to estimate dynamic behavior of the links and related joint torques.

3.1 Inverse Kinematics

The kinematic model concerns relations between different joints on the robot and the position of the individual links. This is done by transforming from joint space to Cartesian space [21]. With this transformation, the position of the links' CoM is calculated from the given rotation of the joints. Required data for the forward kinematic model are the system phase and the angular position of the joints θ_n. Kinematic model output is the global position Gin the global Cartesian coordinates. Mechanical and geometrical data are already extracted from the Nao physical specifications provided by Aldebaran Co. and listed in Appendix. The calculated data for the link vectors are listed in Table 3 together with the accessible CoM vectors.

Denavit–Hartenberg (DH) method and an open loop serial chain system is used in [20–24] for determination of all the joint rotations. The first rotation is assumed in the supporting foot and the last rotation is considered in the non supporting foot. The frames are aligned with the global reference frame in case all the joints are in their zero state because no rotations are needed to describe the system when all the angles of the joints are zero. Furthermore, the position of the ground frame is fixed to the initial frame $\{0\}$ such that the initial frames always contain the global position and all frames have the same direction when the angles are zero. As mentioned above, the kinematic model estimates the global velocity and acceleration vectors as explained in the following paragraphs. The global position of the links CoM, $\{G_n\}$ can be expressed as the following relation [23]:

$$\{G_n\} = \{G_{m,n-1}\} + {}_{n-1}^{\;\;0}R\{b_n\} \tag{6}$$

In which $\{b_n\}$ represents the CoM local position of link n. Position of joint $n-1$ in the global coordinates, $\{G_{m,n-1}\}$ is defined as:

$$\{G_{m,n-1}\} = {}_{n-2}^{\;\;0}R\{a_{n-1}\} + \{G_{m,n-2}\} \tag{7}$$

Local position vector of joints is shown by $\{a_i\}$ in Eq. 7. Controller design and development justifies making use of precise inverse kinematics and dynamic models to satisfy stability and agility requirements in biped robots such as Nao.

Nao robot has 21 degrees of freedom, including 6 in each leg introduced as ankle roll, ankle pitch, knee pitch, hip pitch, hip roll, and hip yaw-pitch. Yaw-pitch joints of the hips are physically bound and driven with one servo motor. The purpose of the inverse kinematic model is to determine the joint angles which yield a specific position of the limbs in Cartesian space. Inverse kinematics model requires expected positions of links which could be attainable by torso and foot trajectories and calculates joint angles corresponding to the determined trajectories. Numerical and closed-form solutions are widely used to solve the inverse kinematics problems.

Goldenberg et al. [25] presented an iterative numerical solution to transform the desired position of a limb into joint angles by solving the pseudo inverse Jacobian to reach accurate results. One of the drawbacks of this method is this iterative method is a time consuming process for both legs solution and does not satisfy the time response requirements of the control loop model [22].

Closed-form solution of the non-linear equations of the robot manipulators with limited degrees of freedom is introduced as an alternative method of solving inverse kinematics equations. This method is also employed to carry out the solution of MRL team Nao robot's kinematic equations with utilization of Pieper's solution [23] for developing transformation and rotation matrices. This method and its modification is described in this section as an approach for solving related equations which enables the approach to compute the same hip Yaw-Pitch joint variable for both legs.

DH frame assignment and related coordinate sequences are shown in Fig. 1, and associated parameters are mentioned in Table 1. DH parameters for the right leg is assumed the same as the left one except α_6 which is $\pi/4$.

Homogeneous transformation ${}^0_6T$ which describes position and orientation of frame $\{6\}$ relative to frame $\{0\}$ is introduced as Eq. 8

$$ {}^0_6T = \begin{bmatrix} {}^0_6R & {}^0p_{6ORG} \\ 0 & 1 \end{bmatrix} \tag{8} $$

Position vector ${}^0p_{6ORG}$ expresses coordination of frame $\{6\}$ origin relative to frame $\{0\}$ in the homogeneous transformation matrix. Joint angles $\theta_1, \theta_2, \theta_3$ could also be computed by using ${}^0p_{6ORG}$, then the last three joint angles of each leg, $\theta_4, \theta_5, \theta_6$ are approximated using ${}^0_6R$ which is rotation matrix explaining orientation of frame $\{6\}$ relative to $\{0\}$. Since origins of frames $\{4\}$, $\{5\}$, $\{6\}$ are coincident at a point, ${}^0p_{6ORG}$ is equal to ${}^0p_{4ORG}$ and mathematically shown in Eq. 9. Calculation of ${}^0p_{6ORG}$ is vital because it results in determination of $\theta_1, \theta_2, \theta_3$ in Pieper's solution.

$$ \begin{aligned} {}^0p_{6ORG} = {}^0p_{4ORG} &= {}^0_3T \times {}^3p_{4ORG} = {}^0_1T \times {}^1_2T \times \begin{bmatrix} f_1(\theta_3) \\ f_2(\theta_3) \\ f_3(\theta_3) \\ 1 \end{bmatrix} \\ &= \begin{bmatrix} c_1g_1(\theta_2,\theta_3) - s_1g_2(\theta_2,\theta_3) \\ s_1g_1(\theta_2,\theta_3) + c_1g_2(\theta_2,\theta_3) \\ g_3(\theta_2,\theta_3) \\ 1 \end{bmatrix} = \begin{bmatrix} x \\ y \\ z \\ 1 \end{bmatrix} \end{aligned} \tag{9} $$

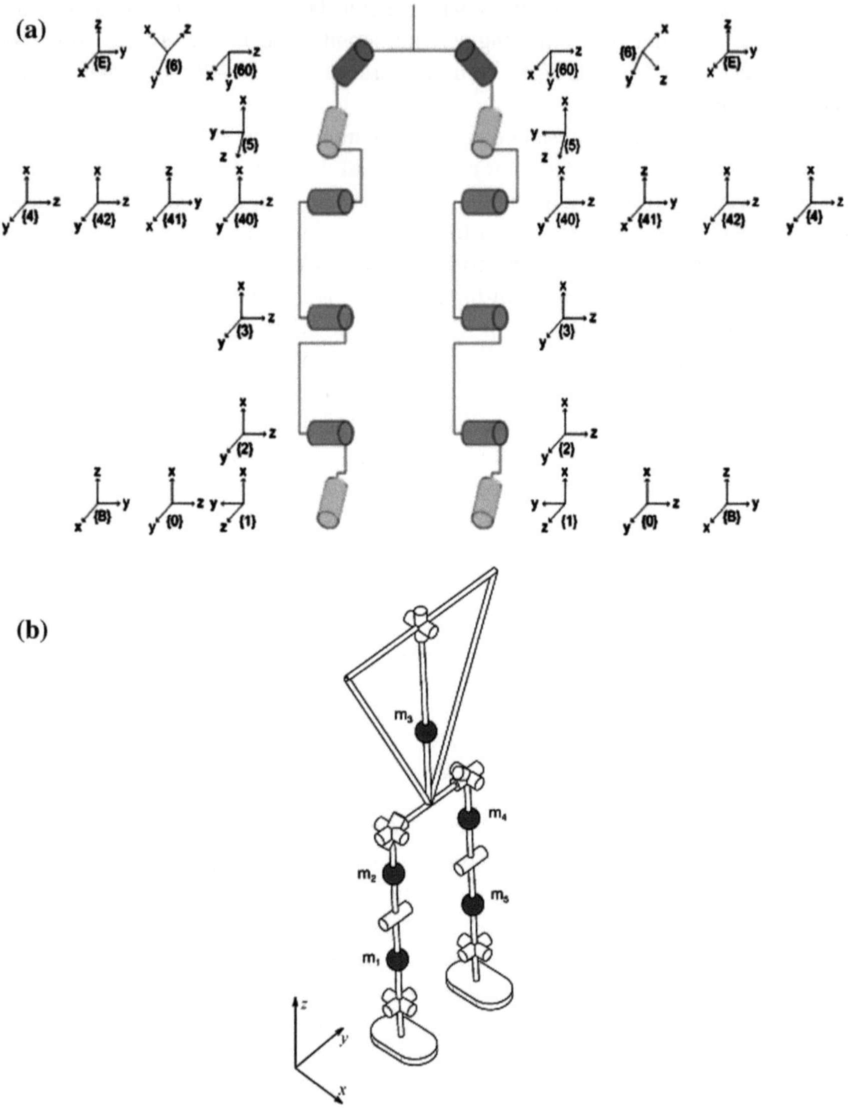

Fig. 1 **a** DH representation of legs for the Nao robot kinematic simulation; **b** mass and geometrical model for the Nao robot

In which c_1 and s_1 represent $Cos\,\theta_1$ and $Sin\,\theta_1$, and approximation functions f and g are given in the Appendix section of this chapter. The first three joint angles θ_1, θ_2 and θ_3 are estimated by position vector components x,y and z and solving Eqs. 10 and 11 simultaneously.

$$z = g_3(\theta_2, \theta_3) \tag{10}$$

Table 1 Dedicated DH parameter values for the left leg

Frame (joint)	α_i	a_i	d_i	θ_i
1	$-\pi/2$	0	0	θ_1
2	$\pi/2$	0	0	θ_2
3	0	Tibia length	0	θ_3
40	0	Thigh length	0	0
4	0	0	0	θ_4
5	$-\pi/2$	0	0	θ_5
6	$-\pi/4$	0	0	$\theta_6 - 3\pi/4$
41		$Rot_z(\pi/2) \times Rot_x(\pi/2)$		
42		$Rot_x(-\pi/2) \times Rot_z(-\pi/2)$		
60		$Rot_x(\pi/2) \times Rot_z(\pi/2)$		

$$r = x^2 + y^2 + z^2 = [g_1(\theta_2, \theta_3)]^2 + [g_2(\theta_2, \theta_3)]^2 + [g_3(\theta_2, \theta_3)]^2 \tag{11}$$

Variables a_1 and $s\alpha_1$ are influencing parameters in solution of Eqs. 10 and 11 and can be extracted from the geometric conditions of the left or right legs. The first joint angle θ_1 is estimated after analytic solution of the second and the third joint angles and replacing these two values in Eq. 9. The following paragraph presents calculation of θ_4, θ_5 and θ_6 in Pieper's solution using ${}^0_6R$ and the estimated first three joint angles. Computation of θ_1, θ_2 and θ_3 leads to calculation of ${}^0_3R$ using Eq. 12 which simply describes the relation between rotation matrices in frames $\{0\}$, $\{3\}$ and $\{6\}$.

$${}^3_6R = \left({}^0_3R\right)^{-1} \times {}^0_6R \tag{12}$$

The cubic spline described in the previous section to generate trajectories for steady state walk is utilized to present the kinematic model. Validation of formulas for both left and right legs is performed through getting joints' values in walking state and generating the hip trajectories relative to the foot with forward kinematic model. Inverse kinematic module employs forward kinematic outputs as a trajectory and produces all joint angles according to the particular trajectory input. This verification is straightforward because the calculated joint angles are expected to be the same as the input ones for the forward kinematic module. Complete description of kinematic equations and model validation is available in [20] which comprehensively focused on joint parameters and closed form solution of inverse kinematics.

3.2 Dynamic Model

Position of estimated center of mass which represents the links, drivers and connection effects shall be calculated for the next stages of the dynamic simulation.

Table 2 Joints axes of rotation

Name	Axis	Joint
Roll	X	1,5,8,12
Pitch	Y	2,34,9,10,11
Yaw-Pitch	Y–Z	6,7

This position will be computed from the newly introduced homogenous transformation matrix in Eq. 13

$$T_i = \begin{bmatrix} R_{i,3\times3} & L_{i,3\times1} \\ O_{1\times3} & 1 \end{bmatrix} \tag{13}$$

R_i s are the basic rotation matrix for all 12 joints about three axes and defined in Eq. 14. In addition, L_i s are joints' vectors as shown in Table 3, Appendix. Basic rotation matrix for three-dimensional transformation in xyz coordinates are as follows:

$$R_{x,\theta} = \begin{bmatrix} 1 & 0 & 0 \\ 0 & \cos\theta & -\sin\theta \\ 0 & \sin\theta & \cos\theta \end{bmatrix}, \quad R_{y,\theta} = \begin{bmatrix} \cos\theta & 0 & \sin\theta \\ 0 & 1 & 0 \\ -\sin\theta & 0 & \cos\theta \end{bmatrix},$$
$$R_{z,\theta} = \begin{bmatrix} \cos\theta & -\sin\theta & 0 \\ \sin\theta & \cos\theta & 0 \\ 0 & 0 & 1 \end{bmatrix} \tag{14}$$

Table 2 represents all 12 joint numbers and their axis of rotation are as depicted in Fig. 1a.

Global position of links' center of mass vectors are defined from the successive multiplication of transformation matrix in Eq. 13 and the local position vectors tabulated in Table 3, Appendix I as:

$$P_{mj} = \left(\prod_{i=1}^{n} T_i\right) L_{mj} \tag{15}$$

$$P_{mi} = \langle x_{mi} \quad y_{mi} \quad z_{mi} \quad 1 \rangle^T \tag{16}$$

where i starts from 1 and ends to the number of joint n before mass i. A dynamic model is developed to determine the joints' angular acceleration for which the mass and diagonal elements of the inertia tensors of each Nao's link are listed in Table 4 and schematically illustrated in Fig. 1b. A movement study of the MRL team Nao robot and ZMP estimation will be done after determination of joint values. The presented dynamic model is for the SSP and will be used as a fundamental calculation for the DSP which is the authors' next research subjects. SSP dynamic equations are derived by the assumption of the Lagrange-d'Alembert Eq. 17 which relates the external force F_i on the Nao links for the single support phase and variations of Lagrangian L relative to the system state r_i.

Table 3 Link vectors and CoM vectors extracted from the Nao physical specifications provided by Aldebaran Co. and geometrical modeling by MRL-Nao team

Center of mass vector (mm)
$L_{m1} = [3.66, -1.52, 38.72]^T$
$L_{m2} = [-3.68, 0.03, 66.79]^T$
$L_{m3} = [-4.80, 50.06, 127.27]^T$
$L_{m4} = [-3.38, -0.03, -33.21]^T$
$L_{m5} = [3.66, 1.52, -64.03]^T$
Joint vector (mm)
$L_1 = [0,0,0]^T$
$L_2 = [0,0,0]^T$
$L_3 = [0,0,102.75]^T$
$L_4 = [0,0,100]^T$
$L_5 = [0,0,0]^T$
$L_6 = [0,0,0]^T$
$L_7 = [0,100,0]^T$
$L_8 = [0,0,0]^T$
$L_9 = [0,0,0]^T$
$L_{10} = [0,0,-100]^T$
$L_{11} = [0,0,-102.75]^T$
$L_{12} = [0,0,0]^T$

Table 4 The mass and diagonal elements of the inertia tensors of each Nao's link

Mass (g)	Inertia (kg × m^2)
$m_1 = 435.98$	$[I_{xx1}, I_{yy1}, I_{zz1}] = [0.0012, 0.0012, 0.0006]$
$m_2 = 605.72$	$[I_{xx2}, I_{yy2}, I_{zz2}] = [0.0018, 0.0018, 0.0010]$
$m_3 = 1026.28$	$[I_{xx3}, I_{yy3}, I_{zz3}] = [0.0049, 0.0047, 0.0016]$
$m_4 = 605.72$	$[I_{xx4}, I_{yy4}, I_{zz4}] = [0.0018, 0.0018, 0.0010]$
$m_6 = 435.98$	$[I_{xx5}, I_{yy5}, I_{zz5}] = [0.0012, 0.0012, 0.0006]$

$$\frac{d}{dt}\left(\frac{\partial L}{\partial \dot{r}_{sn}}\right) - \frac{\partial L}{\partial r_{sn}} = F_n \tag{17}$$

The state vector $\{r_s\}$ consists of the position vectors and the angles as:

$$\{r_s\} = \langle x_1\, y_1\, z_1 \cdots x_5\, y_5\, z_5\, \theta_1 \cdots \theta_{12}\rangle^T \tag{18}$$

The Lagrangian L is defined as the difference between a system's kinetic energy and potential energy and could be written as follows:

$$L = \sum_{n=1}^{Number\ of\ Links} \frac{1}{2} m_n(\dot{x}_n^2 + \dot{y}_n^2 + \dot{z}_n^2 - 2gz_n) + \frac{1}{2}\vec{\omega}_n^T I_n \vec{\omega}_n \tag{19}$$

In which m_n is the mass of nth link, I_n is the inertia tensor of nth link around base frame, and $\vec{\omega}_n$ is the angular velocity vector of the link n around base frame. j is the total number of link masses which is 5. Partial differentiation of Eq. 19 with regards to $\vec{r}_s$ and $\dot{\vec{r}}_s$ yields:

$$\frac{\partial L}{\partial \vec{r}_s} = \left\langle \frac{\partial L}{\partial x_1}\, \frac{\partial L}{\partial y_1}\, \frac{\partial L}{\partial z_1} \cdots \frac{\partial L}{\partial x_5}\, \frac{\partial L}{\partial y_5}\, \frac{\partial L}{\partial z_5}\, \frac{\partial L}{\partial \theta_1} \cdots \frac{\partial L}{\partial \theta_{12}} \right\rangle^T \tag{20}$$

$$\frac{\partial L}{\partial \dot{\vec{r}}_s} = \left\langle \frac{\partial L}{\partial \dot{x}_1} \frac{\partial L}{\partial \dot{y}_1} \frac{\partial L}{\partial \dot{z}_1} \cdots \frac{\partial L}{\partial \dot{x}_5} \frac{\partial L}{\partial \dot{y}_5} \frac{\partial L}{\partial \dot{z}_5} \frac{\partial L}{\partial \dot{\theta}_1} \cdots \frac{\partial L}{\partial \dot{\theta}_{12}} \right\rangle^T \tag{21}$$

Differentiating Eq. 21 results in the determination of the left side of Eq. 17. This vector contains all links' linear and angular accelerations.

The global angular velocities $\vec{\omega}_n$ are calculated in Eq. 22 as described in [23] with consideration of the previous link position and rotation matrix. $\{\zeta\}$ vector represents the axis about which $\dot{\theta}_n$ rotation takes place.

$$\vec{\omega}_n = \vec{\omega}_{n-1} + {}^0_nR\,\dot{\theta}_n\,\{\zeta\} \tag{22}$$

The torque exerted on the links by electrical drivers could be calculated after mapping the Lagrange-d'Alembert equation to the actuators with implementation of Jacobian as:

$$\{T\} = J_{LJ}(\{\theta\})^T \left(\frac{d}{dt}\left(\frac{\partial L}{\partial \dot{\vec{r}}}\right) - \frac{\partial L}{\partial \vec{r}} \right) \tag{23}$$

In which $J_{LJ}(\{\theta\})$ is the Jacobian for mapping the Lagrange to the joints and mathematically described as:

$$J_{LJ}(\{\theta\}) = \frac{\partial\{r\}}{\partial\{\theta\}} = \begin{bmatrix} \frac{\partial r_{(1,1)}}{\partial \theta_1} & \cdots & \frac{\partial r_{(1,1)}}{\partial \theta_{12}} \\ \vdots & \vdots & \vdots \\ \frac{\partial r_{(27,1)}}{\partial \theta_1} & \cdots & \frac{\partial r_{(27,1)}}{\partial \theta_{12}} \end{bmatrix} \tag{24}$$

The generalized equation of motion of each actuator with consideration of $M(\{\theta\})$ for inertia effect, $C(\{\theta\},\{\dot{\theta}\})$ for coriolis and centrifugal effects, and $B(\theta)$ for gravitational effect is described in Eq. 25.

$$\{T\} = M(\{\theta\})\{\ddot{\theta}\} + C(\{\theta\},\{\dot{\theta}\}) + B(\{\theta\}) \tag{25}$$

This yields to determination of $\{\ddot{\theta}\}$ in Eq. 26 with the presence of the total exerted torque vector $\{T\}$ by actuators.

$$\left\{\ddot{\theta}\right\} = M^{-1}(\{\theta\})\Big(\{T\} - C(\{\theta\},\{\dot{\theta}\}) - B(\{\theta\})\Big) \tag{26}$$

Equation 25 requires the predicted joint angles according to the trajectory requirements, but Eq. 26 results in the real joint angles with the robots inertia, coriolis, and centrifugal effects and exerted torques on joints. The main idea of developing control methods on MRL-Nao robots such as torque control or adaptive controllers on joint positions necessitates a comparison between the predefined and feedback joint values.

4 Results and Discussion

The generated trajectory performs as an input for the inverse kinematic model to produce the required joints' angles, velocities and accelerations $\theta_r, \dot{\theta}_r, \ddot{\theta}_r$. The following curves depict joint angles θ_1 to θ_{12} for the predefined trajectory. θ_1 to θ_6 is for the right leg and θ_7 to θ_{12} is contributed to the left leg.

The input torso trajectory could be a cubic type presented in Eqs. 1–5 or real sensory data on the torso and foot provided by experimental results of biped robots. Experimental data of 12 joint angles regarding the Nao trajectory is shown in Figs. 2, 3 and 4. $\theta_5, \theta_6, \theta_7$ and θ_8 symbolize the right leg's hip roll and hip yaw-pitch and left leg's hip yaw-pitch and hip roll angles in Fig. 3 respectively.

Numerical differentiation is employed to determine the joints' angular velocities and accelerations in which time intervals play a significant role. These time intervals are well regulated form the joints sensory position data in this research.

Calculation of joint torques needs joint position specifications as it is stated in Eq. 21 and graphically illustrated in Figs. 5, 6 and 7 in which torque values are shown for all 12 joints.

Drivers' torques are plotted with the supposition of right leg as the supporting leg on the ground and left leg as the non-supporting leg.

Assumption of left leg as the non-supporting leg justifies small values of calculated T_9 and T_{10} values in comparison with the corresponding joints on the right leg and zero values for T_{11} and T_{12}.

Phase estimator calculates the phase parameters for SSP and DSP states, and then joints' reference angles are used in the forward kinematic model together with the phase estimator to produce all links' state position, velocity and acceleration vectors $\{r_s\}, \{\dot{r}_s\}, \{\ddot{r}_s\}$.

Generalized position vectors of links and five masses shown in Fig. 1b are estimated by Eq. 15, then velocity and acceleration values are achievable with numerical differentiation. Figures 8 and 9 demonstrate acceleration of two assumed mass on the torso and the right leg shown as m_2 and m_3 in Fig. 1b.

State position vectors and actuator torques are executed in the dynamic model to generate real joints' angles and velocities. The reference joint angles are compared with the estimated ones to build up a reliable controller such as adaptive PID or fuzzy logic controller. Torque control is this group's next research topic in order to reach a smooth Nao bipedal motion on the slopes.

All plotted curves are based on single support assumption of the right leg, but the transient mode and successive leg changes have not been considered in this simulation. Torque increase at specified times shows this change and may result into some errors in the calculated joint torques.

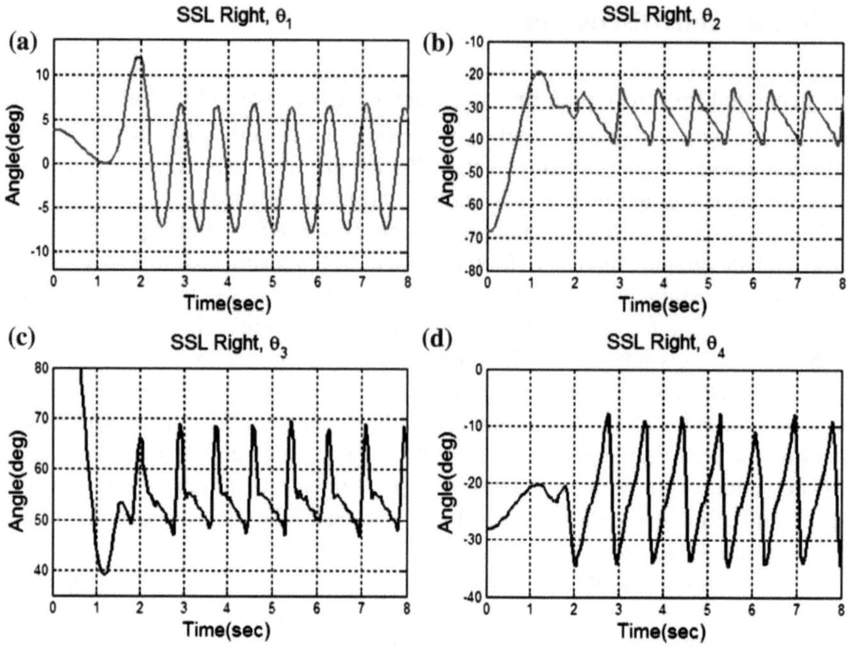

Fig. 2 **a** ankle roll, **b** ankle pitch, **c** knee pitch, and **d** hip pitch joint angles on the right leg extracted from real sensory data

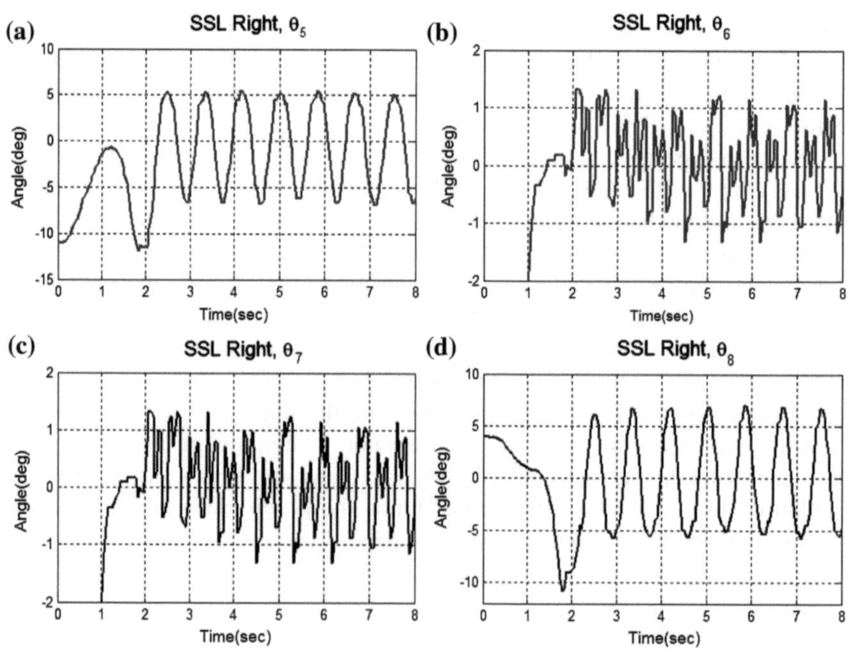

Fig. 3 Real sensory data of (**a**) right hip roll, (**b**) right hip yaw-pitch, (**c**) left hip yaw-pitch, (**d**) left hip roll

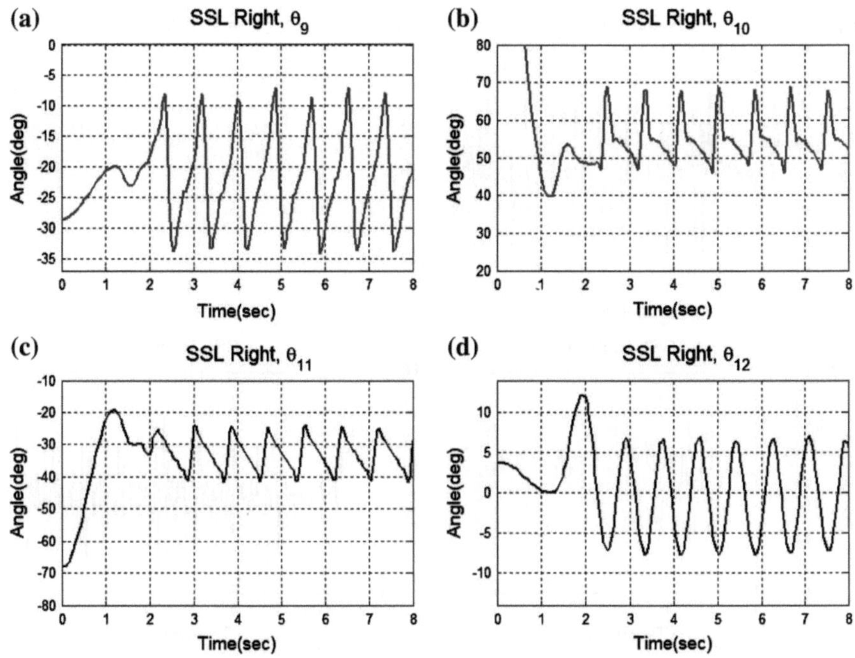

Fig. 4 **a** Hip pitch, **b** knee pitch, **c** ankle pitch, and **d** ankle roll joints angles on the left leg extracted from real sensory data

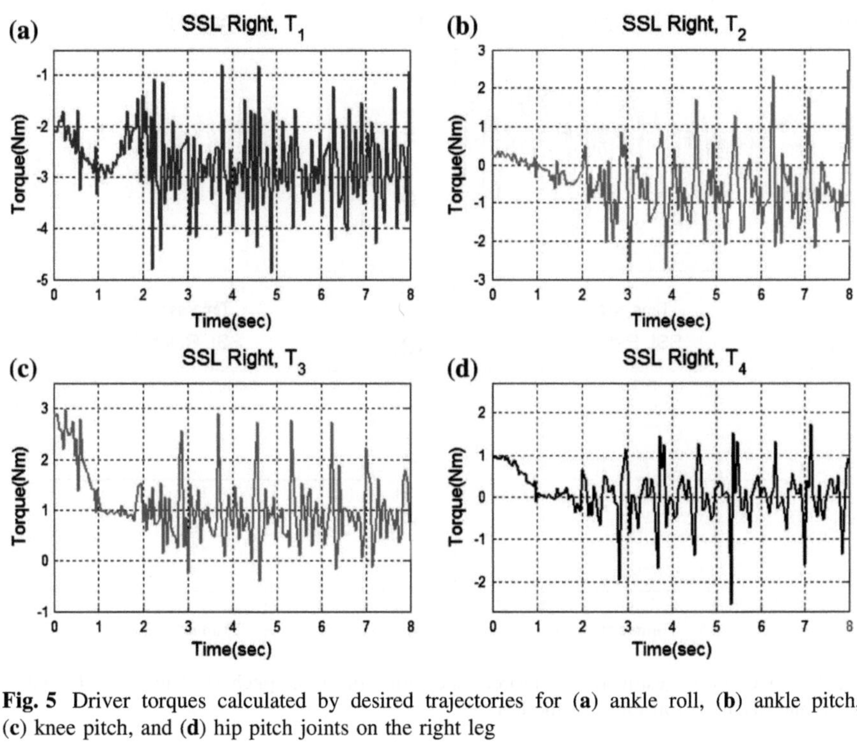

Fig. 5 Driver torques calculated by desired trajectories for (**a**) ankle roll, (**b**) ankle pitch, (**c**) knee pitch, and (**d**) hip pitch joints on the right leg

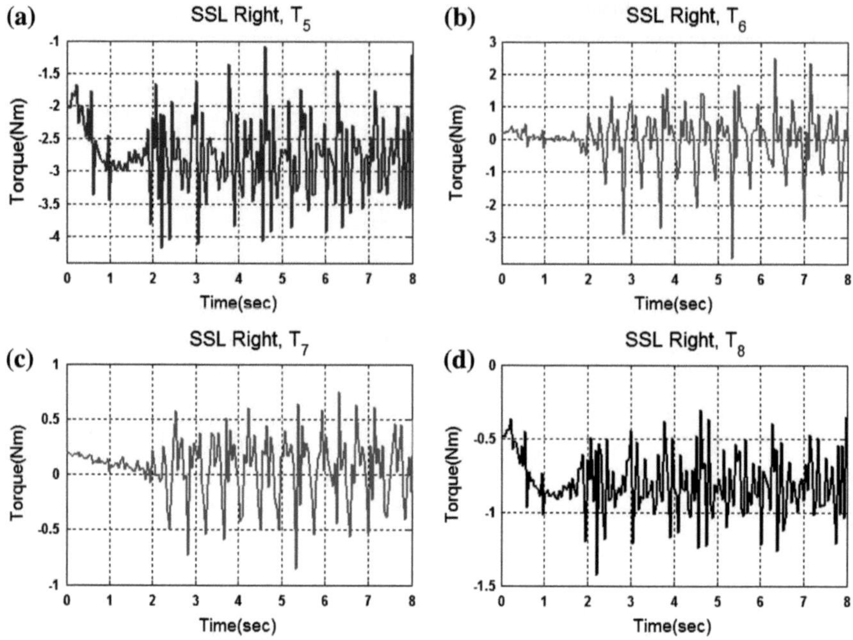

Fig. 6 Driver torques calculated by desired trajectories for (**a**) right hip roll, (**b**) right hip yaw-pitch, (**c**) left hip yaw-pitch, (**d**) left hip roll joints

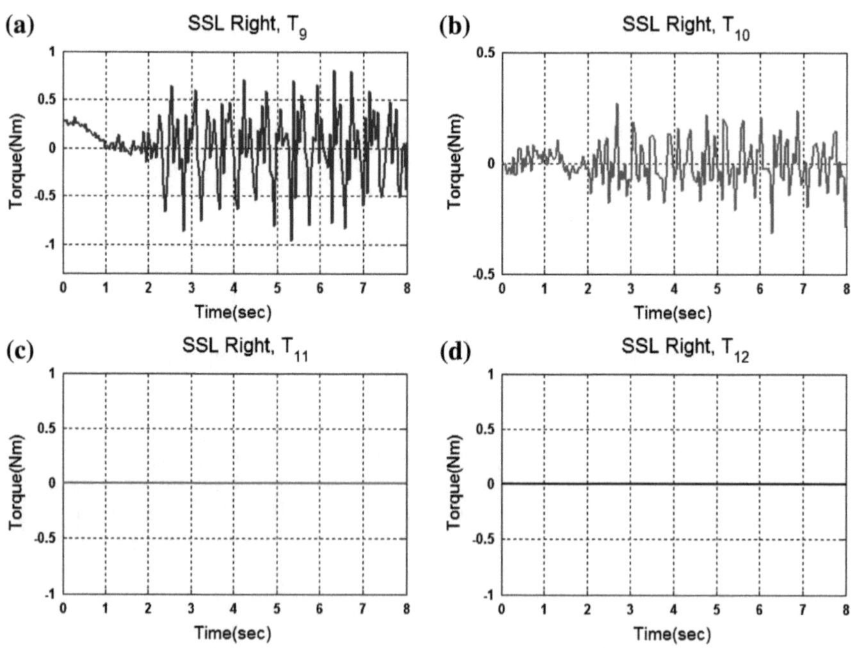

Fig. 7 Driver torques estimated by desired trajectories for (**a**) hip pitch, (**b**) knee pitch, (**c**) ankle pitch, and (**d**) ankle roll joints of the left leg

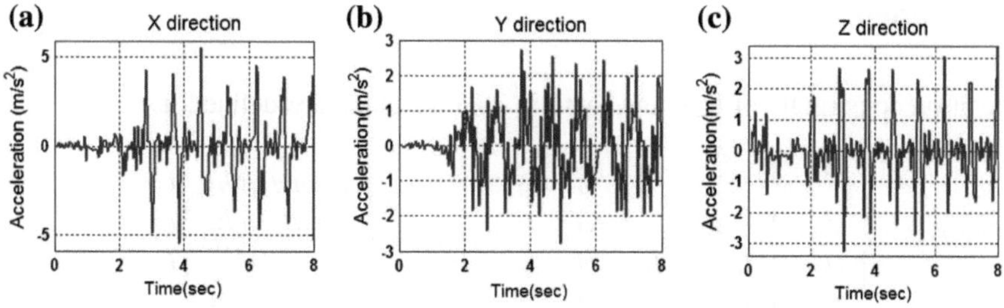

Fig. 8 Acceleration of mass 2 on the right leg computed by generalized position vector

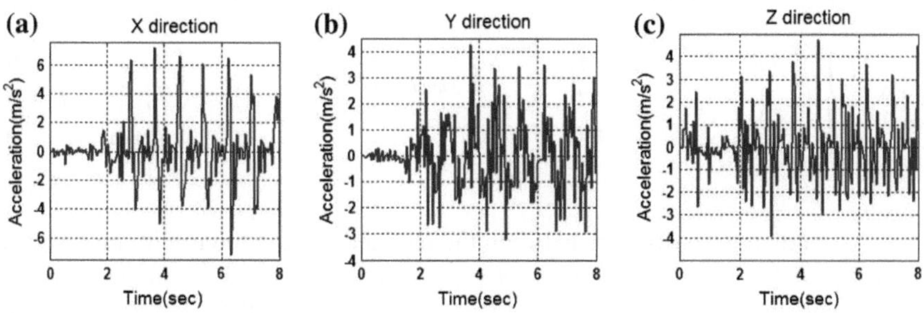

Fig. 9 Acceleration of dedicated mass on torso, m^3, obtained by generalized position vector

5 Conclusions

The proposed dynamic model and related analysis in this chapter simulates the joint values including torque and angular acceleration for the recorded trajectories as the real sensory data. The observed fluctuations on the torque curves are produced during the change support leg behavior in which acceleration values of joint oscillate due to transient phase between SSP and DSP.

The proposed transformation matrix in Eq. 13 contains rotation and position vectors of the joints and reduces the total number of calculation as a combined transformation.

Optimization in trajectories and parameters regulation could improve performance of the robot and lead to minimum energy consumption. Authors' future research topics are focused on a) torso trajectory optimization by genetic algorithm and particle swarm optimization, b) adding the double support phase and the transient mode, and c) torque control for performing stable walk on slopes.

Acknowledgments Authors gratefully acknowledge Qazvin Islamic Azad University, Young Researchers Club (YRC), and technical support of Mechatronics Research Lab. Nao team members.

6 Appendix

Position constraints of foot and torso in *sagittal* plane are defined as:

$$x_f(t) = \begin{cases} (k-1)L_{step} & ,t=t_1 \\ (k-1)L_{step} & ,t=t_2 \\ (k-1)L_{step}+L_{\max} & ,t=t_3 \\ (k+1)L_{step} & ,t=t_4 \end{cases} \quad z_f(t) = \begin{cases} FootHeight & t=t_1 \\ FootHeight & t=t_2 \\ h_{f\max} & t=t_3 \\ FootHeight & t=t_4 \end{cases} \tag{A.1}$$

In which L_{step} is the step length, $L_{\max}$ is the maximum horizontal distance of the ankle from the start point in $T_{\max}$, $h_{f\max}$ is the maximum ankle height during T_{step}. Constraints in torso position for specified times are as below:

$$x_t(t) = \begin{cases} kL_{step}-1.3x_{ts} & ,t=t_1 \\ kL_{step}-x_{ts} & ,t=t_2 \\ kL_{step}+x_{te} & ,t=t_4 \end{cases}, \; y_t = \begin{cases} y_{tmid} & ,t=t_1 \\ y_{t\min} & ,t=t_3 \end{cases}, \; z_t(t) = \begin{cases} h_{t\min} & ,t=t_1 \\ h_{t\max} & ,t=t_3 \\ h_{t\min} & ,t=t_4 \end{cases} \tag{A.2}$$

y_{tmid} stands for the distance between the feet and $y_{t\min}$ is the minimum distance from the ankle of the supporting foot to the spinal column. Experimental results substantiate the margin of $y_{t\min}$ between $-0.2\,y_{tmid}$ and $0.4\,y_{tmid}$. Furthermore, $h_{t\max}$ and $h_{t\min}$ symbolizes maximum and minimum torso height (Tables 3, 4.

Approximation functions of Eq. 9 in text are described as:

$$f_1(\theta_3) = a_3(c\theta_3) + d_4 s\alpha_3(s\theta_3) + a_2 \tag{A.3}$$

$$f_2(\theta_3) = a_3 c\alpha_2(s\theta_3) - d_4 s\alpha_3 c\alpha_2(c\theta_3) - d_4 s\alpha_2 c\alpha_3 - d_3 s\alpha_2 \tag{A.4}$$

$$f_3(\theta_3) = a_3 s\alpha_2(s\theta_3) - d_4 s\alpha_3 s\alpha_2(c\theta_3) + d_4 c\alpha_2 c\alpha_3 + d_3 c\alpha_2 \tag{A.5}$$

$$g_1(\theta_2,\theta_3) = [c\theta_2 f_1(\theta_3)] - [s\theta_2 f_2(\theta_3)] + a_1 \tag{A.6}$$

$$g_2(\theta_2,\theta_3) = c\alpha_1[s\theta_2 f_1(\theta_3)] + c\alpha_1[c\theta_2 f_2(\theta_3)] - s\alpha_1[f_3(\theta_3)] - d_2 s\alpha_1 \tag{A.7}$$

$$g_3(\theta_2,\theta_3) = s\alpha_1[s\theta_2 f_1(\theta_3)] + s\alpha_1[c\theta_2 f_2(\theta_3)] + c\alpha_1[f_3(\theta_3)] + d_2 c\alpha_1 \tag{A.8}$$

References

1. Goswami, A. Kinematic and dynamic analogies between planar biped robots and the reaction mass pendulum (RMP) model. In: Proceedings of the 8th IEEE-RAS International Conference on Humanoid Robots, pp. 182–188 (2008)
2. Kagami, S., Mochimaru, M., Ehara, Y., Miyata, N., Nishiwaki, K., Kanade, T., Inoue, H.: Measurement and comparison of humanoid H7 walking with human being. J. Robot. Auton. Syst. **48**, 177–187 (2003)

3. Wollherr, D., Buss, M., Hardt, M., Stryk, O.V.: Research and development towards an autonomous biped walking robot. In: Proceedings of the IEEE/ASME International Conference on Advanced Intelligent Mechatronics, AIM (2003)
4. Fujimoto, Y.:Trajectory generation of biped running robot with minimum energy consumption. In: Proceedings of the IEEE International Conference on Robotics & Automation, pp. 3803–3808 (2004)
5. Huang, O., Yokoi, K., Kajita, S., Kaneko, K., Arai, H., Koyachi, N., Tanie, K.: Planning walking patterns for a biped robot. IEEE Trans. Robot. Autom. **17**, 280–289 (2001)
6. Sugihara, T., Nakamura, Y., Inoue, H.: Real time humanoid motion generation through ZMP manipulation based on inverted pendulum control. In: The Proceedings of the IEEE International Conference on Robotics & Automation, pp. 1404–1409 (2002)
7. McGee, T.G., Spong, M.W.: Trajectory planning and control of a novel walking biped. IEEE Conf. Control Appl, 1099–1104 (2001)
8. Kajita, S., Morisawa, M., Harada, K., Kaneko, K., Kanehiro, F., Fujiwara, K., Hirukawa, H.:Biped walking pattern generator allowing auxiliary ZMP control. In: Proceedings of the IEEE/RSJ International Conference on Intelligent Robots and Systems, pp. 2993–2999 (2006)
9. Kajita, S., Kanehiro, F., Kaneko, K., Fujiwara, K., Harada, K., Yokoi, K., Hirukawa, H.:Biped walking pattern generation by using preview control of zero-moment point. In: Proceedings of the IEEE International Conference on Robotics and Automation (2003)
10. Hurmuzlua, Y., Genot, F., Brogliatoc, B.: Modeling, stability and control of biped robots-a general framework. Automatica **40**, 1647–1664 (2004)
11. Colbaugh, R., Glass, K., Seraji, H.: An adaptive inverse kinematics algorithm for robot manipulators. Int. J. Model. Simul **11**(2), 33–38 (1991)
12. Azevedo, C., Andreff, N., Arias, S.: Bipedal walking: from gait design to experimental analysis. Mechatron Elsevier **14**(6), 639–665 (2004)
13. Wieber, P. B.: Trajectory free linear model predictive control for stable walking in the presence of strong perturbations. In: Proceedings of the IEEE International Conference on Humanoids, pp. 137–142 (2006)
14. Vukobratović, M., Borovac, B., otkonjak, V. :Contribution to the synthesis of biped gait. In: Proceedings of the IFAC Symposium on Technical and Biological Problem and Control (1969)
15. Huang, Q., Yokoi, K., Kajita, S., Kaneko, K., Arai, H., Koyachi, N., Tanie, K.: Planning walking patterns for a biped robot. IEEE Trans Robot Autom **17**(3), 874–879 (2001). June
16. Choi, Y., You, B. J., Oh, S. R.: On the stability of indirect ZMP controller for biped robot systems. In: Proceedings of International Conference on Intelligent Robots and Systems, 1966–1971 (2004)
17. Huang, Q., Kaneko, K., Yokoi, K., Kajita, S., Kotoku, T., Koyachi, N., Arai, H., Imamura, N., Komoriya, K., Tanie, K.: Balance control of a biped robot combining off-line pattern with real-time modification. In: Proceedings of the 2000 IEEE International Conference on Robotics and Automation San Francisco, pp. 3346–3352, April 2000
18. Huang, O., Kajita, S., Koyachi, N., Kaneko, K., Yokoi, K., Arai, H., Komoriya, K., ane, K.: A high stability, smooth walking pattern for a biped robot. In: Proceedings of the IEEE International Conference on Robotics and Automation, pp. 65–71 (1999)
19. Jensen, B.T., Niss M.O.: Modeling, simulation, and control of biped robot AAU-BOT1. Master's thesis, Aalborg University (2009)
20. Ghaffari Jadidi, M., Hashemi, E., Zakeri Harandi, M.A., Sadjadian, H.: Kinematic Modeling Improvement and Trajectory Planning of the Nao Biped Robot. In proceedings of the Joint International Conference on Multibody System Dynamics, Finland, May 2010
21. Zannatha, J.I., Limon, R.C.: Forward and inverse kinematics for a small-sized humanoid robot. In: Proceedings of the 19th IEEE International Conference on Electrical. Communications and Computers, pp. 111–118 (2009)
22. Christensen, J., Nielsen, J.L., Svendsen, M.S., Ørts, P.F.: Development, modeling and control of a humanoid robot, Master's thesis, Aalborg University (2007)

23. John, J.: Craig, Introduction to Robotics: Mechanics and Control, Pearson Prentice Hall, 3rd edn (2005). ISBN: 0-13-123629-6
24. Mu, X., Wu, Q.: A complete dynamic model of five-link bipedal walking. Proc. Am. Control Conf. **6**, 4926–4931 (2003)
25. Goldenberg, A.A., Benhabib, B., Fenton, R.G.: A complete generalized solution to the inverse kinematics of robots. IEEE J. Robot. Autom. (RA) **1**(1), 14–20 (1985)

Comparison Between BEM Analysis and SONAH Measurements Using an Acoustic Camera

Christian Maier, Rolf Winkler, Wolfram Pannert and Markus Merkel

Abstract There are several methods available to determine Eigenmodes of structural objects. These can be calculated by modal analysis. The Eigenmodes cause maximum of displacements in the material at different positions. This effect is similar to a speaker. So the sound pressure is maximum at the points with maximal surface-velocity. With an acoustic camera these sound sources can be visualized with an appropriate signal processing method. A suitable algorithm is the statistically optimal near-field acoustical holography (SONAH). This algorithm shows its power at low frequencies in comparison to other (Beamforming) algorithms. Beamforming is a signal processing method to locate sound sources using an acoustic camera. In this chapter a comparison between the boundary element method (BEM) calculation and the SONAH sound processing method is performed. The BEM method is used to calculate the sound pressure field, which is caused by a vibrating surface of a structure. In a further step, a BEM software calculates with help of a structural harmonic analysis, which contains the displacement, velocity or acceleration of the elements, the sound pressure in the vicinity of the vibrating surface, especially on the surface. The harmonic analysis is given from a linear finite element method (FEM) software in this case. The measurement setup consists of a sawing blade with a diameter of 0.65 m and a

C. Maier (✉) · R. Winkler · W. Pannert · M. Merkel
Department of Mechanical Engineering, University of Applied Sciences Aalen, Aalen, Germany
e-mail: Christian.Maier@HTW-Aalen.de

R. Winkler
e-mail: Rolf.Winkler@HTW-Aalen.de

W. Pannert
e-mail: Wolfram.Pannert@HTW-Aalen.de

M. Merkel
e-mail: Markus.Merkel@HTW-Aalen.de

A. Öchsner et al. (eds.), *Materials with Complex Behaviour II*,
Advanced Structured Materials 16, DOI: 10.1007/978-3-642-22700-4_43,

thickness of about 0.005 m. In the middle, it is fixed by a clamping spigot. At a small drilled hole at the outer side, the sawing blade is connected to a shaker which impinges the sawing blade with different frequencies.

Keywords Acoustic camera · Beamforming · Eigenmodes · BEM · SONAH

1 Introduction

Achieved as a project at the University of Applied Sciences of Aalen, a comparison between the theoretical BEM method with an experimental method was carried out. The statistically optimal near-field acoustic holography SONAH algorithm was realized as a Labview source code in the Noise Inspector® Software package.

The sawing blade is excited by a shaker with different frequencies over a certain frequency range. At the Eigenmodes, the oscillation and, therefore, the surface velocity is maximal and a regular pattern of a standing wave develops. The regular pattern corresponds to sound sources which are located at the positions of maximal velocity. This pattern is reproduced with a numerical acoustic BEM calculation. As an input, this software needs a result file with the displacement of the elements. This result file can be delivered from a FEM software.

The near-field acoustic holography (NAH) is an experimental technique that makes it possible to reconstruct three-dimensional sound fields from measurements on two-dimensional surfaces.

This can be extremely useful, and NAH is a well-established tool for visualising and analyzing sound fields near sources of noise [1, 2]. Conventional planar NAH is based on discrete spatial Fourier transforms of sound pressure data measured with a microphone array. However, to avoid serious truncation errors caused by the finite two-dimensional spatial transform ('leakage' in the wave number domain), the array must extend well beyond the source so that the sound pressure has decayed to an insignificant level near the edges of the array [2].

SONAH is an interesting variant of NAH developed a few years ago by Steiner and Hald [3]. It has the great advantage of avoiding spatial transforms and thus the mentioned truncation effects; therefore, the measurement array can be smaller than the source [3].

2 Theory

In planar SONAH the 'propagator' that transforms data from one plane to another is a transfer matrix that works directly on the measured data, that is, the sound pressure p at an arbitrary position above the source $r = (x, y, z)$ (where $z > 0$), can be expressed as a weighted sum of sound pressures measured at N positions $(r_{h,n})$ in the hologram plane ($z = z_h$).

$$p(r) \cong \sum_{n=1}^{N} c_n(r)p(r_{h,n}) = p^T(r_h)c(r) \tag{1}$$

With a development coefficient c_n and the measured sound pressure p. Governed by the Helmholtz equation the radial propagation of acoustic pressure p in a linear isotropic acoustic domain is given by:

$$\nabla^2 p + k^2 p = 0 \tag{2}$$

where ∇^2 is the Laplacian operator for the three dimensional flow field and $k^2 = \left(\frac{\omega}{c}\right)^2$ the square of the wave number vector [4].

In the frequency domain, a special solution consists of two parts, which can be plane waves or the evanescent wave, depending on the value of K_z.

$$p(\vec{K}, \omega) = \hat{p}(\omega) \cdot e^{-jK_z z} \cdot e^{j(K_x x + K_y y)} \tag{3}$$

With the three spatial directions of the wave number (K_x, K_y, K_z) we can notice from the formula 3 that the amplitude decays exponentially with the distance z from the surface, if K_z is imaginary. So it is necessary that the measurement distance be very close to the sound source. Otherwise information about the evanescent modes gets lost and the enhanced resolution at low frequencies cannot be realised. The dispersions equation (Eq. (4)) with the k value is defined by the wave number vector and the local wave number K_x, K_y, K_z.

$$k^2 = K_x^2 + K_y^2 + K_z^2 \tag{4}$$

The value K_z is the z component of the wave vector and it can be computed for arbitrary z values (real or complex).

Figure 1 shows the set-up of a SONAH measurement. The sound pressure is measured in the x-y plane in the near-field for the $z = 0$ value. The distance from the microphone array and sound source is d.

The BEM method uses the Green's formula which gives an analytical connection between the values of the sound field on the limiting surface (r_q) and the values somewhere in space (r_n). The integral is carried out on a surface grid which is split in small so-called "boundary elements". The boundary is defined by Γ which contains rigid boundary, vibrating boundary and absorption boundary. The index p describes the observation point and q the source point on the boundary.

$$\frac{1}{2}p(\mathrm{r_p}) = \int_{\Gamma} \left(p(\mathrm{r_q}) \frac{\partial G(\mathrm{r}, \mathrm{r_q})}{\partial n_\mathrm{q}} - \frac{\partial p(\mathrm{r_q})}{\partial n_\mathrm{q}} G(\mathrm{r_p}, \mathrm{r_q}) \right) dS + p_\mathrm{d}(\mathrm{r_p}) \tag{5}$$

where $\partial/\partial n_\mathrm{q}$ specifies an inner normal derivative. At the point p on the boundary, p_d is an external sound pressure to which objects other than the boundaries

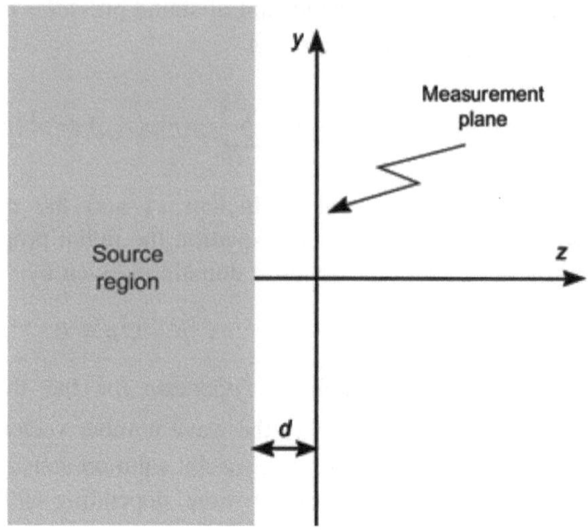

Fig. 1 SONAH measurement plane [5]

contribute. G represents the Green's function (fundamental solution) of the three-dimensional wave equation [6].

The advantage in comparison to a FEM is obvious, if one has to calculate the sound field in a large region—only a surface mesh has to be generated and not a mesh into a infinite distance.

3 Measurements

Acoustic methods based on microphone arrays are able to locate sound sources using an appropriate signal processing. Finally it is possible to superimpose an optical picture with the acoustic picture (see Figs. 3, 4, 5). In this combination it is possible to locate sound sources directly on the object under investigation.

There are many algorithms available to process the microphone signals. Beside the standard Beamforming which is commonly used for far-field measurements the near-field acoustical holography (NAH) is an effective method for near-field measurements. A fundamental criterion for the different methods is the ability to separate closely located sound sources. This ability to separate sound sources is called resolution. Often the resolution is a function of the frequency. In the standard delay-and-sum Beamforming method, a measure of resolution is

$$\Theta_{3dB} \approx \frac{\lambda}{D} \text{ and } R = a\frac{z}{D}\lambda. \tag{6}$$

With the wavelength λ and array dimension D. The resolution in distance z is R with the array geometry-factor a. A steered arrangement of microphones is

Fig. 2 Test set-up

characterised by a so-called main lobe.[1] The width of the main lobe 3 dB below of the maximum is the Θ_{3dB} width [7].

Beside information about sound pressure, the NAH method provides information about velocity potential and sound intensity close to the emitting surface. In contrast to the standard Beamforming advantages, like better resolution and the ability to analyse lower frequencies (< 800 Hz) prevail, SONAH works with regular and irregular microphone arrays. The programmed SONAH method is available as a Plug-In for the Noise Inspector® software[2] package.

For this chapter the Eigenmodes of the above described sawing blade are depicted and the measurements at the frequencies of the Eigenmodes are carried out. For the measurements, the microphone array is positioned at 0.15 m distanced from the sawing blade in the z coordinate and is centered to it (see Fig. 1). To control the results of the SONAH measurements, a simultaneous measurement with a laser vibrometer was carried out. In the following figures, comparisons between laser vibrometer, SONAH method and acoustic BEM calculation is carried out for some frequencies. The SONAH result shows the sound pressure level (SPL) coded in terms of colours directly on the vibrating surface. Furthermore, the SONAH measurements are validated with a sound level meter on the surface to control the SPL. Among other things, the laser vibrometer measurements deliver information about vibrating velocity orthogonal to the surface.

Figures 3, 4, 5 shows the SONAH method on the left and the laser vibrometer measurements on the right side. Number and location of the sound sources are in

[1] The characteristic is more weighted in this direction than in other directions.

[2] The software used to operate the acoustic camera.

SONAH-Method Laser-Vibrometer Acoustic BEM surface pressure plot

Fig. 3 Mode 2 at 73 Hz, four maxima Eigenmode 2: 73 Hz SONAH-method laser-vibrometer acoustic BEM surface pressure plot

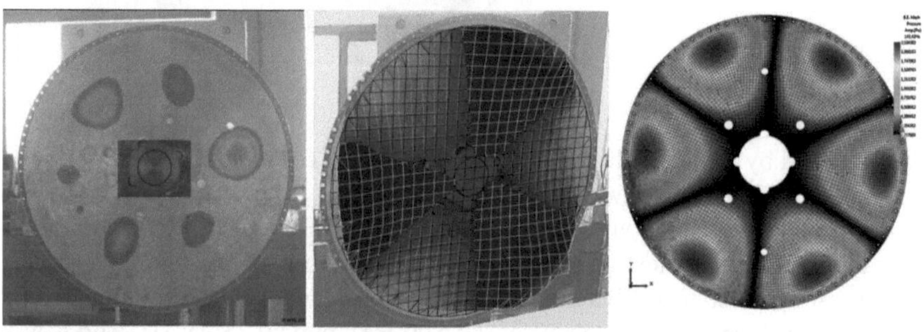

SONAH-Method Laser-Vibrometer Acoustic BEM surface pressure plot

Fig. 4 Mode 3 at 145 Hz, six maxima Eigenmode 3: 145 Hz SONAH-method laser-vibrometer acoustic BEM surface pressure plot

agreement with the measurements with the laser vibrometer and therefore with the maxima of the sound source. In this chapter, the lowest analysed frequency for SONAH measurements is 73 Hz and it works well for that frequency. Frequencies at these low values cannot be handled with the standard Beamforming method. Beamforming yields good solutions for frequencies above 800 Hz. So, the advantage of SONAH at low frequencies in contrast to Beamforming is demonstrated.

Based on simplification of the sawing blade (no modelling of the shaker and adapter), the acoustic BEM calculation may show minimal deviation to the laser vibrometer and SONAH measurements Fig. 2.

The recalculated Eigenmodes with the FE software ANSYS® are close to the actual Eigenmodes. The second Eigenmode is at 73.75 Hz. This frequency is recalculated with the FE software ANSYS® and the acoustic BE software WAON® for 72.5 Hz. The next Eigenmode, the third, is at 145 Hz and

SONAH-Method Laser-Vibrometer Acoustic BEM surface pressure plot

Fig. 5 Mode 4 at 245 Hz, eight maxima Eigenmode 4: 245 Hz SONAH-method laser-vibrometer acoustic BEM surface pressure plot

recalculated for 143.42 Hz. Mode 4 at 245 Hz is recalculated with ANSYS® for 247.65 Hz.

The SONAH plots on the left side of Figs. 3, 4, 5 show the sound pressure level with 6 dB dynamic range on the surface. In the middle, the laser-vibrometer shows the maximum acceleration and the acoustic BEM calculation on the right side shows the pressure on the surface of the structural body.

4 Conclusion

SONAH and BEM are effective methods to show sound fields which are caused by vibrations of structural bodies like a sawing blade. Real situations can be reproduced by these methods. Indeed there is an ANSYS result file for an acoustic BEM calculation necessary, but in combination with this result file WAON is able to calculate the sound pressure level on the surface of the sound source object using Green's formula. The BEM calculation can be compared with SONAH measurement in the case of the Eigenmodes and, concerning the mode-pattern good agreement is achieved.

References

1. Maynard, J.D., Williams, E.G., Lee, Y.: Nearfield acoustic holography. I. Theory of generalized holography and the development of NAH. J. Acoust. Soc. Am. **78**, 1395–1413 (1985)
2. Williams, E.G.: Fourier Acoustics—Sound Radiation and Nearfield Acoustical Holography. Academic Press, San Diego (1999)
3. Steiner, R., Hald, J.: Near-field acoustical holography without the errors and limitations caused by the use of spatial DFT. Int. J. Acoust. Vib. **6**, 83–89 (2001)

4. Zhang, Y.K., Lee, M.R., Trevelyan, J., Adey, R.A., Niku, S.M., Brebbia, C.A.: Acoustic diagnostic analysis of automobile passenger compartments. Pub ID: 103-471-746, internoise 95, pp. 1313–1316 (1995)
5. Hald, J.: NAH and beamforming using the same Array-SONAH, ISSN 0007–2621 DK-2850 Nærum, Denmark (2005)
6. Cybernet Systems Co., Ltd.: WAON theoretical manual, Chiyoda-ku, Tokyo 101-0022 (2009)
7. Christensen, J.J., Hald, J.: Beamforming, ISSN 0007-2621, DK-2850 Nærum, Denmark (2004)

Simulation of Thin Metal Laser Lap Welding

Welding Behavior of Heat Conduction Welding and Keyhole Welding

Takeji ARAI

Abstract The majority of laser welding carried out today is related to the processing of thin metals. Although many papers have been published on thin sheet laser lap welding, few researchers have looked at the heat transfer characteristics during the process and the effect of local deformation of the material during welding and its influence on the welding properties. In lap welding, the gap at the interface is known to have a major influence on the welding performance. The welding phenomenon in keyhole type welding of thin sheet has hitherto not been completely clarified. In the present study, a more detailed understanding of the process when two of thin sheets with a variable gap are welded by high power laser was been established.

Keyhole type welding using a high power density and heat conduction welding using a low power density were compared. The presence of a gap between the sheets and the influence of the sheet deformation were investigated. In addition to welding trials computational simulations were carried out using finite element analysis (FEA). The welding deformation and behavior of the molten metal during the stages of the welding process were analyzed and the deformation of the materials and the angular distortion were calculated.

The analysis was carried out using the best available information regarding simulation of the welding process. Further refinements to the analysis are planned in the future, in particular using the results of X-ray fluoroscopy. However, reasonably good correlation was found between the experimental and analytical results.

Keywords Simulation of laser materials processing · Laser lap welding · Penetration welding · Keyhole welding · Effects of welding gap · Laser processing of metal

---

T. ARAI (✉)
Chuo University R.D.I, 1-13-27 Kasuga Bunkyo-ku, Tokyo, 112-8551, Japan
e-mail: tarai@tamacc.chuo-u.ac.jp

A. Öchsner et al. (eds.), *Materials with Complex Behaviour II*,
Advanced Structured Materials 16, DOI: 10.1007/978-3-642-22700-4_44,

1 Introduction

In the previous study [1], the welding behavior of the thermal conduction type welding was simulated for thin metal butt welding. In this chapter, the welding mechanisms and the molten metal behavior in thin metal lap welding is discussed. There are two types of welding mechanisms in the laser lap welding, a thermal conduction type welding and a keyhole type welding. The thermal conduction type welding is the phenomenon where the melting and joining processes are mainly caused by thermal conduction. This type of welding occurs when the power density of the laser beam is comparatively low. On the other hand, when the power density is high, it is known that the keyhole type welding occurs because a keyhole is generated in the molten pool immediately beneath the heat source, which causes a deep joint penetration. This type of welding occurs in most thicker section welds with a high density laser beam [2–11].

In conventional research into keyhole welding, plates thicker than 10 mm have been processed using a laser power higher than 10 kW. Many of these made investigations into X-ray transmission experiments that used ‘bead on plate’ welds. In those researches, the generation of a molten pool and the porosity behaviors were studied [6–11]. However, bead on plate is not the same as actual welding. Moreover, most of the actual laser welding in industry is either butt welding or lap welding for thin plates. From a content point of view, they have mainly discussed the welding properties such as the melting geometry, defects, metallographic structure of the joining surface and processing conditions. There are few reports and documents on the actual distortion caused while welding [12, 13].

In the course of researches on the phenomenon of thin metal welding, the following is generally known. When lap welding is carried out with high laser power, a keyhole is generated. In lap welding, there is always a gap at the interface of the two plates. It is qualitatively known that the gap between the two plates and the surface condition of the upper plate influence the welding performance. However, there are extremely few researches that discuss the welding process of the lap welding. Especially, the welding phenomena in the keyhole type welding have not been fully understood, even through observations with an X-ray fluoroscopy. Moreover, due to a lack of accurate data and information, simulations that have dealt with the keyhole formation in the welding process are extremely rare. In this study, keyhole welding has been simulated to establish a series of prediction techniques for laser welding. The simulation was based on the latest information available at the present time, although it still contained some assumptions. In the future, if a more accurate understanding of the keyhole in lap welding by X-ray observation method can be specified, the accuracy of calculation of this simulation will be improved further.

2 Experimental Conditions

In the actual processing experiment, a lap welding was carried out, using two SUS304 plates of 1 mm thickness with $100 \times 50\ mm^2$ size. A laser diode (LD) excited yttrium aluminum garnet (YAG) laser with 4 kW power (3.4 kW at the top of the nozzle) was used. The spot diameter for the fiber transmission was $\varphi = 0.6$ mm and the welding speed was 5 m/min. As shown in Fig. 1, a shim with the same thickness as the gap was inserted between the lapped plates and the plates were held with a jig from the upper side while welding. In the simulation, a calculation model was built based on these conditions. The welding process and the amount of distortion were calculated, varying the gap between the two lapped plates from 0.005 mm (which means there is little gap) to 0.25 mm (which means a significantly large gap, considering the plate thickness).

3 The Welding Mechanisms

3.1 Heat Analysis Simulation

When a high power laser is used, the power density is greater at the focused irradiation area so that a keyhole is generated on the material surface, which is known as the so called "keyhole welding". However, there is no simulation model yet for the keyhole type lap welding process for thin plates. In lap welding, the material of the upper plate melts and thermally expands during welding due to the temperature rise. In addition, there is an increase in mass due to oxidization. Because of these phenomena, the gap is closed up and the heat transfers to the surface of the lower plate so that the upper and the lower plates are in contact. The presence of a keyhole makes this phenomenon complicated. The following points are presumed based on several research reports for X-ray observations [10] as well as on discussions among colleagues who were involved in our study, although not all of the phenomenon have been clarified in the present circumstances.

In the laser welding with high energy density, the material is heated and melts to form a molten pool first. As the temperature increases, a keyhole appears in the upper plate immediately beneath the nozzle. Because of this keyhole, the heat conducts in a downward direction. As the material temperature rises, it melts further and the keyhole grows downward. The temperature inside the keyhole is kept higher than its boiling point. As shown above, the distribution of the generated energy was taken into consideration to establish the model for the keyhole welding. The amount of distortion caused by the temperature increase and the volume expansion was obtained quantitatively for the whole material.

This simulation was developed for high power laser welding of a thin metal as a new method. The laser beam is continuously irradiated on the material as the heat source moving at a constant speed, and behavior of the deformation of the material

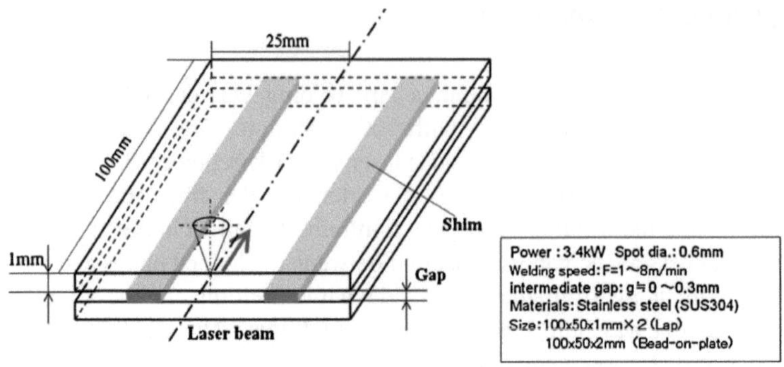

Fig. 1 Schematic diagram of the lap welding experiment which intermediate gap is managed by using shim (slip gauge)

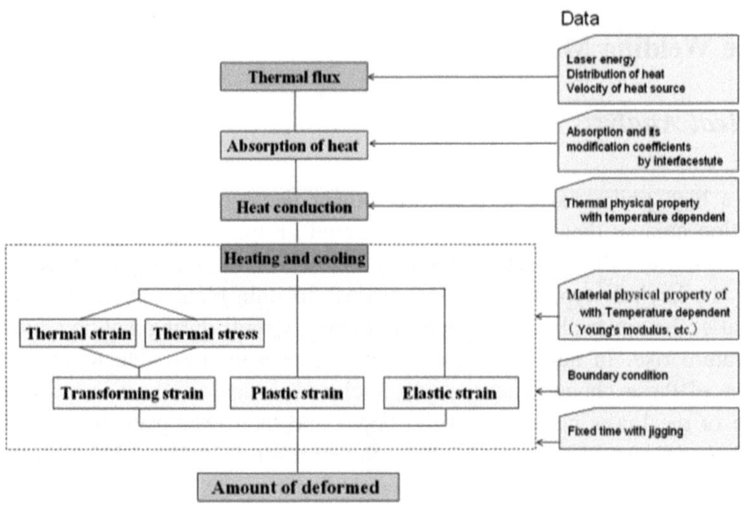

Fig. 2 Flow chart of elastoplasticity stress transformation analysis

was analyzed by 3D Nonstationary elastoplasticity analysis. In this calculation, radiation and convection are taken into account for contact heat transfer at the nodal points of element.

The flow chart is similar to the former report [1]. Figure 2 shows the flow chart of the calculation. This simulation is basically a thermal analysis simulation using the finite element method but it obtains the temperature field by using heat transfer equation which takes radiation, convection and contact heat transfer into account, and then elastic–plastic stress distortion analysis is carried out using the thermal strain obtained from the temperature. In this simulation, the laser heat source is replaced by heat flux which is converted to the quantity of heat, and the temperature is calculated considering the ratio of beam absorption and the thermal properties of the material.

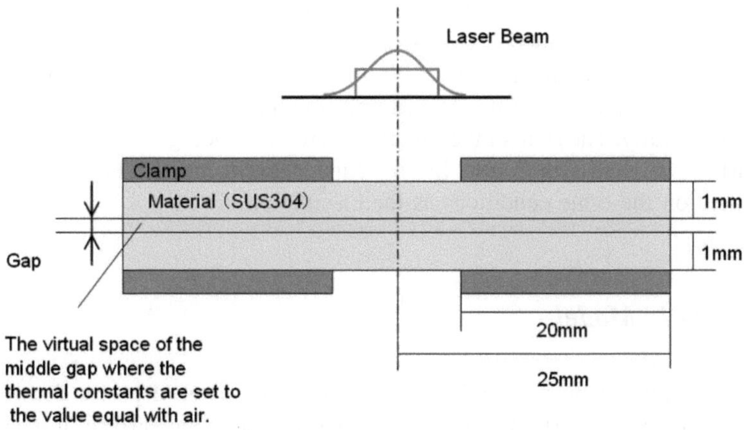

Fig. 3 Simulation model for the calculation

Moreover, the stress, thermal strain, phase transformation strain, plastic strain and elastic strain were determined by using the physical properties of the material such as Young's modulus and Poisson's ratio, and finally the distortion of the entire plate was calculated. Also, temperature dependence was taken into account for the physical properties of the materials and latent heat was also considered in the molten state.

3.2 Computing Model

The calculation model for lap welding simulation assumed two plates of stainless steel (SUS304) 1 mm thick. The laser beam moved to one direction on the material, and the size of the steel was $100 \times 100\ \mathrm{mm}^2$. The laser output at this time was 3.4 kW on the material, and a spot diameter of $\varphi = 0.6$ mm was employed. In the simulation, the material was fixed by fixed clamping at both ends, and the welding jig was released ten seconds after the laser irradiation ended. The volumetric expansion of the metal caused during the melt of the metal by the irradiation of the laser heat source and the increase of mass by the oxidation were calculated, and finally the deformation of the overall sheet metal was evaluated over time, and the residual stress and the amount of deformation were obtained in the calculation process as a result though the method for analyzing used to calculate is the same as a former report [1]. In the calculation for lap welding, the gap in the middle of the plates was varied from 0.05 to 0.25 mm under these conditions.

An original computing model that enabled the calculation by inserting virtual space for the gap layer that would be caused by overlapping two sheets in the model calculation as a layer was used. Figure 3 shows the computing model.

The amount of deformation of the welding plates at various welding speeds was simulated based on this model. On the other hand, as well as the computing model, the size of the material of 100 × 100 mm was investigated with thickness of 1 mm stainless steel materials. The sample for the actual laser experiment was selected carefully. The state of the sample before the welding has been adjusted for the surface-roughness of 5 μm. In an actual experiment, the lap welding was performed on the same conditions as the calculation.

3.3 Keyhole Model

In keyhole welding mode, if a laser above a certain level of power density (for example, higher than 10^5 W/cm^2) is irradiated on the metal surface, it is considered that a keyhole is generated, which is called as keyhole welding. The observation with a high speed X-ray fluoroscopy camera reported that the keyhole generation starts more or less some milliseconds (ms) after the laser irradiation, although the time varies depending on the power density of the laser and the penetration depth. For example, according to the Tsukamoto [8] and Seto [9] group, the X-ray observation indicates that a keyhole starts to be formed in 2–3 ms. If the power density is high, the difference in power accelerated the rate growth of the depth of the keyhole rather than on the time taken to form a keyhole. In the simulation where the power was set as 4 kW, it was assumed that the keyhole generation would start in 2 ms after the form of molten pool.

The diameter of a keyhole is roughly the same as the spot diameter, although the center of the diameter is known to change (in width and length) periodically. Therefore, it is assumed that a keyhole whose diameter is nearly equivalent to the laser spot diameter is generated in the material and the heat source with the temperature higher than the evaporating temperature contributes to warm the material from inside the keyhole. As described above, the keyhole diameter is assumed to be constant in the calculation this time but when the changing cycle becomes clear and if the diameter variation needs to be considered, they can be dealt with in the calculation. The coordinate system for the computation used for this study is shown in Fig. 4.

4 Calculation Results

4.1 Welding Behavior of the Heat Conduction Type Welding

In lap welding where two plates are joined vertically, the size of the gap between the two plates determines the welding quality. Figure 5 shows the simulation results in time-series under the condition where the heat conduction type welding occurs in lap welding, using stainless plates (SUS304) of 1 mm thickness with

Fig. 4 The coordinate system for the computation

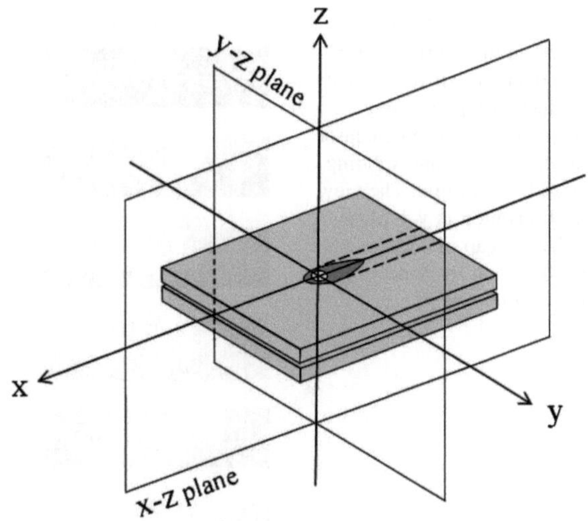

1 kW power. Figure 5a) shows the result for narrow gap (g = 0.05 mm) and Fig. 5b) shows the one for wide gap (g = 0.15 mm). The calculation conditions are as follows: the spot diameter is 0.6 mm, the power density is 8.8×10^4 W/cm^2 and the feeding speed is 1.5 m/min for both simulations. The right circle describes the heat source. The horizontal line in the figure describes the observation point. The welding phenomenon such as heating, expanding, melting and grafting at the cross-section surface were simulated along the heat source movement.

Having an arbitrary cross-section of the welding surface as the observation point, the temperature increases as the heat source approaches and the material expands and then melting starts. The largest expansion is observed when the heat source slightly passed the center on the material and the highest temperature can also be seen in the molten part at that time. Then, as the heat source passes, the natural cooling starts and the material shrinks slightly. The upper plate is heated by the laser heat source and the heat conducts to the intermediate part where the lower plate is located. There is little difference in the heat conduction speed in this process. In lap welding, the heat reaches the gap between the plates and the gap is filled up due to the melting and the expansion. Immediately after the gap is filled up, the heat conducts to the lower plate and its temperature increases. If the gap is small, the heat conducts to the lower plate more quickly so that the penetration bead is generated in the back side of the plate in a short time. Also, a wider interface can be obtained at the gap. On the other hand, if the gap is large, it takes longer time for the heat to reach the bottom of the plate. As the gap is relatively large, more molten metal is needed to fill the gap so that the interface between the upper and the lower plates becomes somewhat small. Consequently, if the gap is large, the interface is decreased and the welding intensity is reduced as well. The time for the penetration bead to be created is also delayed. To compensate this, it will inevitably be required to lower the welding speed and increase the laser power.

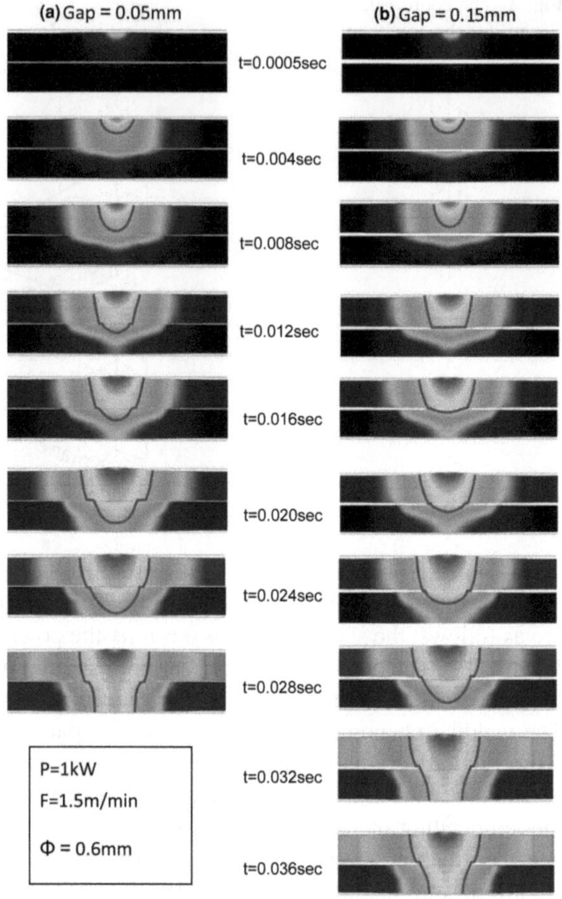

Fig. 5 The simulation results in time-series under the condition where the heat conduction type welding (moving picture). Where laser power is 1 kW, and welding speed is 1.5 m/min. The view of observation is y–z plane. **a** Narrow gap of g = 0.05 mm. **b** Wide gap of g = 0.15 mm

4.2 Welding Behavior of the Keyhole Type Welding

Regarding the keyhole type welding of a thin plate, the study has to contain assumptions for some elements that have not been clarified yet through researches to date, since little researches have been made for the keyhole welding of thin plates. One of the elements that have not been clarified yet is the timing for the keyhole to be generated in the molten pool.

The size of the keyhole diameter is also unknown. Hence, the following assumptions were made for the simulation. The timing for the keyhole to be generated is assumed to be 2 ms after the molten pool appears which was estimated from the studies in the past. The keyhole diameter is assumed to be almost the same as the spot diameter. The keyhole is assumed to appear when the tip temperature immediately beneath the spot diameter exceeds the evaporation temperature, since the keyhole is generated and moves downward when the melting proceeds and the temperature reaches this point. In the actual X-ray fluoroscopy observation for a carbon steel plate which has enough thickness, the downward advance rate of the keyhole is reported as 1.4 mm/ms when processed with 12 kW power [9, 10]. In our research, lap welding is performed with the power output less than half of this and the target material is a thin stainless steel plate of 1 mm so that this result is assumed to be less than the processing rate of previous work. In our research, lap welding is performed with the power output less than half of that used in the referenced work and the target material is a thin stainless steel plate of 1 mm so that this result is assumed to be less than the processing rate of previous study. Under these assumptions, welding was performed at 5 m/min which is the standard speed to process a plate with this thickness. The calculation has been made for the process where the melting is induced by the irradiated laser beam on the material, the keyhole is generated and then the material is welded.

Figure 6 shows the calculation result based on the keyhole generation model. The temperature at the surface of the upper plate increases due to the laser irradiation and just after the molten pool appears (in 2 ms), a keyhole is generated immediately beneath the beam. This keyhole grows and penetrates the upper plate of 1 mm thickness in 5 ms. The molten metal that has reached the intermediate gap expands to fill the gap and the heat starts to conduct to the lower plate. During this period, the heat stays at the intermediate part for a while so that there is a slight delay for the heat shifting to the lower plate. As a result, there is a difference in heat distribution level between the upper and the lower plates. Also, the keyhole in the lower plate is a little smaller than the one in the upper plate. When the middle gap is small ($g = 0.05$ mm), the time required for the two boards to penetrate is 8 ms. On the other hand, when the gap is large ($g = 0.15$ mm), it takes 9 ms. The time to penetrate during welding varies according to the size of the gap.

In this process, it took 9 ms to penetrate the two plates. For justification, actual measurement was performed using a digital video camera (Canon), although it adopts a slightly perfunctory method. The shutter speed is 60 frames per minute, which is 16 ms per frame. In this measurement, the initiation of the irradiation and the penetration splash right under the welding material were confirmed within one frame so that at least it has been proved that the time taken to penetrate the plates is less than 16 ms. Figure 7 show the measurement and observation system for the penetration time with high-speed charge coupled device (CCD) image sensor camera.

In actual laser welding, 15-20 l/min assist gas is blown out coaxially with the laser beam. In the simulation, the gas was blown out toward the molten pool in the same manner. The result shows that this gas flow pushed down the molten metal in the generated molten pool and served as the driving force to move it backward. The molten metal is cooled while going away from molten pool. Also, it was confirmed that the gas flow played the role to coagulate the molten metal moderately.

After a molten pool is generated, the molten metal surrounds the keyhole and then runs backward. Molten metal begins to flow outside of molten pool, and it is cooled soon sequentially and accumulates. This forms a welding nugget and the nugget is cooled from the surrounding base material toward the center. When

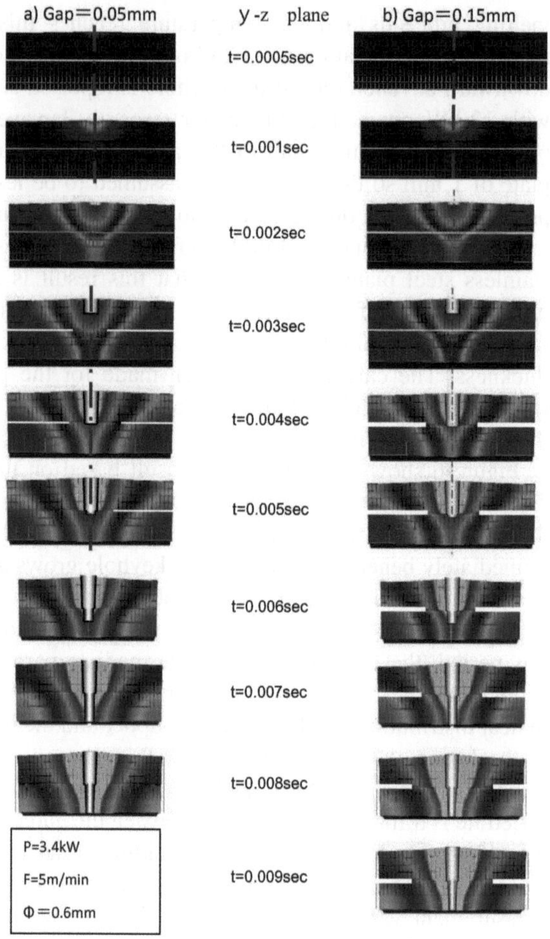

Fig. 6 The simulation results in time-series under the condition where the keyhole type welding (*moving picture*). Where laser power is 3.4 kW, and welding speed is 5 m/min. The view of observation is y–z plane. **a** Narrow gap of g = 0.05 mm **b** Wide gap of g = 0.15 mm

cooling starts, a part of the new molten metal laps to form an "imbricate" nugget. The simulation result is shown in Fig. 8.

In Fig. 8, the lap welding process with keyhole type welding is shown in time-series, observed from the X–Z plane. The process includes the initiation of the irradiation, keyhole generation and penetration. The sections in the Y-Z plane correspond

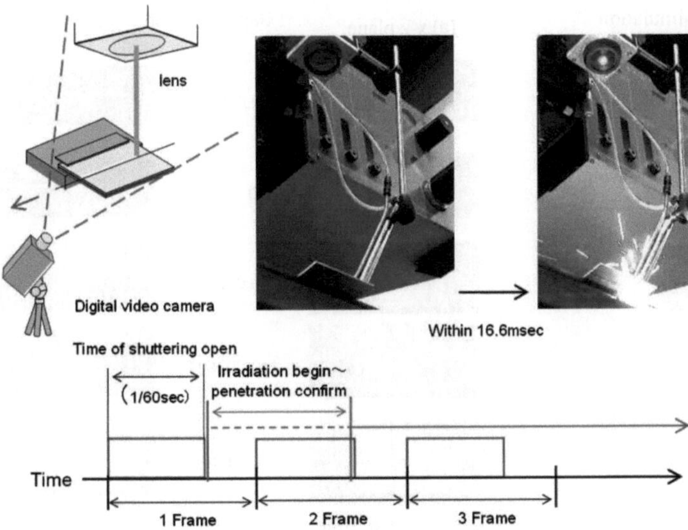

Fig. 7 The measurement and observation system for the penetration time with high-speed CCD camera

to the cross section of the metallographic structure. The actual welding phenomenon was also observed with Simadzu: Hyper Vision, Model HPV-1. Figure 9 shows the surface phenomenon of actual welding behavior observed by high speed CCD camera. It shows that the molten metal is undulating toward the back in the molten pool.

The photo of the cross section showing the structure of the actual weld sample and the melting geometry of the simulated material as well as the temperature distribution were compared. Both of these show the molten cross sections observed from the plane perpendicular to the travel direction (y–z plane) for the same conditions. The result is shown in Fig. 10. It shows the cross section structure of the metal obtained through the lap welding experiment with a gap ($g = 0.05$ mm) and an example of the calculation result obtained through the simulation. From the comparison, it can be seen that similar geometries were obtained. In lap welding with a gap in the middle, the heat stays at the gap interface for a while so that the melting width is assumed to be slightly wider around the gap. By using the keyhole model, the shape that is close to the actual one was obtained.

Figure 11 shows the cross section photos obtained through the actual processing experiments with different gaps. The upper layer shows the material cross section photos for a little gap ($g = 0.005$ mm) processed with 3.4 kW power, changing the welding speed from 2 to 7 m/min for comparison. When the speed is less than 5 m/min, penetration welding is observed, but at a speed of 7 m/min, only partial welding is observed. The lower layer shows the material cross section at a constant welding speed of 5 m/min, and the effect of changing the size of the intermediate gap. It is shown that an under filled condition is observed in the upper plate when the gap is greater than 0.25 mm.

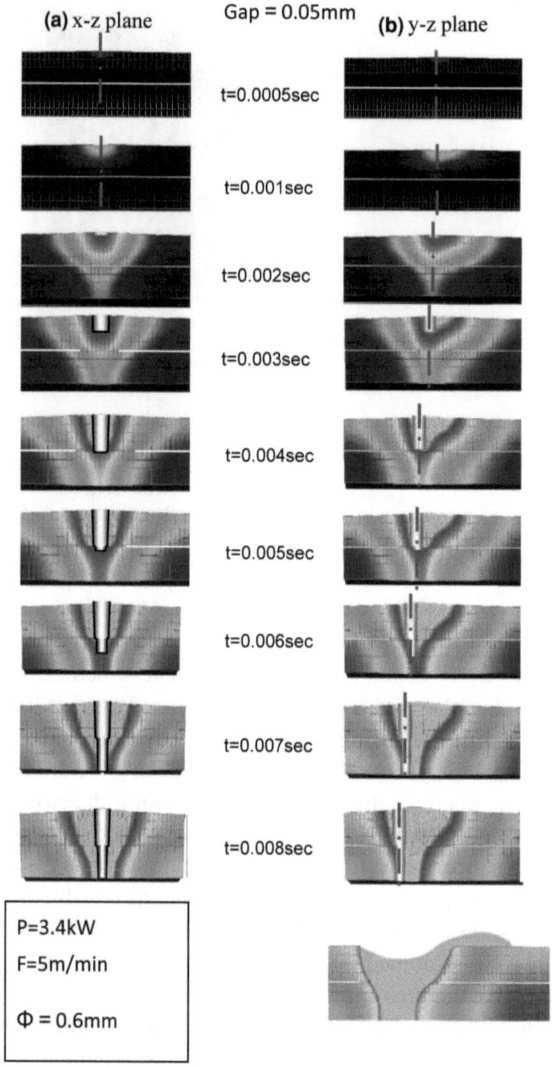

Fig. 8 The simulation results in time-series under the condition where the keyhole type welding (*moving picture*). Where laser power is 3.4 kW, and welding speed is 5m/min, gap is g = 0.05 mm. The views of observation are x–z plane and y–z plane

4.3 Comparison of Heat Conduction Type Welding and Keyhole Type Welding

For comparison purposes, heat conduction type welding was performed on the same material surface with the power density less than 10^5 W/cm^2. As well as keyhole type welding, it is assumed that the two plates with 1 mm thickness are lap welded and the heat reaches to the bottom of the lower plate so that the penetration bead is created for this heat conduction type welding. To achieve

Fig. 9 The actual welding behavior observation by high speed CCD camera. The view of observation is x-y plane. Where laser power is 3.4 kW, and welding speed is 5 m/min

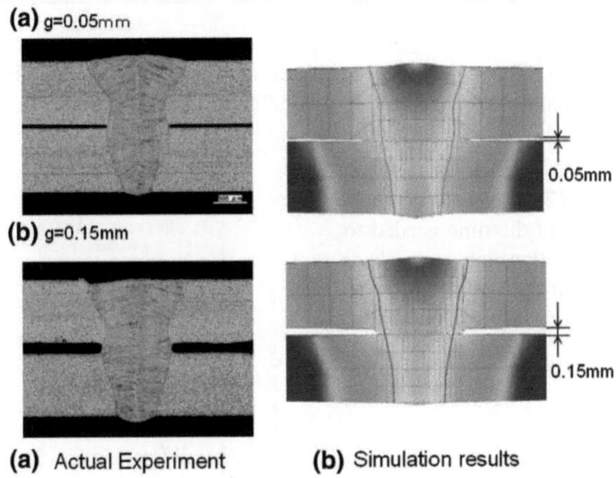

Fig. 10 Comparison between cross section of actual sample and result in simulation in keyhole welding. Where laser power is 3.4 kW, and welding speed is 5 m/min. The view of observation is y–z plane. **a** Narrow gap of g = 0.05 mm **b** Wide gap of g = 0.15 mm

this, the power output should be less than 1 kW. In this comparison, the welding speed was 5 m/min for the keyhole type welding with 3.4 kW power and the power density was 3.0×10^5 W/cm^2, while the welding speed for the heat conduction type welding was 1.5 m/min with 1 kW power and the power density was 8.8×10^4 W/cm^2. Figure 12 shows a comparison of the time needed to penetrate (until the heat is conducted sufficiently to the back side and the penetration bead is created) the lapped plates with 2 mm thickness. The result shows that the time needed for the penetration was 9 μs for keyhole type welding and about 38 μs for heat conduction type welding. This indicates that the presence of a keyhole accelerates the welding speed. From a different point of view, if heat conduction type welding would have been carried out using high power laser welding, it would have taken too long to join the material, which does not reflect industrial processing. Therefore, the result indicates that the keyhole type welding is carried out also in thin plates lap welding with high power.

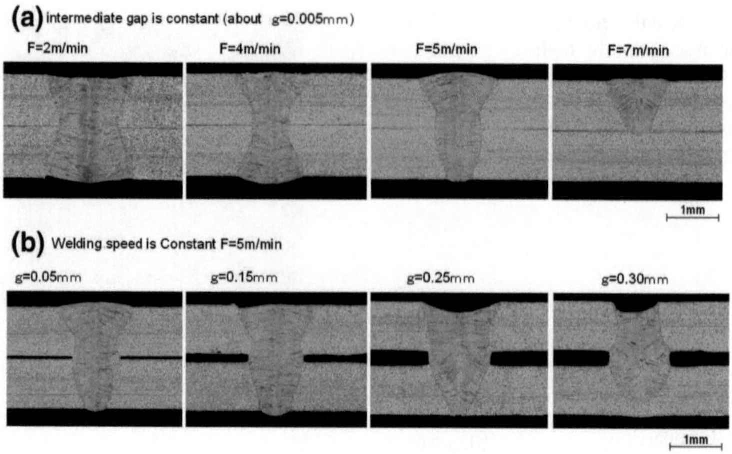

Fig. 11 The cross section photos obtained through the actual processing experiments with **a** different welding speed and **b** different gaps

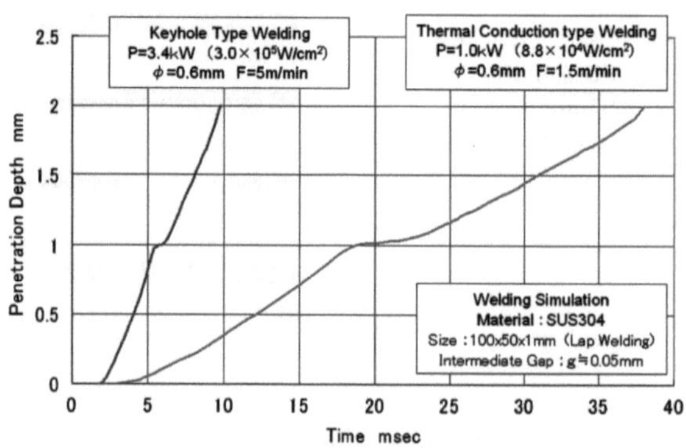

Fig. 12 The comparison result of the time needed to penetration time to the back side in the welding of heat conduction type and keyhole type

4.4 Comparison of the Plate Distortion Through Simulation

Figure 13 shows the amount of distortion along the axis of the welding direction in lap welding with different intermediate gaps, where the vertical axis indicates the amount of distortion and the horizontal axis indicates the distance of the welding axis direction. For comparison purposes, the calculation was made for heat conduction type welding with 1 kW power and 2.5 μm gap in the middle, where the keyhole is not considered to be generated. The speed with which welding can be performed was 2 m/min. This indicates that the processing speed is faster in keyhole type welding than in heat conduction type welding and, what is more, the amount of distortion is smaller for the keyhole type welding, although there is a difference in power and welding speed. In addition, in keyhole type welding, the greater the intermediate gap, the smaller the amount of distortion. It is believed

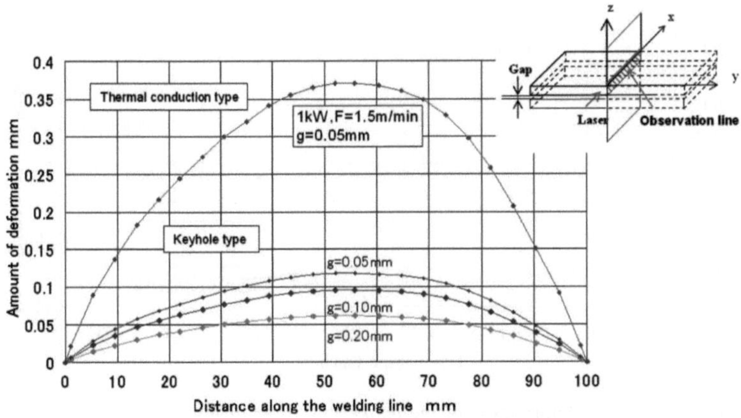

Fig. 13 The amount of distortion along the welding line direction in the lap welding with different intermediate gaps

Fig. 14 The angular distortion process and its distortion behavior in time-series observed from the vertical section (y–z plane)

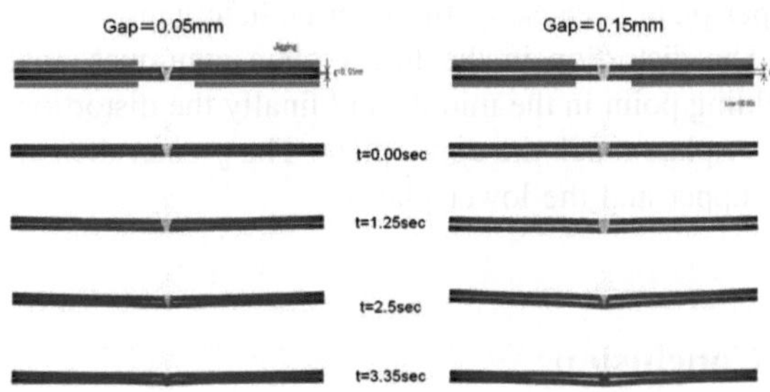

that some of the heat energy is used to fill the intermediate gap so that less heat is conducted into the lower plate. As a result, it is considered that the plate distortion in the whole is reduced.

4.5 Comparison of the Angular Distortion

Figure 14 shows the angular distortion process and its distortion behavior with time observed from the welding surface (y–z plane). The amount of deformation of the upper and lower plates is shown in Fig. 15. It shows the cross section surface where the welding point is located in the center of the plates and the jig to hold the material is arranged on and under the plates. It describes the distortion in the y–z plane, or the angular distortion that starts immediately after the jigs are released when the welding process is completed.

Figure 15 shows that the distortion does not appear in two plates concurrently nor evenly in thin plate lap welding. The upper plate initially deforms toward the beam irradiation direction due to the heating and after the heat source has passed,

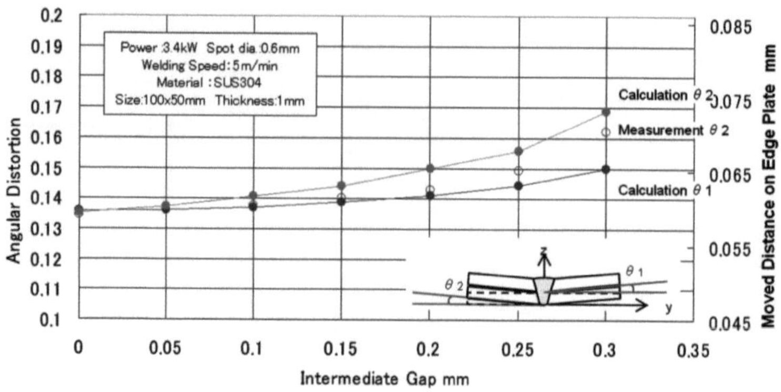

Fig. 15 The amounts of deformation of the *upper* and *lower* plates

the distortion reduces slightly during cooling. During this process, the lower plate deforms in the same manner but the amount of distortion is larger than that in the upper plate because of the intermediate gap.

The distortion in the lower plate continues, maintaining the gap around the welding point in the middle and finally the distortion stops when both edges of the lower plate touch the upper plate. The greater the gap, the wider the angle between the upper and the lower plates.

5 Conclusions

While the simulation process of keyhole welding includes some assumptions, they can be supported by the best experimental observations that are possible at the present time, so that the simulation result is not very different from the experimental results. In addition, the following new knowledge has been obtained through the simulation of high power lap welding.

1) When the intermediate gap is small, the heat transfers more easily between the lapped materials so that sufficient heat is conducted from the upper to the lower plate to increase the molten area. As a result, the overall distortion in the direction of welding becomes greater. On the other hand, if the intermediate gap is wider, less heat is conducted to the lower plate so that the amount of distortion is smaller.
2) When the intermediate gap is large, the molten metal of the upper plate is used to fill the gap so that an under fill condition is generated at the top of the upper plate. The smaller the intermediate gap, the wider interface can be maintained so that the molten cross section surface looks like the bead-on-plate without a gap.
3) In thin plate lap welding, the angular distortion increases as the intermediate gap becomes wider. In addition, the amount of distortion differs between the

upper and the lower plates in lap welding. The angular distortion is larger in the lower plate. The difference is proportional to the size of the intermediate gap both in the actual experiment and the simulation.

4) In lap welding with a gap, the material surface of the upper plate is heated and the heat conducts downward in a short period but the conduction stops at the gap and becomes an adiabatic state. During this adiabatic period, the convective flow of the heat and the volume expansion of the molten metal cause the upper plate to touch the lower plate. Once they touch, the heat conducts rapidly toward the lower plate to form a molten layer.
5) Given that heat conduction type welding would be performed in thin plate lap welding with a high power, the molten area and the time required to join the material does not match the actual processing. Therefore the conclusion can be drawn that keyhole type welding is performed in the high power welding.
6) In thin plate lap welding, there is a difference in the heat distribution between the upper and the lower plates because of the gap between them. In heat conduction welding the laser power is lower, the welding speed is slower and the amount of distortion is greater compared to keyhole welding.

A basic model for keyhole welding has been established. With this simulation, it is assumed that the welding phenomenon observed in thin plate lap welding can be explained. In the future more precise measurement and X-ray fluoroscopy observation data should improve the simulation accuracy.

Acknowledgment The author wants to express gratitude to helpful advice and discussion of Dr Susumu Tsukamoto (National Research Institute for Metals) and Dr. Naoki Seto (National Institute of Advanced Industrial Science and Technology). And also the author want to express thank to Mr. Noritaka Asano of Chuo University for an experimental support of him.

References

1. Arai, T.: The laser butt welding simulation of the thin metal, metals with complex behavior, advanced structured materials 3, modeling, simulation, testing, and applications, pp. 279–296. Springer, Heidelberg (2010)
2. Swift-Hook, D.T., Grick, A.E.F.: Penetration welding with Lasers. Weld. J. **11**, 492–499 (1973)
3. Moon, J., Katayama, S., Matsunawa, A.: Lap Welding Characteristics of thin metals with combined laser beams of Different wavelength. Weld. Int. **17**(6), 442–450 (2003). DO1:10, 1533/wint.2003, 3141
4. Grashchuk, V.P., et al.: Evaluating the permissible inconsistency of process parameters during CO_2 Laser Welding. Avt. Svarka **37**(5), 59–62 (1984)
5. Minamida, K.: High power laser application in nippon steel corporation. In: Proceedings of 1st international symposium on high power laser macroprocessing. Osaka, Japan, SPIE, **4833**, p. 402, 27–31 May 2002
6. Klein, T., Vicanek, M., Kroos, J., Decker, I., Simon, V.: Dynamic behavior of keyhole in laser beam welding. J. Phys. D: Appl. Phys. **27**, 482–486 (1994)

7. Philpp, G., Rudolf, W.: Spatter formation and keyhole observation with high speed cameras–better understanding of the Keyhole Formation. In: Proceedings of ICALEO,'09, **704**, 339-342 (2009)
8. Matsunawa, A., Kim, J.D., Katayama, S., Semak, V.: Experimental and theoretical studies on keyhole dynamics in laser welding. In: Proceedings of ICALEO 96, Vol. 81,. Section B58-67 LIA USA, 17–20 Oct (1996)
9. Tsukamoto, J., et al.: Keyhole behavior in high power laser welding. In: Proceedings of SPIE **4831**, 251–256 (2002)
10. Seto, N., Katayama, S., Mastunawa, A.: High-speed simultaneous observation of plasma and keyhole behavior during high power CO_2 laser welding. Effect of shielding gas on porosity formation. J. Laser **12**(6), 245–250 (2000)
11. Matsunawa, A.: Science of laser welding—mechanisms of keyhole and pool dynamics—(Invited Paper). In: Proceedings of ICALEO 02, LIA, Scottsdale, AZ, USA, (2002)
12. Park, J.U., Lee, H.W., Bang, H.S.: Effects of mechanical constrains on angular distortion of welding joining. Sci. Technol. Weld. Join. **7**(4), 232–239 (2002)
13. Sato, K., Terasaki, T.: Effects of transformation on laser welding residual stress and welding deformations. J.JWS 45–7, pp. 560–566 (1997)

Crack Propagation Analysis in Underwater Laser Drilling

Etsuji Ohmura, Takashi Okazaki, Keiichi Kishi, Toshio Kobayashi, Masahiro Nakamura, Satoshi Kubo and Komei Okatsu

Abstract In recent years, a state-of-the-art method of rock drilling has been developed by Japan Drilling Co., LTD. (JDC) and Japan Oil, Gas and Metals National Corporation (JOGMEC). In that method, a pulsed laser of the wavelength that is absorbed by water is used. Bubbles are generated in the water by the absorbed energy. The rock surface is irradiated by the laser beam through the bubbles. It was reported by the experiments under atmospheric pressure that there are two processing types in this method: melting and spallation. In the former, rock is partially melted and removed by a water stream caused by the bubble formation or the bubble extinction. In the latter, rock is spalled by thermal

E. Ohmura (✉)
Department of Management of Industry and Technology, Graduate School of Engineering, Osaka University, 2-1, Yamada-oka, Suita, Osaka 565-0871, Japan
e-mail: ohmura@mit.eng.osaka-u.ac.jp

T. Okazaki
Division of Materials and Manufacturing Science, Graduate School of Engineering, Osaka University, 2-1, Yamada-oka, Suita, Osaka 565-0871, Japan
e-mail: okazaki@mapse.eng.osaka-u.ac.jp

K. Kishi · T. Kobayashi · M. Nakamura
Japan Drilling Co., LTD, Tokyo, 103-0012, Japan
e-mail: keiichi.kishi@jdc.co.jp

T. Kobayashi
e-mail: toshio.kobayashi@jdc.co.jp

M. Nakamura
e-mail: masahiro.nakamura@jdc.co.jp

S. Kubo · K. Okatsu
Japan Oil, Gas and Metals National Corporation, Chiba, 261-0025, Japan
e-mail: kubo-satoshi@jogmec.go.jp

K. Okatsu
e-mail: okatsu-komei@jogmec.go.jp

A. Öchsner et al. (eds.), *Materials with Complex Behaviour II*,
Advanced Structured Materials 16, DOI: 10.1007/978-3-642-22700-4_45,

stress and removed in the same way as in the former case. In this study, the processing mechanism in the spallation type was clarified first by the thermal stress analysis based on fracture mechanics. Then, the possibility of laser drilling at the sea bottom which is several thousand meters from the sea level was also investigated, because the experiments are very difficult. As a result, the following conclusions were obtained: In the tensile stress field of the rock inside caused by laser irradiation, the rock is spalled by crack propagation in mode I. In the compression field generated in the neighborhood of the rock surface, a rock is spalled by crack propagation in mode II. These spallations occur also under the condition of high pressure that the bottom of the sea was assumed. Contribution of crack propagation in mode II to the spallation increases as the depth becomes large.

Keywords Underwater laser drilling · Crack propagation · Spallation · Granite · Thermal stress · Fracture mechanics

1 Introduction

Rotary drilling is currently the only method used for the digging of oil. The digging of oil using a laser has been studied from the 1970s. In past research, a fused glass component has been generated by laser irradiation, and it disturbs laser drilling. In recent years, a state-of-the-art method of rock drilling has been developed by Japan Drilling Co., LTD. (JDC) and Japan Oil, Gas and Metals National Corporation (JOGMEC) [1]. In that method, a pulsed laser of the wavelength that is absorbed by water is used. Bubbles are generated in the water by the absorbed energy. The rock surface is irradiated by the laser beam through the bubbles. It was reported by the experiments under atmospheric pressure that there are two processing types in this method: melting and spallation. In the former, rock is melted partially and removed by water stream caused by the bubble formation or the bubble extinction. In the latter, rock is spalled by thermal stress and removed in the same way as in the former case.

In this study, the processing mechanism of the spallation type was clarified first by a thermal stress analysis based on fracture mechanics. Granite was used as the rock material. Then, the possibility of laser drilling at the sea bottom which is several thousand meters from the sea level was also investigated, because the experiments are very difficult.

2 Analysis Method

In this study, the commercial software ANSYS was used for analysis. Heat conduction analysis was performed first, and thermal stress analysis was conducted based on the obtained temperature distribution as a plane strain problem. The Analysis model is

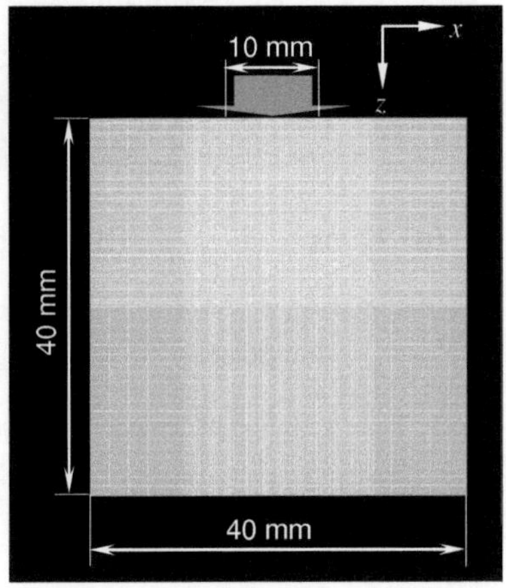

Fig. 1 Analysis Model for heat conduction and thermal stress analyses

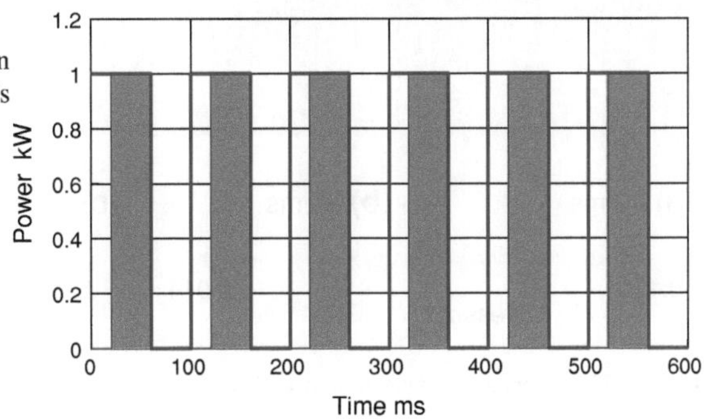

Fig. 2 Laser irradiation condition (real laser irradiation to the rock is shown by halftone dot meshing.)

shown in Fig. 1. The obtained temperature distribution was then applied to the model including precrack, and thermal stress analysis and crack propagation analyses were conducted. In the crack propagation analysis, the mode I stress intensity factor (SIF) and the mode II SIF were calculated at crack tips, and the crack was progressed when the value exceeded each fracture toughness value of rock. Precracks were supposed in the inside of the rock and in the vicinity of the rock surface. The laser irradiation condition was as follows: laser power 1 kW, beam size (width) 10 mm, repetition rate 10 Hz and duty 60%.

Using ANSYS CFX, the velocity of the water stream at the rock surface was analyzed from the formation of the bubble due to laser absorption to the bubble extinction after laser pulse end. As a result, the following results were obtained: among pulse duration which is 60 ms, about 20 ms of the beginning contributes to bubble formation, and laser is irradiated to the rock surface during the remaining pulse duration which is about 40 ms. Bubbles are collapsed after the laser pulse end. Therefore, the real laser irradiation to the rock was assumed, as shown by halftone dot meshing in Fig. 2. In addition, the heat transfer coefficient during

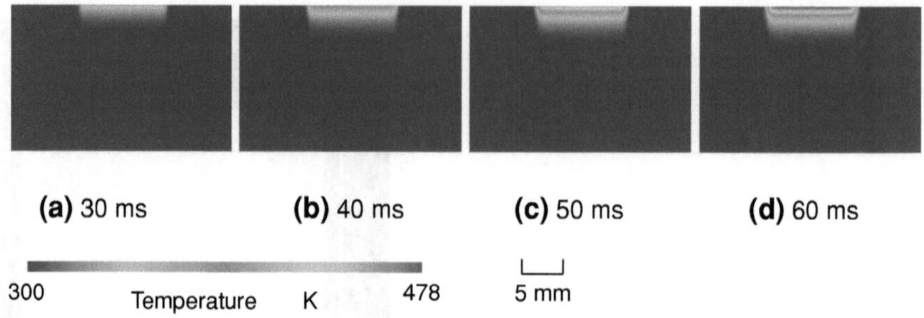

Fig. 3 Time variation of temperature distribution

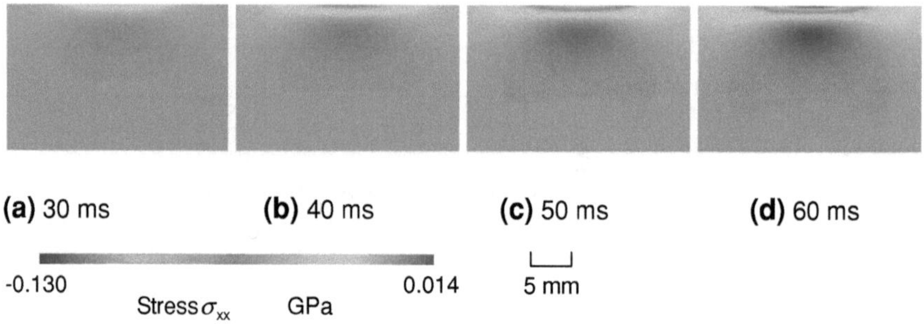

Fig. 4 Time variation of normal stress σ_{xx}

bubble formation and bubble collapse was estimated to be 1.89 kW/m^{2}K from average flow rate on the rock surface [2]. It was assumed that both the initial temperature and the ambient temperature are 300 K.

3 Analysis Results and Discussion

The results of heat conduction analysis and the thermal stress analysis where the obtained temperature distribution was used for are shown in Figs. 3 and 4, respectively. The surface temperature of the granite rises by the laser irradiation. As a result, in the vicinity of the surface, a compressive stress field is present and a tensile stress field is present in the interior.

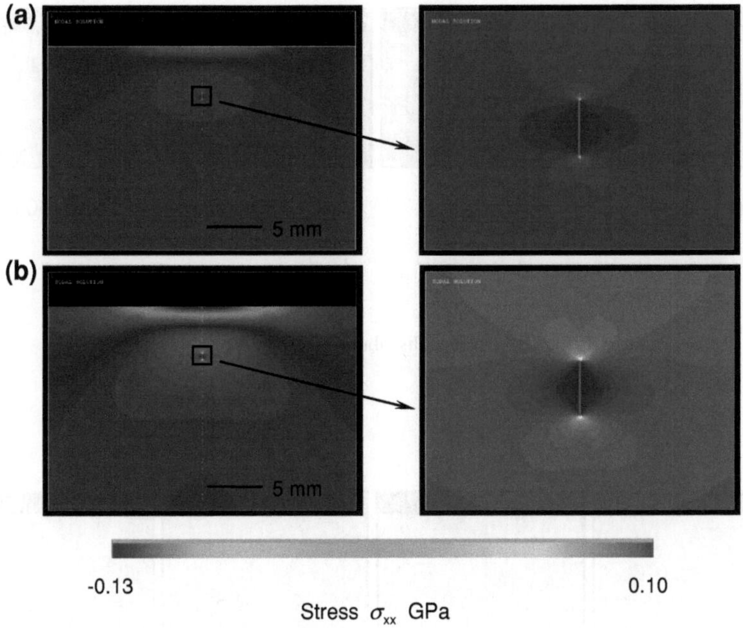

Fig. 5 Normal stress σ_{xx} at 30 and 60 ms

Fig. 6 Time variation of normal stress σ_{xx} at precrack tips

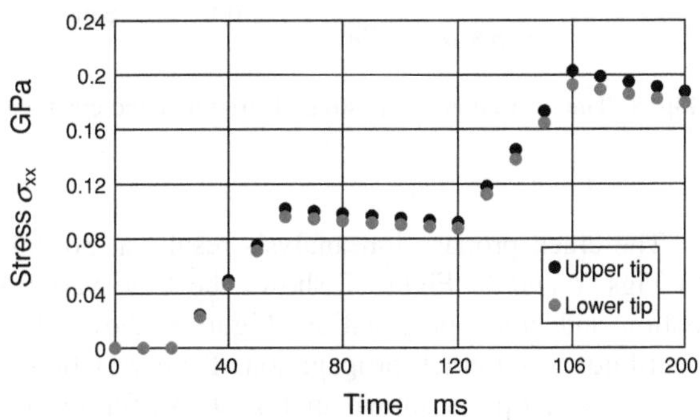

3.1 Case Supposing a Precrack Inside the Rock

A vertical precrack of 500 μm long from a point of 4 mm in depth from rock surface was supposed. Figure 5 shows the normal stress σ_{xx} at 30 and 60 ms. Stress concentrates at the tips of the precrack in the internal tensile stress field at both moments in time. Figure 6 shows the stress concentration at the tips of the precracks. Because such stress concentration occurs at the crack tips, the crack progresses.

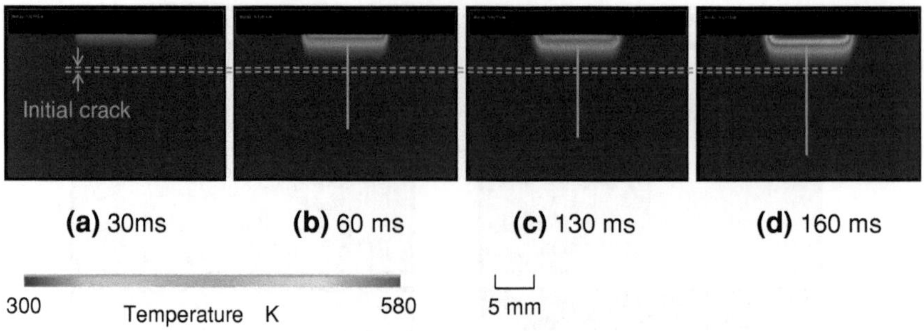

Fig. 7 Time variation of the temperature distribution and crack propagation

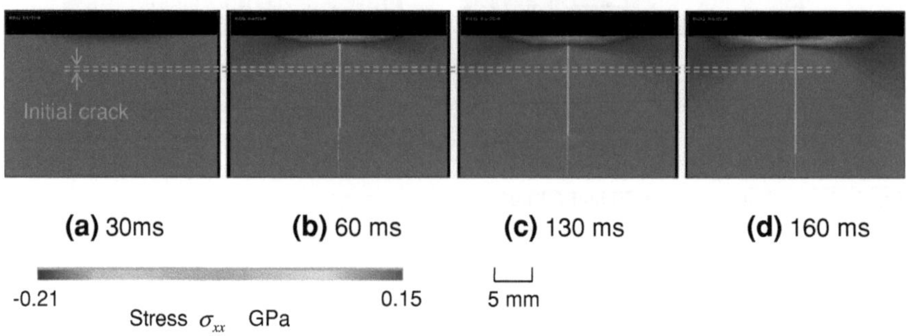

Fig. 8 Time variation of the stress distribution and crack propagation

The crack propagation analysis results at 30, 60, 130 and 160 ms are shown in Figs. 7 and 8. Figure 7 shows the time variation of the temperature distribution and crack propagation. Figure 8 shows the time variation of the stress distribution and crack propagation. Time variation of the depth of the upper and lower crack tips is shown in Fig. 9. As for the upper tip, crack propagates at 60 ms, but stops afterwards. This is because the crack progresses and reaches the compressive stress field. The lower tip of the crack progresses remarkably by the first heating. The tensile stress of the inside of granite decreases after laser pulse ends, and the stress concentration slowly becomes weak. During re-heating by the next laser pulse, the surface is more subjected to a compressive stress field, and the tensile stress field spreads by the counteraction. As a result, larger stress concentration occurs at the lower crack tip, and the crack progresses more.

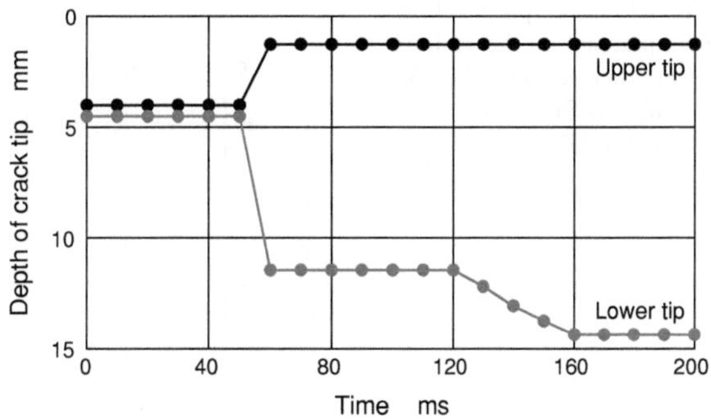

Fig. 9 Time variation of the depth of the upper and lower crack tips

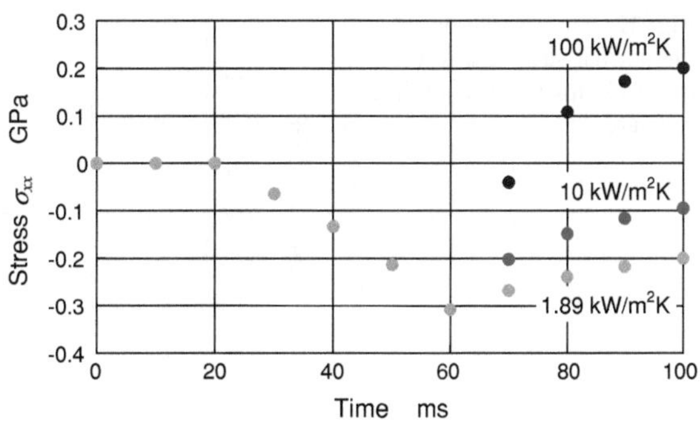

Fig. 10 Influence of heat transfer coefficient on the time variation of normal stress σ_{xx}

3.2 Case Supposing a Precrack in the Vicinity of the Rock Surface

Here a vertical precrack of 500 μm long from the rock surface was supposed. The laser irradiation condition is the same as in the previous section. The influence of heat transfer coefficient on the time variation of the normal stress σ_{xx} at the precrack tip is shown in Fig. 10. The crack does not progress during the heating period under laser irradiation because a compressive stress field is present in the vicinity of the surface. When the heat transfer coefficient is assumed to be 1.89 or 10 kW/m^{2}K, the crack does not progress after laser pulse ends, because the cooling effect is small. On the other hand, a tensile stress acts on the crack tip if a large heat transfer coefficient such as 100 kW/m^{2}K is used, then the crack progresses as shown in Fig. 11. However, its crack propagation is considerably low comparing to the propagation of the above-mentioned interior crack. The heat transfer coefficient 100 kW/m^{2}K is about 50 times larger than 1.89 kW/m^{2}K. Therefore, it is concluded that cracks in the compressive stress field at the vicinity of the rock surface do not almost progress.

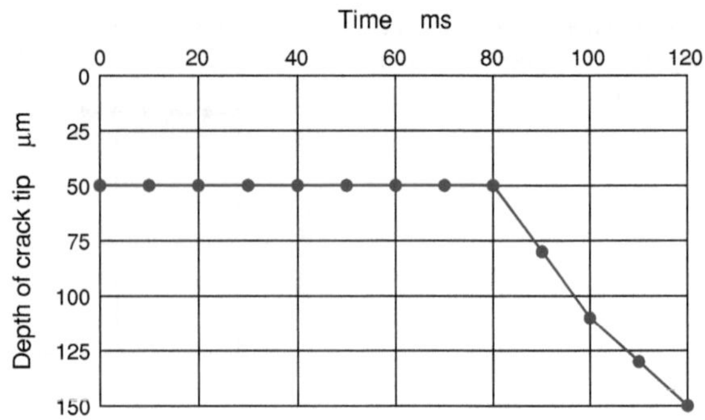

Fig. 11 Crack propagation of a precrack in the vicinity of the rock surface when heat transfer coefficient is 100 kW/m^{2}K

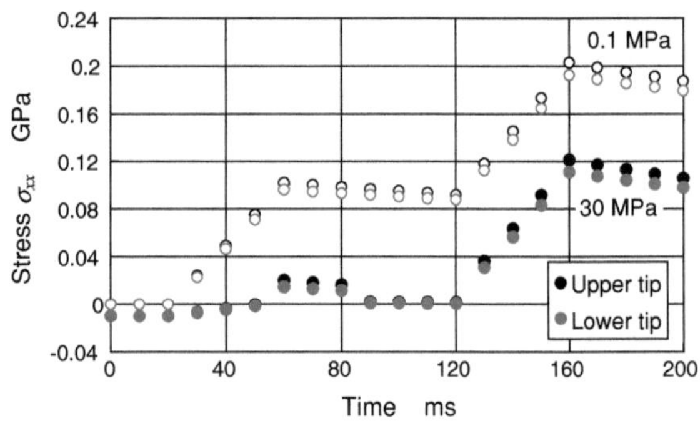

Fig. 12 Time variation of the normal stress σ_{xx} at the precrack tips under confining pressure

3.3 Thermal Stress Analysis Considering a Confining Pressure

The previous analyses were conducted under atmospheric pressure, but the real laser drilling is carried out on the sea bottom whose depth may be several thousand meters. Therefore, the crack propagation analysis was performed assuming constraint pressure on the sea bottom. Here, the depth of sea bottom was assumed to be 3,000 m, where its confining pressure is about 30 MPa. The laser irradiation condition is the same as the above analysis where a precrack of 500 μm long was assumed in the interior of granite. The heat transfer coefficient was also assumed to be 1.89 kW/m^{2}K.

Time variation of the normal stress σ_{xx} at the precrack tips is shown in Fig. 12. For comparison, the analysis result under atmospheric pressure 0.1 MPa is also shown. It is understood that stress is hard to concentrate at precrack tips in comparison to atmospheric pressure. This is because the analysis area is easier to become a compressive stress field under the confining pressure in comparison to atmospheric pressure conditions. Crack propagation began at 60 ms in the analysis under atmospheric pressure, but it can be understood that the SIF does not reach the fracture toughness value 0.35 MPa m$^{1/2}$ of the granite [3] at 60 ms under the

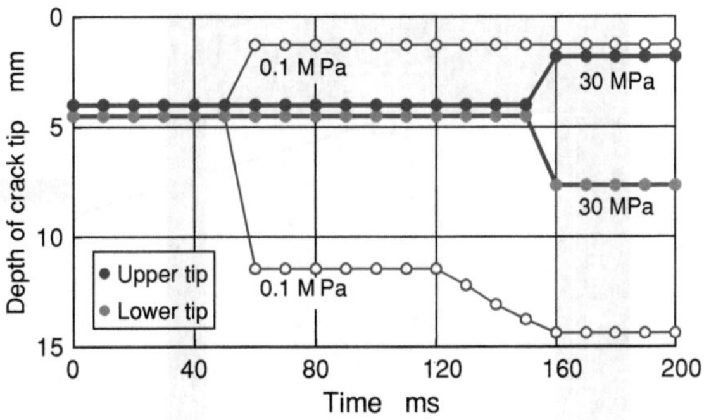

Fig. 13 Time variation of the depth of crack tips under confining pressure

confining pressure of 30 MPa. The tensile stress at the crack tip decreases in the later cooling period, and crack propagation starts at 160 ms by heating of the next laser pulse. Time variation of the depth of crack tips is shown in Fig. 13. It is concluded that while cracks can progress at the sea bottom which is several thousand meters from the sea level, when compared to crack progression under atmospheric pressure, they do not progress as easily.

3.4 Crack Propagation in Mode II

Then, the possibility of crack propagation in mode II, that is, crack propagation by shearing stress in a compressive stress field in the rock surface was examined. The analysis procedure is the same as in the previous sections. Applying the temperature distribution obtained in the heat conduction analysis to the model including a precrack, thermal stress analysis was conducted. The mode II SIF at the crack tips was calculated, and crack progressed when the value exceeded the fracture toughness in mode II of the granite. The laser irradiation condition was the same as above. The mode II fracture toughness value of the granite is 0.46 MPa $m^{1/2}$ [3, 4], and heat transfer coefficient was assumed to be 1.89 kW/m^2K. It was supposed that the precrack is 500 μm long, its center is 500 μm in depth and the inclination angle is 60°. The finite element model is shown in Fig. 14.

The shearing stress at the precrack tips is shown in Fig. 15. Figure 16 shows the time variation of temperature distribution and crack propagation. The time variation of the depth of the crack tips is shown in Fig. 17. The upper crack tip reaches at the surface at 40 ms. The lower crack tip also progresses remarkably at 30–60 ms during laser irradiation time, and crack progresses still slowly after the laser pulse irradiation. It is concluded that cracks in the vicinity of the rock surface, where crack propagation in mode I cannot occur because of the compressive stress field, can progress by the shearing stress.

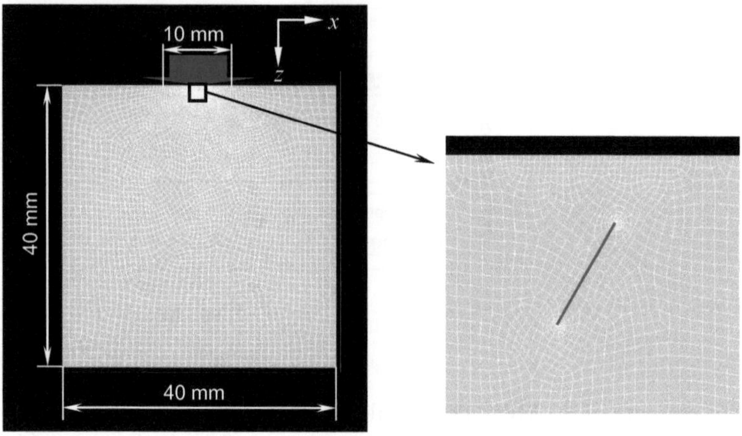

Fig. 14 Finite element model for analysis of crack propagation in mode II

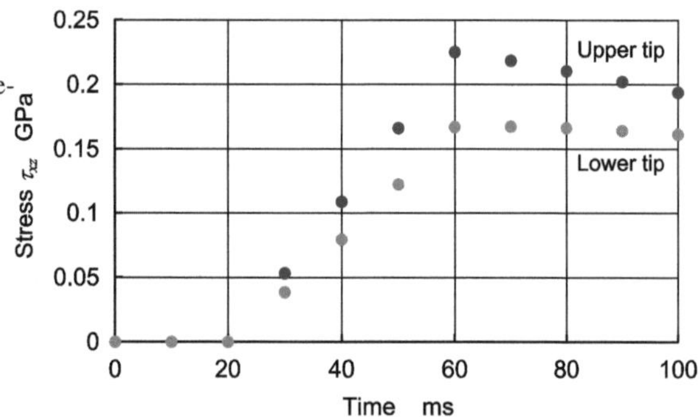

Fig. 15 Shearing stress at the precrack tips in mode II

Confining pressure which acts as hydrostatic pressure at the sea bottom does not affect the shear stress field. Therefore, it can be estimated that the mode II crack propagation at the sea bottom will be almost the same as the propagation under atmospheric pressure. Actually, the analysis result was the same as Fig. 17 when a confining pressure of 30 MPa was given. It is concluded that crack propagation by shear stress becomes dominant in underwater laser drilling as the depth of water increases because the confining pressure becomes increasingly large.

3.5 Multiple Cracks Propagation

In the above-mentioned analysis, only one precrack was supposed in the granite. Here, analysis results where multiple adjacent precracks were supposed are given. At first, the crystal grain boundaries were extracted from a polarizing

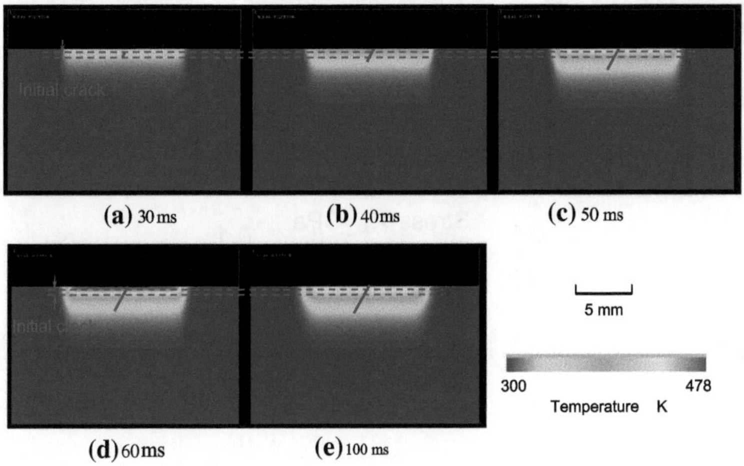

Fig. 16 Time variation of temperature distribution and crack propagation in mode II

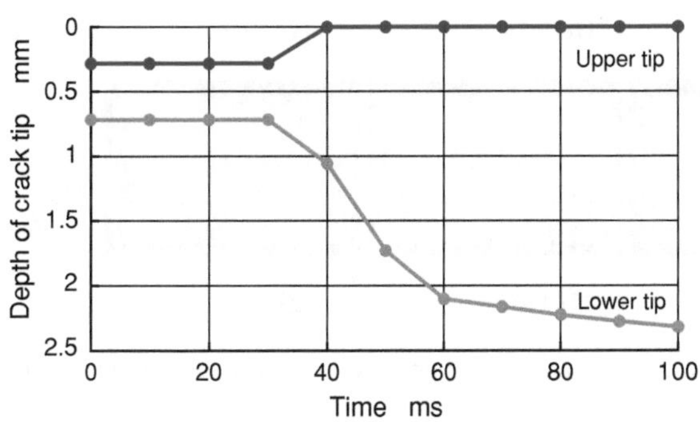

Fig. 17 The time variation of the depth of the crack tips in mode II

microscope image of the granite by an image analysis software Image J. Then, a finite element model where three precracks are in the adjacent of the grain boundaries was made. Each crack was assumed to be a line along the grain boundary. The crack center was defined applying random numbers on the grain boundary, and all of them were assumed to be 500 μm long. Crack propagation analysis was conducted as in the previous sections. The same laser irradiation condition and physical properties of granite were used. The heat transfer coefficient was assumed to be 1.89 kW/m^2K.

Temperature distribution and stress distribution at 130 ms are shown in Fig. 18. It is understood that stress concentration is generated at crack tips.

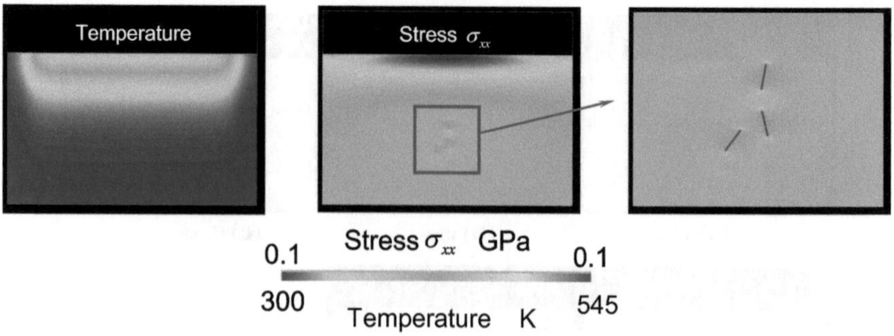

Fig. 18 Temperature distribution and stress distribution at 130 ms for multiple cracks

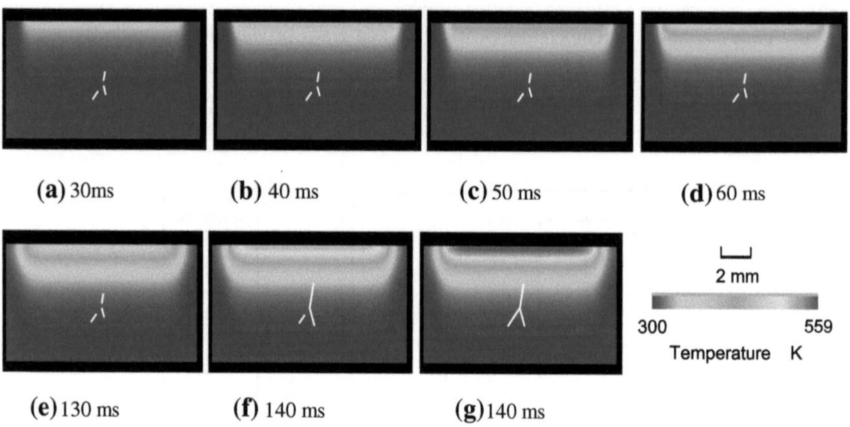

Fig. 19 Time variation of temperature and crack propagation under atmospheric for multiple cracks

For all precracks, mode I is dominant because they are in the tensile stress area. Time variations of temperature and crack propagation under atmospheric pressure and a confining pressure of 30 MPa, are shown in Figs. 19 and 20, respectively. In Fig. 19, two cracks collide at 140 ms at first, then the third crack also collides at 150 ms. Under the confining pressure 30 MPa, cracks are hard to progress comparing to atmospheric pressure, and two cracks finally collide at 560 ms in Fig. 20. The analysis where three precracks on crystal grain boundaries in the granite were supposed was performed here, however in the actual drilling phenomena, this crack propagation occurs at various areas. It is supposed that a rock is spalled by such crack propagation and then pulverized particles are removed by the water jet which is generated by bubble collapse after the end of laser pulse.

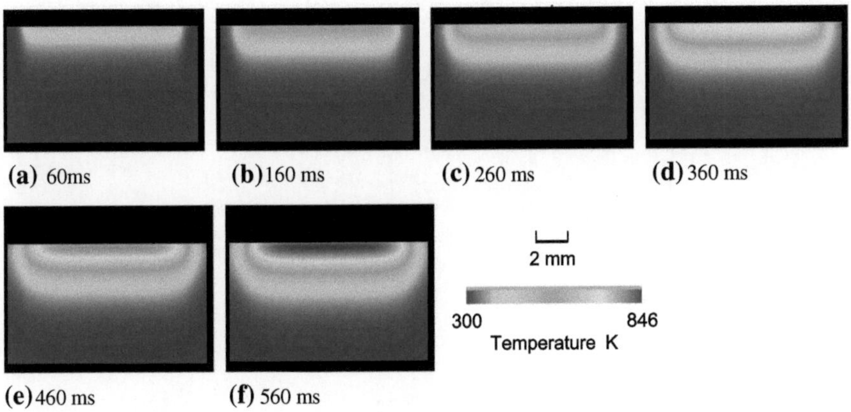

Fig. 20 Time variation of temperature and crack propagation under a confining pressure of 30 MPa for multiple cracks

4 Conclusion

In this chapter, the crack propagation in spallation type, which is one of the processing mechanisms in underwater laser drilling developed by JDC and JOG-MEC, was investigated by a thermal stress analysis based on fracture mechanics. Possibility of laser drilling at the sea bottom which is several thousand meters from the sea level was also investigated, because the experiments are very difficult. As a result, the following conclusions were obtained: In the tensile stress field of the rock inside caused by laser irradiation, the rock is spalled by crack propagation in mode I. In the compressive stress field generated in the neighborhood of the rock surface, a rock is spalled by crack propagation in mode II. These spallations occur also under the condition of high pressure at the bottom of the sea. Contribution of the crack propagation in mode II to the spallation increases as the depth becomes large.

References

1. Kobayashi, T., Otani, K., Takayama, K., Umezu, S., Okatsu, K. In: Congress Proceedings ICALEO 2007, vol. 480 (2007)
2. Lamb, H.: Hydrodynamics. Cambridge University Press, Cambridge (1993)
3. Wada, C., Shoji, T.: Fracture toughness and its relation to initial tangent modulus of granitic rock. J. Soc. Mater. Sci. (Japan) **35**(389), 145–151 (1986) (in Japanese)
4. Chang, S.-H., Lee, C.-I., Jeon, S.: Measurement of rock fracture toughness under modes I and II and mixed-mode conditions by using disc-type specimens. Eng. Geol. **66**, 79–97 (2002)

MIX
Papier aus verantwortungsvollen Quellen
Paper from responsible sources
FSC® C105338

FSC
www.fsc.org

If you have any concerns about our products,
you can contact us on
ProductSafety@springernature.com

In case Publisher is established outside the EU,
the EU authorized representative is:
Springer Nature Customer Service Center GmbH
Europaplatz 3, 69115 Heidelberg, Germany

Printed by Libri Plureos GmbH
in Hamburg, Germany